591.4
H996c
2

Comparative Vertebrate Anatomy

Comparative Vertebrate
ANATOMY

by

LIBBIE HENRIETTA HYMAN
American Museum of Natural History
New York City

1888-

42-21814

THE UNIVERSITY OF CHICAGO PRESS
Chicago · Illinois

THE UNIVERSITY OF CHICAGO PRESS, CHICAGO 37
Cambridge University Press, London, N.W. 1, England
The University of Toronto Press, Toronto 5, Canada

Copyright 1922 and 1942 by The University of Chicago. All rights reserved. *Published February 1922. Second Edition September 1942. Fourteenth Impression 1956. Composed and printed by* THE UNIVERSITY OF CHICAGO PRESS, *Chicago, Illinois, U.S.A.*

PREFACE TO THE FIRST EDITION

Several years ago the method of procedure in the laboratory work in vertebrate zoölogy in this University was changed from the type plan, then in common use, to the comparative plan. No doubt a similar change has been made in many other institutions. A suitable manual for the comparative method has, however, hitherto been lacking; the present publication attempts to supply this need. There can scarcely be any question of the superiority of the comparative method of study of vertebrate anatomy, for by this method the student not only learns all of the anatomical facts brought out by the type method but he also acquires an understanding of vertebrate and human structure which he is quite unable to attain by the older method of study. In view of the fact that the majority of the students taking courses in vertebrate anatomy at the present time are preparing for medicine, it seems obligatory that they be taught the "history of the human body" as revealed by the study of the anatomy of vertebrates. On the other hand, the comparative method is perhaps faulty in that it may not give the student a clear-cut picture of the characteristics of the various classes of vertebrates. Thus, while the student readily learns the history of the aortic arches, for example, he does not readily associate any particular group of vertebrates with a particular condition of the arches, when the comparative method is followed. This defect should probably be remedied in the lecture part of the course. I have attempted to remedy it to a slight extent by introducing the section on the general features of typical chordates.

In this manual I have attempted not only to give the laboratory directions for the dissection of the various systems, but I have also presented in connection with each system a very brief, generalized, and simplified account of the development and evolution of that system. It has seemed to me essential that such an account precede or accompany the laboratory directions in order to give a significance to the facts revealed by the dissection at the time when the student becomes aware of those facts. As the consultation of other texts during a dissection is inconvenient and time-consuming, I have thought it the most practical plan to include such explanatory matter in the laboratory manual. Simple illustrations have been added to clarify further the explanations. I have not, however, in the least intended that these explanations should take the place of reading in the standard texts of comparative anatomy. The students should understand that additional outside reading is expected of them.

21484

In making such brief and generalized explanations as are given here it is impossible to take into account numerous exceptions and variations. I must therefore ask the indulgence of the expert in vertebrate anatomy for the omission of qualifying clauses in the explanatory accounts of the various systems; in some cases, no doubt, I have been unaware of the exceptions; in others I have knowingly omitted them on the grounds that statements of exceptions are more confusing than informative to the beginner.

To avoid confusion the explanatory matter is printed in slightly smaller type than the directions for the dissections.

I have included in the manual such materials as seemed to me to bear most directly on the story of the evolution of the various systems. I have treated the skeleton and the coelom at greater length than is commonly the case. The prevailing neglect of the study of the skeleton in courses in vertebrate anatomy seems to me unjustifiable in view of the importance of this system in the study of phylogenetic and evolutionary questions. It is true that skeletal material is somewhat expensive to purchase and maintain, but a good many of the more important materials can be prepared by the instructor or students. In presenting the vertebral column I have adopted Gadow's conception of the development of the vertebrae from separate arcualia, an idea also adopted by Schauinsland in his account of the development of the vertebrae in Hertwig's *Handbuch.* The conception would appear to be correct in the main, and at least furnishes a simple explanation for the variety of vertebral columns encountered among vertebrates. The difficulties attending the study of the coelom and mesenteries do not seem to me to justify one in disregarding them. I hope the simplified account I have presented after long study and thought on the matter will aid in the understanding of this complex subject. As to the animals to be dissected, the elasmobranch must naturally form the point of departure in the study of comparative anatomy. I have described the skate in addition to the dogfish, since some teachers prefer it; it is certainly more favorable than the dogfish for the study of the nervous and urogenital systems, but less favorable, in my opinion, for the study of the circulation. The not infrequent scarcity of dogfish in recent years makes it desirable that an alternative form be described. The bony fish is omitted because its specialized structure would confuse rather than aid the student in following out the evolution of the systems. The frog is so often used in general and beginning courses that it seems superfluous to consider it here. Further, the urodeles serve our purpose much better. I should like to have included *Cryptobranchus* as alternative to *Necturus,* but the limits of space forbade. The reptile is important for the purposes of a comparative course, and the turtle is perhaps the most readily obtainable form of sufficient size. The bird has been included since it seemed inadvisable to omit altogether an entire vertebrate class. I have described both the cat and rabbit, as the former animal, though perhaps preferable in some respects, is not always readily obtainable.

At the University of Chicago the course consists of a brief glance at the external features of the lower chordates and of representative vertebrates; of all of the work given in the manual on the exoskeleton, endoskeleton, and muscles, with the exception of the muscles of *Necturus;* and of the complete dissection of the dogfish, turtle, and mammal, except the peripheral distribution of the cranial nerves in the mammal. The elasmobranch, urodele, and mammal would make a combination nearly as desirable.

The general statements and explanations given in the manual have been taken from standard works and original papers on comparative anatomy, vertebrate zoölogy and embryology, and human anatomy. I have adopted, for the most part, the usual accounts of the evolution of the various systems and parts, not regarding myself as competent to criticize them. In a few cases I have presented some recent views appearing in the literature. The laboratory directions and descriptions of the anatomy of the several animals have, however, been written entirely from the specimens. Practically the whole of the dissection has been performed twice, some of it more than twice. The dissections have been carried on simultaneously with the writing of the directions. In locating and naming the structures I have been assisted by various texts, manuals, and original papers. In a number of cases I have found it desirable to devise additional names or to change old names. I have employed the terms dorsal, ventral, anterior, and posterior as they are used in vertebrate anatomy, abandoning the usage common in human anatomy. This has involved a change in some familiar names, such as that of the "anterior" abdominal vein.

I have made a particular effort to use all technical words in a very precise and exact manner and to define each such word where it first occurs. One is continually surprised and annoyed in a study of vertebrate structure to note the loose and inexact way in which many terms are employed. It is, in fact, practically impossible to find any precise usage for many terms, such as urethra, bulbus arteriosus, peritoneum, olfactory lobe, perineum, and others. In such cases I have been compelled to adopt such a definition as seems consistent with the majority of the anatomical facts.

I have attempted nothing in this manual but a plain account of the anatomy of the several animals, which account the student follows. This "verification" type of laboratory manual has been recently subjected to much criticism, much of it, in my opinion, undeserved. The critics seem to forget that the student is not in reality engaged in "verifying" the statements in the manual; he is engaged in learning the anatomy of an animal by the shortest and easiest route, a route which the critics themselves would follow if confronted with an animal with which they were not familiar. It is my opinion that human beings in general see chiefly that which is pointed out to them; this has been proved over and over again in the history of biology. The large number and complexity of the anatomical facts to be acquired, the limited time allowed for their acquisition, the large size of the classes, and the limited number of laboratory assistants available seem to me to necessitate that detailed and specific laboratory directions be provided. If the directions are not given in the manual, then the laboratory instructors are compelled to provide them verbally. Personally I am unable to see any pedagogical difference between directions and explanations written in the manual and those given out verbally by the instructor, but I do see a great deal of difference as concerns the time, patience, and energy of instructors and students. Our experience with laboratory manuals of the type in which the burden of discovery is left to the student is that the student becomes highly dissatisfied and that the instructors are brought into a state of irritation and fatigue by the continuous demands for assistance with which they are bombarded. Frankly, I believe in the conservation of instructors, and have written this manual with that end in view. In place of inserting questions in the laboratory manual our method is to hold thorough oral quizzes on the laboratory and textbook work at frequent intervals.

Although a number of drawings are called for in the manual, it is probable that each instructor will prefer to decide for himself what drawings are to be made. Drawings might profitably be omitted altogether, their place being taken by rigorous practical individual quizzes on the dissected specimen.

I am indebted to a number of authors and publishers for permission to reproduce figures from their publications. Due acknowledgment is made in the legends to these figures. I have not listed the numerous original papers to which I have referred, since most of them are given in the bibliographies appended to Kingsley's *Comparative Anatomy of Vertebrates* and Goodrich's account of the fishes in Part IX, first fascicle, of Lankester's *A Treatise on Zoölogy*. I am indebted to Dr. C. R. Moore, Dr. B. H. Willier, and Dr. J. W. Buchanan for calling my attention to errors and omissions in the first draft of the manual which has been used in the laboratory under their direction during the past two years. My thanks are also due to Mr. Kenji Toda for his patience and skill in drawing the illustrations. Finally, I wish to acknowledge the fact that the previous organization of the laboratory work along comparative lines, chiefly through the efforts of Dr. J. W. MacArthur and Mr. J. G. Sinclair, has facilitated the task of preparing this manual.

So laborious has been this task and so great is the number of facts to be considered that I can scarcely hope to have avoided errors, omissions, and statements lacking in clarity. I shall be more than grateful to have my attention called to them.

L. H. HYMAN

UNIVERSITY OF CHICAGO
November 1921

PREFACE TO THE SECOND EDITION

First of all I wish to thank the many users of the first edition of this book. Their generous opinion of its worth has enabled me to live a life of singular freedom and to devote myself exclusively to scientific research and writing.

In this revision not much change has been made in the laboratory directions. These were originally written directly from the dissections, and it has been deemed unnecessary to repeat the dissections. Some changes of terminology have been introduced, and errors have been corrected wherever my attention was called to them by teachers. Some new material has been introduced, especially in the skeletal and muscular systems; and, as before, directions for this have been written from the specimens themselves.

The revision concerns chiefly the textual material, which has been thoroughly revised and considerably expanded with the intention that the book shall now serve as a text as well as a laboratory manual. It is believed that students will not need to purchase any accompanying textbook, although a selection of books on comparative vertebrate anatomy and vertebrate zoölogy should be on hand for reference and consultation. The considerable enlargement of the chapters on general features and external anatomy of chordates was made with the same idea in mind—a more thorough covering of the field of vertebrate zoölogy so that an additional textbook is unnecessary. This material also was written from the specimens, and it is hoped that a variety of vertebrates will be available for study by the student while engaged on these chapters.

In the first edition I was content to adopt and present the standard explanations and stock evolutionary stories found in textbooks of the subject. The years between, however, have taught me to suspect all standardized accounts copied into a succession of college textbooks. These stories of the evolution of the vertebrate organ-systems stem mostly from the great group of German comparative anatomists of the last half of the nineteenth century. These men were pioneers and constructed their evolutionary series on the basis of incomplete and scattered evidence. Their views can no longer be accepted without critical examination. The textual material of the present revision is based wholly on the study of original literature and advanced treatises. In place of the impression of a static subject wherein everything has been worked out, gained from the usual textbooks, it aims to give the student a picture of a vast field full of controversial issues and unsolved problems, depending for their solution on future painstaking embryological and anatomical researches. An army of devoted workers is necessary for elucidating these many questions; but nowadays—alas!—all young biologists want to be experimentalists, and hardly anyone can be found interested in the fields of descriptive embryology and anatomy.

Not only has the textual part of the book been much expanded, but a number of new figures have been added and some old ones, deemed erroneous, dropped out. Some of these figures are original; others are taken from the literature. Text and figures, it is hoped, will give as adequate and modern an account of the vertebrates and the evolution of their systems as can be expected within the limits of an undergraduate college course. It is obviously impossible to cover such a huge subject satisfactorily except in a treatise of several large volumes.

To such a treatise—the great German *Handbuch der vergleichenden Anatomie der Wirbeltiere,* edited by Bolk, Göppert, Kallius, and Lubosch—I wish to express my indebtedness. This work is indispensable to the student of comparative vertebrate anatomy and must serve as a guide in beginning the study of any system. It is a bottomless source of information on any facts one might want to know, although less satisfactory, perhaps, in giving connected accounts of the evolution of the systems. I also here wish to express my admiration for the comparative anatomists of the Russian school headed by the late A. N. Sewertzoff and my indebtedness to their numerous valuable and stimulating publications.

The bibliographies, apart from the references from which figures were taken, are limited to the more recent or more important papers or those dealing directly with the animals dissected. The *Handbuch* mentioned above has extensive bibliographies, as do also Goodrich's valuable book *Studies on the Structure and Development of Vertebrates* and other large works listed in the bibliographies.

I have always believed that zoölogy is best studied and learned not out of books but by actual experience with and handling of material. This book tries to teach comparative vertebrate anatomy by means of real specimens, and it is to be hoped that ample material will be available during the laboratory study. On the same ground I have avoided giving any illustrations of the animals under dissection and, in general, hope that the student will learn the story of the evolution of the various vertebrate organ systems by comparing their condition in the different animals dissected rather than by looking at diagrams in books.

Thanks are expressed to all those who have called my attention to errors in need of correction; to the departments of ichthyology, herpetology, and comparative anatomy of the American Museum of Natural History, especially to Messrs. Nichols, Bogert, and Raven, for unfailing courtesies and assistance and the loan of specimens; to Dr. F. A. Beach, in charge of the department of animal behavior, for generous accommodations in his department and ever ready co-operation; to Mr. Lester Aronson for a critical reading of the last chapter; to the librarians of the splendid library of the American Museum of Natural History for their inexhaustible patience and cheerful aid in the finding of references; and to my friends for encouraging me to complete this almost hopeless task. For I confess, apologetically, that I am not a student of vertebrate anatomy; and more than once I threatened to "bog down" in the vast complexities of a vast subject which was far from my preferred field of work. My task might be described as making molehills out of mountains, of trying to cull from the immense material and literature of vertebrate anatomy a few of the more important facts and concepts. I hope I have not done too badly with them.

LIBBIE H. HYMAN

AMERICAN MUSEUM OF NATURAL HISTORY
March 1942

GENERAL DIRECTIONS

Supplies

1. Dissecting instruments necessary for the course are:
Medium-sized scalpel.
Fine scissors.
Coarse scissors.
Stout probe for dissecting.
Long, slender probe for probing.
Medium-sized forceps with straight points.
A towel and a laboratory coat or gown to protect the clothes are desirable.
Bone scissors or forceps will be provided in the laboratory.

2. Drawing materials necessary for the course are:
Drawing paper, No. 6. This paper must be stiff and hard and have a smooth surface.
Hard drawing pencil, No. 3.
Eraser.
Ruler.
Red, yellow, and blue pencils.
Pad of emery paper to sharpen the hard pencil.

3. Obtain the supplies named above and present yourself with a complete outfit at the first laboratory period. Do not handicap yourself at the start by neglecting to provide yourself with the necessary materials.

Drawings

1. All drawings **must** be made with a hard pencil on good quality drawing paper, unless otherwise specified. Colors are to be used only when specified in the directions. Shading, crosshatching, etc., are undesirable and are to be avoided. Drawings made otherwise than as here specified will not be accepted.

2. Drawings are to be line drawings only—that is, only the outlines of the structures are to be drawn. Every line must represent a structure actually present on the specimen. Lines must be smooth and clean. Correct proportions are of the utmost importance and are to be obtained by use of a ruler. In making a drawing it is best to outline the drawing first with very light lines, correcting these until accurate appearance and proportions are obtained. Then erase the light lines until they are barely visible and go over them with a well-sharpened pencil, making the final lines firm and clear.

3. Drawings are not to be diagrammatized unless so directed in the manual. Many students do not seem to understand the difference between a diagram and a drawing. As an illustration, Figure 14 in this book is a diagram and Figure 13 is a drawing of the upper half of the same structures represented in the diagram. The latter shows what the object actually looks like; the former is for purposes of explanation only.

4. All drawings must be made **directly from the object** with the object before the student and completed in the laboratory. The making of rough sketches in the laboratory to be "improved" elsewhere is unscientific, inaccurate, and absolutely not permitted.

5. Remember that the prime requisite of a drawing is accuracy. A drawing is for the instructor a record of what you have actually seen upon your specimen. If you have not dissected the structures called for, then it is obvious that you cannot draw them accurately. Poor laboratory work invariably reflects itself in the quality of the drawings.

6. Drawings must contain all of the details mentioned in the manual. If, after honest effort and with the aid of the laboratory assistants, you are unable to identify certain structures, omit them from the drawing and make a note to the effect that you were unable to find them. An unreasonable amount of time should not be spent in locating small or unimportant details.

7. All drawings must be thoroughly labeled. Every drawing must be completely labeled regardless of whether the same structures have already been labeled in some preceding drawing. Labels are to be written or printed in hard pencil parallel to the top and bottom edges of the page, and lines drawn with a ruler from the labels to the parts indicated.

8. Draw on the right-hand surface of the page only.

9. Remember that the laboratory instructors are familiar with all of the figures in the various textbooks and that undue resemblance between your drawings and such figures will reflect upon your honesty and raise a suspicion that you have not been exerting yourself in the laboratory.

10. The drawings will be called in at intervals. The dates on which they are due will be announced in advance by the laboratory instructors.

Notes and Quizzes

1. No notes are required in this course. The notebooks will consist of drawings only.

2. Oral and written quizzes upon the subject matter of the laboratory work are to be expected at any time. These quizzes will deal with the anatomy of the animals you are dissecting and with comparative anatomy. You will be expected to know thoroughly the animals and materials which you dissect and study in the laboratory, and to be able to compare them with one another, stating resemblances and differences. You will be required to exhibit your dissections and to be able to identify the structures present on the dissections.

3. An important quiz will follow the completion of each section of the laboratory work and will deal with that section.

4. Reading in the textbooks of comparative anatomy is expected as a part of the laboratory work. Quizzes will include material in such textbooks.

Dissection

1. Dissection does not consist in cutting an animal to pieces. Dissection consists in separating the parts of an animal so that they are more clearly visible, leaving the parts as intact as practicable.

2. In dissecting an animal very little cutting is required. Cleaning away the connective tissue which binds together and conceals structures is the chief process in dissection. In doing this, use blunt instruments, as the probe, forceps, or fingers. Avoid the use of scalpel and scissors. You will probably cut something you will need later on. In short, do not cut; separate the parts.

3. Have the animal firmly fastened. Smaller animals are generally pinned to wax-bottomed dissecting pans. Larger animals, such as are used in the greater part of this course, are tied to screw eyes in the corners of the dissecting pan. Put the particular part you are dissecting on a stretch.

4. Do your own dissecting. Do not watch somebody else do it. Begin at the most easily accessible point of the system you are studying and follow out your structure, cleaning away the tissues that conceal it.

5. Exercise patience and care. Clean the structures by small portions.

6. Follow the directions precisely. Do not cut anything or remove anything unless specifically directed to do so.

7. Your laboratory grade is partly determined by the kind of dissections you make.

Materials

1. But one specimen of each animal is allowed to each student. Each student will be given the necessary specimens and will retain them to the end of the course. Do not discard any animal until the manual so directs.

2. The smaller materials which are provided for the class as a whole should be returned to the bottles or jars from which they came as soon as you have finished studying them.

3. The larger specimens will be kept in large cans. Each table will be allotted the necessary number of cans. Students will attach tags bearing their names to their specimens and will keep them in the cans when they are not in use.

4. Specimens must always be kept moist and must never be allowed to dry up, as this ruins them for dissection. Do not go away and leave your specimens out on the table. When ready to leave the laboratory, wrap the animals in moistened cheesecloth provided in the laboratory and put them into the cans. See that the cans are always covered.

5. Students who, through their own carelessness, render their specimens unfit for further dissection will have to pay for new specimens.

6. The skeletal material provided in this course is expensive. Handle it with care. Be particularly careful with skeletons preserved in fluid.

TABLE OF CONTENTS

I. GENERAL CONSIDERATIONS ON ANIMAL FORM

A. DESCRIPTIVE TERMS

The body of vertebrates is carried in the horizontal position, and the various surfaces are designated as follows with reference to this position:

Dorsal—the back or upper side (posterior in human anatomy).

Ventral—the under side (anterior in human anatomy).

Lateral—the sides, right and left.

Anterior, cephalic, or *cranial*—the head end of the animal (superior in human anatomy).

Posterior or *caudal*—the tail end of the animal (inferior in human anatomy).

Median—the middle.

Adverbs made by substituting *d* for the terminal letter of these words mean "in the direction of," as *craniad,* toward the head; *caudad,* toward the tail; etc.

Other descriptive terms are:

Central—the part of a system nearest the middle of the animal.

Peripheral—the part nearest the surface.

Proximal—near the main mass of the body, as the thigh.

Distal—away from the main mass of the body, as the toes.

Superficial—on or near the surface.

Deep—some distance below the surface.

Superior—above.

Inferior—below.

B. PLANES AND AXES

The structures of most animals are arranged symmetrically with reference to certain imaginary planes and axes.

1. The *median* plane is a vertical longitudinal plane passing from head to tail through the center of the body from dorsal to ventral surfaces. It divides the body into two nearly identical halves, right and left.

2. The *sagittal* plane or section is any vertical longitudinal plane through the body—that is, the median plane or any plane parallel to it. Sagittal planes other than the median plane are sometimes designated as *parasagittal* to avoid misunderstanding.

3. The *horizontal* or *frontal* plane or section is any horizontal longitudinal section through the body—that is, all planes at right angles to the median plane and parallel to the dorsal and ventral surfaces.

4. The *transverse* or *cross* plane or section cuts vertically across the body at right angles to the sagittal and horizontal planes.

5. The *longitudinal* or *anteroposterior* axis is a line in the median sagittal plane extending from head to tail; a *sagittal* or *dorsoventral* axis is any line in the median sagittal plane extending from dorsal to ventral surfaces; a *transverse* or *mediolateral* axis is any line in the transverse plane running from side to side.

C. SYMMETRY

The forms of symmetrical animals are dependent upon the arrangement of their parts with regard to the foregoing axes and planes. There are four fundamental types of animal symmetry—*spherical, radial, biradial,* and *bilateral.* Since all vertebrates are bilaterally symmetrical, the other types of symmetry will not be considered here.

Bilateral symmetry.—The parts of a bilaterally symmetrical animal are arranged symmetrically with reference to three axes—the longitudinal, transverse, and sagittal axes; the two ends of the sagittal axis in any given cross-section are unlike. There is but one plane of symmetry in such an animal—that plane which passes through the longitudinal and sagittal axes—namely, the *median sagittal plane.* It divides the animal into approximately identical right and left halves, which are mirror images of each other. The structures of vertebrates are either cut in half by the median sagittal plane, in which case they are spoken of as *unpaired* structures, or they are placed symmetrically on each side of this plane, equidistant from it, in which case they are *paired* structures. The digestive tract is the only system which does not exhibit a symmetrical relation to the median plane in the adult, although it, too, is bilaterally symmetrical in early embryonic stages.

D. METAMERISM OR SEGMENTATION

Metamerism, or *segmentation,* is the regular repetition of body parts along the anteroposterior axis. The body of segmented animals is composed of a longitudinal series of divisions in each of which all or most of the body systems are represented, either by entire paired organs or parts or by a portion of the median unpaired structures. Each such division of the body is termed a *metamere, segment,* or *somite.* The anterior and posterior boundaries of each segment may or may not be marked externally by a constriction of the body wall. In the former case the animal is said to exhibit both *external* and *internal* segmentation; in the latter case internal metamerism alone is present.

An ideal segmented animal would consist of a series of identical segments; but no such animal exists, since the head and terminal segments must necessarily differ, if only slightly, from the other segments. However, the more primitive ringed worms, such as *Nereis,* closely approach the ideal. The segmentation of the animal body into nearly like segments is spoken of as *homonomous.* The majority of segmented animals have *heteronomous* segmentation, in which the various segments differ from each other to a greater or less extent. In the evolution of segmented animals there has been a continuous progression from the homonomous to the extreme heteronomous condition. Homonomous segmentation is a primitive and generalized state in which the various segments are more or less independent and capable of performing all necessary functions. As heteronomy progresses, the segments become unlike, and different body regions specialize in the performance of different functions. Each segment is then no longer capable of carrying on all functions but becomes dependent upon the other segments, with a resulting unification and harmony of performance lacking in homonomously segmented forms. The heteronomous condition is derived from the homonomous through a number of different processes, such as loss of segments, fusion of adjacent segments, enlargement or reduction of segments, loss of organs or parts from some segments with their retention in others, structural changes among the repeated organs or parts so that those of different segments become unlike, etc.

The segmented groups of animals are the annelids, the arthropods, and the chordates. Whereas relationship between annelids and arthropods is generally acknowledged, most zoölogists are now of the opinion that the chordates stem from a quite different line of evolution and hence that segmentation has arisen in them independently of the metamerism of the annelid-arthropod line. Metamerism in the chordate line apparently began in the hemichordates, where some structures, particularly the gonads, show evidence of serial repetition, a condition termed *pseudometamerism.* Segmentation then progressed among the lower chordates, reaching its climax in *Amphioxus*, and retrogressed among the vertebrates. It appears probable that the primitive chordates never had as fully developed segmentation as do annelids and that no chordate was ever externally segmented. Segmentation in chordates concerns primarily the musculature; that of the endoskeleton and nervous system seems to be secondary to that of the muscles. Among vertebrates there is a rapid loss of segmentation, which can be followed to some extent during embryology. Embryonic stages are much more obviously segmented and more homonomous than adults, and heteronomy progresses during the development of a given vertebrate. Adult vertebrates are thus internally and heteronomously segmented animals.

E. CEPHALIZATION

In the evolution of animals there is a pronounced tendency for the anterior end of the body to become more and more distinctly separated and differentiated from the rest of the body as a *head*. This differentiation of the head consists chiefly of the localization within the head of the main part of the nervous system—i.e., the brain—and of the most important sense organs. Since the brain and the sense organs control, to a very large degree, the activities and responses of the rest of the body, the head thus becomes the dominant part of the organism. This centralization or localization of nervous structures and functions in the head with accompanying dominance of the head is called *cephalization*. Cephalization is more and more marked the higher one ascends in the animal kingdom, and is particularly prominent as a structural and functional feature of the vertebrates.

In segmented animals the advance in cephalization is correlated with the progression of the heteronomous condition. Heteronomy, in fact, appears first in the head region and gradually progresses posteriorly. The anterior end thus retains the least, and the posterior end the most, resemblance to the original homonomous condition. This results in an illusion of a retreat of certain systems toward the posterior regions of the body, whereas the situation in reality arises from the fact that these systems have disappeared from the anterior segments and are retained in the posterior segments. In the case of certain vertebrate organs, as the heart, a real posterior descent occurs during the evolution of the vertebrates. In the vertebrates, as in other heteronomously segmented animals, the head is produced through the fusion of a certain number of the most anterior segments with a loss of some segments or of parts of segments and the disappearance from these head segments of nearly all systems except the nervous system. As cephalization progresses, the head appropriates more and more of the adjacent segments, incorporating them into its structure, so that in general it may be said that the higher the degree of cephalization the greater is the number of segments composing the head. In advanced cephalization, such as is possessed by vertebrates, it is very difficult—indeed, almost impossible—to decipher the number and boundaries of the segments which originally went into the composition of the head; in fact, the problem of the segmentation of the vertebrate head has not been completely solved, although it has received the attention of the foremost vertebrate anatomists.

The vertebrates are, then, animals characterized by the possession of bilateral symmetry, internal and markedly heteronomous segmentation, and a high degree of cephalization. The details of their structure are understandable only in the light of these three broad anatomical conditions.

F. HOMOLOGY AND ANALOGY

The question of the relation of form (anatomy) and function has occupied the attention of comparative anatomists since the dawn of modern zoölogy at the end of the eighteenth century. This question involves the concepts of *homology* and *analogy*, which are understandable only in terms of the principle of evolution. Homology is intrinsic similarity indicative of a common evolutionary origin. Homologous structures may seem unlike superficially but can be proved to be equivalent by any or all of the following criteria: similarity of anatomical construction, similar topographical relations to the animal body, similar course of embryonic development, and similarity or identity of specific physiological function or mechanism. A familiar example of homology is the wing of the bird, the flipper of the seal, and the foreleg of the cat; investigation shows that they have a similar arrangement of bones and muscles, have the same positional relation to the body, develop in the same way from a similar primordium, and work physiologically by the same mechanism. The necessity of including specific physiological function or mechanism as a criterion will become apparent when one considers such structures as the endocrine glands. Although these have a similar anatomical (histological) structure throughout the vertebrates, they frequently differ in position and in the details of their embryonic origin in various vertebrates; but their specific function, which is the ultimate test of their homology, remains the same throughout. Analogy is similarity of general function or of superficial appearance not associated with similarity of intrinsic anatomical construction or of embryonic origin and development. Thus fish and snakes are covered with scales for protective purposes (similar general function), but investigation of the two types of scales shows that they are histologically dissimilar and differ in their mode of embryonic origin. Analogous structures also differ in precise functional mechanism; thus an insect leg and a cat leg serve the same broad general function, that of walking, but the mechanism of walking is quite different in the two cases. When analogous structures present striking similarity of appearance, this is termed *convergence* or *parallelism*. Such correspondences are usually associated with living in a common environment, i.e., they are "adaptations." Not only parts of animals but whole animals may come to resemble each other markedly through living in the same environment, as porpoises and fish. On the other hand, animals closely related by descent may differ greatly in general appearance after long sojourn in different environments, as seals and cats. This phenomenon is termed *divergence.*

II. THE PHYLUM CHORDATA

A. THE CHARACTERISTICS OF THE CHORDATES

While the vertebrates comprise the greater part of the phylum Chordata, two[1] small groups of animals are united with them in this phylum because they possess certain characteristics in common with the vertebrates. These characteristics are:

1. The wall of the pharynx of the embryo or adult is pierced by openings, the *gill slits*, originally probably a food-catching device.

2. A *notochord* is present in embryo or adult. The notochord is a rod lying dorsal to the intestine, extending from anterior to posterior end, and serving as a skeletal support. In vertebrates the notochord is partially or wholly replaced by the skull and vertebral column.

3. The central nervous system is *hollow* (in the tunicates in the embryo only), containing a single continuous cavity, and is situated entirely on the *dorsal* side of the body. In the invertebrates the central nervous system is always solid and lies mainly ventral in the body.

B. THE CHARACTERISTICS OF THE VERTEBRATES

The morphological characters of the vertebrates are the following: animals with bilateral symmetry, internal heteronomous segmentation, and cephalization; with generally two pairs of paired jointed locomotor appendages, in the form of fins or limbs, and sometimes with unpaired appendages in addition; skin separable from the rest of the body wall and commonly producing protective structures, such as scales, feathers, hair, etc., cellular in nature; muscle layer of the body wall decidedly metameric in arrangement; with an internal skeleton, of cartilage or bone, consisting of a skull and gill supports in the head, vertebral column, ribs, and breastbone in the body, and supports for the appendages; vertebral column highly metameric, composed of successive rings around the notochord; central nervous system consisting of a brain, much enlarged, within the skull, and a spinal cord within the vertebral column; nerves highly metameric in arrangement; head with three pairs of sense organs, eyes, ears, and nose; digestive tract giving rise by outgrowth to two digestive glands, the liver and the pancreas; pharynx intimately connected with the respiratory system, either opening to the exterior by openings, the gill slits, in the walls of which the gills are borne, or giving rise by outgrowth to the lungs; heart always ventral in the body; circulatory system closed, always with a median dorsal artery, the aorta, and with one or two portal systems; genital and excretory systems closely related, the excretory ducts generally serving as genital ducts; excretory and genital ducts opening in common with the intestine into a cloaca, or opening separately near the anus; with a well-developed coelom, never segmented, and divided in the adult into two or four compartments; viscera supported by mesenteries.

C. CLASSIFICATION OF THE CHORDATES

Since the enunciation of the principle of evolution by Darwin in 1859 it is known that all animals are related to each other by descent and have become differentiated into distinct groups only through a gradual process of change. Theoretically, therefore, it would be im-

[1] It is customary to include a third group, the Hemichordata (also called Enteropneusta and Branchiotremata), with the vertebrates as a subphylum of the Chordata. Because of their phylogenetic importance, the Hemichordata are treated in the next chapter but, for reasons there given, are regarded as an invertebrate phylum and hence removed from the Chordata.

possible to define any group of animals in such a way as to exclude its nearest relatives. It becomes practical to make such definitions only because of the extinction of intermediate forms in the course of the ages. The poorer the fossil record, the easier it is to erect distinctions between existing animal groups. In the case of the vertebrates where excellent series of fossils have been unearthed, including satisfactory transitional types between the principal groups, the construction of mutually exclusive definitions is very difficult, if not impossible, when the fossil forms are taken into account. Hence the following definitions must not be taken too literally.

Phylum Chordata

Subphylum I. Urochordata or Tunicata, the tunicates or sea squirts.

Subphylum II. Cephalochordata or Acrania, *Amphioxus* and its allies.

Subphylum III. Vertebrata or Craniata, all chordates with an endoskeleton of cartilage or bone or a combination of these.[2]

A. Fish and Fishlike Animals

Cold-blooded aquatic vertebrates with fins as locomotory organs (or, primitively, no locomotory organs) and gills as respiratory organs; vertebral axis ending in a vertical fin; heart with one atrium and one ventricle; embryos without membranes (except the yolk sac).

Class I. Agnatha. Fishlike animals without jaws and with no or poorly developed fins.

Orders 1–3. Osteostraci, Anaspida, Heterostraci, extinct forms commonly called ostracoderms, having an armor of heavy plates.

Order 4. Cyclostomata, the round-mouthed fishes. Skin naked, slimy; round jawless mouth, often employed as a sucker; without paired fins; with a row of from six to fourteen external or concealed gill slits; single median nasal sac and nasohypophyseal aperture and canal.

Suborder 1. Myxinoidea, the hagfishes. Nasohypophyseal aperture terminal, its canal opening into the pharynx, branchial basket vestigial. *Bdellostoma, Myxine.*

Suborder 2. Petromyzontia, the lampreys. Nasohypophyseal aperture dorsal, leading into a blind sac, branchial basket present. *Petromyzon, Lampetra.*

Class II. Placodermi or Aphetohyoidea. Extinct fish with jaws and apparently a full-sized hyoid gill slit; with pectoral or both pectoral and pelvic paired fins; with an armor of bony scales and cartilaginous skeleton, with some ossification. Orders Acanthodii, Arthrodira, Antiarchi, Petalichthyida, Rhenanida, and Palaeospondylia.

Class III. Chondrichthyes, the cartilaginous fishes. Sharklike fishes with jaws and paired fins; skeleton wholly or largely cartilaginous; no

[2] The classification is taken from the publications of Romer; extinct groups in small type.

membrane bones; hyoid gill slit reduced; exoskeleton of small scales homologous with teeth; without lungs or air bladder.

Orders 1–2. Cladoselachii, Pleuracanthodii, extinct sharks.

Order 3. Elasmobranchii or Euselachii, the sharks, skates, and rays. Skeleton wholly cartilaginous, membrane bones lacking, hyostylic type of upper-jaw suspension; armature reduced to small denticles imbedded in the skin; with a spiracle and exposed gill slits; with a spiral valve and conus arteriosus; pelvic fins of male with claspers. *Heptanchus, Squalus, Mustelus, Scyllium, Raja, Squatina, Rhinobatis, Torpedo*, etc.[3]

Order 4. Holocephali, the chimaeras. Fish of curious aspect, with a mostly naked skin; spiracle absent, gill slits concealed under a nonbony skin flap; immovable holostylic type of jaw suspension; notochord large, vertebrae poorly developed; flat bony plates in place of teeth; with spiral valve, conus arteriosus, and claspers. *Chimaera.*

Class IV. Osteichthyes, the bony fishes. Skeleton partly or largely ossified; chondrocranium, jaws, and pectoral girdle incased in dermal bones; body clothed with scales or rhomboid plates; gills covered by a bony operculum; with an air bladder which may function as a lung; nostrils double; no claspers.

Subclass 1. Choanichthyes. Paired fins with a supporting lobe; nostrils opening into the roof of the oral cavity; well-developed air bladder used as lung; no typical branchiostegal rays; clothed with rounded or rhomboid scales.

Superorder 1. Crossopterygii, the lobe-finned fishes.[4] Paired fins with a rounded basal lobe (Fig. 1*A*); two dorsal fins, jaw suspension hyostylic; with a spiracle. *Latimeria, Osteolepis.*

Superorder 2. Dipnoi, the lungfishes. Paired fins with an elongated jointed axis bearing side branches; jaw suspension autostylic; spiracle absent; with tooth plates; with spiral valve and conus arteriosus. *Epiceratodus, Protopterus, Lepidosiren.*

[3] The group Elasmobranchii as here defined is identical with the group Selachii, formerly considered a suborder under Elasmobranchii. Hence, the fishes of this group may be referred to either as selachians or as elasmobranchs. But some consider Elasmobranchii synonymous with Chondrichthyes.

[4] The Crossopterygii were supposed to have been extinct for at least 60,000,000 years, but in 1939 a living crossopterygian (Fig. 1*A*), belonging to the group known as coelacanths, was captured by fishermen off the east coast of South Africa. Unfortunately, the soft parts were too spoiled to be preserved. The fish was named *Latimeria.*

Subclass 2. Actinopterygii or Teleostomi, the ray-finned fishes. Paired fins without axis or basal lobe (Fig. 1*B*), fin rays attached directly to the girdles (except Polypterini); one dorsal fin; branchiostegal rays usually present; no internal openings of nasal sacs; without cloaca; jaw suspension hyostylic.

Order 1. Chondrostei. Endoskeleton largely cartilaginous; with spiral valve and conus arteriosus; spiracle usually present; clothed with ganoid scales; recent forms lack typical branchiostegal rays.

Suborder 1. Polypterini.[5] Pectoral fins with a basal lobe; with vertebral centra; with ventral paired lungs opening into the ventral wall of the esophagus. *Polypterus*, *Calamoichthys*.

Suborder 2. Acipenseroidea. Fins not lobed; with a long snout or rostrum; scales more or less degenerate; notochord large, persistent, centra wanting; lung opening dorsal. *Acipenser*, *Polyodon*.

Two extinct suborders.

Order 2. Holostei. Skeleton moderately ossified; vertebral centra usually fairly well developed; scales ganoid to cycloid; pelvic fins abdominal; spiracle wanting; air bladder single, dorsal, connected to dorsal side of esophagus; with conus arteriosus; spiral valve vestigial.

Suborder 1. Lepidosteoidei. Typically with thick ganoid scales of rhombic shape; tail of shortened heterocercal type. Mostly extinct, *Lepidosteus*.

Suborder 2. Amioidea. Scales tending to thin and round to cycloid type; tail tending to homocercal type. Mostly extinct, *Amia*.

The groups Chondrostei and Holostei are commonly referred to as the *ganoid* fishes.

Order 3. Teleostei, the typical bony fishes. Skeleton almost completely ossified; vertebral centra complete; scales thin, of cycloid or ctenoid type; tail homocercal; pelvic fins often displaced forward; no spiracle; air bladder single, dorsal, often ductless; conus arteriosus vestigial; without spiral valve. All ordinary fish; known as bony fishes.

[5] The Polypterini were formerly considered crossopterygians because of the lobed pectoral fins, but in the skeleton of these lobes and in many other characters they are shown to be somewhat aberrant Actinopterygii.

B. Vertebrates with Limbs (Tetrapods)

Terrestrial or aquatic vertebrates with limbs as locomotory appendages (may be lost) and lungs as respiratory organ (also may be lost); a few with gills in addition; vertebral column terminating in a tail, sometimes absent, or in a horizontal fin; heart with two atria; with internal nares.

Class V. Amphibia. Cold-blooded aquatic or terrestrial vertebrates, naked or with bony dermal scales; with lungs as respiratory organs,

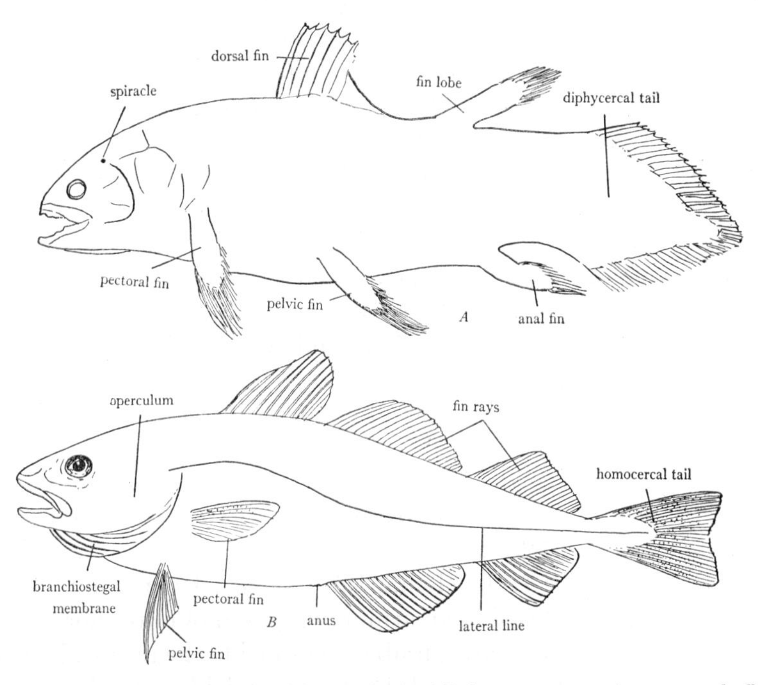

Fig. 1.—Crossopterygian and teleost fish compared. *A*, living crossopterygian captured off the coast of South Africa (sketched from a photograph by J. L. Smith, 1940). *B*, cod, branchiostegal membrane somewhat pulled out. Scales omitted from both fish.

some forms with gills in addition; usually with one sacral vertebra; heart with undivided ventricle; embryos without membranes.

Order 1. Labyrinthodontia. Extinct amphibians, often called Stegocephali, with an armor of small, mostly ventral, scales, and primitive vertebrae, made of separate pieces, without typical centra; important as the first tetrapods. *Eryops.*

Order 2. Lepospondyli. Extinct amphibians with spool-shaped centra.

Order 3. Caudata or Urodela, newts and salamanders. Naked, tailed amphibians, with or without external gills; with two pairs of limbs; vertebrae with solid centra; two occipital condyles. *Siren, Triton, Necturus, Salaman-*

dra, Ambystoma, Triturus, Plethodon, Cryptobranchus, Amphiuma, etc.

Order 4. Salientia or Anura, frogs and toads. Naked, tailless amphibians, without external gills; with typical vertebrae; caudal vertebrae fused into one elongated piece; with two pairs of limbs, tarsus much lengthened; two occipital condyles. *Discoglossus, Pipa, Rana, Hyla, Xenopus, Bufo,* etc.

Order 5. Apoda or Gymnophiona, the coecilians or blind-worms. Vermiform amphibians without limbs or limb girdles; with small scales concealed in skin folds; vertebrae amphicoelous with persistent notochord.

Class VI. Reptilia, the reptiles. Cold-blooded aquatic or terrestrial tetrapods, naked, scaly, or with bony plates, breathing exclusively by lungs; skeleton completely ossified with typical vertebrae; recent forms with one occipital condyle and two sacral vertebrae (to ten in extinct forms); heart with sinus venosus; ventricle partly or completely divided by a partition; embryo provided with embryonic membranes (amnion and allantois).

Subclass 1. Anapsida. Skull completely roofed without temporal vacuities.

Order 1. Cotylosauria. Primitive extinct reptiles, closely resembling the most primitive extinct amphibians and of great phylogenetic importance as links with these. *Seymouria.*

Order 2. Chelonia, the turtles. Trunk short and broad, inclosed in an armor composed of outer horny and inner bony plates; limb girdles inside the armor; jaws without teeth, incased in horny shields. *Chelydra, Testudo, Emys, Chrysemys, Trionyx, Thalassochelys, Chelone.*

Subclass 2. Ichthyopterygia with Order 3, Ichthyosauria, extinct.

Sublcass 3. Synaptosauria with Order 4, Sauropterygia (plesiosaurs, etc.), extinct.

Subclass 4. Lepidosauria. Skull is diapsid with two lateral temporal vacuities or descended from this type; skin with horny scales.

Order 5. Eosuchia, extinct. Primitive diapsid reptiles.

Order 6. Rhynchocephalia. Long-tailed lizard-like reptiles, with weak limbs; skull with two vacuities; teeth fused to jaw, without sockets; vertebrae amphicoelous with persistent intercentra; with abdominal ribs. Mostly extinct, *Sphenodon.*

Order 7. Squamata, lizards and snakes. Long-tailed reptiles covered with horny scales; skull has lost one or both vacuities; quadrate movable; teeth set in sockets; vertebrae usually procoelous; abdominal ribs wanting or rudimentary.

Suborder 1. Lacertilia, the lizards. Limbs usually present; limb girdles always present, reduced in limbless forms; skull with one temporal vacuity (upper one); with mandibular symphysis.

Suborder 2. Ophidia, the snakes. Limbs absent; limb girdles also absent, except remnants of pelvis in a few cases; skull has lost both temporal vacuities; no mandibular symphysis, ligament permits wide gape.

Subclass 5. Archosauria. Skull diapsid with two temporal openings; tending to bipedal gait with associated changes; also includes flying forms. Birds stem from this subclass.

Order 8. Thecodontia, extinct.

Order 9. Crocodilia. Large aquatic or amphibious reptiles with bony plates underneath horny skin thickenings; quadrate fixed; formation of secondary palate displaces internal nares far posteriorly; teeth in sockets; vertebral centra various; ribs double-headed; abdominal ribs present; heart with two ventricles. *Alligator*, *Caiman*, *Crocodilus*.

Orders 10–12. Pterosauria, Saurischia, Ornithischia, extinct; various types of dinosaurs belong here.

Subclass 6. Synapsida. With a single (lower) lateral temporal vacuity; important as showing gradations of skeletal structures in the mammalian direction.

Orders 13, 14. Pelycosauria, Therapsida, extinct. The Therapsida lead to mammals.

Class VII. Aves, birds. Warm-blooded oviparous tetrapods with feathers; one occipital condyle; quadrate free; forelimbs modified into wings; heart with two ventricles, no sinus venosus; embryo with membranes.

Subclass 1. Archaeornithes. Extinct birds with long tail bordered by feathers; metacarpals separate; with teeth. *Archaeopteryx*, *Archaeornis*.

Subclass 2. Neornithes. Tail feathers arranged in a fanlike manner around tail stump; metacarpals fused; teeth present only in extinct forms. All present birds.

Class VIII. Mammalia, the mammals. Warm-blooded, nearly always viviparous aquatic or terrestrial tetrapods clothed with hair, or naked with sparse hairs; skull with zygomatic arch, secondary palate, and two occipital condyles; teeth in sockets; lower jaw of one bone, articulating with the squamosal; quadrate reduced to a middle-ear ossicle; heart as in birds; with muscular diaphragm; young nourished by milk secreted by mammary glands; embryo with membranes.[6]

Subclass 1. Prototheria. Oviparous mammals; mammary glands without nipples; pectoral girdle with separate precoracoid, coracoid, and interclavicle; scapula with spine; oviducts separate; with cloaca.

Order 1. Monotremata, the monotremes. With the characters of the class. *Ornithorhynchus*, the duckbill, and *Echidna*, the spiny anteater.

Subclass 2. Metatheria or Didelphia or Marsupialia. Viviparous mammals with an abdominal skin pouch (marsupium) supported by two marsupial bones, in which the very immature young are kept; teats opening into marsupium; usually without true allantoic placenta; precoracoid and interclavicle absent; coracoid reduced; scapula with spine; clavicle present; four molars on each side; shallow or no cloaca; brain smooth; vagina double.

Order 2. Marsupialia, the marsupials. With the characters of the class. Kangaroos, rat kangaroos, wallabies, wombats, opossums, etc.

Subclass 3. Eutheria or Monodelphia or Placentalia, the placental mammals. Viviparous mammals with an allantoic placenta; without marsupium or marsupial bones; shoulder girdle as in marsupials; mostly three molars on each side; one vagina; no cloaca.

Order 3. Insectivora, the insectivores. Small nocturnal or burrowing mammals with plantigrade clawed feet and often elongated snout; dentition primitive, teeth with sharp cusps; auditory region incompletely ossified, without bulla; uterus bicornuate; with clavicles; brain small, smooth. Moles, shrews, hedgehogs.

Order 4. Chiroptera, the bats. Flying mammals with wing membranes supported by the greatly elongated

[6] Because of the numerous changes and uncertainties in the disposition of the groups of extinct mammals, it has seemed best to omit them, especially as there is little necessity of referring to them in a book of this kind.

metacarpals and fingers; thumb and hind feet with claws; with clavicles; brain smooth; teats thoracic.

Order 5. Dermoptera, the flying lemurs. With a wing membrane as in bats, but fingers not elongated. *Cynocephalus* (*Galeopithecus*).

Order 6. Primates, the primates. Terrestrial or arboreal hairy mammals with five digits on each foot, provided with flat nails; gait plantigrade; forefeet grasping; orbital and temporal depressions partly or completely separated by a bony ridge; with clavicles; stomach mostly simple; teats abdominal, pectoral, or axillary; brain highly convoluted with very large cerebral hemispheres. Lemurs, monkeys, apes, man.

Order 7. Carnivora, the carnivores. Small to large carnivorous mammals; canine teeth prominent; well-developed auditory region usually with expanded bulla; clavicles reduced or absent; stomach simple; brain convoluted.

Suborder 1. Fissipedia, the typical carnivores. Terrestrial carnivores with walking feet, strongly clawed; gait plantigrade to digitigrade; six incisors; anterior cheek teeth sharp, cutting, culminating in a special carnassial tooth, behind which molars are broad or reduced. Cats, lions, tigers, leopards, dogs, wolves, civets, foxes, hyenas, raccoons, bears, otters, weasels, etc.

Suborder 2. Pinnipedia, aquatic carnivores. Aquatic mammals with webbed feet modified into swimming flippers; nails mostly reduced; cheek teeth alike, no carnassial. Seals, sea lions, walruses.

Order 8. Perissodactyla, odd-toed ungulates. Large hoofed herbivorous mammals; third digit forming limb axis; other digits smaller or reduced; gait unguligrade; cheek teeth broad, lophodont (with grinding ridges); clavicles absent; stomach simple; no gall bladder; brain convoluted; teats inguinal.

Suborder 1. Equoidea. Recent forms with only one functional digit (third). Horses, asses, zebras.

Suborder 2. Tapiroidea. Four toes in front, three behind; dentition complete, no horns. Tapirs.

Suborder 3. Rhinocerotoidea. Three or four toes in front, three behind; canines and incisors reduced or wanting; with one or two median horns of epidermal nature. Rhinoceroses.

Order 9. Artiodactyla, even-toed ungulates. Large hoofed herbivorous mammals; third and fourth digits equal and symmetrical, limb axis passing between them; other digits reduced or wanting; broad grinding cheek teeth, bunodont (cusps rounded) or selenodont (cusps crescentic); no clavicles; no gall bladder; brain convoluted; teats inguinal or abdominal.

Suborder 1. Suina. Dentition complete; molars bunodont; feet four-toed; stomach simple to two-chambered; no horns. Pigs, peccaries, hippopotamuses.

Suborder 2. Tylopoda. One upper incisor retained on each side; with ruminating habit; stomach somewhat complex, with water cells. Camels, llamas, alpacas.

Suborder 3. Pecora. Upper incisors wanting; molars selenodont; feet two-toed, rudimentary second and fifth toes rarely present; stomach complex, without water cells, with three or four compartments; with ruminating habit (food after being swallowed is regurgitated and chewed as "cud"); mostly with paired horns with bony cores borne on the frontal bones. Deer, antelopes, cattle, sheep, goats, giraffes, etc.

Order 10. Cetacea, cetaceans. Moderate to gigantic naked fishlike aquatic mammals without external neck; with broad horizontal tail fin (fluke) and often a single dorsal fin; these fins lack skeletal support; hairs limited to few on muzzle; nostrils set far back; forelimbs modified into swimming paddles; hind limbs and girdle vestigial; no clavicles; stomach complex; brain convoluted; teats inguinal. Whales, dolphins, porpoises.

Suborder 1. Odontocoeti, toothed cetaceans. With teeth. Dolphins, porpoises, and some whales.

Suborder 2. Mystacoeti, whalebone whales. Whales without teeth and instead with whalebone, horny fringes hanging along the edge of the upper jaw.

Order 11. Proboscidea, the elephants. Very large herbivores with scanty hair; nose and upper lip drawn out into a proboscis (trunk); two or four incisors enlarged into long tusks; five digits; nails hooflike; legs straight without knee and elbow bend; no clavicles; stomach simple; brain convoluted; teats pectoral.

Order 12. Sirenia, the sea cows. Large aquatic mammals with few hairs; shape seal-like with short neck; forelimbs modified into flippers; hind limbs absent; pelvic girdle vestigial; flattened horizontal tail fin; no clavicles; stomach complex; teats pectoral. Manatee, dugong.

Order 13. Hyracoidea, the coneys. Small, short-tailed herbivorous mammals, with rodent-like upper incisors; ungulate type of molars; gait plantigrade; toes with flat nails, four in front, three behind; no clavicles. Hyrax.

Order 14. Rodentia, the rodents. Small to moderate furry herbivorous mammals with clawed plantigrade feet; with one pair of upper incisors; incisors chisel-like, growing throughout life, with enamel on their front faces only; canines lacking, hence a long space (diastema) between incisors and grinding cheek teeth; with a large caecum; clavicles usually present; brain smooth or slightly convoluted. Squirrels, marmots, prairie dogs, beavers, rats, mice, guinea pigs, agoutis, porcupines, etc.

Order 15. Xenarthra, the edentates. Small to moderate terrestrial or arboreal mammals with strongly clawed feet; teeth without enamel, often reduced or wanting; with clavicles; some have armor of horny and bony scutes, with hair correspondingly reduced; brain smooth. Sloths, armadillos, anteaters.

Order 16. Pholidota, pangolins. Head, body, and tail covered with imbricated horny scales; teeth absent; no clavicles or zygomatic arch; very long tongue. Scaly anteaters or pangolins (*Manis*).

Order 17. Tubulidentata, aardvarks. Scantily haired, ant-eating mammals with a few reduced teeth with perforated dentine and without enamel; long snout and and tongue. Aardvark (*Orycteropus*).

Order 18. Lagomorpha, hares and rabbits. Furry herbivorous mammals of moderate size with short tail and large external ears; with clawed digitigrade feet; with four upper incisors, a small pair behind the larger anterior pair; otherwise similar to rodents.

Order 19. Edentata, the edentates. Small to moderate terrestrial arboreal mammals with strongly clawed feet; teeth without enamel, often reduced or wanting; with clavicles; some have armor of horny scutes and bony plates, with hair correspondingly reduced; brain smooth. Sloths, armadillos, anteaters.

Additional names used for convenience in discussing vertebrates are:

Ichthyopsida, meaning fishlike animals, includes the fishes and amphibians. *Ichthyology* is the study of fish.

Sauropsida, meaning reptile-like animals, includes the reptiles and birds.

Herpetology is the study of amphibians and reptiles.

Anamniota or *Anamnia* includes fishes and amphibians and refers to the fact that their embryos are naked; anglicized to *anamniotes*.

Amniota embraces the reptiles, birds, and mammals and refers to the fact that their embryos are covered by a membrane, the amnion; anglicized to *amniotes*.

Agnathostomata or *agnathostome*, meaning without jaws, refers to the cyclostomes and related extinct forms.

Gnathostomata or *gnathostome*, meaning jawed, refers to all other vertebrates.

The distinction between *teleostome*, meaning both ganoid and bony fishes, and *teleost*, meaning only the bony fishes (Teleostei), should be noted.

III. ESSENTIAL FEATURES OF LOWER TYPES

A. PHYLUM HEMICHORDATA, A PRECHORDATE GROUP

Although the Hemichordata are here separated from the Chordata as a distinct phylum, they possess certain prechordate features and hence are of interest in the study of chordates. The hemichordates are divisible into two groups: the wormlike Enteropneusta, or balanoglossids, and the colonial Pterobranchia, resembling bryozoans. Only the former is considered here, represented by *Saccoglossus*[1] *kowalevskyi* from the New England coast.

1. External features.—Examine the balanoglossid in a dish of water. It is an elongated wormlike animal five or six inches in length. The body is divisible into three regions: *proboscis*, *collar*, and *trunk*. The proboscis is the elongated conical structure at the anterior end, used by the animal in burrowing into the sand; it is much longer in *Saccoglossus* than in other balanoglossid genera. The collar is the band encircling the body just behind the proboscis. The proboscis is attached to the inner surface of the dorsal side of the collar by the slender *proboscis stalk*, below which the collar rim incloses the large, permanently open mouth leading into the *buccal cavity* inside the collar. The greater part of the animal consists of the more or less ruffled trunk having a midventral and middorsal longitudinal ridge. The trunk is also divisible into three regions—*branchiogenital*, *hepatic*, and *abdominal* regions—much better marked in some other species than in *S. kowalevskyi*. The branchiogenital or thoracic region forms the first part of the trunk and contains the gonads, which cause a bulge on each side known as the *genital ridges*. In some genera these project as conspicuous winglike expansions. Between the anterior portions of the genital ridges on the dorsal side will be found two longitudinal rows of small parallel slits, the *gill pores*, which are not the true gill slits (see below). The hepatic region, mostly a little shorter than the genital region, is so named because here the intestine usually bears a paired series of lateral pouches, the so-called *hepatic caeca*. These show externally as bulges in only a few balanoglossids (not in *S. kowalevskyi*, which even lacks definite outpouchings). Although these caeca apparently have some digestive function, there is no evidence that they in any way represent the vertebrate liver. The remainder of the trunk contains a simple tubular intestine opening by a terminal anus. Make a drawing of the animal from the dorsal side.

2. Salient points of the internal anatomy.—The most chordate-like structure of the balanoglossids is the gill apparatus. The gill slits, a series of vertically oriented openings, often numerous, in the wall of the anterior part of the intestine of the genital region, are oval in

[1] *Saccoglossus* Schimkewitsch 1892 has priority over *Dolichoglossus* Spengel 1893.

early stages but become U-shaped by the downgrowth from the dorsal end of each slit of a process called the *tongue bar* (Fig. 2). These tongue bars and the septa between successive gill slits are supported by M-shaped skeletal rods. Tongue bars and rods are identical with those of *Amphioxus*. The gill slits do not open directly to the exterior in most balanoglossids, but instead each leads into a pouch, the *gill sac*, which opens externally by the gill pore (Fig. 2).

From the anterior wall of the buccal cavity a hollow diverticulum projects forward into the proboscis. This diverticulum is believed by some zoölogists to represent the notochord. In some balanoglossids this is said to arise embryologically in the same manner as a notochord, by evagination from the roof of the gut; but in others this is not the case. Altogether the homology of the proboscis diverticulum with the notochord must be regarded as dubious.

The nervous system combines invertebrate and vertebrate characters. It consists of a fibrous layer throughout the entire body surface in the base of the epidermis. This is thickened in the middorsal and midventral ridges into a dorsal and a ventral nerve trunk, respectively, connected by a nerve ring at the junction of collar and trunk. The dorsal nerve continues forward into the collar, where it enlarges into the *collar cord* or central nervous system, separated from the epidermis. The collar cord sometimes has a central lumen but usually contains a number of isolated spaces. Scanty embryological researches indicate that some or all of the collar cord is formed like the vertebrate brain, by the infolding of the middorsal ectoderm.

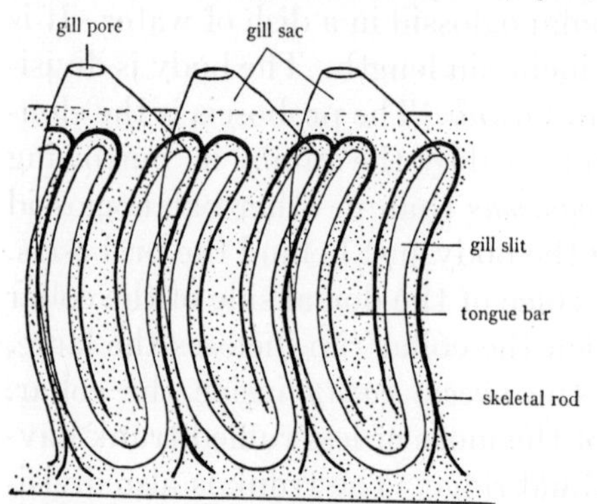

FIG. 2.—Piece of the pharyngeal wall of a balanoglossid, showing three U-shaped gill slits, three tongue bars, gill pores, gill sacs, and skeletal rods.

The blood vascular system is chiefly invertebrate in nature. There is a main ventral blood vessel in which the blood flows posteriorly and a main dorsal vessel in which it flows anteriorly, entering the rear end of the heart, situated in the posterior dorsal part of the proboscis above the proboscis diverticulum. The heart is thus dorsal to the so-called notochord. The heart resembles a half-opened vertebrate heart, consisting of a pericardial sac with a muscular ventral wall acting as propulsatory organ, and a blood channel, the *central blood sinus*, representing the heart lumen, situated between the heart sac and the roof of the diverticulum. From the central blood sinus the blood continues forward and immediately enters the presumed excretory organ, called *glomerulus*, containing a dense network of blood vessels. From the glomerulus branches supply the proboscis and a pair of vessels run downward and backward, embracing the buccal cavity and uniting to form the ventral vessel.

Of great phylogenetic significance is the close resemblance between the *tornaria* larva of balanoglossids and the *bipinnaria* larva of starfish. Other notable similarities of development, especially the number and mode of formation of the coelomic spaces, indicate affinity of balanoglossids with echinoderms. In short, the Hemichordata show as much or even greater relationship to invertebrates as to chordates and hence are best removed from the phylum Chordata.

B. THE LOWER CHORDATES, SUBPHYLUM UROCHORDATA

The adult urochordates or tunicates bear no resemblance to vertebrates, but the tadpole larva exhibits pronounced chordate characteristics.

1. External anatomy.—Place a preserved tunicate in a dish of water. It is an oval saclike creature, scarcely recognizable as an animal. The animal is in life permanently attached to rocks or other objects. The end which was attached is recognizable by its rough and irregular form and by the fragments of wood or other materials which adhere to it. The opposite end, which in the living state extends free into the water, bears two openings, the *siphons.* When the animal is gently squeezed, jets of water are observed to squirt from the siphons; hence the name sea squirt popularly applied to these animals. The upper of the two siphons is the *oral* or *incurrent* siphon and is the degenerate anterior end of the animal; the lower siphon is the *atrial* or *excurrent* siphon and represents the posterior end. The anteroposterior axis of the tunicates is bent into a U-shape. The region of the body between the two siphons is therefore dorsal, and the remaining and much larger part of the surface is ventral. The siphons are operated by circular and longitudinal muscles, which are noticeable in the more transparent species, especially after removing the test. The outer covering of the animal is a thick, tough, sometimes gelatinous membrane, the *tunic* or *test,* secreted by the underlying epithelium and attached to the inclosed body only at the siphons. In some tunicates the test is so transparent that practically all of the internal organs are visible through it.

2. Internal features.—Remove the tunic by making a slit in its base and then peeling it off, noting and severing its attachment to the siphons. Beneath the test is the soft body wall, also called *mantle,* consisting of an outer epithelium and inner connective tissue. It incloses the viscera, to which it adheres considerably.

Fasten the animal in a wax-bottomed dissecting pan by pins through the extreme base and the rims of the siphons; cover with water.

The viscera can usually be seen through the mantle; if necessary to remove the mantle, it must be pulled off gently in small strips. The digestive tract is the most noticeable of the internal features. The oral siphon leads into a large thin-walled bag, the *pharynx,* extending most or all of the body length and looking to the naked eye like mosquito netting. The cavity outside the pharynx, between it and the mantle, is the *atrium,* opening to the exterior by way of the atrial siphon. The lower end of the pharynx narrows abruptly into a short *esophagus,* which opens into the widened *stomach,* situated below or to one side of the base of the pharynx, with its curved axis at right angles to the long axis of the pharynx. The stomach leads into the *intestine,* which immediately doubles back to form a loop parallel to the stomach and then extends straight upward toward the atrial siphon, terminating by an *anus* situated in the atrium.

In the dorsal part of the mantle, between the two siphons, will be found an elongated mass, the *ganglion*, which constitutes the central nervous system of the adult tunicate. Nerves may be seen extending from its ends to the siphons.

Make a longitudinal slit in the side of the pharynx forward through the oral siphon and spread out the pharyngeal walls. In the midventral line of the pharynx will be noticed a conspicuous white cord, the *endostyle* or *hypobranchial groove*, composed of glandular and ciliated cells. Directly opposite the endostyle in the middorsal line of the pharynx is found the *dorsal lamina*, which is either a delicate membranous fold or a fringelike row of projections called *languets*. At the junction of pharynx and oral siphon there generally occurs a circlet of *tentacles* or irregular processes; and posterior to this there is a grooved ridge, the *peripharyngeal band*, in which the anterior ends of endostyle and dorsal lamina terminate. In some tunicates the lateral walls of the pharynx have a number of lengthwise folds. Draw to show above parts.

Cut out a small piece of the pharyngeal wall, mount in water, spread out flat, and examine under low power, or use a prepared slide. The pharyngeal wall is divided into squares by longitudinal bars and crossbars. Each such square incloses several elongated *gill slits*, separated from each other by smaller bars. (The gill slits are curved in *Molgula*.) Finger-like processes or papillae are often present on the longitudinal bars. Draw a portion of the pharyngeal wall.

This whole apparatus is primarily a food-catching device, secondarily a respiratory mechanism. When the animal is undisturbed, a current of water, motivated by the cilia on the gill bars, enters the pharynx through the oral siphon, passes through the gill slits into the atrium, and exits by the atrial siphon. Particles of food are caught in the mucus secreted by the endostyle and passed dorsally by the papillae and cilia of the gill bars to the dorsal lamina, which directs the strands of mucus into the esophagus. At the pharynx entrance tentacles and peripharyngeal groove act similarly, passing food particles to the dorsal lamina.

3. Chordate features of tunicate development.—Tunicates pass through a developmental stage termed the *tadpole*, from a resemblance of shape to the frog tadpole. An early stage of a tunicate tadpole is shown in Figure 3, where the tail especially is diagrammatically chordate. Under the simple one-layered epidermis is seen on the dorsal side the neural tube, widened anteriorly into a brain. Below the neural tube in the tail is located the notochord, composed of large cells; and below that is the entoderm of the future intestine, enlarged anteriorly into the future pharynx.

Whole mounts of mature tunicate tadpoles may be studied if available but are usually not very clear. The larva is sharply marked off into oval body and long slender tail. At the anterior end of the body region are seen the three papillae by which the larva attaches when about to metamorphose. In the body are seen the brain, bearing two black spots, the dorsal one part of an eye, the ventral one part of a statocyst or balancing organ; and below the brain, the mass of entoderm cells constituting the pharynx. In favorable specimens the neural tube may be traced from the brain into the tail, and below this will be seen, by careful focusing, the notochord, appearing as a central rod in the tail. Note that the notochord terminates anteriorly just behind the brain vesicle. Notochord and neural tube are formed embryologically as in vertebrates. There is no trace of segmentation.

The tadpole usually swims about for a few hours and then attaches to some object by the three papillae and undergoes a remarkable metamorphosis. The tail with all its chordate features is partly cast off and partly absorbed, and the brain with its eye and statocyst also disappears; but a portion of the brain earlier constricted off persists and forms the adult ganglion, neural gland and duct. The larval pharynx enlarges and differentiates into the parts noted above, and the atrium arises as a pair of pouches which grow in from the outside and eventually fuse to a single cavity. The *neural gland* is a small mass near the ganglion, opening into the pharynx by a ciliated *duct* whose mouth may be much convoluted or folded into a structure termed the *dorsal tubercle*. This whole apparatus is of nervous origin, and there is some evidence that it is the homologue of the posterior lobe of the pituitary of vertebrates. There is no trace of segmentation in tunicate embryology.

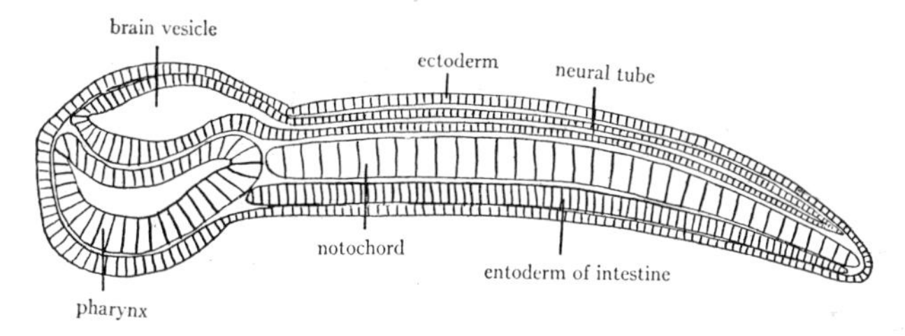

FIG. 3.—Diagram of a young tunicate tadpole. (After Kowalevsky, 1866.)

This developmental history is often taken to indicate that the tunicates are very degenerate remnants of a former advanced chordate group. It is more probable, however, that the tunicate ancestor was an extremely primitive, swimming, tailed chordate without segmentation, in which the notochord originated as a stiffening for the tail.

C. THE LOWER CHORDATES, SUBPHYLUM CEPHALOCHORDATA

The little animal known as *Amphioxus*[2] is of great interest because of its many primitive and generalized chordate features, which also throw light on vertebrate organization. *Amphioxus* lives along ocean shores in coarse sandy or shelly bottom with the oral hood protruding.

1. External anatomy of *Amphioxus*.—Obtain a specimen and place in a dish of water. The body is slender, fishlike, pointed at each end, and compressed laterally. The more blunt end is anterior, the more pointed end, posterior; the dorsal surface is sharp, the ventral surface, for the greater part of its length, flattened. The anterior end represents a poorly developed, somewhat degenerate head. The ventral and greater part of the head consists of an expanded membrane, the *oral hood*, which incloses a cavity, the

[2] The correct generic name is *Branchiostoma* Costa 1834, which has priority over *Amphioxus* Yarrel 1836.

vestibule, at the bottom of which the mouth is located. The borders of the oral hood are extended into a series of stiff *tentacles* or *cirri*.

Turn the animal ventral side up and observe that the flattened portion of the ventral surface is bounded laterally by two membranous folds, the *metapleural folds*, or *lateral fins*, extending posteriorly from the oral hood. These folds meet at a point nearly three-fourths of the distance from anterior to posterior end, behind a median opening, the *atriopore*. From this point a median membranous fold, the *fin*, passes to the posterior end of the body, around to the dorsal side, and forward along the dorsal side to the anterior end. The slightly wider portion of this fin which surrounds the pointed posterior end is the *caudal* fin; that along the dorsal side, the *dorsal* fin. The *anus* is situated to the left of the midventral line near the posterior end, just behind the point where the fin widens.

Along the sides of the body there are clearly visible through the thin, transparent epidermis a longitudinal series of V-shaped *muscle segments* or *myotomes*, separated from each other by connective tissue partitions, termed *myosepta*. Note that the myotomes extend nearly to the anterior tip, diminishing in size above the oral hood. The number of myotomes in *Amphioxus* is definite, about sixty. Immediately below the ventral ends of the myotomes will be seen, in some individuals at least, a row of square white masses, the *gonads* or *sex glands*. As shown by the metameric arrangement of myotomes and gonads, *Amphioxus* is a segmented animal. However, because of certain peculiar features of the embryology, the segments of the two sides do not correspond.

2. Internal anatomy.—The internal anatomy is most easily studied on small, stained, mounted specimens. Examine one under low power and identify: the various parts of the fin, containing rectangular bodies, the *fin rays*, serving as the skeletal support of the fins; the myotomes; and the digestive tract, occupying the ventral half of the body. Study the parts of the digestive tract. Note again the oral hood with its cirri, each supported by an internal skeletal rod. In the posterior part of the oral hood on its inner surface is seen the deeply stained *wheel organ*, consisting of patches of ciliated epithelium, arranged as several finger-like projections from a horseshoe-shaped base. The middorsal and longest finger has a central groove, *Hatschek's groove*, seen to terminate anteriorly in a hollow swelling, *Hatschek's pit*. Just behind the base of the wheel organ is seen a vertical membrane, the *velum*, bearing several short projecting *velar tentacles*. The small mouth opening in the velum (impractical to see) leads at once into the *pharynx*, a wide tube extending nearly half the body length. At the anterior end of the pharynx, just behind and nearly parallel to the velum, note the deeply stained *peripharyngeal band*. The side walls of the pharynx are composed of slender parallel oblique bars, the *gill bars* or *branchial bars*, inclosing between them elongated openings, the *gill slits* or *pharyngeal clefts*. Each gill bar is supported by an internal skeletal branchial rod (difficult to see), and these rods are united into an arcade at the dorsal ends of the gill bars. These skeletal rods are composed of fibrous connective tissue.

The arrangement of gill bars and skeletal rods differs from that of balanoglossids only in that the tongue bars complete their downward growth (see Fig. 2) and fuse with the ventral pharynx wall, so that each of the original larval gill slits of *Amphioxus* becomes divided into two slits. There are, thus, in the adult *Amphioxus* two kinds of gill bars, the *primary* and the *secondary* or *tongue* bars. These are anatomically different in various ways, notably in that the skeletal rod forks at the ventral end of the primary bars, as also in balanoglossids (Fig. 2). Note also fusion of the skeletal rods in the secondary bars in *Amphioxus*. In older specimens of *Amphioxus* numerous cross-unions, called *synapticula*, develop between adjacent gill bars; these also occur in balanoglossids.

Surrounding the pharynx is a large cavity, the *atrium*, the ventral boundary of which is visible below the pharynx as a line which may be traced to the atriopore. Water, propelled by the cilia on the gill bars, enters the pharynx by way of the mouth, passes through the gill slits into the atrium, and exits by the atriopore. The posterodorsal end of the pharynx continues backward as the digestive tube, in which various regions may be recognized. The short *esophagus* behind the pharynx soon widens into the *stomach* or *midgut*, from which a conical outgrowth, the so-called *liver* or *hepatic caecum*, better termed the *midgut caecum*, projects anteriorly along the right side of the posterior part of the pharynx. The midgut narrows posteriorly into a deeper staining region, the *iliocolon ring*, beyond which the narrowed *intestine* or *hindgut* continues to the anus (Fig. 4).

Amphioxus feeds on diatoms, desmids, and other microscopic organisms. In feeding, the cirri are folded over the entrance to the oral hood, excluding large particles. Sufficiently small particles are carried by the main ciliary current into the pharynx, where they become trapped in sheets of mucus secreted by the endostyle. These mucous sheets are passed dorsally by the cilia on the inner faces of the gill bars to the epipharyngeal groove which directs them into the esophagus. Any particles in the oral hood which escape the main current are caught and concentrated by the wheel organ with the aid of mucus from Hatschek's groove and pit; and this material, together with particles in the anterior end of the pharynx, is carried dorsally to the epipharyngeal groove along the peripharyngeal band. During these processes the mucous sheets become rolled into a food cylinder which proceeds along the digestive tube. Upon arriving at the iliocolon ring the food cylinder is thrown into a spiral coil and rotated by the action of the cilia of this ring. Pieces break from the rear end of the rotating strand and pass along the hindgut. All parts of the digestive tube secrete enzymes; intracellular digestion and absorption occur chiefly in the hindgut. The so-called hepatic caecum appears to function chiefly in the secretion of digestive enzymes, and there is no evidence that it has any of the functions of a vertebrate liver. It seems, in fact, to correspond functionally more nearly to the pancreas.

Immediately dorsal to the digestive tube and about as wide as the hindgut is seen a yellowish rod, the *notochord*, extending the body length and running, just above the oral hood, nearly to the extreme anterior tip. Directly above the notochord is situated the much smaller *neural tube* or *central nervous system*, best recognized by the row of black spots which it bears. These spots are very simple *eyes*, each consisting of one ganglion cell and one curved pigment cell, and have been shown to respond to light. The reduced anterior end of the nerve tube seems to represent a brain and may be termed *brain vesicle;* this terminates somewhat behind the anterior tip of the notochord and there bears a black *pigment spot*, shown to have no response to light. Just above the pigment spot there may be seen on the left side a depression, the *flagellated pit*, formerly called olfactory funnel. Although this connects with the cavity of the brain vesicle in the larva, there is no such connection in the adult. The flagellated pit would seem to be a sense organ. Draw the animal from the side, showing as many features as you have been able to see.

Two pairs of nerves arise from the brain vesicle; and the remainder of the nerve tube, which may be termed *spinal cord*, gives off a vertebrate-like metameric succession of *spinal nerves*. These consist of the *dorsal* nerves, chiefly sensory, passing out by way of the myosepta to supply the skin, and the *ventral* nerves, chiefly motor, passing into the myotomes. These nerves alternate on the two sides, because of the asymmetry of the myotomes. Unlike conditions in vertebrates, the dorsal and ventral nerves do not unite outside the cord. There is also an autonomic system, consisting of inner and outer ganglionated plexi in the wall of the digestive tract, connecting with the spinal cord by way of the dorsal nerves.

3. Cross-section through the pharyngeal region.—Examine the cross-section with the low power and identify the following: (*a*) The epidermis, the outer covering of the body composed of a single layer of columnar epithelial cells. (*b*) The dorsal median projection, the dorsal fin, containing an oval mass, the fin ray, which supports it. (*c*) The two ventrolateral projections, the metapleural folds. There are a number of smaller folds in the ventral wall between the two metapleural folds. (*d*) The myotomes, a series of circular masses filling the dorsal and lateral portions of the body wall, and separated from each other by connective tissue partitions. The myotomes are thick dorsally and thin out ventrally. Transverse muscles are present in the ventral body wall, just above the small folds of the epidermis. (*e*) The neural tube, a median dorsal mass, oval or trapezoidal in section, lying between the dorsal portions of the myotomes, below the fin ray. Observe that it contains a central canal, the *neurocoel*. The section may strike one of the eyes described above. (*f*) The notochord, an oval mass much larger than the neural tube and directly ventral to it. (*g*) The atrium, the large cavity occupying the ventral half of the section; it is formed in an advanced larval stage by the invagination of the epidermis. Previous to this occurrence, the larval gill slits open directly to the exterior as in vertebrate embryos. (*h*) The pharynx, occupying the center of the atrium, elongate or heart shaped. As the gill bars run obliquely, a cross-section hits a number of them, so that the pharynx appears to be composed of separate pieces, the gill bars, separated by spaces, the gill slits. The gill bar consists chiefly of a tall ciliated epithelium inclosing at the outer side the skeletal rod and blood vessels and coelomic spaces. In the middorsal line of the pharynx is seen the deep *epibranchial groove;* and in its midventral line, the similar *hypobranchial groove* or *endostyle.* The parts and relations of the pharynx are the same as in tunicates. (*i*) The midgut caecum (liver), an oval, hollow structure lined by a tall epithelium, present to the right side of the pharynx. (*j*) The gonads, if present, are the very large, deeply stained masses at either side, projecting into and nearly filling the atrial cavity; the ovaries consist of cells with large nuclei; the testes present a streaky appearance. (*k*) The nephridia; fragments may be noticed attached to the outer sides of the most dorsal gill bars. Draw the cross-section.

The nephridia are short irregular tubes perched upon the outer surface of the dorsal parts of the secondary gill bars and opening into the atrium. Each nephridium bears several clusters of *solenocytes,* which are flagellated cells drawn out into a long tube in which the flagellum plays. It is a rather puzzling fact that the only other animals known to have solenocytes are certain polychaete annelids, not supposed to be related to the chordate line. Nephridia with solenocytes classify as *protonephridia* and are not homologous to the tubules of vertebrate kidneys.

4. Circulatory system.—The circulatory system lacks a heart and is composed of vessels and tissue channels forming a continuous circuit. The contractile vessels have muscular walls, but the others consist of only a thin membrane; and there is no histological distinction among the smaller vessels between arteries, veins, and capillaries. The blood is colorless. From a junction point, often called *sinus venosus,* situated at the posterior end of the pharynx, a contractile vessel variously called *endostylar artery, ventral aorta,* or *truncus arteriosus,* runs forward beneath the endostyle and branches to both sides into the primary gill bars (Fig. 4). Each such branch bears a contractile swelling (*bulbillus*) and then ascends the gill bar as an *aortic arch* which joins the dorsal aorta of that side. (There also spring from each bulbillus two secondary vessels which ascend the primary gill bar; and there are two similar vessels, connected only indirectly with the ventral aorta, in each secondary gill bar.) Before reaching the dorsal aorta, each aortic arch contributes to a plexus of blood vessels in the corresponding nephridium (which also receives branches from the outer of the two vessels in the secondary gill bar). Thus, before reaching the dorsal aortae, the blood has passed through the gills and the nephridia in part. The noncontractile paired *dorsal aortae* (also called carotids) begin as fine twigs at the anterior end of the animal and run backward above the pharynx, just below the notochord, receiving the aortic arches and other gill vessels of their respective sides. Shortly behind the rear end of the pharynx the two aortae unite into a single dorsal aorta, which continues backward, becoming the *caudal* artery of the tail. From the tail the *caudal* vein runs forward and at the anus forks into the *posterior cardinal* vein and the *subintestinal* vein (Fig. 4). The posterior cardinal vein runs forward along the middle of the body wall just inside the myotomes (Fig. 5) and opposite the sinus venosus joins the similar *anterior cardinal* vein running back from the anterior half of the body. The common *duct of Cuvier,* formed by the union of the two cardinal veins, crosses the atrium (Fig. 5) and enters the sinus venosus. (Several accessory ducts of Cuvier parallel to the main one usually occur.) The subintestinal vein courses along the ventral side of the intestine and breaks up into a network of small ves-

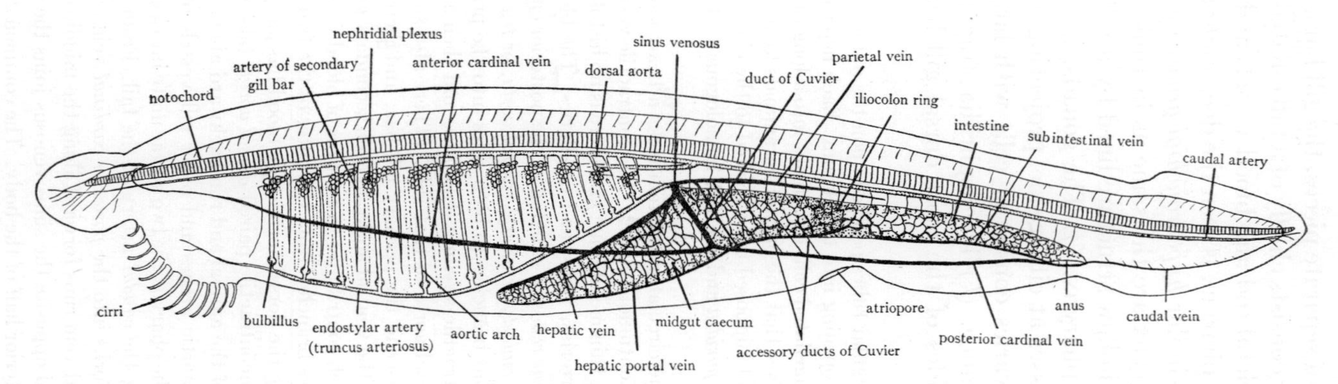

FIG. 4.—Diagram of the circulatory system of *Amphioxus*, showing also the digestive tract. (Altered after Mozejko, 1913.)

sels in the intestinal wall, from which network a vein again forms and runs forward along the ventral surface of the midgut caecum, as the *hepatic portal* vein. This breaks up into a network of vessels in the wall of the caecum, and from this network the contractile *hepatic* vein reforms and, running along the dorsal surface of the caecum, enters the sinus venosus. It is important to note that the arrangement of the circulation of the caecum resembles that of the vertebrate liver, i.e., it appears to be a *hepatic portal system*, and this constitutes the chief grounds for considering the caecum to be a liver. There is a pair of *parietal* veins in the dorsal wall beneath the myotomes (Figs. 4, 5), and all of the main vessels give off or receive metamerically arranged branches from the body wall and digestive tube.

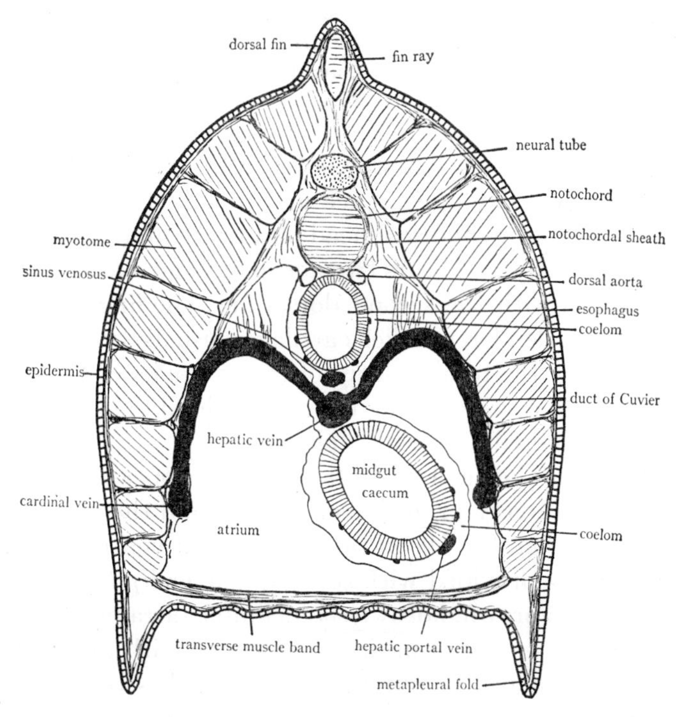

FIG. 5.—Transverse section through *Amphioxus*, at the level of the ducts of Cuvier

The contractions in the circulatory system are slow and irregular. The hepatic vein beats first toward the sinus; then the sinus contracts, followed by the ventral aorta, away from the sinus, and by the bulbilli. The circulation can reverse but rarely does so. The contractile vessels beat only after they have become distended with blood. There is probably not much difference in the oxygen content of the so-called arteries and veins.

5. Relationships.— *Amphioxus* shows affinities with balanoglossids, tunicates, and vertebrates. The mode of formation of the tongue bars and the arrangement of the skeletal rods of the gill bars are identical with those of balanoglossids. There has further been recently discovered in balanoglossids a U-shaped ciliary groove (preoral ciliary organ) on the posterior face of the proboscis which seems to be the homologue of the wheel organ of *Amphioxus*. Hatschek's pit and groove are considered by many a forerunner of the vertebrate hypophysis. Several structures of the *Amphioxus* pharynx (endostyle, peripharyngeal band, epipharyngeal groove) are identical with those of tunicates. Much of the developmental and adult anatomy of *Amphioxus* is typically vertebrate, such as the mode of formation and relations of the myotomes, notochord, neural tube, digestive tract, and main blood vessels. Notable vertebrate features of the last named are the ventral and dorsal vessels of the gill region connected by vessels running through the gill bars, the characteristic body-wall veins (anterior and posterior cardinals) connected to the contractile center by a duct of Cuvier crossing body spaces, relations of caudal and subintestinal veins, and the portal circulation through the midgut caecum.

D. ORDER CYCLOSTOMATA

The cyclostomes or round-mouthed fishes constitute the lowest group of living vertebrates. However, they lead a semiparasitic life with some accompanying specialization and degeneration; but it is now believed that much of their structure is primitive. They are divided into two groups, the Petromyzontia, or lampreys, and the Myxinoidea, or hagfishes. The former occur in streams, lakes, and the ocean but always ascend fresh-water streams to breed; the myxinoids are marine.

1. External anatomy of a lamprey.—Place the specimen in a dissecting tray. Note general eel-like appearance and tough, slimy, naked skin. The body consists of a stout cylindrical head and trunk and a laterally flattened tail. The posterior body half bears two *dorsal* fins and a *tail* fin bordering the tail and continuous with the second dorsal fin. The fins are supported by numerous *fin rays*, slender parallel cartilages usually visible through the skin. Note especially the total absence of the two pairs of ventral fins found in the true fishes and corresponding to fore and hind limbs; their absence in cyclostomes is considered a primitive character.

The anterior end of the head presents a peculiar appearance because of the lack of a lower jaw (a very primitive character) and the presence of a large bowl-shaped depression, the *buccal funnel*, directed ventrally. The edges of the buccal funnel are provided with soft papillae or *lip tentacles*, and its interior is studded with brown, horny *teeth*, definitely arranged. At the bottom of the funnel lies the *tongue*, a projection covered with teeth; the mouth is just dorsal to the tongue. Lampreys attach to fishes by using the buccal funnel as a suction cup and rasp off the flesh of their victims by filing movements of the tongue.

On the dorsal surface of the head there is a median *nasohypophyseal opening*, from which a passage leads to the olfactory sac. This single, dorsally situated entrance into the nasal sac also occurs in other very primitive (extinct) groups of the class Agnatha and contrasts strongly with the paired, laterally or ventrally located nostrils of typical vertebrates. On either side of the head is an *eye*, without eyelids; and behind each eye is seen a row of seven oval openings, the *gill slits*.

The body-wall musculature is segmented into myotomes separated by myosepta, as in *Amphioxus;* these are generally detectable through the skin. At the junction of trunk and tail in the midventral line there is a cloacal pit containing the anal opening in front and the urogenital papilla behind.

2. Sagittal section of the anterior end.—Make a median sagittal section of your specimen to a point about an inch posterior to the last gill slit or study a section so prepared. Examine the cut surface and identify the following:

a) *Digestive tract:* Observe again the buccal funnel with its teeth and tongue. Note the large muscle masses extending posteriorly from the tongue, by means of which the rasping movements of the tongue are brought about. Find the mouth opening above the tongue and follow it into a passage, the *buccal cavity*, which slopes ventrally. The buccal cavity opens at its posterior end into two tubes, an upper smaller one, the *esophagus*, and a larger ventral one, the *pharynx*, the wall of which is pierced by seven oval openings. A fold, the *velum*, is present at the entrance of the buccal cavity into the pharynx. The esophagus leads into the remainder of the digestive tract, but the pharynx, which in the embryo constituted the anterior part of the digestive tract, ends blindly in the adult.

b) *Respiratory system:* The seven openings in the wall of the pharynx are the internal *gill slits*. They open into much-enlarged *gill pouches* which bear the gills on their walls and which communicate with the exterior through the external gill slits already noted. Probe into one of the gill pouches and note the leaves or *gill lamellae* borne on its walls. The esophagus is located between the dorsal parts of the gill pouches. The lamprey respires by pumping water in and out of the gill pouches through the gill slits; when not feeding, it also takes in water through the mouth.

c) *Notochord:* The notochord is the broad, brown rod just dorsal to the esophagus. It is the chief axial skeleton of the animal, as the vertebral column is primitive, consisting only of small separated arches straddling the notochord. There is also a cartilaginous skull inclosing the brain and a complicated lattice termed the *branchial basket* supporting the gills.

d) *Nervous system:* Above the notochord is a narrow canal, the neural canal, containing a slender tube, the spinal cord, and the brain, the enlarged lobed structure just above the anterior extremity of the notochord.

e) *Olfactohypophyseal apparatus:* The section should bisect the nasohypophyseal aperture. From this aperture the *nasohypophyseal canal* slants downward and backward and opens into the large median *olfactory sac*, which probably represents a pair of fused sacs. Note its much-folded wall bearing the olfactory epithelium. Below the olfactory sac the nasohypophyseal canal continues ventrally and posteriorly and then widens into an elongated blind sac, usually called the pituitary sac but better termed the *nasopharyngeal sac*, since it has no hypophyseal function. Glandular tissue corresponding to the anterior lobe of the pituitary of typical vertebrates is situated in the lamprey just behind the olfactory sac in the tissue above the nasohypophyseal canal.

The peculiar dorsal position of the olfactory organ and its relation to the hypophysis result from a shifting of these organs during development. In the lamprey larva the nasal sacs and the hypophysis arise as separate epithelial invaginations on the ventral side of the head anterior to the mouth. The great development of the buccal funnel in front of the mouth brings the nasal and hypophyseal invaginations in contact and shoves them to the dorsal side. They fuse to form the common pore and passage seen in the adult animal.

f) *Pericardial cavity:* Posterior to the last gill pouch is a somewhat conical cavity, the *pericardial cavity,* which contains the heart.

3. External features of a myxinoid.—The myxinoids are probably even more primitive than the lampreys. The common genera are *Myxine* and *Bdellostoma;* whichever is available may be examined. The body is even more slender and eel-like than in lampreys and terminates in a short tail bordered by a continuous tail fin. Note that the anterior end lacks the buccal funnel of the lampreys and instead has a ventral mouth opening bordered by two pairs of tentacles. At the anterior tip of the head is seen the nasohypophyseal opening with two tentacles on either side. (In the myxinoids the nasohypophyseal canal continues beyond the olfactory sac as the *nasopharyngeal canal* opening into the pharynx, and the continuous passage so formed is the respiratory passage by which water for the gill pouches enters. It thus appears that the nasopharyngeal sac of lampreys represents an aborted respiratory passage.) Eyes are wanting. On the sides, some distance behind the head, there is found in *Bdellostoma* a row of from six to fourteen round gill slits. In *Myxine* external gill slits are lacking; instead, canals run backward from the gill pouches under the skin and unite on each side to form an opening, the *branchial opening.* The two branchial openings will be found on the ventral surface about one-third the body length from the anterior end. The left branchial opening is larger because it receives the *pharyngocutaneous* duct, a tube from the pharynx to the exterior, possibly representing a gill slit. This tube also exists in *Bdellostoma,* where it opens in common with the last gill slit of the left side. Along the sides of the body of myxinoids is seen a row of small holes, which are the openings of slime sacs. The slitlike *cloacal aperture* just in front of the beginning of the tail fin receives the anal and urogenital openings.

4. The *Ammocoetes* larva.—The cyclostomes have a larval stage called *Ammocoetes* (because originally supposed to be a distinct genus), which endures from two to five or six years. In the lampreys this larva lives in burrows in the bottoms of streams, feeding on diatoms, desmids, and other minute organisms. The ammocoete larva is of great phylogenetic importance as illustrating primitive vertebrate structure. The following account includes only those details which the student can readily observe; other features will be discussed in later chapters.

a) *Whole mount:* Note general *Amphioxus*-like appearance. The anterior end forms an expanded *oral hood* in the back part of which are seen a number of projections, the *oral papillae;* these serve as strainers in feeding. Behind the oral hood will be noticed the *brain*, above; the *gill* or *branchial* region, below. The brain is an elongated, lobed body showing the characteristic vertebrate divisions and closely related to the three sense organs typical of vertebrates—nose, eye, and ear. The most anterior division of the brain, or *telencephalon*, shows two obvious lobes, the *olfactory bulb* in front and the *olfactory lobe* behind. Directly in front of the olfactory bulb is seen the nasohypophyseal canal ascending to the nasohypophyseal opening, having a raised rim. (A third part of the telencephalon, the *cerebral hemispheres*, is not readily visible from surface view.) Behind the olfactory lobe is a larger division of the brain, the *diencephalon*, which bears an eye on either side. The ventral part of the diencephalon projects forward below the olfactory lobe as the *infundibulum*, most of which represents the posterior lobe of the pituitary. The anterior part of the roof of the diencephalon projects forward over the olfactory lobe as the *pineal body*. The slightly elevated region of the brain above the main mass of the diencephalon and appearing as its dorsal part is really the *mesencephalon* or *midbrain* (*optic lobes*). This is separated by a cleft from the last and largest division of the brain, the hindbrain or *rhombencephalon*, consisting chiefly of the *medulla oblongata*. Along the anterior ventral part of the medulla shortly behind the eye is seen the large oval *ear vesicle*. The medulla oblongata tapers abruptly into the *spinal cord*, which runs as a stained rod nearly to the posterior tip. Immediately below the spinal cord is the much wider notochord, which tapers to an anterior end shortly behind the ear vesicle.

Below the brain and notochord are the parts of the digestive tract and a few other organs. The oral hood leads, by way of the mouth, into the *buccal cavity*, bounded posteriorly by a large flap, the *velum*, situated in front of the gill slits. The velum really consists of a pair of flaps attached dorsally just in front of the ear vesicle and to the sides of the buccal cavity. The flaps first curve forward and then diagonally backward. Rhythmic movements of the velar flaps cause a current of water to pass from the mouth into the pharynx. The pharynx is the large tube behind the velum, having seven gill pouches in its walls. The gill lamellae in the walls of the pouches can be readily seen. Each gill pouch opens to the exterior by a small gill slit. Ventrally below the first five gill pouches there is a conspicuous elongated body, the *subpharyngeal gland*, usually called thyroid or endostyle. Although this gland does arise embryologically as a trough in the pharynx floor, it soon constricts from the pharynx, remaining connected with it during larval life by a duct opening between the third and fourth gill pouches. The gland is of peculiar and complicated construction (see below) and histologically does not closely resemble the endostyle; it further has been shown to have no relation to feeding—in fact, its function is unknown. Hence the usual statement that it is the homologue of the endostyle appears to the author to rest on very insecure grounds. Certain parts of the subpharyngeal gland become the adult thyroid, but experiment has shown that the gland in the *Ammocoetes* has no thyroid function. Hence it does not seem desirable to call it the thyroid gland, as is frequently done. Behind the last gill pouch the digestive tract narrows to a slender tube, the *esophagus*, which extends backward for a short distance and then widens into the *intestine*. Below the esophagus are seen two rounded organs: anteriorly and just behind the last gill pouch is the *heart;* and behind this is the larger *liver*, in which there is a conspicuous vesicle, the *gall bladder*. The indefinite mass above the heart is the *pronephros* or larval kidney. The intestine continues posteriorly to the anus, which marks the boundary between trunk and tail. Immediately behind the anus the tail fin begins and continues posteriorly around the tail and forward in the middorsal line to about the level of the liver. Lateral fin folds, such as the metapleural folds of *Amphioxus*, are lacking in the *Ammocoetes*.

b) *Older larvae:* Older larvae may be examined with a hand lens or under a binocular. Note oral hood; nasohypophyseal opening with a thick rim; slanting row of seven gill slits opening into a groove; and myotomes visible through the skin, anus, and caudal and dorsal fins. On the top and sides of the head will be seen short lines of dots, the organs of the *lateral-line system*, a system of skin sense organs peculiar to aquatic vertebrates. Note that the eyes are not visible; they are deeply sunk and not functional in *Ammocoetes*.

c) *Section through the pharyngeal region:* The section is clothed by a thick epidermis underlain by a thin layer of connective tissue inclosing the layer of myotomes, thicker dorsally than ventrally. Along the sides are the midlateral grooves into which the gill slits open. Middorsally between the myotomes note the large neural canal containing the relatively small flattened spinal cord, and directly below this the large rounded notochord composed of vacuolated cells and inclosed in a thick sheath. To either side of the notochord is seen the cross-section of an anterior cardinal vein, and below the notochord lies the dorsal aorta. The central part of the section is occupied by the large pharynx, whose walls bear the gill lamellae. The appearance of the pharynx depends on whether the section cuts through the anterior or the posterior lamellae of a gill pouch. The slant of the septa which bear the gill lamellae is such that the lamellae on their anterior faces are turned to the outside; those on their posterior faces, to the inside. The gill lamellae are little plates projecting horizontally from the septa, so that a whole set of them from dorsal to ventral side is cut crosswise in a section through the pharynx. Each lamella has lateral ridges and is liberally provided with blood vessels. In the middorsal and midventral line of the pharynx there is a projecting *ciliated ridge* of unknown function. The midventral ridge has a single groove or a pair of grooves, depending on the level of section; these are called the *peripharyngeal grooves* and are continuous around the anterior end of the pharynx with the dorsal ciliated ridge. Below the ventral ridge is the single or paired ventral aorta, and below this the subpharyngeal gland, the appearance of which depends on the level of section. Anterior to its opening into the pharynx floor the gland consists of two chambers separated by a thin partition. In each chamber attached to its floor is seen the cross-section of a longitudinal fold of ciliated epithelium, which at four places is infolded to make four conspicuous columns of tall wedge-shaped cells, termed *cuneiform cells*. Posterior to the duct the central part of the gland makes an upward

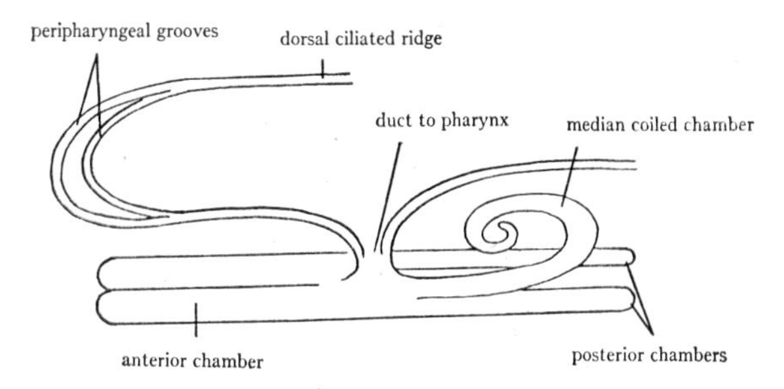

FIG. 6.—Subpharyngeal gland of the *Ammocoetes* with connecting grooves

spiral coil, so that the section cuts through the gland twice and makes a complicated appearance (Fig. 6). The function of this remarkable apparatus is unknown, although the columns of cuneiform cells seem to be secretory. The duct from the subpharyngeal gland is continuous with the peripharyngeal grooves, so that any secretions would be carried along these grooves and so be distributed throughout the pharynx. Food particles in *Ammocoetes* are caught in mucus secreted by the pharynx lining and gill lamellae, and the mucous strands are carried directly backward into the esophagus; direct observation has failed to show any visible participation of the subpharyngeal gland in the process.

d) Section through the trunk: Identify dorsal and ventral fin fold, neural canal, spinal cord, notochord, dorsal aorta below the notochord, posterior cardinal veins to either side of the aorta, and coelomic cavity inclosed by the layer of myotomes. The structures in the coelomic cavity will depend on the level of the section. Shortly behind the pharynx the section shows the esophagus as a small circle of tall epithelium, the paired pronephroi, composed of sections of tubules, situated below the posterior cardinal veins, and the various parts of the heart filled with blood. More posteriorly the liver is encountered, and beneath the cardinal veins are folds of fatty tissue bearing a pronephric duct on each side. Behind the liver the section shows the intestine with a conspicuous infolding, the spiral fold, filled with tissue inclosing the intestinal artery. Beneath the posterior cardinal veins the fatty tissue supporting the pronephric duct continues, or at certain levels the mesonephric tubules may be present. The single unpaired gonad is seen as a small body to the medial side of the kidney tissue.

e) Metamorphosis: After some years of larval life the *Ammocoetes* metamorphoses to the adult condition. In the case of the lamprey the oral hood expands to form the buccal funnel, the eyes move to the surface and become functional, the gall bladder disappears, the subpharyngeal gland undergoes extensive degeneration, and a tube cuts off from the roof of the pharynx and connects with the esophagus, leaving the pharynx and gill pouches as a blind sac having only respiratory functions. In the subpharyngeal gland, the columns of cuneiform cells disappear, but some or all of the ciliated epithelium persists to become the follicles of the thyroid gland. This fact furnishes the basis for the statement usual in books that the thyroid gland of vertebrates is homologous to the endostyle of *Amphioxus* and the tunicates. This statement appears to be unjustified since, first, the homology of the subpharyngeal gland with the endostyle has been shown to be questionable and, second, only a part of the gland becomes the adult thyroid. A more correct statement would be that the thyroid gland evolves from cell elements of the pharynx floor of primitive vertebrates and that these cell elements originally had some other, unknown function. The thyroid of the adult lamprey has typical thyroid properties.

E. SUMMARY

1. Essential chordate features are seen in the tunicate tadpole, *Amphioxus*, and *Ammocoetes*, notably the dorsal tubular central nervous system, underlain by the notochord.

2. A pharyngeal food-catching apparatus, consisting of slits through the pharynx wall and ciliated grooves, occurs in adult tunicates, *Amphioxus*, and *Ammocoetes*. Similarity of anatomical details of this apparatus indicates phylogenetic relationship between these groups and also with balanoglossids. The pharyngeal apparatus secondarily acquires respiratory functions.

3. The thyroid gland evolves from cells in the floor of the pharynx.

4. The oral hood of *Amphioxus* and *Ammocoetes*, without lower jaw, is a primitive chordate feature.

5. Typical segmentation with myotomes is seen in *Amphioxus* and *Ammocoetes*. Its total lack in balanoglossids and tunicates indicates a late origin of this vertebrate feature.

6. The circulatory system of *Amphioxus* (also *Ammocoetes*) displays pronounced vertebrate features, in the presence of dorsal and ventral aortae in the gill region, connecting vessels passing between the gill slits, cardinal veins, duct of Cuvier, and hepatic portal circulation.

REFERENCES

Barrington, E. J. W. 1937. The digestive system of *Amphioxus*. Phil. Trans. Roy. Soc. London, B, 228.

Boeke, J. 1935. The autonomic (enteric) nervous system of *Amphioxus*. Quart. Jour. Mic. Sci., 77.

Bramhall, F. W., and Cole, H. A. 1939. The preoral ciliary organ of the Enteropneusta. Proc. Zoöl. Soc. London, B, **109**, Part 2.

Butcher, E. O. 1930. The pituitary in the ascidians. Jour. Exper. Zoöl, **57**.

De Beer, G. R. 1924. The Evolution of the Pituitary.

Elwyn, A. 1937. Some stages in the development of the neural complex in Ecteinascidia. Bull. Neurol. Inst. New York, **6**.

Franz, V. 1927. Morphologie der Akranier. Ergebn. d. Anat. u. Entw'gesch., **27**.

Horst, C. J. van der. 1927–36. Hemichordata. Bronn's Klassen und Ordnungen des Tierreiches, **4**, Abt. 4, Buch 2, Teil 2.

———. 1932. Enteropneusta. Kükenthal-Krumbach's Handbuch der Zoologie, **3**, Part 2.

Lohmann, H.; Huus, J.; and Ihle, I. E. W. 1933. Tunicata. Kükenthal-Krumbach's Handbuch der Zoologie, **5**, Part 2.

Mozejko, B. 1913. Über das oberflächliche subkutane Gefässsystem von *Amphioxus*. Mitt. zool. Stat. Neapel, **21**.

Orton, J. H. 1913. The ciliary mechanisms on the gill and the mode of feeding in *Amphioxus*, *Ascidia*, and *Solenomya*. Jour. Marine Biol. Assoc., **10**.

Pietschmann, V. 1929. Akrania. Kükenthal-Krumbach's Handbuch der Zoologie, **6**, Part 1.

———. 1929. Cyclostomen. *Ibid.*

Skramlik, E. von. 1938. Über das Blutumlauf bei *Amphioxus*. Pubbl. Staz. Zool. Napoli, **17**.

Weel, P. B. von. 1937. Die Ernahrungsbiologie von *Amphioxus*. Pubbl. Staz. Zool. Napoli, **16**.

Wells, M. M. 1926. Collecting *Amphioxus*. Science, **64**.

Zarnik, B. 1904. Über segmentale Venen bei *Amphioxus* und ihr Verhältnis zum Ductus cuvieri. Anat. Anz., **24**.

IV. EXTERNAL ANATOMY AND ADAPTIVE RADIATION IN GNATHOSTOMES

This chapter will acquaint the student with the external features of typical jawed vertebrates and has further the purpose of showing him the variety of form attained within each vertebrate class. As many representatives of each class as possible should be examined, and visits to museum collections are highly desirable.

A. CLASS CHONDRICHTHYES: EXTERNAL ANATOMY OF THE DOGFISH

The dogfish or dog shark is a representative of the class of cartilaginous or elasmobranch fishes, characterized by their cartilaginous skeleton and exposed gill slits. The dogfish is a very generalized vertebrate, and hence a knowledge of its structure is indispensable for an understanding of vertebrate anatomy. There are two common species of dogfish in use for laboratory dissection: *Squalus acanthias*,[1] the Atlantic spiny dogfish, and *Squalus suckleyi*, the Pacific spiny dogfish. The anatomy of these two species is practically identical.

1. Body and skin.—The body of the dogfish has the shape and proportions which we recognize as most advantageous for free-swimming animals—fusiform (spindle shaped) and pointed at each end, thus offering little resistance to the water. The body is divided into head, trunk, and tail, which are not, however, distinctly bounded from one another. Trunk and tail are provided with fins for purposes of locomotion. The body is clothed with minute *scales*, each of which bears a tiny spine. Pass the hand over the skin of the dogfish and note the rough feeling caused by the spines. A whitish line, the *lateral line*, extends along each side of the body; it overlies the *lateral-line canal* which contains sensory cells serving to detect water vibrations. Through the skin the zigzag muscle segments or myotomes similar to those of *Amphioxus* can often be perceived.

2. The head.—The head is triangular and somewhat flattened; its pointed extremity is known as the *rostrum*. On the ventral side of the head is the narrow crescentic *mouth*, bounded, as in all recent vertebrates except cyclostomes, by the *upper and lower jaws*, both of which bear a number of *teeth* arranged in diagonal rows. The head further bears the three pairs of sense organs characteristic of vertebrates—the olfactory organs, the eyes, and the ears. The *nostrils* are a pair of openings on the ventral side of the rostrum. A little flap of skin extends over the center of each nostril, dividing the opening into two passages, by means of which a current of water circulates through the olfactory sac, a rounded sac into which each nostril leads. The

[1] This is the correct scientific name, although the synonym *Acanthias vulgaris* is often used. The spelling *sucklii* is erroneous.

oval eyes, provided with immovable upper and lower eyelids, are situated on the sides of the head. Behind each eye is a slight prominence, best perceived by feeling with the finger, inside of which the internal ear is located. There is no external ear, ear opening, or drum membrane; but the ears are connected with the surface of the head by two canals, the *endolymphatic ducts*, which open by a pair of small pores in the center of the dorsal surface of the head just back of the level of the eyes. These may be difficult to find in some specimens. The rostrum and adjacent parts of the head are punctured by many pores, which are the openings of long mucous canals imbedded under the skin and of uncertain function.

3. Gill slits.—Just behind each eye is a circular opening, the first *gill slit* or *spiracle;* and a short distance posterior to this, a row of five elongated slits, the second to sixth gill slits. The gill slits communicate with the cavity of the pharynx. In the respiratory movements water enters through the mouth and exits through the gill slits.

4. Fins.—These are of two kinds: the *unpaired* or *median* fins, arising from the median lines of the animal, and the *paired* or *lateral* fins, located on the ventral side of the trunk. The unpaired fins consist of an *anterior* and a *posterior dorsal* fin, arising from the middorsal line and each preceded by a stout spine, and of a tail or *caudal* fin surrounding the posterior end of the body. The caudal fin of elasmobranchs is of an asymmetrical type termed *heterocercal*, consisting of a large dorsal and smaller ventral lobe, with the vertebral column bending upward into the dorsal lobe. There are two pairs of paired fins, corresponding to the fore and hind limbs of tetrapods. The anterior pair, just behind the gill slits, is named the *pectoral* fins; the posterior pair at the junction of trunk and tail, the *pelvic* fins. Note sexual difference in the pelvic fins; in males the medial side of each pelvic fin is modified into a stout process directed backward, called *clasper*, used in mating. Examine the upper surface of the claspers and note sperm-conducting groove formed by the inrolled edges. Near the rear end of the groove there is a spine on the outer side and a hook on the inner side. All the fins are supported by numerous fine parallel flexible rays, the dermal fin rays or *ceratotrichia*, imbedded in the skin.

5. Cloacal aperture.—Between the bases of the pelvic fins is a large opening, the *cloacal aperture* or, more briefly, *anus*. By spreading its walls apart, the projecting *urogenital papilla* will be seen within.

Cloaca is the term applied to the end of the digestive tube when this receives the urinary and genital ducts as well as the intestine. A cloaca is general throughout vertebrates except in placental mammals, where during development a fold divides the cloaca into a ventral urogenital canal and a dorsal intestine. The external opening of the latter in man has been termed anus from ancient times; most terms in comparative anatomy derive from human anatomy. Since the human anus is only the dorsal half of the cloacal aperture, some authors object to using the term anus for the latter. The author considers the point not worth quibbling about, and hence the convenient term anus will be applied throughout the vertebrates to the terminal orifice of the digestive tract.

B. CLASS CHONDRICHTHYES: EXTERNAL ANATOMY OF THE SKATE

The skates or rays are, like the dogfish, elasmobranch fishes but are highly modified as regards external form and proportions in adaptation to bottom-dwelling habits.

1. Body and skin.—The body is divided into a greatly flattened anterior portion comprising head and trunk and a slender posterior portion, the tail. The broad, flat form is characteristic of bottom-feeding fishes and results from a shortening of the dorsoventral axis and an elongation of the transverse axis. The tough skin contains scattered scales with projecting spines; these scales are of the same type as in the dogfish, consisting of an imbedded *basal plate* and a projecting *spine,* but are much larger and fewer than in the dogfish. They have a definite arrangement, differing somewhat in the two sexes. Females have scales over the lateral expansions of the trunk and several rows of scales on the median dorsal part of trunk and tail, whereas in males the lateral expansions are mostly devoid of scales, there are fewer rows along the middle of back and tail, and enlarged scales occur on the head margins. Males further have two rows of curious erectile spines near the margins of the lateral expansions of the trunk. Do scales occur on the ventral surface? Note marked difference in color between the dorsal and ventral surfaces.

2. Fins.—There are median and paired fins, as in the dogfish; but the former are much reduced, consisting of two small dorsal fins on the end of the tail. The enormously enlarged pectoral fins form the lateral trunk expansions already mentioned and continue forward along the sides of the head. The smaller *pelvic* fins are immediately posterior to the pectoral fins, with which they are continuous in some species. They consist of two lobes and in the males bear large stout *claspers,* deeply grooved along their posterior lateral margins, employed in mating.

3. Head.—The head, like the trunk, is greatly flattened and is continuous laterally with the pectoral fins. It terminates in a pointed *rostrum* and bears dorsally a pair of large projecting eyes, without lids. Behind each eye is the large *spiracle* or *first gill slit,* bearing a valve and having on its anterior face parallel ridges representing a rudimentary gill. In skates the respiratory water enters chiefly by way of the spiracle, in correlation with their bottom-dwelling habit. On the ventral side of the head is the mouth, bounded by tooth-bearing jaws. The jaws and teeth are commonly larger in the males than in the females. In front of the mouth are the two nostrils, each provided with a fringed ear-shaped flap. Extending posteriorly from each nostril to the angle of the mouth is a flap with a fringed posterior margin. This is the *nasofrontal process*. On lifting up this process a deep groove, the *oronasal groove*, will be found, extending from the nostril into the mouth cavity. This arrangement foreshadows the appearance of a closed passage from the nostrils into the mouth such as is present in higher vertebrates. Posterior to the mouth are five pairs of gill slits. In skates the pectoral fins have grown forward above the gill slits and fused with the sides of the head, thus shoving the gill slits to the ventral surface.

4. Anus.—The *anus* or *cloacal aperture* is a large opening between the bases of the pelvic fins.

C. ADAPTIVE RADIATION IN ELASMOBRANCHS

Adaptive radiation means the spread of a group into various types of habitat with concomitant alterations of behavior and of morphological features. It is naturally most in evidence in those groups which are represented by large numbers of species, since the competition between the numerous species for food and breeding places forces them to spread into all situations possible to their type of structure. In connection with the following and similar accounts, specimens in museum collections should be examined, or pictures in books may be consulted.

The elasmobranchs, being a rather small group, do not exhibit adaptive radiation as well as the bony fishes; but they do illustrate the characteristic contrast between rapid-swimming predaceous types and sluggish bottom-feeding forms. The spiny dogfishes and various larger sharks, such as the sand shark (*Carcharias*), show the typical shark stream-lined form associated with rapid swimming and predaceous habits. From this type the mackerel sharks diverge in the direction of still greater speed, having more slender bodies, reduced unpaired fins, and a lunate mackerel-like tail; and their sharp teeth add to the picture of a carnivorous hunting type. Here belong several large and voracious sharks, including *Carcharodon*, which may reach a length of over thirty feet, has formidable triangular sharp-edged teeth, and may attack man. Another line of variation is seen in the thresher sharks with their very long flail-like tails. The notodanid sharks, which may have six or seven gill slits instead of the usual five, tend to slender form, culminating in the eel-like frilled shark, *Chlamydoselachus*. More sluggish habits are evidenced by the broadening of the head, flattening of the teeth, and enlargement of the pectoral fins. Beginning stages are seen in the nurse sharks, Port Jackson shark, etc. The broadening of the head reaches a climax in the hammerhead shark, in which the eye-bearing lateral head expansions probably have a rudder action. With increased sluggish, bottom-feeding habits the body flattens dorsoventrally, the pectoral fins enlarge, and the caudal region reduces; transitional stages from the shark to the ray type are seen in the angel ray, *Squatina*, and the guitarfishes, *Rhinobatis*. Finally there are produced the typical skates and rays, extreme examples of bottom-living fishes, with very broad flat bodies, enormously expanded pectoral fins, reduced tail, and crushing teeth for feeding on shelled prey, as crabs and mollusks. The gill slits move to the ventral side, while the spiracle remains dorsal. The reduction of the caudal region reaches its climax in the sting rays, where the tail has become a long slender whip armed with a saw-toothed spine capable of inflicting a severe wound.

In both the sharks and the rays there is found the "saw" type, in which the upper jaw is elongated into a long bladelike snout armed along its sides with teeth, used for damaging small fish used for food.

The torpedoes or electric rays have an electric organ, formed of modified muscle, consisting of a large number of hexagonal cells in the bases of the pectoral fins; these act like a storage battery and can deliver a severe electric shock.

The chimaeras or holocephalans are peculiar-looking fish with smooth, silvery skin, large heads with staring eyes, large pectoral fins, and reduced hinder body, usually tapering to a long slender tail (whence the name of rattails given these fish by fishermen). On the summit of the head of males there is a fleshy projection, the frontal clasper. In addition to the usual claspers of the pelvic fins, males also have accessory protrusible structures in a pocket in front of the pelvic fins. The small gaping mouth has grinding plates. There is but one slit, in front of the pectoral fins, but this is not a gill slit; it is the opercular opening, as the gills are covered with an opercular fold. This differs from the operculum of bony fishes in lacking skeletal support. The lines of sensory organs of the lateral-line system are usually evident on body and trunk. The genus *Hydrolagus* is common on the west coast of North America.

D. CLASS OSTEICHTHYES: EXTERNAL ANATOMY AND ADAPTIVE RADIATION IN TELEOSTS

All common fishes belong to the order Teleostei, characterized by an ossified skeleton and a bony flap, the *operculum*, covering the gill slits. The following description is based upon the perch, but any common bony fish may be used.

1. Body and fins.—The body has the fusiform shape typical of aquatic animals; in the perch and many other teleosts it is somewhat compressed laterally. The body is indistinctly divided into head, trunk, and tail. Trunk and tail are clothed with thin overlapping scales, arranged in diagonal rows and set in pockets in the deeper part of the skin (dermis). Remove some scales and note the pockets. The head is covered by the soft epidermis (which also forms a thin film over the surfaces of the scales) and in some regions bears small scales like those of the body. Beneath the head epidermis will be noted large, thin, flat bones; these, which are the outer bones of the skull, are in reality enlarged scales which have sunk from their original superficial position to a deeper location. A lateral line is present along each side of the body.

2. Head.—The head bears a terminal mouth bounded by well-developed jaws formed of thin, flat bones similar to the skull bones. The terminal position of the mouth is probably more primitive than the ventral position found in the elasmobranchs. On the dorsal side of the anterior end of the head are two pairs of nostrils, a pair to each olfactory sac. This arrangement permits a current of water to circulate through the olfactory sac. The large eyes are without lids. The ears, situated behind the eyes, are invisible externally. The posterior and lateral margins of the head are formed by a large flap, the *gill cover* or *operculum*, which is supported by several *opercular bones*—large, flat, scalelike bones, already noted. It covers a wide slit in the body wall known as the *gill opening*. Attached to the ventral margin of the operculum is a membrane, the *branchiostegal membrane*, supported by seven bony rays, the *branchiostegal rays*. Grasp the membrane with a forceps and spread it out to see the rays. Lift up the operculum and look within the cavity which it covers. Four curved structures, the *gill arches,* which should be separated from each other with a forceps, will be seen. Each bears on its outer surface a *gill,* consisting of a double row of soft filaments, and on its inner margin a series of short toothlike processes, the *gill rakers.* Thrust a probe inward between two gill arches, open the mouth of the fish, and observe that the end of the probe has entered the mouth cavity. The cavity of the pharynx is thus in communication with the exterior through the spaces between the gill arches. These spaces are gill slits corresponding to those which we saw in the dogfish and skate, but here the portions of the body wall between successive gill slits have disappeared, and all open into a common cavity covered by the operculum. This condition is characteristic of all fishes except elasmobranchs. When the fish respires, the mouth opens, the opercula move outward, the branchiostegal membrane unfolds and closes the gill opening; water is thus drawn into the mouth and bathes the gills. The mouth then closes; the opercula move inward; the branchiostegal membrane folds up; and the water passes out through the gill slits and gill opening.

3. Fins.—The body is provided with median and paired fins. Of the former there are an anterior and a posterior dorsal fin, a caudal fin, and a ventral or anal fin. (The number and position of the median fins are very variable in different bony fishes.) The caudal fin is of the bilobed symmetrical type common in bony fishes, termed *homocercal* (actually, as will be seen later, the vertebral column is asymmetrical with regard to the caudal fin). The paired fins comprise the usual pectoral and pelvic fins. The pectoral fins are located just behind the operculum, but the pelvic fins in the perch have moved forward from their normal position at the level of the anus to a position nearly level with the pectoral fins. Such a forward migration of the pelvic fins is common among the teleost fishes (although in many the original position is retained) and is often associated with the diminution and degeneration of these fins. The fins of bony fishes are supported by bony flexible jointed fin rays, termed *lepidotrichia,* the joints of which are modified scales. In many of the higher teleost families, sharp stiff spines replace some or all of the lepidotrichia in certain fins. This is the case with the anterior dorsal fin of the perch and the anterior parts of some of the other fins.

4. Openings.—In the midventral line just in front of the ventral fin is the large anal opening, and behind this there is a depression into which projects a small elevation, the urogenital papilla. The perch has no cloaca, but the intestine and urogenital systems open separately. This is one of the marked differences between elasmobranch and teleost fishes.

5. Adaptive radiation in teleost fishes.—The bony fishes today are a group at the height of their evolution and comprise thousands of species presenting a bewildering range of form, coloration, and habits. The stream-lined, torpedo-like shape of the trout may be taken as the norm. Still more adapted for speed are the scombroid fishes, including the mackerels, tunny, bonito, etc., with pointed head, smooth, small-scaled body, reduced fins, slender tail base, and powerful crescentic tail. A common variation in fishes is lateral compression of the body, leading to forms like the sunfishes of fresh water, and the angelfishes, butterfly fishes, etc., of coral reefs. The high, laterally flattened body appears to be an adaptation for moving among dense growths. This shape reaches an extreme in the *Mola* or ocean sunfish, in which the rear part of the body seems to be missing. The converse variation, dorsoventral flattening, is not so common among teleosts, where the most common bottom-dwelling types, such as the flounders, soles, etc., are really laterally compressed like the foregoing and lie upon one side. These fishes are symmetrical when young but soon adopt the habit of lying on one side, whereupon the head and eyes become twisted to the upper side. The eels and morays illustrate another extreme of form, an elongated snakelike body with numerous vertebrae and a continuous fin around the rear part. One family of fishes, the globefishes, including the puffers, porcupine fish, etc., has the ability to inflate the body to a globular shape, presumably for protective purposes. Of the innumerable variations of head shape, there may be mentioned the pipefishes and sea horses with their long tubular snouts; and the swordfish with its long pointed jaws, used to cut up small fish serving as food. The suckers are a family of fresh-water fishes with an elongated mouth for sucking in bottom food. The catfishes and bullheads are a group of fishes, many living in fresh water, distinguished by their smooth scaleless skins and barbels on the head; these barbels are covered with taste buds and serve to detect food as they are dragged over the bottom.

A large group of fishes called the blennies, mostly of small size with enlarged heads and long dorsal and anal fins, live on rocky shores and try to escape by leaping about when exposed by the turning-over of rocks. Some can clamber about on rocks by the pelvic fins, which are far forward and reduced to a few spines. It is not uncommon for fish to be able to scramble on rocks by fin rays converted into spines. The famous climbing perch (*Anabas*) of India is able to ascend tree trunks by means of spines on the highly movable opercula; this fish and its relatives have an accessory respiratory organ in the form of folds from one of the gill arches as an adaptation for respiring in air.

Suckers for clinging to rocks have been developed in different families of fish. In the lumpsuckers and gobies, the anteriorly placed pelvic fins are united to form a rounded fin which, completed behind by a fold of skin, serves as a sucker. In the cling fishes the sucker consists entirely of skin folds. A different sucking adaptation is seen in the shark suckers or remoras, related to the gobies, where the first dorsal fin is modified into a sucking disk on top of the head, which they use to attach to other fish so as to be carried about. It is related on reliable authority that natives use the remoras to catch fish and turtles; the remora fastened by a line around its tail base is turned loose and attaches to a fish or turtle, which can then be hauled in (Gudger, 1919).

A great enlargement of the pectoral fins for "flying," i.e., planing through the air above the water, has occurred in two different families, the flying fishes and the flying gurnards. After much argumentation it has been established that the sole force of the flight comes from sculling with the tail before the leap from the water. The pectoral fins act merely as planes and are not flapped.

Luminous organs or photophores occur in four unrelated groups of fishes, chiefly deep-sea inhabitants. The photophores consist of luminous spots along the sides of the body or on the head.

Perhaps the most curious of all fishes are the anglers, fishes of bizarre appearance with enlarged froglike heads, and the first spine of the reduced dorsal fin modified into a fishing lure, which curves forward over the head and terminates in an enlargement simulating a bait. This bait is luminous in some of the anglers. It serves to attract other fish, which are then seized as food.

E. SOME TYPES OF LOWER FISHES

The following fish are representatives of groups lower in the taxonomic scale than the teleosts, but the latter have been introduced first to furnish a basis for comparison.

1. **The lungfishes (superorder Dipnoi).**—This group is represented at present by three genera: *Epiceratodus* in Australia, *Lepidosiren* in South America, and *Protopterus* in Africa. The first is less modified than the others and hence more suitable for study. Note rather small head; stout trunk clothed with large, thin, rounded scales (cycloid type) set in pockets and arranged in diagonal overlapping rows, as in teleosts; and pointed tail symmetrically edged with a caudal fin. The tail is of the type termed *diphycercal* (see further, p. 113), i.e., primitively bilaterally symmetrical with a straight vertebral axis along its middle and with the fin equally developed above and below. The most peculiar feature of the fish is the paired fins, which have retained their primitive location. Each consists of a scale-covered lobe edged with fin rays. The skeleton of these lobes is composed of a jointed axis bearing lateral rays. The fin rays of all the fins are soft and flexible and resemble ceratotrichia but apparently are degenerate lepidotrichia. In *Lepidosiren* and *Protopterus* the paired fins are reduced to a slender narrow shape. The anal opening is situated behind the bases of the pelvic fins. Immediately in front of the base of each pectoral fin is a slitlike gill opening, formed by a nonbony operculum. There are no spiracles. Note eyes. The nostrils are located on the under side of the upper lip, and Dipnoi differ from all other living fish[2] and indicate relationship to land vertebrates in that the nasal sacs open into the roof of the oral cavity by internal nares. Dipnoi also have a rather well-developed single or bilobed lung (=air bladder) with a very spongy vascular wall supplied by blood vessels corresponding to the pulmonary arteries of land vertebrates. The lung connects with the esophagus by a duct, and lungfishes can breathe air and so survive drying.

2. ***Polypterus* (order Chondrostei).**—This interesting fish, found in Africa, a member of the great subclass Actinopterygii or ray-finned fishes, presents a number of primitive characters. Note investment of *ganoid* scales, hard shiny scales of rhomboid shape, arranged in diagonal rows. They are continuous with the large scales on the head, which are, in fact, the outer bones of the skull. What type of tail has the animal? There is a small anal

[2] The living crossopterygian fish taken on the African coast (see p. 7 and Fig. 1*A*) surprised its investigators in that it had no internal nares, although these are characteristic of extinct Crossopterygii.

somebody goofed

fin and a long dorsal fin subdivided into finlets. The paired fins retain the primitive position, and each has a basal scaly lobe, as in Dipnoi and Crossopterygii. Note opercula supported by bones dorsally, and joined ventrally across the throat as a membrane containing two *gular* plates. This membrane and its gular plates are the forerunners of the branchiostegal membrane and rays characteristic of teleosts. Find the spiracles by reference to Figure 58*B*. Note pair of double nostrils near the end of the snout; the anterior nostril is mounted on a projecting tube.

3. ***Acipenser,* the sturgeon (order Chondrostei).**—Note ganoid scales arranged in rows on the trunk with areas of apparently naked skin (really bearing small scales) between. Most of the large scales are crested. They pass onto the head, transforming into skull bones. The broad head is extended anteriorly into a *rostrum* clothed dorsally with many small scales, provided ventrally with four branched projections, the *barbels*, used as sense organs for the detection of food. Each nostril is provided with two openings. Note slitlike spiracle above each eye and operculum supported by a single bone. The curious mouth is distensible and bordered with sensory papillae. The degenerate jaws lack teeth. Note paired and median fins and heterocercal tail.

4. ***Polyodon,* the spoonbill (order Chondrostei).**—In this curious fish, native to the Mississippi River and its tributaries, the rostrum is expanded into a broad, thin, spatulate structure, provided with sense organs for the detection of food. In front of the eyes are the double nostrils, and behind them the small spiracles. The operculum lacks opercular bones and continues into a branchiostegal membrane without rays. Lift up the operculum and note the numerous long and fine gill rakers on the gill arches, used by the fish in separating food particles from mud. The jaws open widely and are provided with minute teeth. Note fins and type of tail. Are there any scales? This fish is singularly lacking in external hard parts.

5. ***Lepidosteus,* the gar pike (order Holostei).**—Note complete armor of hard shining ganoid scales of rhomboid shape set in diagonal rows, as in *Polypterus*. They pass onto the head as enlarged plates, which constitute the roof bones of the skull. The very elongated toothed jaws form a snout bearing the small nostrils on its tip. A spiracle is absent, but there is the usual bony operculum. Note median and paired fins with their stout lepidotrichia. The tail is heterocercal but approaches the homocercal type.

6. ***Amia,* the bowfin (order Holostei).**—This fish, found in lakes and streams in the United States, approaches closely the teleost type, as shown by the thin imbricated scales (cycloid type), nearly homocercal tail, bony operculum with attached branchiostegal membrane, and lack of spiracles. The large scales on the head function as roofing bones of the skull. In the

branchiostegal membrane there is a median gular plate in front and a series of lateral gular plates on each side, which are nearly modified into branchiostegal rays. Of the double nostrils, the anterior one is borne on a tube near the anterior end of the head; the posterior one is situated in front of the eye. Note median and paired fins supported by lepidotrichia.

The Polypterini, Chondrostei, and Holostei are often spoken of as *ganoid fishes*, as most of them are covered with the ganoid type of scale. The term is not very exact but has a certain convenience.

F. CLASS AMPHIBIA: EXTERNAL ANATOMY

The amphibians are the lowest land vertebrates, which became adapted to the land habitat through the better utilization of the air bladder for respiratory purposes, the alteration of the lobed fins of the Crossopterygii into walking limbs, and other changes. The members of the order Caudata, or salamanders, resemble in general appearance early land vertebrates, although their internal anatomy is considerably altered from that of primitive amphibians.

1. *Necturus*, example of a salamander.—The body has the form typical of early land vertebrates, carried horizontally close to the ground, with broad flat head, no definite neck, elongate trunk, and relatively large tail, laterally flattened in *Necturus*. The skin is naked and slimy, without scales or other hard parts common in vertebrates. Without such protection, a moist habitat is necessary; in fact, *Necturus* spends its life in water.

The head has a terminal mouth provided with lips and bears the usual three pairs of sense organs. The nostrils or *external nares* are a pair of widely separated openings just back of the upper lip. By probing into them determine that they communicate with the mouth cavity by openings known as the *internal nares*. This arrangement permits air to enter the mouth cavity through the nostrils and contrasts with the condition in most fishes where the olfactory sacs are blind and have no connection with the mouth cavity. In some fish, as in the skate, an external oronasal groove extends from each olfactory sac to the mouth; and by the fusion of the borders of such a groove a closed nasal passage is produced during the embryonic development of land vertebrates. Such a nasal passage and internal nares are already present in those fish groups in or close to the line of ancestry of tetrapods, i.e., the Crossopterygii and the Dipnoi. Note small eyes without eyelids; the ears, as in fish, are internal only and are not detectable on the surface.

From each side of the posterior margin of the head spring three gills, each consisting of a fringe of filaments dependent from a dorsal process. They are *external gills* and do not correspond to the gills of fishes, which are internal. Between the first and second and the second and third gills are the *gill slits*, which open, as in fishes, into the cavity of the pharynx. The animal, however, does not pass water through the gill slits but respires by means of the external gills, kept in constant motion, through the general surface of the body, and to some extent by means of its lungs. A fold of skin, the *gular fold*, passes transversely across the throat.

The trunk bears two pairs of appendages, the walking limbs, which correspond to, and have evolved from, the paired fins of fishes. Each consists of a succession of parts jointed together; these are: *upper arm*, *forearm*, *wrist*, and *hand*, in the case of the forelimb; *thigh*, *shank*, *ankle*, and *foot*, in the hind limb. Both hand and foot bear four *digits* (fingers, toes), although five is the typical vertebrate number; the first digit is the one which is missing. The position of the limbs with reference to the body is very primitive, especially in the case of the hind limb, and should be carefully studied. Note that the hind limb projects out at right angles to the body, all of its parts on a plane parallel to the ground. In this primitive position the limb has an anterior or *preaxial* border, a posterior or *postaxial* border, and dorsal and ventral surfaces. In the forelimb, however, the forearm is bent downward, and the hand is directed slightly forward. This alteration of position is brought about chiefly by an outward torsion of the upper arm, whose former preaxial surface now looks dorsally. The preaxial border of the forearm is turned medially, its postaxial border laterally. As a result of these changes the animal is able to lift itself to a slight extent above the ground.

The flattened tail is bordered by a tail fin which differs from the fins of fishes in that it contains no fin rays. Note large anus at the junction of trunk and tail. Amphibia, like elasmobranchs, have a cloaca.

2. Other salamanders.—Specimens of other salamanders, if available, should be examined and compared with *Necturus*. Note general body proportions; shape of head; presence or absence of external gills; number of gill slits; length of trunk; shape, size, and position of limbs; number of toes; shape and relative size of tail. The larvae of all salamanders have three pairs of external gills, but these are retained in the adult state only in the genera *Necturus*, *Siren*, *Proteus*, and *Pseudobranchus*. The number of gill slits, however, varies from one to three pairs. In some species of *Cryptobranchus* and *Amphiuma* a pair of gill slits remains throughout life after the disappearance of the gills. Most salamanders have neither gills nor gill slits when adult; this is the case in *Ambystoma*, *Triton*, *Triturus*, *Desmognathus*, *Plethodon*, etc. *Siren* lacks hind legs.

3. Comparison with frogs and toads.—These belong to the order Salientia (Anura). Note especially body posture; shape of head; sense organs of head; short, plump trunk; extreme development and specialization of the hind limbs for jumping; number of digits; webs on feet; position of anus; absence of tail. The hind legs appear to have four sections instead of the usual three, because of the extraordinary elongation of the ankle. On each side of the head behind the eye note *eardrum* covering the middle ear. Unlike salamanders, frogs and toads have a middle ear in addition to the inner ear, but the eardrum remains flush with the surface. Scientifically, toads are distinguished from frogs by characters of the skeleton; and the usual conception of a toad as a squatty, warty animal with a pair of large glandular swellings (parotoid glands) on the head behind the eyes applies, in fact, only to some members of the family Bufonidae.

G. CLASS REPTILIA: EXTERNAL ANATOMY

Present-day reptiles are easily distinguished from all other living vertebrates by their dry, scaly skins. Gills are lacking in all stages of the life-history. Many present reptiles are similar in general shape to early land tetrapods, but the class also contains groups highly modified structurally in relation to mode of life.

1. Lizards (order Squamata, suborder Lacertilia).—Any common lizards may be examined—for example, *Anolis* (so-called chameleon of our southern states), *Cnemidophorus* (striped lizard or race runner), or *Sceloporus* (spiny swift) or *Crotaphytus* (collared lizard). The body is characteristically tetrapod in form, divisible into head, neck, trunk, and long tail. We have seen that fishes and amphibians lack a distinct neck, the appearance of which is correlated with the completed adoption of the land habitat. The body is completely clothed with horny scales, plates, or tubercles, which are not separate, detachable structures like the scales of teleost fishes but are merely cornified thickenings of the epidermis and hence form a continuous sheet which is shed at frequent intervals. These horny thickenings occur in transverse or diagonal rows. In many lizards they are crested or spined in at least some body regions, often along the middorsal line, whereas other species present a smooth appearance. Notable among the spiny types are the horned lizards (usually called "toads," genus *Phrynosoma*) of the southwestern United States, armed with conspicuous spiny projections; the head spines or horns are projections of the skull bones covered with a thin horny shell. The scales or plates of lizards are often of different sizes in different body regions and on the head are almost always enlarged to form the *head shields*, much used in identifying lizards and named after the underlying skull bones. In some lizards bony pieces underlie the horny skin thickenings.

Locate external nares (nostrils); do they open into the mouth cavity? The eyes of lizards usually have movable upper and lower eyelids, and there may be present a third eyelid, the nictitating membrane, a thin membrane concealed from view when not in use in the anterior corner of the eye, where it may be sought with a forceps. Halfway between the eye and the base of the forelimb there is in many lizards a conspicuous ear. The visible part of this may consist, as in frogs, of the eardrum flush with the surface or slightly depressed; or, as in *Anolis*, the eardrum may have sunk inward and lies at the bottom of a depression, the *external ear*. In some lizards the eardrum is more or less concealed under projecting scales or may be hidden under the skin. The eardrum belongs to the part of the ear termed the *middle ear*, present in reptiles in general, also in frogs and toads, but lacking in fishes and salamanders; and in some reptiles, as just seen, there is a further advance in the formation of an external ear passage by the insinking of the eardrum.

In many lizards (*Anolis*, *Sceloporus*, *Crotaphytus*, *Phrynosoma*) there occurs in the center of the head, shortly behind the level of the posterior ends of the eyes, a small spot, usually greenish, resembling a lens, often in the center of an enlarged scale. A hand lens or binocular microscope will aid in seeing it. This is a modified, translucent scale which overlies the *pineal eye* (more correctly called *parietal eye*), a third median eye found in a number of vertebrates. It lies in an opening through the roof of the skull, beneath the modified scale. It has the histological structure of an eye and probably serves some optic function.

Skin folds are common in lizards on throat, neck, sides of trunk, or middorsal line. Often these can be erected and are then displayed in courtship or combat. They may be limited to the male sex. In *Anolis* and some other lizards there is such a fold, the *gular or throat fold*, running lengthwise of the throat, which can be extended (with the aid of a special part of the hyoid apparatus) to form a fan or *dewlap*, red in color because the red throat skin shows between the scales; this dewlap is much larger in male than in female *Anolis*. The flying lizards (*Draco*) of the Indo-Malay region parachute by spreading out lateral folds of the trunk, supported by rib extensions.

The paired limbs are similar to those of *Necturus;* they bear five digits, the typical vertebrate number, terminating in horny claws. The limbs depart further from the primitive position described in connection with *Necturus*, thus lifting the animal above the ground to a greater extent. In the hind limb the thigh still extends out at right angles to the body axis but is slightly rotated forward, so that the original dorsal surface is becoming anterior. The shank is directed ventrally and posteriorly and has undergone the same sort of torsion as the thigh, whereas the foot retains the primitive position. In the forelimb the upper arm is rotated in the opposite direction from that observed in the thigh, so that the original preaxial surface is now dorsal. The upper arm is directed posteriorly; the forearm and hand, downward and forward, and so rotated that their original preaxial borders look inward instead of forward. Expanded adhesive digits are common among lizards, enabling them to climb walls and trees.

Along the inner surface of the thighs of many lizards (not *Anolis*), especially in the males, is found a row of pores, the *femoral pores*. These are the openings of sacs which secrete a yellowish waxy material of unknown purpose, probably used in copulation.

The anus has the form of a transverse slit, a shape characteristic of lizards and snakes. In males of lizards there are often a few enlarged scales behind the anus, and these enable one to distinguish the sexes. *Preanal* pores, similar to the femoral pores, may also occur.

2. Snakes (order Squamata, suborder Ophidia).—Snakes in general differ from lizards in lacking limbs and eyelids. However, the absence of fore, hind, or both limbs is not uncommon in lizards; and eyelids are wanting in one whole group of lizards, the geckos. Lizards and snakes are really distinguished by characters of the skeleton. Examine several kinds of snakes; note covering of horny scales or plates, whether smooth or crested, and relative sizes in different body regions. How do you determine boundary of trunk and tail? On the head and jaws note head shields, much used in identification of snakes, position of nostrils, shape of pupil of eye. The middle ear is degenerate in snakes and completely concealed from surface view. On the ventral surface note large *ventral* shields contrasting with small scales of the dorsal side. The *anal* shield, in front of the anus, is divided in many snakes. The ventral tail shields may differ from those anterior to the anus.

A group of poisonous snakes called pit vipers, including the copperheads, moccasins, and rattlesnakes, differs from all other snakes in having a pit on the side of the head behind and below the nostrils. These pits, whose function was long an enigma, are now known to act as temperature-detectors and enable the snake to determine the presence of a warm object. In nature such objects would be warm-blooded animals used as food, such as rabbits. The rattle of rattlesnakes consists of dried rings of cornified skin. One such ring is left behind at each molt; and, as rattlesnakes molt about three times a year, the age of a rattlesnake could be estimated from the rings of the rattle were it not that the terminal rings are apt to break off in time; rattles of more than ten rings are very rare. Another characteristic of pit vipers is the vertically elongated pupil of the eye.

3. *Sphenodon* (order Rhynchocephalia).—The *Sphenodon* of New Zealand, called *tuatara* by the natives, is the sole survivor of the order Rhynchocephalia. It is the most primitive living reptile, but its primitive characters are internal, chiefly skeletal, and externally the animal cannot be distinguished from a large lizard. Compare a preserved specimen with the description of lizards given above; the most noticeable feature is the middorsal row of spines. The parietal eye of *Sphenodon* is better differentiated histologically than in any other living vertebrate, but it is evident externally only in young specimens. In grown ones there is usually no sign of its location present in the skin. *Sphenodon* was in danger of extinction but is now protected in preserves on some small islands of the New Zealand group.

4. Turtles (order Chelonia).—The turtles commonly seen in laboratories are the painted pond turtles (genus *Chrysemys*). The body form of turtles is considerably modified from the typical vertebrate shape. The reptilian head is followed by an unusually long and flexible neck; the trunk, incased in a hard shell, is remarkably broad and flat; and the tail is diminished in diameter and length. The turtle is thus one of those forms in which, as in the skate, the transverse axis is elongated, whereas the other axes are shortened. The skin shows the usual reptilian cornification in the form of large horny shields covering the shell and small scales and thickenings on the legs, tail, and other exposed parts. The skin of the head is usually bare but in some turtles is marked off into large head shields.

The position of the external nares, close together at the tip of the head, permits the animal to breathe air with only a slight exposure above water. The jaws lack teeth and instead are clothed with hard horny beaks. The large eyes have upper and lower eyelids and a nictitating membrane located in the anterior corner of the eye. Just behind the angle of the jaws is seen the circular eardrum flush with the surface, as in frogs.

The shell which incloses the trunk consists of a dorsal arched *carapace*, a ventral flat *plastron*, and lateral *bridges* connecting carapace and plastron. Note—best on a live turtle—how head, neck, limbs, and tail can be neatly folded under the margins of the shell. The shell consists of large *horny* shields overlying *bony* plates, each arranged in a definite pattern described later. The carapace is usually more arched in female than in male turtles. Note parts of the limbs, number of digits on each, claws, webs. The claws are longer in male than in female *Chrysemys* and some other genera. The positions and torsions of the parts of the limbs are similar to those of lizards. Note the rounded anus.

Some variants in external features of turtles may be noted. The sea turtles reach enormous size, up to 1,000 pounds in weight, and have large head shields and limbs modified into swimming flippers. In the snapping turtles (*Chelydra*) carapace and tail are crested, and the plastron is reduced to a narrow cross-shaped form which affords no protection for projecting parts. These turtles defend themselves by their vicious fighting habits. In the mud turtles (*Kinosternon*) the front and rear parts of the plastron are hinged and can be drawn up closely against the carapace. A similar provision is seen in the box turtles (*Terrepene*), where the plastron is hinged across the middle and is not bridged to the highly arched carapace. The box turtles have terrestrial habits. The most curious turtles are the soft-shelled turtles (*Amyda*) with their flat, soft, leathery shells, devoid of skeletal support, their elongated snout, and broad webbed swimming feet.

5. Alligators and crocodiles (order Crocodilia).—The crocodilians are very large reptiles of lizard-like appearance and semiaquatic habits. The order is represented in the Americas by the Mississippi alligator of our southern states and by the American crocodile, found in southern Florida, in Mexico, and southward into South America. The snout of the alligator is broad and blunt, that of the crocodile narrowed and somewhat pointed. Compare the alligator with lizards as to general body form and proportions, attitude of the limbs, number of digits, etc. Note position of the nostrils. The prominent eyes have upper and lower eyelids and a well-developed nictitating membrane; the upper eyelid incloses a bony support. As in some lizards, the alligator has an external ear, opening immediately behind the eye by a longitudinal slit guarded above and below by skin folds.

The external covering of the alligator consists, as in lizards, of squarish and rounded horny thickenings of the skin, which are not molted in large pieces. On the animal's back these thickenings are underlain by bony plates of the same shape and closely adherent to them. Some of the bony plates are crested, and the horny thickenings follow the crests. In larger specimens the horn of the back is often worn away, exposing the bony plates. Alligator leather, like other leather, consists of the tanned inner layer of the skin (dermis, see chap. vi), including the bony plates, and retains nothing of the outer horny thickenings.

H. CLASS AVES: EXTERNAL ANATOMY

The Aves or birds are readily known from all other vertebrates by the covering of feathers and the modification of the forelimbs into wings. The pigeon is convenient as an example; study specimens with and without feathers.

1. Body form and skin.—The proportions of the body, as seen in the plucked specimen, bear a general resemblance to those of the turtle. The head is well developed, the neck long and flexible; the trunk is shorter and considerably plumper than normal, and the tail reduced to a stump, the *uropygium.* The ancestors of birds had long tails like lizards, bearing feathers along their entire length.

The body is clothed with a covering of *feathers,* which conceal its shape. These feathers are called *contour* feathers. The contour feathers are of two kinds: the large, stout feathers of the wings and tail, or *flight* feathers, and the smaller feathers, the *coverts,* which cover the bases of the wings and tail and the general surface of the body. On the plucked bird from which the contour feathers have been removed note the presence of hairlike processes, the *hair feathers* or *filoplumes;* the delicacy of the skin; and the numerous deep pits, the *feather follicles,* into which the contour feathers were set. Observe that the feather follicles are not uniformly distributed on the body but occur in tracts.

2. Head.—The head terminates in the elongated *beak,* which consists of the upper and lower jaws incased in horny sheaths. Teeth are absent on all modern birds, although extinct birds possessed them. The base of the upper beak bears a cushion-like protuberance, the *cere,* a structure occurring only in certain families of birds. Under the anterior margins of the ceres are the slitlike *external nares.* The remarkably large eyes are provided with upper and lower lids and with a nictitating membrane which may be drawn across the eye from the anterior corner. The ear is behind and below the eye and is observable only on the plucked specimen. As in the crocodilians and some lizards, the middle ear has sunk below the surface, with the resulting formation of an external ear. The deep, narrow passage which leads to the middle ear is more accurately called the *external auditory meatus.* The skin around the entrance to the meatus extends to elevate as a fold, and this fold plus the meatus forms the *external ear.*

3. Trunk.—The trunk is very firm and inflexible, owing to a fusion of the bones of the back. Pass your fingers along the back of the plucked bird and feel the skeleton. Feel also in the median ventral line of the trunk the projecting keel of the breastbone, to which are attached the great wing muscles. It is the presence of these muscles, the "breast" of the bird, which produces the plump contour of the trunk. The trunk bears the two pairs of limbs, of which the anterior pair is remarkably modified into *wings* or organs of flight. The hind limbs have also undergone considerable modification as a result of the biped mode of walking used by birds.

The parts of the wing, which are homologous to those of the forelimb of other vertebrates, should be studied on the plucked bird. In the folded condition the sections of the wing make angles with each other like the letter Z. The upper arm is short, directed posteriorly, and slightly twisted on its axis so as to bring the preaxial margin on the dorsal side. The forearm is longer and directed forward. The wrist and hand are fused together, and the whole is considerably elongated and directed caudad. There are but three digits, which are regarded as the first, second, and third. The first digit is the projection found just below the joint between forearm and wrist; the second digit forms the terminal point of the wing; the third cannot be seen externally. When the wing is extended, its parts have nearly the primitive position described under *Necturus*. The great flight feathers of the wings are known as *remiges;* those of the hand are called *primaries;* of the forearm, *secondaries;* and of the upper arm, *tertiaries* or *humerals*. The primaries differ from the others in that the soft part of the feather is wider on the posterior side of the central axis than on the anterior side. The remiges are borne on the postaxial margin of the wing, and the deep large feather follicles exposed by their removal should be noted on the plucked bird.

The hind limb is clothed partially with feathers and partially with horny scales, identical with those found in reptiles. The digits, of which there are but four—the fifth being absent—terminate in claws. The position of the hind limb with reference to the body is greatly altered. The whole limb, instead of extending straight out laterally from the body, is directed ventrally, thus raising the animal high above the ground. In order to achieve this result, it is obvious that the limb must have been rotated 90° from the primitive position, so that the original dorsal surface now faces anteriorly—that is, has become preaxial. The toes are consequently directed forward, with the exception of the first, which, through a secondary modification connected with the perching habit, points posteriorly.

4. Tail.—The tail stump bears a half-circle of large flight feathers, known as *rectrices*. Under the base of the tail is the anus or cloacal aperture, a transverse opening with protruding lips. On the dorsal side, just in front of the base of the tail stump, will be seen on the plucked bird, or by lifting up the tail coverts, a prominent papilla, the opening of the *uropygial gland*, from which the bird obtains oil for preening its feathers.

5. Other birds.—Many thousands of species of birds exist today, exemplifying in a striking way the innumerable possibilities of variation from an anatomical plan, and illustrating notably the principle of adaptive radiation. To gain some idea of adaptive radiation in birds, a visit to the bird collection of some large museum is desirable. Particular attention should be paid to body form and proportions, relative length and shape of wings, and types of bills and feet. It must be borne in mind that similarities in these regards do not necessarily indicate taxonomic relationship but may often represent cases of convergence or parallel evolution, adaptations for similar modes of life. Our conception of normality of body form and appearance among birds is drawn from the multitude of familiar smaller birds, known as the *passerine* or *perching* birds (order Passeriformes), inhabitants of woods, fields, and roadsides—for example, crows, thrushes, sparrows, finches, warblers, flycatchers, etc.; and these should form the basis for comparisons. Among the most striking modifications of birds are those associated with aquatic life. Forms which swim by using the feet as paddles have completely webbed feet, involving all four toes, like the pelicans and cormorants, or webs involving three toes, like the anseriform birds (ducks, geese, and swans), or partially webbed feet, like the grebes and coots. The somewhat squat body with the feet placed far backward is also characteristic of such paddling birds. Birds of wading habits tend to have elongated legs, seen in beginning stages in rails, plovers, and snipe and reaching an extreme of long leggedness in birds like the herons, cranes, storks, and flamingos. Very often such birds also have elongated necks, so that they may reach the ground, and long bills, straight or curved, especially if they seek their food by probing into mud or crevices. Other types of bills among birds of aquatic habits are the strong sharp bills of fish-catching birds like the terns and kingfishers, the pouch on the lower bill of pelicans, used as a scoop in fishing in schools of fish, and the anseriform bill, broad with transverse ridges acting as strainers in grubbing in the bottom mud. The most remarkably adapted aquatic birds are the penguins, with their upright gait, short legs, and reduced wings, used as flippers in diving and swimming under water. These wings can no longer be folded, as in other birds, but hang down extended and are useless for flight. The feathers of penguins are greatly reduced and resemble scales.

Wing shape is naturally correlated with powers and type of flight. Long, slender, pointed wings, sometimes reaching beyond the tail, are seen in birds of great flying powers and soaring habits, as in the gulls and terns and the not closely related birds of prey (falconiform birds, such as eagles, hawks, and vultures). Birds which do not soar but fly by continuous wing strokes have shorter, more rounded wings. Very short, broad wings occur in the gallinaceous birds (fowls, pheasants, grouse, quail, etc.), habitually ground dwellers, with running type of feet, which make short powerful flights by rapid wing strokes as a method of escape. Degeneration of the wings to a flightless condition has occurred in a number of birds: the penguins mentioned above and the ostrich tribe, including the ostrich, emu, cassowary, kiwi, etc. In the ostrich group many other changes accompany the loss of flight, as the disappearance of the keel of the breastbone, which supports the flight muscles; the loss of hooklets on the feathers, making them fluffy; and the development of strong running legs and feet. The *Apteryx* or *kiwi* of New Zealand reaches the extreme in the way of wing reduction.

In addition to aquatic and running modifications of feet already noted, there should be mentioned the powerful grasping feet or talons of birds of prey, with their curved toes and claws, which are used for grasping prey, chiefly smaller birds, while the strong curved beak tears them to pieces. Woodpeckers have highly adaptive feet, with two toes in front, two behind, for clinging to tree trunks; other adaptive features of this family are the strong pointed beak for making holes in trees, the stiff pointed tail feathers used as props, and the long tongue armed at the tip with barbs. In the gallinaceous birds the hind toe (really the first toe) is typically raised, and spurs are often present, whence the name game (i.e., fighting) birds, often applied to this group. Small, weak feet frequently characterize birds which spend most of their time on the wing, as swallows, swifts, and humming birds.

All these variants of structure have evolved by divergence from a common reptilian ancestor; an intermediate stage is represented by the extinct *Archaeopteryx*, of which two specimens have been discovered. This animal had teeth, wings with three separate projecting fingers terminating in claws, and a long tail bordered with feathers, but the feet were birdlike. Claws on the wing digits are also known in some present birds, notably the young of the hoatzin, a cuckoo-like bird of the South American jungles, where they are employed in climbing.

I. CLASS MAMMALIA: EXTERNAL ANATOMY

The mammals are warm-blooded vertebrates provided with hair (very scanty in some forms) and with mammary glands for nourishing the young with milk.

1. External anatomy of the cat (order Carnivora) or the rabbit (order Rodentia).—The body is divisible into head, neck, trunk, and tail. Its proportions depart somewhat from the typical tetrapod form in the larger size of the head and in the reduction of the tail, particularly noticeable in the rabbit. The body is clothed with closely set hairs, forming a covering of *fur*, characteristic of mammals. Around nose and mouth are a number of especially long and stout hairs, the *whiskers* or *vibrissae*, which have a rich nerve supply and serve as important tactile organs. The large size of the head is caused by the great development of the inclosed brain. The head may consequently be divided into an anterior *facial* region, in front of the eyes, and an enlarged posterior *cranial* region. The mouth has well-developed lips; the upper lip is cleft in the center, deeply so in the rabbit (whence the name *harelip* for a similar congenital defect in man), exposing the *incisor* teeth. The external nares are large and elongated, overhung by the mobile nose. The eyes have upper and lower lids and a nictitating membrane, which should be grasped with a forceps and drawn over the eyeball. The ears are provided with a long and flexible external fold, the *pinna*, which springs from the rim of the *external auditory meatus*, a passage descending deeply into the skull. Pinna and meatus constitute the external ear. The trunk is divisible into an anterior *chest* or *thorax*, supported by the ribs, and a posterior *abdomen*. On the ventral surface of the trunk occur four or five pairs of *teats* or *nipples* in female specimens, the openings of the milk glands or *mammary glands*. The trunk bears two pairs of limbs, composed of the same parts as in other vertebrates. The upper section of each limb—thigh or upper arm—is more or less included in the trunk. The limbs terminate in clawed digits—five in front, four behind, as the first hind toe is absent. The claws of the cat and its relatives are retractile.

The limbs have undergone a marked change from the primitive position. Instead of extending laterally from the body, they project ventrally and are elongated, so that the body of the animal is carried high above the ground. This change has involved a rotation of 90° in each limb. The hind limb is rotated forward, so that the original dorsal surface is anterior and the toes point forward. The knee or joint between the thigh and shank is likewise directed anteriorly. The forelimb, on the contrary, has rotated posteriorly, so that the original dorsal surface faces posteriorly and the preaxial surface faces laterally. Consequently, the elbow or joint between upper arm and forearm is directed caudad. As a result of the rotation of the upper arm, the toes would point posteriorly, but by an additional torsion they have been brought around to the front again. This torsion involves a crossing of the two bones of the forearm, the distal end of the preaxial bone being brought internal to the distal end of the postaxial bone (Fig. 7). This position of the forearm is known as the *prone* position and is imitated in the human arm when the arm hangs by the side with the back of the hand directed forward. In this position the crossing of the two long bones of the forearm can be felt. If now the arm is raised sidewise, shoulder high with the palm facing forward, the two bones return to the primitive parallel position, known as the *supine* position. Thus man can change his forearm from the prone to the supine position, but in most mammals the forearm is fixed in the prone position.

The position of the parts of the foot in walking is different in different mammals. The rabbit and the cat walk on the digits, with the remainder of the hand and foot elevated. This type of gait is known as *digitigrade*. Man walks on the whole sole of the foot, the primitive method, known as *plantigrade*. Horses and cattle and other ungulates walk on their nails, which are broadened into hoofs, a mode of walking called *unguligrade*.

In the majority of mammals there is no cloaca, and anal and urogenital openings are separate. The region which includes these openings is termed the *perineum*. The anus is located in the midventral line just in front of the base of the tail. On each side of the anus there is in the rabbit a deep, hairless depression, the *inguinal* or *perineal space*, into which open the *inguinal glands* (not visible externally), source of a characteristic odoriferous secretion. In females the urogenital opening is situated immediately anterior to the anus, which it resembles. It is inclosed by a fold of skin called the *greater lips* or *labia majora*, and opening and labia together constitute the *vulva*. In males the urogenital structures in front of the anus consist of the *penis* or *organ of copulation* (for transmitting sperm to the female) and a double pouch of the body wall, termed the *scrotum* or *scrotal sac*, inside which the testes are located. In cats the scrotum is conspicuous as a pair of rounded eminences anterior to which is a fold of skin, the *prepuce* or *foreskin*. The opening in the center of the prepuce is not, as might be thought, the urogenital opening but is merely the depression left by the withdrawal of the penis into the prepuce. In male cats the penis is generally so far withdrawn as to be invisible externally. In the **rabbit** the prepuce forms a hillock of skin in front of the anus, and the end of the penis is usually visible, projecting from the prepuce.

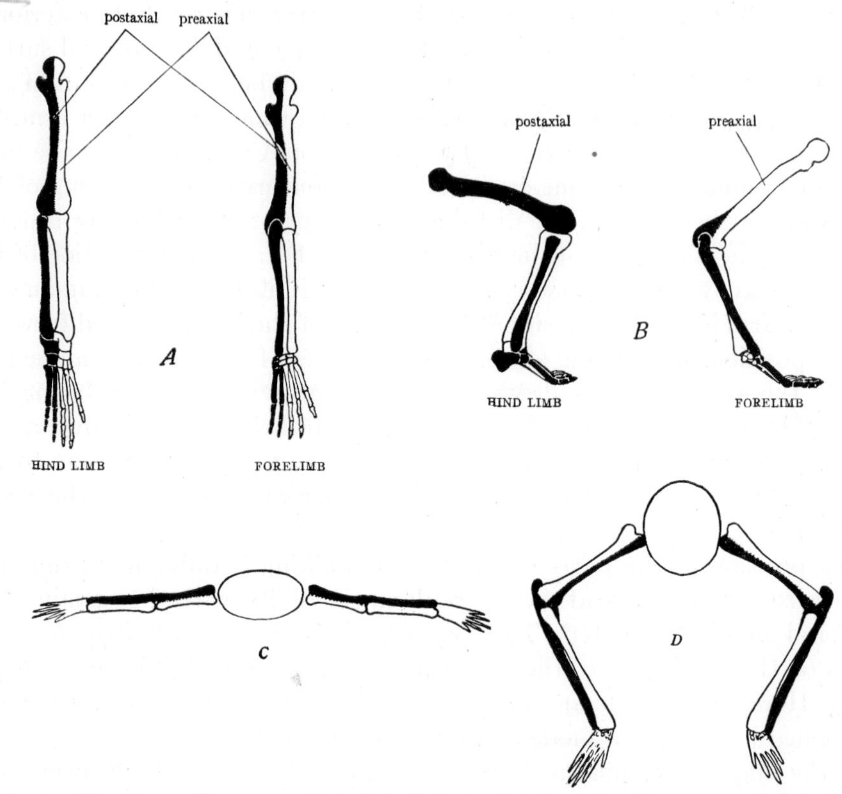

FIG. 7.—Diagrams to illustrate torsion of the limbs; preaxial surface white, postaxial surface black. *A*, primitive position of the limbs seen from above, the preaxial surface facing anteriorly, the postaxial posteriorly. *B*, position of the limbs in mammals, seen from the side; the limbs extend vertically below the body; the upper arm has rotated outward and backward, so that the preaxial surface now faces laterally; the forearm has rotated forward again, resulting in the radius crossing in front of the ulna; the hind limb has rotated outward and forward, so that the postaxial surface faces laterally. *C*, primitive tetrapod, seen from in front, showing position of the limbs. *D*, forelimbs of a mammal seen from in front to show rotation of humerus and crossing of the preaxial forearm bone (radius) in front of the postaxial bone (ulna). (*A* and *B* from Flower's *Osteology of the Mammalia*, 18, 70.)

The scrotum forms an oval swelling to each side of the prepuce and extending forward from it. The urogenital opening of most male mammals is on the tip of the penis, which thus serves a double function, that of copulation and that of discharging the urine to the exterior.

2. Primitive mammals.—The most primitive existing mammals are the *monotremes*, confined to Australia and adjacent islands and including only two types, the duck-billed Platypus (*Ornithorhynchus*) and the spiny anteaters (*Echidna*, *Zaglossus*). Both show, in conjunction with primitive characters, considerable adaptation for special modes of life. The duckbill is a small furry animal which lives in burrows along rivers and ponds and leads a semiaquatic life, evidenced by the webbed swimming feet, short broad steering tail, and ducklike snout, used for grubbing in river bottoms and separating the food from the mud. This snout, however, is not hard and horny like a bird's beak but consists of soft, naked skin, richly supplied with sense organs. The external nares are conspicuous on the top of the snout; from the snout base a fold of skin extends backward. The small eyes have poor vision. A pinna is lacking, but the entrance into the meatus can be tightly closed by horizontal folds of skin. The spiny anteaters are adapted for ant-eating by means of the clawed feet for opening ant nests and the long pointed snout and extensile tongue. Spines are mingled with the fur, and a tail is wanting. Monotremes, especially the males, are armed on the hind legs with a hollow horny spur connected with a poison gland and capable of inflicting a severe wound.

The primitive characters of monotremes are those in which they resemble reptiles. Many such features are skeletal, considered later. Notable is the reptilian type of urogenital system, uniting with the intestine to form a cloaca, as in reptiles, so that there is but one external opening (whence the name Monotremata). Monotremes also lay eggs, which are reptilian in structure, having a large yolk and membranous shell, and reptilian in early development. The duckbill incubates its eggs in a nest in its burrow; the echidnids carry eggs and young in a ventral pouch of skin, which, however, is not homologous to the pouch of the marsupials. Teats are lacking in monotremes, and the milk glands open directly on the skin by several apertures.

Much less primitive than the monotremes are the *marsupials*, which usually have an abdominal pouch of skin in which the young, born in an embryonic condition (the young of a large kangaroo is about the size of one's little finger), are kept for a considerable period. This pouch or *marsupium* is supported by a V-like pair of marsupial bones (epipubes) projecting forward from the anterior end of the pelvic girdle. The marsupials have true teats, which open into the pouch and to which the young adhere by means of modified lips. These peculiarities result from the fact that marsupials lack a placenta (except one genus) and hence cannot nourish the embryos to an advanced state inside the maternal body. The pouch is often evident only during the breeding season. The urogenital system retains some reptilian resemblances, and a shallow cloaca is often present.

The marsupials are confined to Australia and adjacent islands, with the exception of the New World opossums. The group contains arboreal, terrestrial, and burrowing forms and hence presents a considerable variety of appearance. The most familiar marsupials are the kangaroos, notable for their enlarged jumping hind legs and conspicuous forward-opening pouch. The opossums have coarse hair and a prehensile tail associated with arboreal habits. Other types are the arboreal long-tailed phalangers, some of which can parachute by lateral skin expansions; the tailless bearlike koala and wombat; the carnivorous dasyures, including the Tasmanian wolf, Tasmanian devil, and some smaller catlike forms; the bandicoots, somewhat rabbit-like, with backward-directed pouch; and burrowing molelike forms.

3. Adaptive radiation in placental mammals.—The remaining mammals are termed *placental*, because the young, developing to an advanced state in a specialized part of the female tract called the *uterus*, receive food and oxygen by means of a *placenta*, which is a region of the uterine wall intimately associated with, or fused with, certain of the embryonic membranes. The embryonic part of the placenta is connected to the belly of the embryo by the *umbilical cord*, in which run the blood vessels that carry materials between embryo and placenta. A cloaca is always lacking; and there are at least two, sometimes three, openings in the perineum—the anus and one or two urogenital apertures. Placental mammals illustrate adaptive radiation, although not so well as birds, for they have spread into many habitats and have adopted many modes of life, with accompanying modifications of structure. Thus there are flying, arboreal, terrestrial, burrowing, and aquatic types; and mammals of similar appearance and habits may evolve in different orders, furnishing cases of convergence.

True flight is exemplified by the bats (order Chiroptera), in which the mouselike or foxlike body has great lateral skin expansions, the wing membranes, supported by the elongated forearm, especially by the greatly lengthened bones of the hand (metacarpals plus phalanges), which radiate through the outer half of the membranes. The wing membranes can be folded somewhat like an umbrella. The clawed thumb remains free and is employed in climbing and in hanging from objects when the animal is at rest. The hind limbs, except the clawed feet and all or part of the usually abbreviated tail, are also involved in the support of the wing membranes, and for this purpose each ankle bears a spur. Other peculiarities of bats evidence a high sensory equipment, as the relatively large ear pinnae and, in some forms, the skin expansions around the nostrils. Most bats feed on insects and fruits, but it is true that there are blood-sucking bats, found in South America.

Arboreal types able to "plane" by lateral skin expansions between fore and hind legs include the flying lemur (*Galeopithecus*, order Dermoptera), where the expansions involve the tail, and the true flying squirrels (order Rodentia), where the tail is free. Arboreal habits are general throughout the order Primates, including the lemurs, monkeys, and apes, where the arboreal mode of life is reflected in the grasping type of forelimb, with its rotatable forearm, as in man, and with the forefeet modified into a grasping hand, with the thumb opposed to the four fingers. However, the thumb is often reduced or wanting in primates. The hind feet are plantigrade; and, except in man, the first toe is also opposable, as an aid in walking along tree boughs. Commonly in primates, flat nails replace the claws. A long prehensile tail may assist in climbing, although in the higher primates the tail undergoes great reduction. Bare areas on the buttocks (*ischial callosities*) are common among the larger primates. A striking feature of the higher primates is the flattening of the facial part of the head, probably associated with the development of binocular vision. Extreme adaptation to arboreal life is seen in the sloths (order Xenarthra), which spend their entire lives hanging upside down to tree boughs, having for this purpose curved feet terminating in two or three large curved claws.

Dominant among terrestrial mammals are the *ungulates* or hoofed mammals (horses, oxen, sheep, goats, deer, antelope, camels, etc.), with slender running type of legs, lifting the body high above the ground. Ungulates walk on the toenails modified into hoofs. Thigh and upper arm are more or less included with the trunk; and, of the visible part of the legs, the upper half consists of forearm or shank and the lower half of the greatly elongated metacarpals (palm) or metatarsals (sole), fused in each leg into one slender bone, the *cannon* bone. The ungulates are divisible into the odd-toed types (order Perissodactyla: horses, zebras, tapirs, rhinoceros), in which the third toe is dominant and the others reduced or wanting; and the even-toed forms (order Artiodactyla: oxen, deer, etc.), in which the third and fourth toes are equally developed, giving the "cloven-hoof" effect, and the others reduced or wanting. Other external features of ungulates are the elongated neck, reaching an extreme in the giraffes, the large eyes, and often the presence of horns. In general, ungulates are adapted for a life in open country, where they feed chiefly on grass.

A number of remarkable features which do not seem necessitated by any particular mode of life characterize the elephants (order Proboscidea), closely related to the ungulates. Among these features are the huge body, the baggy skin nearly devoid of hairs, the elongation of the snout into a long flexible trunk with the nostrils at its tip, the development of two of the upper teeth (incisors) into tusks, the large flaplike pinnae, and the straight limbs without knee or elbow bend. The feet have five digits, provided with nail-like hoofs.

Another group of mainly terrestrial, sometimes arboreal, habits is the order Carnivora, the carnivores or flesh-eating mammals, including the cats, dogs, wolves, hyenas, otters, weasels, etc. The beast of prey is epitomized by the cat family, with graceful lithe bodies, muscular limbs for leaping, digitigrade gait on padded toes, sharp claws, and tearing teeth with large canines.

Certain structural features, already seen in the Echidna, are correlated with the ant-eating habit, found in placental mammals in three orders: the ordinary anteaters (Xenarthra), the scaly anteaters (*Manis* or pangolins, order Pholidota), and the aardvarks or ant bears (Tubulidentata). All have an elongated head and snout, tubular mouth, slender extensile tongue, degenerate teeth, strongly clawed front feet for opening ant nests, and long tails.

Protective armor has been developed in very few mammals, only by the scaly anteaters just mentioned (order Pholidota) and armadillos (order Xenartha). The former are clothed above with large imbricated horny scales. The armadillos have a dorsal shield composed, like the turtle shell, of outer, thin, horny scales overlying bony plates. In both groups the animals can roll up into a ball, so as to present only the protected surface to an enemy. Some degree of protection is afforded by the conversion of hair into spines, as in the porcupines (Rodentia) and hedgehogs (Insectivora).

The large order Rodentia, or the rodents, typified by the rats and mice, comprises moderate to small furry mammals, with chisel-like gnawing front teeth, short legs, plantigrade gait, and clawed feet. The rodents are predominantly ground forms with a tendency to burrowing habits, although most do not show typical burrowing modifications. Some, however, have altered along the same lines as have the moles (see below). Other adaptive types among rodents are the arboreal squirrels, some of which can plane, as noted above; the aquatic beaver, with its webbed hind feet and flattened scaly tail as aids to swimming; and jumping types with elongated hind limbs, such as various jumping mice and rats, and the hares and rabbits, notable also for their large ears and reduced tails.

Extreme adaptation for a burrowing life is seen in the moles (order Insectivora), evidenced by the small head, degenerate eyes, lack of ear pinnae, sensitive snout, compact body, and shortened limbs with broadened digging forefeet.

Perhaps the most remarkable adaptations of mammals are those for aquatic life, developed independently in three orders: Carnivora, Sirenia, and Cetacea. Seals and walruses form a group of the Carnivora altered for aquatic life, primarily by the conversion of the limbs into short, broad, swimming flippers in which the digits are more or less distinguishable. Other features are the smooth, sleek bodies, reduction or loss of the pinnae, reduction of the tail. Much greater alteration is seen in the Sirenia, or sea cows, and in the Cetacea, the whales and dolphins, all of which are purely aquatic and cannot go upon the land at all. Both groups have a fishlike form; lack pinnae, hind limbs, and (for the most part) hair; have the forelimbs converted into flippers without external trace of the digits; and possess a horizontal tail fin or fluke. The sea cows retain a short thick neck and a normal-looking mammalian head. The whales, however, have become wholly fishlike, with a very large unmammal-like head passing directly into the trunk, and often a dorsal fin. The nostrils open on the top of the head and can be closed by a valve. The dolphins and some porpoises differ from whales chiefly in the long beaklike jaws provided with numerous teeth. The curious narwhal has only two teeth, one of which, usually the left, grows out into a long, spirally twisted tusk. Characteristic of aquatic mammals is the thick layer of fat, or blubber, under the skin to retain body warmth.

J. SUMMARY

Our study of the external features of representative vertebrates may be utilized to direct the student's attention to the following points.

1. The vertebrate body is typically fusiform, that is, moderately thick through the trunk, tapering toward each end. In every group of vertebrates forms may be found which deviate from this typical shape, but such deviations bear no relation to the position of the animal in the vertebrate scale, being rather adaptations to particular modes of life.

2. The body is divided into head, trunk, and tail in the lowest vertebrates. A neck is added in land vertebrates.

3. The head tends to increase relatively in size and the tail to decrease as one ascends the vertebrate series. The former change is associated with increase in the size of the brain; the latter with greater speed and agility of movement.

4. The skin of vertebrates is commonly clothed with protective structures, such as scales, feathers, or hairs. These are more complex in structure in the higher vertebrates and better fitted for keeping the body warm.

5. The head throughout bears three pairs of sense organs—eyes, olfactory sacs, and ears. Of these, the eyes undergo little change throughout the series, except that they degenerate in burrowing and cave-dwelling forms. In most fishes the olfactory sacs open only to the exterior, by way of the nostrils or external nares. As soon, however, as vertebrates adopted the land habitat, internal nares developed, connecting the olfactory sacs with the mouth cavity. This arrangement permits the animal to breathe without opening the mouth, a decided advantage in air-breathing animals, since thereby the drying of the mouth cavity is avoided. In the lower vertebrates an internal ear only is present. To this is added, beginning with the anuran Amphibia, a middle ear, closed externally by the tympanic membrane. In Anura and many reptiles the tympanic membrane is level with the surface of the head, but in some reptiles it begins to sink below the surface. In birds and mammals it has descended deeply into the head, forming a narrow passage, the external auditory meatus. Around the external rim of the meatus in mammals the skin elevates to form a sound-catching device, the pinna. Pinna and meatus constitute the external ear.

6. The gill slits and gills present in the fishes and lower Amphibia disappear in the adults of the higher Amphibia and all forms above them. This is due to the assumption of the air-breathing habit.

7. The trunk bears two pairs of appendages. These are fins in fishes but become limbs in all vertebrates above fishes. Stages in this transformation are found among the fossil lobe-finned fishes (Crossopterygii). The parts of the limbs remain the same throughout the tetrapods, although they are subject to much modification. Among present tetrapods the most primitive limbs, with regard to form, proportions, and position in reference to the body axis, occur in the urodele Amphibia. The somewhat short limbs extend out at right angles to the trunk, and the body almost drags along the ground. In higher vertebrates the position of the limbs is altered by bending and torsion, and there results an elevation of the body above the ground with a correspondingly more rapid progression. As a still further aid to rapid movement the digitigrade or unguligrade mode of walking has been adopted in many cases. Loss of digits is quite common among vertebrates; the missing digits are nearly always the first or last ones, rarely the middle ones.

8. In nearly all vertebrates except mammals the intestine and the urogenital ducts open into a common chamber, the cloaca, which communicates with the exterior by a single opening, the anus or cloacal aperture. In all placental mammals the intestine and the urogenital system open by separate apertures, the urogenital opening being situated always anterior to the anus. The term anus, therefore, does not have the same significance in mammals as in other vertebrates.

REFERENCES

VERTEBRATES

Allen, G. M. 1939. Bats.

Beebe, W. 1906. The Bird: Its Form and Function.

Bolk, L.; Göppert, E.; Kallius, E.; und Lubosch, W. 1931. Handbuch der vergleichenden Anatomie der Wirbeltiere. 6 vols.

Burrell, H. The Platypus.

Cambridge Natural History. 1909: **7**, Fishes, Ascidians, etc.; **8**, Amphibians and Reptiles; **9**, Birds; **10**, Mammalia.

Daniel, J. F. 1934. The Elasmobranch Fishes.

Dickerson, Mary C. 1907. The Frog Book.

Ditmars, R. J. 1936. The Reptiles of North America.

Ellenberger, W., und Baum, H. 1926. Handbuch der vergleichenden Anatomie der Haustiere.

Flower, W. H., and Lydekker, R. 1891. An Introduction to the Study of Mammals, Living and Extinct.

Goodrich, E. S. 1909. Cyclostomes and Fishes. Lankester's A Treatise on Zoölogy, Part IX, Fasc. 1.

———. 1930. Studies on the Structure and Development of Vertebrates.

Groebbels, F. 1932, 1937. Der Vogel. 2 vols.

Gudger, E. W. 1919. Use of sucking fish for catching fish and turtles. Amer. Nat., **53**.

Hamilton, W. J., Jr. 1940. American Mammals.

Howell, A. B. 1936. Aquatic Mammals.

Ihle, J. E. W.; Kampen, P. N. van; Nierstrasz, H. F.; und Versluys, J. 1927. Vergleichende Anatomie der Wirbeltiere (trans. from Dutch into German by G. C. Hirsch).

Jordan, D. S. 1905. Guide to the Study of Fishes. 2 vols.

———. 1907. Fishes.

Kingsley, J. S. 1917. Comparative Anatomy of Vertebrates.

Kyle, H. M. 1926. The Biology of Fishes.

Martin, P. 1912–22. Lehrbuch der Anatomie der Haustiere. 4 vols.

Nelson, E. W. 1916. The larger North American mammals. Natl. Geog. Mag., **30**.

Nelson, E. W., and Fuertes, L. A. 1918. The smaller North American mammals. Natl. Geog. Mag., **33**.

Noble, G. K. 1931. Biology of the Amphibia.

Pope, C. H. 1937. Snakes Alive and How They Live.

Pycraft, W. P. (ed.). 1931. The Standard Natural History.

Reese, A. M. 1915. The Alligator and Its Allies.

Romer, A. S. 1933. Vertebrate Palaeontology.

———. 1941. Man and the Vertebrates. 3d ed.

Sisson, S., and Grossman, J. D. 1938. The Anatomy of the Domestic Animals.

Slijper, E. J. 1936. Die Cetaceen vergleichend, anatomisch, und systematisch. Capita Zoologica, **7**, Part 2.

Smith, J. L. B. 1940. A living coelacanthid fish from South Africa. Trans. Roy. Soc. South Africa, **28**.

Spencer, W. B. 1887. On the presence and structure of the pineal eye in Lacertilia. Quart. Jour. Mic. Sci., **27**.

Stresemann, E. 1927–31. Aves. Kükenthal-Krumbach's Handbuch der Zoologie, **7**, Part 2.

WEBER, M.; BURLET, H. M. DE; and ABEL, O. 1927. Die Säugetiere. 2 vols.
WETTSTEIN, O. VON. 1931–37. Rhynchocephalia. Crocodilia. Kükenthal-Krumbach's Handbuch der Zoologie, 7, Part 1.
WILLISTON, S. W. 1914. Water Reptiles of the Past and Present.
WOODWARD, A. S. 1925. K. A. Zittel's Textbook of Palaeontology: 2, Fishes to Birds; 3, Mammals.

DISSECTION MANUALS

BAUM, H., und ZIETSCHMANN, O. 1936. Handbuch der Anatomie des Hundes.
BENSLEY, B. A. 1918. Practical Anatomy of the Rabbit. 3d ed.
BRADLEY, O. C. 1920–22. Topographical Anatomy of the Horse. 3 parts.
———. 1927. Topographical Anatomy of the Dog.
CAHN, A. R. 1926. The Spiny Dogfish.
DAVISON, A. 1937. Mammalian Anatomy with Special Reference to the Cat. 6th ed.
FIELD, HAZEL. 1939. The Fetal Pig.
FRANCIS, E. T. B. 1934. The Anatomy of the Salamander.
GERHARDT, J. 1909. Das Kaninchen.
GREENE, E. C. 1935. Anatomy of the Rat.
GRIFFIN, L. E. 1937. A Guide for the Dissection of the Dogfish. 6th ed.
M'FADYEAN, J. 1922. The Anatomy of the Horse. 3d ed.
REIGHARD, J., and JENNINGS, H. S. 1901. Anatomy of the Cat.
STUBBS, G. 1938. The Anatomy of the Horse.

V. GENERAL FEATURES OF CHORDATE DEVELOPMENT

Since it is impossible to understand the comparative anatomy of vertebrates, which forms the main subject of this course, without some knowledge of the way in which a vertebrate develops, it is *necessary* that the student learn some of the elementary facts about vertebrate development. Every student must master the contents of this chapter; supplementary reading in reference works is also desirable.

A. THE CHORDATE EGG

The manner of development of the chordate egg is dependent upon the amount of food material, or *yolk*, which it contains. On this basis chordate eggs are classified as follows: (1) *Isolecithal* eggs, with little yolk, evenly distributed, as in *Amphioxus* and mammals. (2) *Telolecithal* eggs, with *total* cleavage; these have a moderate amount of yolk, which accumulates in one-half of the egg and retards its development, as in Amphibia, cyclostomes, and ganoid fishes. (3) *Telolecithal* eggs, with *meroblastic* cleavage, in which there is an enormous amount of yolk and the protoplasm is mostly concentrated in a small disk, the *germinal disk*, which floats on the surface of the yolk. The true egg is the yolk, as the albumen ("white") is merely a nutritive envelope. Meroblastic eggs are characteristic of teleosts, reptiles, and birds, and the mammalian egg has evolved from the reptilian type by loss of the yolk mass. These three types of eggs are illustrated diagrammatically in Figure 8.

B. THE CLEAVAGE OF THE EGG AND THE FORMATION OF THE BLASTULA

Development begins by the division of the egg into two, four, etc., cells until a large number of cells has been produced, a process called *cleavage*. The way in which it occurs depends on the amount of yolk which the egg contains. It should be understood that the yolk is inert material and that the process of development is carried out only by the living protoplasmic portions of the egg.

1. Holoblastic equal cleavage.—In the case of isolecithal eggs the entire egg divides and produces a number of approximately equal cells. Such cleavage is said to be *holoblastic* and *equal*. The cells, as they increase in number, gradually withdraw from the center and arrange themselves in a single layer on the surface, thus producing a hollow ball of cells. This ball is called the *blastula;* its cavity is known as the *segmentation cavity* or *blastocoel*. Such a blastula is produced in the development of *Amphioxus*. Cleavage and formation of the blastula in *Amphioxus* are illustrated in Figure 9*A*. Study also the models of the cleavage of *Amphioxus* provided in the laboratory.

2. Holoblastic unequal cleavage.—This type occurs in those telolecithal eggs which contain a moderate amount of yolk. The half of the egg which contains the majority of the yolk is called the *vegetal hemisphere;* that which contains the majority of the protoplasm is the *animal hemisphere.* The early cleavage planes are shifted toward the animal hemisphere, and, further, the cleavage processes are delayed in the vegetal hemisphere because of the inert yolk. In consequence of these two factors the cells produced in the animal hemisphere are smaller and more numerous than those of the vegetal hemisphere, although the entire egg cleaves. Such cleavage is holoblastic but *unequal.* The cells withdraw from the center, producing a blastula with a somewhat reduced *blastocoel* and a wall several layers of cells thick. The cells of the blastula are of unequal sizes, grading from the smallest at the animal pole to the largest at the vegetal pole.

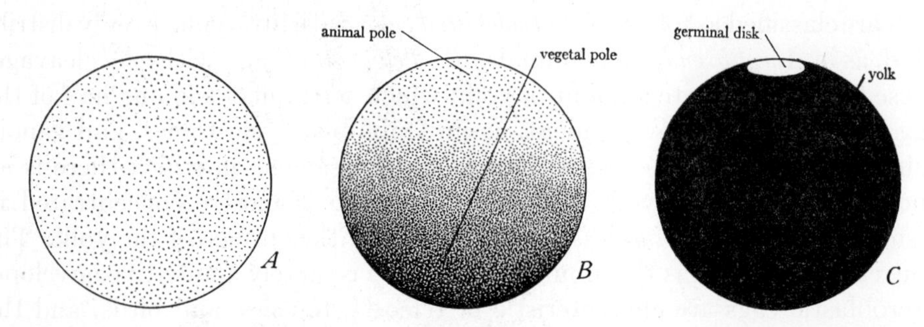

FIG. 8.—Three types of chordate eggs. The black coloring represents the yolk. *A*, isolecithal egg with small and evenly distributed yolk particles. *B*, telolecithal egg with total unequal cleavage, with the yolk more abundant in the vegetal than in the animal half. *C*, telolecithal egg with meroblastic cleavage, consisting completely of yolk except for the small germinal disk of protoplasm at one pole. The size of the germinal disk in the figure is greatly exaggerated with respect to the size of the yolk.

This type of development is characteristic of Amphibia. It is illustrated in Figure 9*B;* study also the models of the cleavage of the amphibian egg, provided in the laboratory. Then obtain a section through an amphibian egg in the blastula stage and examine under the low power of the microscope. The blastula is a hollow sphere whose wall is composed of two or three layers of cells. The wall of the animal hemisphere is thin and consists of small cells; it is the future dorsal side of the embryo. The wall of the vegetal hemisphere is much thicker than that of the animal hemisphere and is composed of large cells, laden with yolk and with indistinct boundaries; it is the future ventral side. The blastocoel is smaller than in the blastula of *Amphioxus* and is displaced dorsally, because of the thickness of the ventral wall. Draw, showing outlines only of the cells.

3. Meroblastic cleavage.—In eggs containing large quantities of yolk only the small germinal disk undergoes cleavage. This kind of cleavage is called *meroblastic.* As a result, a minute disk of cells is produced on the surface of

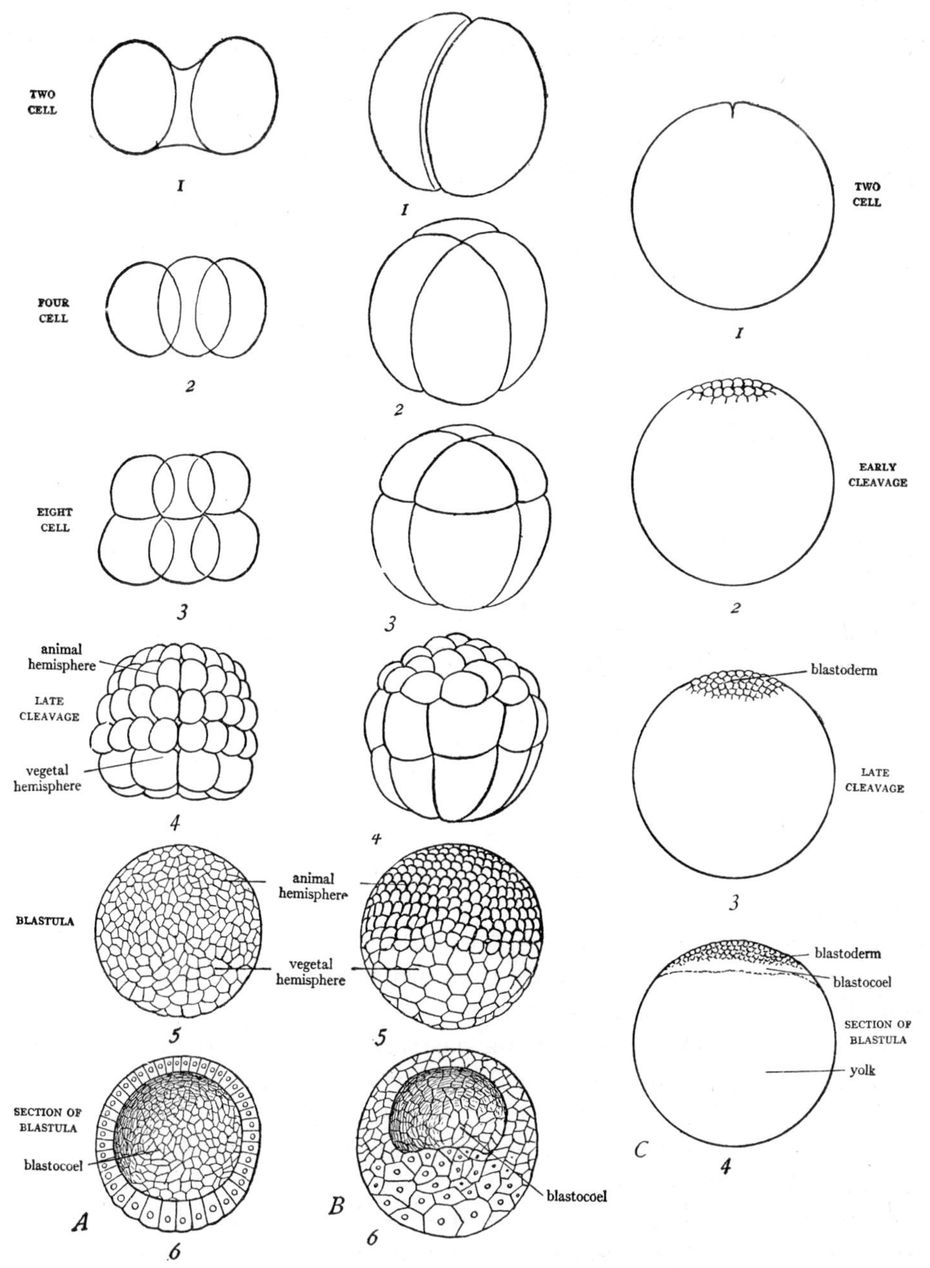

FIG. 9.—Cleavage of the three types of chordate eggs and formation of the blastula. *A. Amphioxus: 1–4*, cleavage; *5*, external view of the blastula; *6*, section of the blastula. Note that the cells of the vegetal hemisphere are but slightly larger than those of the animal hemisphere, the wall of the blastula is one cell layer in thickness, and the blastocoel is large. *B*, amphibian: *1–4*, cleavage; *5*, blastula; *6*, section of the blastula. The cells of the vegetal hemisphere are considerably larger than those of the animal hemisphere, the wall of the blastula is at least two cell layers in thickness, and the blastocoel is smaller and displaced dorsally. *C*, reptile or bird egg with meroblastic cleavage: *1–3*, cleavage; *4*, median sagittal section of the blastula. Only the germinal disk cleaves, forming a disk of cells—the blastoderm—resting on the yolk; a slight slit between this and the yolk represents the blastocoel. (*A* and *B*, *1–5*, from Parker and Haswell's *Textbook of Zoölogy*, after Hatschek, courtesy of the Macmillan Co.; *B6* from Prentiss and Arey's *Textbook of Embryology*, courtesy of the W. B. Saunders Co.)

the relatively enormous yolk. A slight split appears between the disk and the yolk, and this corresponds to the segmentation cavity of other developing eggs; this stage is consequently the blastula stage. Meroblastic cleavage is illustrated in Figure 9*C*. The disk of cells produced by meroblastic cleavage soon begins to expand over the surface of the yolk, being then termed *blastoderm*.

C. FORMATION OF THE GASTRULA

1. In eggs of the *Amphioxus* type.—In such eggs the vegetal hemisphere begins to bend inward and continues this process of *invagination* until its wall comes in contact with the wall of the animal hemisphere. An embryo with a wall two cell layers thick is thus produced, called a *gastrula*. The outer layer is named the *ectoderm*, and the inner layer the *entoderm*. Because of their role in the subsequent development, these layers are referred to as the first two *germ layers*. The hollow tube of entoderm is called the *archenteron* or *primitive intestine;* the cavity of the gastrula is the cavity of the archenteron or *gastrocoel;* and the opening of the archenteron to the exterior is the *blastopore*. Note that the blastocoel is eliminated in the production of this type of gastrula. The formation of the *Amphioxus* gastrula is illustrated in Figure 10*A;* study also the models exhibited in the laboratory.

2. In eggs of the amphibian type.—In these eggs gastrulation is somewhat modified by the presence of inert yolk in the vegetal hemisphere. It is accomplished partly by the invagination of the entoderm, particularly at the dorsal lip of the blastopore, and partly by the expansion of the ectoderm ventrally, pushing the entoderm into the interior. The result is the same as the foregoing; a gastrula is formed. A small portion of the inclosed yolk-bearing cells commonly remains for some time protruding through the blastopore; this portion is called the *yolk plug*.

The formation of the amphibian gastrula is illustrated in Figure 10*B;* study further the models of amphibian development illustrating this stage, noting especially the sagittal section of the gastrula. Then obtain a slide bearing a sagittal section of the gastrula and study it with the low power of the microscope. The gastrula is slightly elongated in the anteroposterior direction. The side with the thinner wall is the dorsal side; that with the thick wall, the ventral side; the end with an opening is the posterior end; the opposite end is anterior. The wall consists of two layers each composed of more than one sheet of cells. The outer and thinner layer is the ectoderm, uniform in width over the whole embryo. The inner layer is the entoderm, separated from the ectoderm by a slight space, and very thick ventrally, where its cells are laden with yolk. The cavity inclosed by the entoderm is the gastrocoel. The opening of the archenteron to the exterior at the posterior end is the blastopore. Ectoderm and entoderm are continuous at the rim of the blastopore. A portion of the entoderm, the *yolk plug*, protrudes through the blastopore and nearly occludes the opening. Make a *diagram* of the section, *coloring ectoderm blue and entoderm yellow.*

3. In meroblastic eggs.—The meroblastic type of development of reptiles and birds is an adaptation for providing large stores of food for the embryo, which thus completes its development inside the egg shell and hatches as a

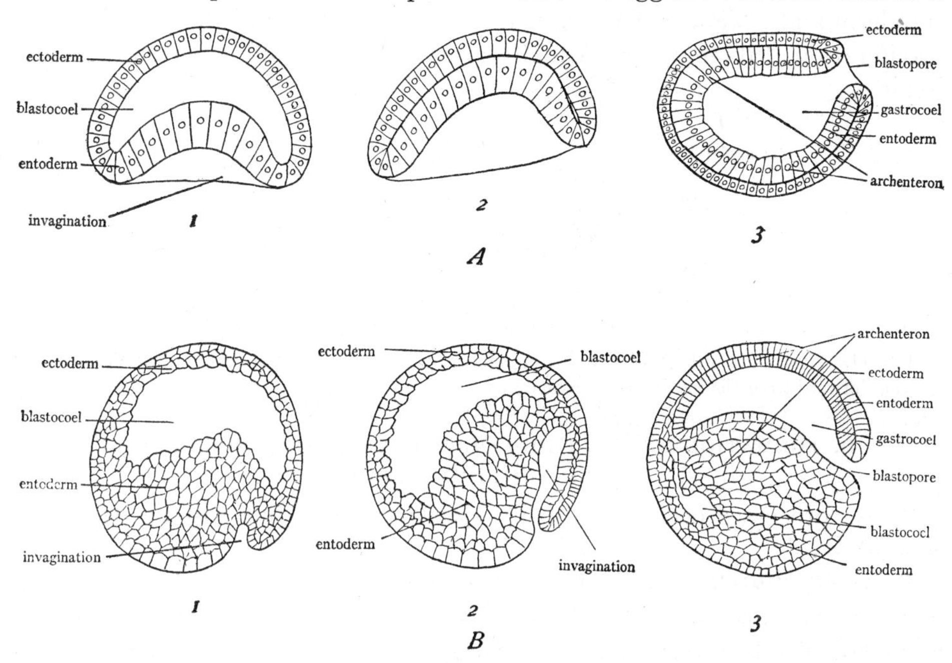

Fig. 10.—Gastrulation and formation of the entoderm in holoblastic chordate eggs. All shown in median sagittal section. *A*, *Amphioxus: 1*, beginning of the invagination; *2* invgaination completed; *3*, completed gastrula having a wall of two layers, ectoderm and entoderm, and an internal cavity, the gastrocoel. *B*, amphibian: *1*, beginning of the invagination; *2*, progress of the invagination accompanied by downward growth of the ectoderm; *3*, completed gastrula, with very thick entoderm ventrally. In *A3* and *B3*, the anterior end of the embryo is to the left, posterior end to the right, dorsal surface above, ventral below. (*A* from Parker and Haswell's *Textbook of Zoölogy*, after Hatschek, courtesy of the Macmillan Co.; *B*, from Kellicott's *General Embryology*, courtesy of Henry Holt & Co.)

small replica of the adult. A larval stage is completely wanting. The change from holoblastic to meroblastic development is a drastic one and naturally involves many alterations, especially of the early embryonic processes. The mode of entoderm formation in the Sauropsida is still under dispute. According to modern researches, the dominant process is delamination, i.e., the downward migration of cells from the under side of the blastoderm to form a new layer, the entoderm, next to the yolk. In reptiles there is also some persistence of entoderm formation by invagination, with the formation of a blastopore and a short archenteron (Fig. 11). The entoderm so formed unites with the delaminated entoderm to become the final entoderm. Archenteron formation is best retained by turtles but among other reptiles is evidently disappearing, and in birds it has been completely lost. In birds the entoderm is formed entirely by delamination, and no traces remain of archenteron or blastopore.[1]

After entoderm formation the blastoderm consists of two strata, the outer ectoderm and the inner entoderm; it lies on the surface of the yolk and by proliferation at its margins gradually spreads over the yolk, eventually inclosing it.

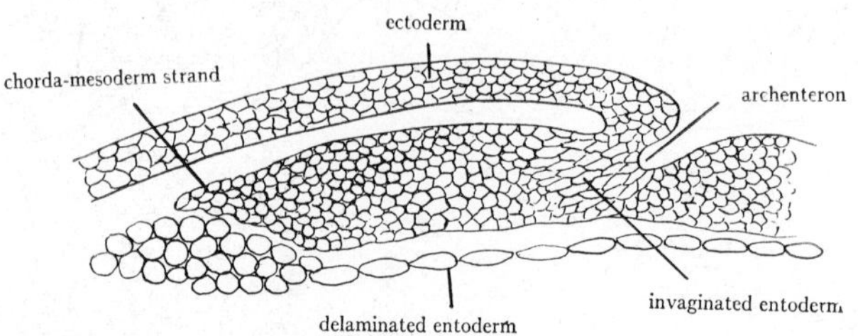

FIG. 11.—Gastrulation in reptiles (based on figures by Peter, 1939). Note chorda-mesoderm strand growing forward from the invaginated archenteron.

D. FORMATION OF THE THIRD GERM LAYER, THE NEURAL TUBE AND THE NOTOCHORD

These processes occur practically simultaneously but will be described separately.

1. In *Amphioxus*.—After the embryo has attained the gastrula stage, it elongates and presents a flattened dorsal surface, a rounded ventral surface, and recognizable anterior and posterior ends (see Fig. 10*A3*). From the dorsolateral regions of the entoderm, which, it will be remembered, forms the "inner tube" of the gastrula, hollow pouches begin to grow out in pairs. These pouches are called the *coelomic sacs* or *mesodermal pouches*. The walls of the pouches constitute the *mesoderm*, or *third germ layer*, which, unlike the ectoderm and entoderm, consists of two walls. The pouches grow laterad and ventrad, filling the space between ectoderm and entoderm. The outer wall of the pouches, in contact with the ectoderm, is called the *somatic* or *parietal* mesoderm; the inner wall, in contact with the entoderm, is the *splanchnic* mesoderm. The cavity of the pouches is the *body cavity* or *coelom*. Eventually the anterior and posterior walls of the pouches break down, so that those of each side unite to form a tube. Thus, the coelom, originally segmented, comes to consist of a pair of continuous cavities, one on each side of the embryo.

[1] This statement is based on the researches of Peter. According to Peter, the much-quoted work of Patterson on gastrulation in the pigeon is wholly erroneous and that of Jacobsen (1938) on the chick is largely so.

Meantime, the ectoderm rises up on either side of the middorsal line as a fold or ridge; and these two folds meet and fuse, forming a tube, the *neural tube,* which is the primordium of the brain and spinal cord. From the mid-dorsal wall of the archenteron a solid rod of cells is elevated and separated off; this is the *notochord* or primitive axial skeleton. These processes are illustrated in Figure 12; see also the models of *Amphioxus* development.

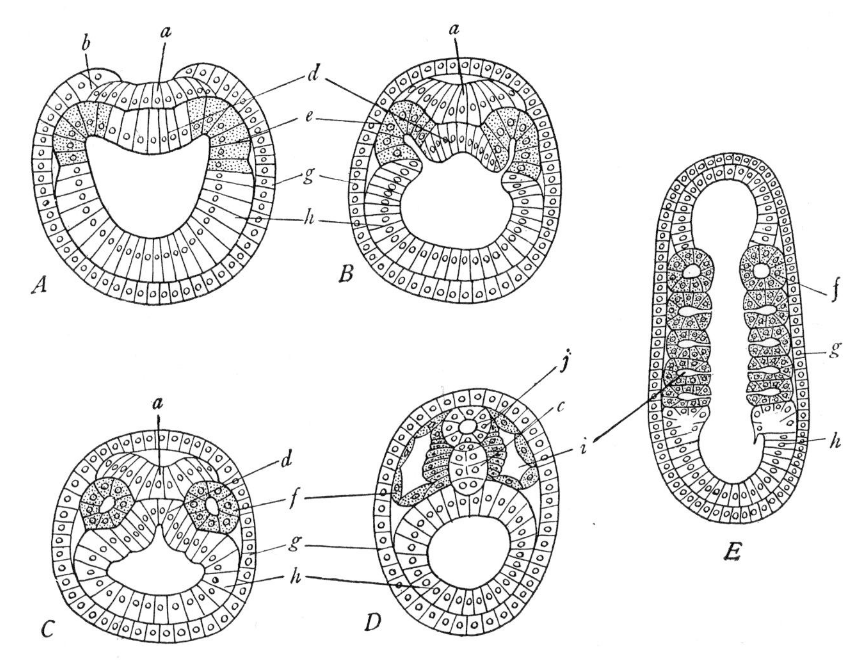

Fig. 12.—Formation of the neural tube, notochord, mesoderm, and coelom in *Amphioxus. A–D,* cross-sections; *E,* frontal section. *A,* differentiation of the medullary plate *a,* the notochordal plate *d,* the neural folds *b,* and the mesodermal pouches *e. B,* the neural folds have closed across above the medullary plate; the mesodermal pouches are further evaginated. *C,* the medullary and notochordal plates are beginning to close; the mesodermal pouches *f* are completely separated from the entoderm. *D,* the neural tube *j* and the notochord *c* are completed; the mesodermal pouches are increasing in size. *E,* frontal section to show the mesodermal pouches *f* originating from the entoderm segmentally. *a,* medullary plate; *b,* neural fold; *c,* notochord; *d,* notochordal plate; *e,* mesoderm; *f,* mesodermal pouches; *g,* ectoderm; *h,* entoderm or archenteron; *i,* coelom. In all figures the mesoderm is stippled. (From Parker and Haswell's *Textbook of Zoölogy,* after Hatschek, courtesy of the Macmillan Co.)

2. In vertebrates.—The outpouching of the archenteron roof to form notochord and mesodermal sacs is usually considered the primitive chordate method, but it does not occur in any vertebrate. In vertebrates with holoblastic unequal cleavage as cyclostomes and amphibians, there split off from the archenteron roof (from the blastopore forward) a median strand, the notochord, and lateral sheets, the mesoderm (Fig. 13). The mesoderm sheets are never segmented at first, nor do they contain a coelom. They spread laterally and ventrally; and a central split, the coelom, appears in them, dividing them into somatic and splanchnic walls.

In meroblastic development the problem is faced of covering the yolk quickly with mesoderm, since the blood vessels are of mesodermal origin and are necessary for carrying the yolk to the embryo as food. Reptiles failed to solve this problem efficiently, as notochord and mesoderm arise much as in amphibians, from a middorsal strand that grows forward from the short invaginated archenteron and therefore represents the archenteron roof (Fig. 11). In birds the problem has been satisfactorily solved. As already noted, birds have no archenteron or entodermal invagination. Along the axis of the future embryo the ectoderm begins to migrate and proliferate into the interior. These processes produce a thickening of the ectoderm, and this thickening is so pronounced that it becomes visible to the naked eye as an opaque streak, the *primitive streak*. The ectodermal material thus sent into the interior forms medially the notochord and laterally the mesoderm. The mesoderm spreads rapidly over the yolk as sheets in which a coelomic split arises as in amphibians. There has long been promulgated in vertebrate embryology the theory that the primitive streak represents an elongated blastopore whose sides have fused together (concrescence theory). This theory now appears definitely erroneous. The primitive streak is a rapid, short-cut method of mesoderm formation. The middorsal ectoderm of the embryo, in place of going through the tedious process of getting into the interior by invagination at the blastopore (a process not very practical, anyway, in meroblastic eggs), passes directly into the interior by migration and proliferation. The primitive streak therefore represents phylogenetically the formation of the roof of the archenteron, since, like that roof, it is the source of the notochord and the mesodermal sheets.

The neural tube or future central nervous system is formed throughout vertebrates in the same manner as in *Amphioxus*. A pair of longitudinal ectodermal folds, the *neural* or *medullary folds*, rise up in middorsal region of the embryo and fuse together, producing a tube (Fig. 14).

These processes will be more clearly understood by reference to the accompanying figures (Figs. 13, 14). Obtain a mounted cross-section through an amphibian embryo at the stage of the formation of the mesoderm, and examine under the low power. The section is oval in form; it is in most cases still surrounded by the delicate *egg membrane*. The outer layer of the embryo is the ectoderm, relatively thin and of the same width over the whole surface. In the median dorsal line the ectoderm is producing, or has already produced, the neural tube. In the former case the ectoderm exhibits a pair of *neural folds*, inclosing a thick plate of ectoderm between them. In the latter case the folds have fused across in the median line, forming a tube, the *neural tube*, which is the oval hollow mass in the median dorsal line, just beneath the ectoderm. The greater part of the section is occupied by the *archenteron* or *primitive intestine*, composed of entoderm. The archenteron has a thin dorsal wall, a thick ventral wall, whose cells contain yolk, and incloses the relatively small *gastrocoel*, which occupies its dorsal part. In the

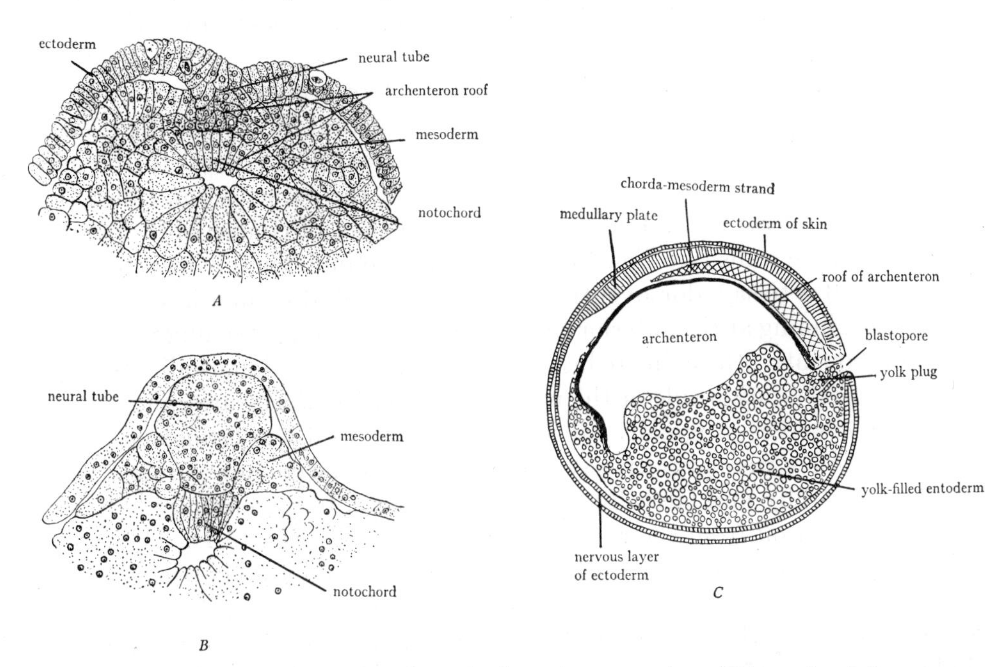

FIG. 13.—Notochord and mesoderm formation in cyclostomes and amphibians. *A* and *B*, cross-sections of the *Petromyzon* embryo, showing separation of notochord and mesoderm from the archenteron roof; neural tube forms as solid invagination (after Selys-Longchamps, 1910). *C*, median sagittal section through the frog embryo, showing chorda-mesoderm strand (diagonal hatching) separating from archenteron roof (black); the ectoderm has separated into two layers, an outer one to become the skin epidermis, and an inner nervous layer; in the latter the thickening of the medullary plate is seen.

median dorsal region of the archenteron a mass of cells will be seen protruding dorsally, or in some slides this mass of cells may have separated from the archenteron and lies between the latter and the neural tube. This mass is the *notochord*. Between the ectoderm and the archenteron on each side is a narrow sheet of cells extending from the sides of the neural tube ventrally. In some slides these sheets will extend only a short distance, while in others they reach nearly to the median ventral line. These sheets are the mesoderm. Make a *diagram* of the section, *coloring ectoderm blue, entoderm yellow, and mesoderm red.*

E. FURTHER HISTORY OF THE MESODERM

The history of the mesoderm is of the utmost importance for the understanding of vertebrate structure. We have already noted that the mesoderm splits into two layers, an outer or somatic layer and an inner or splanchnic layer, and that the space between the two layers is the body cavity or coelom. The mesoderm grows from each side of the embryonic axis ventrally to the median ventral line, or in meroblastic eggs grows out over the yolk, pushing out between ectoderm and entoderm.

The mesoderm next becomes differentiated into three regions: a dorsal region, called the *epimere,* which lies to each side of the neural tube; a middle region, called the *mesomere* or *nephrotome,* situated lateral and ventral to the epimere; and a large ventral region on each side of the archenteron, called the *hypomere* or *lateral plate.* Each of these regions has, of course, both somatic and splanchnic walls (see Fig. 15*A*). The epimere immediately becomes *segmented;* that is to say, dorsoventral clefts appear in it at regular intervals, the process beginning at the anterior end of the embryo and proceeding posteriorly. Consequently, the epimere becomes divided up into a longitudinal row of blocks, a row on each side of the neural tube. These blocks are epimeres, or *mesoblastic somites* (originally called provertebrae, as it was erroneously supposed that they were primitive vertebrae). At first the epimeres are still continuous ventrally and laterally with the mesomere, but eventually they are completely cut off from the rest of the mesoderm (see Fig. 15*B*). The mesomere and the hypomere do not become segmented and remain permanently in close relation to each other. Within the mesomere little tubules appear, which open into the cavity of the hypomere; they are the tubules of the kidney (see Fig. 15*B*). The hypomeres of each side fold around the archenteron, their inner walls coming in contact above and below the archenteron to form double-walled membranes, the *dorsal* and *ventral mesenteries* (see Fig. 15*B*). The cavities of the two hypomeres become the coelom of the adult; the cavity in each epimere disappears; and that of the mesomere remains as the cavities of the tubules of the kidney.

In embryos of the amphibian type the archenteron is a closed tube, and the two hypomeres are closed cavities which meet below the archenteron. In embryos resulting from meroblastic eggs, however, the archenteron is open below and spread out on the yolk, and the hypomeres extend out over the yolk. The differences between the two types of embryos are illustrated in Figure 15*A*, *C*. The embryo in the case of meroblastic development is later constricted from the yolk by the formation of deep grooves on all sides. The yolk then hangs from the ventral surface of the embryo, inclosed in a sac of blastoderm, the *yolk sac,* which is connected with the embryo by a stalk, the *yolk stalk,* as shown in Figure 15*D*. At the time of hatching, the yolk has been practically used up, the remnant is withdrawn into the embryo, and the opening in the body wall where the yolk stalk arises is finally closed over.

Obtain a slide bearing a cross-section through the trunk of a chick embryo of two days' incubation. As explained above, the intestine of the chick embryo is open below on the yolk. In making such sections the embryo is cut off from the yolk, the lines of section being indicated in Figure 15*C* at *x*. After understanding the relation of embryo and yolk, examine the section with the low power. The dorsal boundary of the section is a thin layer, the ectoderm, which is slightly elevated in the median dorsal line; the ventral boundary is

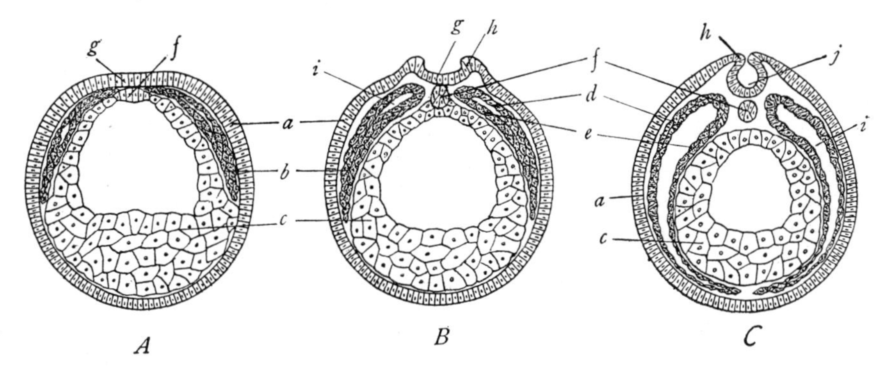

FIG. 14.—Diagrams to show the formation of the neural tube, notochord, mesoderm, and coelom in vertebrates, based on Amphibia. Cross-sections. *A*, differentiation of the notochord (*f*) in the roof of the entoderm; mesodermal plates (*b*) spreading ventrally. *B*, neural folds (*h*) rising at the sides of the medullary plate (*g*); notochord (*f*) separated from the entoderm; mesodermal plates (*b*) extended farther ventrally and developing a central cavity, the coelom (*i*). *C*, neural folds (*h*) nearly closed to form the neural tube (*j*); mesodermal plates have reached the midventral line; coelomic split (*i*) has extended ventrally. *a*, ectoderm; *b*, mesoderm; *c*, entoderm or archenteron; *d*, somatic mesoderm; *e*, splanchnic mesoderm; *f*, notochord; *g*, medullary plate; *h*, neural fold; *i*, coelom; *j*, neural tube. (*A* from Hertwig-Mark's *Textbook of the Embryology of Man and Mammals*, courtesy of the Macmillan Co.)

another thin layer, the entoderm, which makes a slight upward bend in the median ventral line, indicating the future intestine. In the median dorsal line just beneath the ectoderm is the oval hollow section of the *neural tube*. Immediately ventral to this is a small circular mass of cells, the *notochord*. On each side of the neural tube is a squarish mass, its cells radiating from the center. This is the *epimere* or *mesoblastic somite*. Lateral to the epimere and continuous with it is a smaller mass, the *mesomere* or *nephrotome*, in which one or more tubules with central holes are distinguishable. Beyond the mesomere the mesoderm is observed to split into two layers. This region of the mesoderm is the *hypomere* or *lateral plate*. The outer or dorsal layer of the hypomere is the somatic mesoderm. It ascends and comes in contact with the ectoderm, the two together constituting the *somatopleure* or body wall. The lower or ventral wall of the hypomere is the splanchnic mesoderm; it descends and comes in contact with the entoderm, and the double layer thus formed is the *splanchnopleure* or intestinal wall. The cavity between the somatic and splanchnic walls of the hypomere is the coelom. As already explained, the hypomere in such embryos extends far out over the yolk. Observe that the splanchnopleure contains many holes; these are the cross-sections of blood vessels, which convey the food from the yolk sac to the embryo. There is also a large artery in the embryo below each epimere. *Draw the section in diagram*, coloring the three germ layers as before.

F. THE FATE OF THE ECTODERM

As we have seen, the ectoderm gives rise to the neural tube, from which develop the brain, spinal cord, and nerves. The ectoderm also forms the external layer of the skin and all of its derivatives, such as hair, nails, etc. It also gives rise to the sensory part of all the sense organs, the lining membrane of the nasal cavities, the mouth, and anus, the glands and other outgrowths of the nasal and mouth cavities, the glands of the skin, the enamel of the teeth, and the lens of the eye.

G. THE FATE OF THE ENTODERM

The entoderm is, as we have seen, the primitive intestine. This intestine is the inner lining of the adult intestine. The entoderm thus forms the epithelial lining of the intestine and the epithelial lining and epithelial cells of all of the outgrowths of the intestine, which include the gill pouches and gills, the larynx, windpipe, and lungs, the tonsils, the thyroid and thymus glands, the liver, the gall bladder and bile duct, the pancreas, and the urinary bladder and adjacent parts of the urogenital system. The student should note that only the epithelial cells of these structures arise from the entoderm.

H. THE FATE OF THE MESODERM AND THE FORMATION OF MESENCHYME

1. Mesenchyme.—In the further development of the mesoderm, *mesenchyme* plays an important role. Mesenchyme is not a germ layer but a particular type of tissue. It is a primitive kind of connective tissue, consisting of branched cells, whose branches are more or less united to form a network (see Fig. 16). Nearly all of the mesenchyme comes from mesoderm; however, it may arise from the other germ layers also. Hence tissues and structures which arise from mesenchyme may owe their origin to more than one germ layer. Consequently, to avoid inaccuracy it is usually merely stated that they arise from mesenchyme, without specifying the particular germ layer or layers involved. When a germ layer is about to produce mesenchyme, its cells become loose, separating from their fellows, lose their epithelial form, and, taking on a branched irregular shape, wander away by amoeboid movements to more or less definite regions, where they give rise to certain tissues (see Fig. 16). Those parts of the mesoderm which do not become mesenchyme but retain their epithelial characteristics are called *mesothelium.*

2. The fate of the epimeres.—The medial wall of each epimere transforms

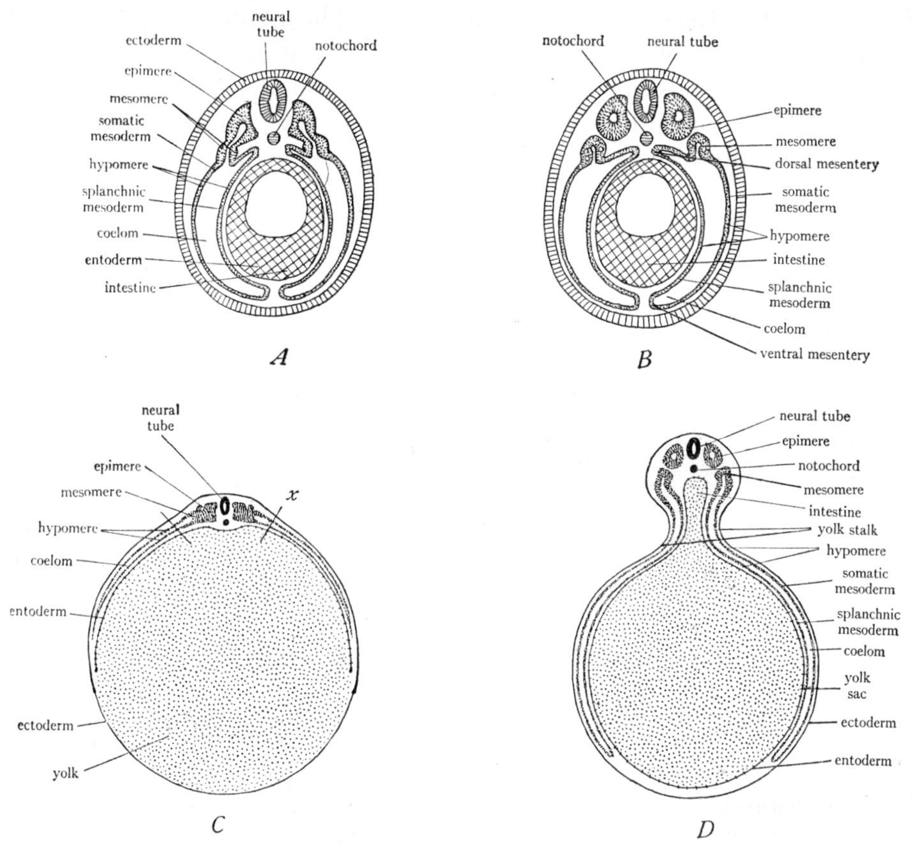

FIG. 15.—Diagrams to show the differentiation of the mesoderm in holoblastic and meroblastic types of development. *A*, and *B*, holoblastic type; *C* and *D*, meroblastic type. *A*, differentiation of the mesoderm into epimere, mesomere, and hypomere. *B*, separation of the epimere from the mesomere, appearance of kidney tubules in the mesomere, and closure of hypomere around the intestine to form dorsal and ventral mesenteries. *C*, similar to *A* but in the meroblastic type, showing entoderm and mesoderm growing around the yolk; note how embryo is spread out on the surface of the yolk; *x*, indicates lines where embryo is cut off from yolk in making sections for miscroscopic study. *D*, similar to *B* but in the meroblastic type; the entoderm has completely surrounded the yolk; the mesoderm has nearly done so; the yolk sac is seen to be a part of the intestine; the embryo is partly constricted from the yolk sac, the constriction being the yolk stalk. (*C* from Wilder's *History of the Human Body*, courtesy of Henry Holt & Co.)

into a mass of mesenchyme cells which migrate to a position around the notochord and there give rise to the vertebral column. This mass of mesenchyme is known as the *sclerotome* (see Fig. 17). The outer wall of each epimere transforms into mesenchyme cells which migrate to the under side of the ectoderm and there give rise to the inner layer (dermis) of the skin. This part of the epimere is called the *dermatome* (Fig. 17). The remainder of the epimere persists in place as mesothelium and is known as a *myotome* or *muscle segment*. Each myotome becomes separated from the adjacent ones by a connective tissue partition, the *myocomma* or *myoseptum*. The myotomes give rise to the voluntary muscles of the body (with certain exceptions). Each grows from its original dorsal position ventrally between the ectoderm and the hypomere to the median ventral line, where it meets its fellow from the opposite side. There is thus produced a complete muscular coat for the body (see Fig. 17).

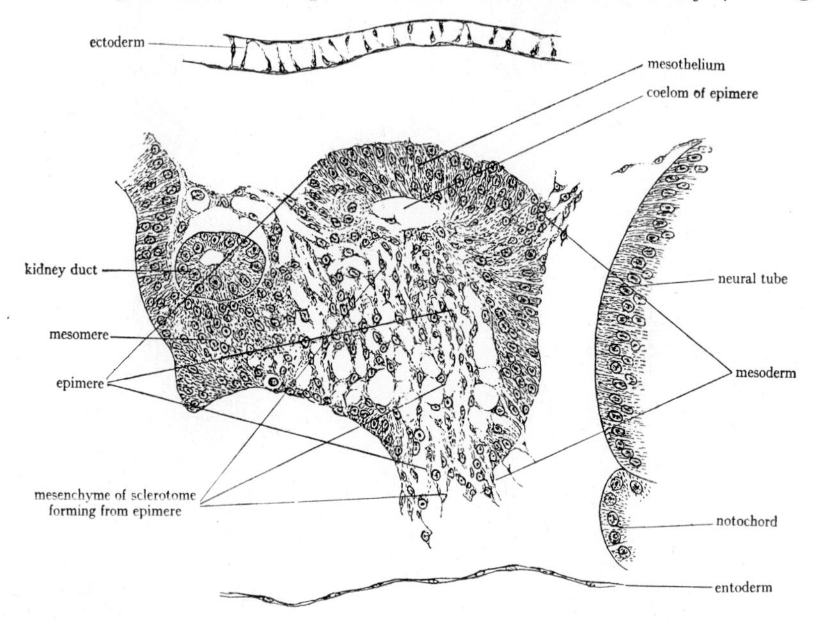

FIG. 16.—Enlarged view of the epimere of a chick embryo of two days' incubation to show the transformation of a portion of the epimere into mesenchyme. These mesenchyme cells constitute the sclerotome, from which the vertebral column arises.

3. The fate of the mesomere.—The mesomere gives rise to the kidneys, the reproductive organs, and their ducts (the terminal portions of the urogenital ducts may have ectodermal or entodermal linings).

4. The fate of the hypomere.—The cavity of the hypomere is the coelom of the adult. The splanchnic walls of the hypomeres of the two sides fold around the archenteron and give rise to mesenchyme, from which are produced the smooth muscle and connective tissue coats of the digestive tract, and also the smooth muscle, connective tissue, cartilage, etc., as the case may be, of all the derivatives of the digestive tract mentioned above. The hypomere also gives rise to the linings of all the coelomic cavities, the serosa of the viscera, and all of the mesenteries. The splanchnic mesoderm of the hypomere produces the heart. In the region of the gill slits the hypomere produces voluntary muscles.

5. The products of the mesenchyme.—The mesenchyme gives rise to all of the connective tissues of the body, including cartilage and bone; to all of the involuntary or smooth muscles; to the blood cells, the blood vessels, the lymph vessels, and lymph glands; and to the voluntary muscles of the appendages. It has already been stated that the vast majority of the mesenchyme is of mesodermal origin, but a small part arises from the other germ layers.

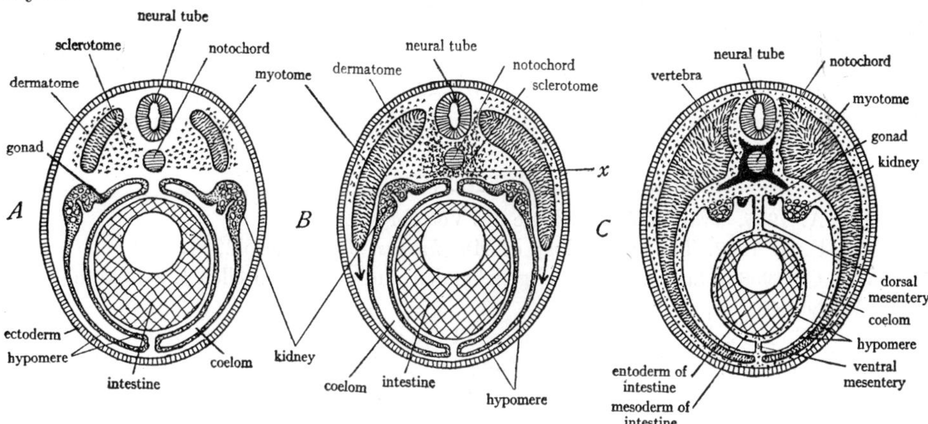

FIG. 17.—Diagram of cross-sections of vertebrate embryos to show the differentiation of the epimere into dermatome (skin-producer), myotome (muscle-producer), and sclerotome (skeleton-producer). Dermatome and sclerotome consist of mesenchyme; myotome, of mesothelium. In *B* and *C* the dermatome is seen spreading beneath the ectoderm to form the dermis of the skin; the myotomes are growing ventrally to form the muscle layer of the body wall; the sclerotome is accumulating around the notochord at *x* to form the vertebrae; and the hypomere incloses the intestine producing the dorsal and ventral mesenteries and the mesoderm of the intestine.

I. THE HOMOLOGY OF THE GERM LAYERS

The foregoing account follows the method standard in embryology books of describing vertebrate development in terms of the germ layers. Such an account tacitly assumes the homology of the germ layers and their products throughout the vertebrate series. In recent years the homology of the germ layers, and consequently the descriptive embryology based on such homology, has been severely criticized in many quarters. The basis of this critical attack is the brilliant work in experimental embryology of the last twenty years under the leadership of Spemann of Freiburg. By grafting pieces of embryos, chiefly amphibians, between the same and other species, these workers have shown that structures normally arising from one germ layer can be caused to form from another germ layer. The author, however, is of the opinion that such results have no bearing at all upon the principle of the homology of the germ layers. Homology is not concerned with the potentialities of cells or the capacities they may possess for alteration under altered conditions. A biologist of a physiological turn of mind will naturally assume that any cell can differentiate into any other kind of cell if taken early enough and placed under the proper conditions. What concerns the student of comparative anatomy is whether the structures of vertebrates normally originate and develop in much the same way from early embryonic regions throughout the vertebrate series. There cannot be any doubt that this is the case; and, consequently, in particular instances the origin and manner of development of a structure can be utilized to determine its homology. At the same time, it must always be borne in mind that adaptability is the outstanding feature of living things and that alterations in the processes of development will occur concomitant with altered conditions of life.

REFERENCES

Brachet, A. 1935. Traité d'émbryologie des Vertébrés. 2d ed.

Conklin, E. G. 1932. The embryology of *Amphioxus*. Jour. Morph., **54**.

Hertwig, O. 1906. Handbuch der vergleichenden und experimentellen Entwicklungslehre der Wirbeltiere.

Jacobsen, W. 1938. The early development of the avian embryo. Jour. Morph., **62**.

Kerr, J. Graham. 1919. Textbook of Embryology: 2, Vertebrata with the Exception of Mammalia.

Pasteels, J. 1940. Un aperçu comparatif de la gastrulation chez les Chordés. Biol. Rev. Cambridge, **15**.

Peter, K. 1938. Gastrulation und Homologie. Anat. Anz., **86**.

———. 1938–39. Untersuchungen über die Entwicklung des Dotterentoderms. Zeitschr. f. mikr.-anat. Forsch., **43**, **46**.

———. 1939. Gastrulation und Chordaentwicklung bei Reptilien. Zeitschr. Anat. u. Entw'gesch., **109**.

Selys-Longchamp, M. de. 1910. Gastrulation et formation des feuillets chez *Petromyzon planeri*. Arch. de Biol., **25**.

VI. THE COMPARATIVE ANATOMY OF THE SKIN AND THE EXOSKELETON

A. GENERAL CONSIDERATIONS ON THE SKELETON

The term *skeleton* includes all of the hardened portions of the bodies of animals. The skeleton of the invertebrates is commonly external, forming a hard covering inclosing the body, while that of the vertebrates is both external and internal. In invertebrates, further, the skeleton is a lifeless secretion, containing no cells, whereas the vertebrate skeleton is almost invariably cellular, either being composed entirely of hardened cells or consisting of cells and intercellular products. There are two distinct kinds of skeleton in vertebrates, different in origin and function: (1) the external skeleton, or *exoskeleton*, derived from the skin, and forming a covering and protective layer on the outside of the body; (2) the internal skeleton, or *endoskeleton*, derived chiefly from the inner wall of the epimere, and constituting a support and framework for the body and a place of attachment of the voluntary muscles.

B. THE STRUCTURE OF THE SKIN

Since the exoskeleton is derived from the skin, the structure of the skin must first be understood. The skin occurs only in vertebrates and may be defined as a surface covering easily separable from the underlying muscular layer of the body wall. Study of the development of the skin and of its microscopic structure (Fig. 18) reveals that it consists of two distinct parts: an outer layer, the *epidermis*, composed of *epithelial cells*, and derived from the *ectoderm* of the embryo; and an inner layer, the *dermis*, *corium*, or *cutis*, composed of *connective tissue* and formed from the mesenchyme of the *dermatome*, which in turn comes from the epimere (Fig. 16).

1. Conditions in the lower chordates.—In *Amphioxus* and the tunicates a skin is lacking. The surface is clothed with a one-layered epidermis similar to the embryonic ectoderm; this is underlain by connective tissue corresponding to the vertebrate dermis. An exoskeleton is lacking, but tunicates are inclosed in a "tunic" secreted by the epidermis. This is a thick gelatinous layer composed chiefly of cellulose and containing loose wandering cells.

2. The vertebrate skin.—The epidermis of vertebrates differs from that of all other animals. It is a *stratified* epithelium, composed of several to many layers of cells, produced by proliferation from the original single layer (Fig. 18). It usually contains *one-celled glands* (Fig. 19) or is invaginated into the dermis as *many-celled glands* (Fig. 18)—hollow or solid, flask-shaped or tubular bodies opening to the surface. The dermis characteristically consists of white, fibrous connective tissue of loose irregular or of dense parallel fibers.

3. The skin of aquatic vertebrates.—In aquatic vertebrates (cyclostomes, fishes, tailed amphibians) the relatively thin epidermis contains numerous mucous gland cells (Fig. 19), whose secretion keeps the skin moist and slimy; otherwise its cells are much alike throughout. The dermis usually presents a loose layer next to the epidermis and a compact layer below (Fig. 19).

4. The skin of land vertebrates, exemplified by the frog.—The secretion of mucus is, in general, an insufficient protection for land vertebrates. Instead, the outer layers of the epidermis undergo *keratinization*, i.e., flattening and hardening into a *horny* stratum (Fig. 18) which is shed continuously in small bits or at intervals as a whole (constituting a molt) and is replenished continuously by proliferation from below. Multicellular glands replace the unicellular glands of aquatic forms.

As an example of the skin of a land vertebrate, a cross-section of the frog's skin may be examined. Identify the following parts:

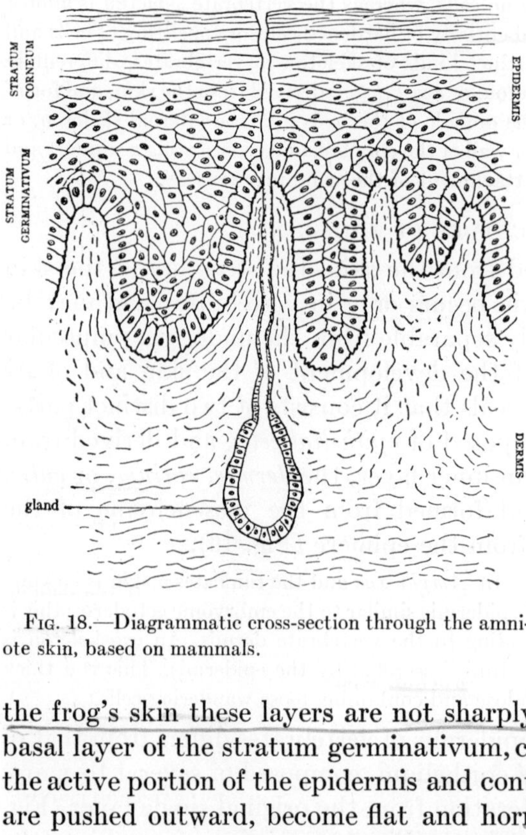

a) *Epidermis:* The outer part of the skin, the epidermis, consists of several layers of epithelial cells. The outermost layers are thin, flattened, and keratinized, i.e., converted into a dead, hard, horny material, protecting against drying. This keratinized stratum is termed the *stratum corneum*. Beneath the stratum corneum the cells gradually change from a flattened to a rounded and finally to a columnar shape. These lowermost layers of rounded to columnar cells constitute the *stratum germinativum* (also called *stratum mucosum*, *profundum*, and *malpighii*).[1] In the frog's skin these layers are not sharply demarcated. The lowermost or basal layer of the stratum germinativum, consisting of tall columnar cells, is the active portion of the epidermis and continuously proliferates cells which are pushed outward, become flat and horny, and finally form part of the stratum corneum, which thus is constantly renewed from below to compensate for surface shedding.

Fig. 18.—Diagrammatic cross-section through the amniote skin, based on mammals.

b) *Dermis or corium:* The dermis is the inner part of the skin, consisting of fibrous connective tissue (Fig. 19). In the frog and other lower vertebrates

[1] In many books only the lowermost row of cells is called stratum germinativum or malpighii, since this row is the active proliferating part of the epidermis. This, however, makes it necessary to invent some other name for the layers between this row and the stratum corneum. It seems best, to the author, to call the whole epidermis below the stratum corneum stratum germinativum; and, where necessary, the lowermost row can be called basal layer of the stratum germinativum.

the dermis next the epidermis consists of loose open connective tissue (*stratum laxum*) and deeper down of layers of dense, parallel, wavy fibers (*stratum compactum*). In addition to the connective tissue fibers the dermis contains: *pigment cells,* dark, irregular branching cells just beneath the epidermis; the *cutaneous glands,* flask-shaped bodies composed of epithelial cells and opening on the surface by a narrowed *neck;* and columns containing smooth muscle cells, blood vessels, and nerves, ascending through the dermis. Of these structures, the most conspicuous are the cutaneous glands, really parts of the stratum germinativum, invaginated into the loose portion of the dermis. They produce mucous and other protective secretions.

Draw a small portion of the skin to show the parts named above.

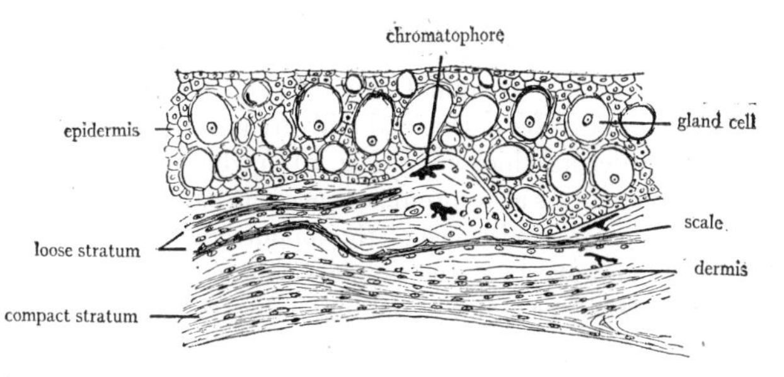

FIG. 19.—Section through the skin of a fish showing unicellular mucous glands in the epidermis (after Klaatsch, 1890). Note loose and compact strata of the dermis and location of scales in the former.

5. The amniote skin.—The amniote skin in correlation with land life has a well-developed stratum corneum, except in the cetaceans, where only a thin, imperfectly keratinized layer is present. Keratinization reaches its height in the reptiles, where among snakes and lizards the stratum corneum is shed *in toto* periodically as a *molt,* preceded by the formation of a new stratum corneum beneath. The skin of birds is notably thin and delicate. There is a paucity of cutaneous glands among reptiles and birds; the most constantly present type is the uropygial gland of birds, opening on a prominent papilla in front of the tail stump (p. 52). The thick skin of mammals has a deep stratum corneum, separated from the stratum germinativum by two thin layers—the transparent *stratum lucidum* and the granular *stratum granulosum.* The mammalian stratum germinativum is characterized by the scalloping of its under surface caused by dermal projections, called *papillae,* which carry blood vessels to the epidermis. The dermis of amniotes generally lacks distinct stratification into loose and compact layers. Leather consists of dermis which has undergone a tanning process.

The mammalian skin is particularly rich in number and types of glands except in Cetacea and Sirenia. Besides various kinds of glands localized in particular regions or limited to particular groups of mammals, there are generally present three types of cutaneous glands: *sebaceous, sweat,* and *mammary* or *milk* glands. The sebaceous or oil glands are solid epidermal ingrowths, usually but not necessarily associated with hair follicles, which secrete an oily substance for lubricating the skin and hairs and protecting against water.[2] The sweat glands are long, hollow tubular glands, usually more or less coiled, which secrete the sweat as a means

[2] The "pores" of the skin so frequently mentioned in advertisements of cosmetics are really the openings of the sebaceous glands; there are no pores through the skin.

of lowering body temperature. The mammary glands are histologically similar to sweat glands, and both are presumably derived from a common type of hollow tubular gland.[3]

The milk glands of monotremes form a pair of oval masses, each composed of numerous lobules, in turn made of branching tubular glands. Each lobule opens directly on the flat surface, in common with a hair follicle; and the numerous milk pores thus occupy paired, oval milk areas on the abdomen. Teats are lacking.[4] Among other mammals the number of milk glands is diminished, and they are clustered about elevations, the *nipples* or *teats*, of which up to twenty-five may be present. At the base of each teat the mammary glands, especially when swollen during lactation, may cause a rounded elevation, the *mamma* or *breast*. Although possibly there were originally in marsupials four rows of teats (producing median teats by fusion of members of the inner rows), the teats in all placental mammals are arranged in two longitudinal rows or derived from such by loss. Loss from different levels explains why the remaining teats may have any position on the ventral surface, as pectoral, abdominal, inguinal, etc. Reduction to a single pair of nipples is seen in man, horse, elephant, sheep, etc. Occurrence of extra nipples is probably a throwback to an ancestral condition. In embryology each row of nipples appears on a longitudinal elevated ridge termed the *milk ridge*, which can be regarded as derived by elongation of the oval milk areas of monotremes. In the milk ridge each nipple begins as an epidermal thickening (Fig. 20*A*) which sinks inward and proliferates the milk glands, and in some forms also sprouts evanescent hair follicles, hairs, and sebaceous glands (Fig. 20*C*), probably a reminiscence of the monotreme condition. In most placental mammals a papilla forms and everts as the nipple, forming an *eversion* type of nipple (Fig. 20*E*). In some mammals, however, notably the ungulates, the nipple arises chiefly by elevation of the adjacent skin, so-called *proliferation* nipples (Fig. 20*H*).[5] The human nipple is intermediate between these extremes.

The marsupial pouch of marsupials probably has no relation to the mammae or teats of placental mammals. It is lacking in the most primitive marsupials and therefore has evolved within the group. The work of Bresslau (1920) indicates that in higher marsupials each teat was sunk in a pocket and that the marsupium arose by fusion of these pockets.

6. Color.—Color in vertebrates is vested in the skin and is caused either by pigment or by structures capable of diffracting light. The pigment may occur as diffuse substance or as granules, or may be located in special branched cells, called *chromatophores*. Epidermal pigment is usually of the diffuse or granular type, whereas dermal pigment is nearly always inside chromatophores, which typically form a layer just beneath the epidermis. Chromatophores are of three general sorts: *melanophores*, containing *melanin*, a brown to black protein pigment; *xanthophores* or *lipophores*, containing yellow to red fatty pigment; and *guanophores* or *iridocytes*, filled with crystals of guanine, which reflect or refract light and produce white, silvery, metallic, or iridescent colors. The pigment can wander to various extents along the branches of the chromatophores or can aggregate into a central ball, so that various shades of color result. Chromatophores are characteristic of fishes, amphibians, and reptiles. The col-

[3] The view originating with Gegenbaur, and widely copied, that the milk glands of monotremes are modified sweat glands and those of other mammals are modified sebaceous glands is untenable; all mammary glands are of the sweat-gland type.

[4] Although it is commonly stated that the milk exudes upon the coarse hairs of the milk areas and is licked from these hairs by the young, it appears that no one has ever actually witnessed the manner of nursing of monotreme babies.

[5] The idea of true and false teats must be abandoned. The ungulate teat is not, as formerly supposed, formed wholly from adjacent skin and hence is not a false teat; as Fig. 20 shows, only its outer surface is so formed; the lumen is really the termination of the milk glands.

ors and color patterns of these groups result from fixed arrangements of the various kinds of chromatophores in definite locations. Thus the green color of frogs is caused by a blue diffraction of the guanophores seen through a layer of yellow xanthophores. Black spots represent aggregations of melanophores. Many of these forms exhibit striking color changes with reference to day and night, background, or physiological state. Such color change results from pigment migrations in the chromatophores, rearrangements of the chromatophores, and sometimes changes in the number of chromatophores, and is usually controlled by hormones mediated through the nervous system, although in some cases nerves may act directly to

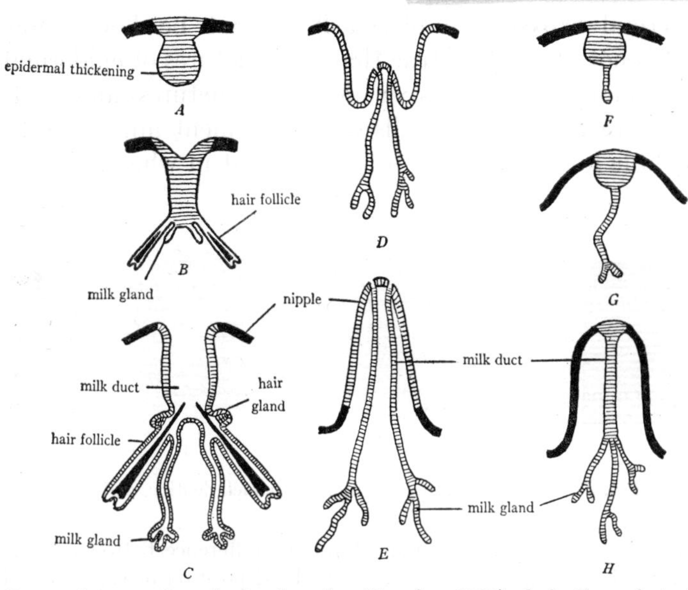

FIG. 20.—Types of mammalian nipples (based on Bresslau, 1920); derivatives of nipple primordium crosshatched, adjacent skin black. *A*, nipple primordium formed as epidermal ingrowth; *B*, primordium sprouting hair follicles and milk glands; *C*, primitive nipple, with hair follicles, milk glands, and sebaceous (hair) glands opening into milk duct; *D*, first stage in the formation of an eversion nipple (hair follicles and sebaceous glands have degenerated); *E*, eversion nipple complete, nipple formed wholly of epidermis of original primordium; *F*, early stage of ungulate nipple; *G*, later stage; *H*, final stage of ungulate nipple; the nipple is formed of adjacent skin, and its lumen is the milk duct, as in the case of other nipples.

disperse or concentrate the pigment granules. The colors of birds and mammals consist chiefly of diffuse or granular pigment in the feathers and hairs, which get them from pigment cells during development; but pigment also occurs in the epidermal cells, and dermal chromatophores may be present. The different skin colors of man are vested in epidermal pigment granules. Colors produced by physical means—e.g., diffraction of light by fine parallel striations—are not uncommon in feathers. Thus the iridescent blue or green wing patch (speculum) seen in the wings of male ducks is physically produced.

C. GENERAL REMARKS ON THE EXOSKELETON

The exoskeleton is derived from the skin by hardening processes in epidermis or dermis or both. Exoskeleton derived from the epidermis is spoken of as *epidermal;* it is produced by the activity of the stratum germinativum and consists of many flat horny cells pressed firmly together to make a hard structure. Epidermal exoskeletal structures are therefore only special portions of the stratum corneum. Exoskeleton derived from the dermis is called *dermal*, consists of bone or substance allied to bone, and is produced by mesenchyme cells originating from the dermatome. Embryologically, epidermal exoskeleton is of ectodermal origin, dermal exoskeleton of mesodermal origin. They thus differ both structurally and embryologically. Epidermal and dermal exoskeleton obviously cannot be homologous to each other; but it is not necessarily true that all structures of epidermal origin are homologous among themselves or dermal structures among themselves. Topographical relations and details of development must be taken into account. Probable homologies will be pointed out in what follows.

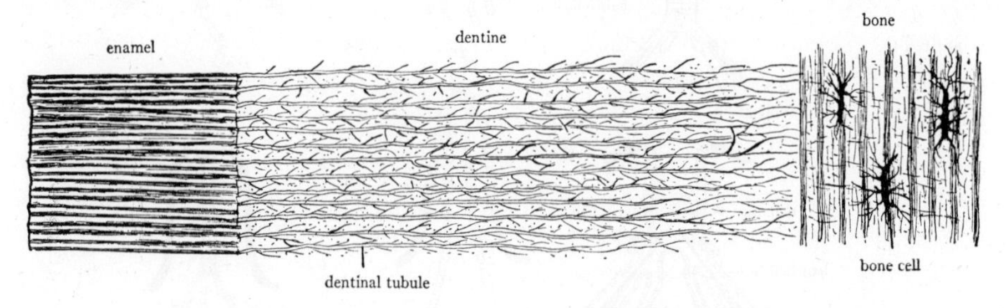

FIG. 21.—Schematic representation of enamel, dentine, and bone

First, however, it is necessary to understand the difference between *bone*, *dentine*, and *enamel*. These hard structures are produced by the deposition of calcium and magnesium salts, chiefly calcium phosphate. Bone consists of fibrous connective tissue mostly arranged in lamellae (thin plates) impregnated with salts and containing numerous branching bone cells (Fig. 21). Dentine, the chief constituent of teeth, is harder than bone, containing a higher percentage of salts, also deposited on a connective tissue matrix; it differs in that the secreting cells are outside and send long processes into the dentine along numerous parallel branching canals, the so-called *dentinal tubules* (Fig. 21). Dentine merges into bone by way of types which lack the characteristic tubules. Enamel, the shiny outer coat of teeth, is extremely hard, containing over 90 per cent of salts, as contrasted with 20 per cent in bone; the salts occur in long prisms (Fig. 21), and very little other structure appears to be present. Enamel is of epidermal origin, secreted on the under side of the basal layer of the stratum germinativum, whose cells are then called *ameloblasts* (Fig. 62).

D. EXOSKELETON OF FISHES

As a rule, fishes are clothed with scales which probably are wholly of dermal origin, for the presence of epidermal contribution has not been satisfactorily established. They are composed of substances similar to, or allied to, bone; and the most primitive types of fish scales consist of or contain dentine. There are six kinds of fish scales.

1. The placoid scale.—This is characteristic of the elasmobranch fishes and consists of a *basal plate* carrying a projecting *spine*. In most elasmobranchs the placoid scales, usually termed *denticles*, are small, almost microscopic, giving a rough texture to the skin; but large ones are obtainable from skates. Cut out from a skate a small piece of skin containing one spine. Clean away the skin so as to expose the complete scale and note toothed basal plate from which arises a shiny curved spine. The spine contains a *pulp cavity*, to be located by probing with a needle point in the center of the under side of the basal plate. Placoid scales consist of dentine[6] and are identical in microscopical structure (Fig. 22) with the dentine of teeth. Draw.

The dogfish is covered with placoid scales, which are much smaller and more closely set than in the skate. The spines of the scales, directed backward, can be felt by passing the hand over the dogfish skin from behind forward. Pieces of sharkskin, called *shagreen*, were formerly used as an abrasive. The scales may be freed by boiling a piece of skin with alkali. Examine such isolated scales under the microscope and note similarity to skate scales. They differ considerably in shape in different regions of the dogfish.

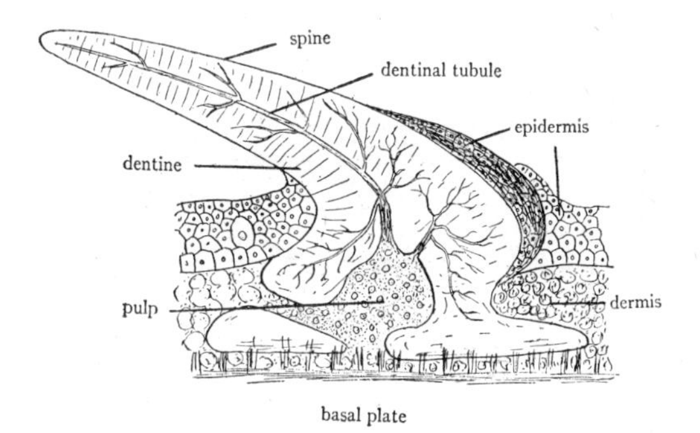

Fig. 22.—Longitudinal section of a placoid scale of a dogfish (*Mustelus*), showing its dentinal nature. (After Hertwig, 1874.)

The placoid scale is best understood by considering its development (Fig. 23). Each scale begins as an aggregation of dermal cells (Fig. 23*A*), which then pushes up against the epidermis as a conical projection, termed a *dermal papilla* (Fig. 23*B*). The layer of epidermis in contact with the dermal papilla alters to a tall columnar form, resembling the ameloblasts of an enamel organ, and probably plays a role in shaping the spine. The surface of the dermal papilla then secretes the placoid scale in the form of dentine, and after finishing the scale the papilla remains as a nutritive pulp occupying the pulp cavity. The epidermis withdraws from around the spine (Fig. 23*D*), which then projects above the skin. The mode of formation of the scale and its dentinal nature indicate that placoid scales are homologous with vertebrate teeth. Although the best researches indicate that enamel, present as a surface layer on the exposed part of teeth and secreted by the under surface of the enamel organ, is lacking from placoid scales, the enamel organ is present as in developing teeth. The homology of teeth and scales may be ascribed to the fact that the mouth lining is inturned skin and hence contains skin structures.

2. The cosmoid scale.—This occurs only in extinct fish of the groups Crossopterygii and Dipnoi. It consists (Fig. 24) of three layers: an outer layer of dentine, a middle layer of very vascular bone, permeated with blood channels, and an inner layer of lamellate bone (so-called

[6] It has long been taught that the shiny surface covering of the spine is enamel and hence that the placoid scale completely resembles a tooth. But all the evidence indicates that the shiny material is merely a particularly hard kind of dentine.

enamel is lacking

isopedine). The dentine, called *cosmine*, is peculiar in that the dentinal tubules occur in clusters (Fig. 24*A*, *B*) with channels between, which open on the surface by pores. The cosmoid scale is the most primitive scale to be found among the Osteichthyes. Its structure suggests that it has arisen by the fusion of a number of placoid scales (each represented by a cluster of dentinal tubules in the cosmine) to a bony dermal plate, and this theory is generally accepted as the mode of origin of the scales of bony fishes. Evanescent spines representing placoid scales occur in the development of some bony fishes—e.g., *Lepidosteus*.

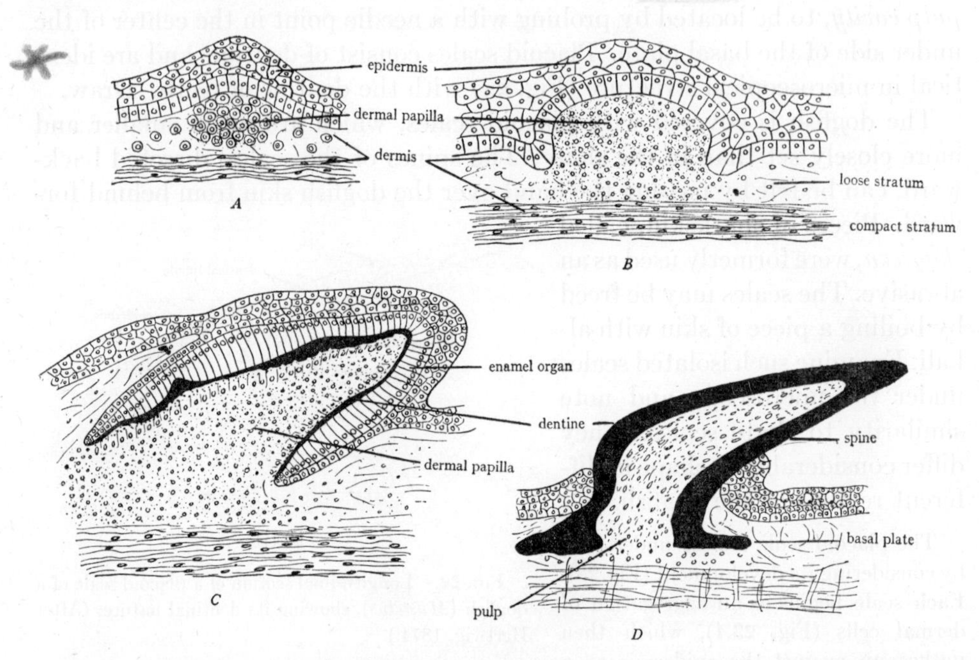

FIG. 23.—Stages in the development of a placoid scale (based on Klaatsch, 1890). *A*, early stage, aggregation of dermal mesenchyme cells; *B*, next stage, aggregation has formed a papilla over which the basal layer of the epidermis becomes columnar, forming an enamel organ; *C*, papilla has lengthened and is secreting dentine on its outer surface under the still more columnar enamel organ; *D*, completed scale, epidermis has withdrawn, papilla becomes the pulp, dentine black.

3. The palaeoniscoid scale.—This type, found in the order Chondrostei (extinct palaeoniscids and *Polypterus*), is a ganoid scale, which, however, is intermediate between the cosmoid and typical ganoid scale. It has (Fig. 24*B*) a lower layer of lamellate bone (isopedine), a middle reduced cosmine layer, and an upper layer of lamellae of *ganoin*, a very hard shiny substance secreted by the dermis. Recall the scales of *Polypterus* (p. 43) or re-examine the specimen. Palaeoniscoid scales cannot be distinguished externally from the next type.

4. The lepidosteoid or true ganoid scale.—The cosmine layer has disappeared, so that the true ganoid scale consists of lamellae of ganoin deposited on a layer of lamellate bone (Fig. 24*C*). It is characteristic of gar pikes, sturgeons, and their allies. Examine a gar pike (*Lepidosteus*) and note its complete investment of hard, shiny, rhomboid plates, fitted closely together in diagonal rows. Cut out a piece of skin containing several scales and note that each diagonal row is movable on the adjacent rows along the line of junction or hinge line, but the members of each row are immovably joined to each other by a peg-and-socket arrangement (visible only on the under side of scales from large specimens). Draw a few scales.

In the sturgeon note similar ganoid scales—bony rhombic crested plates,

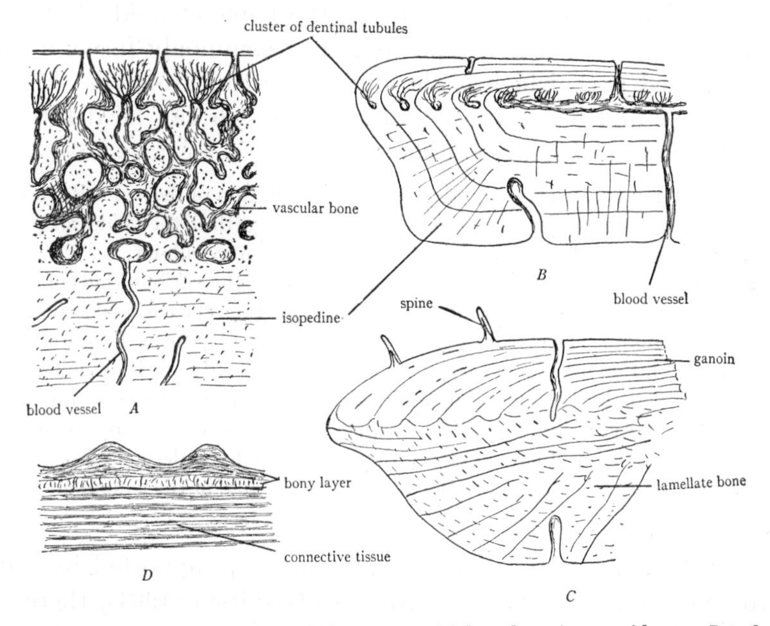

Fig. 24.—Diagrammatic sections through four types of fish scales. *A*, cosmoid type; *B*, palaeoniscoid type; *C*, lepidosteoid or true ganoid type; *D*, ctenoid type. (*A–C*, based on Goodrich, 1907; *D*, based on Hase, 1907.)

arranged in five longitudinal rows with areas of apparently naked skin between the rows. In some sturgeons (*Scaphirhynchus*) the tail is completely invested with ganoid scales.

5. The cycloid scale.—The stiff, hard ganoid scales obviously hinder movement; hence they are replaced in modern types of fishes by thin flexible scales of which there are two sorts, *cycloid* and *ctenoid*, differing only in small details. They consist of an under layer of fibrous connective tissue, which lends flexibility, and an outer hard bony layer (Fig. 24*D*),[7] making the scale firm. Cycloid scales are the earlier type and occur in Holostei, present Dipnoi, and some Teleostei. Examine the bowfin (*Amia*) and note the thin,

[7] Called hyalodentine in many books, but it has no microscopic resemblance to dentine and is more nearly like bone.

rounded scales, set in pockets of the skin, in overlapping rows, shingle fashion (imbricated). How does the imbricated arrangement assist lateral movements? Remove a scale and examine under the microscope, noting concentric ridges, fluted inner half, and smooth free half. Draw.

6. The ctenoid scale.—This type of scale occurs in about half of the bony fishes. On the perch or other common fish compare scale arrangement with *Amia*. Examine a ctenoid scale with the microscope and note fluting and concentric ridges; it differs from the cycloid scale only in that the free part is covered with small teeth (whence the name ctenoid), but these are poorly defined in some fishes. The epidermis commonly forms a thin film over the toothed part, which is the part projecting from the skin pocket. Draw. The age and hence the growth rate of fish can be determined by studying the concentric ridges of the scales.

Cycloid and ctenoid scales are secreted by an accumulation of dermal cells to both sides. The posterior edge of the scale grows backward, carrying with it the epidermis, which may persist as a thin coat over the exposed part of the scale or may rub off.

There appears to be no accepted version of the relation of the various kinds of fish scales to each other. If the placoid scale is the most primitive type, then the cosmoid scale was derived by the fusion of a number of placoid scales to an independently formed bony dermal plate; but the reverse evolution may have occurred. The palaeoniscoid scale is obviously descended from the cosmoid type by the reduction of the cosmine layer and the appearance of a layer of ganoin on top of the cosmine. Complete disappearance of the cosmine results in the typical ganoid scale, with the layer of ganoin placed directly on the bony dermal plate. Ctenoid and cycloid scales seem to have come from the ganoid type by loss of the ganoin and thinning of the bony dermal plate.

7. The dermal fin rays.—The fins of fishes are supported by stiff rays, found to be of dermal origin. According to Goodrich (1904), there are four kinds of fin rays.

a) *Ceratotrichia:* These are the slender, flexible, unjointed fin rays found in elasmobranchs and Holocephali; they consist of a fibrous material. Reexamine them in the dogfish.

b) *Actinotrichia:* These occur during the development of bony fish and persist in a few fins and near the growing edge of fins, but they are usually replaced by the next type. They are identical in structure and mode of formation with ceratotrichia and appear to be a recapitulation of the latter during development.

c) *Lepidotrichia:* These are characteristic of the Osteichthyes in general. They are branched, jointed rays composed of bone; the joints have been shown by Goodrich to be homologous with the scales of the same fish. Reexamine them in the perch or other bony fish. Fin spines arise by alteration and fusion of the joints of the lepidotrichia.

d) Camptotrichia: This type of fin ray is limited to present Dipnoi and is intermediate between *a* and *c*.

The dermal fin rays are in turn supported through ligamentous connections with the endoskeletal fin rays.

E. EXOSKELETON OF AMPHIBIA

The vast majority of present-day Amphibia have naked skins, devoid of exoskeleton. Minute concealed dermal scales are found in the Gymnophiona. The extinct orders of Amphibia, formerly termed Stegocephala, usually had a ventral armor of bony dermal plates, probably derived from the cosmoid scales of crossopterygian fish.

F. EXOSKELETON OF REPTILES

The bodies of reptiles are characteristically clothed in a horny exoskeleton marked off into scales or scalelike areas. These are of epidermal origin, representing thickened areas of the stratum corneum. In the formation of such a thickening a dermal papilla first appears to furnish nutriment for the process, and the stratum germinativum over the dermal papilla begins to proliferate rapidly, producing cells which flatten, cornify, and compress into a scale or scalelike area. Such scales are never separable, like the scales of fishes, but are simply thickened parts of a continuous horny layer. The scales of reptiles and of fishes are thus wholly different structures (Fig. 25). In addition, many reptiles possess bony plates of dermal origin beneath the epidermal scales. To avoid confusion it is well to refer to the epidermal thickenings as *scales* or *scutes*, to the dermal structures as *plates*.

1. Exoskeleton of lizards and snakes.—Recall the condition in these animals (pp. 47, 49) or re-examine the specimens. In a few lizards there are bony plates under the scales; on this point see also what was said about the Crocodilia (p. 50).

2. Exoskeleton of turtles.—Carapaces and plastrons which have been separated by sawing through the bridges will be provided; cooking them renders the sutures more distinct. The following description is based upon our common pond turtles; as already noted (p. 50), the numbers and shapes of scutes and plates differ in different families of turtles.

a) Carapace: The dorsal surface of the carapace consists of large, thin, horny epidermal scutes, whose boundaries are marked by grooves. They are arranged in three groups: a median row of five *neural* scutes, a lateral row of four *costal* scutes on each side, and a set of small *marginal* scutes (how many?) along the entire margin. The anterior narrow median unpaired marginal scute is called the *nuchal* scute; and two median posterior ones, behind the fifth neural scute, are often called *pygal* scutes. Observe continuation of the marginal scutes to cover the margin of the under side. Draw half the dorsal side of the carapace, filling in the scutes accurately.

Study the ventral surface of the carapace. It is composed of heavy bony plates of dermal origin, bounded by jagged *sutures* and fused to the vertebrae and ribs. Like the scutes, they occur in three groups. The median row of plates, fused to the vertebrae, consists of a single large anterior *nuchal* plate, followed by eight small *vertebral* or *neural* plates, followed by two postneural or precaudal plates, not attached to vertebrae. On each side is a row of eight elongated costal plates, each attached to a rib, giving the appearance of an expanded rib. The margins are formed of a circle of marginal plates, paired, except for the median posterior *pygal* plate. Make an accurate drawing of the under side of the carapace, showing bony plates and their relation to the vertebrae and ribs.

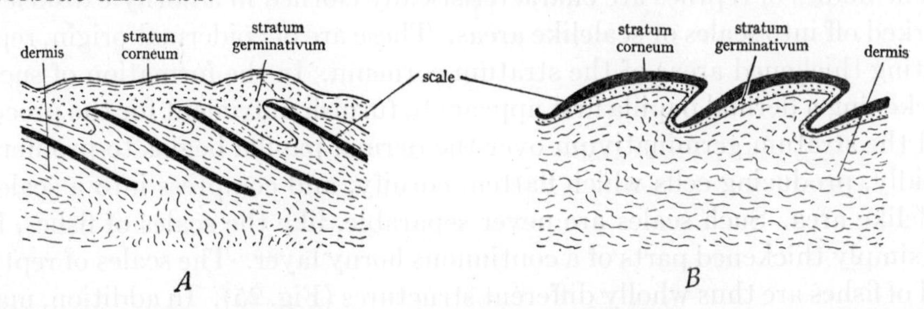

FIG. 25.—Diagrammatic longitudinal sections through the skin of *A*, a teleostome fish, and *B*, a reptile, to show the locations of the scales. In *A* the scales are in the dermis, while in *B* they represent thickened portions of the stratum corneum. (From Wiedersheim's *Comparative Anatomy of Vertebrates*, courtesy of the Macmillan Co.)

The best available evidence favors the view that the neural and costal plates are independent formations, secondarily attached to the vertebrae and ribs, and are not, as formerly thought, expanded vertebrae and ribs. Hence, all of the plates of the carapace are of uniform origin and comprise one exoskeleton, sometimes called the thecal exoskeleton. There is also considerable evidence that at one time turtles had an additional dermal carapace, the epithecal exoskeleton, between the present one and the epidermal scutes. The epidermal scutes correspond to these lost epithecal plates; hence they do not correspond to the thecal plates.

b) *Plastron:* This, like the carapace, consists of a set of horny epidermal scutes covering bony dermal plates. Study the external (ventral) surface of the plastron. It has six pairs of scutes, named from in front backward: *gular*, *humeral*, *pectoral*, *abdominal*, *femoral*, and *anal*. Irregular *inframarginal* scutes cover the bridges. Study the internal (dorsal) surface, noting the large bony plates, united by jagged sutures, which compose it. They comprise: a small anterior pair, the *epiplastra;* a single median plate between them with posteriorly projecting point, the *entoplastron;* and behind these three pairs of squarish plates, the *hyoplastra*, *hypoplastra*, and *xiphiplastra*, respectively. Draw the plastron, showing outlines of scutes and plates.

c) *Other exoskeletal structures:* The exoskeleton also includes the claws and the horny beaks which incase the jaws. Examine these beaks in a demonstration specimen in which they have been loosened from the underlying bones. Turtles also possess scales or thickened scalelike areas on the legs and tail, and in some turtles there are enlarged head shields. All these structures are epidermal, consisting of special portions of the stratum corneum.

G. EXOSKELETON OF BIRDS

Birds are clothed in an exoskeleton consisting of feathers on the greater part of the body, scales and claws on the feet, and horny beaks, all of epi-

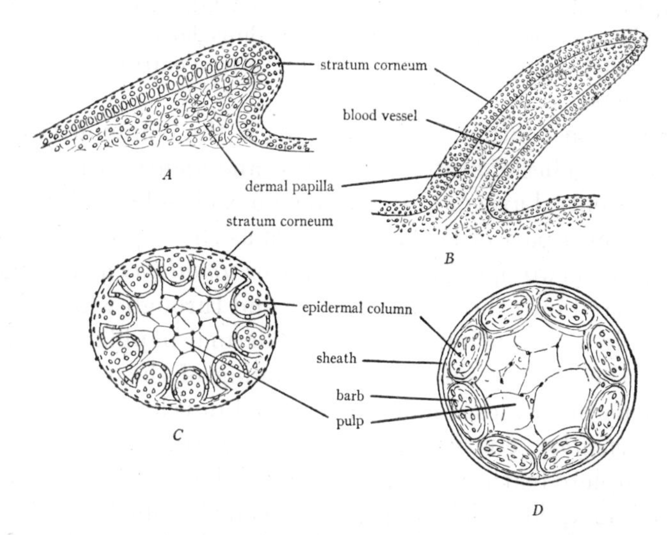

FIG. 26.—Development of a plumule (based on Davies, 1889). *A*, early stage of feather papilla; *B*, papilla elongated, pulp formed; *C*, cross-section of stage similar to *B*, showing epidermal columns forming; *D*, cross-section, showing epidermal columns separated to become the barbs.

dermal origin, formed from the stratum corneum. The scales, claws, and beaks are obviously indicative of reptilian descent. Feathers are believed to be homologous to reptilian scales but resemble them only in the very earliest stages of formation. No stages intermediate between scales and feathers are known. Dermal exoskeleton is lacking in birds.

There are three kinds of adult feathers: *down* feathers or *plumules*, *contour* feathers, and *hair* feathers or *filoplumes*.

1. Structure of the down feather.—Down feathers (better called plumules) constitute the fluffy covering of young birds and also occur in many adult birds (especially aquatic birds and birds of prey) between the bases of the contour feathers. Obtain a plumule or a prepared slide of one. Identify the short stem or *quill*, the soft rays or *barbs* which spring in a circle from the top of the quill, and the minute side rays or *barbules* on the barbs. Draw.

2. Development of the plumule.—A plumule arises from a papilla of the skin, the *feather papilla*, consisting of a dermal core, the *dermal papilla*, covered by epidermis (Fig. 26*A*, *B*). The dermal papilla is richly vascular and furnishes nutrition for the developing feather. Later the papilla sinks into a pit in the skin, the *feather follicle*. The stratum germinativum at the base of the papilla begins to proliferate upward, forming a number of longitudinal columns which project into the dermis (Fig. 26*C*). These columns cornify and separate, each being a barb (Fig. 26*D*). In the meantime the stratum corneum of the developing feather has become a definite *sheath* which covers the circlet of barbs. When the feather is finished, the sheath splits open, releasing the barbs. Meantime the whole papilla has been elongating and comes to project above the surface. The basal part of the feather does not split into barbs but remains undivided as the quill, within which the nutritive material, termed the *pulp*, elaborated from the dermal papilla, dries up to form a pith. The feather papilla remains alive at the bottom of the follicle and continues to form feathers throughout life. The development of a feather resembles that of a reptilian scale in that it involves a nutritive dermal papilla and an epidermal thickening.

3. Structure of a contour feather.—This is the common type of feather which covers the bodies of birds. Obtain one and identify the following parts. It consists of a central axis or *quill* bearing on each side a weblike expansion, the *vane*. The lower part of the quill, known as the *calamus*, is bare and hollow. The quill has two openings: the *inferior umbilicus* at its proximal end, which was inserted into the feather follicle and through which during development the pulp was continuous with the dermal papilla; and the *superior umbilicus*, on the ventral side at the beginning of the vane. From the superior umbilicus protrudes a more or less well-developed accessory feather, called the *afterfeather*, consisting of only a few tufts in some birds but in others of a complete feather, nearly or quite as large as the primary feather.

The part of the quill which supports the vane is called the *shaft* or *rachis;* it has on its ventral surface (side next the bird's body when the feather is in place) an *umbilical groove*. The vane is divided into the *outer* vane, exposed when the feather is in place, and the *inner* vane, covered by the adjacent feather, and often narrower.[8] The vane is obviously composed of a large number of parallel, obliquely placed rays, adhering to each other. These rays are the *barbs*, and each barb bears side rays or *barbules*, as in the down feather. The barbules interlock, causing the barbs to adhere to produce an unbroken surface. To see the method of interlocking of the barbules, microscopic examination is necessary. It is then seen that the barbules of the upper side (side toward the feather tip of each barb) are placed diagonally across the barbules of the lower side (side toward the calamus) of the adjoining barb, and that the former are provided with hooklike projections which fit over and catch flangelike extensions of the latter. The diagonal arrangement permits one hooked barbule to catch a number of flanged barbules.

[8] The definitions of quill, shaft, vane, etc., here employed follow Chandler (1916). Other usages are current; e.g., quill and calamus are often regarded as synonymous, and the term vane may include the rachis.

By pulling on the barbs one can unhook the barbules, and by stroking the barbs one can hook them up again. The latter process is the chief object of the frequent preening of the feathers habitual with birds. In the rectrices and remiges the hooklets are well developed throughout, so that the barbs are interlocked over the entire feather; but in the coverts the lower barbules lack the hooklets, and the lower barbs are loose and fluffy. In some birds, as the ostrich, all of the barbules lack hooks, and the whole feather is fluffy.

Draw a contour feather, showing its parts.

4. Development of the contour feather.—It is seen that in plumules the barbs spring in a circle from the top of the quill (Fig. 27*A*), whereas in contour feathers they spring from the sides of the quill. The development of the two kinds of feathers is similar to a certain stage; then the middorsal region of the growing zone in the base of the feather papilla begins to push upward in the case of contour feathers carrying the barbs with it (Fig. 27*B*) and becoming the

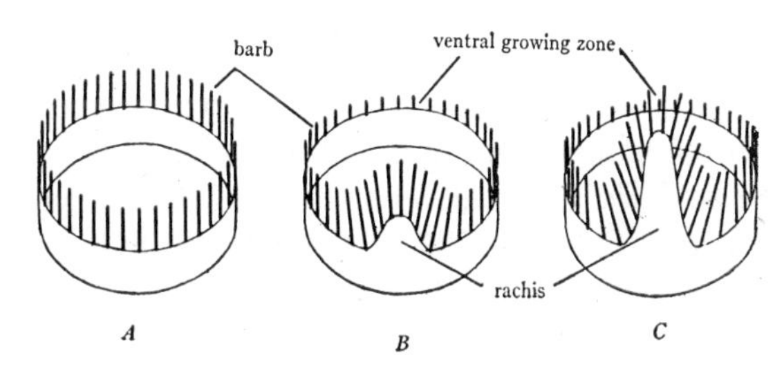

FIG. 27.—Diagrammatic comparisons of developing down and contour feathers. *A*, diagram of a plumule, equal parallel barbs springing from the ringlike basal growing zone of the feather papilla; *B*, early stage of a contour feather, barbs originating in a ventral growing zone and enlarging dorsally, where a median projection throws them into a diagonal position; *C*, later stage of the contour feather. (*B* and *C* based on Lillie and Juhn, 1937.)

quill. New barbs form in the ventral side of the growing region and gradually swing dorsally into an oblique position. The afterfeather is formed in the midventral region in exactly the same way (not shown in the figures) and hence is a mirror-image twin of the primary feather. The young contour feather is thus rolled up inside the sheath, forming the *pin feathers* of popular terminology. The future dorsal or outer surface of the feather is next to the sheath, and the future ventral or inner surface next to the pulp. When the sheath splits and falls away, the feather flattens out. The calamus is the base of the feather which failed to split into barbs, and the white flaky material noticeable inside it is the dried pulp. Contour feathers grow from the same follicles as the plumules and, in fact, are continuous with them.

5. Feather tracts.—Birds appear to be completely covered by contour feathers, but actually the feathers are borne only by certain areas of the skin called feather tracts or *pterylae*, with featherless areas, *apterylae*, between them. Observe the feather tracts on a demonstration specimen of a young bird. Plumules usually occur evenly distributed over the body but may be limited to either the pterylae or the apterylae.

6. Structure of the filoplume.—The filoplumes are the "hairs" visible on the plucked bird. Examine a prepared slide, or mount a filoplume from a plucked bird in a drop of water. It consists of a main axis bearing a few terminal barbs and seems to be a miniature degenerated contour feather. Draw.

H. EXOSKELETON OF MAMMALS

The exoskeleton of mammals consists primarily of *hair*, found in no other vertebrates; however, a number of mammals are secondarily almost devoid of hair, as Cetacea, Sirenia, elephants, hippopotamus, rhinoceros, although the embryos of some of these possess a complete coat of hairs. In some mammals scales are also present in addition to hairs.

1. Structure and development of hair.—On your own skin determine that each hair springs from a pit in the skin, the *hair follicle;* the part of the hair inside the follicle is termed the *root*, the exposed part the *shaft*. Examine a light or white human hair or hair of any other mammal with the high powers of the microscope. By focusing on the surface of the hair observe the irregular wavy lines of the scalelike cells which form the surface layer of the hair, termed *cuticle*. On focusing down, one will note a central strand of degenerated material, the *medulla*, and between this and the cuticle a layer of elongated cells, the *cortex*. The medulla is often absent, especially in smaller hairs. Color, when present, is located chiefly in the cortex. Threads from clothing may also be examined; genuine wool is recognized by the wavy cuticular cells. Study a microscopic section showing growing hair follicles in longitudinal view. They are deep pits lined by inturned epidermis. At the bottom of the follicle the dermis forms a bulb-shaped papilla over which the stratum germinativum is seen proliferating a conical heap of cells; this undergoes keratinization and becomes the hair. This cone extends up the follicle, and after a short distance a split is seen separating the hair from the wall of the follicle. The hair continues to be pushed outward by proliferation from below. The transition from the ordinary epithelial cells of the proliferating area to the horny cells of the hair is readily observable. The lining of the follicle, called the *inner root sheath*, is also somewhat cornified; it is the white coat which clings to the roots of hairs when they are pulled from the follicles. From the foregoing account it follows that hairs are epidermal structures. Hairs grow only from the base of the root next the dermal papilla and, so far as known, grow at an even rate regardless of cutting. Hairs of mammals are continually being shed and replaced by new ones from the same follicles. In many mammals the pelt as a whole is shed at a definite season and may be replaced by a pelt of a different color. Color changes in mammalian fur can occur only by the shedding of the old hairs and the formation of new ones.

The hairs of mammals are by no means of uniform morphology. They are generally cylindrical but may be more or less flattened; they may be straight or wavy, taper quickly or gradually to a point, or may have enlarged thickened or spoonlike ends. The following types are found generally distributed among mammals: *facial vibrissae*, also called *sinus* hairs, the large tactile hairs found on the chin, lips, cheeks, snout, and around the eyes; spinelike hairs or bristles, like those of the spiny anteaters; *guard* hairs, especially long, stiff, straight, pointed, coarse hairs scattered in the fur; *aristate* hairs, smaller and fairly numerous with a thickened or flattened end; and the ordinary wool or fur, composed of fine, short, usually wavy hairs.

A follicle does not necessarily form a single hair at a time; in fact, in many mammals a bundle of hairs grows from each follicle by a budding process of the papilla. In such bundles there is often a central larger hair, surrounded by shorter hairs.

2. Scales of mammals.—A number of mammals possess horny scales like those of reptiles. These may cover the body in imbricated fashion, as in the scaly anteaters (p. 59), but commonly occur only on the tail or legs. Such scaly parts are also provided with scanty hairs. Examine the tail of a rat or beaver and observe the scales and hairs upon it.

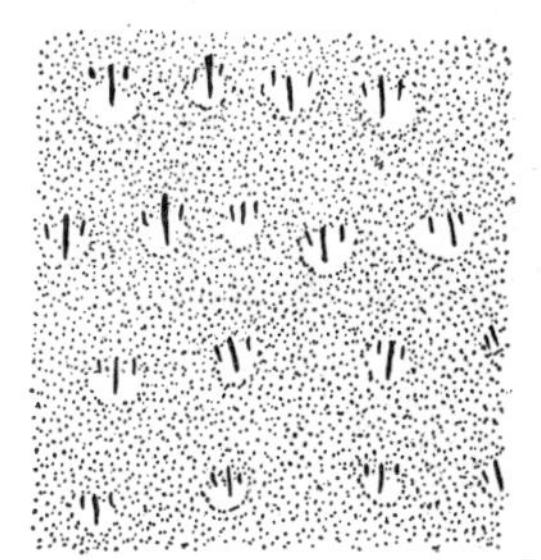

FIG. 28.—Hairs arising in groups of three in connection with elevations suggestive of scales in the embryo of *Centetes* (order Insectivora). (After Emery, 1893.)

The armadillos are the only living mammals which possess, like the turtle, an armor composed of both epidermal scutes and dermal plates. Obtain a dried armadillo armor and examine the external surface. It is made up of an *anterior shield* of small polygonal scales, a middle region composed of nine movable *bands* with bare areas of skin between them, and a *posterior shield* similar to the anterior shield. The outer surface of the armor consists of thin, horny epidermal scales or scutes, polygonal on the shields, triangular on the bands. The triangular scutes are alternately reversed in position, so that in half of them the apex points anteriorly, and in half posteriorly. The former bear hairs at their posterior margins. How many hairs to each scute? Draw some scutes and hairs to show their relation. Turn the armor over and study the internal surface. It is composed of bony plates of dermal origin, polygonal on the shields, rectangular on the bands. With a knife point scrape off some of the epidermal scutes; note relative thickness of scutes and plates, differences in the materials of which they are composed—one of horn, the other of bone—and the impressions left on the plates by the scutes.

3. The relation of hairs to scales and the homology of hair.—It is probable that hairs are not homologous to reptilian scales. For one thing, their development differs considerably, since a hair follicle begins as a solid epidermal ingrowth which only later acquires a dermal papilla, whereas the formation of a large dermal papilla is the first stage in scale production. Furthermore, as just seen, scales and hairs occur simultaneously on the same body parts in many mammals. In such cases a definite arrangement of the hairs with respect to the scales can usually be noticed. The hairs occur in groups of three to five behind each scale. Such hair groupings have been shown to be of frequent occurrence among mammals, whether scales are present or not (Fig. 28). It is therefore plausible to suppose that hairs originated in relation to scales, not from them, and that subsequent loss of the scales permitted the hair to spread over the skin. As to what the reptilian forerunner of hair was remains very uncertain. The most accepted hypothesis is that hairs are derived from skin sense organs.

4. Claws, nails, hoofs, and horns.—Claws, nails, and hoofs are essentially composed of compressed layers of stratum corneum. Claws are practically identical in reptiles, birds, and mammals. Each is a curved horny sheath covering the last joint of the digit and therefore inclosing the terminal skeletal joint (phalanx). Generally the ventral side of the claw, termed sole, is formed of less compact horn than the dorsal and lateral sides, and hence the junction with the sole often forms a sharp edge or angle (Fig. 29*A*). The base of the claw is usually

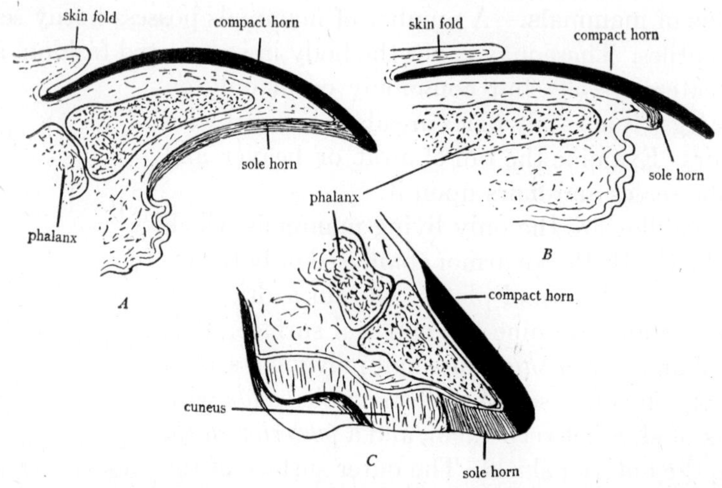

Fig. 29.—Claw, nail, and horse's hoof compared (based on figures of Boas, 1884). *A*, longitudinal view of a claw, note compact and sole horn about equal in extent; *B*, primate nail, note great reduction of sole horn; *C*, longitudinal view of horse's hoof; the surface of the hoof consists of compact horn; inside this is a crescentic zone of sole horn; and the median part forming a wedge fitting into the crescent of the sole horn is the so-called "frog" or cuneus, made of horny tissue and corresponding to the foot pads of other mammals.

imbedded in a projecting fold of skin and flesh. The claw grows mainly at the base, but generally also new horn similar to that of the sole is continually deposited under the tip to strengthen it. The nails of primates differ from claws in the great reduction of the sole, which is limited to a brief zone on the under side of the nail just back of the free edge (Fig. 29*B*). In hoofs, on the other hand, the sole is greatly enlarged and thickened and forms an important part of the hoof (Fig. 29*C*).

The horns of mammals are of several different sorts. Those of the rhinoceros are made of bundles of hair matted together and mounted on bony bosses of the skull bones. The short horns of the giraffe consist of bony cores covered by ordinary skin. The true horns of cattle, sheep, goats, and antelopes are hollow horn sheaths covering hollow bony cores which are outgrowths of the frontal bone of the skull. Such horns occur in both sexes, do not branch, and are never shed. The horns of the pronghorn antelope of our western plains differ from those of other antelopes, although like them structurally, in that they branch and are shed annually, as in deer. The horns of the deer family (Cervidae), better called antlers, are solid bony outgrowths of the frontal bone. When young, they are covered with a modified skin, termed the "velvet," which soon wears off. Antlers are limited to the male sex, except in reindeer, usually branch, and are shed annually, generally in late winter; new ones then grow from the old stumps.

I. SUMMARY

1. The vertebrate skin consists of an outer epidermis, ectodermal in origin and composed of epithelial cells, and an inner dermis, mesodermal in origin and composed of connective tissue.

2. Exoskeletal structures are produced by the skin, either by the epidermis or by the dermis or by both.

3. Epidermal exoskeletal structures consist of horn and are composed of numerous flattened dead cornified cells pressed together. These cells arise by proliferation of the lower layers of the epidermis (stratum germinativum). Of such nature are the superficial scales of reptiles, birds, and mammals, and feathers, hair, claws, nails, beaks, hoofs, and horns such as those of cattle.

4. Dermal exoskeleton structures consist of bone or of substances allied to bone, such as dentine, and are secreted by the mesenchyme cells of the dermis. Of such nature are the scales and fin rays of fishes, the bony plates of reptiles and mammals, the plates of recent and extinct Amphibia, and the antlers of deer.

5. The placoid scales of elasmobranch fishes consist of dentine; and this fact, together with their manner of development, indicates a homology to teeth.

6. In the formation of exoskeleton a dermal papilla is involved which furnishes nutrition for the proliferating or secreting cells. Dermal structures arise directly from the cells of the papilla. In the case of epidermal exoskeleton the papilla takes no part in the structure itself but simply brings a blood supply to the developing part, which arises solely by proliferation and subsequent keratinization of epidermal cells.

7. Epidermal structures exhibit more modifications and a greater complexity of structure than do dermal parts.

8. There has been no definite progressive evolution of the exoskeleton throughout the vertebrate groups, but each group shows a correlation between exoskeletal structures and habits and mode of life. There is, however, a certain tendency for the higher vertebrates to develop complicated epidermal structures and to discard dermal ones.

REFERENCES

ANTHONY, H. C. 1928. Horns and antlers, their evolution, occurrence, and function in the Mammalia. Bull. N.Y. Zoöl. Soc., **31.**

BARGMANN, W. 1937. Zur Frage der Homologisierung von Schmelz und Vitrodentin. Zeitschr. f. Zellforsch. u. mikr. Anat., **27.**

BOAS, J. E. V. 1884. Ein Beitrag zur Morphologie der Nägel, Krallen, Hufe, und Klauen der Säugethiere. Morph. Jahrb., **9.**

———. 1894. Zur Morphologie der Wirbelthierkralle. *Ibid.*, **21.**

BRESSLAU, E. 1904. Zur Entwicklung des Beutels der Marsupialier. Verhandl. d. deutsch. zool. Gesellsch.

———. 1920. The Mammary Apparatus of the Mammalia.

CHANDLER, A. C. 1916. A study of the structure of feathers with reference to their taxonomic significance. Univ. Calif. Pub. Zoöl., **13.**

DAVIES, H. R. 1889. Die Entwicklung der Feder und ihre Beziehungen zu anderen Integumentgebilden. Morph. Jahrb., **15.**

EMORY, C. 1893. Über die Verhältnisse der Säugetierhaare zu schuppenartigen Hautgebilde. Anat. Anz., **8.**

Goodrich, E. S. 1904. On the dermal fin rays of fishes. Quart. Jour. Mic. Sci., 47.

———. 1907. On the scales of fish, living and extinct, and their importance in classification. Proc. Zoöl. Soc. London, Part 2.

Hase, A. 1907. Über das Schuppenkleid der Teleosteer. Jena Zeitschr. Naturwiss., 42.

Hay, O. P. 1928. Further consideration of the shell of *Chelys* and the constitution of the armor of turtles in general. Proc. U.S. Natl. Mus., 73.

Hertwig, O. 1874. Über Bau und Entwicklung der Placoidschuppen und der Zähnen der Selachier. Jena Zeitschr. Naturwiss., 8.

Hosker, A. 1936. Studies on the epidermal structures of birds. Phil. Trans. Roy. Soc. London, B, 226.

Klaatsch, H. 1890. Zur Morphologie der Fischschuppen und zur Geschichte der Hartsubstanzgewebe. Morph. Jahrb., 16.

Lillie, F. R., and Juhn, Mary. 1932, 1937. The physiology of development of feathers. I, II. Physiol. Zoöl., 5, 11.

Meijere, J. C. H. 1894. Über die Haare der Säugetiere besonders ihre Anordnung. Morph. Jahrb., 21.

Neaves, F. 1936. On the development of the scales of *Salmo*. Trans. Roy. Soc. Canada, 3d ser., 30.

———. 1940. On the histology and regeneration of the teleost scale. Quart. Jour. Mic. Sci., 81.

Nickerson, W. S. 1893. Development of the scales of *Lepidosteus*. Bull. Mus. Comp. Zoöl. Harvard, 24.

Parker, G. H. 1940. Neurohumors as chromatophore activators. Proc. Amer. Acad. Arts and Sci., 73.

Pocock, R. I. 1914. On the facial vibrissae of Mammalia. Proc. Zoöl. Soc. London, Part 2.

Procter, J. 1922. A study of the remarkable tortoise *Testudo loveridgii* Blgr. and the morphogeny of the chelonian carapace. Proc. Zoöl. Soc. London, Part 2.

Sayles, L. P., and Hershkowitz, S. G. 1937. Placoid scale types and their distribution in *Squalus acanthias*. Biol. Bull., 73.

Schmidt, W. J. 1940. Das porsellanartige Dentin (Durodentin) der Selachier. Zeitschr. f. Zellforsch. u. mikr. Anat., 30.

Toldt, K., Jr. 1910. Über eine beachtenwerte Haarsorte und über das Haarformensystem der Säugetiere. Ann. naturhist. Hofmuseums, Wien, 24.

Turner, C. W. 1939. The Comparative Anatomy of the Mammary Gland.

Zangerl, R. 1939. The homology of the shell elements in turtles. Jour. Morph., 65.

VII. THE ENDOSKELETON: THE COMPARATIVE ANATOMY OF THE VERTEBRAL COLUMN AND RIBS

A. GENERAL CONSIDERATIONS ON THE ENDOSKELETON

1. The endoskeleton and its parts.—The endoskeleton is defined as an internal supporting system of hardened material. It is highly characteristic of vertebrates, since among invertebrates hardened supporting structures are usually exoskeletal. The endoskeleton of vertebrates consists of cartilage or bone or mixtures of these two types of tissue and is of mesodermal origin in general. The parts of the endoskeleton of vertebrates are: the *skull*, in the head; the *branchial skeleton*, composed of *gill arches* supporting the gills; the *vertebral column*, occupying the median dorsal region; the *ribs*, projecting from the vertebrae, one pair to each vertebra primitively; the *sternum*, occupying the median ventral region of the anterior part of the trunk; the *pectoral girdle*, supporting the anterior paired appendages; the *pelvic girdle*, supporting the posterior paired appendages; and the *skeleton* of the *appendages*. The four parts first named constitute the *axial* skeleton, while the other parts constitute the *appendicular* skeleton.

2. The notochord.—The notochord, as already learned, is a stiffened rod running longitudinally in the middorsal region of the chordate body, just beneath the central nervous system. It probably originated as a support for an elongated body swimming by lateral undulations. It is formed embryologically from the roof of the archenteron or its equivalent. It is first seen in tunicates, where it forms a supporting rod for the tail of the tadpole larva, disappearing at metamorphosis in most tunicates. In *Amphioxus* it constitutes the axial support of the body. The notochord appears in the embryos of all vertebrates, but in them it extends anteriorly only to the hypophysis. Its extension in *Amphioxus* nearly to the anterior tip appears to be a secondary, rather than the primitive, condition. In the vast majority of vertebrates the notochord is replaced as an axial support by the vertebral column. The vertebral column forms around the notochord and gradually squeezes it out of existence, or nearly so, although remnants often remain between the vertebrae. Early in development the vertebrate notochord becomes invested with a connective tissue sheath, which strengthens it in forms with a slight development of the vertebral column and in others plays a role in the formation of the vertebrae. The sheath (usually divisible into an inner thicker sheath of fibrous connective tissue and an outer thinner one of elastic fibers) appears to be laid down by the notochord.

3. The skeletogenous regions.—The endoskeleton develops from mesenchyme (p. 74). The mesenchyme for the vertebral column and ribs comes chiefly from the sclerotomes, which, as we have learned (p. 76), are formed by the breaking-down of the medial sides of the epimeres (somites) into mesenchyme; but contributions from other mesodermal sources no doubt occur. The skeletogenous mesenchyme accumulates around the notochord and neural tube and in certain other *skeletogenous regions*. The arrangement of the latter is largely dependent on the disposition of the myotomes. As already explained, the myotomes or muscle segments, which are those portions of the epimeres remaining after the giving-off of mesenchyme, grow down between the skin and the digestive tract, so as to form the muscular layer of the body wall. (Review pp. 76–77.) Each myotome is separated from the adjacent ones by a transverse partition of mesenchyme, the *myoseptum* or *myocomma*. Each myotome (except in cyclostomes) is further divided into a dorsal (*epaxial*) and a ventral (*hypaxial*) half by a horizontal partition, the *horizontal skeletogenous septum*, which extends from the notochord to the level of the lateral line on the sides of the body. The mesenchyme surrounding the notochord (*perichordal* mesenchyme) and neural tube continues to the median dorsal line as the *dorsal skeletogenous septum*, and similarly from the notochord to the median ventral line (in the tail) as the *ventral skeletogenous septum*. In the trunk region the ventral skeletogenous septum is naturally split into two *ventrolateral* septa by the intervention of the coelom (Fig. 30). The horizontal, dorsal, and ventral septa are, it is to be understood, continuous longitudinal septa, running the length of the body. The skeletogenous septa are illustrated in Figure 30, which should be thoroughly mastered. As their name implies, the skeletogenous septa are regions of skeleton formation. *At the intersection of every myoseptum with the dorsal, ventral, and horizontal septa and with the perichordal mesenchyme a vertebra arises.* As the

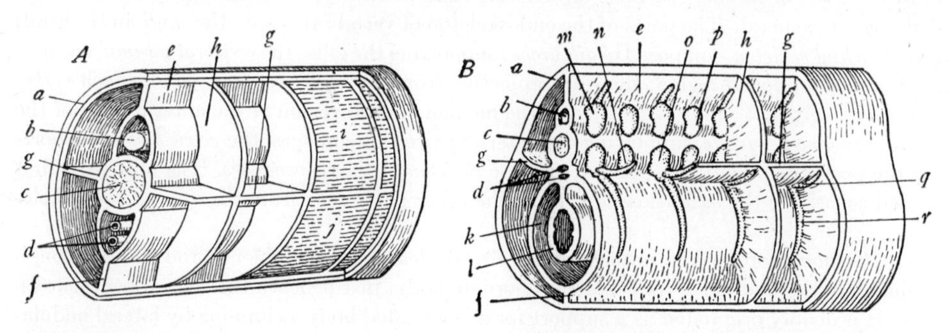

FIG. 30—Diagrams to show the skeleton-forming septa in *A*, the tail region, and *B*, the trunk region, of a vertebrate. *a*, skin; *b*, neural tube; *c*, notochord; *d*, blood vessel; *e*, dorsal skeletogenous septum; *f*, ventral skeletogenous septum; *g*, horizontal skeletogenous septum; *h*, myoseptum; *i*, epaxial part of the myotome; *j*, hypaxial part of the myotome; *k*, coelom; *l*, intestine; *m–p*, cartilages from which the vertebrae are formed (*m*, basidorsal; *n*, interventral; *o*, basiventral; *p*, interdorsal) *q*, intermuscular rib; *r*, subperitoneal rib. In *B* note positions of the vertebral cartilages and ribs with respect to the skeletogenous septa. (*A* after Kingsley's *Comparative Anatomy of Vertebrates*, copyright by P. Blakiston's Son & Co.; *B* from Goodrich in Part IX of Lankester's *Treatise on Zoölogy*, courtesy of the Macmillan Co.)

myosepta are segmentally repeated, because of the primary segmentation of the myotomes, it follows that the vertebrae are also *segmentally repeated* and that the vertebrae *alternate with the myotomes*.

4. Cartilage and membrane bones.—In its development the endoskeleton typically passes through three stages: mesenchyme, cartilage, and bone. The mesenchymal, also called *membranous*, stage is that indicated above, namely, the accumulation of mesenchyme in the skeletogenous regions. This is followed by the laying-down of cartilage in the mesenchyme, and this cartilage gradually takes on the shape of the parts of the definitive endoskeleton. In the lower vertebrates the endoskeleton may remain wholly or partly cartilaginous, and in that case the cartilage may be stiffened by the deposition in it of calcium salts. Such cartilage is said to be *calcified*. In most vertebrates, however, the cartilage is more or less replaced during development by bone, deposited by bone-forming cells. The skeleton is then said to be *ossified*. Bone produced in this manner by the replacement of pre-existing cartilage is known as *cartilage bone*.

Investigation of the development of the vertebrate skeleton shows that not all of the bones arise in this manner but that some of them develop directly from the mesenchyme without passing through a cartilage stage. Such bones are called *dermal*, *membrane*, or *investing* bones. They are really *derived from the dermis of the skin* and are therefore dermal plates homologous to the scales of ganoid and bony fishes and with the plates of the turtle's shell. Consequently, they are actually *parts of the exoskeleton* which have sunk inward from their original position in the skin and have attached themselves so closely to the true endoskeleton that for convenience they are always studied with it. Dermal bones occur in connection with the skull, jaws, and pectoral girdle (these parts also contain cartilage bones, of course).

The student should clearly understand that cartilage bones and dermal bones look alike, have the same histological structure and chemical composition, and cannot be distinguished by examination. *It is only their manner of origin that is different.* To determine which bones of the endoskeleton are cartilage bones and which membrane bones, their embryonic development must be studied. Knowledge of this sort is essential in tracing the homology of the parts of the skeleton, and the necessary information will be given in the following pages.

However, it has become clear in recent years that the embryonic history of bones is not so reliable for distinguishing true endoskeleton from added dermal elements as formerly supposed, for it is now known that bones unquestionably of true endoskeletal nature may omit the cartilage stage and develop directly from mesenchyme. For instance, with regard to a number of teleosts it has been found that the first few vertebrae are preformed in cartilage, whereas the rest of the vertebrae mostly develop directly from mesenchyme. Such omission of the cartilage stage is one of those short cuts common in embryonic development and illustrative of the wonderful adaptability of organisms. The occurrence of the cartilage stage in the development of bone is, on the whole, an ancestral reminiscence, a phylogenetic recapitulation. In many cases it can now serve no useful purpose, and hence a tendency toward its omission may be expected. Such omission in no way diminishes the theoretical importance of the distinction between true endoskeleton and superimposed dermal bones, but it adds considerably to the difficulties of following the evolutionary history of particular bones. A bone preformed in cartilage is necessarily endoskeletal, but one ossifying directly may be exo- or endoskeletal.

It is a general rule that the oldest parts of the endoskeleton retain the cartilage stage, and the omission of this stage is limited to phylogenetically younger elements.

B. ANATOMY AND EMBRYONIC ORIGIN OF VERTEBRAE AND RIBS

1. Parts of a typical vertebra.—The axis of the vertebrate skeleton is the *vertebral column* or backbone, composed of a longitudinal series of similar bones, the *vertebrae.* A typical vertebra consists of a central cylindrical mass, the *body* or *centrum,* which incloses the notochord, a dorsal *neural arch* inclosing the neural tube, and a ventral *haemal arch,* inclosing blood vessels (Fig. 31*A*). Neural and haemal arches are commonly prolonged dorsally and ventrally, respectively, into *neural* and *haemal* spines. In the trunk region the haemal arch is missing or may be represented by basal stumps, the *basapophyses* (Fig. 31*B*). A vertebra also commonly has a variety of projecting processes, termed *apophyses,* serving for articulation with adjoining vertebrae or with ribs or for muscle attachment. The most common apophyses are:

a) *Zygapophyses,* articulations between successive vertebrae; they are divisible into *prezygapophyses,* anterior projections, and *postzygapophyses,* posterior projections, of the basal region of the neural arch.

b) *Basapophyses,* also called haemapophyses and basal stumps, are a pair of ventral projections of the centrum, which represent the remains of the haemal arch and serve for rib attachment.

c) *Diapophyses,* lateral projections of the centrum for the attachment of the upper head of two-headed ribs (Fig. 31*C*).

d) *Parapophyses,* lateral centrum projections for the attachment of the lower head of two-headed ribs (Fig. 31*C*).

e) *Pleurapophyses*, lateral projections representing the rib attachments of the vertebra plus the fused rib.

f) *Hypapophyses*, midventral projections from the centrum.

The expression *transverse process* for any lateral projections of vertebrae is useful, although it must be understood that it has no exact morphological meaning.

2. Shapes of centra.—The vertebral column is formed and functions as an axial support by the end-to-end placing of the centra of the vertebrae. The shape of the ends of the centra is thus of importance in the mechanics of the vertebral column; there are five principal types of ends:

a) *Amphicoelous*, both ends concave.

b) *Procoelous*, anterior end concave, posterior end convex.

c) *Opisthocoelous*, anterior end convex, posterior end concave.

d) *Heterocoelous*, both ends shaped like the seat of a saddle placed transversely.

e) *Amphiplatyan*, both ends flat.

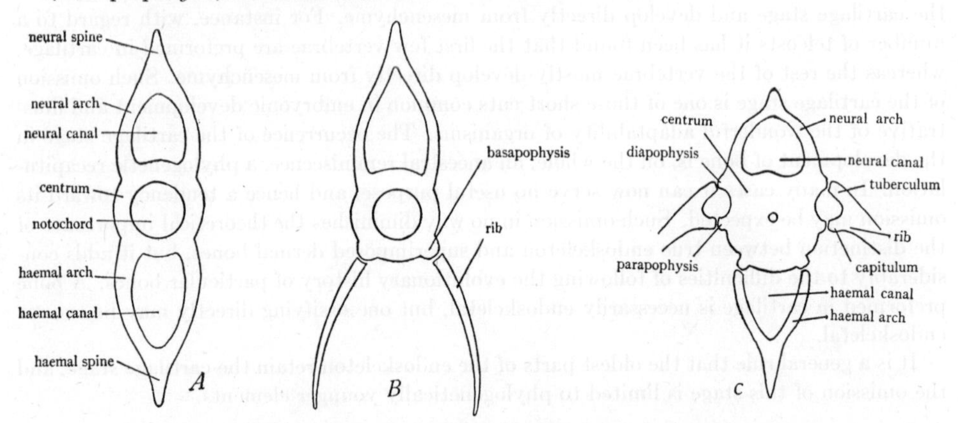

FIG. 31.—Diagrams of typical vertebrae. *A*, tail vertebra of a teleost fish. *B*, trunk vertebra of the same, showing opening of the haemal arch to form the basapophyses. *C*, vertebra of a land vertebrate, showing particularly relation of ribs to the centrum. (After Kingsley's *Comparative Anatomy of Vertebrates*, copyright by P. Blakiston's Son & Co.)

Infrequent types are those convex on both ends (*biconvex*), flat in front and concave behind (*platycoelous*), and concave in front and flat behind (*coeloplatyan*).

Between the ends of the centra there may occur the remains of the notochord or intervertebral disks of fibrocartilage.

3. Development of the sclerotome.—The vertebrae arise from the sclerotomes, which, as already stated, migrate medially, surround the notochord and neural tube, and spread along the skeletogenous septa as mesenchyme. Cartilages then appear in the sclerotomal mesenchyme of the skeletogenous regions, and these may remain in the cartilage stage, often stiffened by calcification, or may undergo partial or complete ossification. In most vertebrates direct ossification of the sclerotomal mesenchyme also occurs, and the vertebra is compounded of parts preformed in cartilage with parts ossifying directly. The student must clearly understand that the vertebral column does not come from the notochord but is formed *around* it. The notochord is thus inclosed inside the centra, where it is readily located in the more primitive vertebrates; but in most vertebrates it is gradually squeezed down and more or less disappears.

A pair of sclerotomes does not, as might be supposed, produce a vertebra as one piece. Not only is each vertebra formed from a number of pieces (see below), but the sclerotomes of two adjacent segments co-operate in building a vertebra. This happens as follows: Each sclerotome soon becomes divided by a vertical split into a loose anterior half and a denser posterior half (Fig. 32*A*). A vertebra arises from the material of the posterior half of one pair of sclerotomes plus the material of the anterior half of the succeeding pair of sclerotomes (Fig. 31*B*). The posterior halves of a pair of sclerotomes become the anterior part of a vertebra, and the anterior halves of the next pair of sclerotomes become the posterior part of the same vertebra. It thus happens that the centrum of a vertebra intersects a myoseptum and that the vertebrae alternate with the myotomes; this arrangement is necessary because muscles must run from one vertebra to another in order to bend the vertebral column.

4. Vertebral components.—As already intimated, each vertebra is formed by the fusion of a number of originally separate components, some or all of which appear first as cartilaginous pieces and may remain in the cartilaginous stage or may ossify later. These components are of three sorts as to origin.

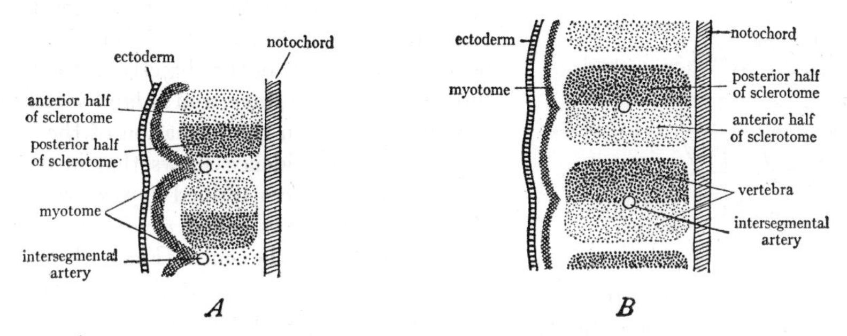

FIG. 32.—Two stages in the development of the vertebrae from the sclerotomes, only one side of the body being shown. *A*, division of the sclerotome into anterior and posterior halves. *B*, union of the posterior half of one sclerotome with the anterior half of the succeeding sclerotome to form a vertebra. (From Arey's *Developmental Anatomy*, courtesy of the W. B. Saunders Co.)

a) *Arch components or arcualia:* Definite paired cartilages, termed *arcualia,* arise in the sclerotomes. Typically, there are four pairs of arcualia to each vertebra: an anterior dorsal pair, the *basidorsals;* an anterior ventral pair, the *basiventrals;* a posterior dorsal pair, the *interdorsals;* and a posterior ventral pair, the *interventrals* (Fig. 30*B*). Basidorsals and basiventrals are usually larger than the interdorsals and interventrals. Basidorsals and basiventrals arise in the original posterior halves of the pair of sclerotomes which become the anterior half of the vertebra; and interdorsals and interventrals arise in the original anterior halves of the pair of sclerotomes which become the posterior half of the vertebra. These four pairs of arcualia rest upon the sheath of the notochord (Fig. 30*B*, 33). There may, in addition, be other elements above and below these, such as *supradorsals* and *infraventrals*, but these are not of general occurrence. In the further development the basidorsals extend dorsally around the neural tube and fuse to form the neural arch, and the basiventrals extend ventrally around the blood vessels and fuse to become the haemal arch (in the tail region) (Fig. 33). Similar fusion of the interdorsals may result in an *interneural* or *intercalary* arch, but the interventrals usually do not form an arch. The arcualia also usually spread around the notochord and so contribute to the centrum.

b) *Sheath component:* Among the Holocephali, Elasmobranchii, Chondrostei, and Dipnoi, cells from the basidorsals and basiventrals invade the sheath of the notochord at the four points in contact with these four arcualia and spread throughout the sheath, which may then greatly increase in thickness. In elasmobranchs these cells then chondrify (i.e., produce cartilage), forming a ring-shaped centrum. A centrum formed in this manner is termed a *chordal* centrum, in contrast with a centrum formed outside the notochordal sheath, termed a *perichordal* or *arch* centrum. It was formerly believed that the distinction between chordal and perichordal centra was absolute—the one, characteristic of elasmobranchs; the other, of all other vertebrates. But work on the development of vertebrae within the last twenty years, notably that of Ridewood (1921) on elasmobranch vertebrae, has destroyed this distinction, for pure chordal centra are now known to occur in only a very few elasmobranchs, and in most elasmobranchs the centra are formed by a fusion of the sheath component with other components (Fig. 36*G*). It is probable that perichordal centra are the more primitive.

c) *Perichordal component or intermedialia:* Mesenchyme of sclerotomal origin accumulates around the notochord to the inner side of the main sclerotome blocks. This perichordal mesenchyme also contributes to the centra, also sometimes to the arches, combining with arch elements. The perichordal components frequently ossify directly without passing through a cartilage stage, and in land vertebrates are a major element in the formation of the vertebrae. Paleontological evidence shows that in tetrapods a pair of ventral pieces termed *hypocentra* and a pair of dorsolateral pieces called *pleurocentra* are important components of the centra (Fig. 36*A–D*). Because of a lack of embryological material among fossil forms it has been difficult to relate these pieces to the embryological story of vertebra development, but it seems best to regard them as representing perichordal components in large part. Hypocentra have the same location as the basiventrals; and pleurocentra, the same as the interdorsals. Although some authorities identify the hypocentra as basiventrals and the pleurocentra as interdorsals, the probable better view is to regard hypo- and pleurocentra as perichordal elements which develop around and incorporate the basiventrals and interdorsals, respectively. The terms hypo- and pleurocentra will then be understood to include the basiventrals and interdorsals, respectively, but not to be identical with them, being larger pieces developed largely from the perichordal mesenchyme.

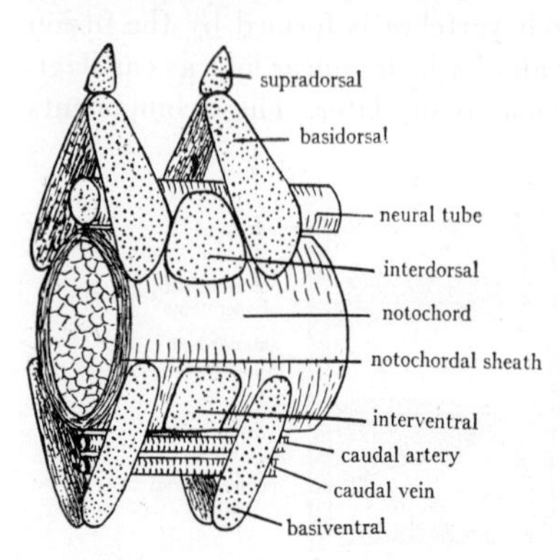

FIG. 33.—Diagram showing the arcualia in the tail region and their relation to neural tube, notochord, and caudal blood vessels.

From the foregoing account it is evident that the composition of the centrum is a very complex matter, not yet completely clarified, and that many different combinations of the various elements are possible. Only the most careful embryological researches can reveal what elements enter into the centra in different vertebrates. Many of the older studies of the development of vertebrae were inaccurate. Modern work indicates that the composition of the centra is or may be different in different groups of vertebrates or even within the same group. The available information as to the particular combination of elements involved in each case will be given below in connection with the descriptions of the vertebrae of various vertebrates. It must be realized, however, that the whole subject of vertebra development and homology of components is still in an unsatisfactory state.

5. Diplospondyly.—One consequence of the fact that vertebrae arise by the fusion of a number of components is that these may fuse in such a way as to produce two centra to the segment (but only one neural and haemal arch). This is termed *diplospondyly*. Diplospondyly could theoretically occur in several ways, but it usually happens by the fusion and enlargement of the hypocentra to produce one centrum and of the pleurocentra to form the other centrum (Fig. 36*C*). Diplospondyly of this type occurs in the tail region of *Amia* and other fishes and in extinct amphibians (Fig. 36). Other types of diplospondyly also occur as in the tail region of elasmobranchs (p. 110) and in the tails of many lizards and of *Sphenodon.* In these reptiles the tail vertebrae are split along the plane of union of the anterior and posterior halves of the sclerotomes; and the tails, when grasped, are apt to break at the splits, subsequently regenerating. The usual condition of one vertebra per segment (but alternating with it), termed *monospondyly,* commonly arises by the expansion of some elements and the reduction or suppression of others.

6. The ribs.—Each vertebra is theoretically provided with a pair of *ribs,* which articulate to various projections of the centrum, extend out into the body wall, and serve to strengthen the latter and provide muscle attachments. The ribs, like other parts of the axial skeleton,

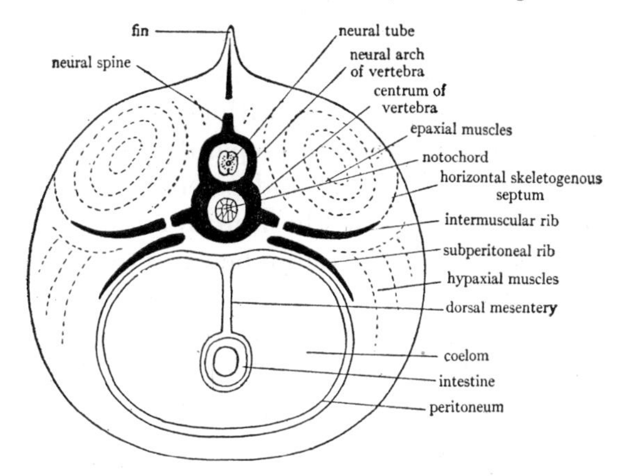

Fig. 34.—Diagrammatic cross-section through the trunk of a vertebrate showing the relation of the ribs to the skeletogenous regions and the positions of the two kinds of ribs. (From Parker and Haswell's *Textbook of Zoölogy,* courtesy of the Macmillan Co.)

arise in the skeletogenous septa. There are two kinds of ribs, both of which are situated in the myosepta and hence are segmental in arrangement. One type of rib is formed at the intersection of each myoseptum with the horizontal skeletogenous septum. Since the horizontal septum divides the myotomes into dorsal and ventral halves (p. 99), such ribs lie between the muscles and hence are called *intermuscular* ribs, also *dorsal* or *upper* ribs (Fig. 34). The second type of rib arises at the points of intersection of the myosepta with the ventral skeletogenous septa or its derivatives. It will be recalled that in the trunk region the ventral septum is split into two lateral septa because of the intervention of the coelom and its contents on the ventral side of the body (Fig. 30*B*). The second type of rib appears at the points of intersection of the myosepta with these ventrolateral septa. They typically lie just outside the coelomic lining between the coelomic wall and the muscle layer (Fig. 34). They are therefore called *subperitoneal* ribs, also *ventral, lower,* or *pleural* ribs. Both kinds of ribs may occur simultaneously on a vertebra, and, in fact, some fishes may have additional ribs of the category of dorsal ribs.

The ventral or pleural ribs are phylogenetically the older. It was formerly supposed that the ribs of most vertebrates were of the dorsal type, but the investigations of Emelianov indicate that they are of the pleural type. According to this worker, the ribs may shift their position with regard to the muscles during development, and hence the type of rib cannot be determined by its location in the adult.

Whether the ribs are independent formations or represent cutoff extensions of the vertebrae is still an open question. It appears definitely established by the work of Emelianov that the dorsal ribs are independent formations, since they begin some distance from the vertebrae and grow toward them. The pleural ribs are probably best regarded as extensions of the basiventrals, although it appears that they are no longer continuous with the basiventrals even in embryonic stages. They are first seen very near the vertebrae and grow outward from them.

The pleural ribs are always preformed in cartilage and may subsequently ossify. The dorsal ribs may pass through a cartilage stage or may omit this. This conforms to the general principle that the omission of the cartilage stage is confined to the phylogenetically more recent parts of the endoskeleton.

C. SOME PRIMITIVE VERTEBRAL COLUMNS

Primitive vertebrae are those in which the various components have not yet combined to form typical vertebrae with centra. Such vertebrae (called *acentrous* vertebrae) therefore consist of separate pieces resting on the large persistent notochord. Embryological and paleontological evidence testifies that the dorsal arcualia appear first, the ventral ones next, and the centrum last. Corresponding stages may therefore be expected among primitive and extinct vertebrates.

1. Vertebral column of cyclostomes.—In cyclostomes there is a fully developed notochord provided with a thick sheath. Vertebral elements are present only in the lampreys, where they consist of two pairs of dorsal pieces (Fig. 35*A*) per segment, resting on the notochord, one pair behind the other. These two pairs are believed to represent the basidorsal and interdorsal arcualia, respectively, but it is not clear which pair is which. Apparently, the posterior pair in the segment is the basidorsals, the anterior pair the interdorsals.

2. Vertebral column of the sturgeon.—The vertebrae of this fish are very primitive, acentrous, and composed of separate cartilaginous arcualia. Examine a demonstration specimen and compare it with Figure 35*B*. The large unconstricted notochord traverses the center of the vertebral column and is covered dorsally and ventrally by cartilage pieces, leaving its middle bare. On the dorsal side are the large *basidorsals*, united to form a *neural arch* topped by a *neural spine* (*supradorsal*). Between the bases of the basidorsals the small *interdorsals* (sometimes subdivided) are situated. The ventrolateral regions of the notochord are covered on each side by the large *basiventrals*, each bearing a projecting *basapophysis* for articulation with a rib. Note that the basiventrals of the two sides do not meet below the notochord; hence the haemal arch has not yet been formed. The small *interventrals* lie between the ventral portions of the basiventrals.

3. Vertebrae of extinct amphibians.—Fossil amphibians of the order Labyrinthodontia furnish important evidence on the composition of vertebrae. One group of these (suborder Rhachitomi) had vertebrae composed of several pieces, hence called *rhachitomous*, without typical centra. Usually (Fig. 36*A*), such vertebrae consist of a large neural arch bearing a neural spine and representing the fused basidorsals; a crescentic ventral piece, the hypocentrum, serving as centrum (composed of the fused basiventrals plus probably some perichordal contribution); and a pair of ventrolateral pieces, the pleurocentra (representing the interdorsals or the interdorsals plus the interventrals plus a probable perichordal contribution). In the tail region of these rhachitomous amphibians the hypocentrum bears a haemai arch. In a few cases separate interventrals are present (Fig. 36*B*).

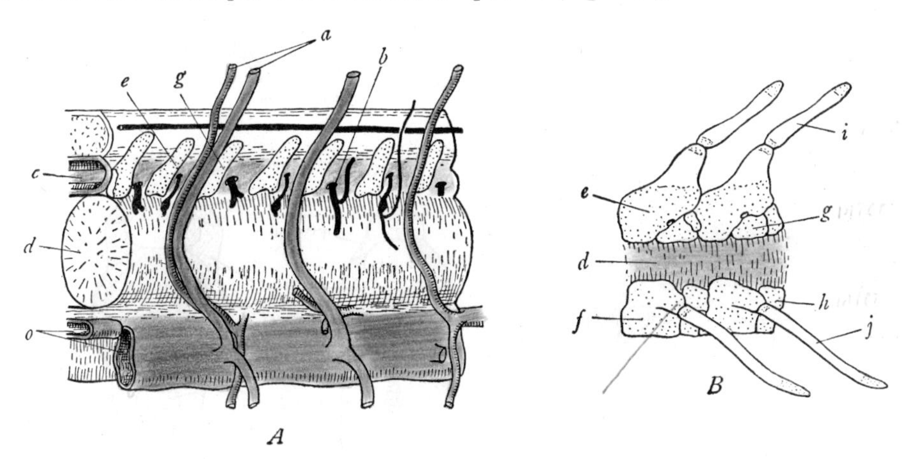

FIG. 35.—Two primitive vertebral columns. *A*, cyclostomes; the vertebral column consists of two pairs of pieces per segment (only one of each pair shown), resting on the dorsal surface of the notochord. *B*, sturgeon; separate arcualia partially surround the notochord. (After Goodrich, 1909, courtesy of the Macmillan Co.) *a*, intersegmental blood vessels; *b*, nerve; *c*, neural tube; *d*, notochord; *e*, basidorsal or neural arch; *f*, basiventral; *g*, interdorsal; *h*, interventral; *i*, neural spine or supradorsal; *j*, rib; *o*, blood vessels.

4. Diplospondyly in the tail of *Amia* and in embolomerous amphibians.—Examine a demonstration specimen of the skeleton of the tail of the bowfin and compare it with Figure 36*E*. Note that the notochord is not visible, being inclosed by the vertebrae, which thus have formed complete centra. Note the alternation of a centrum bearing neural and haemal arches with one devoid of arches, so that there are two centra to each body segment. Study of fossil fish related to *Amia* (Fig. 36*F*) indicates that the posterior centrum (without the arches) represents the fused pleurocentra extended to form a ring and that the anterior centrum (with the arches) is the extended hypocentra. The monospondylous trunk vertebrae of *Amia* arise by the fusion of the two centra.

Diplospondyly similar to that of *Amia* occurs in extinct amphibians of the order Labyrinthodontia, suborder Embolomeri (Fig. 36*C*). In these embolomerous vertebrae the large neural arch rests upon two disklike centra, an anterior one formed by the upward extension of the fused hypocentra, and a posterior, often larger, one formed by the downward extension of the fused pleurocentra. In the tail region the hypocentral centrum carries a haemal arch.

The vertebrae of these extinct amphibians lead to those of amniote vertebrates in general by the emphasis of the pleurocentral element in the formation of the centrum and the decline in the hypocentral contribution.

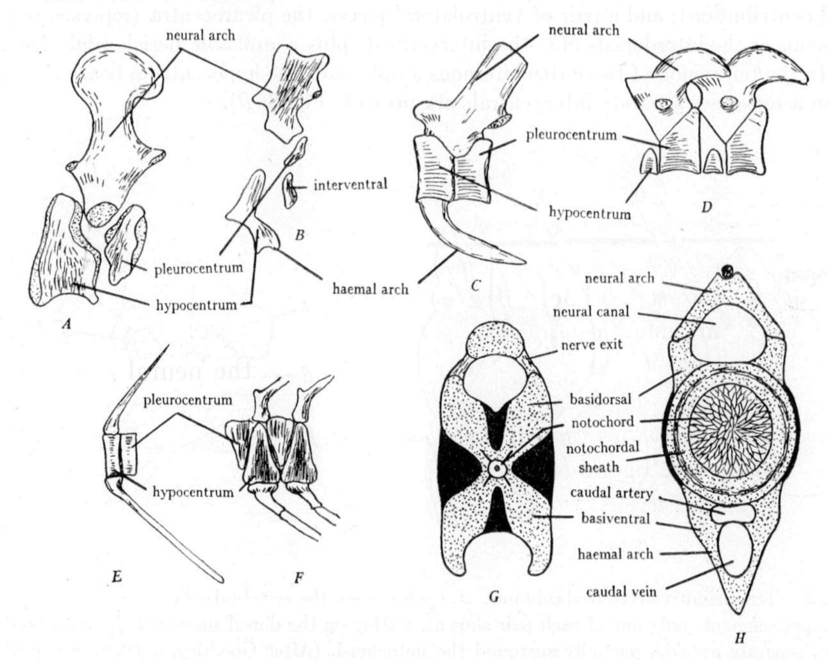

FIG. 36.—Vertebral components. *A*, trunk vertebra, and *B*, tail vertebra, of an extinct rachitimous amphibian, *Archegosaurus* (after Jaeckel, 1896); the vertebrae are made of separate pieces. *C*, tail vertebra of an extinct embolomerous amphibian, showing typical diplospondyly. *D*, trunk vertebrae of *Seymouria*, showing separate components. *E*, tail vertebrae of *Amia*, showing diplospondyly (after Shufeldt, 1885). *F*, tail vertebrae of an extinct amioid fish, showing stage toward the *Amia* condition (after Zittel). *G*, section through vertebra of the dogfish *Mustelus*, showing centrum composed of alternate arch (stippled) and perichordal (black) components (after Ridewood, 1921). *H*, tail vertebra of *Squalus*, showing nearly pure arch centrum, chondrified in the sheath of the notochord; small black strips are perichordal components (after Goette, 1878; altered according to Ridewood, 1921). (*C* and *D*, after Williston, *Osteology of the Reptiles;* courtesy Harvard University Press.)

D. VERTEBRAL COLUMN OF THE DOGFISH

1. Cross-section of the tail.—Obtain a cross-section through the tail of the dogfish and study the cut surface, being sure that the section passes through the junction between vertebrae and not through the center of a vertebra.[1] The center of the section contains the vertebra, composed of clear, relatively soft cartilage. Between the vertebra and the skin is a thick layer of voluntary muscles, composed of a number of leaves, the myotomes or muscle seg-

[1] When the section passes through the center of the vertebra, areas of calcification in the form of rings or rays will be seen.

ments, separated from each other by plates of connective tissue, the myosepta. The myotomes appear in whorls because they are zigzag in form like those of *Amphioxus*, and hence a number will be cut across in any cross-section. The myotomes are somewhat indistinctly divided into dorsal and ventral portions by a connective tissue partition, the horizontal skeletogenous septum, which extends from the centrum of the vertebra to the skin, where it meets the lateral line. The muscles above the septum are the dorsal or *epaxial* muscles; those below it, the ventral or *hypaxial* muscles.

Turning now to the vertebra itself, we note that it consists of a central circular concave portion, the *centrum* or *body;* dorsal to this, of an arch, the *neural arch,* which incloses a cavity, the *neural canal,* in which the spinal cord, a soft white body, is located; ventral to the centrum, of another arch, the *haemal arch,* which incloses a cavity, the *haemal canal,* containing the *caudal artery* and *vein.* The neural arch terminates in a point, the *neural spine,* and the haemal arch similarly terminates in the *haemal spine.* Observe the connective tissue partitions which extend from the neural spine to the median dorsal line and from the haemal spine to the median ventral line. These are the *dorsal and ventral skeletogenous septa;* and they, together with the horizontal skeletogenous septum already mentioned, mark the chief sites of skeleton formation.

Draw the section, labeling all parts.

2. Sagittal section of the tail.—Obtain or make a median sagittal section through a piece of the tail of a dogfish. The vertebrae form a row in the center of the section. Identify the centra in the section. Each consists of two somewhat triangular pieces, apparently separate, the rounded apexes of the triangles directed toward each other, the whole shaped somewhat like an hourglass. The two ends of the centra are concave, so that diamond-shaped spaces are present between successive centra. These spaces are filled with a soft, gelatinous material, the notochord, which also fills the canal which runs through the center of the centrum. Hence the notochord is constricted by the centra but expands to nearly its embryonic size in the space left between the concave ends of adjacent centra. The centra of elasmobranchs are, as just seen, biconcave or *amphicoelous* (p. 102). Above each centrum identify the neural arch, arching over the neural canal. Between successive neural arches, and lying therefore dorsal to the diamond-shaped spaces between the centra, observe an extra arch, inverted, however, with apex inclosing the neural canal. This is the *intercalary arch.* Below the centrum is the haemal canal, its sides formed by the haemal arches, rectangular in section. Draw the section.

3. Cross-section of the trunk region.—In a cross-section of the anterior part of the trunk region of the dogfish identify the following parts. The muscle segments are arranged as in the tail region, their division into dorsal and ventral masses being well marked by the horizontal skeletogenous septum. The dorsal or *epaxial* muscles above the septum are thick masses, but the ventral or *hypaxial* muscles below the septum form a thin layer inclosing a large cavity, the *body cavity* or *coelom*, lined by a smooth membrane, the *pleuroperitoneum*. The coelom incloses the viscera, some of which will be observed to be suspended by a delicate membrane, the *dorsal mesentery*, from the median dorsal line of the coelomic wall. Study the vertebra, which appears in the middle of the myotomes. It consists of centrum and neural arch, similar in appearance to those of tail vertebrae; but the haemal arch appears to be absent. It is represented by a pair of small cartilages at the sides of the ventral part of the centrum. These are the *basapophyses* or *basal stumps*, the remains of the haemal arch, which may be regarded as having opened out and shifted to a more lateral position. Examine the horizontal skeletogenous septum carefully and find within it, by picking away the muscles if necessary, a slender cartilage on each side, articulating with the basapophyses. These cartilages are the ribs. From their location in the horizontal septum it has usually been supposed that they are dorsal ribs; but, according to Emelianov, they are pleural ribs which have migrated into the musculature. Draw the section.

4. Caudal diplospondyly of elasmobranchs.—In the tail region of elasmobranchs the occurrence of two centra per segment is very common and apparently lends flexibility to a region important in swimming. Not only are the centra doubled, but, unlike typical diplospondyly (p. 104), all other parts of the vertebrae are also doubled. The origin of this diplospondyly is uncertain, but apparently the centra divide in two and the smaller pieces arise *de novo*.

5. Composition of elasmobranch vertebrae.—In the development of elasmobranch vertebrae the usual four pairs of arcualia appear in the sclerotomes and chondrify to form the neural and haemal arches, the intercalary pieces, and sometimes also interventral pieces; minor arcualia forming additional pieces are often present. Cells from the arcualia, particularly the arch bases, invade the notochordal sheath and there chondrify to a cartilaginous cylinder which breaks up into centra. It was formerly supposed that elasmobranch centra were formed wholly in this manner and hence are of pure sheath (chordal) origin. But Ridewood (1921) has shown that the centra include small to large amounts of cartilage derived from the perichordal mesenchyme. This perichordal or intermedial component often takes the form of large wedge-shaped areas (Fig. 36*G*), alternating with cartilage of arch origin. Hence the vertebrae of elasmobranchs are compounded from arch, sheath, and perichordal (intermedial) components. There is no ossification in the elasmobranch skeleton, which consists wholly of cartilage; but this is probably a secondary condition, since it is known that bone was present in the most ancient vertebrates (ostracoderms). The vertebrae are usually strengthened by calcification, however, in the form of rings or radiations in the centra.

E. VERTEBRAL COLUMN OF TELEOSTS

1. The tail vertebrae.—Examine a separate, dried tail vertebra of any bony fish. Note that the vertebra is very much harder and more opaque than the dogfish vertebrae, owing to the fact that it is composed of bone. Identify the same parts as already seen in the dogfish vertebrae: the biconcave or amphicoelous centrum, bearing a minute canal in its center for the notochord; the neural arch, terminating in a very long sharp neural spine; the haemal arch, terminating in a similar haemal spine. The neural canal or space inclosed within the neural arch is generally smaller than the haemal canal, inclosed by the haemal arch; in this way the dorsal and ventral sides of the vertebrae may be distinguished. The spines are directed posteriorly. In some fish there are two neural spines to each vertebra, an anterior and a posterior one; the second one probably corresponds to the intercalary arch of the dogfish. Draw a vertebra.

2. The trunk vertebrae.—Obtain a separate dried trunk vertebra of a bony fish. Identify, as before, the centrum and the neural arch and neural spine. The haemal arch and spine appear to be missing. Instead there is a pair of projections at the sides of the base of the centrum to each of which a long slender *rib* is articulated. These projections are termed basal stumps or basapophyses and represent the opened bases of the haemal arch. Draw a vertebra with ribs.

3. Section through the trunk of a bony fish.—In such a section identify the parts already described for a similar cross-section of the dogfish. Note the muscle segments, the centrum and neural arch and spine of the vertebra, and the coelom with its lining. Find the ribs located just outside of the coelomic lining; this situation, together with the facts of their development, show that the ribs of teleosts are pleural (subperitoneal) ribs.

4. Composition of teleost vertebrae.—In the development of teleost vertebrae, basidorsal and basiventral arcualia appear and form the neural and haemal arches, but the other arcualia are generally absent. Perichordal contributions to the arches have also been described. The arcualia may contribute to the centra, but the centra are formed chiefly or wholly from the perichordal mesenchyme (which ossifies directly without passing through a cartilage stage) plus ossification in the notochordal sheaths. Teleost centra are thus compounded of perichordal and sheath elements.

5. Further study of ribs.—Some fishes have two (or more) pairs of ribs simultaneously on each vertebra. Examples are *Polypterus* (p. 43) and many teleosts, including members of the salmon, herring, and pike families. Examine the skeleton of *Polypterus* and note two pairs of ribs attached to each vertebra. The dorsal ribs are articulated to projecting processes of the centrum; the lower or pleural ribs are loosely attached to the ventral surface of the centrum. In the intact fish the dorsal ribs are located in the horizontal septum, the pleural ribs next the peritoneum; these locations, together with the details of the development, show that *Polypterus* does, in fact, possess the two kinds of ribs characteristic of vertebrates. Vertebrae of fishes like the salmon may also be examined, or sections through the trunk of such fishes.

Note the pleural ribs; also the dorsal ribs located in the horizontal septum. Additional ribs may also be present dorsal to these, articulated with centrum or neural arch and extending out into the myosepta between myotomes. Teleost ribs may, in fact, be formed at any level of the myosepta; these extra ribs are of the nature of dorsal ribs, according to Emelianov. Make a diagram showing a vertebra and its ribs.

6. The vertebral column as a whole.—Study the entire mounted skeleton of a bony fish, noting its very complete ossification. Observe that the vertebral column is formed by the end-to-end placing of the centra, held together in life by ligaments and muscles. The ends of the centra of fish in general are amphicoelous, and the considerable space left by the concavities of the ends is occupied in life by the remains of the notochord. The vertebral column is divisible into *trunk* and *tail* regions. In the former the haemal arches are reduced to basapophyses supporting the long slender ribs; in the tail region, haemal arches replace the ribs. Observe the transition between trunk and tail regions, noting gradual elongation of the basapophyses and reduction of the ribs toward the posterior end of the trunk. At the beginning of the tail region the reduced ribs finally vanish, and the elongated basapophyses fuse to form the haemal arches. This transition in teleost fishes is the best evidence that the pleural ribs represent cutoff pieces of the haemal arches.

Note that the neural arches of successive vertebrae together inclose a continuous neural canal which in life contains the spinal cord. Similarly, the haemal arches of the tail vertebrae inclose a continuous haemal canal in which blood vessels run in life.

F. ENDOSKELETAL FIN SUPPORTS OF FISHES

As already seen (p. 88), the fins of fishes are supported by dermal rays; these, however, articulate with endoskeletal supports, the *pterygiophores*, more or less concealed in the animal's flesh. Here only the unpaired fins will be considered, as the paired fins are treated in the next chapter.

1. Median fins.—On skeletons of elasmobranch and teleost fish note the endoskeletal supports between the dermal rays of the median fins and the neural and haemal spines of the vertebrae. In elasmobranchs these pterygiophores comprise one or more rows of cartilaginous pieces or rods, which are usually larger and fewer in number next the vertebrae. In teleosts a row of slender bony rods, sometimes flattened, is articulated with the vertebral spines at one end, and the dermal fin rays at the other. There may be one such pterygiophore to each vertebral spine or more. These pterygiophores seem to be cutoff portions of the vertebral spines.

2. Tail fin.—The dermal rays of the tail fin of fishes are articulated directly to the arches of the vertebrae. Much attention has been paid to the form of this terminal part of the vertebral column, and several types are recognized (Fig. 37):

a) Protocercal tail: Hypothetical primitive type, perfectly bilaterally symmetrical with fin developed equally above and below the straight vertebral (or notochordal) axis; found in cyclostomes and passed through in the development of teleosts.

b) Heterocercal tail: Asymmetrical with vertebral axis or notochord bent upward in the tail, so that the external fin is larger below than above the axis; characteristic of elasmobranchs and most other lower fishes. Examine the skeleton of an elasmobranch and note asymmetrical tail fin with vertebral column turning upward in it. Observe larger fin expanse below than above the column, and particularly note that the anterior part of the lower

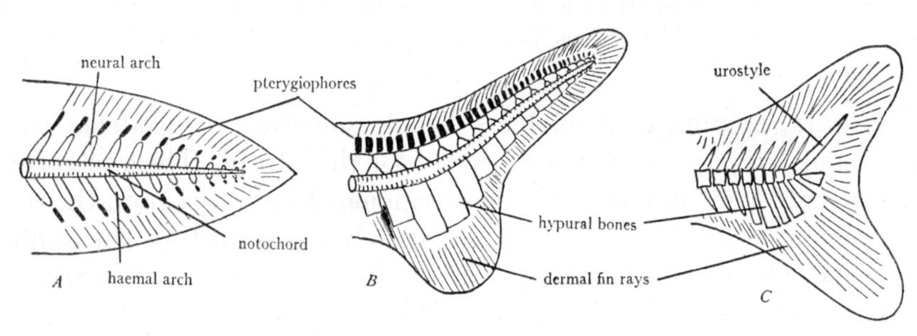

Fig. 37.—Diagrams of tail types of fishes. *A*, protocercal type, bilaterally symmetrical. *B*, heterocercal type, asymmetrical with lobe below. *C*, homocercal, externally symmetrical, internally of the heterocercal type. Pterygiophores in black, vertebral arches blank.

fin is expanded into a lobe, which probably represents an originally separate fin now fused to the true tail fin. Note how dermal fin rays articulate to the neural and haemal arches of the vertebrae and that the haemal arches are larger and more expanded than the neural ones. A reversed heterocercal (*hypocercal*) tail, with vertebral column bending downward and lobe on the upper side, is characteristic of certain ostracoderms.

c) Diphycercal tail: Secondarily symmetrical tails, derived by modification from the heterocercal type; seen in present Dipnoi and coelacanth Crossopterygii (Fig. 1*A*). They are indistinguishable from the protocercal type, but intermediate fossil forms show their derivation from the heterocercal condition.

d) Homocercal tail: Externally more or less symmetrical but internally like a shortened heterocercal type; found in all higher fishes and derived during embryology from the heterocercal type. Examine the tail of a skeleton of any bony fish. Note upturned end of the vertebral column with last centrum (probably consisting of several fused centra), forming an elongated *urostyle* which turns sharply upward. The haemal arches accompanying this upturned part of the vertebral column are enlarged and flattened, forming the *hypural* bones; and a few corresponding enlarged neural arches, or *epural* bones, may be present. It is seen that the tail fin is formed largely or wholly of the ventral part of the fin of the heterocercal tail with a great reduction or loss of the dorsal part. This ventral fin has become secondarily bilaterally symmetrical, so that the homocercal tail appears symmetrical externally.

G. VERTEBRAL COLUMN OF AMPHIBIA

1. Extinct types.—The importance of the extinct Labyrinthodontia for the understanding of the evolution of tetrapod vertebrae was already mentioned (p. 107), especially the significance of the embolomerous and rachitomous types (Fig. 36*A*, *C*). These vertebrae consist of neural arch, haemal arch, hypocentrum, and pleurocentrum as separate pieces. The embolomerous type with hypocentrum and pleurocentrum equal or nearly so is now considered the more primitive from which the rachitomous type was derived by reduction of the hypocentrum and pleurocentrum.

2. Vertebral column of urodeles.—Whole skeletons of urodeles, such as *Necturus* and *Cryptobranchus*, are needed for this study. The vertebral column consists of four, not very distinct, regions. The first region, the *cervical* or *neck* region, consists of one cervical vertebra, supporting the skull and lacking ribs. There follows a long *trunk* region of similar rib-bearing vertebrae. The next or *sacral* region, also called *sacrum*, consists of one vertebra, whose ribs, the *sacral ribs*, support the hind legs. Behind the sacrum is the *caudal* or *tail* region, composed of vertebrae lacking ribs and usually having haemal arches. By moving the vertebrae apart, note that the ends are amphicoelous, as in fishes.

Study individual vertebrae. The tail vertebrae are similar to those of fishes, having neural and haemal arches and a centrum bearing projecting lateral processes. The neural arches are notably low and elongated. In the trunk region the haemal arch is absent, and ribs are articulated to the lateral projections of the centrum. The vertebrae are articulated to each other by *zygapophyses*, consisting of a pair of projections on the neural spine fitting over a similar pair on the anterior end of the succeeding vertebra. Thus each vertebra has a pair of *prezygapophyses* on its anterior end whose articulating surfaces face upward and a pair of *postzygapophyses* on its posterior end whose articular surfaces face downward; these zygapophyses yoke the vertebrae together.[2]

The ribs of *Necturus* are *bicipital;* i.e., their attached ends are forked into two *heads*, a dorsal *tuberculum* and a ventral *capitulum*, which articulate with similar but less marked dorsal and ventral projections of the vertebra called *diapophysis* and *parapophysis*, respectively. The ribs are dorsal ribs.

[2] Zygapophyses are poorly developed or absent in fishes.

Embryological studies on urodele vertebrae show that basidorsals appear and contribute to the neural arch, which is completed above by a supradorsal element; in the tail region basiventrals contribute to the haemal arch, also completed by an infraventral. The centrum is formed wholly from perichordal mesenchyme, and this mesenchyme also contributes extensively to the neural and haemal arches. Large intervertebral disks are formed which in genera with opisthocoelous centra divide in two to form the ball-and-socket ends of adjacent centra. Paleontological evidence suggests that the urodele centrum corresponds to the hypocentrum of extinct forms (Fig. 38). A peculiarity of urodeles is that the rib-bearing part of the vertebra originates as a separate piece (rib-bearer), which subsequently fuses to the centrum, and this forms the dia- and parapophyses, which therefore are not homologous to the structures of the same name in amniotes. According to Emelianov, this rib-bearer represents the ventral rib. A projection of basiventral origin which grows out and joins the rib-bearer apparently represents the true parapophysis.

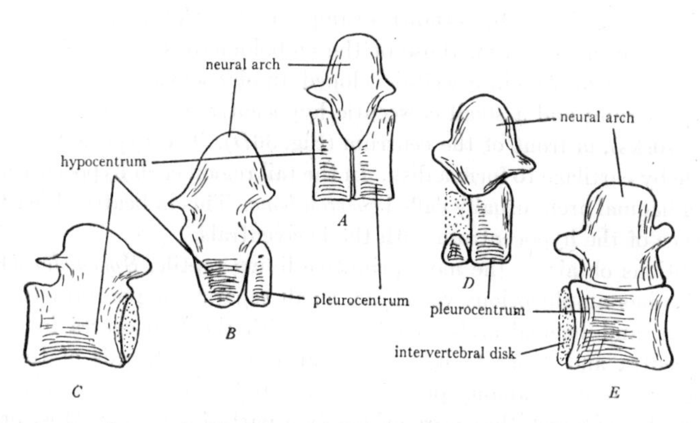

FIG. 38.—Diagrams to show probable derivation of amphibian and amniote centra from the embolomerous type. *A*, embolomerous vertebra, pleurocentrum and hypocentrum equal. *B*, *C*, evolution toward present amphibians, pleurocentrum reduced. *D*, *E*, evolution toward amniotes, hypocentrum reduced, becomes intervertebral cartilage.

3. Vertebral column of an anuran amphibian, the frog.—In frogs there is one cervical vertebra, followed by seven trunk vertebrae, succeeded by one sacral vertebra supporting the hind limbs. Behind the sacrum a long piece, the *urostyle*, completes the vertebral column; a tail is lacking. The trunk vertebrae have low neural arches, no haemal arches, zygapophyses, and centra bearing long transverse processes which possibly correspond to diapophyses. Embryology indicates that the ends of these processes originate separately and hence are ribs. In single frog vertebrae determine that the centra are procoelous; opisthocoelous centra also frequently occur among Anura.

The development of anuran vertebrae is similar to that of urodeles; basiventrals are even more reduced except for the urostyle, which seems to be formed largely from them. Although it is usually supposed that the urostyle represents several fused caudal vertebrae, these are not evidenced embryologically, as the urostyle ossifies from a continuous cartilaginous rod. The nature of the large intervertebral disks, which in Anura, as in Urodela, divide to form the ball and socket ends of adjacent centra, is disputed. Older authors regarded them as representing the missing interdorsal and interventral arcualia, but recent work indicates their perichordal nature, and they may be regarded as the reduced pleurocentra. Consequently, in recent Amphibia, the centra represent the hypocentra, the intervertebral disks the pleurocentra (Fig. 38), just the opposite of the conditions in amniotes.

H. VERTEBRAL COLUMN OF REPTILES AND AMNIOTES IN GENERAL

1. Primitive reptilian and amniote vertebrae.—Among amniotes the vertebral centra are monospondylous, and this condition has arisen by the enlargement of the pleurocentra to become the centrum and by the reduction of the hypocentra to a small crescentic piece ("intercentrum") wedged between the lower parts of the centra or to an intervertebral disk (Fig. 38). Stages in this evolution are found among primitive extinct and living reptiles.

The most primitive reptiles are the members of the extinct order Cotylosauria which are almost indistinguishable from their ancestors, the labyrinthodont Amphibia. The important fossil reptile *Seymouria*[3] is the best-known example of the Cotylosauria. The vertebrae of *Seymouria*, directly derivable from those of the embolomerous labyrinthodonts, consist of three pieces: a large neural arch; a well-developed amphicoelous centrum, representing the enlarged pleurocentrum; and a small crescentic hypocentrum (often called intercentrum in paleontological works), in front of the centrum (Fig. 36*D*). The hypocentrum was probably completed in life by cartilage to form a disk. In the tail region each hypocentrum is extended ventrally into a haemal arch, usually called *chevron bone*. This indicates at least some degree of correspondence of the hypocentrum with the basiventrals.

Similar conditions obtain in the most primitive living reptile, *Sphenodon* (Fig. 39*B*, *C*). This animal also has amphicoelous vertebrae, small separate hypocentra (intercentra) between the centra, and haemal arches continuous with the hypocentra in the tail region. Amphicoelous centra and separate hypocentra also occur in lizards of the gecko group. Chevron bones are common among present reptiles but usually are simply forked pieces, since the hypocentra of which they were originally a part are lacking. Most present reptiles have procoelous centra.

2. Evolution of the atlas and the axis.—In amniotes the first two vertebrae, termed *atlas*[4] and *axis* (or *epistropheus*), respectively, are usually modified for the support and movements of the skull. The atlas is usually a ring-shaped piece with one or two concavities articulating with skull projections (*condyles*), and the axis has an anteriorly projecting *odontoid process* acting as a pivot in the turning of the head. Paleontological and embryological evidence combine to show that the atlas consists of the neural arch above and hypocentrum below and that its pleurocentrum is incorporated into the odontoid process of the axis. Hence the axis has two pleurocentra as well as a neural arch and hypocentrum. Other pieces may also contribute, especially a piece called the *proatlas*, which appears to be the remains of a vertebra originally intercalated between the present axis and the skull. Stages in the formation of the amniote atlas and axis are seen in *Seymouria* and other primitive reptiles. In *Seymouria* (Fig. 39*A*) atlas and axis are similar to the other vertebrae; and the pleurocentrum of the atlas, which is the future odontoid process, is only slightly modified and displaced. The Cotylosauria probably also had a proatlas. A single or paired proatlas and a composite condition of the atlas and axis are also seen in various extinct reptiles and in *Sphenodon*, Crocodilia, and other living members of the class. In *Sphenodon* the atlas consists of proatlas, neural

[3] *Conodectes* Cope 1896 has priority over *Seymouria* Broili 1904 and hence is the correct name of this fossil, according to nomenclatorial rules.

[4] The first vertebra of Amphibia is often called atlas but appears not to be a true atlas, simply an ordinary vertebra.

arch, and hypocentrum (Fig. 39*B*); but its pleurocentrum has already joined the axis as the odontoid process, and the axis has developed its characteristic large neural arch. Embryological studies (Hayek, 1923, 1924) indicate that in Squamata and in mammals the proatlas is attached to the anterior end of the odontoid process of the axis (Fig. 39*E*).

3. The tetrapod rib.—Land vertebrates have but one pair of ribs per vertebra. It was formerly believed that these are dorsal ribs, but it is now considered more probable that they

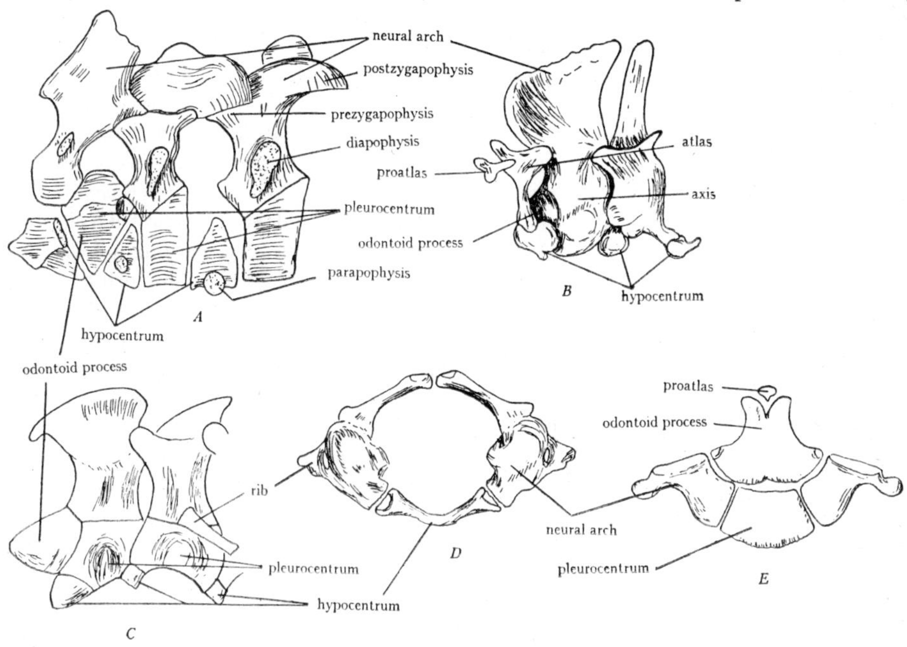

FIG. 39.—Primitive and embryonic stages of amniote vertebrae, showing composition. *A*, first three cervical vertebrae of *Seymouria*, showing composition of atlas and axis, also rib articulations (after Watson, 1915). *B*, first three cervical vertebrae of *Sphenodon*, showing proatlas, separate hypocentra (after Albrecht, 1883). *C*, axis and next cervical vertebra of young *Sphenodon*, showing separate odontoid process, neural arches, centra, and hypocentra (after Schauinsland, 1906). *D*, developmental stage of human atlas, showing the three components (paired neural arches, hypocentrum below). *E*, developmental stage of human axis, seen from the ventral side, showing proatlas, odontoid process (=pleurocentrum of atlas), pleurocentrum, and paired neural arches, all as separate elements. (*D* and *E*, after Flower, 1885.)

are pleural ribs. Some anatomists think they have been formed by the fusion of the dorsal and pleural ribs, and in this way they account for the two-headed condition. Tetrapod ribs contrast with fish ribs in that they are typically two-headed or bicipital, as already seen in urodeles, with an upper head, the tuberculum, articulated to the diapophysis and a lower head, the capitulum, articulated to the parapophysis. The bicipital condition is probably primitive among tetrapods, since the oldest amphibians and *Seymouria* already possessed it. The fusion theory of its origin is not much accepted; a better view is that the capitulum is the original head and that the tuberculum is simply an outgrowth from it. In tetrapods with a single-headed rib the single head may be either the tuberculum or the capitulum or a fusion of both. The bicipital rib presumably functions to strengthen the trunk in relation to the habit of walking on land.

Primitively in tetrapods the tubercular head of the rib articulates with the diapophysis borne on the neural arch, and the capitular head articulates with the hypocentrum (Fig. 40). With the reduction of the hypocentrum and its transformation into the intervertebral disk the capitulum comes to articulate between successive centra on half-facets. The capitulum may, however, shift to the centrum, where a parapophysis may develop to articulate with it. However, both diapophysis and parapophysis may undergo so many shifts of position in different amniotes that it is hardly profitable to attempt to follow the changes.

Primitively, each tetrapod vertebra bears a pair of ribs articulated with it and hence movable. Thus *Seymouria* had a complete set of ribs along the whole vertebral column except the last few tail vertebrae. In most amniotes, however, the ribs tend to reduce and to fuse immovably with the vertebrae except in the anterior part of the trunk (*thoracic* region), where the ribs attain their best development in relation to the air-breathing habit. These thoracic ribs, playing an important role in lung respiration, are imbedded in the body-wall musculature and curve around to the midventral line, where they articulate with the sternum. They become divided into two or three sections for greater flexibility, usually an upper bony *vertebral* section and a lower cartilaginous *sternal* or *costal* section (Fig. 40*B*). Elsewhere along the vertebral column reduced ribs are usually detectable, at least during embryology.

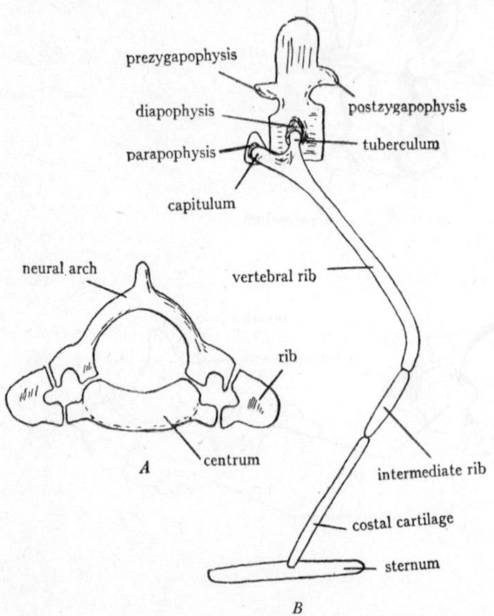

FIG. 40.—Relations of ribs. *A*, *cervical vertebra* of young *Echidna*, showing cervical ribs still separate from neural arch and centrum (after Flower, 1885). *B*, diagram of typical amniote rib.

4. Abdominal ribs.— *Ventral* or *abdominal* ribs, also called *gastralia*, occur imbedded in the ventral wall of the trunk in many fossil and recent amphibians and reptiles. In most cases these represent dermal elements which are homologous to the ventral armor of labyrinthodont amphibians (p. 89) and which have sunk into the interior. They are usually V-shaped with the apex directed forward. A complete set of such dermal bones occurred along the whole ventral wall of *Seymouria*, and they are also present in *Sphenodon*, the Crocodilia, and other present reptiles. Some reptiles, however, have similar-looking ventral ribs which are cartilage bones and hence endoskeletal. It is desirable to call the dermal ventral ribs gastralia, and the endoskeletal ones abdominal ribs; the latter probably evolve into the sternum (see next chapter).

5. Development of reptilian vertebrae.—Embryological studies confirm the paleontological evidence. In amniotes the division of the sclerotomes into cranial and caudal halves and the union of caudal halves with the succeeding cranial halves in the production of vertebrae is very pronounced. Reptilian vertebrae form embryologically by the fusion of originally separate elements (Fig. 41*A*). A pair of basidorsals appears in the original caudal sclerotomic halves and becomes the greater part of the neural arch. Interdorsals appear in the cranial halves and contribute to the neural arch; both pairs of dorsal arcualia also spread ventrally to contribute to the centrum. The main part of the centrum is of perichordal origin and hence corresponds to the pleurocentra; it becomes covered over by the arch contributions to the centra. The hypocentra originate as independent ventral cross-pieces (hypochordal bars). Basiventral elements also appear to be concerned in the formation of the anterior ventral part of the centrum. In the tail region haemal arches appear as outgrowths of the hypocentra (or basiventrals). In *Sphenodon* two sets of hypocentra occur during embryology; the first set is preformed in cartilage and gives rise to the haemal arches; the second set replaces most of the first set and ossifies directly in the perichordal mesenchyme. This may be a recapitulation of the replacement of the basiventral arcualia by the perichordal hypocentra in the evolution of tetrapod vertebrae. In reptiles (*Sphenodon*, alligator), the neural arches form first, the haemal arches next, and the pleuro- and hypocentra last. All parts go through a cartilage stage. Amniote vertebrae are thus seen to be compounded of arch and perichordal components.

6. Vertebral column of the alligator.—On the whole-mounted skeleton of the alligator note division of the vertebral column into five regions: cervical, much longer than in Amphibia; thoracic, bearing long ribs; lumbar, without

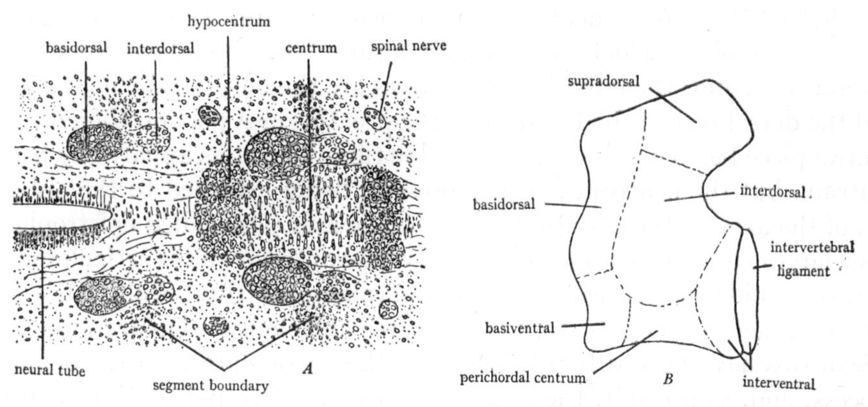

FIG. 41.—Development of amniote vertebrae, showing components. *A*, embryo of *Sphenodon*, horizontal section, slightly curved, showing arcualia, distinct from centrum and hypocentra (=hypochordal bars) (after Schauinsland, 1906); anterior end to the left. *B*, components of bird vertebra (after Piiper, 1928), based on an embryological study.

ribs; sacral, composed of two fused vertebrae, supporting the hind limbs; and caudal. The caudal or tail vertebrae are the most primitive in form and will be studied first. Each consists of a large centrum, a neural arch with a high neural spine, prominent transverse processes directed straight out from the centrum, and a haemal arch, missing on the first caudal vertebra. The haemal arches, often called chevron bones or chevrons, are relatively small and articulate nearly or quite between successive centra (an indication that they originally articulated with the missing hypocentra). Toward the end of the tail the various vertebral projections tend to reduce and finally vanish altogether, so that the last vertebrae consist of centra only. The sacral region is composed of two vertebrae (one sacral vertebra is probably the primitive condition in tetrapods), each bearing a stout sacral rib, to which the supports of the hind limbs are articulated. The sacral vertebrae have high neural spines but lack chevrons. The five lumbar vertebrae, anterior to the sacrum, possess high neural spines and broad transverse processes. The ten thoracic vertebrae are similar to the lumbars but bear long ribs reaching the midventral region. Of the nine cervical vertebrae the first two are somewhat different from the other seven, which have strong neural arches, long neural spines, and short transverse processes bearing ribs. Most of these cervical ribs are of peculiar form, being V-shaped with the two ends of the V applied to the vertebra. The centra of the cervical and anterior thoracic vertebrae possess short midventral projections or hypapophyses. All the vertebrae are provided with well-developed pre- and postzygapophyses. By moving some vertebrae apart note that the centra are procoelous.

The first two cervical vertebrae, the atlas and the axis, are highly instructive, in that their components have remained separate. The atlas consists of four separate pieces which together form a ring. The ventral piece represents the hypocentrum; the lateral curved pieces are the bases of the neural arch; and the dorsal piece is believed to be the proatlas or a portion thereof. The ventral piece bears a pair of long movable ribs. The atlas apparently has no centrum, but the centrum is really present, being attached to the anterior end of the axis, as the odontoid process. The axis (also called epistropheus) has a large centrum (pleurocentrum) to whose anterior end is articulated the large odontoid process, really the centrum (pleurocentrum) of the atlas. The axis is characterized by its strong neural arch and elongated neural spine. Its pair of ribs have moved forward, so that they articulate with the odontoid process; and, as a result, the atlas seems to have two pairs of ribs and the axis none.

7. Ribs of the alligator.—On the cervical or first thoracic vertebrae note the bicipital ribs. The upper tuberculum is attached to a short blunt diapophysis, the lower capitulum articulates to the side of the anterior end of the centrum. A large opening, the *transverse foramen*, is naturally inclosed by the two heads of the ribs. The successive openings form the *vertebrarterial canal* in which blood vessels to the head are located. Observe gradual increase in the length of the ribs as one proceeds backward and the shifting of the capitular head dorsally, so that it finally articulates to the enlarged transverse process, which here appears to be a diapophysis. The thoracic vertebrae differ from the cervical ones in that their ribs reach around to the ventral side. The little smooth places on the vertebrae to which the ribs articulate are called *facets*, and rib-bearing vertebrae can always be recognized by such facets.

The thoracic ribs are segmented into three parts: an upper bony part, the *vertebral* rib, which includes the heads; a middle, partly cartilaginous *intermediate* part; and a lower, partly cartilaginous *sternal* part or *costal* cartilage.

The last two thoracic ribs consist of vertebral ribs only. Note *uncinate processes* on the vertebral ribs.

On the ventral side of the lumbar region will be noted a series of riblike bones, the *gastralia*, not attached to any other part of the skeleton. These are dermal bones, part of the exoskeleton, homologous to the dermal plates of the turtle's plastron. They are likely to occur in long-bellied reptiles.

Draw a cervical or anterior thoracic vertebra of the alligator with all of its parts, showing particularly the relation of the ribs.

8. The vertebral column of the turtle.—In our study of the carapace of the turtle we already noted certain peculiarities of the vertebral column of these animals. These peculiarities arise from the circumstance that part of the vertebral column is fused to the exoskeleton. The vertebral column of the turtle consists of cervical, trunk, sacral, and caudal regions. As usual, the caudal region is the most primitive. The caudal vertebrae have neural and haemal arches and transverse processes, of which the most anterior are ribs, as shown by the suture at their bases. The first caudal vertebra is fused to the sacrum, which consists of two sacral vertebrae bearing sacral ribs. The first caudal, the two sacral, and the ten trunk vertebrae are all fused to the carapace. As already explained in connection with the exoskeleton, the neural arches of these vertebrae are inseparably fused to the vertebral plates of the carapace, and the ribs of the trunk vertebrae are expanded and fused to the inner surface of the costal plates of the carapace. Turtle ribs have a single head, the capitulum, articulating at or near the boundary between successive centra. There are eight ribless cervical vertebrae which are very flexibly articulated by a variety of centrum ends, as may be observed if separate vertebrae are available. The first two cervical vertebrae are differentiated into atlas and axis, having a construction similar to those of the alligator.

I. VERTICAL COLUMN OF BIRDS

The vertebral column of birds is very similar throughout the class and, like their entire structure, is modified in relation to the flight habit.

1. Vertebral column of the chicken or pigeon.—Note differentiation of the column into the usual regions, but all except the cervical vertebrae are fused together to strengthen the back. The numerous cervical vertebrae (sixteen in the chicken, thirteen to fourteen in the pigeon) have very flexible articulations, permitting birds to turn their heads halfway around. This flexibility results from the saddle-shaped or *heterocoelous* centrum ends, which should be examined on isolated neck vertebrae. The atlas is a small ring-shaped bone; the axis bears the usual anteriorly projecting odontoid process. The typical cervical vertebrae behind the atlas have low neural arches and spines, well-developed zygapophyses, often midventral projections, the hypapophyses, and ribs. The ribs are the lateral masses having below a sharp posteriorly directed spine; as in reptiles, they are two-headed, forming a vertebrarterial canal, and their heads are fused to the diapophyses and parapophyses of the vertebrae. The thoracic vertebrae whose ribs reach the ventral side have high neural spines and well-developed transverse processes; spines, centra, transverse processes, and zygapophyses are immovably fused together. The last thoracic, the lumbar, the sacral, and the first few caudal vertebrae are united into one continuous piece, the *synsacrum*, separated by a suture from the broad hip bones which extend along each side of it. On the ventral side of the synsacrum the individual vertebrae which compose it can be distinguished by their separate transverse processes, which apparently represent the diapophyses; the fused hypapophyses form a midventral ridge. The number and regions of vertebrae entering into the synsacrum vary in different birds. Only a few of the vertebrae in the synsacrum are true sacral vertebrae corresponding to those of reptiles. Posterior to the synsacrum are a few free caudal vertebrae ending in an enlarged piece, the *pygostyle*, which represents several fused vertebrae. Recall that a long tail composed of regular caudal vertebrae was present in the extinct bird *Archaeopteryx;* in present birds the tail is reduced to a stump. Haemal arches are absent throughout, although embryology indicates that they participate in the pygostyle.

The ribs of birds are divisible into the usual vertebral and sternal sections. The vertebral ribs bear backwardly directed *uncinate* processes, characteristic of birds and serving apparently to strengthen the ribs. Uncinate processes also occur in *Sphenodon* and other reptiles.

2. Development of bird vertebrae.—The available accounts vary considerably; the most recent (that of Piiper, 1928) shows much resemblance to the development of vertebrae in *Sphenodon.* All four pairs of arcualia appear, as well as supradorsals. The neural arch is formed by the fusion of the basidorsals, interdorsals, and supradorsal (Fig. 41*B*). The primary centra arise as rings in the perichordal mesenchyme but become covered over above and at the ends by the spreading of all four pairs of arcualia, so that the final centrum is a combination of perichordal and arch elements (Fig. 41*B*). A complete series of hypocentra appears, contributes to atlas, axis, and pygostyle, but disappears elsewhere along the vertebral column; the hypocentra are said to arise from contributions from basiventrals and interventrals (thus differing from the story in reptiles). The ribs and parapophyses are derivatives of the basiventrals; this supports the belief that the ribs of amniotes are pleural or ventral ribs.

J. VERTEBRAL COLUMN OF MAMMALS

The vertebral column of mammals is markedly differentiated into the usual five regions, and the typical vertebrae of these regions are so well characterized as to be readily identifiable when isolated. Study whole-mounted skeletons of the cat or rabbit and isolated vertebrae.

1. The cervical vertebrae.—There are seven cervical vertebrae, of which the first two are differentiated as the atlas and the axis. The atlas is ring-shaped with wide winglike lateral projections which represent ribs and hence are perforated by an opening, the vertebrarterial canal. The low, flat neural arch of the atlas is also perforated by a pair of holes for the passage of nerves. The anterior surface of the atlas has a pair of large curved concavities which articulate with the condyles of the skull. As in reptiles, the atlas consists mainly of the bases of the neural arch (basidorsals), and the median ventral region is the hypocentrum (Fig. 39*D*). Draw the atlas from in front.

The axis has a very large elongated neural arch having a neural spine which projects forward over the atlas and bears posteriorly a pair of postzygapophyses. The centrum of the axis has at its anterior end a median pointed projection, the odontoid process, which fits into the ring of the atlas, allowing the turning of the head; and on each side of this has a large rounded articulating surface for the atlas. From the sides of the centrum project the transverse processes, which consist in part of a fused rib (hence should be called pleurapophyses), as shown by the presence of a vertebrarterial canal at their base. As in the case of reptiles, the odontoid process is the centrum (pleurocentrum) of the atlas; embryology shows that its anterior tip is the proatlas (Fig. 39*E*). Draw the axis from the side.

The remaining cervical vertebrae are more or less similar, with well-developed neural arches and spines, pre- and postzygapophyses, and transverse processes, really pleurapophyses, consisting in part of a fused-on rib, as shown by the presence of the vertebrarterial canal. Draw a cervical vertebra from in front.[5]

There are seven cervical vertebrae in nearly all mammals, regardless of length of neck; sloths have six or nine and Sirenia six. In aquatic mammals with short necks, as in whales and dolphins, the vertebral centra are also very short and often more or less fused together, and a similar fusion is seen in armadillos. In monotremes the sutures between the cervical ribs and the processes of the centrum are evident in young animals (Fig. 40*A*). A separate hypocentrum is seen in the atlas of some marsupials, but in others it is missing and its place is taken by a ligament.

2. The thoracic vertebrae.—These vertebrae bear long ribs, reaching the ventral side, and hence can be recognized when isolated by the smooth costal half-facets where the ribs were attached. There are thirteen thoracic vertebrae in the cat, generally twelve in the rabbit, various numbers in other mammals (twelve in man). The majority of the thoracic vertebrae have very tall neural spines, directed caudad, short centra, small pre- and postzygapophyses, and short stout transverse processes (diapophyses) to whose outer ends the tubercula of the ribs articulate. At the ends of the centra are half-facets for the capitula of the ribs, so that the lower rib heads really articulate between the centra (i.e., on the missing hypocentra). The last

[5] The anterior and posterior ends of vertebrae can be recognized by the fact that the facets of the prezygapophyses face upward or forward; those of the postzygapophyses, downward or backward.

thoracic vertebrae differ somewhat from the others, having reduced neural spines and transverse processes, more prominent zygapophyses, a single rib facet, and an extra process from the prezygapophyses, called *metapophyses* or *mammillary processes*.

3. The ribs.—The ribs of mammals usually consist of a bony vertebral rib and a cartilaginous sternal rib or costal cartilage. The vertebral rib is usually bicipital, with the reduced tuberculum engaging the costal facet on the under side of the transverse process, and the capitulum attached between two centra to the demifacets. The tuberculum diminishes posteriorly, so that the last three ribs have only capitular heads. The narrowest part of the rib between the two heads is termed the *neck;* the remainder of the rib, the *shaft;* and the point of greatest curvature of the shaft, the *angle*. Those ribs which reach the ventral side and are independently attached to the breastbone are termed *true* ribs; those which join the preceding ribs or are unattached below are called *false* ribs; and the unattached false ribs are known as *floating* ribs. The cat has nine true and four false ribs, of which the last one is floating; the rabbit has seven true and five false ribs, of which three are floating; and man is the same as the rabbit but with only two floating ribs. Draw, from in front, a typical thoracic vertebra with its ribs.

There are commonly in mammals twelve to fifteen thoracic vertebrae; but eighteen to twenty are seen in the horse, tapir, rhinoceros, and elephant; the largest number occurs among the sloths (up to twenty-five), the smallest among Cetacea (nine). In the Sirenia and the whalebone whales all but the first one to three ribs are floating, and this seems to be related to the respiratory needs of these large aquatic mammals. The division of the ribs into three sections is seen in monotremes and some sloths, but it is not clear that these correspond to the divisions in reptilian ribs. Uncinate processes are lacking in mammals.

4. The lumbar vertebrae.—The cat and rabbit have seven lumbar vertebrae; four to seven is the usual number among mammals, but up to twenty-one are seen among Cetacea, and two to four in monotremes and some edentates. The lumbar vertebrae are large and stout with prominent neural spines and long transverse processes directed forward; the latter are probably pleurapophyses, i.e., composed of fused diapophysis and rib. A prominent metapophysis projects above the prezygapophyses, and a spinelike *anapophysis* (*accessory process*) is seen below the postzygapophyses. Draw a lumbar vertebra from the side.

5. The sacrum.—The sacrum is composed of a variable number of vertebrae fused together for articulation with the hind limbs. There are three sacral vertebrae in the cat, generally four in the rabbit (of which only the first two really contribute to the attachment), five in man. The boundaries between the fused sacral vertebrae are readily made out by means of the openings between them through which the spinal nerves pass out, and by means of the number of neural spines, zygapophyses, etc. The first sacral vertebra assumes the greater part of the task of transmitting the support of the hind limbs to the vertebral column; for this purpose it has large lateral expansions bearing articular surfaces for the insertion of the bony structure which supports the hind limb. These lateral expansions consist in part of transverse processes and in part of sacral ribs, indistinguishably fused to the vertebra. In most mammals there are three to five sacral vertebrae, but six to eight occur in some perissodactyl ungulates, and up to thirteen are found among edentates.

6. The caudal vertebrae.—These are very variable in number among mammals, as shown by the varying tail lengths, and run from a few up to fifty. Neural arches, transverse processes, and zygapophyses diminish caudally, so that the last vertebrae consist only of centra. Chevron bones (=hypocentra of reptiles) are of common occurrence in the tails of mammals; very small ones occur in the cat tail but are usually lost in preparing the skeleton. Man has three to five tail vertebrae fused into a single piece, the urostyle or *coccyx*.

7. General remarks on the whole column.—Observe, by inspection of isolated vertebrae, that both centrum ends are flat, a condition termed amphiplatyan. However, opisthocoelous centra occur in the cervical vertebrae of ungulates. Between the centrum ends of mammals are found in life cartilaginous disks, the *intervertebral cartilages*, which are homologous to the hypocentra of reptiles. On the mounted skeleton observe on each side between successive centra a series of openings, the *intervertebral foramina*, through which the spinal nerves pass out from the spinal cord. The spinal cord occupies the continuous neural canal formed by the neural arches. No trace of the notochord remains except in the intervertebral disks, of which it forms a central area.

8. Development.—Although the development of mammalian vertebrae has been described by a number of workers—most recently by Dawes (1930) and Bochmann (1937)—there is much disagreement in regard to the number of distinct elements which can be observed entering into the composition of a vertebra. Whereas Dawes recognizes all four pairs of arcualia, Bochmann does not find any of them in the same animal (mouse). In general, it may be said that the neural arch arises from a pair of elements which represent the basidorsals and that the centrum comes from one or a pair of chondrifications (pleurocentra?) which form a ring-shaped mass. Ventral crossbars (the hypochordal bars) are recognizable throughout the column but persist as vertebral elements only in the atlas and the tail; they are evidently the hypocentra. They usually contribute to the intervertebral disks. Evidence of the interdorsals is seen in at least some forms. Peculiar to mammalian vertebrae are disks, the *epiphyses*, one at each end of each centrum, which ossify separately from the main mass of the centrum and fuse with the latter only in postembryonic life or sometimes not at all (whales). Their phylogenetic significance is unknown.

K. SUMMARY

1. The vertebral column forms the axial endoskeletal support of the vertebrate body; it consists of a series of bones, the vertebrae, laid end to end. Typically, each vertebra consists of a spool-shaped body or centrum inclosing the notochord; a dorsal arch, the neural arch, inclosing the neural tube; and a ventral arch, the haemal arch, inclosing blood vessels. The haemal arch is limited to the tail region.

2. The vertebral column replaces the notochord as the axial support; it does not come from the notochord but forms around it, squeezing it more or less out of existence.

3. In development the vertebrae arise at the intersections of the myosepta, with the sclerotomal mesenchyme surrounding the notochord and neural tube. Each vertebra is produced by the union of the posterior halves of the pair of sclerotomes of one segment with the anterior halves of the pair of sclerotomes of the following segment. The vertebrae consequently alternate with the myotomes; and this is necessary for muscle action on the vertebral column.

4. In its development the vertebral column (and the endoskeleton in general) passes through three stages: mesenchymatous, cartilaginous, bony. It may remain in the cartilage stage or may ossify in part or whole. Ossification may take place directly in the mesenchyme or in the cartilage already formed.

5. A vertebra does not arise as a single piece but is formed by the fusion of various originally separate components, as is proved by embryological studies and by the condition of the vertebrae in primitive fish, amphibians, and reptiles.

6. The components of a vertebra come from three sources: the arcualia, the notochordal sheath, and the perichordal mesenchyme.

7. The arcualia are paired cartilages which form in the sclerotomes and rest upon the notochord. There are four main pairs: the basidorsals and interdorsals above, and the basiventrals and interventrals below; minor arcualia may also occur. The basidorsals and basiventrals come from the original posterior halves of the sclerotomes and hence are situated in the anterior halves of the future vertebra. They form the neural and haemal arches, respectively, and may contribute to the centrum. The interdorsals and interventrals form in the original anterior halves of the sclerotomes. The interdorsals may form an intercalary arch behind the neural arch or may contribute to the neural arch or the centrum. The interventrals are of less importance.

8. In some vertebrates, chiefly elasmobranchs, cells from the arcualia invade the sheath of the notochord, and the sheath then chondrifies and becomes part of the centrum.

9. Mesenchyme from the sclerotomes accumulates around the notochord to the inner side of the main masses of the sclerotomes, to form the perichordal mesenchyme. This contributes to the centrum either by direct ossification or by way of cartilage in most vertebrates, particularly in amniotes.

10. Because the vertebrae are thus formed from a number of components, primitive vertebrae consist of separate pieces, since they represent a stage before the components have fused.

11. Among primitive land vertebrates, two pieces, termed the pleurocentrum and the hypocentrum, are important in the final evolution of the completed vertebra. The pleurocentrum more or less corresponds to the interdorsal arcualia; the hypocentrum, to the basiventrals. In amniotes the pleurocentrum forms the main mass of the centrum, and the hypocentrum remains as haemal arches (chevron bones) in the tail and also contributes to the atlas and sometimes the axis.

12. Primitively, the ends of the centra are concave—amphicoelous type of centrum. Other types of centrum ends are derived from this.

13. In fishes the vertebral column is differentiated into trunk and tail regions. In amphibians a short cervical and a sacral region are added. In reptiles, birds, and mammals the cervical region is longer; and the trunk region is differentiated into an anterior thoracic region, bearing long ribs, and a posterior lumbar region with reduced or no ribs. The differentiation of the vertebral column into regions is more marked the higher one ascends in the vertebrate scale.

14. Originally, each vertebra was provided with a pair of movable ribs; but in the higher vertebrates these are reduced or absent except in the trunk or thoracic regions. Reduced ribs are generally present on the cervical vertebrae and always present on the sacral vertebrae.

15. Ribs are of two kinds: those that arise at the intersection of the myosepta with the horizontal skeletogenous septum, known as dorsal or intermuscular ribs; and those that arise at the intersection of myosepta with the ventral skeletogenous septa, known as ventral, pleural, or subperitoneal ribs. The former are probably independent formations; the latter probably represent cutoff extensions of the haemal arches (basiventral arcualia). The ribs of the majority of vertebrates are of the ventral type. Some fishes may have both kinds simultaneously and may also develop additional ribs (of the dorsal type) at other levels of the myosepta.

16. The ribs of land vertebrates primitively articulate with the vertebrae by two heads, and the space between the heads forms in the cervical region a vertebrarterial canal for the passage of blood vessels.

REFERENCES

Bochmann, G. 1937. Die Entwicklung der Säugetierwirbel der hinteren Körperregionen. Morph. Jahrb., **79.**

Dawes, B. 1930. The development of the vertebral column in mammals. Phil. Trans. Roy. Soc. London, B, **218.**

Emelianov, S. W. 1925–28. Die Entwicklung der Rippen und ihr Verhältnis zum Wirbelsäule. Rev. zool. russe, **5, 6, 8.**

———. 1933. Die Entwicklung der Rippen der Tetrapoden. Trav. lab. morph. evolution, **1.**

———. 1935. Die Morphologie der Fischrippen. Zool. Jahrb., Abt. Anat., **60.**

———. 1936. Die Morphologie der Tetrapodenrippen. *Ibid.*, **62.**

———. 1940. Omission of cartilaginous stages in the development of chondral ossifications in teleosts. Compt. rend. Acad. sci. U.S.S.R., **26.**

Evans, F. G. 1939. The morphology and functional evolution of the atlas-axis complex from fish to mammals. Ann. N.Y. Acad. Sci., **39.**

Gamble, D. L. 1922. The morphology of the ribs and transverse processes in *Necturus*. Jour. Morph., **36.**

Goette, A. 1878. Beiträge zur Morphologie des Skeletsystems der Wirbeltiere. Arch. f. mikr. Anat., **15.**

Graham-Smith, W. 1936. The tail of fishes. Proc. Zoöl. Soc. London.

Hayek, H. 1923. Über den Proatlas und die Entwickelung der Kopfgelenke beim Menschen und bei Säugetieren. Sitzungsb. Akad. Wiss. Wien, **130–131,** Abt. 3.

———. 1924. Über das Schicksal der Proatlas und über die Entwicklung der Kopfgelenke bei Reptilien und Vögeln. Morph. Jahrb., **53.**

Higgins, G. M. 1923. Development of the primitive reptilian vertebral column. Amer. Jour. Anat., **31.**

HOWES, G. B., and SWINNERTON, H. H. 1901. Development of the skeleton of the tuatara, *Sphenodon*. Trans. Zoöl. Soc. London, **16**.

JAECKEL, O. 1896. Die Organisation von *Archegosaurus*. Zeitschr. f. deutsch. geol. Gesellsch., **48**.

MACBRIDE, E. C. 1932. Recent work on the development of the vertebral column. Biol. Rev. (Cambridge), **7**.

MOOK, C. C. 1921. Postcranial skeleton in the Crocodilia. Bull. Amer. Mus. Nat. Hist., **44**.

MOOKERJIE, H. K. 1930. On the development of the vertebral column of Urodela. Phil. Trans. Roy. Soc. London, B, **218**.

———. 1930. On the development of the vertebral column of Anura. *Ibid.*, **219**. See also Jour. Morph, **64**, **67**.

PIIPER, J. 1928. On the evolution of the vertebral column in birds. Phil. Trans. Roy. Soc. London, B, **216**.

RIDEWOOD, W. G. 1899. Caudal diplospondyly of sharks. Jour. Linn. Soc. London, Zoöl., **27**.

———. 1921. On the calcification of the vertebral centra in sharks and rays. Phil. Trans. Roy. Soc. London, B, **210**.

SCHAUINSLAND, H. 1903. Beiträge zur Entwickelungsgeschichte der Wirbeltiere. I. *Sphenodon, Callorhynchus, Chamaeleo*. Zoologica, **16**.

———. 1906. Die Entwickelung der Wirbelsäule nebst Rippen und Brustbein. Hertwig's Handbuch der vergleichenden und experimentellen Entwickelungslehre der Wirbeltiere, **3**, Part 2.

SHUFELDT, R. W. 1885. Osteology of *Amia calva*. Rept. U.S. Comm. Fish. for 1883.

WATSON, D. M. S. 1918. On *Seymouria*, the most primitive known reptile. Proc. Zoöl. Soc. London.

WHITE, T. E. 1939. Osteology of *Seymouria*. Bull. Mus. Comp. Zoöl. Harvard, **85**.

WILDER, H. H. 1903. The skeletal system of *Necturus*. Mem. Boston Soc. Nat. Hist., **5**.

WILLISTON, S. W. 1908. The oldest known reptile. Jour. Geol., **16**.

———. 1911. American Permian Vertebrates.

———. 1925. The Osteology of the Reptiles.

VIII. THE ENDOSKELETON: THE COMPARATIVE ANATOMY OF THE GIRDLES, THE STERNUM, AND THE PAIRED APPENDAGES

A. GENERAL CONSIDERATIONS

1. Definitions.—The *girdles* are crescent-shaped or arch-shaped portions of the endoskeleton which function for the support of the paired appendages. The main mass of the girdles occupies a ventral position, and the ends of the girdles extend dorsally. In the lower fishes the girdles are composed of cartilage, but in bony fishes and land vertebrates they are more or less ossified. The *pectoral* girdle supports the anterior appendages; the *pelvic* girdle, the posterior ones.

The *sternum* or *breastbone* is an elongated structure lying in the midventral region of the anterior part of the trunk. It is lacking in fishes and is first seen among amphibians. Primitively, it is a simple cartilaginous plate, but in later forms usually consists of a chain of cartilages or bones or both; it articulates with the pectoral girdle and in amniotes also with the ribs. The ribs fail to reach the sternum in all present amphibians, but it is possible that they did so in extinct forms. The combination of pectoral girdle, ribs, and sternum serves to strengthen the anterior trunk in relation to the air-breathing habit and the presence of lungs.

The paired appendages consist of fins (*ichthyopterygium*) in fishes and limbs (*cheiropterygium*) in land vertebrates and are supported by an endoskeleton. The anterior appendages, termed *pectoral fins* or *forelimbs*, articulate with the pectoral girdle; and their support is transmitted to the body by way of this girdle, which, however, rarely is directly jointed to the vertebral column. The posterior appendages, called *pelvic fins* or *hind limbs*, similarly articulate to the pelvic girdle and transmit their support to this girdle. As the support of the hind appendages is commonly the more important, the pelvic girdle is usually larger and more massive than the pectoral girdle, especially in biped vertebrates, and is strongly articulated to the vertebral column by way of the sacral ribs. As already learned, that region of the vertebral column which bears the sacral ribs and hence supports the pelvic girdle is called the sacrum. A sacrum is lacking in fishes, since the fins usually do not support the body but serve only for locomotion. Primitively, the pectoral fins are situated just behind the gill region, and the pelvic fins are located immediately in front of the anus.

2. Origin of the paired appendages.—Since the paired limbs of land vertebrates are unanimously acknowledged to have evolved from the paired fins of fishes, the question of origin becomes limited to the origin of the paired fins. According to the most probable theory, the *fin-fold theory*, the ancestral vertebrate possessed a pair of continuous fin folds, one along each side of the trunk, which fused behind the anus to a single median fin, extending around the tail and along the middorsal line (Fig. 42). This hypothetical condition resembles that actually obtaining in *Amphioxus*, where the paired metapleural folds and median caudal and dorsal fin may be recalled. The paired fins of present fishes are supposed to have arisen through the persistence of certain regions of the paired fin folds; and the median fins through the persistence of portions of the median fin fold (Fig. 42*B*). This theory is supported by a number of facts: the development and early structure of the unpaired and paired fins of fishes are identical; in elasmobranchs the muscle buds and nerve branches to the embryonic paired

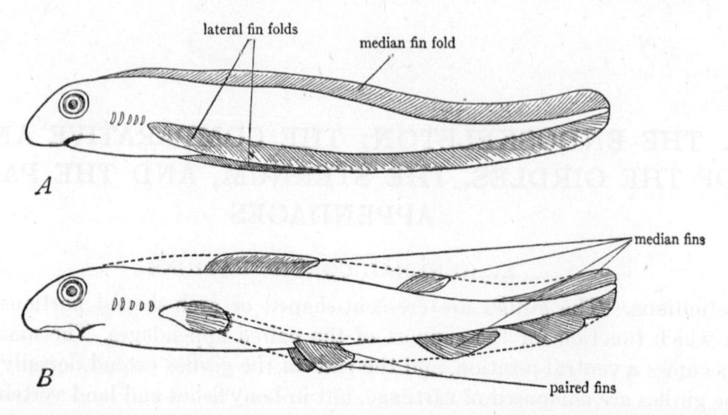

FIG. 42.—Diagrams to illustrate the theory of the origin of the median and paired fins through the persistence of certain regions of originally continuous median and lateral fin folds. *A*, early stage, showing the median dorsal fin fold and the two lateral fin folds uniting at the anus. *B*, later stage, illustrating the persistence of certain regions of the fin folds as the median and paired fins and the disappearance of the remainder of the fin folds, as indicated by dotted lines. (From Wilder's *History of the Human Body*, courtesy of Henry Holt & Co.)

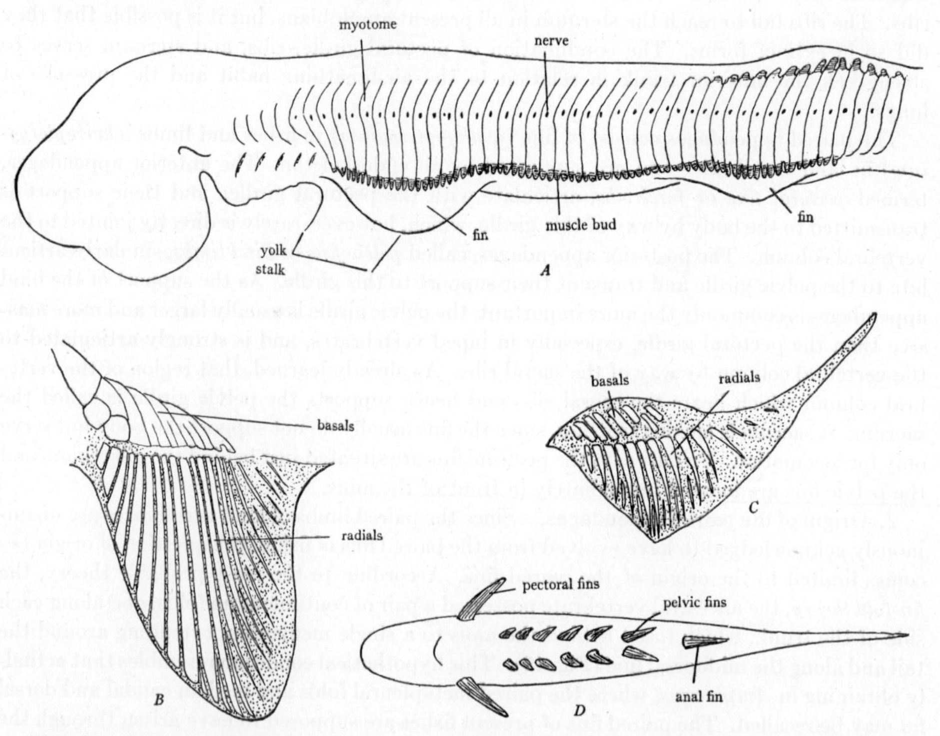

FIG. 43.—Evidence for the fin-fold theory of the origin of the paired appendages. *A*, embryo of the shark *Scyllium*, showing complete set of muscle buds along the side of the body (after Goodrich, 1906). *B*, pectoral fin, and *C*, pelvic fin, of *Cladoselache*, extinct shark, showing lappet shape and parallel pterygiophores (after Dean, 1894). *D*, an acanthodian shark, from below, showing paired row of fins between pectoral and pelvic fins (after Watson, 1937).

fins involve many more segments than in the adult (Fig. 43*A*); in elasmobranch embryos and the extinct shark *Cladoselache* (Fig. 43*B*) the shape of the fins and the parallelism of the fin rays strongly suggest origin from a continuous fold;[1] and finally, in the very primitive extinct acanthodian sharks, called by Watson (1937) the earliest-known jawed vertebrates, there was a row of fins on each side between the pectoral and pelvic fins (Fig. 43*D*).

3. Origin of the limb girdles.—As previously learned, the fins are supported by a series of endoskeletal rods, the pterygiophores, to which are attached muscles for moving the fins. (These muscles originate from the adjacent myotomes as the muscle buds mentioned above and seen in Fig. 43*A*.) Primitively, there is one pterygiophore (and one muscle bud) per segment. The pterygiophores early become subdivided into three or four pieces, or possibly this was the primitive condition. The distal pieces next the dermal fin rays are termed *radials;* the proximal pieces in the fin base are called *basals* (Fig. 43*B*). The basals tend to fuse into larger pieces, and this condition of a few large basals and one or more rows of small radials is seen in the paired fins of present elasmobranch and ganoid fishes. By the extension of the

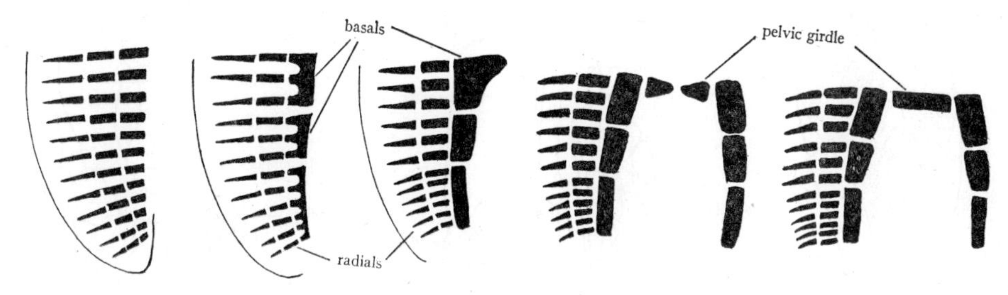

FIG. 44.—Diagrams of the origin of the limb girdles by the concentration, cutting off, and fusion of basal pterygiophores. (Based on Thacher, 1877.)

large, most anterior basal toward the midventral line and its fusion with its fellow of the opposite side, a primitive girdle in the form of a cartilaginous bar arises (Fig. 44). This is the generally accepted explanation of the origin of the girdles.

4. Evolution of the fish fin into the tetrapod limb.—As ancestral to the tetrapods, only the fish groups Dipnoi and Crossopterygii need be considered, as they alone have internal nares, functional lungs (swim bladder) connected by a duct to the ventral pharyngeal wall, and lobed fins containing a series of endoskeletal elements transmutable into limb bones. In the past the idea of a dipnoan ancestry for tetrapods was urged by a number of prominent anatomists, but at present the crossopterygian ancestry has been universally accepted. The skeleton of the lobed fins of both recent and ancient Dipnoi consists of an axial series of elements with rays to both sides (Fig. 45*A*). This monaxial type of fin was called *archipterygium* by Gegenbaur and was considered by him the prototype of all other kinds of fish fins and of the tetrapod limb. This view has now been abandoned, since the Dipnoi are no longer regarded as ancestral to tetrapods, and instead attention has been directed to the endoskeleton of the lobed fins of extinct Crossopterygii. Unfortunately, only two or three such fin skeletons are known; that of *Sauripterus* (Fig. 45*B*) with its radiating or dichotomous arrangement of endoskeletal basals and radials is generally accepted as illustrating a stage in the transformation of fish fin into amphibian limb. In Figure 45 it is shown compared to the skeleton of the

[1] It now appears that *Cladoselache* is not as primitive a shark as was formerly supposed; but this need not necessarily detract from the phylogenetic significance of its lappet-like fins with their parallel rays, although Gregory regards the fin rays of this fish as an example of polymerism (i.e., repetition of parts) rather than as primitive in nature.

most ancient tetrapod limb known, that of the embolomerous amphibian *Eryops*. No agreement has been reached between vertebrate paleontologists as to the identification of the wrist elements in *Eryops*, and hence these are left unlabeled.

Holmgren (1933, 1939), after an extensive embryological study, has concluded that the limbs of urodeles are derived from the dipnoan archipterygium and the limbs of all other tetrapods from the crossopterygian fin. If true, this necessitates removing the urodeles from the Amphibia as a separate class of vertebrates. It also indicates the evolution of fish into land vertebrates twice independently.

5. Parts of the primitive limb.—The tetrapod limb (or cheiropterygium) is divisible into three sections or segments: *stylopodium, zeugopodium*, and *autopodium*. The stylopodium is the *upper arm* in the forelimb, the *thigh* in the hind limb, and contains a single bone—*humerus*

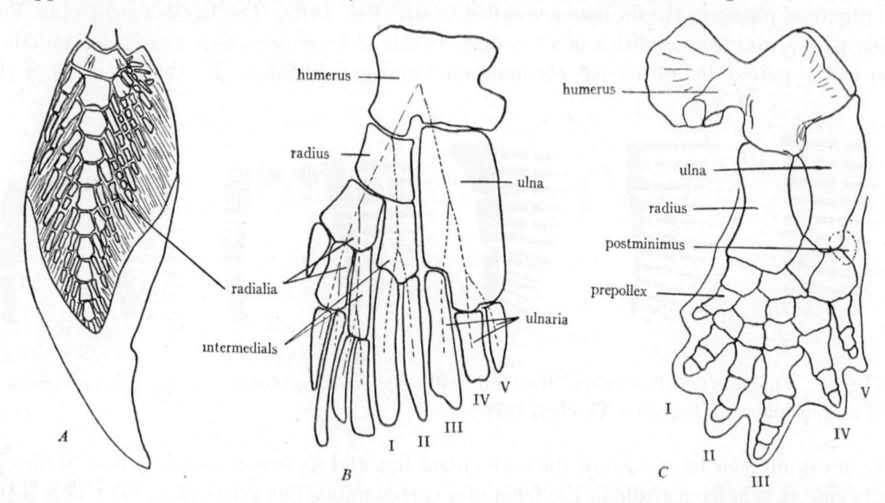

FIG. 45.—Types of early appendage skeletons. *A*, pectoral fin of the lungfish *Ceratodus*, showing archipterygium, monaxial with lateral rays (after Holmgren, 1933). *B*, endoskeletal elements of the pectoral fin of the extinct crossopterygian *Sauripterus*, showing dichotomous arrangement (after Holmgren, 1939). *C*, endoskeleton of the most ancient tetrapod limb known, the forelimb of the extinct embolomerous amphibian *Eryops* (after Gregory, 1935); note diagonal arrangement of the bones of the carpus. Preaxial side to left.

or *femur*, respectively. The zeugopodium is the *forearm* in the forelimb, the *shank* in the hind limb, and contains in each case two parallel bones—*radius* and *ulna* in the former, *tibia* and *fibula* in the latter. Radius and tibia are on the preaxial side, ulna and fibula on the postaxial side. The stylopodium forms with the zeugopodium an important joint, the *elbow* joint in the forelimb, the *knee* joint in the hind limb. The autopodium also consists of three divisions: *carpus* (wrist), *metacarpus* (palm), and *digits* (fingers) in the forelimb; and *tarsus* (ankle), *metatarsus* (sole), and *digits* (toes) for the hind limb. Generalized names for these three divisions are *basipodium, metapodium*, and *acropodium*, respectively. The basipodium (wrist, ankle) is composed of a number of small bones, which originally seem to have had the following arrangement: a proximal row of three bones—*radiale* or *tibiale, intermedium, ulnare* or *fibulare;* a central group of four bones—the *centralia;* and a distal group of five *carpalia* or *tarsalia* (Fig. 46). The metapodium (palm, sole) consists of five bones in a row, the *metacarpals* or *metatarsals*. The acropodium typically is pentadactyl, i.e., composed of five digits, numbered from one on the preaxial side to five on the postaxial side. Each digit consists of a linear row of bones termed *phalanges*, and the number of phalanges is primitively from first to last digit: 2, 3, 4, 5, 4. There is some evidence that the first tetrapods had seven digits, an extra one on the preaxial side called *prehallex* or *prepollex*, respectively, and a *postminimus* on the postaxial side (Fig. 46). However, it is more probable that these were additional wrist or ankle bones. There is evidence that the first metacarpal is really one of the carpalia and that the fifth metacarpal is really a phalanx. This indicates a marked asymmetry of the primitive hand, corresponding to the asymmetry of the crossopterygian fin (Fig. 5*B*), with extension along the preaxial side, crowding together along the postaxial side. In existing and most extinct tetrapods reduction has occurred in the number of phalanges and of wrist and ankle bones, and loss or reduction of the end digits is also very common.

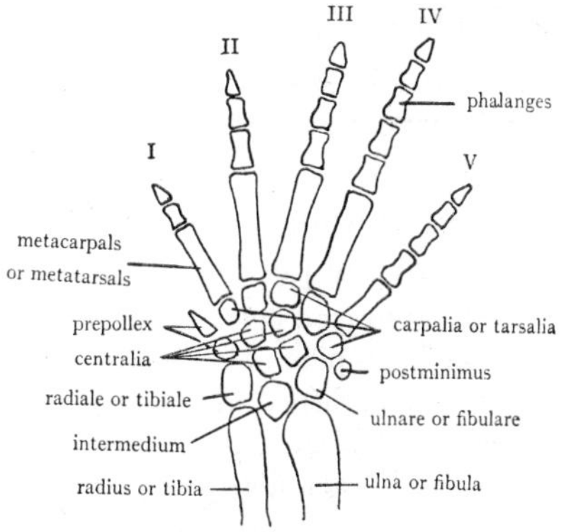

Fig. 46.—Endoskeletal elements of the primitive autopodium.

B. THE PELVIC GIRDLE AND THE POSTERIOR APPENDAGES

1. The primitive pelvic girdle.—As explained above, the primitive pelvic girdle consists on each side of the enlarged, most proximal cartilage of the basal series. Such a girdle, composed of a pair of rods or plates, is seen in various extinct and present fish, as in *Cladoselache* (Fig. 43*B*, *C*), *Eusthenopteron* (Fig. 47), and Holocephali, etc. Examine the skeleton of *Amia*, *Polypterus*, or *Acipenser* and note the pair of pelvic bones or cartilages. The paired pelvic cartilages of *Acipenser* are particularly instructive, as they show signs of having been formed by the fusion of basal cartilages, and a row of radials is attached to each.

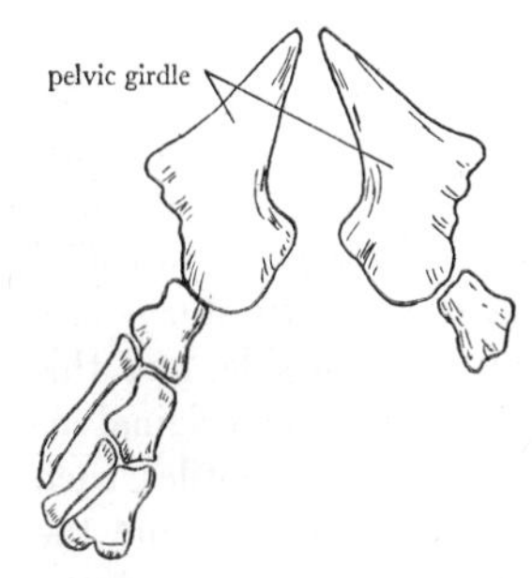

Fig. 47.—Pelvic fin and girdle of the extinct crossopterygian fish *Eusthenopteron*, showing primitive paired condition of the girdle. (After Goodrich, 1901.)

2. Pelvic girdle and pelvic fin of elasmobranchs.—The next stage of the pelvic girdle is seen in elasmobranchs, in which the pair of pelvic rods or plates has fused across in the midventral line to form the cartilaginous pelvic girdle. Study preserved skeletons of elasmobranchs. The pelvic girdle is the bar of cartilage across the ventral side at the end of the trunk region. It consists of the *puboischiac* bar ventrally and slightly projecting dorsal ends termed the *iliac processes*. Articulating to the puboischiac bar, at the base of the iliac processes, are the pelvic fins, which are seen to have an endoskeleton of a number of cartilaginous pieces, the pterygiophores or cartilaginous fin rays. These continue up to the beginnings of the dermal fin rays, considered previously. The pterygiophores are arranged in a medial series consisting of one or two (in some forms three or five) much enlarged cartilages, the *basals*, and an outer series of one or more rows of small, rodlike cartilages, the *radials*. There is usually a single basal, the *metapterygium*, a long, curved cartilage extending along the whole medial border of the fin; a small *propterygium* may be present anterior to the metapterygium. The basals have probably arisen by the fusion of a number of pieces similar to the radials. In males large radials support the claspers.

Draw the pelvic girdle and fin of an elasmobranch.

Observe complete lack of connection of the pelvic girdle with the vertebral column; this obtains throughout fishes. In teleosts the pelvic girdle is often reduced and may occupy an anterior position, in which case it may be attached to the pectoral girdle.

3. The tetrapod pelvic girdle.—The original form of the tetrapod pelvic girdle is evidently a cartilaginous puboischiac plate with dorsally extending iliac processes. This must soon have become strengthened for the support of walking legs and their musculature by ossification processes. Typically, two paired centers of ossification appear in the puboischiac plate, and one center in each iliac process. The two pairs of bones so formed in the puboischiac plate are the *pubis* in front and the *ischium* behind; the bone formed in the iliac process is called the *ilium*. The bones of the pelvic girdle are thus all cartilage bones. Usually the pubes and ischia of the two sides are united in the midventral line by way of cartilage; such union is termed a *symphysis*. A cup-shaped depression, the *acetabulum*, receives the head of the femur; and typically all three bones of the pelvic girdle meet at the acetabulum and contribute to the depression. The ilium is attached along its upper edge to the sacral ribs, so that sacrum and pelvic girdle together form a skeletal circle through which the exits of the urogenital and digestive system must pass to gain the exterior.

Such a primitive ossified tetrapod girdle is seen in extinct amphibians and reptiles, such as *Cacops* (Fig. 48*A*). The paired pubes and ischia form an unbroken shield below, from which the paired ilia extend dorsally; the sutures between the three bones occur in the acetabulum. Primitively, the only opening through the girdle was the small *pubic* foramen (also called diazonal and obturator foramen) through each pubis for the passage of the obturator nerve. Soon, however, an opening formed between the pubis and ischium on each side, and this *puboischiac* foramen may become of large size, extending to the symphysis. The original independence of the pubic and puboischiac foramina is seen in various primitive reptiles, such as *Sphenodon* (Fig. 48*B*). In later forms and in mammals the two become confluent, and the resulting foramen may be termed the *obturator* foramen.[2] An *acetabular* foramen through the acetabulum is seen in the Crocodilia, birds, and a few urodeles; it is obviously a secondary formation.

Minor elements of very uncertain homology occur among various tetrapods. A single or paired median forward projection from the pubis is termed *epipubis* and may be cartilaginous or ossified; lateral forward pubic projections are called *prepubic* or *pectineal* processes. A single or paired backward projection from the ischium, termed *hypoischium*, is found among reptiles (Fig. 48*B*). A separate acetabular bone in the acetabulum occurs among many mammals.

[2] The name obturator foramen has certainly often been loosely used and applied to pelvic openings that were not homologous. Strictly speaking, it is another name for pubic foramen, since originally the obturator nerve went through this foramen. However, the fact that the term obturator has become familiar for the pelvic opening of mammals, which represents the fusion of the pubic and puboischiac foramina (as shown by embryological studies) and through which several nerves pass, makes it desirable to limit the name to such a composite opening. It should be understood that in life such large foramina are closed by muscle and connective tissue.

4. Pelvic girdle and hind limb of urodeles.—The urodele girdle has remained in a primitive, largely unossified condition. Study preserved skeletons of *Necturus* or *Cryptobranchus*. The girdle has the form of a flattened puboischiac or pelvic plate, the anterior part of which, representing the pubic region, has remained wholly cartilaginous. In the posterior part of the plate is seen a pair of rounded ossification centers, representing the ischia. From each side of the pelvic plate a rod of bone, the ilium, extends dorsally and is firmly articulated to the end of the sacral rib. At the point of junction

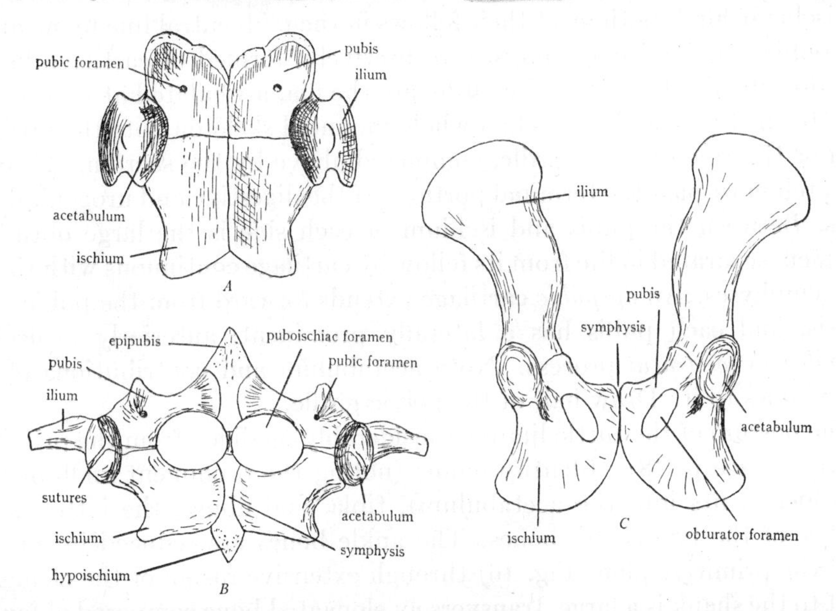

Fig. 48.—Representative tetrapod pelvic girdles. *A*, extinct embolomerous amphibian *Cacops*, showing platelike form with only the small pubic foramina (after Williston, 1925). *B*, *Sphenodon*, puboischiac foramina have formed, pubic foramina still present. *C*, pelvis of a mammal (ape), with obturator foramen; note also shape of ilium and resulting tilt.

of pubic cartilage, ischium, and ilium there is a depression, the acetabulum, into which the proximal end of the femur is inserted by a ball-and-socket joint.

The posture of the hind limbs of urodeles is rather primitive (see pp. 46, 56). The limb is divided into a proximal segment, the thigh, consisting of a single bone, the femur; a middle segment, the shank, composed of two parallel bones, a preaxial tibia and a postaxial fibula; and a distal segment, made of ankle and foot. The ankle or tarsus consists of several small bones, usually impossible to make out; it is followed by four elongated bones, the metatarsals, composing the sole; and these bear the toes, each made of a row of two or three small bones, the phalanges. *Necturus* and some other urodeles have but four toes on the hind feet, although most amphibians have five; it is probably the fifth which is missing. Draw pelvic girdle and hind limb from above.

5. Pelvic girdle and hind limb of reptiles.—In reptiles the ossification of the pelvic girdle is complete, and not much cartilage remains. In primitive fossil reptiles, as noted above, the pelvic girdle is platelike (Fig. 48*A*), but among present reptiles a more open construction prevails. The pelvic girdle of the turtle is representative. This consists of three pairs of stout bones—two pairs ventral, one pair dorsal. The ventral pairs are the pubes in front, the ischia behind; both meet their fellows in the midventral line by means of the pubic and ischiac symphyses, respectively, formed of cartilage in life. The dorsal pair of bones of the girdle are the ilia, articulated at their dorsal ends to the two sacral ribs. On a whole-mounted skeleton note the inverted arch or U formed by the girdle, completed above by the sacrum. Through this arch must pass the terminal portions of the digestive and urogenital systems. Between the pubis and ischium of each side is the large obturator foramen, separated in life from its fellow by cartilage continuous with that of the symphyses. An *epipubic* cartilage extends forward from the pubic symphysis, and each pubis has a laterally prominent, anteriorly projecting *prepubic* or *pectineal* process. Note acetabulum and contributions of the three bones to it. Draw half of the pelvic girdle.

The posture of the turtle limb is considerably modified from the primitive condition (see p. 48). Identify femur (noting the prominent knob or *head* by which it fits into the acetabulum), tibia, and fibula; the latter is the smaller of the two shank bones. The ankle bones are reduced in number from the primitive plan (Fig. 46) through extensive fusion of the elements. Next to the shank is a large, transversely elongated bone composed of four or more fused pieces (tibiale, intermedium, fibulare, and one or more centralia). In some turtles the fibulare, at the distal end of the fibula, is separate. Distal to this compound bone is a row of four tarsalia, of which the fourth consists of the fourth and fifth tarsalia, fused, plus possibly one or more centralia. Distal to the tarsalia are the five metatarsals, followed by the phalanges, of which the terminal ones are provided with horny claws. The number of phalanges in the digits is that regarded as typical for later tetrapods, namely, 2, 3, 3, 3, 3, but reduced from the original formula (Fig. 46). It is worthy of note that, in turtles and reptiles in general, the ankle joint occurs between the two rows of ankle bones, i.e., it is an *intratarsal* joint.

The pelvic girdle of lizards resembles that of *Sphenodon* (Fig. 48*B*), having separate pubic and puboischiac foramina; but that of the alligator (and of other Crocodilia) has several peculiarities. Ilia and ischia are normal, but the so-called pubes are movably jointed to the ischia and take no part in the acetabulum. This has led some authorities to consider them prepubes and to regard the true pubis as reduced to the anterior projection of the ischium seen in front of the acetabulum. Present evidence does not suffice to settle the question. There is an acetabular foramen through the acetabulum, and the obturator foramina between the prepubes and ischia form a very large opening, divided in two in life by a median ligament. In the snakes the pelvic girdle and hind limbs are wanting, except for some remnants in pythons and in a few other snakes.

6. Pelvic girdle and hind limb of birds.—These structures, like the remainder of the skeleton, are highly modified in birds, although consisting of the same parts as in other vertebrates. Examine isolated backbones with the pelvic girdles attached, or study the whole mounted specimens. The pelvic girdle consists of three pairs of bones as in reptiles, i.e., *ilium, ischium,* and *pubis.* All three are fused on each side to form a continuous broad bone, the *innominate bone.* The ilium is the largest and most dorsal part of the innominate bone, forming a thin elongated plate—concave in front, convex behind—extending from the last thoracic vertebra to the tail region, fused along its entire length with the synsacrum, from which it is bounded by a suture. (In the bird embryo the ilium is articulated to only two vertebrae, the true sacral vertebrae.) The ischium is the part of the innominate bone below the rear half of the ilium, separated from the latter by the large, oval *ilioischiac* foramen. The pubis is the long, slender bone along the ventral border of the ischium, from which it is separated by a more or less distinct suture and by the slitlike obturator foramen, sometimes subdivided. Ilium, ischium, and pubis contribute to the formation of the acetabulum, which is perforated by an acetabular foramen, as in the Crocodilia. The anterior projection of the pubis, in front of the acetabulum, is homologous to the pectineal process of reptiles. In bird development, ilium, ischium, and pubis originate as separate bones, and the pubis starts out anterior to the acetabulum but later grows posteriorly. Observe the absence of pubic and ischiac symphyses, so that the two innominate bones are widely separated ventrally, probably an adaptation for the laying of large eggs.

The hind limb offers several peculiarities. The femur has a large *head* fitting into the acetabulum and a prominent projection lateral to the head, called the *great trochanter.* The distal end of the femur is shaped like a pulley, consisting of a central depression with curved ridges—the *condyles*—on either side. Over the joint between thigh and shank is an extra small bone, the *patella* or *kneecap,* not found in the lower vertebrates. The patella is a *sesamoid* bone, i.e., a bone developed in a tendon. Such sesamoid bones are quite common in the limbs of higher vertebrates. The shank is composed of two bones, a medial large one and a lateral short rudimentary bone. The large bone is the *tibiotarsus.* It consists of the tibia fused at its distal end with the proximal tarsal bones. The proximal end of the tibiotarsus has two *condyles* for articulation with the condyles of the femur, and bears in front two diverging elevations or *crests* for muscle attachments. The small bone of the shank is the *fibula*, whose distal portion is atrophied. The distal end of the tibiotarsus has a pulley-like surface for articulation with the succeeding bone, formed by raised articular surfaces named *malleoli*. Beyond the tibiotarsus is a long, stout bone, the *tarsometatarsus*, evidently formed by the fusion of three bones, as shown by the three ridges on its distal end. The three fused bones are the metatarsals, and the distal ankle bones are also included. Hence the ankle bones do not exist separately in adult birds; instead the proximal ankle bones are fused to the lower end of the tibia, and the distal ones to the upper ends of the metatarsals, so that the ankle joint is intratarsal, as in reptiles. The three metatarsals participating in the tarsometatarsus are the second, third, and fourth; a remnant of the first is present as a small projection on the medial side of the distal end of the tarsometatarsus. Each metatarsal articulates with its respective clawed digit; fifth metatarsal and digit are absent. The gait of birds is digitigrade. Draw pelvic girdle and hind limb from the side.

The peculiarities of the bird pelvic girdle, especially the backward-turning of the pubes, are already seen among dinosaurs, with which birds are believed to have had a common ancestry. *Archaeopteryx* has a pubic symphysis, well-developed fibula, and separated metatarsals.

7. Pelvic girdle and hind limb of mammals.—The mammalian pelvic girdle is derivable from that of reptiles; it consists of the usual three bones indistinguishably fused into an *innominate* or *hip* bone on each side. The ilium, the most dorsal and largest of the three components of the hip bone, articulates with the sacrum and terminates anteriorly and dorsally in a curved border, termed the *crest* of the ilium. The ischium is that part of the dorsal region of the girdle posterior to the acetabulum. The posterior end of the two ischia form prominent projections in the rabbit or a rough curved surface in the cat, called the *ischiac tuberosity*, and extend toward the midventral line as the *rami* of the ischia, which meet to form the ischiac symphysis. The anterior ventral part of the innominate bone is the pubes, and the two pubes also have *rami* meeting to form the pubic symphysis. Both symphyses in life consist of cartilage. Between the rami of the ischium and pubis is the large obturator foramen. Ilium, ischium, and pubis meet at the acetabulum and participate in its walls. Draw the pelvic girdle from in front.

On a demonstration specimen of a kitten or other young mammal note complete separation and boundaries of ilium, ischium, and pubis. Note further in the acetabulum a small acetabular bone which forms that part of the acetabulum which would otherwise be occupied by the pubis; its phylogenetic significance is uncertain. The three bones of the pelvic girdle are separate in the young of all mammals (Fig. 49), which thus recapitulate the condition of adult reptiles.

The hind limb is fairly typical. The femur has a head, a greater trochanter lateral to the head (continuing in the rabbit to a small posterior projection, the third trochanter), and a lesser trochanter, below the head. These trochanters serve for muscle attachments. The large articulating surfaces at the distal end of the femur are condyles (medial and lateral); and they bear additional elevations or roughened areas, the epicondyles. At the knee joint a patella is present. The shank is composed of a stout tibia and slender fibula, the latter in the rabbit fused with the tibia for the greater part of its length. The anterior face of the tibia presents a crest; its proximal articulating surfaces are known as condyles; its distal ones as malleoli. The bones of the ankle are identical with those of the human ankle and are designated by the same names, which are, unfortunately, somewhat fanciful and not based upon comparative anatomy. The name derived from comparative anatomy is given in parenthesis after the name derived from human anatomy. The ankle consists of seven bones (cat) or six (rabbit). The largest and most conspicuous of these, which projects backward as the heel, is the *calcaneus* (*fibulare*). Articulating with the malleoli of the tibia and fibula is the *astragalus* or *talus* (believed to represent the *tibiale* plus the *intermedium*). Directly in front of the astragalus is found the *navicular* or *scaphoid* (representing one or two *centralia*), a curved bone reaching to the medial side of the foot. Directly in front of the calcaneus is the *cuboid* (*fourth* and *fifth tarsalia* fused), articulating with the fourth and fifth metatarsals. Medial to the cuboid is the *third* or *lateral cuneiform* (*third tarsale*), articulating with the third metatarsal. Medial to this is the second or *intermediate cuneiform* (*second tarsale*), articulating with the second metatarsal. In the cat there is a first or *medial cuneiform* (*first tarsale*) along the medial border of the anterior part of the ankle in front of the navicular. It articulates with the small rudimentary first metatarsal, which lies directly in front of it. In the rabbit the first cuneiform is fused to the proximal end of the second metatarsal. The homology of these ankle bones with those given in the primitive vertebrate plan (Fig. 46) is quite evident. The sole consists of four long *metatarsals* and one rudimentary one (the first) on the medial or ventral side of the proximal end of the second metatarsal. The terminal phalanges of the digits are curiously beak shaped, for the support of the horny

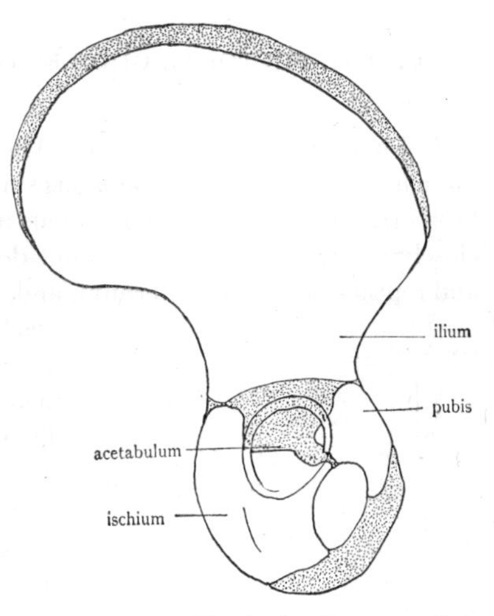

FIG. 49.—Half of the human pelvic girdle at birth, showing the three bones of which it is composed. Stippled regions represent cartilage, which later ossifies, obliterating the boundaries between the bones.

claws, and in the cat have sheaths at their bases into which the bases of the horny claws fit.

The joint between foot and shank in mammals is between the ankle bones and the malleoli of the tibia and fibula, unlike the condition seen in birds and reptiles, where the joint lies between the distal and proximal ankle bones. The gait of the cat and rabbit is chiefly digitigrade, although the hind legs assume the plantigrade posture when the animal sits down.

Review pages 54–55 on the torsion of the mammalian hind limbs.

C. THE PECTORAL GIRDLE, THE STERNUM, AND THE ANTERIOR PAIRED APPENDAGES

1. Phylogeny of the sternum.—The sternum is limited to land vertebrates. It tends to remain in a more or less cartilaginous state among amphibians and reptiles. It appears to have been originally a simple cartilaginous plate level with the rear part of the pectoral girdle and closely associated with it. It is not definitely known whether primitive extinct amphibians and reptiles possessed a sternum, and, if so, whether ribs were articulated to it. Ribs do not reach the sternum in any present amphibians, but this condition may have been the result of degeneration. No agreement has been reached concerning the origin of the sternum. There are three main theories, deriving the sternum from the ribs, the pectoral girdle, and the true abdominal ribs, respectively. The rib theory, based on embryology, is that the ventral ends of the ribs fuse to form on each side a rod-shaped cartilage and that these two rods then unite to become the sternum. This meets the difficulty that in amphibians the ribs do not reach the sternum and have no relation to the sternum during development. Some have met this objection by declaring the amphibian to be of different origin from the amniote sternum, the former derived from the pectoral girdle (archisternum), the latter from the rib ends (neosternum). The pectoral-girdle theory postulates that the sternum is the separated median region of the primitive pectoral girdle. This theory relates the sternum to conditions in fishes but has little supporting evidence. The third theory derives the sternum from cartilages forming in the ventral myosepta. Such cartilages are the same as the true abdominal ribs (also called *parasternals*), previously mentioned. Fusion of the most anterior parasternals would result in a sternum, independent at the start of ribs or pectoral girdle. This theory is regarded at present as the most probable. It is further supported by the fact that in certain lizards the series of parasternals is directly continuous with the sternum (Fig. 50).

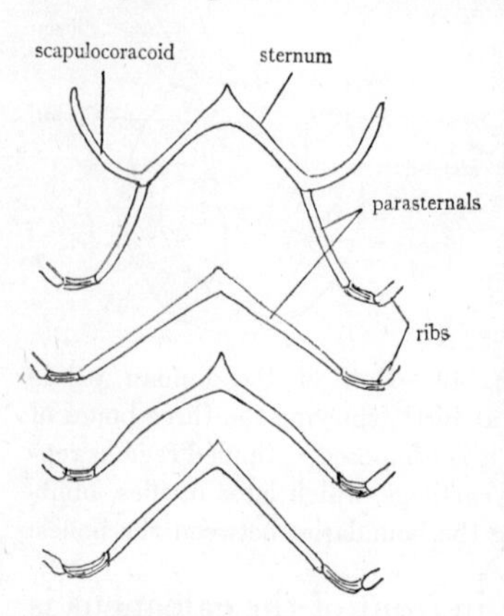

Fig. 50.—Transition of parasternals into sternum in a lizard. (After Camp, 1923.)

2. Evolution of the pectoral girdle.—The pectoral girdle originates like the pelvic girdle, i.e., by the enlargement and midventral fusion of a pair of basal pterygiophores (Fig. 44), and is seen in a simple condition in elasmobranchs, where it consists of a cartilaginous bar corresponding to the pelvic bar. In bony fishes, however, a set of dermal bones is added to this original endoskeletal girdle. These bones are really enlarged skin scales which have sunk into the interior and applied themselves to the ventroanterior surface of the pectoral girdle, adding strength. A number of such dermal bones may be present in fishes, as in *Polypterus* (Fig. 51); the principal ones from the ventral side dorsally are on each side: *clavicle, cleithrum, supracleithrum,* and *posttemporal;* the posttemporal in nearly all Osteichthyes attaches the pectoral girdle to the skull. This dermal component of the pectoral girdle persists in all tetrapods, and its history is easily followed. In the earliest embolomerous amphibians, as in *Eogyrinus* (Fig. 52*A*), the dermal pectoral girdle is practically identical with that of fishes, including even a probable skull attachment by way of the posttemporal; but from the start in tetrapods there was an added element, an unpaired midventral *interclavicle.* In the further evolution of tetrapods there has been a gradual *reduction and loss of the dermal elements of the pectoral girdle* (Fig.

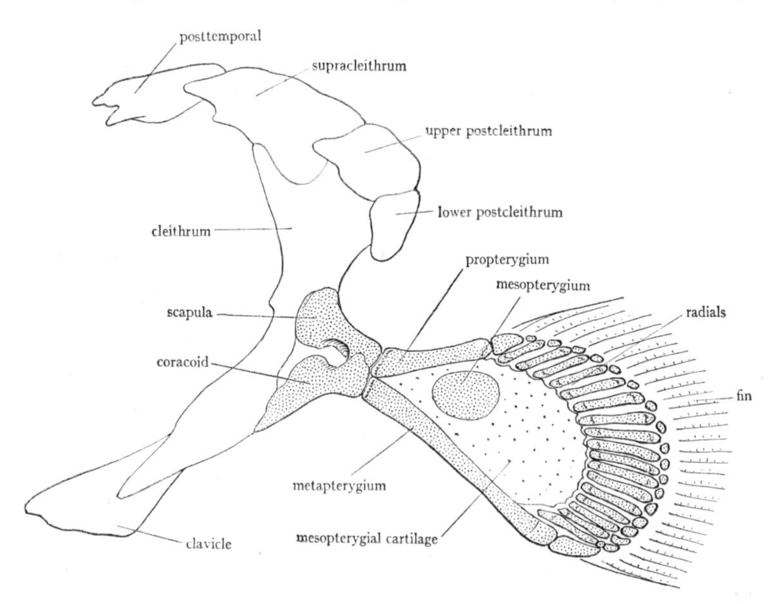

FIG. 51.—Pectoral girdle of *Polypterus*, to show the large number of membrane bones occurring in the pectoral girdle of teleostome fishes. Viewed from the inside (dorsal view). Membrane bones, blank; cartilage bones, close stippling; cartilage, open stippling. Note also the arrangement of the basals and radials in the "stalk" of the fin. (After Goodrich in Part IX of Lankester's *Treatise on Zoölogy*, courtesy of the Macmillan Co.)

52). Interclavicle and paired clavicles and cleithra are seen among many extinct amphibians and reptiles, but the cleithrum is soon lost and does not occur in any living amniote. Interclavicle and clavicles persist among present reptiles and birds; but the interclavicle is lost in mammals, being present only in monotremes (Fig. 52). Clavicles are also frequently absent in mammals.

The history of the true endoskeletal part of the pectoral girdle is less clear, and homologies are still under dispute. The original single pectoral bar or plate became paired, forming on each side a *scapulocoracoid* cartilage, having near its middle a depression, the *glenoid fossa,* for articulation with the humerus. The part above the glenoid fossa is the *scapular* region; that below, the *coracoprecoracoid* region. Among early fossil amphibians the endoskeletal girdle ossifies on each side into a single scapulocoracoid bone; presumably considerable areas of unossified cartilage remained, but these are not preserved as fossils. In early reptiles, also, the endoskeletal girdle consists of a scapulocoracoid on each side; but this becomes divided by a

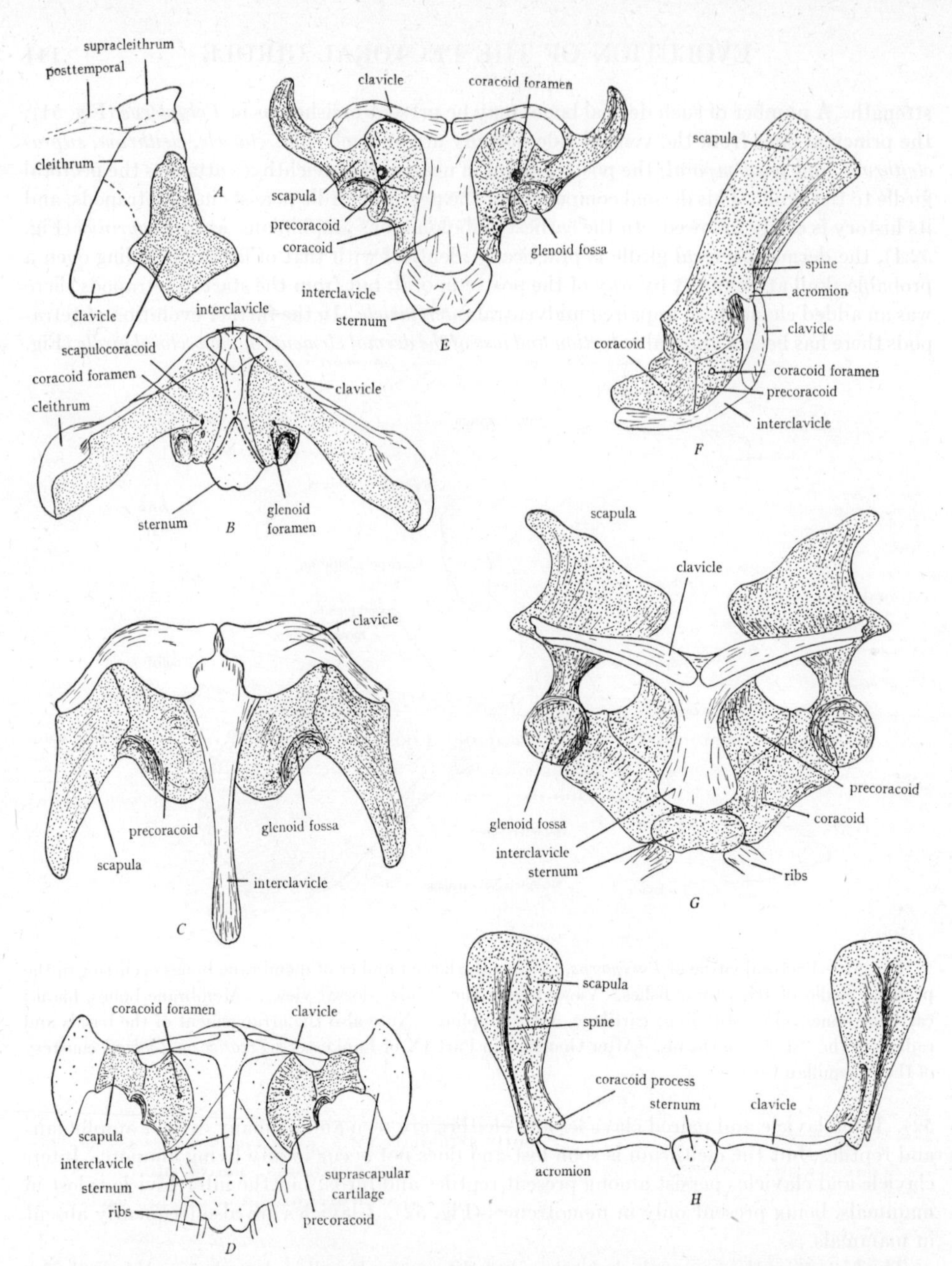

Fig. 52.—Series of tetrapod pectoral girdles. *A*, half of the pectoral girdle of the embolomerous amphibian *Eogyrinus*, seen from the side, showing fishlike series of membrane bones (after Watson, 1926). *B*, pectoral girdle of the rachitomous amphibian *Eryops*, cleithrum still present, ossified scapulocoracoid in one piece (after Miner, 1925). *C*, pectoral girdle of *Seymouria*, cleithrum lost, suture separating scapula and precoracoid (after Williston, 1925). *D*, pectoral girdle of *Sphenodon*, retains same elements as *Seymouria*, clavicles reduced. *E*, pectoral girdle of extinct mammal-like reptile, showing two bones in the coracoid region (after Broom, 1932); note that coracoid foramen is through the anterior bone, indicating this to be the single coracoid element of present reptiles. *F*, same type of girdle as *E*, viewed from the side in place (after Pearson, 1924); note incipient mammalian features, spine and acromion on scapula. *G*, pectoral girdle of the monotreme *Echidna* (after Broom, 1932); note identity with reptilian girdle of Figures *E* and *F*, except that precoracoid is completely excluded from the glenoid fossa. *H*, pectoral girdle of a placental mammal (muskrat); precoracoid completely lost, coracoid reduced to process.

suture into two bones—a *scapula* or shoulder blade above the glenoid fossa, and a *precoracoid* below it, as in *Seymouria* (Fig. 53*A*); both bones contribute to the glenoid fossa. This primitive girdle is perforated by three foramina—the *supraglenoid* through the scapula above the glenoid cavity, and the *glenoid* and *coracoid* foramina through the precoracoid (Fig. 53*A*). The coracoid foramen transmits the supracoracoid nerve (hence also called supracoracoid foramen), but the function of the other two openings is uncertain. The endoskeletal girdle remains in the *Seymouria* condition, consisting of a scapula and a precoracoid (usually called coracoid) on each side in present reptiles and in birds, and the precoracoid appears to be homologous throughout these forms. But in some primitive reptiles, and especially in the extinct reptilian groups ancestral to mammals, there occurred two ossifications in the coracoprecoracoid cartilage instead of one, resulting in two bones below, an anterior precoracoid and a posterior *coracoid* (also called *metacoracoid*); the scapula, of course, was also present. Among the mammal-like reptiles the anterior bone, the precoracoid, gradually became reduced in importance and excluded from the glenoid fossa, whereas the posterior bone or true coracoid came to surpass it in size (Fig. 52*F*). A continuance of this process has resulted in the loss of the precoracoid in all mammals except monotremes (where it is usually called *epicoracoid*). The coracoid is fully developed in monotremes (Fig. 52*G*) but in other mammals is reduced to a projection of the scapula, the *coracoid process* (Fig. 52*H*).

According, then, to the above account (which follows the views of Broom, Watson, and Romer), there was in early tetrapods one ossification in the coracoprecoracoid cartilage, and this formed the precoracoid bone, which is the single coracoid element of present reptiles and of birds; but in certain extinct reptiles ancestral to mammals two ossifications occurred in the cartilage, resulting in an anterior precoracoid and a posterior coracoid. Both bones are seen in the reptiles in question and in monotremes, but in other mammals only the posterior one survives. Thus the so-called coracoid of reptiles and birds is not homologous to the bone called coracoid in monotremes; the former is really the precoracoid and hence will be so termed in what follows. It must be mentioned that this view is not accepted by all vertebrate anatomists. On the alternative view, the original single coracoid ossification is equivalent to the two later ones; i.e., the reptilian and avian coracoid element represents both precoracoid and coracoid. This conclusion would contradict the rule that separate centers of ossification represent originally distinct bones.

It is thus seen that throughout the tetrapods the pectoral girdle is compounded of dermal and cartilage bones. It was previously mentioned that dermal bones, really part of the exoskeleton, may participate in the endoskeleton; and we here meet them for the first time, associated with the pectoral girdle. The dermal bones never take part in the articulation with the forelimb, whereas the cartilage bones (scapula, precoracoid, and coracoid) regularly do so. The coracoid elements of the girdle frequently articulate with the sternum, whereas the scapula rarely has any connection with the vertebral column (exceptions: skates, pterosaurs). In this respect the pectoral girdle contrasts with the pelvic girdle, which, as already seen, is always in land vertebrates firmly articulated with the sacrum.

Some of the original cartilage of the endoskeletal part of the pectoral girdle is likely to persist unossified. Various names have been given to such cartilages of the girdle, but no particular importance can at present be attached to such names.

3. Pectoral girdle and pectoral fin of elasmobranchs.—In these fish the pectoral girdle is purely endoskeletal and in the same primitive condition as the pelvic girdle, consisting simply of a curved cartilaginous bar nearly encircling the anterior part of the trunk. The median ventral portion between the bases of the two fins is called the *coracoid bar;* the long processes extending dorsally beyond the articulations of the fins are the *scapular processes;* the ends of the scapular processes may consist of separate pieces, the *suprascapular cartilages.* The pectoral fins are similar to the pelvic. They are supported at their bases by several series of cartilaginous fin rays and externally by dermal fin rays. The cartilaginous fin rays consist of a proximal row of enlarged *basals* and several distal rows of smaller *radials.* There are generally three basals: an inner one, the longest, the *metapterygium;* a middle one, the *mesopterygium;* and an outer one, the *propterygium.* There is no trace of a sternum in *Squalus,* but in some other sharks the midventral part of the coracoid bar is a separate piece; this piece has been considered the beginning of a sternum by some, who cite its occurrence as evidence for the pectoral girdle theory of the origin of the sternum (p. 140). Draw the pectoral girdle and one fin.

4. Pectoral girdle of bony fishes.—In bony fishes the dermal components are added to the endoskeletal girdle and, in fact, constitute the major and most conspicuous part of the girdle. On a skeleton of a teleost note a series of dermal bones extending from the ventral to the dorsal side in front of the pectoral fins. The uppermost bone of the series, the posttemporal, usually a forked bone, is attached by one fork to the rear end of the skull. The endoskeletal part of the girdle, more or less concealed by the dermal bones, consists on each side of a scapulocoracoid, usually more or less divisible into a lower coracoid and an upper scapular region (Fig. 51). These may be cartilaginous or in many bony fishes are partially ossified into two bones, coracoid and scapula. The homology of these to the bones of the same names in tetrapods is, however, uncertain. In those teleosts in which the pelvic fins have moved far forward, note their attachment by a bony rod to the pectoral girdle.

5. Pectoral girdle and forelimb of amphibians.—The importance of the extinct embolomerous and rachitomous amphibians in the story of the evolution of the pectoral girdle was noted above (p. 141). They originally had a series of membrane bones similar to that of bony fishes plus a midventral interclavicle (lacking in all fishes), and the posttemporal was attached to the skull in at least the earliest forms (Fig. 52*A*). But some of the membrane bones were very soon lost, so that the dermal girdle comes to be composed among primitive tetrapods of interclavicle, clavicles, and cleithra (Fig. 52*B*). The endoskeletal girdle of these early amphibians consists of a single scapulocoracoid on each side (Fig. 52*B*).

In urodele amphibians, as an adaptation to aquatic life, where less support is needed, the dermal pectoral girdle has been wholly lost and the endoskeletal part has remained in or has reverted to a simple, largely cartilaginous condition. Study *Necturus* or *Cryptobranchus*. The girdle consists of two halves, the scapulocoracoids. The ventral coracoid region forms a flat plate of cartilage, often overlapping its fellow. Dorsally above the glenoid fossa is seen the scapula, the only ossified part of the girdle; its dorsal border bears the *suprascapular cartilage*. In *Cryptobranchus* and a number of other urodeles the sternum is present as a cartilaginous piece between the rear parts of the coracoid cartilages. In *Necturus* the sternum is represented by two or three transverse cartilages in the ventral myosepta.

Note, again, primitive posture of the limbs (p. 46). Identify the bones of the forelimb; the small wrist elements are impractical to distinguish (there are six or seven cartilages in three rows).

In the anuran amphibians the girdle is better ossified, and a more definite sternum is often present. Primitive types of frogs present conditions similar to urodeles in that the cartilaginous portions of the coracoid region overlap and the sternum is limited to a piece between the rear parts of these. In common frogs (*Rana*) note the firm construction and the union of girdle and sternum. The girdle consists ventrally on each side of an anterior clavicle and a posterior coracoid; the clavicle covers the precoracoid, which remains cartilaginous.[3] The sternum consists of two parts: a *prezonal* part in front of the girdle, termed *episternum* and composed of an ossified and a cartilaginous section; and a *postzonal* part or *xiphisternum* behind the girdle, also partly ossified and terminating in a rounded cartilage. The dorsal part of the girdle is composed of the scapula, participating in the glenoid fossa, and the suprascapular cartilage above the scapula; this cartilage is partially covered by a thin, flat bone, now believed to represent the cleithrum.

Note lack of attachment of the pectoral girdle of amphibians to the vertebral column. There are always four fingers in present amphibians; it is now accepted that the fifth is lacking. A prepollex is present in Anura.

6. Pectoral girdle, sternum, and forelimb of reptiles.—Primitively, the girdle resembles that of amphibians, having a dermal part, composed of interclavicle, clavicles, and often traces of cleithra, and an endoskeletal part, consisting of scapula and precoracoid in the lines of reptiles which led to present reptiles and birds, and of scapula, coracoid, and precoracoid in the lines that led to mammals (Fig. 52). Among present reptiles *Sphenodon* has a representative girdle (Fig. 52*D*); the dermal part is composed of the T-shaped interclavicle and slender clavicles; the endoskeletal girdle is largely cartilaginous; and the sternum between the coracoprecoracoid cartilages is

[3] By some the clavicle is believed to be compounded of a dermal element and of an ossified part of the precoracoid. The dermal element is termed *thoracale*. Supporters of this view think that the name thoracale should replace clavicle wherever the clavicle is a pure dermal bone, as in early tetrapods, present reptiles and birds, and monotremes, and that the name clavicle should be reserved for a bone compounded of the thoracale and the precoracoid, such as the clavicle of anurans and placental mammals.

wholly so. In the coracoprecoracoid cartilages there is a single bone on each side, the precoracoid, having a coracoid foramen; the scapula is also ossified and bears on its upper end a large suprascapular cartilage. A similar pectoral girdle is found among lizards, except that the ossified portions are scalloped to form openings.

The pectoral girdle of the alligator consists of the midventral dagger-shaped interclavicle, a stout ventral bone on each side, the precoracoid, and a stout dorsal bone, the scapula. Clavicles are lacking. The sternum is a plate of cartilage between the ventral ends of the coracoids, underlain by the interclavicle. The sternum is drawn out posteriorly into long, curved cartilages, the *xiphisternal horns,* behind which is the series of gastralia. Note attachment of the sternal ribs to the sternum and xiphisternal horns.

The turtle lacks a sternum—no doubt because of the presence of the plastron; and its pectoral girdle is peculiar in that it has moved to the inside of the ribs. The dermal part of the turtle's girdle is in the plastron, where the entoplastron and epiplastra represent the interclavicle and clavicles, respectively; the other elements of the plastron may be gastralia. The endoskeletal part of the girdle is a tripartite bony structure. There is a ventral precoracoid, an elongated dorsal scapula, reaching the carapace, and an anterior projection from the scapula, the *prescapular process,* believed to represent the acromion process seen in mammals. Identify the bones of the forelimb of the turtle. The carpus of turtles is in a fairly primitive condition, which should be studied if good specimens are available. At the distal end of the ulna are two bones, an outer ulnare and an inner intermedium. The center of the carpus is occupied by a long bone, composed of fused radiale and centrale and probably other centralia, fused with the ulnare and intermedium. A row of five carpalia articulates with the metacarpals. Compare with the carpus of *Sphenodon* (Fig. 53*B*).

7. Pectoral girdle, sternum, and forelimb of birds.—The pectoral girdle of birds is similar to that of reptiles, consisting of the scapula, a long swordlike bone lying above the ribs; the precoracoid, stout bone reaching the sternum; and the *wishbone* or *furcula,* in front of the precoracoids. The wishbone is the dermal part of the girdle and really consists of the clavicles, forming the two forks and united posteriorly to a rounded median piece, the interclavicle. Precoracoid and scapula participate in the glenoid fossa. The sternum is an elongated bone bearing a strong ventral projection, the *keel* or *carina,* serving for the attachment of the powerful flight muscles. As in reptiles, the ribs join the sternum by way of their costal cartilages. The anterior end of the sternum has short *costal* processes, each side two long *xiphisternal* processes. Draw from the side, showing girdle and sternum.

On a demonstration specimen of the sternum of an ostrich or other flightless bird note smooth convex sternum lacking a carina. Such birds are called *ratite* birds, whereas flying birds which have the keel are called *carinate* birds.

The forelimb of birds is adapted for flight by the concentration of its distal elements. The stout humerus fits into the glenoid by a convex head, to either side of which are prominent projections, the greater and lesser tuberosities, for muscle attachment. The lesser tuberosity continues distally as a sharp ridge, the *deltoid ridge*, which marks the preaxial or radial side of the limb and so indicates the dorsal rotation which the humerus has undergone. The greater tuberosity is postaxial and bears on its under surface a large hole, the *pneumatic foramen*, the entrance into the air space in the humerus. The radius is more slender than the ulna, which has at its proximal end a projection,

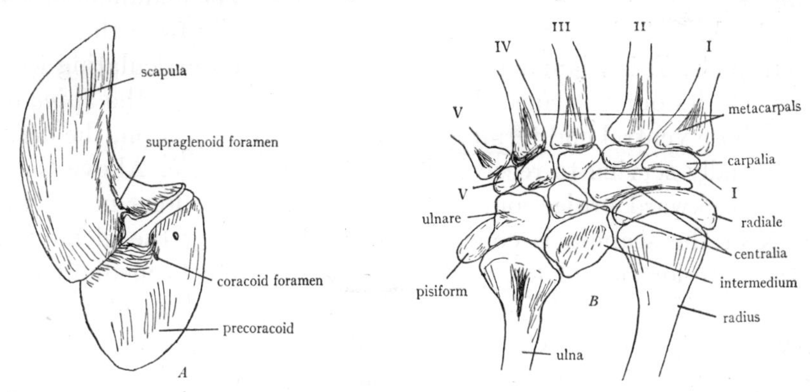

FIG. 53.—*A*, scapula and precoracoid of *Seymouria*, to show the foramina (after White, 1939); the unlabeled opening is a nutrient foramen for the passage of a blood vessel into the bone. *B*, wrist bones of *Sphenodon*.

the *olecranon process* or *elbow*, here met for the first time. The wrist is greatly altered, consisting of but two separate bones, the homologies of which are unclear. The remaining wrist bones are fused to the metacarpals to form the *carpometacarpus*, which includes three metacarpals—two long ones and a short hump on the radial side of these. These three metacarpals are now generally considered to be the first, second, and third;[4] they bear the corresponding digits. A claw is frequently present on the first finger. *Archaeopteryx* had separate metacarpals and claws on all three fingers.

8. Pectoral girdle, sternum, and forelimb of mammals.—As already explained, the reptilian ancestors of mammals had two bones in the coracoid region, precoracoid and coracoid. This reptilian condition of the pectoral girdle is retained in monotremes (Fig. 52*G*), which have a dermal girdle of interclavicle and clavicles and an endoskeletal girdle of precoracoid (called

[4] Formerly thought to be the second, third, and fourth. Holmgren (1933) in an exhaustive embryological study again maintains that they are the second, third, and fourth, so that the question must be regarded as still open.

epicoracoid in older accounts), coracoid, and scapula on each side. In all placental mammals the interclavicle and the precoracoid have vanished and the coracoid is reduced to a projection on the scapula. Clavicles also are frequently reduced or wanting, so that the scapula becomes the most important part of the mammalian pectoral girdle.

Study the pectoral girdle of the rabbit or cat. It consists, on each side, of a clavicle and a scapula. The clavicles are small, slender bones imbedded in muscle; and, as they are not articulated to any other part of the skeleton, they generally fall off in prepared skeletons. They will be seen in place later, when the muscles are dissected. The scapulae or shoulder blades are the large, flat, triangular bones above the anterior ribs. The mammalian scapula has certain characteristics which differentiate it readily from the scapula of other tetrapods. It is triangular in form, with the apex articulating with the humerus. On its outer surface there is a prominent ridge, the *spine* of the scapula, whose ventral end terminates in a pointed projection, the *acromion* process, usually serving to receive one end of the clavicle. Above the acromion is a backward projection, the *metacromion*, very long in the rabbit. The apex of the scapula is concavely curved, forming the glenoid fossa. From the anterior side of the glenoid rim there projects medially a small beaklike *coracoid process*, the vestige of the coracoid bone, which embryologically forms by a separate center of ossification. For facilitating the description of muscle attachments the various surfaces and borders of the scapula are named as follows: the part of the external surface anterior to the spine is the *supraspinous fossa;* the part posterior to the spine, the *infraspinous fossa;* the whole of the internal surface is the *subscapular fossa;* the dorsal border is the *vertebral border;* the anterior margin, the *anterior border;* the posterior margin, the *axillary border*. Draw the scapula from the outer surface.

The sternum consists of a longitudinal series of pieces, the *sternebrae*—eight in the cat, six in the rabbit. The first sternebra is called the *manubrium* and articulates with the first thoracic rib at its center. The next six (cat) or four (rabbit) sternebrae constitute the *body* of the sternum. The last piece is called the *xiphisternum* and terminates in a *xiphoid* or *ensiform* cartilage. Note points of articulation of the sternebrae with the ribs.

The forelimb is fairly typical. The humerus has a large, rounded *head* fitting into the glenoid fossa and *greater* and *lesser tuberosities* at the sides of the head. The anterior surface of the humerus below the tuberosities is slightly elevated into *ridges* or *crests* (two in the cat, one in the rabbit), which serve as points of muscle attachment. The lower end of the humerus is rounded for articulation with the bones of the forearm and is divided into two portions—an outer mass, the *capitulum*, and a medial mass, the *trochlea*. Above the capitulum is a projecting ridge, the *lateral epicondyle;* and a similar *medial epicondyle* is situated above the trochlea. Near the medial epicondyle the bone is pierced by an opening, the *supracondyloid foramen*, absent in the rabbit.

The forearm consists of the smaller radius and larger ulna; the proximal end of the ulna forms a prominent projection, the *olecranon* or *elbow*. Distal to this is a deep semicircular concavity, the *semilunar notch*, which articulates with the trochlea of the humerus. The distal border of the notch forms another projection, the coronoid process. Observe that the radius crosses obliquely in front of the ulna, and review pages 54–55 regarding the cause of this condition.

The wrist is composed of a number of small bones, arranged in two transverse rows. The proximal row consists of four pieces in the rabbit, three in the cat; the distal row, of five in the rabbit, four in the cat. Articulating with the distal end of the radius is the large *scapholunar* in the cat, separated in the rabbit into a medial *navicular* (*radiale*) and a lateral *lunate bone* (*intermedium*). Lateral to the lunate portion or bone and articulating with the ulna is the *triquetral bone* (*ulnare*). The *pisiform* is the element projecting prominently lateral to the triquetral bone in the cat or behind it in the rabbit; the pisiform is a sesamoid bone, i.e., a bone formed in a tendon. The distal row of pieces beginning at the medial side and proceeding laterally are: the *greater multangular* (*first carpale*), the *lesser multangular* (*second carpale*), the *central* (*centrale*, missing in the cat), the *capitate* (*third carpale*), and the *hamate* (*fourth and fifth carpales* fused). These carpales are situated at or near the proximal ends of their respective metacarpals. These are five *metacarpals*, of which the first is very much reduced, and five digits whose terminal phalanges support the horny claws.

D. SOME VARIATIONS OF STERNUM, GIRDLES, AND LIMBS AMONG MAMMALS

1. Monotremes and marsupials.—In monotremes and most marsupials a pair of *marsupial* or *epipubic* bones supporting the marsupial pouch project forward from the pubes to which they are movably articulated. It is probable that these are ossifications of either the epipubic or prepubic cartilaginous processes of the pubis, most likely the latter. Peculiar to monotremes are projections of the fibulae resembling elbows. Among kangaroos and some other marsupials there is commonly a great slenderizing of the second and third metatarsals and digits, and these may become inclosed in a common skin.

2. Bats.—Bats are notable for the great elongation of the clavicles, coracoid processes, and all the sections of the forelimb. The greatly reduced ulna is fused to the radius. The carpus is reduced; but the metacarpals and phalanges are immensely lengthened to support the flying web, except the thumb, which is excluded from the web. It is interesting to compare bats with the flying reptiles known as pterosaurs, where there is the same reduction of the wrist and tremendous elongation of the bones supporting the flying web.

3. Ungulates.—The limbs of ungulates show extreme adaptation for running. Clavicles are absent, the scapula is generally narrow, and the ilia are often broad and expanded. Instructive stages in the modifications of the limbs and elongation and reduction of the distal sections are seen in the elephant, tapir, and rhinoceros. In the most modified ungulates, humerus and femur are strong and short and included in the body musculature, so that the level of the legs, where ordinarily the knee and elbow would be located, is really the wrist and ankle. Ulna and fibula remain distinct in the pig, hippopotamus, tapir, and rhinoceros but in other ungulates are reduced. The lower part of the ulna is diminished or lost, and its upper part fuses to the radius. Wrist and ankle bones are about as usual, but metacarpals and metatarsals become more and more elongated. The first is always lacking; among perissodactyls the third becomes dominant, and the others reduce gradually to mere splints, as in the horse. Among the artiodactyls the third and fourth metacarpals, metatarsals, and digits remain of equal importance as the supporting bones, but the second and fifth gradually reduce. They are present in pigs and hippopotamuses; in ruminants the third and fourth metacarpals and metatarsals are more or less fused, forming the cannon bone, while the second and fifth reduce greatly and may disappear, although their corresponding digits may remain. Note hoof shape of terminal phalanx fitting into the horn part of the hoof.

4. **Aquatic mammals.**—These lack clavicles. Seals do not show much modification from ordinary carnivores except that the limb bones tend to become short and stocky. Great alterations of the limb skeleton are seen in the two orders of strictly aquatic mammals, Sirenia and Cetacea. In these the pelvic girdle is reduced to a pair of bones, and the hind limbs are completely wanting or represented only by bony rudiments. In Cetacea the scapula is typically short and fanlike, and in both groups the bones of the forelimbs tend to be short, broad, and flattened. Among whales, especially the whalebone whales, a large part of the wrist and hand remain cartilaginous, so that the bones are set in cartilage and hence are but little movable. The outer digits tend to reduce, but the central ones are remarkable in that the number of phalanges may increase much beyond the usual numbers—up to twelve or thirteen. The skeleton of the flippers of whales should be compared with that of the limbs of extinct aquatic reptiles, such as mosasaurs, plesiosaurs, and ichthyosaurs; there is seen the same shortening of the limb bones to a flattened rounded form and the same increase in the number of phalanges.

Among the Cetacea and Sirenia the sternum is frequently reduced to a single piece, with which only the most anterior ribs articulate.

E. SUMMARY

1. The paired appendages probably represent remnants of a pair of continuous lateral fin folds, supported by cartilaginous fin rays.

2. The two girdles probably arose through the fusion in the median line of some of these fin rays.

3. The primitive girdles are bars or plates of cartilage in which subsequently ossification occurred with the formation of cartilage bones.

4. In tetrapods three pairs of bones ossified in the pelvic girdle: pubis, ischium, and ilium. At first these are closely attached, but later a large opening arises between pubis and ischium. There are never any dermal bones associated with the pelvic girdle.

5. The pelvic girdle is free in fishes (or may have moved anteriorly and then be attached to the pectoral girdle), but in all other vertebrates it is immovably fused (ankylosed) to the sacral ribs.

6. The pectoral girdle has a more complicated history and consists of a combination of cartilage and dermal bones.

7. The endoskeletal part of the girdle, consisting in early tetrapods of a scapulocoracoid

cartilage or bone on each side, may give rise to either two or three cartilage bones: one dorsal bone, the scapula; and one or two ventral bones, an anterior precoracoid and a posterior coracoid. Only one ventral bone, believed to be the precoracoid, ossifies in early tetrapods and in present reptiles and birds. Two bones, precoracoid and coracoid, are present in extinct mammal-like reptiles and in monotremes. In placental mammals the precoracoid is lost and the coracoid reduces to a projection of the scapula.

8. The dermal part of the pectoral girdle originates in fishes as a series of scalelike bones overlying the endoskeletal part and attached to the skull. Among tetrapods some of these and the skull attachment are lost. The typical dermal girdle of primitive tetrapods consists of median unpaired interclavicle and paired clavicles and cleithra. The cleithrum is soon lost; the interclavicle persists in present reptiles, in birds, and in monotremes but is lost in placental mammals.

9. The mammalian scapula is distinguished by the presence of a spine and of the coracoid process mentioned in paragraph 7.

10. The pectoral girdle is very rarely connected to the vertebral column.

11. The sternum is limited to tetrapods and originally consists of a cartilaginous plate; later this has one to several ossifications in a linear arrangement.

12. The phylogenetic origin of the sternum is disputed; the theory that it arises by the fusion of the anterior members of a series of true (endoskeletal) abdominal ribs (parasternals) is considered most acceptable at present.

13. Ribs articulate with the sternum in most amniotes.

14. The bones of the limbs are probably derived from the cartilaginous fin supports (pterygiophores) of crossopterygian fish. They are very similar in all tetrapods but are subject to many modifications, especially loss and fusion of distal parts.

15. Each limb is divided into three segments: stylopodium, containing one bone, femur or humerus; zeugopodium, containing two parallel bones, tibia and fibula, or radius and ulna; and autopodium, including ankle or wrist, sole or palm, and digits. Ankle or wrist consists primitively of twelve small bones; a proximal row of three bones: tibiale or radiale, intermedium, and fibulare or ulnare; a middle group of four centralia; and a distal row of five tarsalia or carpalia. Sole or palm consists of five metatarsals or metacarpals. There are typically five digits, with some evidence of an original number of seven. Each digit has a skeletal axis of a row of phalanges; the original number of phalanges for the five digits from the preaxial to the postaxial side was 2, 3, 4, 5, 4.

16. Loss or fusion of tarsals or carpals, metatarsals or metacarpals, and digits and phalanges is common. These losses begin from both preaxial and axial sides and proceed toward the middle. Increase in the number of phalanges (hyperphalangy) occurs in aquatic (extinct) reptiles and aquatic mammals.

REFERENCES

Broom, R. 1899. On the development and morphology of the marsupial shoulder girdle. Trans. Roy. Soc. Edinburgh, **39**.

———. 1932. The Mammal-like Reptiles of South Africa and the Origin of Mammals.

Bryant, W. L. 1919. On the structure of *Eusthenopteron*. Bull. Buffalo Soc. Nat. Sci., 13.

Camp, C. L. 1923. Classification of the lizards. Bull. Amer. Mus. Nat. Hist., **48**.

Dean, B. 1894. Contributions to the morphology of *Cladoselache*. Jour. Morph., **9**.

———. 1895. Fishes, Living and Fossil.

Fuchs, H. 1922. Mitteilungen über den Schultergürtel der Amphibia Anura. I. Suprascapulare und Cleithrum, Procoracoid und Thoracale. Zeitschr. f. Morph. u. Anthropol., **22**.

Goodrich, E. S. 1901. Shoulder girdle and fin of *Eusthenopteron*. Quart. Jour. Micr. Sci., **45**.

———. 1906. Development, structure, and origin of the median and paired fins of fish. *Ibid.*, **50**.

Gregory, W. K. 1935. Further observations on the pectoral girdle and fin of *Sauripterus*. Proc. Amer. Philos. Soc., **75**.

Gregory, W. K.; Miner, R. W.; and Noble, G. K. 1923. The carpus of *Eryops* and the structure of the primitive cheiropterygium. Bull. Amer. Mus. Nat. Hist., **48**.

Holmgren, J. 1933. On the origin of the tetrapod limb. Acta Zool., **14**.

———. 1939. Contribution to the question of the origin of the tetrapod limb. *Ibid.*, **20**.

Howell, A. B. 1935. The primitive carpus. Jour. Morph., **57**.

Miner, R. W. 1925. The pectoral limb of *Eryops* and other primitive tetrapods. Bull. Amer. Mus. Nat. Hist., **51**.

Moy-Thomas, J. A. 1936. The Evolution of the Pectoral Fins of Fishes and the Tetrapod Fore Limb.

Nilsson, T. 1939. Cleithrum und Humerus der Stegocephalen und rezenten Amphibien. Acta Univ. Lund, **35**.

Pearson, H. 1924. A dicynodont reptile reconstructed. Proc. Zoöl. Soc. London, Part 2.

Romer, A. S. 1924. Comparison of mammalian and reptilian coracoids. Anat. Rec., **24**.

Romer A. S., and Byrne, F. 1931. The pes of Didectes: notes on the primitive tetrapod limb. Palaeobiology, **4**.

Sewertzoff, A. N. 1926. Die Morphologie der Brustflossen der Fische. Jena Zeitschr. Naturwiss., **62**.

———. 1926. Development of the pelvic fins of *Acipenser*, new data for the theory of the paired fins of fishes. Jour. Morph., **41**.

Steiner, H. 1922. Die ontogenetische und phylogenetische Entwicklung des Vogelflügelskelettes. Acta zool., **3**.

———. 1935. Beiträge zur Gliedmassentheorie: die Entwicklung des Chiropterygium aus dem Ichthyopterygium. Rev. suisse zool., **42**.

Vos, C. M. de. 1938. The inscriptional ribs of *Liopelma* and their bearing upon the problem of abdominal ribs in Vertebrata. Anat. Anz., **87**.

Watson, D. M. 1917. The evolution of the tetrapod shoulder-girdle and fore-limb. Jour. Anat., **52**.

———. 1926. The evolution and origin of the Amphibia. Phil. Trans. Roy. Soc. London, B, **214**.

———. 1937. The acanthodian fishes. *Ibid.*, **228**.

Wettstein, O. von. 1931. Rhynchocephalia. Kükenthal-Krumbach, Handbuch der Zoologie, **7**, Part 1.

IX. THE ENDOSKELETON: THE COMPARATIVE ANATOMY OF THE SKULL AND THE VISCERAL SKELETON

The *skull* or *cranium* is that part of the axial endoskeleton found inside the head where it incloses and protects the brain. The skull is the most complex of all the parts of the endoskeleton, chiefly because it is derived from several different sources, It consists, in fact, of three components: the *chondrocranium*, the original cartilaginous brain case which ossifies as cartilage bone and is then termed *neurocranium;* the *dermatocranium*, a set of dermal bones that become attached to the surface of the chondrocranium; and the *splanchnocranium*, a series of endoskeletal arches that originally acted as gill supports. Because of this complex history and the large number of bones participating in the finally evolved skull, many questions of homology remain unsettled. The circumstance that a bone is called by the same name in different vertebrate groups cannot be taken to establish its homology. Only exact paleontological and embryological investigations can decide questions of the homology of the bones of the skull.

In what follows, an attempt is made to facilitate the understanding of the skull by studying its three components separately.

A. THE CARTILAGE STAGE OF THE SKULL

1. Development of the chondrocranium.—The formation of the chondrocranium has been studied in a large number of vertebrate embryos and found to differ considerably in detail among different groups (see De Beer, 1937); here only a generalized account can be given. In the head mesenchyme there appear two pairs of elongated cartilages: the *parachordals*, alongside the notochord; and the *trabeculae*, anterior to these (Fig. 54*A*).[1] In a number of vertebrates a pair of *polar* cartilages is present between parachordals and trabeculae; in others they seem to be represented by the rear ends of the trabeculae. Cartilaginous capsules also develop around the chief sense organs of the head: *nasal capsules* around the olfactory sacs, *optic capsules* around the eyes, and *otic* or *auditory capsules* around the internal ears. The optic capsules do not participate in the formation of the chondrocranium, since the eyes must remain freely movable; but the two other sense capsules become immovably incorporated into it. The parachordals fuse medially to form the *basal plate*, which incloses the notochord, and the basal plate unites on each side with the otic capsules. The anterior ends of the parachordals usually become connected by a transverse bar, the *acrochordal* cartilage; and in this process an opening, the *basicranial fenestra*, is usually left behind the acrochordal bar (Fig. 54*B*). Polar cartilages (when present) and trabeculae unite with each other and with the anterior ends of the parachordals. At their anterior ends the trabeculae fuse to form the *intertrabecular plate* (also called *ethmoid plate*), and this leaves a large *hypophyseal fenestra* between this plate and the acrochordal bar, serving to transmit the hypophysis and the internal carotid arteries. The intertrabecular plate fuses with the nasal capsules, or more commonly the nasal capsules are formed by lateral extensions of the trabeculae plus various independently arising cartilages. An anterior median prolongation of the intertrabecular plate, termed the *rostrum* (Fig. 54*B*), usually gives rise to the septum between the nasal capsules, the *internasal septum*. Lateral projections of the intertrabecular plate, termed the *trabecular horns* (*cornua trabeculae*), typi-

[1] Trabeculae are absent in cyclostomes; in mammals they are fused, from the start, to a median cartilage.

cally contribute to the nasal floor (Fig. 64). Other components of the nasal capsules are median *paraseptal* cartilages and lateral *paranasal* cartilages (Fig. 55*B*). The side and hind walls of the nasal capsules, separating them from the orbits (spaces lodging the eyes), are typically formed by the *orbitonasal lamina* (Fig. 55*B*, *C*), also called *planum antorbitale*, which fuses to the preoptic pillar of the orbital cartilage (Fig. 55*C*), so as to leave between them an *orbitonasal fissure* or *canal*.

In regard to the trabeculae, there are two types of skull: the more primitive *platybasic* type with widely spaced trabeculae, hence a large hypophyseal fenestra and no interorbital septum; and the *tropibasic* type with closely placed trabeculae, hence a small hypophyseal fenestra and an interorbital septum. The former occurs among the lower fishes and in amphibians; the latter, in higher fishes and in most tetrapods.

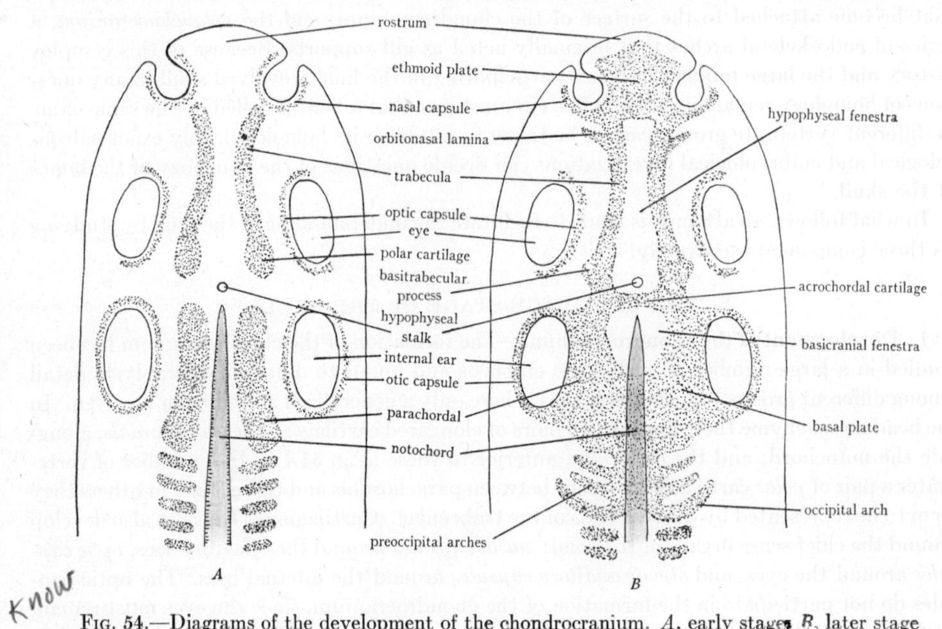

FIG. 54.—Diagrams of the development of the chondrocranium. *A*, early stage; *B*, later stage

The trabeculae are regarded by some as a pair of gill arches anterior to the first typical pair, but this view has not been generally accepted. A variable number of vertebrae incorporate into the rear parts of the parachordals (Fig. 54*B*).

The fusions described above result in an incomplete cartilaginous trough beneath the brain. The side walls in the orbital region arise chiefly from a pair of *orbital* cartilages (Fig. 55) which connect with the trough by three main outgrowths: the *antotic pillars* (or *roots*) in front of the auditory capsule, fusing with the acrochordal region; the *metoptic* or *postoptic pillars*, fusing to the trabeculae behind the optic nerve; and the *preoptic pillars*, fusing to the trabeculae in front of the optic nerve. Of these the first is the most important and constant, but among mammals it is retained only by the monotremes, being one of the many primitive characters seen in this group. The upper edge of the orbital cartilages also sends out a slender process anteriorly to join the nasal capsules (*anterior marginal taenia*, also called *sphenethmoid commissure*), and another posteriorly to join the auditory capsules (*posterior marginal taenia*) (Fig. 55*B*). However, the side walls remain very incomplete and pierced by numerous openings. The fusions of the preoptic pillars with the trabeculae may produce an *interorbital septum* between the orbits continuous anteriorly with the internasal septum; above the interorbital septum the anterior parts of the orbital cartilages (anterior marginal taenia) may fuse to form a flat median expansion termed the *planum supraseptale,* characteristic of reptiles and birds (Fig. 55*B*). There is usually no roof to the chondrocranium except in those vertebrates, such as the elasmo-

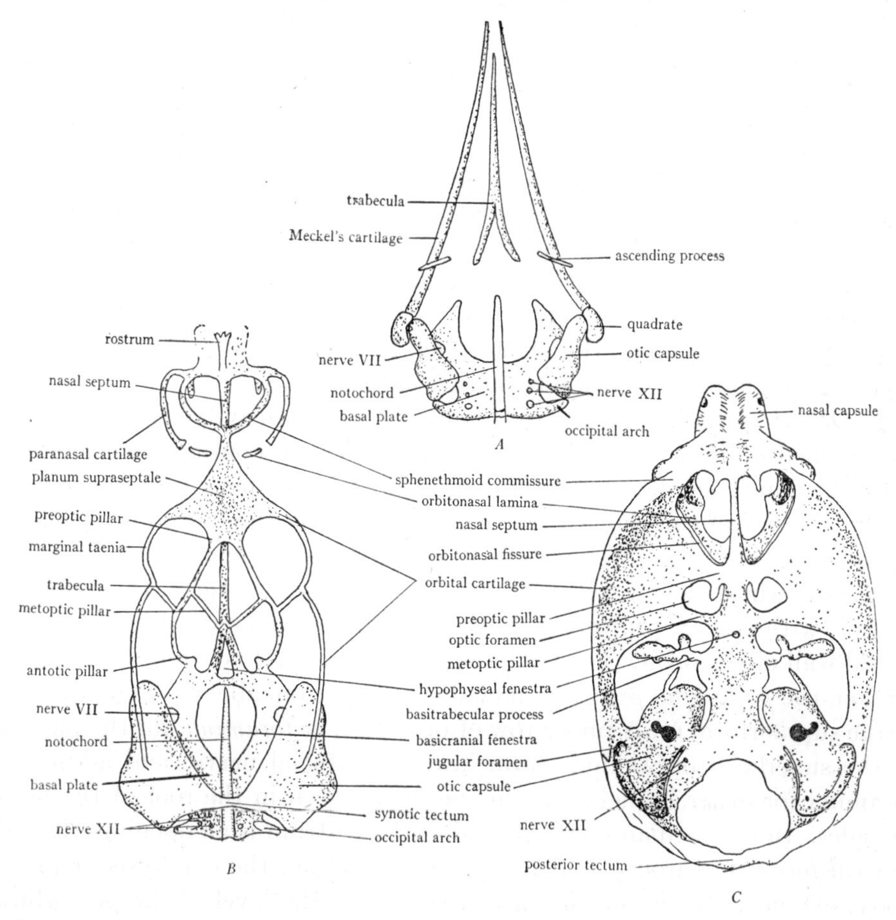

FIG. 55.—Examples of vertebrate chondrocrania. *A*, early stage, and *B*, later stage of the chondrocranium of a lizard (*Lacerta*) (after De Beer, 1930). *C*, chondrocranium of the rabbit (after De Beer and Woodger, 1930).

branchs, where the chondrocranium remains as the adult skull, in which case upgrowths of cartilage complete the chondrocranium above. This is presumably a secondary condition.

As already mentioned, the parts of the parachordals behind the otic capsules are segmental in nature, and in the development of the skull they usually show a series of upward projections on each side which are really the bases of neural arches (Fig. 54*B*). The last one is termed the *occipital arch*, the ones in front of this the *preoccipital arches;* in addition, other vertebrae may be added from behind to the skull, and these are called the *occipitospinal arches*. Some of these neural arches may become completed above by a roof (*tectum*); the most constant are the *synotic tectum*, an arch connecting the otic capsules (Fig. 64), and the *posterior tectum*, an arch behind this (Fig. 55*C*). One or the other of these two may be present, or they may be fused to one; in any case, preoccipital, occipital, and occipitospinal arches, together with the tecta, become the *occipital arch* of the definitive skull, and the large opening so bounded is the *foramen magnum*, through which the brain is continuous with the spinal cord.

The cartilaginous chondrocranium resulting from the fusion of originally separate cartilages, as described above, occurs in the embryo of all vertebrates. It typically has no roof except at the rear end, and the side walls remain imperfect. Above the lower fishes it becomes partly or completely ossified by the usual formation of centers of ossification and is completed above by dermal bones; dermal bones also attach to its ventral surface.

2. Chondrocranium of the dogfish.—Study preserved specimens in fluid. The chondrocranium is a cartilaginous mass without divisions or sutures, broad and flat above, narrower and more irregular below. Its anterior troughlike region is termed the *rostrum*. On each side at the base of the rostrum is a *nasal capsule*. The middle of the sides of the chondrocranium forms a large depression, the *orbit*, which in life holds the eye. Behind the orbits the region on each side is the *otic capsule*.

a) Dorsal surface: The rostrum contains a large cavity, the *precerebral cavity*, filled in life with a gelatinous material; this is continuous posteriorly with the much smaller *precerebral fenestra*, which opens into the *cranial cavity* (interior of the chondrocranium, occupied by the brain in life) but in life is closed by a membrane.[2] The nasal capsules are rounded, thin-walled structures whose dorsal wall is continuous by way of a ridge, the *antorbital process*, with a thick projecting shelf, the *supraorbital crest*, which forms the dorsal wall of the orbit. The posterior end of the crest continues into a projection, the *postorbital process*, triangular as seen from above. Along the base of the supraorbital crest runs a row of openings, or *foramina*, for the passage of the superficial ophthalmic branches of the facial nerve; and additional foramina for other branches of this nerve are seen in the roof of the nasal capsules. In the median line just behind the rostrum is an opening, the *epiphyseal foramen*, through which a brain projection, the epiphysis or pineal body, extends. In the median line just behind the level of the postorbital processes is a rounded depression, the *endolymphatic fossa*, in which are situated two pairs of openings, foramina for the passage of the *endolymph* (smaller holes) and *perilymph ducts* of the internal ear. These ducts connect the fluid-filled channels of the ear with the skull surface. The massive region to each side of the endolymphatic fossa is the *auditory* or *otic capsule* incorporated into the chondrocranium. The terminal opening shortly behind the

[2] The precerebral cavity and fenestra were formerly called the anterior or prefrontal fontanelle. A fontanelle is an opening in the roof of the skull. It has been shown that the precerebral fenestra is not in the roof but is at the anterior end of the embryonic chondrocranium and that the precerebral cavity is merely a depression.

endolymphatic fossa is the *foramen magnum,* through which the brain continues into the spinal cord. **Draw the chondrocranium from the dorsal side.**

b) Ventral and lateral surfaces: The rostrum bears a midventral keel, the *rostral carina;* to each side of the rear end of this keel is an oval fenestra, through which one may look into the cavity of the chondrocranium, or cranial cavity. The nasal capsules are situated lateral to these fenestrae; because of their thin walls they are liable to breakage; but, when complete, they are nearly spherical, with a ventrally directed opening (*external naris*) divided in two by a cartilaginous bar. Behind each nasal capsule the antorbital process already mentioned forms the anterior wall of the orbit. The walls of the orbit are pierced by nerve foramina, of which the largest, ventrally located, is the optic foramen for the passage of the optic nerve.[3] The narrowed ventral wall between the two orbits furnishes articulation for the orbital processes of the upper jaw; these articulations are called the *basitrabecular processes.* Behind this narrow region the ventral surface of the chondrocranium expands to a broad flat region, the *basal plate,* continuous on each side with the otic capsules. The median streak in the basal plate is the notochord; its anterior end turns dorsally into the cartilage at about the level of a small median foramen through which the united internal carotid arteries pass. The posterior end of the basal plate projects slightly on either side of the notochord as an *occipital condyle,* which articulates with the first vertebra. Draw a ventral view of the chondrocranium.

B. THE SPLANCHNOCRANIUM

The *splanchnocranium* or *visceral skeleton* is that part of the endoskeleton which supports the gills and furnishes attachment for the respiratory muscles. It is called visceral skeleton because the gills, as will be seen later, are part of the wall of the digestive tract, and the mesoderm in which the visceral skeleton develops is the splanchnic mesoderm, not the sclerotomal mesoderm. The visceral skeleton consists of a longitudinal series of crescentic elements situated in the pharyngeal wall between the gill slits. Each such element is termed a *gill arch* or skeletal visceral arch; it is usually subdivided into a number of separate pieces in fishes. There are typically seven gill arches in vertebrates (Fig. 56), although some elasmobranchs have nine, and the primitive number was probably greater. Although the gill arches arise in complete independence of the chondrocranium, they early become associated with it, and in the course of evolution the most anterior ones become incorporated into the skull.

1. The visceral skeleton of the dogfish.—Obtain a specimen in which the visceral skeleton has been left attached to the chondrocranium, and study it carefully. The seven gill arches form a series of curved cartilages ventral to the posterior part of the chondrocranium and extending posteriorly to the pectoral girdle. The first gill arch, the *mandibular arch,* is the largest and

[3] Projecting into the orbit may be present a mushroom-shaped structure, the *optic pedicel,* which in life supports the eyeball.

most modified of the series. It is seen, when viewed from below, to consist of dorsal and ventral halves. Each side of the dorsal half is called the *palatoquadrate* or *pterygoquadrate* cartilage; in profile view it will be seen that this cartilage is closely applied to the ventral surface of the chondrocranium, to which in life it is attached by a ligament. It sends up a well-developed orbital process into the orbit. The pterygoquadrate cartilages bear teeth and, in fact, *constitute the upper jaw of the animal.* The ventral half of the mandibular arch consists of two halves, each of which is known as *Meckel's cartilage.* These bear teeth and together *constitute the lower jaw of the dogfish.* The wide gap between the two jaws is the mouth opening. At their posterior ends the pterygoquadrate and Meckel's cartilages join at an acute angle called the *angle* of the jaw, forming a hinge joint, permitting opening and closing of the mouth.

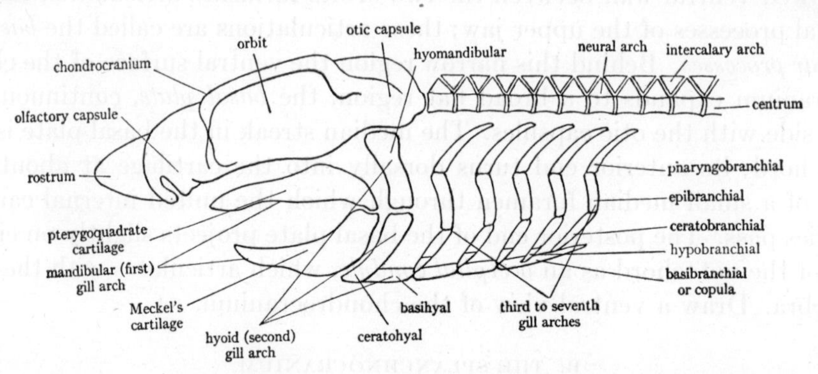

FIG. 56.—Diagram of the chondrocranium, vertebral column, and gill arches of an elasmobranch to show particularly the parts and relations of the seven gill arches. (Slightly modified from Vialleton's *Eléments de morphologie des Vertébrés.*)

The second or *hyoid* arch is more slender than the mandibular arch, to the posterior face of which it is closely applied. It consists of a ventral median piece, the *basihyal;* a slender bar, the *ceratohyal,* on each side of the basihyal; and a stout piece, the *hyomandibular*, dorsal to each ceratohyal. The hyomandibular articulates to the otic region of the skull and thus acts as a *suspensor* of the lower jaw. The hyomandibular bears on its posterior margin some slender projections, the *gill rays,* which in life support the gills.

The remaining arches, known simply as *gill* or *branchial* arches, are similar to each other. Each consists typically of five pieces, named from the dorsal side ventrally: *pharyngobranchial,* the most dorsal piece, elongated and directed posteriorly; *epibranchial,* the succeeding much shorter piece; *ceratobranchial,* another elongated piece; *hypobranchial,* curved ventral pieces, of which there are but three pairs to the five branchial arches; and the *basibranchials,* two in number—an anterior small one situated between the medial ends of the first and second pairs of hypobranchials and a large posterior piece between the bases of the fifth ceratobranchials and terminating in a caudally directed point. Epi- and ceratobranchials bear gill rays. Note that the gill arches are not attached to the vertebral column.

Draw from the side, showing chondrocranium and visceral skeleton.

2. Palatoquadrate attachments and jaw suspension.—The association of the visceral skeleton with the chondrocranium results from the tendency of the jaws to attach to the skull for support and strength. However, this whole subject of jaw attachments is very complicated and confused. According to Goodrich (1930) and De Beer (1937), there may be up to four attachments of the palatoquadrates to the skull (Fig. 57):

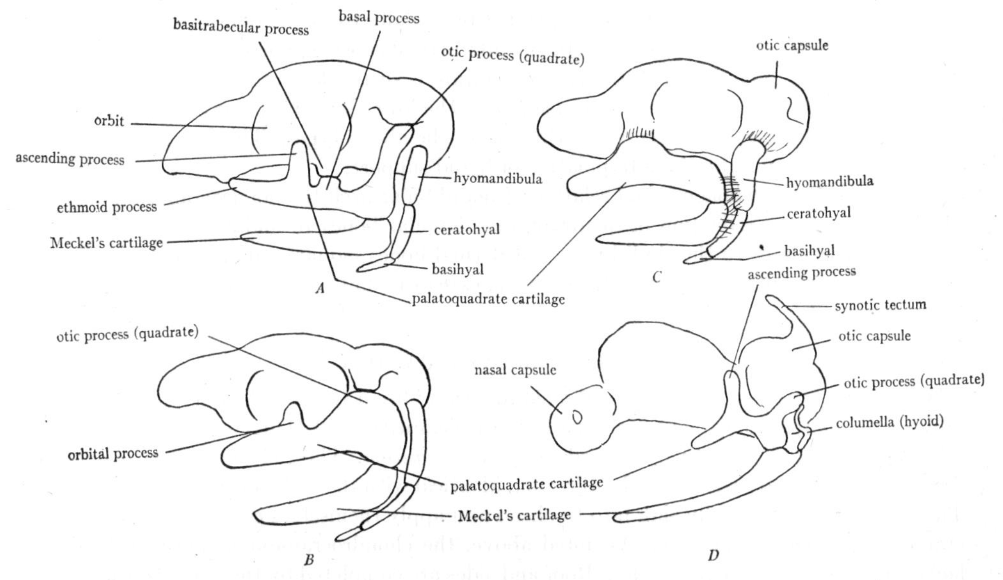

FIG. 57.—Diagram of types of jaw suspension. *A*, diagram to show primitive articulating processes of the palatoquadrate. *B*, amphistylic type of jaw suspension. *C*, hyostylic type. *D*, autostylic type, based on the embryo of *Sphenodon.*

a) Basal process: Identical with the orbital process of elasmobranchs, articulated to projections of the chondrocranial base termed *basitrabecular processes.* This *palatobasal* articulation appears to be the most important of the palatoquadrate attachments and is found in elasmobranchs, Dipnoi, primitive teleostomes, and tetrapods.

b) Otic process: Upward projection of the rear end of the palatoquadrate for attachment to the otic region; this becomes the quadrate bone in tetrapods and is also present in most fishes.

c) Pterygoid or ethmoid process: Projection of the anterior end of the palatoquadrate making contact with or fused with the ethmoid region of the skull; found in fishes and amphibians.

d) Ascending process: In front of the otic process and lateral to the basal process; found in Dipnoi and tetrapods, where it becomes the epipterygoid bone.

The question of the types of jaw suspension has also been in a very confused state but has been clarified by Gregory (1904) and De Beer (1937); the following omits some of their categories (Fig. 57):

a) *Amphistylic:* Orbital (= basal) and otic processes of the palatoquadrates attached by ligaments to the chondrocranium; hyoid arch attached to chondrocranium and lower jaw by ligaments, more or less contributing to the suspension of both jaws; primitive elasmobranchs (Fig. 57*B*).

b) *Hyostylic:* Both jaws suspended from the chondrocranium by way of ligamentous attachments to the hyomandibular, which is attached to the otic region; attachment by basal or other processes of the palatoquadrate may also be present; most elasmobranchs (Fig. 57*C*).

c) *Autostylic:* Processes of the palatoquadrate articulated to or fused with the chondrocranium; hyoid arch does not participate in jaw suspension; most vertebrates (Fig. 57*D*).

d) *Holostylic:* Variety of autostylic found only in Holocephali, in which the palatoquadrates are indistinguishably fused to the chondrocranium and support the lower jaw in the quadrate region.

e) *Methyostylic:* Palatoquadrates suspended principally from the otic capsule by way of hyoid derivatives; ethmoid or other attachments may also occur; teleostomes.

The autostylic type, in which the palatoquadrate is suspended from the chondrocranium solely by means of its own processes and supports the lower jaw at its rear end (quadrate region), appears to have been the most primitive of the above types. This has been modified in most elasmobranchs by the interpolation of hyoid support.

In mammals the whole original relation of mandibular arch to chondrocranium has been altered through the reduction of quadrate and articular bones to middle-ear ossicles. The lower jaw articulates to the skull by way of dermal bones (squamosal and dentary); hence mammals do not fall under any of the foregoing categories but are undoubtedly derived from the autostylic type.

C. THE DERMATOCRANIUM

1. Origin of the dermal bones of the skull.—In addition to the chondrocranium and gill arches, still another component participates in the definitive skull. This consists of the dermal or membrane bones, collectively termed the *dermatocranium.* These elements appear in bony fishes as the enlarged scales of the head region, identical with the scales of the rest of the body. These large scales gradually sink into the head and apply themselves closely to the chondrocranium and mandibular arch. As noted above, the chondrocranium in most vertebrates lacks a roof and has imperfect sides. Roof and sides are completed by these fish scales, which thereupon become the *superficial bones of the skull.* Similarly, scales incase the palatoquadrate and Meckelian cartilages, which are the primitive upper and lower jaws, and become the superficial bones of the jaws. In this fashion the chondrocranium and jaws become sheathed in dermal scales, originating in the skin and hence really belonging to the exoskeleton. These incasing bones are known as dermal, membrane, or investing bones; they develop directly in the mesenchyme. Early stages of the dermal bones of the skull are best seen in primitive fish, such as extinct crossopterygians, but a good idea of them may be gained from examining ganoids.

2. Dermal bones of some ganoid fish.—Study the head scales of any one of the following: *Polypterus*, sturgeon (*Acipenser*), gar pike (*Lepidosteus*), or bowfin (*Amia*), and compare with Figure 58. Note general resemblance of arrangement of these scales with the superficial bones of the skull of land vertebrates by comparison with Figure 59. The larger head scales have received the same names as the main bones of roof of tetrapod skulls and are probably really homologous to them; but many of the fish scales, especially those of the operculum, have been lost in the transition to land life. Understand that in these fish the head scales are in the skin and that a typical cartilaginous or partially ossified chondrocranium is present inside the covering of scales.

a) *Polypterus* (Fig. 58*B*): Identify the two small nasal openings near the anterior tip and the larger oval orbits posterior to them. In front of the nasal openings is a pair of bones bearing teeth, the *premaxillae;* behind the nasal openings is a pair of *nasals;* between the two nasals is situated a small triangular membrane bone, the *dermal mesethmoid.* Posterior to the nasals are two large *frontal* bones, and posterior to them, two smaller *parietal* bones. The several small bones posterior to the parietals are called *temporals.* A row of small bones extends from each orbit posteriorly; these are *postorbitals.* Below the orbit is an elongated bone bearing teeth, the *maxilla.* The sides of the head behind the orbit are covered by large flat bones, the *operculars.* The lower jaw is similarly clothed in dermal bones, consisting of a tooth-bearing *dentary* in front and a toothless *angular* behind.

b) *Sturgeon* (Fig. 58*C*): The anterior end of the sturgeon's head is extended into a snout or rostrum, covered by many small *rostral* scales. At the sides of the base of the snout are the two pairs of nasal openings, and just posterior to them, the orbit. On the dorsal side between the two orbits are two large scales, the *frontals,* and posterior to them, two *parietals.* Numerous other small bones need not be considered. The jaws and visceral skeleton of the sturgeon are degenerate because of its method of feeding.

c) *Gar pike* (Fig. 58*D*): The head is prolonged into a long snout having the nasal openings at its extremity. The small *nasal* bones surround the nasal openings. Posterior to these, occupying the center of the dorsal surface of the snout, are the elongated *ethmonasal* bones. Posterior to them are the *frontals,* whose anterior ends inclose the posterior ends of the ethmonasals. Posterior to the frontals are the large *parietals,* and behind them a number of *temporals.* The edges of the upper jaw are composed of the *maxillae,* bearing teeth and each consisting of a series of squarish bones. The lower jaws consist chiefly of the long *dentary* bones, bearing teeth. The sides of the head behind the orbit are covered by a large number of small *cheek plates* and, posterior to them, by the larger *operculars.*

d) *Bowfin* (Fig. 58*A*): Identify on the skull anterior stalked and posterior nostrils and the orbits. Between the anterior nostrils projects the small triangular dermal mesethmoid; anterior to this are the teeth-bearing premaxillae. Covering the space between the four nostrils are the nasals, posterior to them in the median line the large frontals, behind them the parietals, and at the posterior end of the skull the four temporals. Below and anterior to the orbit is the large lacrimal, below the orbits the two suborbitals,

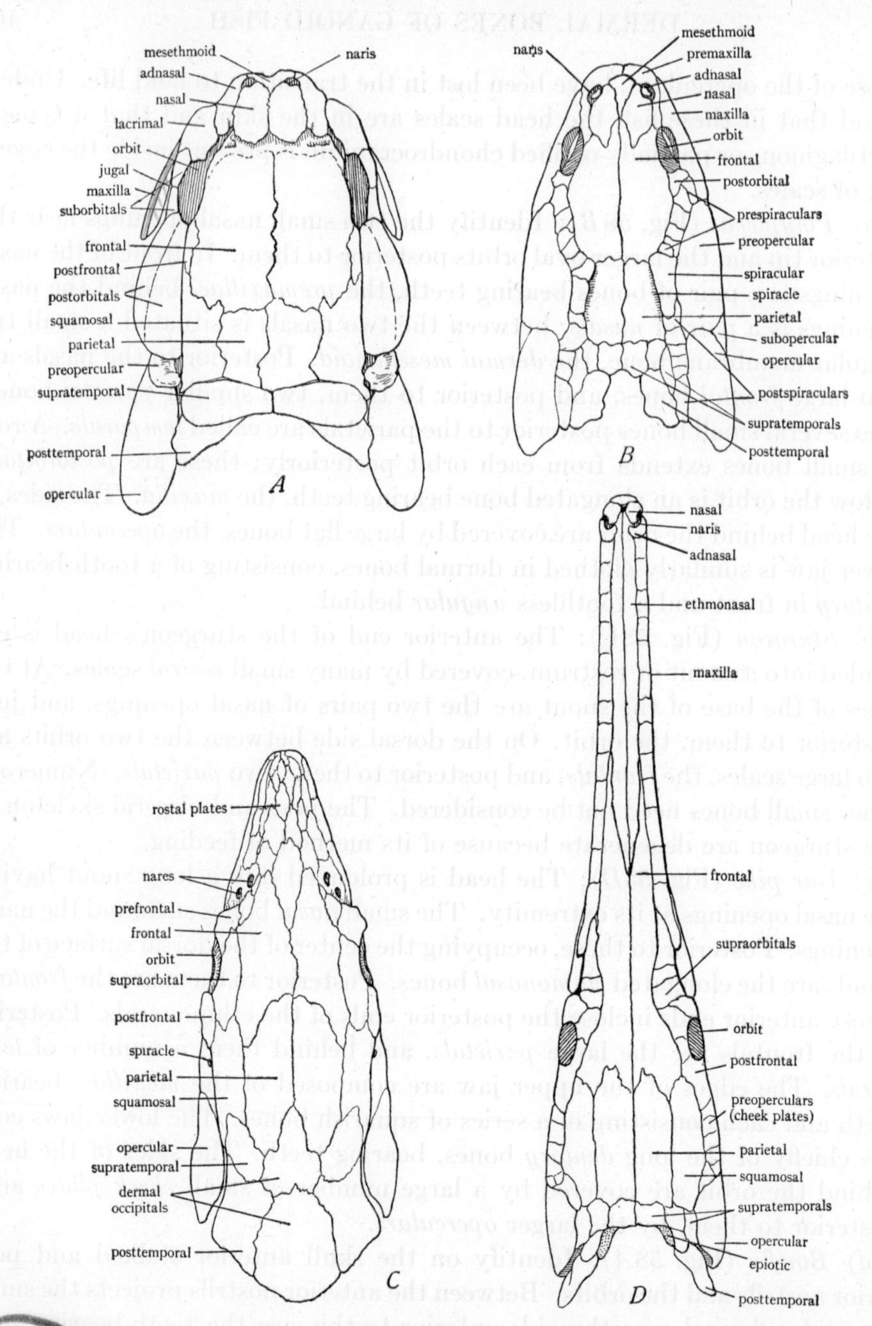

Fig. 58.—Dermal scales (superficial bones of the skull) of the head of some ganoid fishes. *A*, bowfin (*Amia*). *B*, *Polypterus*. *C*, sturgeon (*Acipenser*). *D*, gar pike (*Lepidosteus*). The arrangement of some of these scales, which in teleosts become the bones of the dorsal surface of the skull, greatly resembles that of the dorsal skull bones of the land vertebrates, and these scales have received the same names as the skull bones, although the strict homology is doubtful. The supratemporals are also called postparietals. (After Goodrich in Lankester's *Treatise on Zoölogy*, courtesy of the Macmillan Co.)

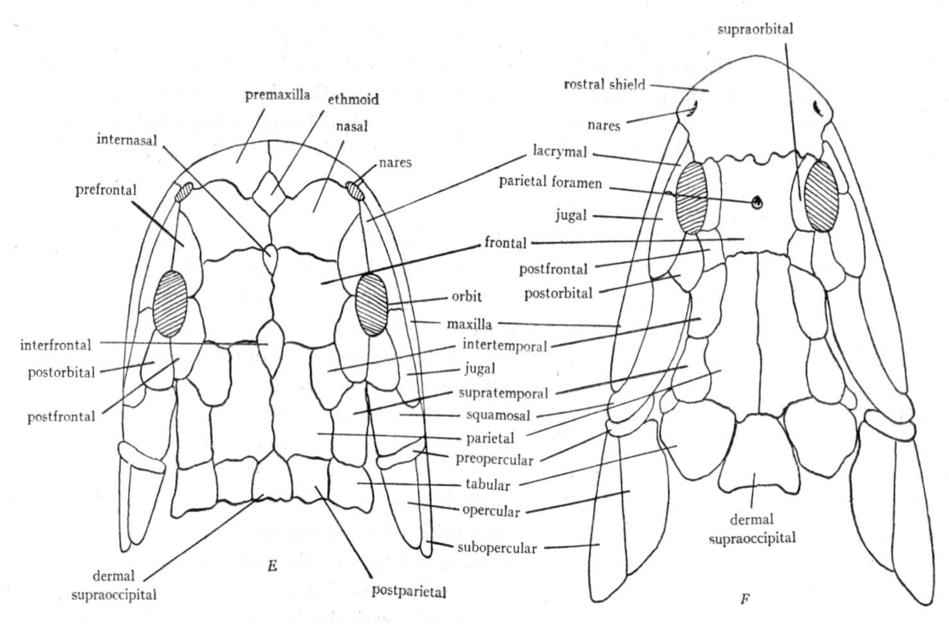

FIG. 58, *continued.*—*E*, plan of the dermal bones of a generalized fish leading to tetrapods (after Goodrich, 1925). *F*, skull of *Osteolepis*, an extinct crossopterygian fish, showing similarity of dermal bones to primitive tetrapods (after Goodrich, 1918). In *F* the far forward position of the parietal foramen, in what is usually considered to be the frontal bone, is noteworthy. It has recently been suggested that this is really the parietal bone (the bone where one would expect the parietal foramen to occur) and that in the evolution of the dermal bones of the skull there has been a backward progression of the parietal bone and a lengthening of the skull region in front of it.

and posterior to the orbit two large postorbitals. The tooth-bearing maxillae form the sides of the upper jaw. Between the parietal and the upper postorbital is found the squamosal. The operculum is supported by several opercular bones.

3. Demonstration of the separate origin of chondrocranium and dermal bones.—Examine the demonstration specimen of the head of a ganoid fish in which the incasing dermal bones have been loosened from the underlying chondrocranium. Remove the sheath of dermal bones, noting that they are situated in the skin, and note the chondrocranium, similar to that of the dogfish, lying within the sheath. Such a specimen shows clearly the origin of the skull from two separate sources, the cartilaginous chondrocranium and the dermal scales.

D. THE FORMATION OF THE CARTILAGE BONES OF THE SKULL AND THE COMPOSITION OF THE COMPLETE SKULL

The next and last step in the formation of the skull is the production of cartilage bones in the chondrocranium, including the sense capsules, and in the mandibular and hyoid arches. In definite regions of these structures *centers of ossification*—that is, centers of bone formation—arise. Each of these centers transforms a certain area of cartilage into bone, and each such bone is primitively a cartilage bone of the skull. These centers of ossification do not, however, correspond to the cartilages whose fusion produces the chondrocranium. Since, as already emphasized, the chondrocranium consists of a ventral trough completed above only in the occipital region, so also the cartilage bones of the skull are limited to the floor and rear end of the cranial cavity (*basicranial* region). The roof and sides of the skull are formed of dermal bones which represent the scales of fishes. Similarly, cartilage bones arise in the mandibular and hyoid arches and become sheathed in dermal bones. The complete vertebrate skull, therefore, consists of cartilage bones ossified in the chondrocranium, sense capsules, and mandibular and hyoid arches, and of dermal bones covering the cartilage bones everywhere except on the midventral surface and posterior end of the skull. In general, the degree of ossification of the chondrocranium is greater in the higher members of each group, but sometimes highly cartilaginous skulls are the result of retrogressive processes.

The following attempts to tabulate the bones of the tetrapod skull with reference to origin. It must always be borne in mind that homologies are not as certainly established as the names imply.

1. Cartilage bones ossified in the chondrocranium proper.—These may be grouped by regions into the occipital, posterior sphenoid, and orbitosphenethmoid groups.

a) *Occipital group:* This consists typically of four bones at the rear end of the skull, encircling the foramen magnum, ossified in the parachordals and tecta. They are the *supraoccipital* above; the two *exoccipitals*, one on each side; and the *basioccipital* below. Modern amphibians lack supra- and basioccipital, but they were present in extinct amphibians. A prominent projection of the occipital region by which the skull articulates with the atlas is termed the *occipital condyle.* Fishes and primitive tetrapods have a single occipital condyle formed chiefly by the basioccipital with some contributions from the exoccipitals. Through the reduction of the basioccipital part and enlargement of the exoccipital projections there come to be two occipital condyles, as in present amphibians and mammals.

b) *Posterior sphenoid group:* This consists of a median bone and a pair of lateral bones ossified in the rear parts of the orbital cartilages and trabeculae. In the cranial floor in front of the basioccipital is found the unpaired *basisphenoid.* To either side of the basisphenoid, forming the rear walls of the orbit, there is found in reptiles and birds a *pleurosphenoid* bone (also called *laterosphenoid*), ossified in the antotic pillars. This pair of bones (formerly confused with the mammalian alisphenoids) is absent in mammals (which do not have antotic pillars [p. 154]), and it appears to have been lacking in primitive tetrapods also. In both groups this region of the skull is occupied by a pair of bones (*epipterygoids*) derived from the palatoquadrate cartilages (see below).

c) *Orbitosphenethmoid region:* This region remains more or less unossified in tetrapods, and homologies are not well worked out. According to Broom (1926), the following bones may ossify in this mass: presphenoid, orbitosphenoids, and mesethmoid. In mammals there is typically in front of the basisphenoid a median *presphenoid,* flanked on each side by an *orbitosphenoid* bone, ossified in the orbital cartilages. These three bones form the walls of that part of the cranial cavity containing the olfactory lobes. The presphenoid arises from paired ossifications in most mammals and, although having separate boundaries in young mammals, is usually indistinctly bounded in adults. In some mammals there is an unpaired ossification, the *mesethmoid,* in the internasal septum, in front of the presphenoid. The basicranial axis in such mammals therefore consists of a chain of four cartilage bones (Fig. 65*B*): basioccipital, basisphenoid, presphenoid, mesethmoid. But this story cannot, as formerly supposed, be applied to the tetrapod or even to the mammalian skull in general, for in many mammals and also in the therapsid ancestors of mammals a mesethmoid is lacking and the nasal septum is invaded by the presphenoid (Fig. 65*A*), so that there are only three bones in the basicranial axis. Below mammals homologies are not clear. There are no ossifications at all anterior to the posterior sphenoid group in most present reptiles. In extinct reptiles and amphibians and in present frogs there is a single ossification here, termed the *sphenethmoid* bone. This apparently corresponds to the presphenoid and paired orbitosphenoids of mammals.

The ethmoid or nasal region of the chondrocranium always remains more or less cartilaginous. In addition to the mesethmoid mentioned above (found only in some mammals), there may be lateral ossifications in the orbitonasal lamina.

2. Cartilage bones ossified in the sense capsules.—

a) *In the otic or auditory capsules:* A number of otic bones arise in the walls of the otic capsules. Since the latter are fused with the parachordal region of the skull, the otic bones are naturally closely associated with the occipital group of bones and often fused with them. There may be as many as five otic bones (in teleosts): the *prootic*, *epiotic*, *opisthotic*, *pterotic*, and *sphenotic;* but they are commonly fused together or fused with near-by bones. The three first named are the ones most constant in the higher vertebrates. When fused into one bone, they are designated the *periotic* or *petromastoid* bone.

b) *In the optic capsules:* As already explained, the optic capsules do not fuse with the skull because of the necessity for retaining free movement in the eyes. The optic capsule becomes the *sclerotic cartilage* in most vertebrates (not in mammals, except monotremes). In some vertebrates, particularly reptiles and birds, a ring of *sclerotic* bones occurs around the pupil.

c) *In the nasal capsules:* The principal products of the lateral walls of the olfactory capsules are the *turbinals* or *conchae*, scrolled bones in the side walls of the nasal passages. They are best developed among mammals having a good sense of smell. Homologizing the turbinals in the different vertebrate classes is impractical at present.

3. Cartilage bones ossified in the gill arches.—

a) *In the palatoquadrate cartilages:* These cartilages are, as we have seen, the dorsal halves of the first or mandibular pair of gill arches and form the primitive upper jaw, which makes contact with the chondrocranium proper by three or four dorsally directed processes on each side (p. 159, Fig. 57*A*). Whereas in primitive fishes, such as early crossopterygians, the entire palatoquadrate cartilage ossifies, forming a series of *pterygoid* bones and a terminal *quadrate* bone, in tetrapods in general most of the palatoquadrate remains unossified and degenerates, because its function is taken over by a covering of dermal bones. The two principal cartilage bones produced from each palatoquadrate in tetrapods are the *epipterygoid* (=so-called alisphenoid of mammals), ossified in the ascending process, and the *quadrate*, ossified in the otic process. Except in mammals the quadrate has the function of bearing the articulating surface for the lower jaw (autostylic type of jaw suspension). The quadrate usually becomes an integral, immovable part of the auditory region of the skull, a condition termed *monimostylic*. But in some vertebrates, notably lizards, snakes, and birds, the quadrate is movable, a condition termed *streptostylic*, permitting a wide gape for swallowing large prey.

b) *In Meckel's cartilages:* These are the lower halves of the mandibular arches and function as the lower jaw in elasmobranchs. Their history is similar to that of the palatoquadrates, i.e., several bones may ossify in each in primitive fishes; but in higher fishes and in tetrapods dermal bones take over their function, and they reduce and disappear or remain as a cartilaginous core inside the dermal bones. The principal bone ossified in Meckel's cartilage in vertebrates is the *articular* bone in its rear end; an anterior *mentomeckelian* bone is common in amphibians. The articular articulates with the quadrate to form the suspension of the lower jaw, and this quadrate-articular type of suspension (autostylic) is characteristic of tetrapods except mammals, where both bones are reduced and have become bones of the middle ear.

c) In the hyoid and other gill arches: The dorsal part of the hyoid arch is the hyomandibular cartilage of elasmobranchs and ossifies into a bone, the *hyomandibula* in bony fishes. This element may function as a suspensor of the lower jaw (hyostylic type). In tetrapods the hyomandibula, at least in part, becomes the columella of the ear. The rest of the hyoid arch in tetrapods associates with the remaining gill arches, much reduced and simplified, to form the *hyobranchial skeleton.* This consists of the *hyoid apparatus* and the cartilages of the *larynx.* The hyoid apparatus is a plate or rod of cartilage or bone, situated in the throat and having projecting processes extending to the otic region. It acts as a support for the tongue and larynx, serves for muscle attachment, and in amphibians plays an important role in buccal respiration. The larynx or voice box is a chamber at the top of the windpipe whose walls are supported by cartilages derived from the gill arches. The exact composition of the hyoid apparatus and the laryngeal cartilages varies in different vertebrates.

4. The dermal bones of the skull and jaws.—

a) Skull roof: As the chondrocranium is typically roofless, there are never any cartilage bones in the skull roof, which is formed of dermal bones. The general arrangement of these roofing bones was already seen in the study of the head scales of ganoid fishes (Fig. 58), but the exact correspondence of these with the dermal bones of the tetrapod skull remains uncertain, despite the identity of the names applied to them. The original arrangement of the dermal bones of the dorsal surface of the tetrapod skull is seen in early extinct forms, such as the embolomerous amphibians and cotylosaurian reptiles and is shown in Figure 59*A*, *B*. Compare also with fish (Fig. 58*E*, *F*). Note that there was practically no difference in the skull roof between early amphibians and early reptiles. The skull is completely roofed over (anapsid type) with openings only for the nostrils (*nares*), eyes (*orbits*), and the pineal body (*epiphyseal* or *parietal foramen*). At the back on each side there is an indentation, the *otic notch*, shown but not labeled in Figures 59*B* and 58*E*. In the median region from the nostrils backward are found the paired *nasals, frontals, parietals*, and *postparietals* (also called *dermosupraoccipitals*). An unpaired *interfrontal* bone, seen in Dipnoi, also occurs in *Eryops* and in a few other earliest amphibians. A dermal bone termed the *septomaxilla* is present in the interior of the nose in many vertebrates, and in some extinct reptiles is exposed on the surface; it protects Jacobson's organ. Around each orbit there is a set of bones: the *lacrimal, prefrontal, postfrontal*, and *postorbital.* Alongside the parietals are found on each side the *intertemporal*, the *supratemporal*, and the *tabular;* and on the outer side of the otic notch is the *squamosal.*

In the course of the evolution of tetrapods many of the original dermal bones of the roof of the skull are lost, so that in mammals only a few remain (see Fig. 59). There is also a tendency for the formation of vacuities in the posterior part of the roof. The chief of these, making their appearance in reptiles, are an upper *supratemporal fossa* between the parietal and the squamosal and a lower *infratemporal* or *lateral temporal* fossa between the squamosal and the jugal. Reptiles with only the lower fossa are termed *synapsid* and include those reptiles ancestral to mammals; mammals inherit the single temporal fossa, which enlarges and becomes confluent with the orbit. *Diapsid* groups of reptiles with two temporal fossae include *Sphenodon*, the crocodiles (Fig. 59*C*), and extinct groups ancestral to birds (also diapsid). When only the upper fossa is present, the condition is termed *parapsid* (few forms, possibly lizards). A pair of *posttemporal fossae* in the rear of the skull above the exoccipitals occurs in many reptiles—for example, in *Sphenodon*—and traces of them are seen in monotremes.

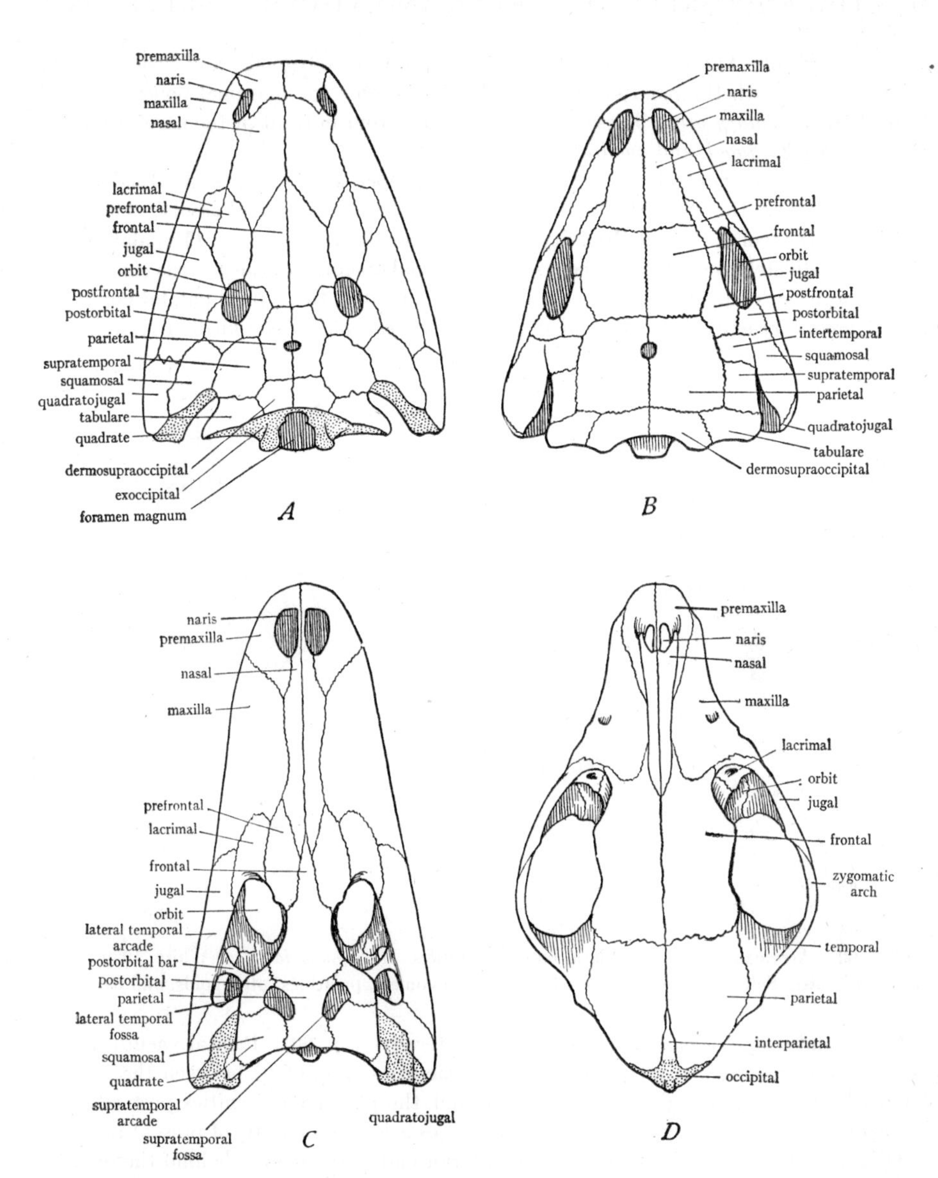

Fig. 59.—Dorsal view of the skulls of four representative vertebrates to show the reduction of the dermal bones of the roof in the course of evolution. *A*, skull of an extinct amphibian, *Capitosaurus*, belonging to the Stegocephali; note the large number of membrane bones completely roofing the skull. *B*, skull of one of the most ancient reptiles, *Seymouria*, belonging to the Cotylosauria; the dermal bones are nearly as numerous as in the amphibian, are similarly arranged, and completely roof the skull. *C*, skull of a modern reptile, the alligator; several of the dermal bones present in the extinct forms have been lost, and the roof bears several openings. *D*, skull of a modern mammal, the dog, showing still greater loss of dermal bones. Dermal bones blank; cartilage bones stippled. (*A* from Reynolds' *The Vertebrate Skeleton*, courtesy of the Macmillan Co.; *B*, from Williston's *Water Reptiles of the Past and Present*, University of Chicago Press.)

b) Upper jaw: The upper jaw is fused to the skull in tetrapods, and hence the dermal bones which sheath the palatoquadrate cartilages become the marginal bones of the skull, termed the *maxillary arch.* These dermal bones are from in front of the nares caudad: *premaxilla, maxilla, jugal,* and *quadratojugal* (Fig. 59*C*).

c) Palate: The ventral surface of the skull proper also becomes sheathed in dermal bones concealing the cartilage bones of the true floor of the cranial cavity. These dermal bones form the roof of the mouth cavity and hence constitute the *palate.* The palate originally also was completely roofed over except for the internal openings of the nostrils (*internal nares*) near its anterior end. Beginning behind the internal nares, the typical palatal bones are the paired *prevomers* and *pterygoids* and unpaired *parasphenoid* in the median region, and the

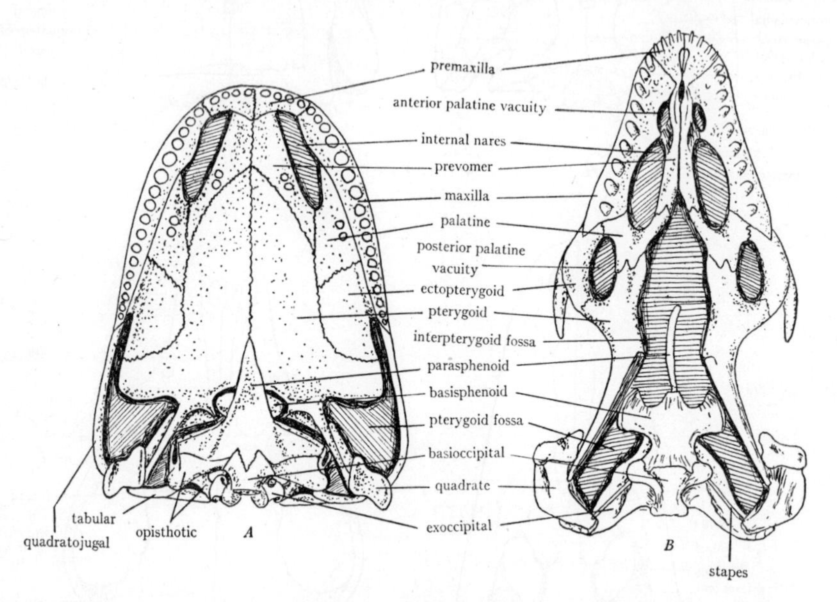

FIG. 60.—Arrangement of the palatal bones of reptiles. *A*, *Seymouria* (after White, 1939). *B*, lizard (*Varanus*) (after Schmidt, 1919). Note relations of prevomers, palatines, pterygoids, and parasphenoid.

palatines and *ectopterygoids* (=*transpalatines*) laterally (Fig. 60). Palatal vacuities soon develop, however, principally the *interpterygoid* vacuities (Fig. 60*B*) between the pterygoids, often spanned by the median parasphenoid, and the pterygoid vacuities posteriorly, as in *Seymouria* (Fig. 60*A*). Very large palatal vacuities are characteristic of present amphibians.

Originally the internal nares are at the anterior end of the palate, behind the premaxillae or between the prevomers. In certain reptiles, particularly those ancestral to mammals, there begins to form the *secondary* or *hard* palate, by means of which nasal passages are formed in the maxillary and palatine bones and the internal nares are shoved backward. This process is complete in mammals, where the fleshy "soft" palate, not supported by bones, continues the secondary palate, so that the nasal passages then open very far back, near the entrance into the windpipe. The hard palate arises in the mammalian embryo from horizontal plates which form on the maxillae and palatines, grow toward the median line, and fuse with a median vertical plate of bone. This last is part of a bone characteristic of the mammalian skull termed the *vomer* and is now generally believed to represent the parasphenoid of lower forms.

d) *Lower jaw:* This, too, becomes sheathed in membrane bones, with the one cartilage bone, the *articular*, exposed at the rear end. In primitive tetrapods there was a large number of dermal bones in the lower jaw: *dentary, splenial, postsplenial, angular*, and *surangular* (or *supra-angular*) on the outer surface, and a *prearticular* and three *coronoids* (*coronoid, inter-*

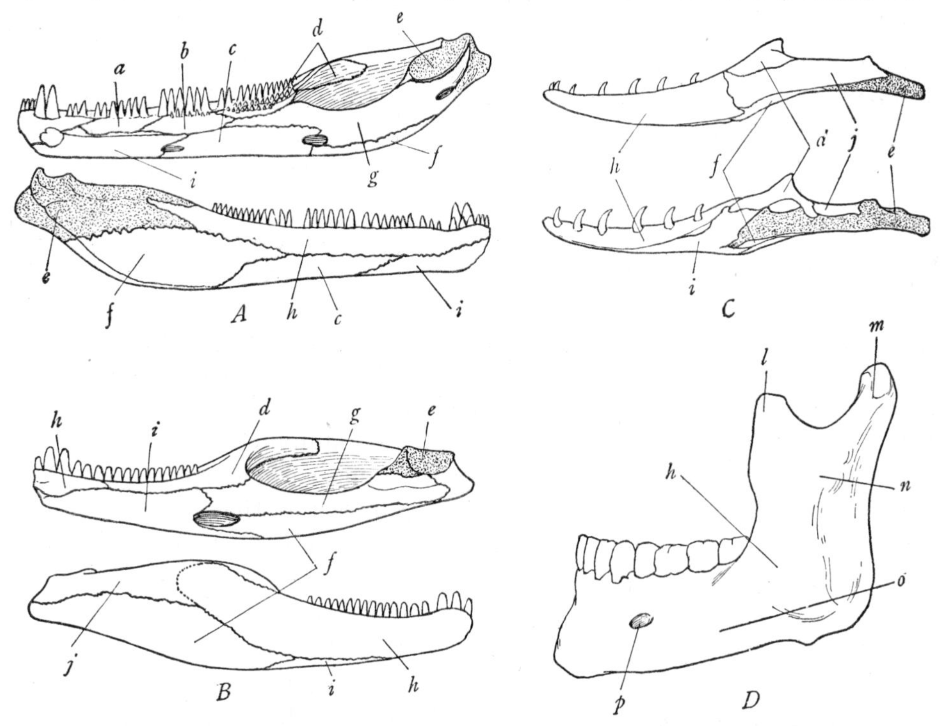

FIG. 61.—Lower jaws of four vertebrates to show the reduction in the number of bones in the course of evolution. *A*, lower jaw of an extinct amphibian, *Trimerorhachis*, belonging to the Stegocephali; inner surface above, outer surface below; note large number of dermal bones. *B*, lower jaw of an extinct reptile, *Labidosaurus*, belonging to the Cotylosauria; inner surface above, outer surface below; note reduction in the number of dermal bones and increased size of the dentary (*h*) and the splenial (*i*). *C*, lower jaw of a modern reptile, a lizard, *Varanus*, showing still further reduction in the number of bones; outer surface above, inner surface below. *D*, half of the lower jaw of man, seen from the outer surface; it consists of but one bone, the dentary, all other bones having vanished. Dermal bones blank; cartilage bones stippled. *a*, precoronoid; *b*, intercoronoid; *c*, postsplenial; *d*, coronoid; *e*, articular; *f*, angular; *g*, prearticular; *h*, dentary; *i*, splenial; *j*, supra-angular or surangular; *l*, coronoid process; *m*, condyloid process; *n*, ramus; *o*, body; *p*, mental foramen. (*A* and *B* from Williston's *Water Reptiles of the Past and Present*, University of Chicago Press; *C* from Reynolds, *The Vertebrate Skeleton*, courtesy of the Macmillan Co.)

coronoid, precoronoid) visible only on the inner surface (Fig. 61*A*). In the evolution of tetrapods there is a gradual loss of the dermal bones of the lower jaw until finally in mammals only the *dentary* remains (Fig. 61*D*). As both the articular and the quadrate have become reduced to bones of the middle ear (p. 185), the articulation of the lower jaw with the skull is in mammals between the dentary and the squamosal.

e) *Other gill arches:* No dermal bones occur in connection with the hyoid or branchial arches.

E. THE TEETH

Although teeth are usually considered part of the digestive system, it seems more convenient and logical to treat them with the skeleton. As already learned, teeth are homologous to placoid scales and become located in the mouth cavity through the inturning of the skin to line this cavity.

1. Teeth of elasmobranchs.—Examine a variety of available specimens. In many elasmobranchs the biting surface of the jaws resembles the external skin, having a mosaic of similar small or minute, rounded, diamond-shaped, or hexagonal teeth. Other specimens show a great enlargement of the teeth in those regions which bear the brunt of food-chewing, especially when hard food, such as mollusks or crustaceans, is ingested. Examples are the Port Jackson shark (*Cestracion*), which has small pointed teeth in front and very large flattened ones farther back, and certain skates, in which the middle teeth are elongated, the lateral teeth small polygons. Among the larger predaceous sharks the teeth are enlarged formidable cutting weapons with one or more sharp points and often serrated edges. They are seen to be arranged in rows, of which only one or a few rows are in use, standing upright on the jaw edges, whereas the others lie flat along the inner surface of the jaws. It is generally believed that shark teeth can be replaced by the moving-up of the back rows; but some authorities doubt this, and, in fact, direct observation on the matter appears to be scanty.

The teeth of elasmobranchs are not fastened directly to the jaw cartilages but are held in place only by strong connective tissue bands of the dermis.

2. Teeth of bony fishes.—When the jaw cartilages and palate become sheathed in dermal bones, the teeth naturally are borne upon the dermal bones. Consequently, in bony fishes teeth may occur on any of the palate or jaw bones of dermal origin. The teeth of bony fishes vary greatly in shape and arrangement; available materials may be examined. Fish teeth may be fastened by fibrous bands or are sometimes movably jointed to the bones, but usually they are ankylosed, i.e., immovably soldered to the underlying bones by way of intervening bone.

A type of tooth structure termed *labyrinthine* occurred among extinct crossopterygians; the enamel is infolded along longitudinal grooves, often making a complicated pattern in the interior. Similar teeth are inherited from crossopterygian ancestors by early amphibians, whence the name Labyrinthodontia applied to the latter.

3. Development, structure, and arrangement of tetrapod teeth.—As already indicated, the development of the teeth of bony fishes and of tetrapods is similar to that of a placoid scale. The stratum germinativum of the lining of the mouth cavity proliferates a solid ingrowth, the *enamel organ*, whose epithelium, one cell thick, takes on a cup shape (Fig. 62). The interior of the cup is filled with dermis, forming the *dental papilla*. The epithelial cells of the enamel organ, termed *ameloblasts*, secrete enamel on their under surface, and the dermal papilla secretes dentine on its outer surface. The combined secretions become a tooth, and the remains of the dental papilla persist as the pulp which nourishes the tooth.

Obtain a longitudinal section of a simple tetrapod tooth and identify the following parts: the *crown*, or shiny upper part of the tooth, which in life projects above the gum; the *root*, or dull lower part, which sets into the jaw; the *pulp cavity*, the interior space, filled in life with the *pulp*, consisting of connective tissue, nerves, and blood vessels; the *neck*, or junction of crown and root; the *dentine*, the bonelike material of which most of the tooth is composed; the *enamel*, the thin, shiny layer covering the dentine of the crown; the *cement*, a thin layer of bone which coats the dentine of the root.

In amphibians and reptiles the teeth are usually of a simple conical form, are generally numerous, and often occur on bones of the palate as well as along the jaw margins. Among the reptiles ancestral to mammals (order Therapsida) palatal teeth disappear, the teeth become limited to a single row along the jaws, and they begin to differentiate into several kinds, as in mammals. These changes lead to the condition in mammals where the teeth form a single row along the jaw margins, are usually limited in number, and are of four different kinds (see later). The teeth of tetrapods are immovably ankylosed to the skull and jaw bones.

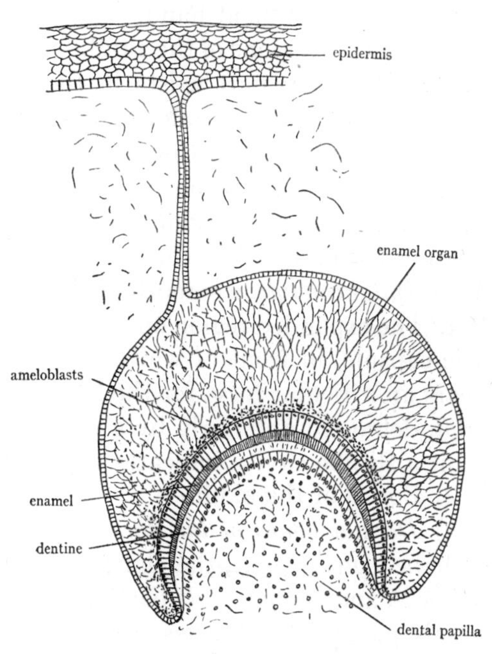

FIG. 62.—Diagram of the development of a tetrapod tooth, showing the enamel organ.

4. Terminology of teeth.—When all the teeth of an animal are alike in form, the dentition is said to be *homodont;* when they vary in form in different parts of the mouth, the term *heterodont* is applied. Heterodonty is characteristic of mammals but, as noted above, is also seen in many fishes. Animals in which the teeth cannot be replaced when lost are spoken of as *monophyodont;* those in which the first evanescent set of teeth is replaced by a second permanent set, as in most mammals, are *diphyodont;* and those with a continuous succession of teeth, as fishes, amphibians, and reptiles, are termed *polyphyodont.* When the teeth are ankylosed to the side of the supporting bones, the term *pleurodont* is applied; when to the summit of the bones, the teeth are *acrodont;* and when set in sockets, they are *thecodont.* Socketing is characteristic of mammals and some reptiles.

F. THE SKULL OF NECTURUS, A PARTIALLY OSSIFIED SKULL

In the skull of Amphibia the ossification of the chondrocranium has proceeded to but a limited extent, so that a partially cartilaginous chondrocranium is present from which the incasing dermal bones can be readily separated, as in fishes. The palatoquadrate cartilages are inseparably fused to the ventral and lateral sides of the skull proper and are partially ossified.

In the lower jaw Meckel's cartilages are persistent as cores inclosed by dermal bones. The number of dermal bones in the skulls of present-day Amphibia is considerably less than that of extinct forms, such as shown in Figure 59*A*.

For the study of the skull of *Necturus*, complete skulls, preferably preserved in fluid, and chondrocrania, from which the dermal bones have been removed, should be at hand.

1. General regions of the skull.—The skull is partly bony, partly cartilaginous. The bony part exists in the form of distinct areas or bones, separated from one another along wavy or jagged lines, the *sutures*. The dermal bones are somewhat distinguishable from the cartilage bones by their more superficial positions. The skull is divisible into a median portion, the skull proper, and marginal regions formed of the upper jaws and palatal bones and bearing teeth. Identify the foramen magnum and, below this on each side, an *occipital condyle*, a smooth projection for articulation with the first vertebra. The expanded region on each side of the posterior part of the skull is the otic capsule, in front of which a projection, formed of the quadrate bone, bears an articular fossa for the lower jaw. Anterior to the otic capsule is the *orbit*, delimited anteriorly by a projecting *antorbital process*. In front of the orbit a slitlike cavity indicates the position of the nasal capsules.

2. Bones of the skull.—

a) *Dermal bones of the dorsal side:* From front to rear are found the V-shaped *premaxillae*, the elongated *frontals* covered anteriorly by one limb of the V, and the *parietals* partially covered by the frontals. Note absence of nasals, maxillae, and bones around the orbit. On each side of the parietal is a slender *squamosal* bone, extending from the quadrate along the side of the otic capsule.

b) *Dermal bones of the ventral side:* The chief bone here is the large median *parasphenoid* extending from the occipital condyles anteriorly between a pair of tooth-bearing bones, the *prevomers*, which form the floor of the nasal capsules. Between the prevomers, at the anterior end of the parasphenoid, there is exposed a bit of the cartilage of the chondrocranium, the *ethmoid plate*. At the sides behind the prevomers are the *pterygoids*, bearing teeth on their anterior ends and articulating posteriorly with the quadrate.

c) *The teeth:* The teeth are arranged in two rows—a short row on the premaxillae, a longer row on the prevomers and pterygoids. The teeth of the lower jaw fit into the space between these two rows when the jaws close. The teeth are pleurodont.

d) *Cartilage bones ossified in the chondrocranium and otic capsules:* Only the rear end of the chondrocranium is ossified, forming the exoccipital bones, which bear the occipital condyles and constitute the lateral walls of the foramen magnum, completed above by a strip of cartilage, the *synotic tectum*. The otic capsules are partially ossified, each containing two cartilage bones, the *opisthotic* and the *prootic*. The opisthotic, lateral to the exoccipital, is a cone-shaped bone composing the projecting angle of the skull noticeable to either side of the occipital region. The dorsal portion of the opisthotic makes contact with the parietal and squamosal above and the parasphenoid below. In front of the opisthotic on the dorsal side is an area of cartilage; and in front of this is the prootic bone, wedged between the squamosal and the parietal. Examine the otic capsule from the side and note a small rounded bone, the *columella*, in its wall between the opisthotic and the prootic. A projection of the columella, termed the *stylus*, meets a columellar projection from the middle of the squamosal. The rounded base of the columella, called the *foot plate* or *fenestral plate*, fits into an opening, the *oval window* or *fenestral ovalis*, in the cartilage between the prootic and the opisthotic. This window leads to the internal ear, situated inside the prootic bone. The columella is a sound-transmitting apparatus, part of the middle ear, here degenerate.

e) Middle ear of urodeles: The middle ear originates in amphibians but is degenerate in urodeles. It is a chamber interpolated between the internal ear (lodged in the otic capsule) and the skin. It is closed externally by the *eardrum* or *tympanic membrane*, a double-layered membrane of which the outer layer consists of skin and which primitively is flush with the surface of the head. Across the *tympanic cavity* (=cavity of the middle ear) there stretches a sound-transmitting apparatus called in nonmammalian tetrapods the *columella auris*, which is a persisting remnant of the upper part of the hyoid arch, hence more or less homologous to the hyomandibular bone of fishes. The columella auris usually consists of more than one piece and may be incompletely ossified. The distal end of the columella abuts against the eardrum; its proximal end fits into the fenestra ovalis and is expanded into the foot plate, fenestral plate, or stapedial plate, to fit the window exactly.

According to Reed (1920), the columella auris of urodeles consists of two pieces, a distal *plectrum*, and a proximal *operculum* (the name is unfortunate; no relation to the fish operculum implied) (Fig. 63*A*). It seems to be generally attested that the operculum originates in the wall of the otic capsule and hence is not a hyoid-arch derivative. The plectrum typically has a stylus which primitively articulates with the squamosal but may shift its connection onto the quadrate. In *Necturus*, plectrum and operculum are fused (Fig. 63*B*).

f) Cartilage bones ossified in the palatoquadrate cartilages: The only bone so formed is the quadrate, already noted as occupying the projection between squamosal and pterygoid and bearing an articulating fossa for the lower jaw. Medial to the quadrate, a considerable area of cartilage represents unossified portions of the palatoquadrates.

Draw dorsal and ventral views of the skull, outlining the bones accurately.

3. Chondrocranium.—When the skull of *Necturus* is soaked in warm soap solution, the dermal bones can easily be lifted off, revealing the chondrocranium beneath. This consists largely of cartilage with a few cartilage bones already noted. The otic capsules are inseparably incorporated into the chondrocranium, but the nasal capsules have very thin walls and are not firmly attached to the chondrocranium, so that they are usually lost in preparing the chondrocranium.

Compare the chondrocranium with Figure 64. Its posterior part consists chiefly of the rounded otic capsules connected dorsally by the narrow *synotic tectum* and ventrally by a still narrower bridge between the exoccipitals,

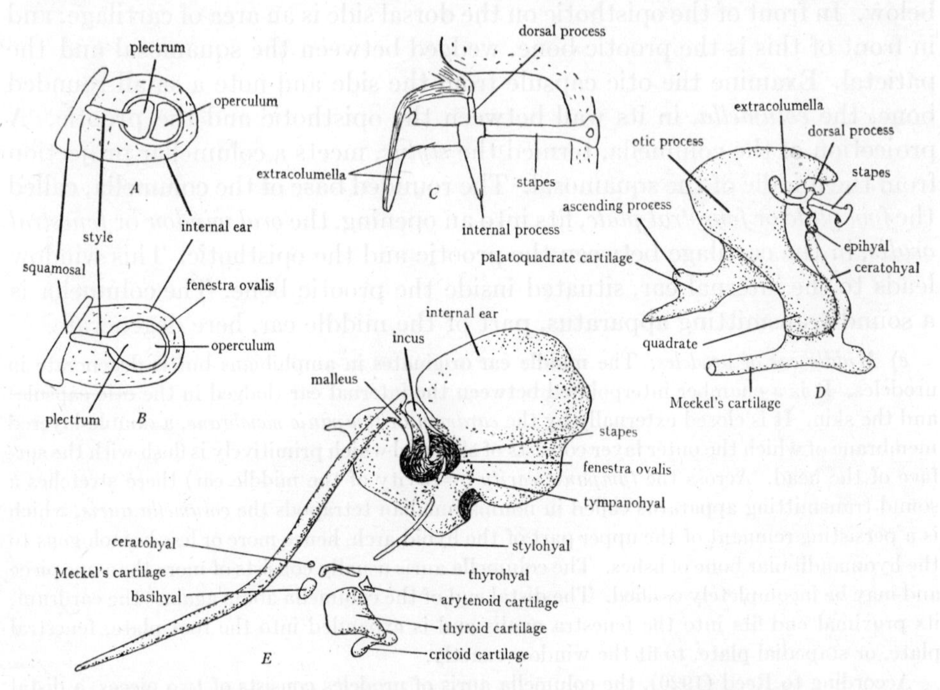

Fig. 63.—Middle-ear ossicles of tetrapods. *A*, scheme for urodeles. *B*, condition in *Necturus*, plectrum and operculum fused. (*A* and *B* after Reed, 1920.) *C*, scheme for lizards (after Versluys, 1903). *D*, embryo of the crocodile (after Parker, 1883); the fusion of the hyoid arch with Meckel's cartilage is peculiar to the Crocodilia. *E*, developing ear ossicles of the rabbit, showing malleus forming from Meckel's cartilage (articular) and stapes in line with chain of elements representing the hyoid arch; cartilages of the larynx below (after De Beer and Woodger, 1930).

which apparently represents all there is of the basal plate. There is thus very little fusion of the parachordals in *Necturus*. From the anterior end of each otic capsule a slender curved *trabecular* cartilage extends forward and fuses with its fellow to form the ethmoid plate already noted. The trabeculae inclose between them a very large fenestra, composed of the fusion of the basicranial and hypophyseal fenestrae. (In the complete skull the fenestra is covered above by the frontals and parietals and below by the parasphenoid.) From the ethmoid plate a slender process continues forward on each side, the *trabecular horn* or *cornua trabeculae*. (In front of the otic capsule is the quadrate bone and accompanying cartilage; the latter sends processes to the otic capsule and trabecula. These are not, of course, parts of the chondrocranium proper.) In front of the orbit is the projecting *antorbital* cartilage; but, as it is attached only by a ligament, it may be missing. The boundaries of prootic, opisthotic, and columella can be seen better than in the complete skull.

4. Lower jaw.—The lower jaw consists of a pair of Meckel's cartilages, united in front and sheathed for the greater part of their course in dermal bones. The outer surface of each half of the lower jaw consists of the *dentary*, a dermal bone bearing teeth. The inner surface is formed of two dermal bones, the *splenial* and the *angular*. The former is a small bone situated at about the middle of the inner surface, bearing the last group of teeth, consisting of five or six teeth. The angular covers the remainder of the inner surface and passes onto the outer surface at the extreme posterior end of the jaw, below the posterior end of the dentary. The articulating surfaces of the lower jaw are composed of cartilage, which is the posterior end of Meckel's cartilage. This articulates with the quadrate of the upper jaw. Meckel's cartilage runs almost the entire length of the jaw, concealed between the dentary and the angular. It can be revealed by removing these dermal bones.

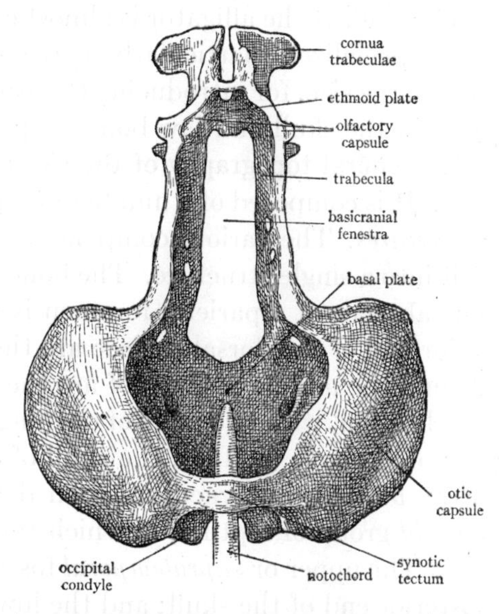

FIG. 64.—Chondrocranium of a urodele larva. (After Gaupp in Hertwig's *Handbuch der vergleichenden und experimentellen Entwickelungslehre der Wirbeltiere.*)

5. The remaining gill arches.—The hyoid and three succeeding gill arches are present in *Necturus* and are almost completely cartilaginous. Preserved material is necessary for their study. The hyoid arch is a broad, somewhat V-shaped cartilage situated just posterior to the lower jaw in the floor of the mouth cavity. On each side it is divisible into two cartilages—a small anterior *hypohyal* and a much larger posterior *ceratohyal*. The third gill arch (first true gill-bearing arch) is more elongated and slender than the hyoid arch and is likewise divisible into two pieces—an anterior *ceratobranchial* and a posterior, slightly longer, *epibranchial*. Between the median ventral portions of hyoid and third arch is a triangular *copula* representing a *basibranchial* piece. The fourth and fifth gill arches are short, curved rods of cartilage on each side, composed of *epibranchials*. At the anterior end of the epibranchial of the fourth arch is a small *ceratobranchial.* In the median ventral line is a slender bone, the second *basibranchial.* It will be seen that the gill arches of *Necturus* are reduced both in number and as regards the pieces of which they are composed, as compared with the gill arches of the elasmobranchs. For the typical condition of the gill arches in elasmobranchs see Figure 56, page 158, and compare with the condition in *Necturus.*

G. THE REPTILIAN SKULL EXEMPLIFIED BY THE ALLIGATOR

The skull of the alligator is almost completely ossified and, although somewhat specialized (particularly as regards the palate), is convenient, because of its large size, for introducing the student to the names, locations, and relations of the skull and jaw bones of present tetrapods.

1. General topography of the skull.—Obtain a skull and study its structure. It is composed of a number of separate bones united along jagged lines, the *sutures.* The various components of the vertebrate skull are here closely knit into a single structure. The bones of the skull roof are rough and pitted. The absence of a parietal foramen is peculiar to the Crocodilia. At the anterior end of the dorsal surface are the two nasal openings or *external nares.* Behind the middle of the roof are the two orbits; and behind them the temporal region, with two pairs of openings, termed temporal fossae or vacuities. As noted above (p. 166), the skull of early tetrapods was completely roofed (Figs. 58, 59), but vacuities soon developed. The alligator belongs to the *diapsid* group of reptiles, in which there are two temporal vacuities on each side. The upper or *supratemporal* fossae are the openings in the roof near the posterior end of the skull; and the lower or *infratemporal* or *lateral* temporal fossae are situated immediately behind the orbits, from which they are separated by a bony rod, the *postorbital bar* (Fig. 59*C*). The projecting ledge to the side of the supratemporal fossae is called the *supratemporal arcade;* and that below the infratemporal fossae, the *infratemporal arcade.* Beneath the supratemporal arcade is an elongated space which in life lodges the external auditory meatus. The middle oval opening, seen on looking deep into this space, is covered in life by the eardrum, and consequently the cavity to the interior of this opening is the cavity of the middle ear or tympanic cavity. Dorsally in the tympanic cavity will be seen a canal leading to the tympanic cavity of the other side. The opening behind that which bears the eardrum, and somewhat confluent with it, is a foramen for the passage of the seventh cranial nerve. The small ear bones which occupy the tympanic cavity are generally missing (see below). Turn the skull so as to view its rear end. Note: the two small slitlike *posttemporal fossae* just beneath the rear edge of the roof; the foramen magnum; the single occipital condyle characteristic of reptiles below the foramen; the quadrate bones forming the side projections and bearing an articular surface for the lower jaw; and the enormous pterygoid fossae below the quadrates, continuous above with the orbits and infratemporal fossae. On looking into the foramen magnum a pair of bulges will be seen; these are the otic capsules containing the internal ears. The brain in life occupies the relatively small cavity, seen around and in front of the otic capsules.

Study the ventral surface and identify: a small opening near the tip, the *anterior palatine vacuity*, also called the *premaxillary vacuity;* a pair of larger oval openings below the orbits, the *posterior palatine* or *suborbital vacuities;* and the *internal nares*, a pair of small openings in the median region near the posterior end of the skull. Note that the internal nares are very far posterior, compared to their position in amphibians; this is caused by the development of a secondary palate, as in mammals.

Most of the small openings in the rear part of the skull are nerve foramina; some serve for the passage of blood vessels.

2. Dermal bones of the dorsal side.—These are: the *premaxillae*, in front of, and at the sides of, the external nares; the *nasals*, the pair of long median bones behind the external nares; the *maxillae*, the large bones to the sides of the nasals; the *prefrontals*, to the sides of the rear ends of the nasals and contributing to the anterior wall of the orbits; the *lacrimals*, lateral to the prefrontals, also bounding the orbit anteriorly; the *frontal* bone, the median bone between the orbits, single in the adult but paired in the embryo; the *parietal* bone, behind the frontal and between the supratemporal fossae, also single in the adult but paired in the embryo; the *postorbital*,[4] forming the anterior part of the supratemporal arcade and sending down a process which constitutes the upper half of the postorbital bar; the *squamosal*, directly behind the postorbital and forming the rest of the supratemporal arcade; the *jugal* or *malar*, the elongated bone below the orbit and bearing a projection which forms the lower half of the postorbital bar; and the *quadratojugal*, a slender, obliquely placed bone behind the infratemporal fossa. The *quadrate*, a cartilage bone, is wedged between the quadratojugal, which it parallels, and the squamosal; its outer end has a concave surface for articulation with the lower jaw, and its inner end forms a bony curve for the eardrum. A small bone, often lost in prepared skulls, occurs in the eyelid; this has been called *adlacrimal* and *supraorbital*, but *palpebral* (=eyelid) bone appears the best name for it.

Premaxillae, maxillae, jugal, quadratojugal, and quadrate form the *maxillary arch* and are really the upper jaw inseparably incorporated into the skull. All are dermal bones except the quadrate, which is ossified in the otic process of the palatoquadrate cartilage.

[4] Also called postfrontal; which one it is does not seem to be definitely settled.

Compare the bones of the roof of the alligator skull with those of primitive tetrapods (Fig. 59*A*, *B*). Which are missing?

3. Palate and ventral view of the maxillary arch.—Study the ventral surface and identify premaxillae and maxillae, both bearing teeth. Behind the maxillae are the *palatines*, forming the inner boundaries of the posterior palatine vacuities. Behind the palatines are the broad *pterygoids*, inclosing the internal nares. The *ectopterygoids* (or *transpalatines*) extend from the pterygoids laterally to meet the maxillae and jugals and form the posterior boundaries of the posterior palatine vacuities. Identify also quadratojugals and quadrates.

In the Crocodilia the maxillae and palatine bones have put out a horizontal shelf which, growing medially, has met in the midventral line. This produces a long nasal passage and shoves the prevomers into the interior, away from the surface. It is more typical of reptiles that the prevomers contribute to the palatal region (Fig. 60). By the formation of this secondary palate the internal nares are moved far back, and the animal is able to remain under water with only the tip of the snout exposed for respiration. A parasphenoid bone, also a typical palate bone of amphibians and reptiles, is lacking in the Crocodilia, although represented by a vestige in the embryo. A more typical reptilian palate is shown in Figure 60.

4. Occipital region.—This region forms the posterior end of the skull and consists of four cartilage bones ossified in the basal plate and synotic tectum. Examine the rear end of the skull. The foramen magnum is bounded by three bones, the *exoccipital* on each side, the *basioccipital* below. The latter forms most of the large rounded occipital condyle, but processes from the exoccipitals participate. Above and between the exoccipitals is the triangular *supraoccipital*, which articulates above with the parietal; between the supraoccipital and the squamosals are the posttemporal fossae, already mentioned. Between the exoccipital and the quadrate on each side is a slanting passage for the seventh nerve, the inner end of which was already seen close to the opening into the tympanic cavity. In the exoccipitals, on the side of the region taking part in the condyle, are foramina for the ninth, tenth, and eleventh nerves; and below these, near the margin of the exoccipital, is a passage for the internal carotid artery.

5. Posterior sphenoid region.—This region lies in front of the occipital region when the skull is oriented ventral side up with the posterior end toward you; it forms the floor of the cranial cavity. The *basisphenoid* appears as a V-shaped bone in the median region between the basioccipital and the pterygoids. Most of the basisphenoid is concealed from view by the pterygoids and quadrates, but its anterior end projects as a thin plate, the *rostrum* (considered by some to represent a presphenoid), into the space between the orbits. This space in life is closed by the interorbital septum, which remains cartilaginous except for the rostrum of the basisphenoid. The rostrum is best seen by looking from the side into the pterygoid fossa. Above the rostrum on each side is a *pleurosphenoid* bone (also called *laterosphenoid*), which ascends to the skull roof, where it articulates with the frontal, postorbital, and parietal.[5] Between the base of the pleurosphenoid and the quadrate a conspicuous oval opening, the *foramen ovale*, provides passage for the fifth nerve. Above this is the supratemporal fossa seen from the inside. The two pleurosphenoids fail to meet in the median line, leaving a cleft through which pass the optic nerves and the olfactory tracts.

6. Orbitosphenethmoid region.—There are no ossifications in the chondrocranial floor anterior to the basisphenoid; the cartilaginous interorbital septum is continuous anteriorly with the cartilaginous internasal septum.

7. Otic capsule and median sagittal section of the skull.—The otic capsules, as already noted, can be seen by looking into the foramen magnum but can be studied only in sagittal sections of the skull. In such a section locate the capsule as the rounded swelling in the side wall shortly in front of the foramen magnum. Embryology shows that three bones ossify in the otic capsule: *epiotic*, *prootic*, and *opisthotic*. A vertical suture in the capsule marks the boundary between opisthotic and prootic. The opisthotic is behind the suture, fused to the exoccipital; the prootic, in front of the suture, remains as an independent bone. Through its lower part passes the foramen ovale, noted above; and the prootic can be seen in external view of the skull as the posterior boundary of this foramen. The dorsal part of the otic capsule above the suture represents the epiotic fused to the supraoccipital. The row of foramina below the capsule beginning behind the foramen ovale serves for the exit of the seventh to the twelfth nerves; the slitlike foramen below the capsule gives passage to the ninth to eleventh; and those behind this are for the twelfth. Below the foramen ovale is the foramen for the sixth nerve.

In the parietal bone note the large canal connecting the tympanic cavities of the two sides. Identify pleurosphenoid and basisphenoid in the section. The rounded depression in the basisphenoid between pleurosphenoid and rostrum is the *sella turcica*, which lodges the pituitary. The brain occupies the cavity from the foramen magnum to the anterior limits of the pleurosphenoids. The upright bar in front of the interorbital space is formed chiefly of the prefrontal. Observe formation of the long nasal passage by the pterygoid, palatine, and maxilla. In front of the long partition belonging to the pterygoid is seen the thin *prevomer*.

8. The middle ear of reptiles.—In reptiles, as in amphibians, the columella auris stretches from the eardrum to the fenestra ovalis and is derived wholly from the hyoid arch. It typically consists of the following parts (Fig. 63*B*): the stapedial part, having an expanded

[5] The pleurosphenoid was formerly called alisphenoid and was believed to be homologous to the bone called alisphenoid in mammals, but much study of this region by several authorities has shown that the reptilian and mammalian element are not homologous. The pleurosphenoid is a true bone of the cranial floor, ossified in the antotic pillars.

stapedial plate fitting into the fenestra ovalis; and the extra-stapedial part, also called *extra-columella*, having an expanded end butting against the eardrum and bearing a *dorsal* and an *internal* process (Fig. 63*C*). There appears to be no exact correspondence between these parts and the operculum and plectrum of amphibians. From the dorsal process a ligament extends between the eardrum and the expanded end of the extra-columella. The middle ear of the Crocodilia (Fig. 63*D*) is similar to the typical plan except that the internal process is wanting. The hyoid arch (horn of the hyoid) of the Crocodilia persists throughout life as a tendon which extends from the extra-columella through a passage in the quadrate bone into the articular bone of the lower jaw, where it becomes continuous with Meckel's cartilage.

9. The lower jaw.—The lower jaw or mandible is composed of two halves or *rami*, united in front by a symphysis. Each ramus consists of six separate bones. Near the posterior end of the jaw is a large vacuity, the *external mandibular foramen*, and in front of this on the inner side a smaller *internal mandibular foramen*. The bones of each ramus are: the *dentary*, bearing teeth and forming the outer surface of the anterior two-thirds of the ramus; the *splenial*, of about the same shape and size as the dentary and in the same position on the inner surface; the *angular*, below the external mandibular foramen, and separated from the splenial by the internal mandibular foramen; the *supra-angular* or *surangular*, above the external mandibular foramen; the *coronoid*, a small bone on the inner surface, between the anterior ends of the angular and the supraangular; and the *articular*, above the posterior end of the angular and bearing a concavity for articulation with the quadrate. A cavity exists in the interior of the ramus between the dentary and the splenial. This cavity is occupied in life by Meckel's cartilage, which, it may be recalled, is the original lower half of the first gill arch. The posterior end of this cartilage has ossified into the articular, which is therefore the only cartilage bone in the lower jaw. The articulation between upper and lower jaw is by way of the articular and the quadrate, a condition found in the majority of land vertebrates. Draw the lower jaw from both inner and outer views.

10. The hyoid apparatus.—As has already been explained, this is derived from the hyoid arch and remaining gill arches. The hyoid apparatus is generally missing on dried skeletons, and for its study preserved specimens are necessary. It consists of a broad cartilaginous plate, the *body* of the *hyoid*, and a pair of processes or *horns* (*cornua*) extending posteriorly and dorsally from the body, one on either side. The cornua are partially ossified. The body of the hyoid apparatus is derived from the bases of the second and other arches, while the horns are remnants of the third arch. The second (hyoid) arch, as noted above, persists as a tendon in the quadrate and articular bones. In lieu of alligator material the hyoid apparatus may also be studied on turtle skeletons. In turtles the hyoid apparatus consists of a median ventral plate, the body of the hyoid, located in the floor of the mouth cavity, and two pairs of posteriorly extending processes, the anterior and posterior horns of the hyoid; they are portions of the third and fourth gill arches.

11. The teeth.—The teeth of the alligator occur in a single row on the margin of the premaxillae, maxillae, and dentary bones; but among other reptiles teeth may also be borne on the palatine and pterygoid bones. Note that the teeth are of simple form and are all alike (homodont) except that certain ones, the grasping teeth, are considerably larger than the others. The teeth of the lower jaw fit between those of the upper jaw when the jaws are closed, but some upper teeth overlap the lower ones. The teeth are thecodont, set into sockets in the jaw; other present reptiles have mostly acrodont or pleurodont teeth. On an isolated tooth note large pulp cavity widely open at the base; this condition is related to the ability of reptiles to replace the teeth an indefinite number of times. Numerous developing teeth occur in the jaw alongside the roots of the old ones.

In poisonous snakes there are special teeth termed *poison fangs*, provided with a groove or a canal (formed by the closing of the groove) for the conduction of the venom. The venom comes from glands of the nature of salivary glands. The vipers and rattlesnakes have the canal type of poison fang. There is a pair of poison fangs, large pointed teeth, one ankylosed to each maxilla and folded back against the roof of the mouth when not in use. By means of a movable chain of bones, including the maxilla, palatine, pterygoid, ectopterygoid, and quadrate, the poison fangs can be "erected," i.e., brought to a vertical position for striking into the prey. In another group of poisonous snakes, the elapine snakes, which includes the cobras, the poison fangs are immovable and permanently erect and are provided along their front faces with a poison groove.

H. THE MAMMALIAN SKULL

1. Distinguishing features of the mammalian skull.—The mammalian skull is completely ossified except for a small part of the ethmoid region. It is notable for the setting-off of a *facial* region or snout from the enlarged *cranial* region containing the brain. The number of bones in the mammalian skull is greatly reduced, as compared with the number in primitive tetrapods (Figs. 58, 59), partly through loss of dermal bones, partly through extensive fusions, especially among the cartilage bones. Thus, as Williston says: "The most primitive reptiles had no less than seventy-two separate bones in the skull; the human skull has but twenty-eight inclusive of the (six) ear bones." Other characteristic features are: the formation of a secondary palate, causing a posterior displacement of the internal nares; the presence of a temporal fossa more or less confluent with the orbit, so that certain bones of the maxillary arch are forced away from the skull to form the zygomatic arch; the greater differentiation of the internal ear; the reduction of the quadrate and articular to middle ear bones; the jaw articulation between the squamosal and the dentary; the loss of all lower jaw bones except the dentary; and the marked heterodonty. All the typical features of the mammalian skull are already evidenced in the extinct mammal-like reptiles (order Therapsida). These reptiles were of the synapsid type, and the mammalian temporal fossa is an expansion of the reptilian infratemporal fossa.

2. General features and regions of the skull.—The cat and the rabbit are here presented as examples of the mammalian skull. They are typical of the orders Carnivora and Rodentia, respectively. The student must learn the bones of the mammalian skull and be able to state to what skull component each bone belongs.

The skull is a hard, bony case composed of separate bones, immovably jointed together in dovetail fashion along the sutures. The *facial* region supporting the nose and eyes is distinguishable from the expanded posterior *cranial* region inclosing the brain and including the middle and internal ears. At the anterior end of the facial region are seen the two external nares, separated in life by a cartilaginous partition, which constitutes part of the *septum of the nose.* At the side of the facial region, the large, nearly circular orbit is partially separated by bony projections from the *temporal fossa* behind it, filled in life with muscles; the temporal fossa is very small in the rabbit. The lower boundary of these fossae is formed by a projecting arch, the *zygomatic arch,* a very characteristic feature of the mammalian skull. In the cat the orbit is bounded posteriorly by two *postorbital* processes, one projecting dorsally from the middle of the zygomatic arch and the other descending from the roof of the skull. In the rabbit a postorbital process extends downward and backward from the roof of the orbit and is connected in life to the zygomatic arch by a ligament which thus marks off a small temporal fossa. Above the orbit is a projecting margin, the *supraorbital arch,* the posterior end of which forms the postorbital process already mentioned; in the rabbit its anterior end also forms a projection.

The cranial portion of the skull presents the following features. At the posterior end is the large *foramen magnum;* on each side of this is a projection, the *occipital condyle,* which articulates with the atlas. Lateral and slightly anterior to each occipital condyle is a conspicuous hollow expansion, the *tympanic bulla,* which contains the middle ear. On the posterior surface of each bulla are two processes, an anterior *mastoid* process and a posterior *jugular* process. The mastoid process will be seen to be part of a bone, differing in its rough and pitted surface from the other bones of the skull; this bone, the *petromastoid* bone, contains the internal ear. The tympanic bulla opens laterally by a large opening, the *external auditory meatus,* which in the rabbit is bounded by a bony tube. The meatus leads into the cavity of the bulla, which is the *tympanic cavity* or cavity of the middle ear. From the dorsal side of the bulla a ridge begins which proceeds across the posterior part of the skull to the other bulla; this ridge is the *superior nuchal line* (or *lambdoidal ridge*) and is the most anterior point of attachment of the muscles of the vertebral column. From the middle of this ridge there projects posteriorly the *external occipital protuberance,* slight in the cat but forming a rectangular prominence in the rabbit.

On the ventral surface the anterior part of the skull is occupied by the *hard palate.* Dorsal to this are the nasal passages, which open at the posterior end of the hard palate by the *internal nares* or *choanae.* The hard palate contains a pair of openings, the *incisive* foramina or *anterior palatine* foramina; these are small and at the anterior extremity of the palate in the cat but much longer and more prominent in the rabbit. They lead into the nasal cavities. At the posterior end of the zygomatic arch on its ventral side is a depression, the *mandibular fossa,* for the reception of the lower jaw. Medial to this is the *pterygoid fossa,* for the attachment of certain muscles. The pterygoid fossa in the cat is continuous with the temporal and orbital fossae, while in the rabbit it is included between two projecting plates of bone which point toward the tympanic bulla.

3. Dermal bones of the roof of the skull.—Beginning just behind the external nares are the paired *nasals,* roofing the nasal cavities; next posterior, the paired *frontals;* and last, the paired *parietals,* terminating at the superior nuchal line. Between the posterior ends of the parietals there is generally a small triangular *interparietal* bone, the boundaries of which are not always distinct. This interparietal bone is believed to be homologous to the postparietals (=dermosupraoccipitals) of primitive tetrapods and already exists as an unpaired bone in therapsid reptiles. The frontal bones form the supraorbital arches and postorbital processes already noted as bounding the orbit above and behind. In the anterior wall of the orbit is the small *lacrimal* bone, which has at its anterior end the opening of the *nasolacrimal duct,* by which the tears drain into the nasal cavity.

4. The bones of the upper jaw and palate.—The maxillary arch consists, on each side, of the following elements: *premaxilla,* ventral to the anterior nares and bearing teeth; *maxilla,* forming the side of the facial region of the skull and also bearing teeth; *malar* or *jugal,* forming most of the zygomatic arch; and *temporal,* completing the zygomatic arch and covering the side of the cranial part of the skull, including the tympanic bulla. The premaxillae send *frontal processes* dorsally alongside the nasal bones; these are very pronounced in the rabbit. The premaxillae also form the anterior part of the hard palate by means of their *palatine processes,* which meet in the median ventral line and include the incisive foramina. The maxilla is the main bone of the facial region; in the rabbit it is much fenestrated. It forms part of the anterior wall of the orbit, ventral to the lacrimal bone, by its *orbital process;* it extends to the frontal bone above by its *frontal process;* its *palatine process* meets its fellow in the median ventral line continuing the hard palate; its *alveolar process* bears teeth; and its *zygomatic process* constitutes the beginning of the zygomatic arch. The malar or jugal bone is distinct in the cat but in the adult rabbit is fused to the zygomatic process of the maxilla. The temporal bone is a compound bone characteristic of mammals. It consists of a *squamous portion*, which by its zygomatic process completes the zygomatic arch and which also contributes to the cranial wall, ventral to the parietal; of the *tympanic bulla*, composed of a tympanic bone of uncertain homology; and of the *periotic* or *petromastoid* bone, only slightly visible on the surface and consisting of the fused otic bones of lower vertebrates. The squamous part of the temporal bone is homologous to the squamosal bone of the lower vertebrates.

The palatal region is distinguished, as already noted, by the formation of the secondary (or "false") palate. The anterior part of this is composed of the palatine processes of the premaxillae and maxillae; its rear part is formed of the palatine bones which embrace the internal nares. Above the palatines the extreme posterior tip of the *vomer* is visible in the roof of the internal nares. The vomer forms the base of the nasal septum and is best seen later in the study of the sagittal section of the skull. The rear parts of the palatines extend up into the orbit, and in the rabbit there is a very deep cleft between them in the midventral line.

Observe that the quadrate is wanting. The lower jaw consequently articulates with the squamosal (temporal) by means of a depression, the mandibular fossa, on the under surface of the zygomatic process of the temporal bone. This feature distinguishes mammalian skulls from those of all other vertebrates. Because of the absence of the quadrate, all of the bones of the upper jaw are membrane bones.

5. The occipital region.—This region surrounds the foramen magnum and consists of a single *occipital* bone extending from the superior nuchal line to a point between the anterior ends of the tympanic bullae. It is really composed of the four occipital bones present in reptiles; these are distinct in embryonic and young mammals (Fig. 66) but fuse later. The occipital bone bears the two occipital condyles, derived from the single one of reptiles by the retrogression of the basioccipital contribution; two condyles are characteristic of mammalian and amphibian skulls, whereas birds and reptiles have a single condyle. The dorsal part of the occipital may appear to extend anteriorly between the posterior ends of the two parietals because the interparietal bone is commonly fused to the occipital; this is the case in man. The occipital bone includes the lambdoidal ridge and the jugular process resting on the bulla.

6. The otic capsules and middle ear.—As already explained, the bones of the otic capsules are all fused together and fused with the squamosal and the tympanic bulla to form the temporal bone, a bone very characteristic of mammals (Fig. 66*C*, p. 188). That part of the temporal bone which is composed of the otic bones is named the *periotic* or *petromastoid* bone. The separate otic bones of which the petromastoid is composed can be seen only in early embryonic stages. The mastoid portion of this bone is visible on the external surface of the skull between the bulla and the occipital; it projects over the bulla as the mastoid process, very prominent in man as the bump behind the pinna. The petrous portion of the petromastoid bone incloses the internal ear and is visible only from the inside of the skull. The tympanic part of the temporal bone is termed the *tympanum* and forms the wall of the *tympanic cavity*, which contains the middle ear and its three little bones, the *auditory ossicles*. The latter can be seen in well-cleaned cat skulls as a delicate chain of bones extending across the anterior part of the tympanic cavity. If specimens are available, identify them as follows: the *malleus* or *hammer*, having an enlarged knob or *head* at one end and a slender "handle" or *manubrium*, the end of which abuts against the eardrum; the *incus* or *anvil*, a smaller bone with two pointed processes; and the *stapes* or *stirrup*, shaped like a stirrup.

The tympanum is not necessarily inflated to form a bulla; in fact, all degrees of this condition may be found among mammals. The composition of the tympanum varies greatly, and surrounding bones often participate in its walls. Also, in mammals it usually contains two bones of its own—the *ectotympanic*, often called simply *tympanic*, and the *entotympanic*. The former is a dermal bone and forms the partial or complete ring which supports the eardrum; it also may contribute to the bony external meatus. The entotympanum is a cartilage bone which appears to have no homologue in lower vertebrates. It generally forms all or part of the bulla.

The history of the auditory ossicles of mammals has been studied by a long list of comparative anatomists; and general agreement has been reached that the malleus represents the reduced articular bone, the incus the reduced quadrate, and that the stapes is the remnant of the columella auris of reptiles and hence derived from the hyoid arch. It is also believed by many that the ectotympanic is one of the reduced dermal bones of the lower jaw, probably the angular. Stages in the reduction of the quadrate and lower jaw bones to auditory ossicles can be followed in a series of fossil reptiles ancestral to mammals, and embryological evidence also supports the foregoing conclusions (Fig. 63*E*). The joint between the head of the malleus and the incus represents the articulation of lower jaw with the quadrate seen in nonmammalian vertebrates; with this joint removed into the middle ear, it is necessary for what remains of the lower jaw (namely, the dentary bone) to form an articulation with the skull bone adjacent to the quadrate (namely, the squamosal). This dentary-squamosal articulation is, as already noted, a distinguishing feature of the mammalian skull.

7. The posterior sphenoid region.—On the ventral surface of the skull in front of the occipital is the basisphenoid. This extends laterally in front of the tympanic bullae as processes, the wings of the basisphenoid, or *alisphenoids*, separate in young stages. The alisphenoids meet the squamous part of the temporal bones dorsally and extend forward as the *pterygoid processes*, which meet the posterior ends of the palatines. In the rabbit the pterygoid process presents two backwardly projecting thin plates of bone, the *lateral* and *medial lamellae*, which inclose between them the pterygoid fossa. The medial lamella in the rabbit and the pterygoid process itself in the cat bear a pointed process, the *hamulus*. The pterygoid processes ossify from two centers embryologically (Fig. 66*B*).

Much argument has occurred concerning the homologues of the mammalian alisphenoids and pterygoids (pterygoid processes) in reptiles. It is now generally accepted that the mammalian alisphenoid is homologous to the epipterygoid bone of reptiles (absent in the Crocodilia, present in *Sphenodon*). This was already stated to be a derivative of the palatoquadrate cartilage and not a true bone of the cranial floor (p. 165). The homology of the pterygoid is

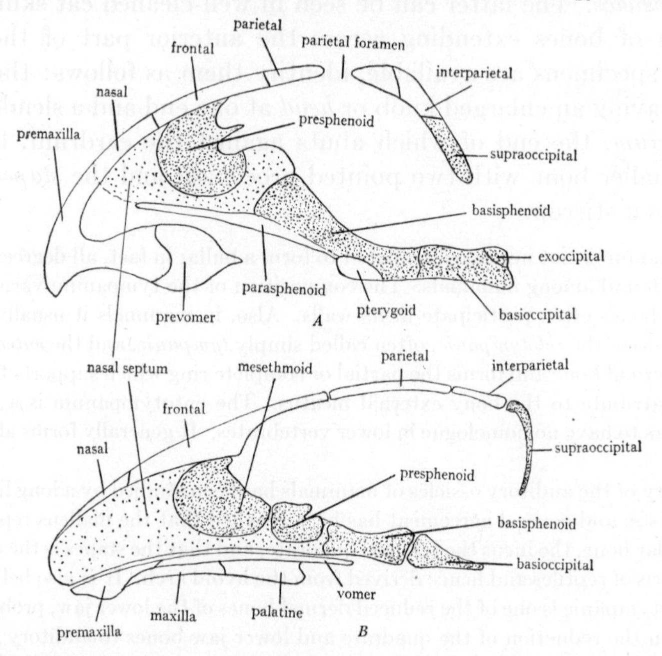

FIG. 65.—Median sagittal section to show arrangement of skull bones in *A*, a therapsid reptile (*Dicyonodon*) (after Broom, 1926); and *B*, in a developing mammal (*Procavia* = *Hyrax*) (after Broom, 1927). Note especially the cartilage bones of the basicranial axis. Cartilage bones, close stippling; cartilage, open stippling; dermal bones, blank.

much more confused. This bone ossifies in most mammals from two centers; and in *Echidna* also there are two pterygoids, a dorsal and a ventral; the latter forms the hamulus and is separate in adult monotremes. According to De Beer (1937), the dorsal element is homologous to the lateral parts of the reptilian parasphenoid, and the ventral one is the reptilian pterygoid. Parrington and Westoll (1940) agree that the ventral one or hamulus represents the reptilian pterygoid, but they take the view that the dorsal element corresponds to the reptilian ectopterygoid. The pterygoid processes of the mammalian skull would then represent the combined pterygoids and ectopterygoids of reptiles, and the alisphenoids are the epipterygoids. Other views about the pterygoids are listed by De Beer (1937), and it seems impossible at present to reach a conclusion.

8. The anterior sphenoid region.—In the median region in front of the basisphenoid is the slender *presphenoid;* this in the rabbit is in the deep cleft between the palatines. The presphenoid sends wings into the orbit which meet the frontal bones above and contain the large optic foramen for the passage of the optic nerve; in the cat this is the most anterior of a row of four foramina. The wings of the presphenoid are best seen in a lateral view of the skull. They are separate in young stages and are believed to be homologous to the *orbitosphenoids* of lower tetrapods. The presphenoid ossifies from two centers (Fig. 66*B*); its history was discussed above (p. 164). In man, basisphenoid, alisphenoids, presphenoid, and orbitosphenoids are fused into one bone, termed the sphenoid (Fig. 66*B*).

Draw a ventral view of the skull, showing sutures accurately.

9. The ethmoid region and the sagittal section of the skull.—The ethmoid region can be studied only in sagittal sections. There should be available sections cut slightly to one side of the median line, so that one-half retains the septum of the nose and the other half exposes the turbinals. The section shows that the interior of the cranial portion of the skull is occupied by a large cranial cavity, divisible into three regions of unequal size. The most posterior cavity, inclosed by the occipital and temporal bones, is the *posterior* or *cerebellar fossa* of the skull. Its anterior boundary is marked by a prominent (cat) or slight (rabbit) ridge or shelf of bone, the *tentorium*, which in life is completed by a membrane. In the lateroventral wall of the cerebellar fossa is a rounded area of very hard, compact bone bearing two openings; this is the *petrous* part of the petromastoid bone and incloses the internal ear. The greater part of the cranial cavity comprises the *middle* or *cerebral fossa*, extending forward from the tentorium. Its roof and walls are formed by the frontal, parietal, and temporal bones; its floor by the sphenoids. In the floor of the cerebral fossa, located in the basisphenoid bone, is a marked saddle-shaped depression, the *sella turcica*, in which in life the pituitary body is lodged. The presphenoid bone contains a cavity, the *sphenoidal sinus*. In the anterior part of the frontal bone, cavities, the *frontal sinuses*, are also present. The anterior end of the cranial cavity is the small *anterior* or *olfactory fossa* located between the anterior parts of the two frontal bones. The olfactory fossa is separated from the nasal cavities, which lie in front of it, by a nearly vertical plate of bone, perforated by numerous holes, the *cribriform plate* of the ethmoid. This plate is best seen in the intact skull of the cat by looking through the foramen magnum. The plate pierced by holes like a sieve is then seen closing the anterior end of the cranial cavity. Our study of the sagittal section shows that the floor of the skull is composed of a chain of cartilage bones—occipital, sphenoids, and ethmoid—on which the brain rests. These bones, as has already been explained, are derived from the chondrocranium.

The nasal cavities or nasal fossae are inclosed partly by cartilage bones and partly by dermal bones. The roof of the cavities consists of the nasal bones and a small part of the frontals. The two cavities are separated by a median, vertical, bony partition, the *perpendicular plate* of the ethmoid; in the living state this is continued to the anterior nares by a cartilaginous plate. The two together constitute the *septum* of the nose. The bony part of the septum, i.e., the perpendicular plate of the ethmoid, is small in the rabbit. Dorsally the septum meets the nasal and (cat) frontal bones; ventrally it meets the *vomer*, an elongated bone dorsal to the maxillae and palatines. The posterior end of the septum meets the cribriform plate.

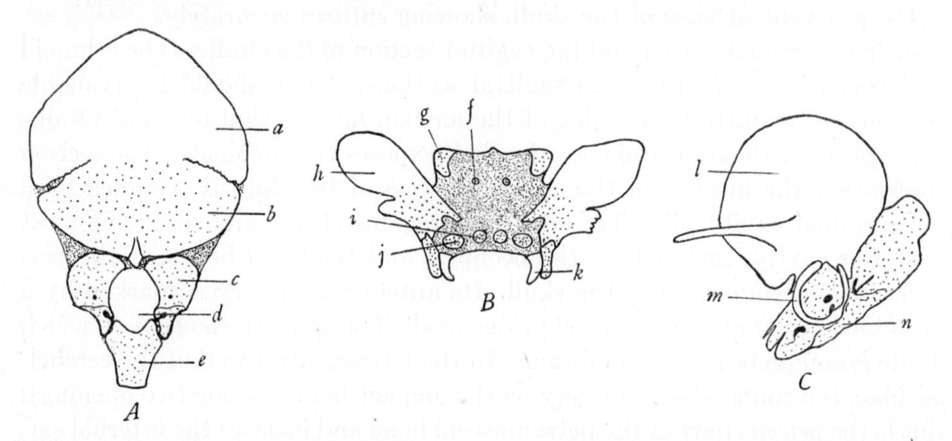

FIG. 66.—Some bones of the human skull at an early age, showing their compound nature. *A*, the occipital bone at birth, showing the five elements of which it is composed: *a*, interparietal; *b*, supraoccipital; *c*, exoccipital; *d*, the foramen magnum; *e*, basioccipital. *B*, the sphenoid bone in an embryo of four months, showing its components: *f*, center for the presphenoid; *g*, center for the orbitosphenoid; *h*, alisphenoid; *i*, center for the basisphenoid; *j*, center for the lingula; *k*, two centers for the pterygoid. *C*, the temporal bone at birth, showing its three components: *l*, squamosal; *m*, tympanic; *n*, petromastoid or periotic. Dermal bones blank; cartilage bones open stippling; cartilage, close stippling. The subsequent ossification of the cartilage obliterates the boundaries between the components. (*A* and *C* from specimens; *B* after McMurrich's *Development of the Human Body*, copyright by P. Blakiston's Son & Co.)

The homology of the mammalian vomer, an unpaired dermal bone, is also still under dispute. It was formerly supposed to represent the fused prevomers of reptiles, which were then called vomers. This was succeeded by the present view that the vomer is homologous to the parasphenoid bone of lower tetrapods; when this idea became prevalent, the name of the paired bones previously called vomers in lower tetrapods was changed to prevomers, and this usage has been followed in this book. But the parasphenoid homology has recently been challenged by Parrington and Westoll (1940), who believe the parasphenoid has been wholly lost in mammals and who return to the original and more sensible view that reptilian prevomers and mammalian vomer are identical. It will be noted that the vomer in the section of the cat skull occupies the same position as the prevomer of the alligator skull.

On the half of the skull where the septum is missing, the *turbinated bones* or *conchae* may be studied. They are peculiar, delicate, grooved, and folded bones which occupy the lateral walls of the nasal cavities and fill most of the interior. The most posterior of these bones is the *ethmoturbinal* or *ethmoid labyrinth*, situated just in front of the cribriform plate and composed of folds which inclose spaces known as the ethmoid cells. In front of the ethmoturbinal is the smaller *maxilloturbinal*, borne on the inner surface of the maxilla. A single elongated bony ledge, the *nasoturbinal*, lying above the uppermost scrolls of the ethmoturbinals, occurs on the inner surface of each nasal bone. The function of the turbinals is to increase the respiratory and olfactory surface of the nose. The ethmoturbinals are covered by olfactory epithelium, whereas the maxillo- and nasoturbinals serve to moisten and strain the inspired air. Turbinals occur in reptiles and birds and present much variation among mammals; homologizing them in different groups has so far proved impractical.

The bony part of the septum of the nose appears to represent the mesethmoid in some mammals and to be developed from the presphenoid (interorbital septum) in others. The ethmoturbinals of mammals develop from the orbitonasal lamina. Their union with the septum between the branches of the olfactory nerve forms the cribriform plate; and all three structures—septum, cribriform plate, and ethmoturbinals—together constitute the ethmoid bone of the mammalian skull. A small bit of the ethmoid bone is visible on the surface in the cat between the lacrimal and frontal bones.

Draw the sagittal section.

10. The foramina of the skull.—The skull is pierced by numerous openings for the passage of nerves and blood vessels and sometimes other structures. These are listed below for convenient reference.

Cat:

a) *Incisive* foramina. Anterior end of ventral side of maxillae; connect roof of mouth with nasal cavities.

b) *Infraorbital* foramen. Large opening in the maxilla at the beginning of the zygomatic arch; for the passage of certain branches of the fifth cranial nerve and blood vessels.

c) *Nasolacrimal* canal. Anterior end of the lacrimal bone and passing through the maxilla into the nasal cavities; for the draining of the tears.

d) *Sphenopalatine* foramen. In that part of the palatine bone which extends into the orbit, posterior to the lacrimal; for the passage of branches of the fifth nerve into the nasal cavity.

e) *Posterior palatine* canal. The posterior end of this is immediately ventral to the sphenopalatine foramen; its anterior end is in about the middle of the palatine process of the maxilla; for the passage of a branch of the fifth nerve to the palate.

f) Optic foramen. In the orbitosphenoid part of the presphenoid, in the posterior part of the orbit, most anterior of a row of four foramina; for the passage of the optic nerve.

g) Orbital fissure. Second and largest of the row of four; through it pass the third, fourth, and sixth nerves to the muscles of the eyeball, and a part of the fifth.

h) Foramen rotundum. Third of the row; in the alisphenoid bone; transmits part of the fifth nerve.

i) Foramen ovale. Last of the four; in the alisphenoid; transmits part of the fifth nerve.

j) Canal for the auditory tube. In the anterior wall of the bulla, its roof formed by the alisphenoid; for the passage of the auditory tube from the pharynx into the bulla.

k) Pterygoid canal. Each bulla terminates anteriorly in a point (styliform process) lying on the basisphenoid; this point is directed to a minute opening, the pterygoid canal, lying in the suture between the basisphenoid and the pterygoid process; for the passage of a branch of the fifth nerve into the bulla.

l) Jugular foramen. Large foramen on the medial side of the posterior end of the bulla; for the passage of the ninth, tenth, and eleventh nerves.

m) Hypoglossal foramen. In the medial side of the preceding foramen; for the passage of the twelfth nerve.

n) Stylomastoid foramen. At the ventral tip of the mastoid process; for the passage of the seventh nerve.

o) Internal auditory meatus. Opening in the center of the petromastoid bone, as seen in the sagittal section; for the passage of the auditory nerve into the brain. Behind it in a diagonal ventral direction are two foramina leading to the jugular foramen.

Rabbit:

a) Incisive foramina. As in the cat but larger.

b) Infraorbital foramen. As in the cat but more slitlike and elongated, forming an *infraorbital canal,* opening into the orbit above the expanded part of the maxilla.

c) Nasolacrimal canal. As in the cat, situated under the pointed anterior end of the supraorbital arch.

d) Posterior palatine foramen. On the ventral side, in the suture between the palatine process of the maxilla and the palatines; forms the anterior opening of the *palatine canal.* The posterior opening is at the posterior end of the expanded mass of the maxilla located in the orbit. This canal is for the passage of a branch of the fifth nerve.

e) Sphenopalatine foramen. In common with the posterior end of the palatine canal just described; for the passage of branches of the fifth nerve.

f) *Anterior* and *posterior supraorbital* foramina. The projecting anterior and posterior ends of the supraorbital arch are continued in life by ligaments, thus forming foramina, through which branches of the fifth nerve pass.

g) *Optic* foramen. Large opening in the center of the orbit; for the passage of the optic nerve.

h) *Orbital fissure.* Posterior and ventral to the preceding, and including the foramen rotundum of other mammals; for the third, fourth, and sixth nerves to the eyeball, and the greater part of the fifth.

i) *Anterior, middle,* and *posterior sphenoidal* foramina. Three foramina in a row in the lateral lamella of the pterygoid process at the place where this is continuous with the alisphenoid; for the passage of part of the fifth nerve.

j) *Foramen lacerum.* In front of the tympanic bulla on the ventral surface and including the foramen ovale of other mammals; for the passage of part of the fifth nerve and an artery.

k) *External carotid* foramen. A small foramen in the middle of the medial surface of the bulla; for the passage of the internal carotid artery, the other end of the canal lying in the foramen lacerum.

l) *Jugular* foramen. Just posterior to the preceding, in the depression between occipital condyle and bulla; for the ninth, tenth, and eleventh cranial nerves, and a vein.

m) *Hypoglossal* canal. Including small apertures posterior to the preceding; for the passage of the twelfth (hypoglossal) nerve.

n) *Stylomastoid* foramen. In front of the middle of the mastoid process; for the passage of the seventh nerve.

o) *Internal auditory meatus.* In the center of the petromastoid bone, seen in the sagittal section, below a larger depression; for the passage of the auditory nerve from the internal ear to the brain; the seventh nerve also begins its exit in common with this foramen and appears on the outside of the skull through the stylomastoid foramen.

11. The lower jaw.—The lower jaw or *mandible* consists of a single pair of bones, the *dentaries,* fused in front by a *symphysis.* All other bones seen in the lower jaw of the alligator have vanished (except the articular, which has become the malleus of the middle ear). The horizontal part of the mandible is named the *body;* the vertical part, the *ramus.* (In the lower vertebrates each half of the lower jaw is also named ramus.) The posterior end of the mandible (cat) extends dorsally into a strong *coronoid* process, which in the natural position projects into the temporal fossa. In the rabbit this is reduced to a slight projection, which forms the lateral boundary of a deep groove. The articulating surface of the mandible is borne on the *condyloid* process. The depressed areas in the posterior part of the mandible are for the insertion of the muscles of mastication. Near the anterior tip of the mandible on the outer surface is the *mental* foramen (or two in the cat), through which the nerve of the lower jaw exits. Near the caudal end of the inner surface is the *mandibular* foramen, through which the nerve enters and pursues a course in the interior of the mandible to the mental foramen. In the rabbit there is an additional foramen just above the mandibular foramen, for the passage of a vein.

In the absence of the quadrate and of all of the bones of the lower jaw except the dentary, the articulation of the lower jaw to the skull is *between the dentary and the squamosal.* This feature distinguishes mammals from all other vertebrates, for in the latter the articulation is between the articular and the quadrate. The condition found in mammals is, however, approached by those reptiles directly ancestral to mammals.

12. The teeth.—The teeth of mammals are thecodont, heterodont, and diphyodont; some of them are of complicated form, having several cusps or ridges and more than one root. They are important in working out taxonomic and phylogenetic relationships. The cat serves as an example of the teeth of a carnivorous mammal; the rabbit, of a gnawing type, with chisel-like front teeth and grinding back teeth.

Cat: At the tip of the jaws are six small, simple teeth, named *incisors.* On either side of the incisors is a *canine,* a long, sharp but simple tooth. Back of the canine on each side are four teeth in the upper jaw, three in the lower. These teeth are mostly more complicated than the preceding, having more than one *cusp* or pointed projection and more than one root. These teeth are known as *premolars* and *molars.* In the upper jaw the first three on each side are premolars, and the first two in the lower jaw; the last tooth on each side in each jaw is a molar. Note that the upper incisors are borne on the premaxillae; the other teeth of the upper jaw on the maxilla. Between the canines and the premolars is more or less of a gap, known as a *diastema.*

Rabbit: At the tip of the premaxillae are borne four chisel-like *incisor* teeth, a small pair behind a large anterior pair. The chisel-like form of the incisors is characteristic of rodents and is due to the fact that the enamel is present on the front face of the tooth only; the posterior face, being composed of the softer dentine, wears away, leaving a sharp edge to the enamel. Furthermore, the incisors of rodents continue to grow indefinitely, so that the loss from use at the tip is replaced by growth at the root. At the tip of the lower jaw are two similar incisors. Posterior to the incisors is a very wide gap or *diastema.* The canine teeth, found in most mammals posterior to the incisors, are missing in the rabbit and rodents in general. On the alveolar process of the maxilla are borne six teeth in the upper jaw and five in the lower, on each side. Of these the first three in the upper and first two in the lower jaw are *premolars;* the last three, *molars.* The teeth are ridged crosswise, an adaptation for the grinding of vegetable food; the ridges consist of enamel with dentine between them.

From the foregoing it is seen that the teeth of mammals are differentiated into four kinds: incisors, canines, premolars, and molars. The incisors are usually simple teeth with a cutting edge; they are always borne in mammals on the premaxillae. The canine is usually a simple conical tooth and is often greatly enlarged, as in the tusks of the walrus and boars and other pigs. The premolars and molars are often spoken of together as *cheek teeth* and are not exactly definable from each other, although in general the cheek teeth increase in size and complexity from before backward. In man and many other mammals the premolars are those cheek teeth which occur in the milk dentition and are replaced by permanent premolars, whereas the molars have no milk predecessors and are not replaceable. Most mammals have two sets of teeth: the juvenile or *milk* dentition, which falls out early in life; and the permanent dentition which replaces the milk teeth and includes, in addition, the molars not represented in the milk dentition.

Mammalian heterodonty is often described by means of a *dental formula,* which expresses the number of each different kind of tooth in each half-jaw from the anterior median line to the posterior end of the jaw. The upper teeth are placed in the numerator of the formula; the lower teeth, in the denominator. The complete original mammalian dentition is believed to have consisted of three incisors, one canine, four premolars, and three molars in each half-jaw, or a total of 44, and is written thus: $\frac{3\cdot1\cdot4\cdot3}{3\cdot1\cdot4\cdot3}$. In the cat the formula is $\frac{3\cdot1\cdot3\cdot1}{3\cdot1\cdot2\cdot1}$; in the rabbit, $\frac{2\cdot0\cdot3\cdot3}{1\cdot0\cdot2\cdot3}$; in man, $\frac{2\cdot1\cdot2\cdot3}{2\cdot1\cdot2\cdot3}$.

Since the primitive vertebrate tooth is believed to have been of simple conical shape with a single root, it remains to explain the complex mammalian molar tooth with several cusps or ridges and two or three roots, and several theories have been advanced. The idea that the mammalian molar arose through the fusion of simple teeth has now been generally abandoned; and the contrary theory, of a process of cusp formation on a simple tooth, is now generally accepted. Such additional cusps are supposed to have arisen as upgrowths from the *cingulum,* which is a ridge around the tooth at the neck. The formation of two such secondary cusps leads to the *trituberculate* tooth, believed by many to be the primitive form of the mammalian cheek tooth from which all more complicated types are derived. The trituberculate tooth has three cusps in the form of a triangle. These cusps are called *protocone, paracone,* and *metacone* in the upper jaw; *protoconid, paraconid,* and *metaconid* in the lower jaw (Fig. 67). The protocone is at the apex of the triangle facing inside; the protoconid, at the apex facing outward. Embryological evidence indicates that paracone and protoconid represent the original reptilian cusp. In the lower jaw the tooth soon develops a posterior extension, termed the *heel* or *talonid;* and this often bears additional cusps. The teeth are so placed that the lower triangles fit between the upper triangles when the jaw is closed and the protocone rests upon the talonid, which means that each upper tooth is slightly behind its mate of the lower jaw. Among herbivorous mammals the cheek teeth tend to enlarge, broaden, and develop many cusps, or the cusps alter into crescents (*selenodont* type of grinding molar) or fuse into long and complicated ridges (*lophodont* type of molar).

Alternative theories to the trituberculate theory are the *multituberculate* theory, that the early mammalian cheek teeth had a number of cusps, and Bolk's *dimer* theory, according to which each mammalian tooth corresponds to two fused tricuspidate reptilian teeth.

13. The hyoid apparatus.—This is, as already explained, the remnant of the hyoid and other gill arches. It is generally absent on prepared skeletons, and isolated specimens will be provided for its study.

The hyoid of the **cat** consists of a bony bar placed in the root of the tongue just in front of the larynx; this bar is called the *body* of the hyoid, or *basihyal.* From it extend two pairs of processes or horns. The anterior horns consist of a chain of four pieces, called *hypohyal, ceratohyal, stylohyal,* and *tympanohyal,* of which the last is attached to the tympanic bulla just ventral to the stylomastoid foramen. The groove which it occupies can generally be seen on the side of the bulla. The posterior horn consists of a single piece (*thyrohyal*) attached to the larynx. The body and anterior horns belong to the hyoid arch, of which an additional piece, as already learned, forms the stapes in the

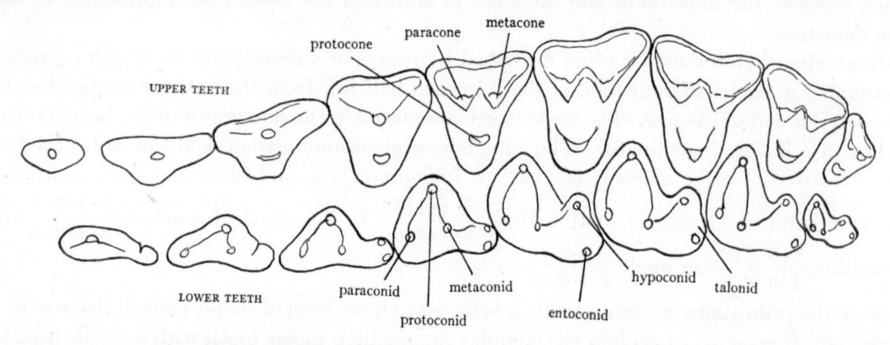

FIG. 67.—Cheek teeth of a generalized primitive mammal (after Gregory, 1916). Premolars at left gradually change into molars; upper jaw above shows three-cusped or tritubercular condition of molars; lower jaw below shows also the talonid or heel with additional cusps; note alternation of upper and lower teeth but meeting of protocone and talonid.

middle ear; the posterior horns are remnants of the third gill arch (Fig. 63*E*). The remaining gill arches are represented in the cartilages of the larynx, studied in a later chapter.

The hyoid apparatus of the **rabbit** consists of a stout bone, the *body* of the hyoid, situated at the base of the tongue in front of the larynx. It bears two pairs of processes or *horns.* The anterior horn is a short piece connected by a muscle with the jugular process of the skull. The posterior horn is a longer piece connected by ligament with the larynx and by muscle with the jugular process. The body and anterior horn are remnants of the hyoid arch, the posterior horn of the third gill arch.

14. Some variants of the mammalian skull and teeth.—

a) *Monotremes:* Both *Echidna* and *Ornithorhynchus* have a long facial region. The latter is remarkable for the forceps-like form of the premaxillae and maxillae which support the ducklike beak. In each half of each jaw at about the middle is a broad, horny grinding plate which incloses the true teeth. In *Echidna* the face is extended into a long, pointed beak which lacks teeth.

b) Edentates: This group, including the anteaters, sloths, and armadillos, displays a variety of skull form. Notable is the very elongated face of the anteaters, formed of the long nasals and maxillae, with the nares at the tip. The name edentate means "without teeth," and, in fact, a few edentates are totally devoid of teeth, such as the South American anteaters; but most edentates possess simple cheek teeth, although incisors are always lacking. The teeth have no enamel, and most edentates lack a milk dentition.

c) Ungulates: The horse is notable for the very large and elongated face, as compared to the relatively small cranial region; the large and long nasal passages have well-developed turbinals. The orbit is completely inclosed by a ring of bone (postorbital bar), beneath which, however, the orbital and temporal fossae are confluent. There is a strong lambdoidal ridge and marked *paroccipital* processes (projections of the exoccipitals behind the tympanic region). The horse has a complete dentition, although the canines are small and often lacking in females; there is a long diastema in front of the cheek teeth, used for insertion of the bit. The cheek teeth, six in each half of each jaw, are high and broad, with a complicated pattern of enamel ridges—hence of the *lophodont* type. The evolutionary history of the horse, which has been very well worked out, shows that these ridges are formed by the confluence of originally separate cusps.

In the pig tribe the cranial part of the skull has a high median crest (occipital crest) and strong lambdoidal ridge; the orbit is open behind. There is usually a complete dentition, and all four canines may form tusks, growing from persistent pulps and often making an upward curve. The cheek teeth are *bunodont* (with separated rounded cusps).

In the ruminants (sheep, oxen, antelopes, etc.) there are large nasal cavities with well-developed turbinals, the orbit is closed by a postorbital bar, the frontals bear the horn cores, there are often long paroccipital processes, and there is usually an evident bulla with a bony meatus. The upper jaw lacks incisors and instead has a horny pad for occlusion with the lower incisors. The cheek teeth, adapted for grinding, are of the *selenodont* type; i.e., the cusps are altered into crescents, formed of enamel, mostly four in number.

d) Elephants: The skull of the elephants is remarkable for its short, high form, very thick walls permeated with air spaces, the high placement of the large external nares, and the prolongations of the premaxillae in front of the nares for the support of the tusks. The tusks represent the only incisors and grow throughout life from persistent pulps. They form the ivory of commerce, really a fine-tubed dentine. The only other teeth are the cheek teeth, of which there are six—three milk premolars and three molars in each jaw-half; but these are so large that only one is present at a time in each jaw-half, and the six teeth succeed each other, in place of occurring simultaneously as in other mammals. Each tooth is made of transverse plates of dentine cemented together and topped by enamel ridges. The number of ridges increases from the first to the last cheek tooth.

e) Carnivores: The skull of the fissipede carnivores is very similar to that of the cat except that the facial region is often more elongated. Various degrees of inflation of the tympanic bulla occur. The teeth are notably adapted for tearing flesh, as shown by the large, sharply pointed canines and the progression in size and cutting edges of the cheek teeth, culminating in the large *carnassial* tooth. In the pinnipede carnivores (seals, walruses) the skull is similar to that of the fissipedes, but the cheek teeth are simpler and all alike, for there is no carnassial tooth. The walrus has enormous upper canines, which form the tusks.

f) Rodents: The rodents are divisible into the Duplicidentata, including the hares and rabbits, in which there is a small second pair of upper incisors behind the principal pair; and the Simplicidentata, including all other rodents (rats, mice, squirrels, beavers, guinea pigs, etc.), in which this second pair is lacking. Canines and the first premolars are missing, so that the rodent skull is easily recognized by the gnawing incisors, the long diastema between them and the cheek teeth, and the closely set grinding series of cheek teeth. The remarkable chisel-like incisors extend far back into the jaws, behind the cheek teeth in the case of the lower incisors, and grow throughout life from persistent pulps.

(*g*) *Cetaceans:* Perhaps the most remarkable variant of the mammalian skull is seen among the whales and dolphins. The cranial region is short, broad, and high; the external nares are moved far back and open upward, by way of the single blowhole; the very reduced nasal bones lie behind them; the parietals have been pushed to the sides; the enlarged supraoccipital spreads forward in the roof, almost meeting the nasals; the frontals extend laterally as broad supraorbital plates; the squamosal is also curiously expanded; and the jugal is often reduced to a slender bar. The nasal passages are small and without turbinals. The face forms a beak or rostrum, caused by the great elongation of premaxillae and maxillae. Teeth, when present, are often limited to the lower jaw; they are homodont, of simple conical shape, and monophyodont and may be very numerous. The whalebone whales are devoid of teeth, although tooth germs occur in the young; instead there is along each side of the upper jaw a series of frayed horny plates hanging vertically and acting to sieve out food from the water. They are folded back when the mouth is closed. These constitute the whalebone or baleen of commerce. The narwhal has only two teeth, upper incisors; and in the male, the left one grows as a tusk to a length of 10–12 feet.

The great sperm whale has a curious boxlike head, caused by the presence of a huge oil cavity which may hold as much as 20 barrels of oil. This cavity is above the skull, formed by a transverse crest behind the nares plus the concave surface of the premaxillae and maxillae.

(*h*) *Primates:* The skull of man and the anthropoid apes is distinguished by the large, rounded cranial portion, the flattened facial portion, gradually taking on a vertical orientation, the complete separation of the orbits from the temporal fossae, the reduction of the nasal cavities and the turbinals, the large mastoid process, the absence of a tympanic bulla, and the extensive fusions of the skull bones. The unspecialized teeth indicate an omnivorous diet. There are only two, instead of the usual three, incisors; the canines are slightly enlarged; and the cheek teeth are bunodont.

I. SUMMARY

1. The vertebrate skull or cranium is formed by the fusion of three originally separate components: the cartilaginous cranium or chondrocranium, the dermatocranium, and the splanchnocranium.

2. The chondrocranium arises through the fusion of originally separate cartilages—the parachordals alongside the notochord and the trabeculae anterior to these—with the nasal and otic capsules and with lateral cartilages, the orbital cartilages. Its roof is open and its side walls imperfect, but one or more arches occur at the rear end.

3. The ossification of the chondrocranium produces the cartilage bones of the skull, which are limited to its ventral surface and rear end. The principal cartilage bones are: the supraoccipital, exoccipital, basioccipital, basisphenoid, pleurosphenoid, orbitosphenoid (or sphenethmoid), the otic bones, and some bones of the nasal capsules. As many as five otic bones may ossify in the otic capsules, but the usual number is three; in mammals they are fused to form a periotic or petromastoid bone. In mammals there may be one or two additional bones in the basicranial axis, the presphenoid and mesethmoid.

4. The dermatocranium consists of dermal bones (fish scales) which incase the chondrocranium and jaws and become part of the definitive skull. In early (extinct) tetrapods they form a complete roof for the skull, but in the evolution of higher tetrapods they reduce greatly

in number, chiefly through loss. Vacuities also tend to arise in the posterior part of the roof, and these temporal fossae are of importance in the evolution of the various reptilian groups and of birds and mammals.

5. The number of cartilage bones also tends to reduce, especially in mammals, through the fusion of originally separate bones.

6. The splanchnocranium consists of the gill arches, typically seven in number. Of these the first or mandibular arch forms the upper and lower jaws, of which the upper jaw becomes incorporated into the skull, while the lower jaw forms a movable joint with it. The second or hyoid arch and the remaining five arches contribute to the hyoid apparatus and the laryngeal cartilages in land vertebrates.

7. The upper jaw (palatoquadrate cartilages) ossifies to form cartilage bones and also becomes ensheathed in dermal bones which incorporate into the skull as bones of the palate and maxillary arch. Only one cartilage bone formed in the palatoquadrate cartilage persists through the vertebrates; this is the quadrate bone. The lower jaw cartilages, termed Meckel's cartilages, give rise to one main cartilage bone, the articular, at the rear end of the lower jaw and also become covered with dermal bones. The upper end of the hyoid arch may form the hyomandibula, typical of fishes.

8. In all tetrapods except mammals the lower jaw articulates with the skull by way of the articular and quadrate bones. In mammals, because of the reduction of these two bones, the articulation is by way of dermal bones, the dentary and the squamosal.

9. In primitive tetrapods there is a large number of dermal bones in the lower jaw; these become progressively lost or reduced until in mammals the lower jaw consists of a single pair of bones, the dentaries.

10. The middle ear contains a sound-transmitting apparatus which in all tetrapods except mammals consists of one element or chain of elements, termed the columella auris, which is a remnant of the hyoid arch. This persists in mammals as the stapes. In addition the mammalian middle ear contains two more little bones, the malleus and the incus, believed to represent the greatly reduced articular and quadrate, respectively.

11. The teeth of vertebrates are homologous to the placoid scales of elasmobranchs. Usually of simple form and all alike among lower tetrapods, they become heterodont, i.e., differentiated into several kinds in mammals, where they are also thecodont, set in sockets in the jaw bones. Borne in lower tetrapods on various jaw and palatal bones, they become limited in higher ones to the jaw margins. Of the several theories advanced to explain the complicated form of mammalian molar teeth, the tritubercular theory (three-cusped tooth the primitive type) is now most accepted.

REFERENCES

BEER, G. R. DE. 1926. The orbito-temporal region of the skull. Quart. Jour. Micr. Sci., **70**.

———. 1930. The early development of the chondrocranium of the lizard. *Ibid.*, **73**.

———. 1937. The Development of the Vertebrate Skull.

BEER, G. R. DE, and WOODGER, J. H. 1930. The early development of the skull of the rabbit. Phil. Trans. Roy. Soc. London, B, **218**.

BROOM, R. 1911. On the structure of the skull in cynodont reptiles. Proc. Zoöl. Soc. London.

———. 1926. On the mammalian presphenoid and mesethmoid bones. *Ibid.*, Part 1.

———. 1927. Some further points on the structure of the mammalian basicranial axis. *Ibid.*, Part 1.

———. 1935. The vomer-parasphenoid question. Ann. Transvaal Mus., **18**.

EATON, T. H., JR. 1939. The crossopterygian hyomandibular and the tetrapod stapes. Jour. Wash. Acad. Sci., **29**.

EDGEWORTH, F. H. 1925. On the autostylism of Dipnoi and Amphibia. Jour. Anat., **59**.

———. 1926. On the hyomandibula of Selachii, Teleostomi, and *Ceratodus*. *Ibid.*, **60**.

GAUPP, E. 1900. Das Chondrocranium von *Lacerta*. Anat. Hefte, **15**.

GOLDBY, F. 1925. The development of the columella auris in the Crocodilia. Jour. Anat., **59**.

GOODRICH, E. S. 1918. Restorations of the head of *Osteolepis*. Jour. Linn. Soc. London, Zoöl., **34**.

———. 1925. On the cranial roofing bones of the Dipnoi. *Ibid.*, **36**.

GREGORY, W. K. 1910. The orders of mammals. Bull. Amer. Mus. Nat. Hist., **27**.

———. 1916. The Cope-Osborn theory of trituberculy and the ancestral molar patterns of the primates. *Ibid.*, **35**.

———. 1922. The Origin and Evolution of the Human Dentition.

GREGORY, W. K., and NOBLE, G. K. 1924. The origin of the mammalian alisphenoid bone. Jour. Morph., **39**.

KLAAUW, C. J. VAN DER. 1931. The auditory bulla in some fossil mammals. Bull. Amer. Mus. Nat. Hist., **62**.

MUMMERY, J. H. 1924. The Microscopic and General Anatomy of the Teeth, Human, and Comparative. 2d ed.

OSBORN, H. F. 1907. Evolution of Mammalian Molar Teeth.

PARKER, W. K. 1883. On the Structure and Development of the Skull of the Crocodilia. Trans. Zoöl. Soc. London, **11**.

PARRINGTON, F. R., and WESTOLL, T. S. 1940. On the evolution of the mammalian palate. Phil. Trans. Roy. Soc. London, B, **230**.

PLATT, JULIA B. 1897. The development of the cartilaginous skull in *Necturus*. Morph. Jahrb., **25**.

REED, H. D. 1920. The morphology of the sound-transmitting apparatus in caudate Amphibia and its morphological significance. Jour. Morph., **33**.

SCHMIDT, K. P. 1919. Herpetology of the Belgian Congo. Bull. Amer. Mus. Nat. Hist., **39**.

VERSLUYS, J. 1903. Entwicklung der Columella auris bei den Lacertilien. Zool. Jahrb., Abt. Anat., **19**.

WELLS, GRACE. 1917. The skull of *Acanthias vulgaris*. Jour. Morph., **28**.

WILDER, H. H. 1903. The skeletal system of *Necturus*. Mem. Boston Soc. Nat. Hist., **5**.

WILLISTON, S. W. 1904. The temporal arches of the Reptilia. Biol. Bull., **7**.

X. THE COMPARATIVE ANATOMY OF THE MUSCULAR SYSTEM

A. INTRODUCTION

1. Kinds and origin of muscle.—The muscles of the vertebrate body may be divided into two general classes, the *involuntary* and the *voluntary*. The involuntary or smooth muscles occur in the walls of the digestive tract and other viscera and in the skin and certain derivatives thereof. They originate through the transformation of mesenchyme cells, which may be of various origins; but typically they come from the splanchnic mesoderm of the hypomere, since in development the hypomere closes around the archenteron and its derivatives. The voluntary or striated muscles, on the other hand, arise from the myotomes (also called muscle plates), with certain exceptions noted below. The myotomes, it will be recalled, are those portions of the epimeres remaining after the sclerotomes and dermatomes have been given off. From their original dorsal position the myotomes grow down between the hypomere and the skin, and those of opposite sides meet in the midventral line. Review pages 76–77. In this way there is produced a complete coat of voluntary muscles, lying beneath the skin. This muscle coat is divided into dorsal and ventral parts by the horizontal skeletogenous partition (also called lateral septum) which extends from the vertebral column to the lateral line. The muscles dorsal to the septum are called the *epaxial;* those below the septum the *hypaxial* muscles. It should be noted, however, that the horizontal septum is absent in *Amphioxus* and cyclostomes and hence apparently is not a primitive chordate feature.

The muscles originating from the myotomes are called the *parietal* or *somatic* muscles. Not all of the voluntary muscles are, however, of myotomic origin. In the gill region of vertebrates a system of voluntary muscles is developed for moving the gill arches. Since the gills and related parts are of entodermal origin, the muscles in the walls of the gill region are homologous to the gut musculature, and embryology shows that they come from the mesoderm of the hypomeres.[1] For this reason the gill arch musculature is frequently called the *visceral* musculature, although, unlike the musculature of the viscera, it is voluntary and striated. A better term, and one that avoids confusion, is *branchial* or *branchiomeric* musculature; this will be used here. There are consequently as to embryonic origin two kinds of voluntary muscle: the parietal or somatic muscles, derived from the myotomes and covering most of the body; and the branchial muscles, derived from the hypomere, limited to the gill region, and innervated by cranial nerves.

As a result of the presence of the gill region the musculature for the midventral region of the head and neck comes from the myotomes behind the gills; these myotomes send out downgrowths, which then turn forward and grow anteriorly to the jaw. The muscles so originating are termed the *hypobranchial* musculature and are innervated by spinal nerves, like the rest of the myotomic musculature. There is also an *epibranchial* musculature above the gill region, but this is of slight importance, being limited to elasmobranchs.

The terms muscle and muscular system refer only to the voluntary muscles, as only these are studied in vertebrate dissection; the study of the involuntary muscles belongs properly to histology. It is assumed that the student understands the histological difference between smooth and striated muscle.

[1] Edgeworth in his scholarly book *The Cranial Muscles of Vertebrates* takes the view that the gill arch musculature is also of myotomic origin, but this is contrary to the findings of embryology.

2. Terminology of muscles.—The function of the muscular system is to move the skeleton. The muscles are therefore attached to the bones directly or indirectly, and these attachments usually consist of connective tissue. When the connective tissue forms flat sheets, these are termed *fasciae* or *aponeuroses;* when concentrated into a band, this is termed a *tendon.* A muscle is attached at its two ends and is more or less free in the middle. The place of attachment which in any particular movement remains fixed when the muscle contracts is called the *origin;* that which is caused to move is the *insertion;* and the in-between free part of the muscle is the *belly.* When a muscle has more than one origin, these are called *heads.* When there are several points of attachment segmentally arranged, these are termed *slips.* Persistent myosepta in the course of a muscle which extends over more than one body segment are called *inscriptions.* The specific movement produced by a muscle is called its *action* and is brought about by the shortening of the muscle.

The best method of naming muscles is to combine the origin and the insertion into a compound word in which the origin precedes, as *pubofemoralis,* which would mean a muscle originating on the pubis, inserting on the femur, and causing the femur to move toward the pubis on contraction. Unfortunately, many of the muscle names in use in vertebrate anatomy have been carried over from human anatomy, where they were based on shape, action, fanciful resemblances, etc. In many muscles origin and insertion are interchangeable, i.e., either end may be held fixed and the other end moved.

The voluntary muscles are so arranged that the action of one muscle or group is offset by the antagonistic action of another muscle or group. The chief actions of muscles are: *flexion, extension, adduction, abduction, rotation, elevation, depression,* and *constriction.* Flexion is the bending of a part, and the antagonistic action of extension is the straightening of a part; these terms are applied chiefly to limbs, and muscles with such action are named *flexors* and *extensors,* respectively. Adduction is the drawing of a limb toward the ventral surface; abduction, away from the ventral surface; and the corresponding muscles are called *adductors* and *abductors.* Muscles with a rotating action are termed *rotators, pronators,* and *supinators.* Elevation and depression are illustrated by the closing and opening of the lower jaw, and the muscles concerned are called *levators* and *depressors,* respectively. Constrictor muscles act by compressing a part; when encircling an aperture, such as the mouth or anus, they are termed *sphincters.*

3. Relation of muscle and nerve and the question of muscle homology.—A muscle is the end organ of a motor nerve and is caused to contract by a nervous impulse originating in the central nervous system and transmitted to the muscle along its nerve. If the nerve is cut across, the animal is no longer able to make the muscle contract; but contraction can be elicited by artificially stimulating the nerve stump attached to the muscle. However, denervation of a muscle eventually results in its atrophy.

Primitively, the chordate muscular system consists of an axial series of myotomes from anterior to posterior end. The number of myotomes and vertebrae is equal, and each myotome is supplied by a corresponding nerve issuing from between two adjacent vertebrae, for, as already learned, vertebrae and myotomes alternate. The evidence indicates that this primitive relation of a nerve of a given body segment to the myotome of that segment is very constant and that the same nerve continues to supply the muscles derived from that myotome through all their evolutionary changes. Hence, nerve supply is the most valuable criterion for determining muscle homology throughout the vertebrate classes. However, this principle, which seems simple enough in itself, is not so easy to apply in practice because of the branching and anastomoses of nerves and the difficulty of tracing fine branches. Other criteria of muscle homology are the origins and insertions, and the topographical relations to other muscles. The former are subject to considerable shift, especially in case of loss of the bones to which they were originally attached; e.g., the disappearance of the cleithrum forced muscles that were fastened to it to find new attachments.

Thus the task of homologizing muscles among the vertebrates is one of extreme difficulty, and most of the progress in this matter is relatively recent. Muscles are very plastic structures and change their shape and attachments with adaptive structural modifications of the animal. Generalized connecting groups are extinct and are known chiefly by their skeletons; but it is possible to reconstruct probable muscle arrangements by means of the marks left on bones by muscle attachments. Much progress has been made in such reconstructions of limb muscles of extinct amphibians and reptiles by Romer and others. But the whole subject of muscle homology is too difficult for adequate presentation here. To the natural difficulties of the subject has been added an immense confusion of terminology; the same muscle in the same animal has been called different names by different investigators and has received different names in different animals; nor has complete uniformity been reached by modern investigators attempting to produce a terminology that will express homologies among vertebrate groups.

It is not the purpose of the following account to give all the details of muscle origins and insertions or to describe every muscle of the animals dissected; nor is it desired that the student should memorize the details or mutilate the specimen by trying to dissect out all the muscle attachments. The intention is rather to give an idea of the primitive arrangement of the vertebrate musculature and to follow some of the larger evolutionary changes.

B. GENERALIZED VERTEBRATE MUSCULATURE AS EXEMPLIFIED BY THE DOGFISH

The musculature of the dogfish serves as a convenient starting-point for the study of vertebrate musculature, although, of course, it has undergone some modification from the primitive chordate plan. The latter is presumably exemplified by *Amphioxus*, where, it will be recalled, there is a complete set of myotomes from anterior to posterior end. These myotomes in *Amphioxus* nearly cover the gill region, which has an intrinsic musculature of hypomeric origin. A horizontal septum is absent in primitive chordates.

The musculature of the dogfish will be considered under the heads of somatic, fin, branchial, and hypobranchial musculature.

1. The parietal or somatic muscles.—Strip off the skin from the dogfish at the base of the tail, in the neighborhood of the pelvic fins. In doing this, make a cut through the skin, grasp the cut edge with the fingers, and strip off a piece of skin without the further use of the knife. Trying to cut the skin from the body usually results in cutting into the muscles. After removing a considerable area of skin, note that the body is completely sheathed in a coat of muscles, the parietal muscles. They consist of a series of zigzag myotomes separated from one another by white connective tissue partitions, the myosepta. In the middle of the side of the body there is a white longitudinal line, the outer edge of the horizontal skeletogenous septum, more briefly termed lateral septum. This divides the myotomes into dorsal or epaxial portions and ventral or hypaxial portions. The epaxial muscles form the *dorsal longitudinal bundles*, usually two in number, as can be seen in cross-sections. The hypaxial muscles are also divisible into two longitudinal bundles: a *lateral* one, just below the lateral septum and easily distinguishable by its darker color; and a *ventral* longitudinal bundle, which in cross-sections can be seen to be subdivided into two bundles. Thus in most elasmobranchs the fibers of the myotomes are arranged into five longitudinal bundles.

In the midventral line note a white partition, the *linea alba*, which separates the myotomes of the two sides of the body. Muscles do not cross the middorsal or midventral lines and hence are always paired, even when appearing single.

Draw from the side a portion of the body to show the myotomes.

2. Fin musculature.—Strip off the skin from both surfaces of a pelvic fin, preferably in a female, as the muscles of the clasper introduce complications. On the dorsal side of the fin a mass of muscles will be seen extending into the fin; it originates on the fascia (connective tissue) of the myotomes and on the iliac process of the pelvic girdle and may be regarded as an abductor, levator, or extensor of the fin. On cutting through this mass the myotomes will be revealed beneath it; this suggests that the fin muscles are outgrowths of the myotomes, as, in fact, is the case. The cut also reveals a deeper part of the muscle which originates on the metapterygial cartilage and radiates into the fin. Both parts of the muscle insert by way of connective tissue on the pterygiophores and ceratotrichia. On the ventral side of the pelvic fin, the muscle mass, which acts as a depressor, adductor, or flexor, is divisible into two parts, one extending from the linea alba and puboischiac bar to the metapterygium, and the other from the latter into the fin, inserting like the above.

Similarly skin the base of the pectoral fin. On the dorsal side of the fin is seen the fanlike extensor or levator muscle originating on the scapular process of the pectoral girdle and adjacent fascia. A similar ventral flexor or depressor mass originates on the coracoid bar of the pectoral girdle. Both muscles radiate into the fin in bundles and insert on the pterygiophores.

Thus the fin musculature consists essentially of a dorsal extensor mass and a ventral flexor mass. Both masses have an obviously segmental appearance, and embryology shows that the fin musculature originates by budding from the myotomes. Typically there are two buds from each myotome (Fig. 43*A*); each of these again subdivides, and the upper two become dorsal fin musculature, the lower two ventral fin musculature.

3. The branchial musculature.—Make a median ventral incision in the skin of the head from the pectoral girdle forward and strip off the skin in an upward direction over the gill slits and up to the middorsal line. Note that the epaxial part of the myotomes (dorsal longitudinal bundles) continues unaltered above the gill region and attaches to the rear part of the chondrocranium. The lateral longitudinal bundle is seen to terminate at the scapular process. Laterally and ventrally the ventral longitudinal bundles continue unaltered to the pectoral girdle, to which they are attached. The region in front of the pectoral fin is seen to be occupied by a set of muscles separate from the myotomes. These are the branchial muscles, which operate the gill arches and jaws. In the evolution of vertebrates, with the loss of gill respiration, the branchial muscles undergo striking changes but continue to serve the jaws and remains of the gill arches as well as spreading into new territory, and continue to be supplied by the same motor nerves (cranial) which go to them in the dogfish.

a) *Constrictor series:* The musculature covering the gill region from the pectoral fin to the eye and mouth is termed the *superficial constrictor* musculature and consists of six *dorsal* constrictors above, and six *ventral* constrictors below, the external gill slits. Only the most posterior part ("gill-hood" of Lightoller) of each of the last four constrictors is exposed on the surface, and these exposed parts are separated from one another by vertical lines of connective tissue (*raphe*). Most of each constrictor is concealed under the preceding constrictor, as can be seen by slitting along a raphe upward and downward from a gill slit. The innermost part of the constrictor supports the wall bearing the gill plates (branchial part of Lightoller, interbranchial muscle of other authors).

The second dorsal constrictor, belonging to the hyoid arch, is very large, extending from the raphe above the second gill slit forward to the muscle mass at the jaw angle; then continuing above this mass as the *epihyoidean* muscle which originates on fascia and the otic capsule and inserts on the hyomandibula. (From behind the spiracle there is seen emerging the hyomandibular trunk of the seventh cranial nerve [facial], which slants downward across the lower part of the epihyoidean muscle.) Immediately in front of the epihyoidean is the small *dorsal constrictor* (*craniomaxillaris*) muscle, passing in front of the spiracle, originating on the otic capsule and inserting on the palatoquadrate cartilage. (The muscle immediately in front of the dorsal constrictor belongs to the levator series [see below].) The large muscle mass at the jaw angle is termed the *quadratomandibularis;* it originates on the rear part of the palatoquadrate cartilage and inserts on the mandible (=Meckel's cartilage). The *preorbitalis* (also called *suborbitalis* and *levator labialis superioris*) will be found by dissecting deeply between the upper jaw and the eye.[2] It is a cylindrical muscle originating on the midventral surface of the chondrocranium, running along each side of the upper jaw, and narrowing posteriorly to a tendon which joins the quadratomandibularis. The latter plus the preorbitalis forms the *adductor mandibulae*, which acts to close

[2] In so dissecting, there will be uncovered the slender *labial* cartilage which bends around the mouth angle; there are also similar *extrabranchial* cartilages under the skin over the gill pouches; the significance of these cartilages external to the regular gill arches is unknown.

the lower jaw. The adductor mandibulae plus the dorsal constrictor are part of the first constrictor, that belonging to the mandibular arch.

In ventral view, the ventral constrictors of the last four gill pouches are seen to be similar to the dorsal ones. The first ventral constrictor is the very broad sheet originating on a midventral raphe and slanting forward on both sides to insert on the mandible as the *intermandibularis* muscle. The most posterior fibers attach to the fascia of the quadratomandibularis muscle, and this part is termed *mandibularis* by Lightroller. By carefully cutting through the middle of the intermandibularis of one side, another thin sheet will be found immediately above it; this is the rest of the second ventral constrictor, belonging to the hyoid arch, and properly termed *interhyoideus.*

The constrictors function in general to compress the gill pouches, forcing water out and closing the gill slits, also closing the mouth. These actions constitute the expiratory phase of respiration.

b) *Levator series:* On turning again to the dorsal side of the animal, there will be found directly in front of the dorsal constrictor, but a little deeper, the first levator or *levator maxillae superioris,* which originates on the otic capsule and inserts on the palatoquadrate cartilage next to the quadratomandibularis. It acts to raise the upper jaw. The second or hyoid levator is under the epihyoideus, to which it is fused. The remaining levators (3–8) are, according to Lightroller (1939), represented by the *cucullaris,* an elongated muscle mass lying between the dorsal longitudinal bundles and the upper ends of the last gill pouches. It appears like a continuation of the lateral longitudinal bundle in front of the scapular process. It originates on the fascia of the dorsal longitudinal bundle and inserts on the epibranchial cartilage of the last gill arch and also on the scapular process of the pectoral girdle, hence acting as a levator scapulae. The anterior end of the cucullaris thins out to a point. The whole levator series, termed *levatores arcuum,* acts to raise the gill arches.

c) *Interarcual series:* The upper ends of the gill pouches can be easily separated from the dorsal longitudinal bundles and the cucullaris (the space above the gill pouches so exposed is the anterior cardinal sinus [see p. 316]). By spreading these structures apart as much as possible, the gill cartilages will be seen; and running between their upper ends, slanting forward, are the short *interarcual* muscles. They extend chiefly between the pharyngobranchial cartilages of the gill arches and act to draw the arches together and so expand the pharynx.

4. The hypobranchial musculature.—This group of muscles occupies the region between the coracoid bar and the mandible, passing dorsal to the mandibularis and intermandibularis muscles. They strengthen and elevate the floor of the mouth cavity, strengthen the walls of the pericardial cavity, and assist in opening the mouth and expanding the gill pouches in inspiration of water. Immediately in front of the coracoid bar in the midventral region is a triangular area containing a pair of diverging muscles, the *common coracoarcuals,* which originate on the coracoid bar and insert on fascia just above the rear edge of the mandibularis. Slit up the midventral raphe forward to the middle of the mandible and note just above the constrictor layer a median (really paired) muscle, the *coracomandibular* (also called *geniocoracoid* and *geniohyoid*), extending forward to the mandible. Dorsal to this is a pair of strong muscles, the *coracohyoids* (=*rectus cervicis* of tetrapods, according to Edgeworth), running obliquely forward to insert on the basihyal. Just behind the center of the lower jaw, between the anterior parts of the coracomandibular and coracohyoids, a flat dark-colored mass is the *thyroid gland.* On cutting through the middle of the coracohyoids and deflecting them, there will be seen, dorsal to them, the *coracobranchials,* a muscle sheet extending obliquely laterally, with slips of insertion on the ceratohyal cartilage and on various parts of the last five gill arches.[3] To avoid cutting important blood vessels, the details of these muscles should not be traced at present.

The hypobranchial musculature arises by forward growth of the ventral ends of the myotomes behind the branchial region and may be regarded as a forward continuation of the ventral longitudinal bundles. The hypobranchial musculature, like the body myotomes, is innervated by spinal nerves, whereas the branchial musculature is innervated by cranial nerves. The coracobranchial series is in dispute, in that some think it belongs to the branchial musculature; but the majority opinion seems to be that this series (in elasmobranchs) is hypobranchial.

C. EARLY TETRAPOD MUSCULATURE, EXEMPLIFIED CHIEFLY BY NECTURUS

When vertebrates evolved from an aquatic to a land mode of life, the musculature underwent many accompanying alterations. With the gradual decline of gill respiration, much of the branchial apparatus and its musculature was lost, but some of the gill arches and muscles persist with altered functions. More complex movements are required of the jaws, with greater specialization of mandibular arch muscles. Walking limbs require a heavier musculature with broader attachments than fins, and their greater mobility and varieties of movement necessitate the splitting of the simple muscle masses of fish fins into a number of specific muscles. The trunk musculature undergoes less change but becomes more or less covered by the expanded musculature of the limb girdles.

The muscles of urodeles serve to illustrate the transition from aquatic to land vertebrates and have been supplemented by recent work on the musculature of extinct primitive tetrapods, such as *Eryops.*

Specimens of *Necturus* preserved in formalin are preferable for the study of the muscles. Make a middorsal incision extending the length of head and trunk. Loosen the cut edges of skin with the fingers, noting in the cut sur-

[3] Throughout this book the typical gill arches will always be referred to as the third to seventh arches; many books call them the first to fifth.

faces the flask-shaped cutaneous glands which secrete slime. Then with the fingers separate the skin from the muscles in a ventral direction until you have removed the skin from head, trunk, and appendages, leaving the gills in place. The white fibrous material between the skin and muscles is termed the subcutaneous connective tissue or superficial fascia. Confine the dissection of the muscles to one side of the animal.

1. The muscles of the trunk and tail.—These muscles preserve the generalized arrangement typical of primitive and embryonic vertebrates. They consist, as in the dogfish, of a series of myotomes, separated by myosepta. The myotomes are long, nearly rectangular blocks extending from the middorsal to the midventral line. Their narrowed dorsal ends slant forward. Note their division into *epaxial* and *hypaxial* portions by the horizontal septum. Although the trunk muscles appear to be unmodified, they are, in reality, already separating into layers. On cutting into the epaxial muscles they will be seen to constitute a mass whose fibers are all directed forward; this is the dorsal muscle mass, most of which forms the *dorsalis trunci,* whose fibers are attached to the myosepta and, deeper down, to the transverse processes of the vertebrae. A deeper portion of the dorsal muscle mass runs between the vertebrae (*interspinalis*). The general function of the dorsal muscle mass is lateral flexion of the body.

The hypaxial musculature is divisible into the *rectus* and the *oblique* groups. The oblique muscles form the sides of the body wall and, although seemingly one mass below the horizontal septum, are really split into three layers. The outer layer, or *external oblique* muscle, has fibers directed obliquely caudad and ventrad. On cutting through this rather thick layer, there will be found internal to it the thin *internal oblique* muscle, with fibers directed obliquely craniad and caudad. When this is cut through, the thin *transverse* muscle is revealed next the body cavity; it is formed of circular fibers. These three layers, typical of tetrapods in general, serve to compress the viscera and have other subsidiary actions according to their attachments.

On either side of the linea alba the hypaxial musculature is, as in tetrapods in general, differentiated into the *rectus abdominis* muscle, a narrow band of longitudinal fibers interrupted by regular segmental inscriptions (=myosepta). The rectus extends from the anterior edge of the pubis to the sternum and hyobranchial apparatus, and acts to compress the viscera, curve the body ventrally, and retract the hyobranchial apparatus.

The epaxial and hypáxial muscle masses continue into the tail and act to bend the tail from side to side. On skinning the tail, note the numerous mucus-secreting cutaneous glands in the tail fin.

2. General features of the limb musculature.—Note, particularly for the forelimb, the fanlike muscle sheets which originate on trunk and girdle and converge toward the limb, on both dorsal and ventral sides; these sheets illustrate the spread of limb musculature over the myotomes. A similar arrangement is seen on the ventral side of the hind limb but is here less evident dorsally. Infer actions of the dorsal and ventral muscle sheets on the limb as a whole. Within the limb itself the general arrangement comprises a set of muscles from each part of the limb to the next more distal part—e.g., from upper arm to forearm, forearm to wrist and hand, etc. The muscles of the

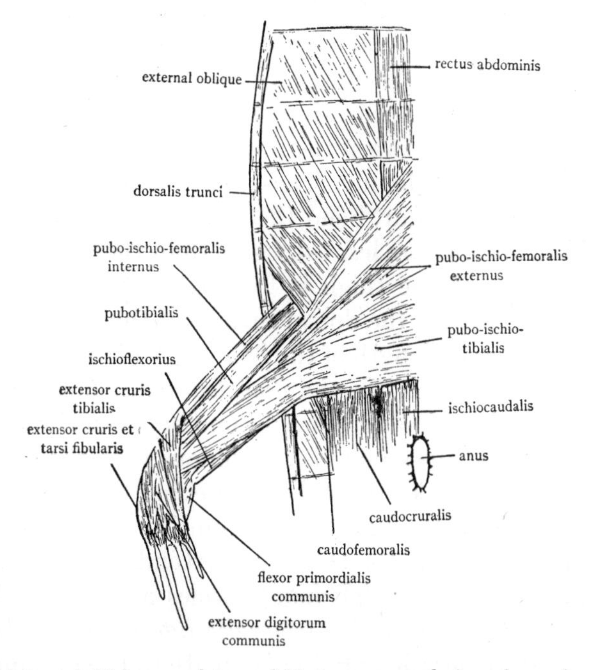

FIG. 68.—Pelvic and thigh musculature of *Necturus*, ventral view, drawn from a dissection

original upper surface of the limb sections have an extensor action; those of the lower surface, a flexor action. Note great similarity of the muscles of the distal parts of both limbs, less similarity of the proximal parts.

3. The muscles of the pelvic girdle and hind limb.—Here, as in the dogfish, the continuity of the muscles of the pelvic girdle and hind limbs with the body myotomes is more or less evident. The limb musculature obviously consists of dorsal abductor and ventral adductor masses which have become split up into a number of separate muscles.

The ventral surface of the pelvic girdle is covered by a flat, fan-shaped sheet of muscle divisible into two parts along a transverse line of connective tissue about the middle of the mass (Fig. 68). The most anterior muscle is the *pubo-ischio-femoralis externus*, originating on the median line of the girdle and inserting on the femur. It covers the rear end of the rectus abdominis; and, on lifting up its anterior edge, the latter muscle can be traced to its attachment on the anterior end of the pubic cartilage. The posterior part of the pubo-ischio-femoralis externus is covered by the second of the two muscles mentioned above, the *pubo-ischio-tibialis*, which originates on the median region of the pelvic girdle and passes to the tibia.

Around the anus there is a glandular mass, the *anal gland;* and, on scraping this away, there is seen a set of muscles running between the tail and the pelvic girdle. The median one is the *ischiocaudalis*, from the posterior border of the ischium into the tail; lateral to this is the *caudocruralis* (or *caudo-pubo-ischio-tibialis*), originating on the fascia of the pubo-ischio-tibialis. On turning the animal sidewise and separating the hypaxial mass from the caudocruralis, there will be seen between them a slightly deeper muscle, the *caudofemoralis*, originating on the femur. These three muscles, through their fixation on pelvic girdle and femur, bend the tail in various ways.

The dorsal girdle muscles are less extensive than the ventral ones. Dorsally the preaxial border of the thigh is formed of the *pubo-ischio-femoralis internus*, which originates on the dorsal surface of the puboischiac plate and inserts along a considerable length of the femur. Behind this is the *iliotibialis* and immediately posterior to this the *ilioextensorius;* both originate on the base of the ilium (noticeable as a curved bone between the myotomes) and are inserted on the tibia; the second is continuous with the aponeurosis over the knee and along the shank. Immediately behind the ilioextensorius is the slender *iliofibularis*, from the base of the ilium to the proximal end of the fibula. These muscles abduct the thigh and extend the shank.

On the ventral side of the thigh four muscles are visible (Fig. 68). The most anterior one is the *pubo-ischio-femoralis internus* seen above. Next to this is the *pubotibialis* from the puboischiac plate to the proximal part of the tibia. Next comes the distal part of the pubo-ischio-tibialis already noted (this part is sometimes called *gracilis*); and the most posterior muscle is the slender *ischioflexorius*. This blends with the pubo-ischio-tibialis at its origin but is distinct distally, where it forms the posterior edge of the thigh and inserts on fascia of the shank. The foregoing muscles serve chiefly to flex the shank; they also adduct the thigh.

The shank musculature comprises a dorsal extensor mass and a ventral flexor mass, both operating the feet. Twist the shank so that it is in line with the thigh. The extensor mass, on the dorsal side, is divisible into three muscles: a preaxial *extensor tibialis;* a median and largest mass, the *extensor digitorum communis*, which sends three conspicuous slips to the digits; and the postaxial *extensor fibularis*. Note the short extensors (*extensores breves*) on the proximal phalanges. On the ventral or sole surface of the foot the main muscle mass is the *flexor primordialis communis*, which continues distally into an aponeurosis, the *plantar fascia*, having tendinous slips to the digits.

4. The muscles of the pectoral girdle and forelimb.—On the ventral surface, the broad oblique sheet from the linea alba to the humerus is the *pectoralis;* note continuity of its fibers with the rectus abdominis and external oblique muscles. The anterior part of the pectoralis, where the fibers run transversely to the upper arm, is somewhat separable as the *supracoracoideus;* it overlies the rounded coracoid part of the pectoral girdle. In front of this and extending anteriorly on each side of the neck is the elongated

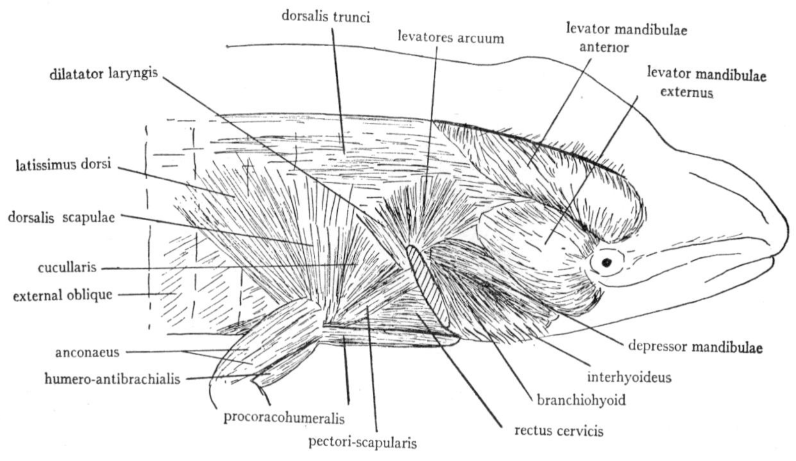

FIG. 69.—Shoulder and branchial musculature of *Necturus*, dorsolateral view, drawn from a dissection. The crosshatched area is the place where the gills have been cut away.

procoracoid cartilage; and this is covered by the *procoracohumeralis* muscle, which proceeds from the surface of the cartilage to the humerus, lateral to the insertion of the supracoracoideus. The procoracohumerales inclose between them a large area of neck, covered by the *sternohyoid* muscle. This is segmented by inscriptions, is obviously a forward continuation (above the pectoralis) of the rectus abdominis, and hence is more properly termed *rectus cervicis*.

Turn the animal sidewise, push the gills forward, and note the series of shoulder muscles from the fascia, suprascapula, and scapula to the humerus (Fig. 69). The most posterior one is the *latissimus dorsi* (or *dorsohumeralis*), originating by separate slips on the fascia of the epaxial muscle mass and inserting on the shoulder joint and humerus. In front of this is the *dorsalis scapulae* (representing the later *deltoid*) from the suprascapular cartilage to the humerus. Anterior to this is the *cucullaris*, originating on fascia and inserted on the girdle at the shoulder joint. The origin of the cucullaris is partly covered by the most posterior of a series of muscles above the gills (see below). Next anterior to the cucullaris is the *pectoriscapularis* (formerly *omohyoid*). This muscle slants obliquely backward to its origin on the scapula next to the insertion of the cucullaris. Between the pectoriscapularis and the procoracohumerales, the lateral part of the rectus cervicis is exposed on the side of the neck. Some separate strands of the rectus cervicis attached to the dorsal surface of the procoracoid cartilage are sometimes regarded as a distinct muscle (*omoarcual*).

Compare the shoulder muscles of *Necturus* with the more typical arrangement in the primitive extinct amphibian *Eryops* (Fig. 70).

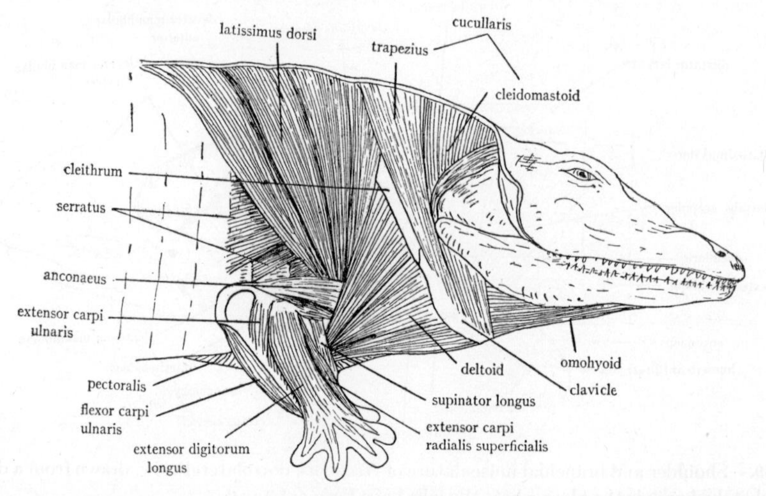

Fig. 70.—Shoulder musculature of the extinct amphibian *Eryops*, reconstruction (after Miner, 1925). Note muscle attachments to clavicle and cleithrum; with loss of the cleithrum, these shift elsewhere.

In ventral view the upper arm has a median slender muscle, the *humero-antibrachialis*,[4] which extends from the humerus to the radius. Behind this on the inner surface of the upper arm is the larger *coracobrachialis*. These muscles are flexors of the arm. The outer or back side of the arm is covered by a large extensor mass, the *anconeus*, divisible into four parts; it originates on shoulder girdle and humerus and inserts on the fascia over the elbow.

The musculature of the forearm closely parallels that of the shank. On the dorsal or upper surface there is a median *extensor digitorum communis*, and this is flanked on the preaxial side by the *extensor antibrachii et carpi radialis* and on the postaxial side by the *extensor antibrachii et carpi ulnaris*. On the ventral surface of the forearm there is a similar arrangement of a median *flexor primordialis communis*, a preaxial *flexor antibrachii et carpi radialis*,

[4] Often called *biceps;* but a true biceps, i.e., a flexor muscle from the pectoral girdle to the forearm, is lacking in present amphibians.

and a postaxial *flexor carpi ulnaris*, of which the antibrachial part is deeper down.

5. Branchial and hypobranchial musculature.—As already learned, these are the muscles of the head, jaws, gill arches, floor of the mouth cavity, and throat. As *Necturus* retains three pairs of gills and the corresponding arches, the arrangement of the branchial musculature bears considerable resemblance to that of the dogfish.

Above the gills is seen a series of thin, flat muscles, originating on the skin and fascia and passing to the gill arches; these are the *levatores arcuum* or *branchiarum*, the gill levators. The thin, narrow muscle behind them and in series with them, covering the origin of the cucullaris, is the *dilatator laryngis*, which acts to dilate the larynx and open the glottis. Although this muscle seems to be a derivative of the levator series, Edgeworth states that this is not the case.

On the dorsal surface of the head the dorsalis trunci is seen to terminate by attaching to the skull. The large muscle mass which covers the roof of the skull is the *levator* (or *adductor*) *mandibulae*, functioning to close the jaw (Fig. 69). It is divisible into two main portions: the elongated mass to either side of the middorsal line, termed the *levator mandibulae anterior;* and the large mass lateral to this, lying behind the eye, called the *levator mandibulae externus*. The large mass between the latter and the gills is divisible into two muscles, of which the anterior one is the *depressor mandibulae*, or opener of the jaw, and the posterior one, next the gills, the *branchiohyoideus*. The latter is traced further below. The levator mandibulae inserts on about the middle of the mandible, below the eye, and the depressor mandibulae inserts on the rear end of the mandible at its articulation with the quadrate region of the skull.

The ventral side of the head and the throat are covered with a sheet of transverse fibers, as in the dogfish. The anterior part of this sheet is the *intermandibularis*, extending from the median raphe to the mandible on each side. Behind this is the *interhyoideus*, which also inserts on the median raphe and originates chiefly on the fascia of the depressor mandibulae and branchiohyoideus. The interhyoideus is divisible into anterior and posterior parts; the latter is also called *sphincter colli* and *gularis*. Some of its fibers attach to the skin of the gular fold.

Cut through the intermandibularis and interhyoideus slightly to one side of the median raphe and deflect them. The branchiohyoideus can now be followed forward as it curves along the cartilage of the hyoid arch on which it inserts. To either side of the median line is seen the narrow *geniohyoid* muscle, which extends from the mandible to the second basibranchial cartilage (=urobranchial). The rectus cervicis also inserts on this cartilage;

and, in fact, the geniohyoids may be regarded as a forward continuation of the median part of the rectus cervicis. At the sides of the urobranchial cartilage and somewhat under the branchiohyoid muscle is a small glandular mass, the thyroid gland. On cutting through and deflecting the geniohyoids and pressing the branchiohyoid to the side, the gill arches are exposed. The *subarcual* muscles are now seen—short straplike muscles running longitudinally from one gill arch to the next; the first one from the hyoid to the third arch and the second from this to the fourth arch are easily seen. The *depressores arcuum* or gill depressors are muscles which run along the gill cartilages into the gills and can be seen lateral to the subarcuals. The attachment of the rectus cervicis to the hyobranchial skeleton is now exposed. Deep down between the rectus cervicis and the last gill cartilage will be seen the transverse fibers of the *transversi ventrales* series.

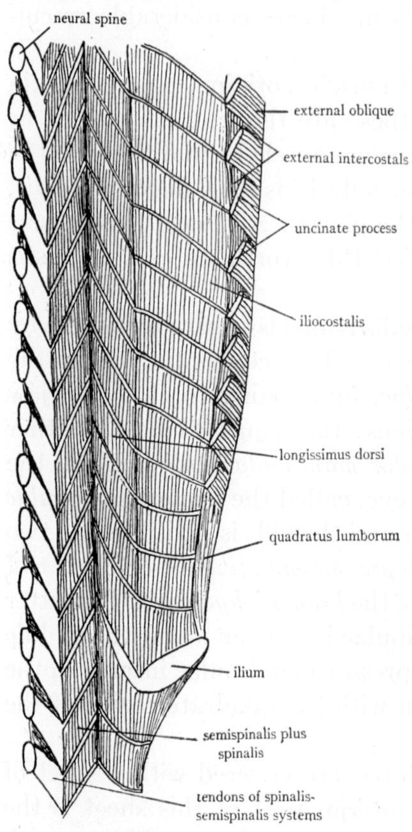

FIG. 71.—Epaxial musculature of *Sphenodon*, viewed from the dorsal side. (After Nishi, 1916.)

Drawings may be made of views not shown in Figures 68 and 69.

The levator mandibulae and the intermandibularis are mandibular arch muscles, homologous to the first constrictor set of elasmobranchs, and are innervated by the fifth cranial nerve (trigeminus). The depressor mandibulae and the interhyoideus are hyoid arch muscles, homologous to the hyoid levator and ventral hyoid constrictor, respectively, of elasmobranchs and are innervated by the seventh cranial nerve (facial). The levators and depressors of the gills, the branchiohyoid, the dilatator laryngis, the subarcuals, the transversi ventrales, and the cucullaris belong to the remaining gill arches and are innervated by the ninth (glossopharyngeal) and tenth (vagus) cranial nerves. The geniohyoids, rectus cervicis, and pectoriscapularis are hypobranchial muscles with a spinal innervation.

D. REPTILIAN MUSCULATURE

In reptiles the muscular system is well on the road to its final mammalian form. The greater development of the limbs, associated with higher elevation above the ground and need of greater variety of movements, entails greater splitting-up of the intrinsic limb masses into individual muscles and greater spread of shoulder and hip musculature over the trunk muscles. The primitive segmental arrangement of the latter is still evident, but the epaxial mass differentiates into a number of longitudinal series of muscles operating the spine and ribs and moving the head. A horizontal septum is lacking in amniotes—even in their embryos, according to Emelianov (1935).

Because of the losses and alterations of the turtle musculature, on account of the presence of the carapace, as well as the lack of a modern study of turtle muscles, only a generalized account of reptilian musculature will be given, with examples drawn mainly from *Sphenodon*, crocodilians, and extinct forms.

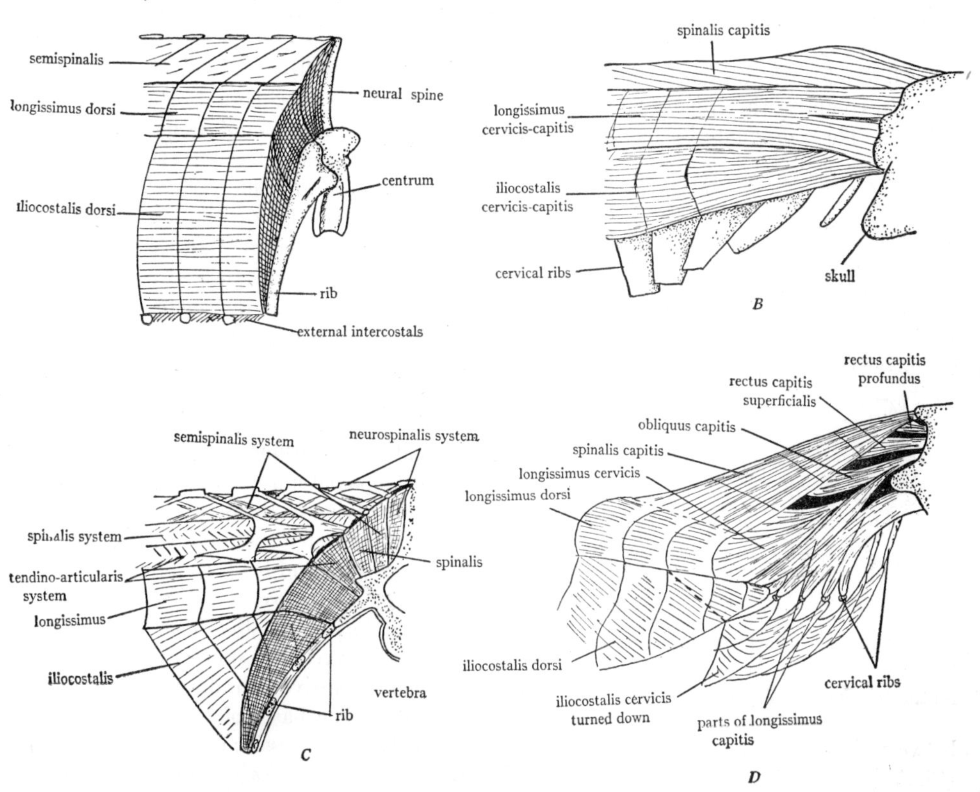

Fig. 72.—Epaxial musculature (side view) of the extinct cotylosaurian reptile *Diadectes* (*A* and *B*) (after Olson, 1936), and of the crocodile (*C* and *D*) (after Vallois, 1922). *A* and *C*, arrangement in the trunk; *B* and *D*, arrangement in the neck; note subdivisions of the longissimus, and, in *D*, the occipital muscles.

1. Reptilian trunk musculature.—The epaxial muscle mass, which in *Necturus* was seen to remain as an undivided bundle, is in reptiles subdivided into several longitudinal systems—some long, others segmental. These systems may again be divided into lateral and median groups. The lateral systems are of the long type and consist of two bundles, the *iliocostalis* and the *longissimus*, both often showing evidences of segmentation (Figs. 71, 72). The iliocostalis lies just above where the horizontal septum would be if present, and meets the external oblique below; it originates on the ilium and fascia of the longissimus and inserts on the ribs; in *Sphenodon*, where the muscle has segmental inscriptions, it inserts on the uncinate processes of the ribs (Fig. 71). The trunk part of this muscle is called *iliocostalis dorsi;* the neck part, which inserts on the atlas and occipital region of the skull, is called *iliocostalis cervicocapitis*. The longissimus muscle lies directly above the iliocostalis; it originates on ilium, sacrum, and vertebrae and is inserted on the ribs, transverse processes of the vertebrae, and adjacent fasciae. It is segmentally divided by inscriptions in *Sphenodon*, Crocodilia, and some lizards. In the trunk region it is called *longissimus dorsi* and in the neck region *longissimus cervicocapitis*. The latter, on approaching the skull, divides into two to five, usually four portions—two superficial ones (Fig. 72*B*) and two deeper; all insert on the rear parts of the skull.

The median mass of the epaxial system lies next to the vertebral column above the longissimus and consists of long and short systems attached to various parts of the vertebrae and often intermingled in a complicated way. The long systems, collectively termed *transversospinalis*, comprise the *spinalis* nearest the vertebrae and the more lateral or deeper *semispinalis*. These originate by tendons on the vertebral spines and neural arches, often in a criss-cross anastomosing fashion (Figs. 71, 72*C*) and insert on more anterior vertebrae and the skull. They are termed *dorsi* in the trunk region, *cervicis* and *capitis* anteriorly. In the Crocodilia there are four long systems instead of two (Fig. 72*C*); the extra ones are the *tendinoarticularis*, just above the longissimus, and the *neurospinalis*, next to the neural arches with complicated tendons; both of these intermingle anteriorly with the spinalis semispinalis systems, of which they are presumably special differentiations.

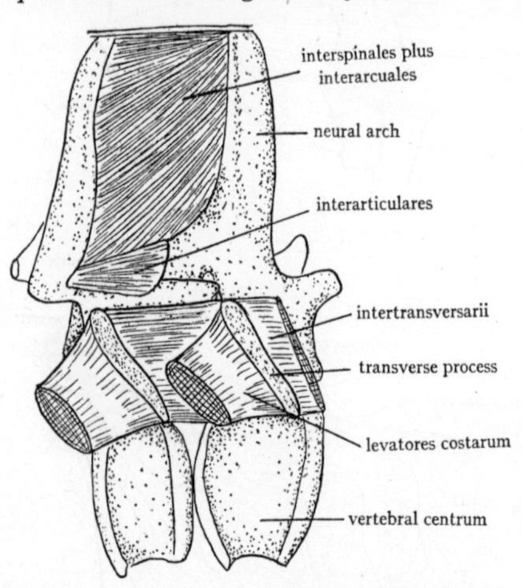

FIG. 73.—The short segmental muscles of the epaxial series in the extinct cotylosaurian reptile *Diadectes*. (After Olson, 1936.)

The short epaxial muscles are a group of segmental muscles lying to the inner side of the spinalis semispinalis systems and often intermingled with them; they extend between adjacent vertebrae or ribs or run between vertebrae and ribs. The principal series are: the *interspinales*, between successive vertebral spines; the *interarcuales*, between successive neural arches; the *interarticulars*, along the bases of the neural arches, between the zygapophyses; and the *intertransversarii*, between successive transverse processes (Fig. 73). Belonging to the last system are the *scalenes* (or *costocervicals*), from the foremost ribs to the cervical ribs or transverse processes of the cervical vertebrae.

In addition to the anterior ends of the long systems already described, there is, in the neck region, deep to these, a set of muscles, termed the *occipital* muscles, which extend from the foremost cervical vertebrae to the atlas and skull and assist in turning the head. They comprise a medial group termed *rectus capitis* and a lateral group termed *obliquus capitis;* each usually occurs as two or more distinct muscles (Figs. 72*D*, 74).

The hypaxial musculature of the general body wall is similar to that of *Necturus*, comprising the oblique series on the sides and the rectus series in the midventral region. The oblique series consists of the external oblique, internal oblique, and transverse; but each of these is frequently split into subsidiary layers. In the region of the ribs there is interposed between the external and internal oblique an *intercostal* layer, of short muscles running between adjacent ribs. This is typically composed of the *external* intercostals with fibers slanting forward and the *internal* intercostals with fibers slanting backward. Posteriorly the external intercostal layer is represented by the *quadratus lumborum* (Fig. 71). The intercostal layer appears to be a derivative of the external oblique layer of urodeles. The intercostal muscles, although belonging to the hypaxial series, are more or less shoved dorsally under the epaxial muscles and may be underlain here by a part of the transverse layer called *subvertebral* and *subcostal*. A deep set of small segmental muscles, the *levatores costarum*, which run between the transverse processes and the heads of the adjacent ribs (Fig. 73), also belongs to the hypaxial musculature.

2. Pelvic and hind-limb musculature.—It should be mentioned now that the limb musculature of tetrapods does not originate embryologically by budding from the myotomes (as it does in fishes) but develops *in situ* from formless mesenchyme cells. Nevertheless, a myotomic origin may be accepted phylogenetically for the limb and girdle musculature of all tetrapods.

According to Romer (1922), the ilium served originally primarily to anchor the long lateral systems of the epaxial trunk musculature, and only its basal part was occupied by pelvic musculature. In the evolution of tetrapods the pelvic musculature came to occupy a larger and larger surface of the ilium, and the epaxial muscles were pushed dorsally and eventually carried to the inner surface of the ilium.

The pelvic musculature is divisible into long muscles, which extend from the girdle to the shank, and short muscles, which extend from the girdle to the femur. On the dorsal (outer) or extensor side the long muscles comprise a large extensor mass, variously termed *extensor femoris* or *triceps* or *quadriceps*, inserted on the tibia and divisible into the *iliotibialis*, the *ambiens*, and the *femorotibialis;* and a smaller mass attached to the fibula, the *iliofibularis* (Fig. 75). The short extensor muscles include the *pubo-ischio-femoralis internus*, the *iliofemoralis*, the *pubo-ischio-femoralis posterior* (or *ischiotrochantericus*). On the ventral or flexor side there is a varied series of long flexors from the pubis and ischium to the tibia or shank fascia, including the *pubotibialis*, *pubo-ischio-tibialis*, *flexores tibiales internus* and *externus*, and *ischioflexorius*. The short flexors, which are more deeply located than the long flexors, are of the adductor type, consisting chiefly of the *pubo-ischio-femoralis externus*.

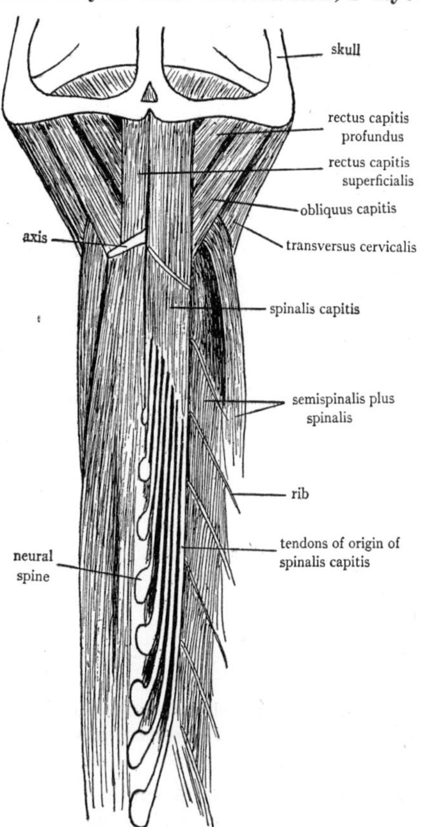

FIG. 74.—The spinalis semispinalis system and the occipital muscles of *Sphenodon* (after Nishi, 1916), viewed from above. The longissimus has been removed on both sides, and the spinalis capitis on the left side.

The shank or lower-leg musculature consists of an extensor series on the upper or outer surface, flexor series on the inner side. The extensors form three groups: on the fibular side, the *peronaeus* muscles from the femur and fibula to the last metatarsal and sole of the foot; a middle *extensor digitorum communis longus* with tendons to the metatarsals; and on the tibial side, the *tibialis anticus* from the tibia to the first metatarsal. The flexor muscles are in three layers, of which the outer one consists of the *gastrocnemius*, with two or more heads, from the femur and tibia to the sole of the foot; under this the *flexor digitorum communis longus* (or *flexor primordialis communis*), also with tendons along the sole; and an innermost group, of variable composition, including the *tibialis posticus*, a *popliteus* muscle, and in some reptiles an *interosseus* muscle.

3. Shoulder and forelimb musculature.—The series of shoulder muscles from the trunk and pectoral girdle to the pectoral girdle and humerus is similar in higher tetrapods to the condition seen in *Necturus*, but the muscles are stronger and with a wider spread. The most posterior member is the broad fan-shaped *latissimus dorsi* (Fig. 76), originating on the dorsal fascia and inserting on the humerus. A *teres major* muscle, found in the Crocodilia and some lizards, appears to be a derivative of the latissimus which became attached to the scapula. Anterior to the latissimus is the *cucullaris*, which in amniotes is generally divided into two parts—a posterior and larger *trapezius* and an anterior straplike *sternomastoid*, *cleidomastoid*,

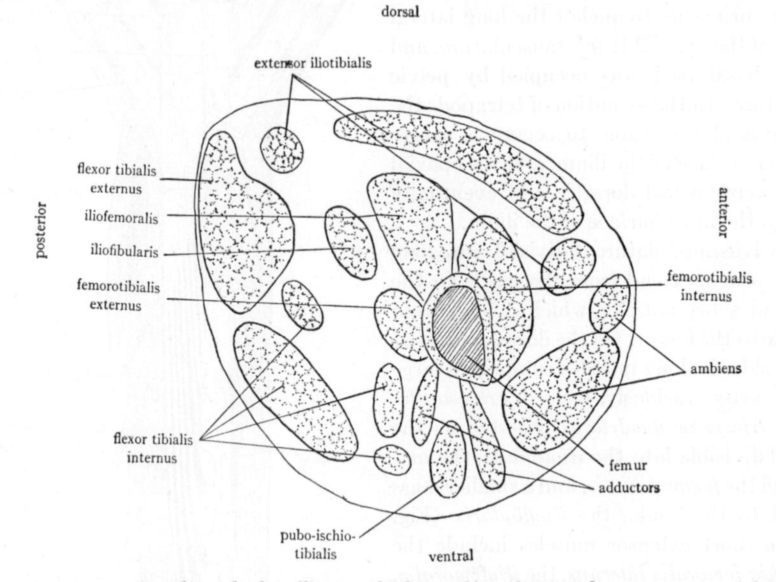

Fig. 75.—Cross-section through the alligator thigh, to show the muscle arrangement (after Romer, 1923). Compare with Figure 82*A*, *B*, with the aid of the table of homologies on p. 248.

or *sternocleidomastoid*, depending on its attachments. The trapezius originates on the dorsal fascia, often extending forward to the skull, and inserts on the anterior border of the scapula. The sternocleidomastoid part originates on the skull and inserts on scapula, clavicle, or sternum, as the case may be. If the clavicle is reduced or lacking, the muscle becomes a *cephalohumeralis*, from skull to humerus.

Between the latissimus and the trapezius there is exposed the *deltoid* group of shoulder muscles (Figs. 76, 77), very characteristic of amniotes. In reptiles there are usually two: the *deltoides scapularis*, from the scapula to the humerus; and the more ventral *deltoides clavicularis*, from the clavicle to the humerus. The first appears to represent the *dorsalis scapulae* of *Necturus* and is often still so labeled; the second, often called cleidohumeralis, seems to correspond to the procoracohumeralis.

Beneath these superficial shoulder muscles is a series of muscles acting on the girdle which belong to the epaxial musculature, probably its lateral bundles. These comprise one or more *levatores scapulae* running under the trapezius between the anterior part of the scapula and the ribs or transverse processes of the anterior cervical vertebrae; and the *serratus* muscle, often in several parts, originating by slips on a number of ribs and inserting on the inner surface of the scapula (Fig. 77). The *rhomboideus* muscle is part of this series but is found, among reptiles, only in the Crocodilia. It originates on the dorsal fascia and extends to the inner surface of the scapula (Fig. 77*B*).

On the ventral side the large *pectoralis* muscle forms the external layer of the musculature acting on the limb; it originates on the girdle, sternum, gastralia when present, and surface of the hypaxial musculature and inserts on the humerus, acting as a powerful adductor of the humerus. Beneath (dorsal to) the pectoralis is a layer of muscles from the girdle to the humerus. These include a *coracobrachialis* series originating on the coracoid and a *supracoracoideus* anterior to this (Fig. 77*B*). Continuing this layer around the front of the shoulder are the *scapulohumeralis* muscles.

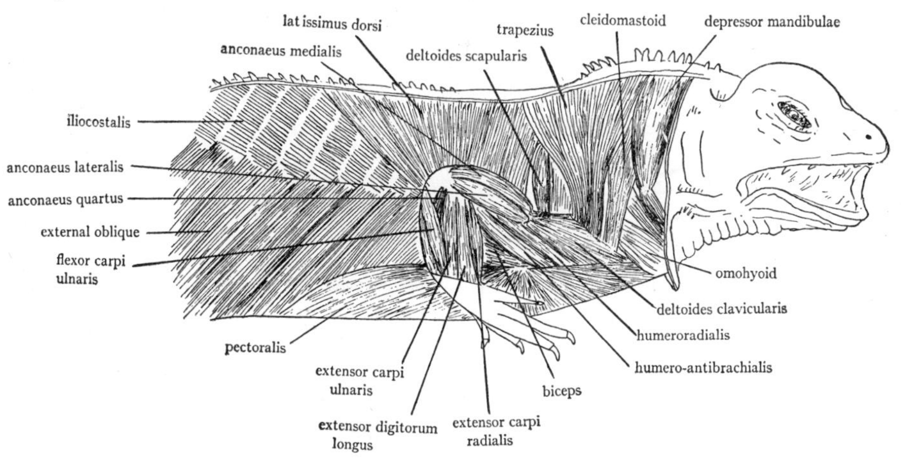

Fig. 76.—Shoulder and arm musculature of *Sphenodon* (after Miner, 1925). Compare with Figure 70, noting changes due to loss of cleithrum.

Muscles on the upper arm proper include on the extensor side the *anconeus*, usually with four parts in reptiles, and on the flexor side the *brachialis* and *biceps*.

The arrangement of the forearm musculature (Fig. 78) is similar to that seen in *Necturus*. On the flexor side there is a median mass, the *flexor digitorum communis*, with tendons to most or all of the fingers, one or more *radial flexors* on the preaxial side, and one or more *ulnar flexors* on the postaxial side of the common flexor. In addition, there are often one or more *pronators*, oblique muscles which originate on the proximal part of the ulna and insert on the radial side of the carpus, having a rotating action (also found in urodeles). The extensors also have an arrangement similar to that of urodeles. There is a median *extensor digitorum communis* (also called *humerodorsalis*) and *radial* and *ulnar* groups of extensors to either side of this. The rotator muscle of the extensor side is termed *supinator*. The forearm muscles are inserted on carpus and metacarpals, on which they act.

Howell (1936) has proposed the following terminology for the primitive tetrapod arrangement of the forearm musculature: extensor humeroradialis, extensor humerodorsalis, and extensor humero-ulnaris for the three main extensor masses; and flexor humeroradialis, flexor humeropalmaris, and flexor humero-ulnaris for the three main flexor masses. These become subdivided into various muscles in different tetrapods.

4. Branchial and hypobranchial musculature.—In reptiles the first gill arch, as in all vertebrates, forms the upper and lower jaws, and its muscles, together with contributions from the hyoid arch, are the muscles of mastication. With the complete change to lung respiration the hyoid and remaining gill arches are reduced to the hyoid apparatus and the cartilages of the larynx; and such of their musculature as persists operates the tongue, hyoid apparatus, and larynx. Despite the considerable change in the general anatomy of this region resulting from the loss of gills, the musculature is not unlike that of *Necturus* (Fig. 79).

The *adductor mandibulae* (called by some authors *capitimandibularis*) forms the chief muscle of mastication and in reptiles is usually divided into three parts—superficial, medial, and

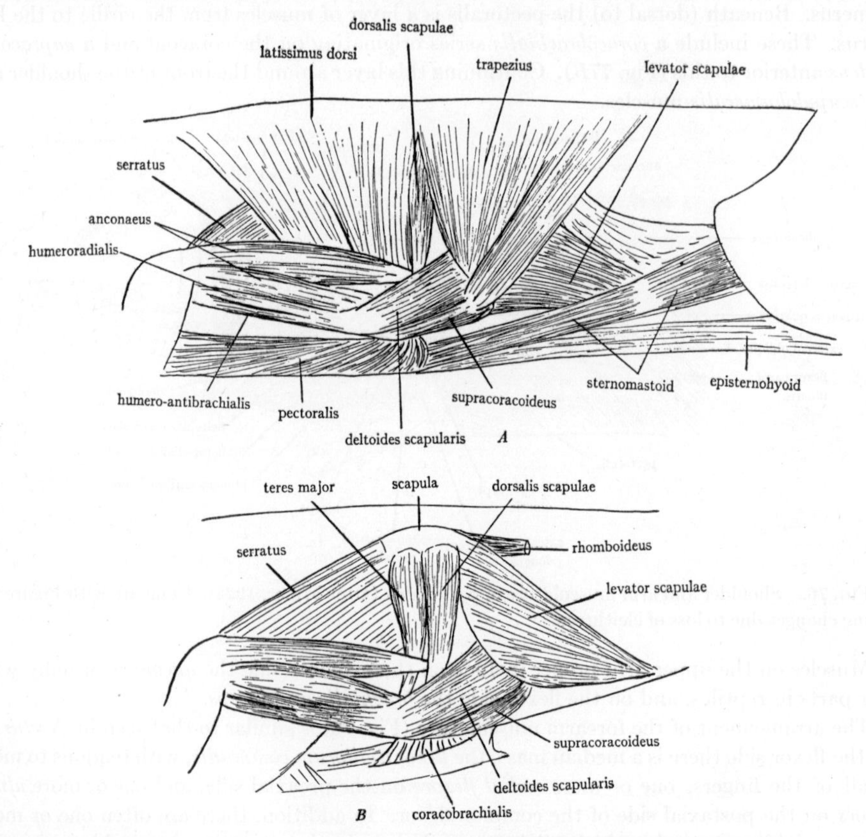

FIG. 77.—Shoulder and arm musculature of the crocodile (after Fürbringer, 1876). *A*, superficial layer; *B*, deep layer.

deep. The adductor mandibulae is also often referred to as the *temporomassetericus*, for it is represented in mammals by the *temporal* and *masseter* muscles. More internal portions of the adductor mass are usually called *pterygoideus*, of which there may be one or more in reptiles (four in the alligator). The *depressor mandibulae*, belonging to the hyoid arch, is typically present in reptiles, passing from the rear part of the skull to the articular bone of the lower jaw.

The floor of the mouth cavity and the throat are covered with a constrictor sheet, as in *Necturus*. This comprises anteriorly the *intermandibularis* (usually called *mylohyoid*) and more posteriorly the *constrictor colli* or *cervical constrictor*, which presumably includes the interhyoideus. The constrictor colli is very long in long-necked forms such as turtles. Dorsal to the constrictor sheet are the muscles of the tongue and hyoid apparatus, such as the *geniohyoid, genioglossus,* and *hyoglossus.* There are also in reptiles persistent gill arch muscles, such as the *branchiomandibularis* and *branchiohyoid,* which are said to be homologous to the first subarcual muscle of the urodeles. This muscle originally passed between the two horns of the hyoid, then called branchiohyoid; but in some reptiles it shifts one end to the mandible and so becomes a branchiomandibularis.

The ventral surface and sides of the neck are covered by the *rectus cervicis,* which in reptiles usually furnishes two muscles: the *sternohyoid* or some variant thereof, according to its origin from sternum, interclavicle, clavicle, coracoid, etc.; and the more lateral *omohyoid,* from shoulder to the hyoid apparatus (Fig. 76).

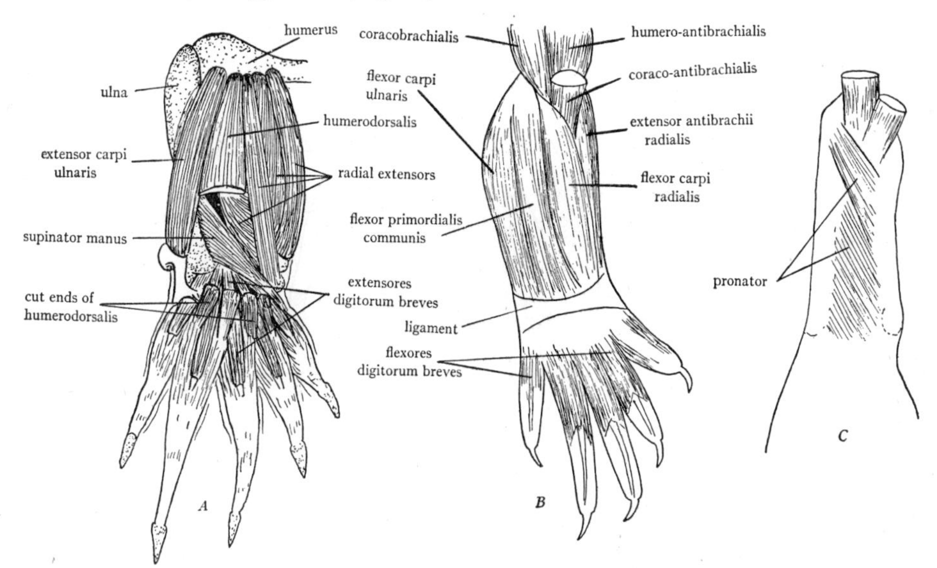

FIG. 78.—Forearm musculature of *Sphenodon. A,* extensor side (after Haines, 1939); the humerodorsalis has been cut across and a piece removed to reveal the supinator. *B,* superficial, and *C,* deep flexor, side (after Osawa, 1898).

E. MAMMALIAN MUSCULATURE

The muscular system of mammals derives directly from that of reptiles, and its general organization can be understood by reference to the foregoing account of reptilian musculature. The adult musculature of mammals exhibits only slight evidence of the original segmentation. With complete elevation of the trunk above the ground there is further differentiation of the limb musculature into individual muscles, permitting a variety of movements, and still greater spread of the shoulder and hip musculature over the trunk. The branchial musculature is not greatly altered from its reptilian state.

The following directions apply to the cat and the rabbit; they do not profess to give a complete account of the muscular system of these two animals; and, if this is desired, more detailed manuals should be consulted. For the terminology of muscles see page 200.

1. **The dermal or integumental muscles.**—These are muscle sheets which

have acquired insertion on the skin and which act to move the skin or skin structures such as hairs, scales, bristles, etc. Although present in all amniotes, they are particularly developed in mammals, where they accomplish such movements as shivering of the skin, erection of the fur, movements of vibrissae, changes of facial expression, etc. They are extensions of trunk musculature, especially of the pectoral muscle, or of branchial muscles, such as the trapezius and hyoid constrictor. The last named is the source of the

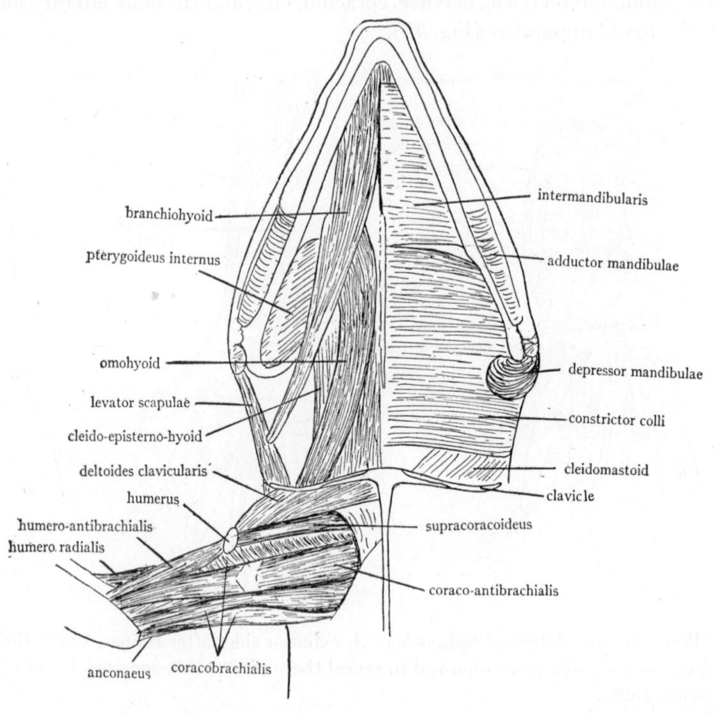

Fig. 79.—Branchial, hypobranchial, and upper-arm musculature of *Sphenodon* (combined from figures by Fürbringer, 1900). Superficial throat layer on left has been removed on the right; the pectoral muscle has been removed also.

platysma, the dermal muscle of the head, with numerous subdivisions acting to move the lips, cheeks, eyelids, ears, etc.

Skin the animal and, in doing so, note the structures named below. (If the animal has already been skinned, this part of the work will have to be omitted.) In skinning, make a middorsal incision from the base of the tail to the back of the head, being sure to cut only through the skin. Make an incision through the skin around the throat, around ankles and wrists, and incisions along the outer surface of the limbs. Connect these with the middorsal incision. Loosen the skin along the incisions and gradually work the skin loose from the muscles, using fingers and back of the scalpel. Avoid, as far as possible, the method of cutting the skin from the body, as one is liable to cut into the muscles by this procedure. Work from the dorsal side toward the ventral side. Leave the skin on the head and on the perineal region for the present.

The following points should be noted during the skinning. The skin is connected with the underlying muscles by a loose weblike material, the *subcutaneous connective tissue* or *superficial fascia*, often impregnated with fat. Below this is the much firmer and tougher connective tissue on the surface of the muscles, forming the *deep fascia.* Passing from among the muscles into the skin will be seen at regular intervals, which represent the segments of the body, a slender cord, composed of an artery, a vein, and a sensory nerve. These may be severed. Other blood vessels, not segmentally arranged, will also be seen passing onto the under surface of the skin, from anterior and posterior regions toward the middle. The arteries are colored by an injection mass and are readily recognized. The veins are usually of a very dark reddish-brown color. All vessels to the skin should be severed.

When the skin has been loosened from the sides of the animal, there will be noted a thin layer of muscle fibers on its under surface, appearing like a fine striping. Toward the chest and shoulder region this assumes the form of a thin sheet. This muscle is a dermal or skin muscle, the *panniculus carnosus* or *cutaneous maximus.* It covers the entire lateral surface of the thorax and abdomen, being more prominent anteriorly. On continuing to skin forward and ventrally, the muscle will be found to take its origin from the outer surface of a muscle (latissimus dorsi) situated posterior to the shoulder, and from the axilla in the cat, the medial side of the humerus (rabbit), and from the linea alba and various points on the ventral side of the thorax in both animals. These points of origin should be cut through, and the cutaneous maximus removed with the skin to which it generally adheres. The muscle is inserted on the skin and serves to shake the skin; it is degenerate in man. The cutaneous maximus is chiefly an outgrowth of the pectoral muscles and has various parts in different mammals.

The other chief dermal muscle, the *platysma,* will be found on the under surface of the skin of the neck and head, and consists of many different parts, bearing separate names. Some of these will be seen later, but the study of the parts of the platysma is almost impossible in any but freshly killed specimens. The platysma muscle is inserted on the skin of the ears, eyelids, lips, etc., and serves to move them; in man it constitutes the muscles of facial expression. The platysma is a branchial muscle, derived from the muscles of the hyoid arch by extension.

In females on the under surface of the skin of the ventral side the *mammary* or *milk glands* will be noted spread out as a thin, irregular layer.

The skin having been removed and discarded, clean away fascia and fat from the surface of the muscles. There is generally a large mass of fat at the base of the hind legs. It will now be seen that the exposed surface in part consists of muscles, pinkish masses composed of parallel fibers, and in part of the deep fascia which forms very strong white sheets. The posterior half of the back is covered by such a sheet, known as the *lumbodorsal fascia.* In the median ventral line is the linea alba. The angle between the base of the thigh and the abdominal wall is known as the *inguinal* region. At the bottom of this will be found in the rabbit a stout white shining cord, the *inguinal ligament*, which stretches from the pubic symphysis to the crest of the ilium. It is absent in the cat. The angle between the upper arm and chest is called the *axilla*, or *axillary fossa.*

In studying the muscles it is necessary to separate each muscle from its neighbors. This is done by searching carefully for the white lines of connective tissue which mark the boundaries of muscles and slitting along these lines with the point of the scalpel. Observing the direction in which the fibers run will also aid in separating muscles, since the fibers in one muscle generally run in the same direction, which is usually different from that of the neighboring muscles. After freeing the margins of a muscle the fingers should be worked under the muscle until it is separated from its fellows. As each muscle is inclosed in a connective tissue sheath, it will separate smoothly from its neighbors; the presence of rough edges indicates that the muscle itself has been cut. Avoid the use of sharp instruments in freeing the muscles. In case it is necessary to cut through a muscle in order to reveal another muscle beneath it, *always cut through the center of the muscle.*

In the following dissection of the muscles the dissection is to be confined *strictly to the left side*, leaving the right side intact for the dissection of other systems.

2. The muscles of the abdominal wall.—The abdominal wall is composed of three layers of muscles with their aponeuroses. The aponeuroses of these muscles are quite extensive. The three layers are: an external layer, the *external oblique;* a middle *internal oblique;* and an internal layer, the *transversus abdominus.*

a) *External oblique:* This is the large muscle constituting the outermost layer of the abdominal wall; clean away superficial fascia from its surface. Its anterior part is concealed under a large flat muscle, the latissimus dorsi, which covers the anterior part of the back and slopes toward the upper arm. The posterior boundary of this muscle should be located and slit, and it should then be lifted from the surface of the external oblique by thrusting the fingers between. Any fat present between the two muscles should be cleaned away. The most posterior chest muscles also cover the anterior part of the external oblique and should be lifted off in a similar manner. The external oblique originates from the lumbodorsal fascia and from the posterior ribs by separate slips. Its fibers pass obliquely downward and backward, and in the rabbit the more dorsal ones are directed straight caudad. The external oblique is inserted on an extensive neurosis which passes to the linea alba; in the rabbit also on the inguinal ligament; in the cat also on the pubis. Action, constrictor of the abdomen.

b) Internal oblique: Very carefully cut through the middle of the belly of the external oblique, in a longitudinal direction, and separate it from the underlying muscle, which is the *internal oblique.* This separation is often difficult. The internal oblique is a short muscle lying beneath the more dorsal portion of the external oblique. Its fibers are directed obliquely downward and forward, and are continued by a very broad aponeurosis. Origin, rabbit—second sheet of the lumbodorsal fascia, posterior ribs, and inguinal ligament; cat—second sheet of the lumbodorsal fascia and border of the pelvic girdle. Insertion, on the linea alba by the extensive aponeurosis; action, compressor of the abdomen.

c) Transverse: On cutting through the preceding and separating the edges, a third muscle layer, very thin, will be found. This is the *transversus abdominis.* Its fibers are directed ventrally and slightly posteriorly. Origin, insertion, and function similar to the preceding.

d) Rectus abdominis: This is a long slender muscle on each side of the linea alba, extending from the pubic symphysis to the anterior part of the thorax. It is found inside of and between the aponeuroses of the preceding muscles. Slit open these aponeuroses along each side of the linea alba and expose the rectus abdominis. Its fibers run longitudinally and in the cat are crossed at regular intervals by white lines, the myosepta or inscriptions. Origin, anterior end of pubic symphysis; insertion, sternum and costal cartilages; action, retracts ribs and sternum and compresses abdomen. Internal to the transverse and rectus muscles is the peritoneal membrane or lining of the coelom.

The foregoing muscles are hypaxial muscles, formed into sheets by the sidewise fusion of myotomes. The rectus abdominis comes from the ventral ends of the myotomes. The three layers of the sides of the abdomen have resulted from a splitting process. The arrangement of the hypaxial trunk musculature of the mammal is seen to be similar to that of *Necturus* and reptiles.

3. The epaxial muscles.—Remove the lumbodorsal fascia over the extreme posterior part of the back, finding beneath it a great thick mass of muscle inclosed in a tough shining fascia; this is the epaxial muscle mass, sharply separated from the hypaxial muscles in the abdominal region by a furrow corresponding to the position of the horizontal skeletogenous septum of lower vertebrates. The epaxial mass in the lumbar region is divisible into a slender, narrow median portion, corresponding to the transversospinalis system of reptiles but called in mammals the multifidus spinae, and a very thick

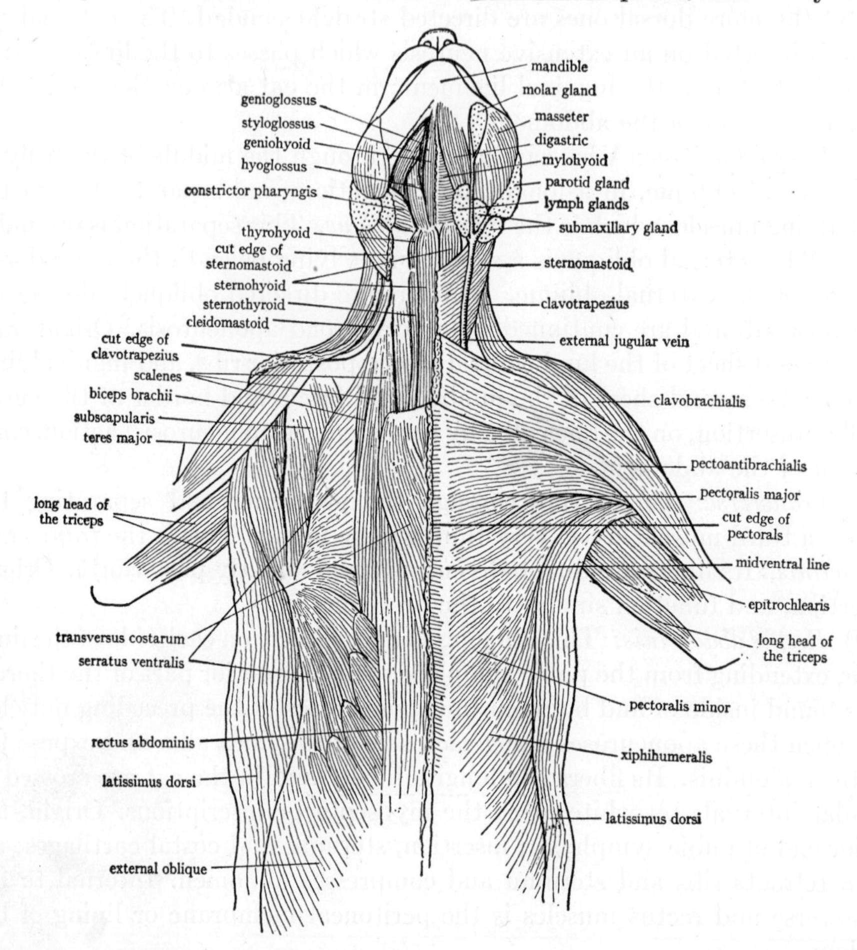

Fig. 80.—Ventral view of the anterior part of the cat to show the muscles. All dermal muscles have been removed. Superficial muscles on the right side, deeper layer of muscles on the left side, after removal of the pectoral muscles, sternomastoid, mylohyoid, and digastric. The nerves and blood vessels which cross the axilla have been omitted. The view of the axilla is different from that revealed in the dissection given in the text.

lateral portion, the sacrospinalis, divisible anteriorly into three longitudinal parts—semispinalis, longissimus, and iliocostalis—which will be followed later. The resemblance of the epaxial systems to those of reptiles is evident.

4. The muscles of the chest (Fig. 80).—Turn the animal on its back and expose the chest by spreading and fastening the forelimbs. The great muscles covering the chest or ventral side of the thorax are the *pectoral* muscles.

They are divisible into several portions which are not very definitely separable from each other. Some authorities list five parts in the rabbit and four parts in the cat; of these we may consider the following:

Rabbit:

a) *Pectoralis major.* This is a large muscle originating from the whole length of the sternum and inserted on the humerus. It covers most of the surface of the chest, but the insertion is concealed by a muscle coming down from the head (clavodeltoid). Action, draws the arm toward the chest.

b) *Pectoralis primus.* A slender muscle at the anterior end of the preceding and covering its anterior fibers. Origin, manubrium of the sternum; insertion, humerus; action, like the preceding. Its anterior border is in contact with the clavodeltoid, which also partly covers its insertion.

c) *Pectoralis minor.* Cut through the middle of the belly of the pectoralis major, and upon deflecting the cut edges note, internal to it, a similar muscle, the *pectoralis minor.* Origin, manubrium; insertion, clavicle and spine of the scapula. To find the insertion loosen up the clavodeltoid and locate in its fibers at the shoulder a small, slender bone, the clavicle. The clavicle is on the inner surface of the muscle. Then loosen the muscle next lateral to the clavodeltoid, a long muscle coming down from the back of the neck (anterior trapezius). The pectoralis minor is inserted by some fibers on the clavicle, but most of its fibers sweep over the shoulder internal to the clavodeltoid and anterior trapezius and are inserted on the spine of the scapula. Action, draws arm and shoulder toward the chest.

Cat:

a) *Pectoantibrachialis.* Anterior and most superficial of the chest muscles. Origin, manubrium; insertion, by a flat tendon on the fascia of the forearm; action, draws the arm toward the chest.

b) *Pectoralis major.* Next posterior to the preceding and extending anteriorly dorsal to the preceding, which should be cut across; originating on the sternum and median ventral raphe and inserted on the humerus. Action, like the preceding.

c) *Pectoralis minor.* Next posterior to the preceding and covered in large part by the pectoralis major. The latter should be cut through and the extent of the pectoralis minor noted. The pectoralis minor is divisible into several parts. Origin, sternum; insertion, humerus; action, like the preceding. The insertion cannot be fully traced at the present stage of the dissection.

d) *Xiphihumeralis.* The last of the chest muscles. A thin, flat, long muscle, passing from the xiphoid process of the sternum, its anterior part passing dorsal to the posterior part of the pectoralis minor, and inserted on the humerus. The insertion is covered by a mass of fat in the axilla. Remains of the cutaneous maximus are probably present on its surface. Action like the preceding.

5. The muscles of the neck and throat (Fig. 80 or 81).—Slit the skin up the center of the throat to the tip of the lower jaw and loosen it so as to expose fully the lower jaw. Note, in doing so, parts of the *platysma* muscle on the under side of the skin. It sweeps from the median dorsal line of the neck around the sides of the head to face and ears, and portions of it generally are attached near the anterior end of the sternum. In dissecting the throat muscles work on one side only, leaving the other intact for the study of other parts. Avoid cutting any blood vessels. A large vein, the *external jugular vein*, runs in the superficial muscles of the throat. At the angle of the jaw is a rounded pinkish body, the *submaxillary salivary gland*, crossed by the posterior facial vein. Other small bodies are *lymph glands*.

Rabbit:

a) *Special portion of the platysma*. A broad, thin sheet of dermal muscle extends from the manubrium of the sternum forward, forking like the letter V, each half inserting at the base of the ear. This is the *depressor conchae posterior* and is the most superficial muscle on the ventral surface of the neck. The external jugular vein runs in it. It is a part of the platysma. It should be well separated from the underlying muscles and the posterior end severed and turned forward, without, however, injuring the vein.

b) *Sternohyoid*. This is the long muscle in the median line of the neck, the two members of the pair being closely fused in the median ventral line. Origin, manubrium of the sternum; insertion, anterior horn of the hyoid. Follow the muscle up to the throat and feel with the fingers the bony hyoid on which the muscle is inserted. Action, draws the hyoid posteriorly or raises the sternum.

c) *Sternomastoid*. The long muscle on each side of the preceding, the two members of the pair converging toward the manubrium of the sternum from which they originate ventral to the origin of the preceding. Insertion, mastoid process of the skull; action, singly turn the head, together depress the head on the neck.

d) *Cleidomastoid* and *basioclavicularis*. These two long, strap-shaped muscles are next lateral to the preceding and unite at the clavicle with the clavodeltoid. The cleidomastoid is the more medial one and lies lateral and somewhat dorsal to the sternomastoid. Origin, mastoid region of the skull; insertion, clavicle; action, elevates clavicle or turns the head. The basioclavicularis is slightly lateral to the preceding at its cranial end but crosses ventral to it caudally, so that its insertion on the clavicle is medial to that of the cleidomastoid. Origin, occipital bone; insertion, clavicle; function, like preceding. The origins of these muscles cannot be followed out conveniently.

e) *Clavodeltoid.* Continuation of the two preceding muscles. Origin, clavicle; insertion, humerus; action, raises the humerus.

f) *Masseter.* The great mass of muscle covering the angle of the jaws, its outer surface with a very tough shining fascia. Origin, zygomatic arch; insertion, outer surface of the posterior end of the mandible; action, closes the lower jaw (elevator of the jaw.)

g) *Digastric.* The muscle along the ventral surface of each half of the jaw bone, terminating in a slender tendon. Origin, occipital bone; insertion, ventral surface of the mandible; action, opens the jaw (depressor of the jaw).

h) *Mylohyoid.* The thin sheet of muscle crossing transversely between and dorsal to the two digastrics. Origin, mandible; insertion, median ventral line (raphe) and the hyoid; action, raises the floor of the mouth and brings the hyoid forward.

i) *Sternothyroid.* Divide the two sternohyoids in the median ventral line. This exposes the *trachea* or *windpipe,* a tube stiffened by rings of cartilage. At the top of the trachea at a level about between the two submaxillary glands there is an enlarged chamber, the *larynx* or Adam's apple, whose walls are supported by cartilage. The chief cartilage of the larnyx is the large, shield-shaped cartilage, which forms the ventral wall; this is called the *thyroid* cartilage. The sternothyroid muscle will be found, one on each side of the trachea, dorsal to the sternohyoid, originating on the sternum and inserted on the thyroid cartilage. Action, pulls the larynx posteriorly.

j) *Thyrohyoid.* A thin muscle at each side of the larynx, extending from the thyroid cartilage to the hyoid. Action, raises the larynx.

Cat:

a) *Sternomastoid.* This is the superficial muscle of the ventral side of the neck. A large vein, the *external jugular* vein, crosses its surface at an angle to the direction of its fibers. Origin by two parts, from the median raphe and the manubrium of the sternum, the first-named origin lying ventral to the second, so that the muscle appears divisible into two muscles. From the origins the muscle passes obliquely away from the median ventral line around the sides of the neck and is inserted on the skull from the lambdoidal ridge onto the mastoid process. The muscle passes internal to the submaxillary gland and the *parotid* gland; the latter is a mass at the base of the ear. The insertion on the mastoid process is by means of a thick tendon. Action, singly turn the head, together depress head on neck.

b) *Sternohyoid.* The anterior ends of these muscles are visible between the two sternomastoids, as the latter diverge from the median raphe. Slit the raphe of the sternomastoids to the manubrium of the sternum, thus exposing the full length of the sternohyoids. They extend in the median ventral line from the first costal cartilage to the body of the hyoid bone, the two being closely united in the median line. Action, draw the hyoid posteriorly.

c) *Cleidomastoid.* Lateral to the sternomastoid is a long muscle passing from the head to the upper arm. Loosen this up and find, internal to it, a narrow, flat muscle, the cleidomastoid. Extends from the clavicle, which will

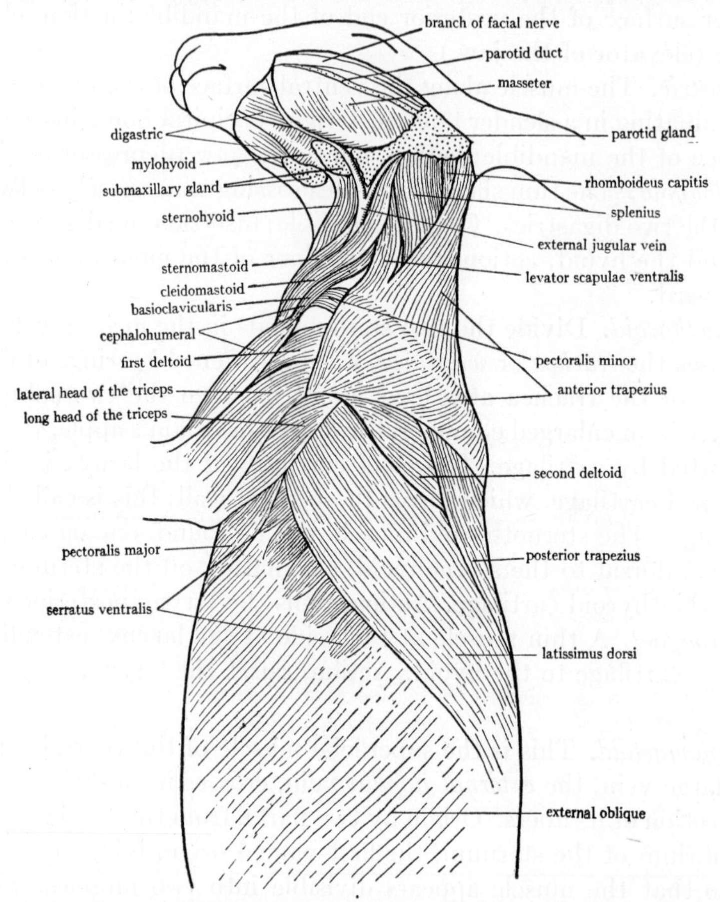

Fig. 81.—Lateral view of the anterior part of the rabbit to show the muscles. The head is turned slightly to give a ventral view of the throat. All dermal muscles have been removed.

be found as a slender bone on the internal surface of the long muscle just mentioned at the level of the shoulder, to the mastoid process, dorsal to the insertion of the sternomastoid. Action, pulls clavicle craniad or turns head, acting singly, or lowers head on neck. Origin and insertion are thus interchangeable.

d) *Clavotrapezius* and *clavobrachial.* The long muscle on the side of the neck and passing over the ventral surface of the shoulder to the forearm is sometimes considered as one muscle, the *cephalobrachial,* or as two. In the latter case the upper part is known as the *clavotrapezius* and extends from the skull to the clavicle. It will be considered later. The lower part from the clavicle to the forearm is the *clavobrachialis*. Origin, clavicle and fibers of the clavotrapezius; insertion, ulna; action, flexor of the forearm. The clavicle will be found on the inner surface of the muscle in the shoulder region, imbedded in the muscle.

e) *Masseter.* The great thick muscle covering the angle of the jaws, situated in front of the submaxillary and parotid glands. It is covered by a very tough shining fascia. Origin, zygomatic arch; insertion, posterior half of the lateral surface of the mandible; action, elevator of the lower jaw.

f) *Temporal.* Remove the skin from the side of the head up to the median dorsal line. A great mass of muscle covered by a strong shining fascia will be seen occupying the temporal fossa of the skull, dorsal to the ear. Origin, from the side of the skull from the superior nuchal line to the zygomatic process of the frontal bone, and from part of the zygomatic arch; insertion, coronoid process of the mandible; action, elevator of the jaw, in common with the masseter.

g) *Digastric.* The muscle lying along the medial surface of each half of the mandible. It extends posteriorly internal to the submaxillary gland. Origin, jugular and mastoid processes of the skull; insertion, mandible; action, depressor of the lower jaw.

h) *Mylohyoid.* The thin transverse sheet passing across between the two digastrics from one half of the mandible to the other. Origin, mandible, the origin concealed by the digastrics; insertion, median raphe; action, raises floor of the mouth and brings hyoid forward.

i) *Geniohyoid.* Cut through the median raphe of the mylohyoid. This exposes a pair of long, slender muscles, the geniohyoids, lying in the median line. Origin, mandible near the symphysis; insertion, body of the hyoid; action, draws the hyoid forward.

j) *Sternothyroid.* Separate the two sternohyoids in the median line. This exposes the *trachea* or *windpipe*, a tube stiffened by rings of cartilage. At the top of the *trachea* is a chamber with cartilaginous walls, the *larynx*. The chief cartilage of the larynx is the large, shield-shaped *thyroid* cartilage, forming the ventral walls of the larynx. Just in front of the thyroid cartilage the *body* of the *hyoid* is felt as a bony bar. The sternothyroid muscles are located, one on each side of the trachea, dorsal to the sternohyoids. Origin, sternum in common with the sternohyoid; insertion, thyroid cartilage of the larynx; action pulls the larynx posteriorly.

k) *Thyrohyoid.* Short, narrow muscle on each side of the thyroid cartilage, from which it takes its origin; insertion, posterior horn of the hyoid; action, raises the larynx.

The foregoing muscles of the neck and throat belong chiefly to the branchial and hypobranchial groups of muscles. The temporal and the masseter are homologous to the adductor mandibulae of nonmammalian tetrapods but do not exactly correspond to any of the parts of the latter. There are also in the cat and rabbit two pterygoideus muscles found along the inner surface of the mandible which belong to the general adductor mandibulae mass. The mylohyoid is homologous to the intermandibularis. All of these muscles are branchial muscles derived from the original musculature of the first or mandibular arch and innervated by the fifth cranial nerve. The digastric muscle is so called because in most mammals it is a compound muscle with a tendon dividing the belly into two parts; the anterior part is innervated by the fifth nerve and hence is a mandibular arch muscle; the posterior belly is innervated by the seventh nerve and hence is a derivative of the musculature of the hyoid arch. The posterior belly is not identical with the depressor mandibulae of reptiles and amphibians but comes from the same embryonic muscle mass as this muscle. Other mammalian muscles derived from the mandibular arch musculature are the tensor palati and tensor tympani (not seen in the dissection).

The posterior belly of the digastric and the various parts of the platysma muscle of mammals are derivatives of the hyoid arch musculature. An interhyoideus muscle is present in monotremes.

The cucullaris muscle of lower tetrapods is subdivided in mammals into one or more trapezius muscles and the sternomastoid, sternocleidoid, etc. There are also branchial muscles. A ceratohyoid muscle, homologous to the first subarcual muscle of *Necturus*, is found in many mammals, including the cat; it runs between the two horns of the hyoid. Monotremes have, in addition, an interthyroid muscle homologous to the third subarcual. The striated pharyngeal and laryngeal muscles are also remnants of branchial arch musculature.

The hypobranchial musculature, formed from growths of somatic myotomes into the region below the branchial musculature and innervated by the eleventh cranial nerve, includes the geniohyoid and various other muscles of the tongue and hyoid apparatus. The rectus cervicis of lower tetrapods is split up in mammals into the sternohyoids, sternothyroids, and thyrohyoids, with the addition of omohyoids in some mammals.

6. The muscles of the upper back and shoulder and back of the neck (Fig. 81).—

Rabbit:

a) *Latissimus dorsi.* Turn the animal on one side, so that the side on which the muscles have already been dissected will be uppermost. The large, flat muscle extending obliquely from the middle of the back to the forelimb is the latissimus dorsi. Origin, lumbodorsal fascia and posterior ribs; insertion, on the crest on the medial side of the humerus, the insertion covered by the chest muscles; action, draws the arm caudad and dorsad.

b) *Anterior* and *posterior trapezius.* These two muscles are the flat, thin muscles covering the upper back and back of the neck anterior to the latissimus dorsi. The posterior trapezius originates from the lumbodorsal fascia and the neural spines of the thoracic vertebrae and is inserted on the spine of the scapula. Action, draws the scapula dorsally. The anterior trapezius originates on the external occipital protuberance of the skull and ligament in the middorsal line and is inserted on the metacromion process (which is very long in the rabbit) and near-by muscles and fascia; action, draws the scapula and limb upward and forward. The space between the two trapezius muscles is filled by a stout fascia.

c) *Levator scapulae ventralis* or *major*. This long, slender muscle runs along the ventral border of the anterior trapezius near its insertion, then diverges to its origin from the ventral surface of the skull at the suture between occipital and basisphenoid; insertion, metacromion process in common with the anterior trapezius; action, pulls the scapula anteriorly.

d) *Rhomboideus*. Cut across the middle of the bellies of the two trapezius muscles and the latissimus dorsi. The large, thick muscle extending from the vertebral border of the scapula to the middorsal line is the rhomboideus. Origin, middorsal ligament of the neck and succeeding neural spines; insertion, vertebral border of the scapula; action, draws scapula toward vertebral column.

e) *Splenius*. A fairly broad but thin muscle on the back of the anterior part of the neck under the anterior trapezius. (Running along its external surface is a narrow straplike muscle, the *rhomboideus capitis;* see below.) Origin of splenius, middorsal line of neck and adjacent fascia; insertion, occipital region of the skull and atlas; action, singly turns the head, together raise the head.

Under the splenius are the epaxial muscles, continuations of those already noted in the lumbar region.

f) *Supraspinatus*. The superficial muscular layer of that part of the scapula anterior to the spine consists of the pectoralis minor, which sweeps over the scapula to be inserted on the spine and vertebral border. Lift up its anterior border and separate it from the muscle beneath it. This muscle is the *supraspinatus*, filling the supraspinous fossa of the scapula. Origin, supraspinous fossa; insertion, greater tuberosity of the humerus; action, extends the humerus.

g) *Deltoids*. There are three deltoids in the rabbit, of which one, the clavodeltoid, has already been considered. The second deltoid is a small triangular muscle lateral to the clavodeltoid. Origin, acromion process; insertion, humerus; action, raises the humerus. The third deltoid is lateral to the second and is a longer muscle. It passes under the long metacromion process and takes its origin from the fascia of the muscle which fills the infraspinous fossa. Insertion and action like the preceding.

h) *Infraspinatus*. The muscle partly covered by the third deltoid, which is attached to its surface. The deltoid may be removed to see it. Origin, infraspinous fossa and spine; insertion and action like the supraspinatus.

i) *Teres major.* The stout muscle along the axillary border of the scapula behind the preceding. Origin, dorsal half of the axillary border of the scapula; insertion, on the humerus in common with the latissimus dorsi; action, draws humerus against body and rotates it.

j) *Teres minor.* Separate the teres major well from the infraspinatus and look in between them. On the inner surface of the latter will be found a small but stout muscle. Origin, ventral half of the axillary border of the scapula; insertion, greater tuberosity of the humerus; action, like the preceding.

k) *Rhomboideus capitis* (or *levator scapulae minor*). Cut through the rhomboideus. A slender, bandlike muscle lies in contact with the inner surface of the rhomboideus and passes along the external surface of the splenius to be connected with the skull. Origin, side of the skull above the tympanic bulla; insertion, posterior end of the vertebral border on the medial side; action, draws scapula craniad and rotates it.

l) *Subscapularis.* Lift the scapula. A large muscle, the subscapularis, completely covers the medial or inner surface of the scapula, its fibers disposed in several directions. The muscle has a shining fascia; its posterior end is more or less continuous with teres major. Origin, medial surface of the scapula; insertion, lesser tuberosity of the humerus; action, pulls the humerus toward the median ventral line.

m) *Serratus ventralis.* On raising with the finger the vertebral border of the scapula a large, fan-shaped muscle will be seen extending anteriorly and posteriorly from the scapula to the walls of the thorax. This is the serratus ventralis; it is readily divisible into anterior and posterior portions. The anterior or *cervical* portion originates on the transverse processes of the cervical vertebrae by separate slips and on the first two ribs. The posterior or *thoracic* portion takes its origin by seven slips from the ribs. Insertion, vertebral border of the scapula above the subscapularis; action, draws scapula forward, backward, or against the body.

n) *Scalenes.* The scalenes are several long, flat muscles extending from the transverse processes of the cervical vertebrae to the ribs. They will be seen by lifting up the scapula and looking on the ventral side of the origin of the serratus ventralis. They lie internal to the sternomastoid, cleidomastoid, etc., previously identified; and farther posteriorly they form the layer next internal to the pectoralis muscles. Action, raise the ribs and bend the neck.

o) *Serratus dorsalis.* The dorsal half of the thorax underneath the latissimus dorsi, trapezius, and rhomboideus muscles is covered by a strong aponeurosis (part of the lumbodorsal fascia), in the ventral part of which muscle fibers are present which are inserted on the ribs by slips. The foremost slips are quite fleshy and take their origin by a tendon from the median dorsal line of the neck. This muscle is the serratus dorsalis. Action, raises the ribs craniad.

p) Intercostals. On the sides of the chest a series of muscles will be seen running from one rib to the next one. They are the external intercostals. They extend on the chest wall ventral to the insertion of the serratus dorsalis, which muscle, in fact, covers their most dorsal portions. Origin, posterior margins of the vertebral ribs; insertion, anterior margins of the succeeding vertebral ribs; action, pull the ribs forward. Observe that the fibers of the external intercostals are directed obliquely backward. On carefully cutting through any of the external intercostals, a layer of internal intercostals will be found inside of them, their fibers being directed obliquely forward. The internal intercostals are best seen in the ventral thoracic wall, internal to the scalenes, which may be cut through. Here between the costal cartilages the internal intercostals are not covered by the external intercostals. Origin and insertion, margins of the vertebral and sternal ribs; action, lower the ribs. The intercostals are the chief muscles concerned in the respiratory movements of the thorax. The scalenes, serratus, and other muscles assist.

q) Epaxial muscles of the thorax. The mass of epaxial muscles is conspicuous, running along the dorsal part of the thorax. This mass passes internal to the serratus dorsalis, which should be cut through; it lies upon the dorsal portions of the ribs and thus conceals the dorsal portions of the intercostal muscles. It is covered by the tough, shining lumbodorsal fascia, which should be removed. The epaxial mass is easily divisible into a narrow median portion next to the median dorsal line, the *semispinalis dorsi*, and a very thick lateral portion, the *longissimus cervicis.* The latter gives off on its ventral margin the narrow *iliocostalis*, lying on the ribs to which it is inserted, as is also the longissimus.

The epaxial mass may be followed along the neck by cutting through the splenius. There is a longissimus capitis, a semispinalis cervicis, and a semispinalis capitis. The occipital muscles comprise five muscles of the rectus capitis group and two of the obliquus capitis.

Cat:

a) Latissimus dorsi. Turn the animal on one side, so that the side on which the muscles have already been dissected will be uppermost. The large flat muscle extending obliquely forward from the middle of the back to the upper arm is the latissimus dorsi. Origin, from the neural spines of the last thoracic and most of the lumbar vertebrae and from the lumbodorsal fascia; insertion, by a tendon on the medial surface of the humerus; action, pulls the forelimb dorsally and caudally.

b) Trapezius muscles. There are three trapezius muscles in the cat; they are the thin, flat muscles covering the back and neck anterior to the preceding. The posterior trapezius or *spinotrapezius* takes origin from the spines of the thoracic vertebrae and passes obliquely forward, covering part of the latissimus to be inserted on the fascia of the scapula; action, draws the scapula dorsad and caudad. In front of this is the middle trapezius or *acromiotrapezius*. Origin, neural spines of cervical and first thoracic vertebrae; insertion, metacromion process and spine of the scapula and fascia of the preceding muscle; action, draws scapula dorsad and holds the two scapulae together. The anterior trapezius or *clavotrapezius* is the anterior part of the long muscle already described as cephalobrachial. Origin, superior nuchal line and median dorsal line of neck; passes obliquely ventrally to be inserted on the clavicle, which is imbedded on its inside surface; it is continuous with the clavobrachial muscle. Action, draws the clavicle dorsad and craniad.

c) Levator scapulae ventralis. Carefully free the three trapezius muscles. Along the ventral border of the acromiotrapezius and apparently continuous with it is seen a flat, bandlike muscle which passes anteriorly, diverging from the acromiotrapezius and passing internal to the clavotrapezius, which should be cut across. Origin, transverse process of the atlas and occipital bone; insertion, metacromion process and neighboring fascia; action, draws the scapula craniad.

d) Rhomboideus. Cut across the middle of the bellies of the spino- and acromiotrapezius muscles. A thick muscle will be seen beneath them extending from the vertebral border of the scapula to the middorsal line; this is the rhomboideus. Origin, neural spines of the vertebrae and adjacent ligaments; insertion, vertebral border of the scapula; action, draws scapula dorsad. The most ventral portion of this muscle is a practically separate muscle, the *rhomboideus capitis,* which extends as a slender band forward to originate from the superior nuchal line; insertion, scapula; action, draws the scapula craniad and rotates it.

e) Splenius. This is the large sheet of muscle covering the back of the neck in front of the rhomboideus, internal to the trapezii and crossed by the rhomboideus capitis. Origin, from the middorsal line and fascia; insertion, superior nuchal line; action, raises or turns the head.

f) Supraspinatus. On turning back the ventral half of the cut acromiotrapezius a stout muscle is seen occupying the supraspinous fossa of the scapula. Origin, whole surface of the supraspinous fossa; insertion, greater tuberosity of the humerus, next to the insertion of pectoralis minor; action, extends the humerus.

g) Deltoids. There are two deltoids in the cat. The clavobrachialis, already described, is sometimes also considered as a deltoid and is called the *clavodeltoid* by some authorities. The *acromiodeltoid* is a short, thick muscle passing ventrally from the acromion process; it is inserted on the surface of other muscles of the shoulder. It lies lateral to the clavobrachial. Action, with the next. Turn the flap of the acromiotrapezius forward again. Posterior to the line marking the insertion of the acromiotrapezius and the levator scapulae ventralis is a muscle which passes to the upper arm, across the upper ends of the muscles of the upper arm. This is the spinodeltoid. Origin, spine of the scapula; insertion, ridge (deltoid ridge) of the humerus, the insertion being concealed by the acromiodeltoid; action, the deltoids raise and rotate the humerus.

h) *Infraspinatus*. Cut across the belly of the latissimus dorsi. Bring the anterior parts of the latissimus and the spinotrapezius forward so as to expose the posterior part of the scapula. Two large muscles are here seen. The anterior one fills the infraspinous fossa of the scapula, from whose surface it takes its origin, and is inserted on the greater tuberosity of the humerus, the insertion being concealed by the deltoids, which may be cut across to see it. This muscle is the infraspinatus. Action, rotates the humerus.

i) *Teres major*. The stout muscle immediately behind the preceding, its fibers running in the same direction. Origin, axillary border of the scapula and fascia of neighboring muscles; insertion, in common with latissimus dorsi on the medial surface of the humerus; action, rotates the humerus and lowers it.

j) *Teres minor*. Carefully separate the infraspinatus from teres major and separate the former from the deltoids and the muscles of the upper arm. On the posterior border of the infraspinatus and somewhat covered by it is the small teres minor. Origin, axillary border of the scapula; insertion, greater tuberosity; action, assists the infraspinatus.

k) *Subscapularis*. Place the finger under the anterior border of the scapula and clear away connective tissue from the under surface of the scapula. The subscapular fossa is seen to be occupied by a muscle, the subscapularis, which covers the inner or medial surface of the scapula. Origin, subscapular fossa; insertion, lesser tuberosity of the humerus; action, pulls the humerus medially. Posterior to the subscapularis will be found part of teres major, which extends onto the medial surface of the scapula.

l) *Serratus ventralis*. Cut through the rhomboideus close to the vertebral column. The scapula then swings loose. On raising the scapula, a large, fan-shaped muscle is seen extending from the ventral border of the scapula to the sides of the thorax and neck. Origin, by slips from the first nine or ten ribs and the anterior part (sometimes called levator scapulae) from the transverse processes of the last five cervical vertebrae; insertion, scapula near the vertebral border; action, draws the scapula craniad, ventrad, and against the thoracic wall.

m) *Serratus dorsalis*. The anterior part of this muscle arises by a number of fleshy slips from the ribs near their angles. The short slips soon pass into a thin aponeurosis which overlies the epaxial muscles of the thorax and which is fastened to the median dorsal line. The posterior part of this muscle consists of a few slips lying under the latissimus dorsi and appearing like a forward continuation of the internal oblique. These slips insert on the last ribs and originate by means of an aponeurosis from the median dorsal line. Action, draw the ribs forward.

n) *Scalenes.* Raise up the pectoralis muscles from the chest wall by passing the fingers under them. If necessary, their posterior parts may be cut into. Several long muscles will be seen in the chest wall ventral to the origin of the serratus ventralis and in front of the anterior boundary of the external oblique. These muscles are scalenes; they originate on the ribs and pass forward in a nearly straight course to be inserted on the transverse processes of the cervical vertebrae, uniting anteriorly into one band, which will readily be seen by looking immediately ventral to the origin of the anterior part of the serratus ventralis. Insertion, transverse processes of the cervical vertebrae; action, draw the ribs forward and bend the neck.

o) *Intercostals.* The intercostals are a set of muscles extending from one rib to the next. The external layer is called the *external intercostals* and will be seen in part by looking at the chest wall between the origins of the serratus ventralis and dorsalis. Their fibers run obliquely backward and downward. On cutting through some of them, another layer, the *internal intercostals,* will be seen inside of the external layer. The fibers of the internal intercostals run obliquely forward and downward. Near the median ventral line the external intercostals are lacking, so that the internal ones are exposed by cutting through the scalenes. Action, external intercostals bring the ribs forward, internal intercostals draw them back again. The intercostals are the chief respiratory muscles of the thoracic wall.

p) *Epaxial muscles of the thorax.* On cutting through the aponeurosis of the serratus dorsalis, the thick mass of epaxial muscle is exposed. This may be followed up into the neck by cutting through the splenius. In the thoracic region the muscle is divisible into three parts of about equal width, a dorsal part next the median dorsal line, the *semispinalis dorsi;* a median part, the *longissimus;* and a ventral part, the *iliocostalis.* The latter is composed of a number of separate bundles with prominent tendons between them and lies on the ribs to which it is attached. The longissimus may be traced forward to its insertion on the transverse processes of the vertebrae in contact with the origin of the serratus ventralis; this part is the longissimus capitis. In the neck the semispinalis becomes the semispinalis cervicis et capitis and is divisible into a medial part, the biventer cervicis, and a lateral part, the complexus.

In the cat and other mammals there are also the short segmental muscles described for reptiles, comprising the interspinales and the intertransversarii. The occipital muscles consist of three muscles of the rectus capitis series and two of the obliquus capitis set.

The arrangement of the shoulder and dorsal neck musculature of mammals is similar to that of reptiles. The latissimus dorsi and trapezius muscles form the superficial layer of shoulder muscles, with the deltoid groups more lateral and slightly deeper. The levator scapulae, rhomboideus, and serratus muscles belong to the epaxial series. The splenius appears to be a muscle peculiar to mammals, having no homologue among reptiles. The teres major is regarded by Romer as a separated part of the latissimus dorsi. The supra- and infraspinatus have moved up onto the scapula from the chest and probably have a common origin with the supracoracoideus of reptiles; teres minor and the subscapularis belong to the scapulohumeralis group of reptiles. The intercostals are a layer of the hypaxial trunk musculature, and the scalenes are part of the cervical epaxial system.

7. The muscles of the upper arm.—Separate these muscles from each other and identify them.

Rabbit:

a) *Triceps brachii.* The great extensor mass on the back of the upper arm, called anconeous in lower tetrapods, is named triceps in mammals, because it usually has three heads. These are practically distinct muscles in the rabbit. Free the heads. The *long* head is the large mass on the back of the upper arm; origin, scapula, from the axillary border; insertion, olecranon. The *lateral* head is on the lateral surface of the upper arm, ventral to the preceding; origin, greater tuberosity of the humerus. The *medial* head is in contact with the humerus. To see it, spread the other two heads apart and look deep between them or cut through the middle of the belly of the lateral head; origin, along the dorsal surface of the humerus. All three heads insert on the olecranon and are the great extensors of the forearm.

b) *Epitrochlearis* or *extensor antibrachii.* On the medial surface of the long head of the triceps is a fascia from the lower part of which this muscle originates. Insertion, olecranon; action, in common with the triceps.

c) *Biceps brachii.* On the anterior surface of the upper arm found by removing the insertions of the pectoral muscles is a spindle-shaped muscle, the biceps. Origin, glenoid fossa; insertion, ulna and radius; action, flexor of the forearm.

d) *Brachialis.* This muscle is next lateral to the biceps and in contact with the humerus. Origin, ventral and lateral surface of the humerus; insertion and action, in common with the biceps.

Cat:

a) *Triceps brachii* (or *anconeus*). As above, under rabbit. There is also in the cat a small fourth part of the triceps,[5] a small triangular muscle at the elbow joint, covered by the distal end of the lateral head, which should be deflected. Origin, distal end of the humerus; insertion, lateral surface of the ulna; action, strengthens the elbow joint, which it covers, and possibly rotates the ulna.

b) *Epitrochlearis* or *extensor antibrachii*. On the medial side of the long head, a thin sheet, taking origin from the latissimus doris and inserted on the olecranon; action, in common with the triceps, tending also to rotate the ulna.

c) *Biceps brachii*. As in the rabbit; visible only after cutting the insertions of the pectoral muscles. Origin, glenoid fossa; insertion, radius; action, flexor of the forearm.

d) *Brachialis*. Lateral to the biceps, in contact with the lateral head of the triceps. Origin, lateral surface of the humerus; insertion, ulna; action, with the biceps.

8. The muscles of the forearm.—These consist of long, straplike extensors on the outer or lateral surface of the arm and similar flexors on the inner or medial surface. They are here described for the cat only. Remove the tough fascia from the surface of the muscles as you proceed. At the wrist the tendons to the metacarpals and fingers pass under tough ligaments of the wrist, which must be removed if the tendons are to be traced.

a) *Extensor carpi ulnaris:* Begin on the outer surface of the forearm next to the ulna. The olecranon process of the ulna is exposed at and below the elbow, and the first muscle preaxial to the ulna is the extensor carpi ulnaris. Origin, lateral epicondyle of the humerus, and semilunar notch of the ulna; insertion, proximal end of fifth metacarpal; action, extends fifth digit and ulnar side of wrist.

b) *Extensor digitorum lateralis:* This is next to the preceding, going toward the preaxial side. Origin, lateral surface of humerus above the lateral epicondyle; insertion, tendon passes internal to wrist ligaments and then splits into three or four tendons somewhat underlying the tendons of the next muscle; these tendons go to three or four digits, which they extend.

c) *Extensor digitorum communis:* Next to the preceding muscle. Origin, just above the preceding; tendon and action like the preceding.

d) *Brachioradialis or supinator longus:* Next to the preceding on the preaxial border of the forearm, but loose and standing away from the under-

[5] Called anconeus in works on cat anatomy, but the name is obviously inadmissible.

lying extensor (*e*). Origin, about middle of humerus; insertion, lower end of radius and adjacent ligaments; action, rotates hand to supine position.

e) *Extensor carpi radialis:* This underlies the brachioradialis and extends onto the inner or medial surface of the arm. The forearm should now be turned so that the medial side faces the student. The extensor carpi radialis is divisible into a long part (longus) and a shorter part (brevis); the former is the part seen on the medial surface. Origin, humerus near other extensors; insertion, second and third metacarpals; action, extends hand.

f) *Pronator teres:* Proceeding on the medial surface of the forearm toward the ulnar side, the pronator teres is next to the extensor carpi radialis longus and somewhat under it. Its broader upper end slants proximally toward the radius, to which the edge of its narrower lower part is fastened. This muscle is continuous with a strong fascia overlying the tendon of the next muscle (*g*). Origin, medial epicondyle of the humerus; insertion, radius; action, rotates radius to prone position.

g) *Flexor carpi radialis:* Lies next to, and mostly under, the preceding. Origin, medial epicondyle of humerus; muscle narrows to a tendon which is much bound to adjacent ligaments and finally passes through a deep groove between the capitate (third carpale) and the first metacarpal, inserting on second and third metacarpals; action, flexes these metacarpals.

h) *Palmaris longus:* Flat muscle forming outer surface of forearm next to preceding. Origin, medial epicondyle of humerus; insertion, flat tendon passes through wrist ligaments and divides into four or five tendons, which pass to pads of palm and phalanges; action, flexor of digits.

i) *Flexor profundus digitorum:* This is the complex muscle which lies under the preceding and projects to the radial side of it. It has five parts, originating on ulna and humerus and converging to a broad flat tendon under the tendon of palmaris longus. This tendon divides into five, inserted on the basal phalanges. Action, general flexor of the fingers.

j) *Flexor carpi ulnaris:* This consists of two nearly separate muscles which form the ulnar border of the forearm from the medial view. One head, somewhat under palmaris longus, originates on the medial epicondyle of the humerus in common with one of the heads of flexor profundus digitorum. The other head lies alongside the exposed olecranon process of the ulna, from which it originates. Near the wrist both heads join to a tendon which inserts on the pisiform bone of the wrist. Action, flexes ulnar side of wrist.

The muscle along the ulna between flexor carpi ulnaris and extensor carpi ulnaris is part of the flexor profundus digitorum.

9. The muscles of the thigh (Fig. 82).—

Rabbit:

a) *Tensor fasciae latae.* The anterior half of the lateral surface of the thigh is covered by a tough fascia or aponeurosis called the *fascia lata.* In the dorsal part of this will be found a short muscle, the tensor fasciae latae, which is more or less continuous with adjacent muscles. Origin, ilium; insertion, in the fascia; action, tightens the fascia.

b) *Biceps femoris.* Slit open the fascia lata. Posterior to the fascia occupying the middle portion of the lateral surface of the thigh is the large biceps femoris. Separate it from the muscle posterior to it which forms the posterior margin of the thigh (semimembranosus). On removing the surface fascia from the biceps this will be found divisible into two heads. Follow these heads dorsally to their origins, clearing away the tough surface fascia. Origin of the smaller anterior head, neural spines of adjacent vertebrae, and of the larger posterior head, ischial tuberosity; insertion, by a tendon on the kneecap and on the fascia of the shank; action, abductor of the thigh and flexor of the shank.

c) *Gluteus maximus.* Remove the fascia over the sacral region. The gluteus maximus is under the fascia, a thin muscle whose posterior part is partly covered by the biceps femoris. The muscle also has an anterior part which is continuous with the tensor fasciae latae. The two parts of the gluteus maximus are united by an aponeurosis. Origin, fascia of the sacrum and part of the ilium; insertion, third trochanter, the small projection below the great trochanter; action, abducts the thigh.

d) *Gluteus medius.* The larger, thicker muscle anterior to the preceding and partly covered by it. Origin, crest and anterior part of the ilium; insertion, greater trochanter; action, like the preceding. When the two gluteus muscles have been well separated, the origin of the tensor fasciae latae will be found on the ventral edge of the gluteus medius dipping deep down to the ilium.

e) *Vastus lateralis.* Cut through the fascia lata and its tensor by a longitudinal slit extending to the knee. Under the fascia will now be seen the stout vastus lateralis. Origin, greater trochanter dorsal to the insertion of the gluteus medius.

f) *Vastus intermedius.* This is the name given to what appears to be the posterior part of the preceding muscle. It is partly separable from the vastus lateralis and lies between this and the anterior head of the biceps femoris. Origin, greater trochanter and surface of the femur.

g) *Rectus femoris.* This is the thin muscle which is folded over the anterior margin of the thigh; it extends on both lateral and medial surfaces of the thigh. It originates on the ilium and the fascia lata and is fused with the fibers of the tensor fasciae latae. The rectus femoris consists of two parts, the part just described and a second part, which may be located as follows: Separate the rectus femoris from the vastus lateralis, to which it will be found slightly fused, continuing this separation to the medial side of the thigh, spreading the legs apart. The first part of the rectus femoris may be cut through at the middle. On the medial side of the thigh internal to the first part of the rectus femoris and ventral to the vastus lateralis is the second part of the rectus femoris, a cylindrical muscle. Origin, ilium in front of the acetabulum.

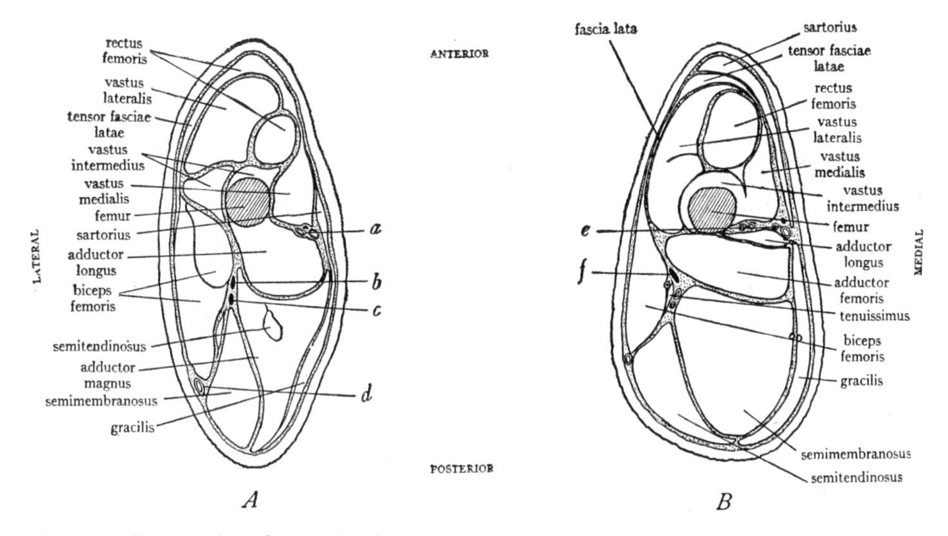

FIG. 82.—Cross-sections through the thigh of *A*, rabbit, and *B*, cat, to show the location of the muscles. Black spots are nerves; small circles, blood vessels. *a*, greater saphenous nerve, artery, and vein; *b*, peroneal nerve; *c*, tibial nerve; *d*, sciatic vein; *e*, femoral nerve, artery, and vein; *f*, sciatic nerve. (*A* from Bensley's *Practical Anatomy of the Rabbit*, University of Toronto Press.)

h) *Vastus medialis.* On the medial side of the thigh posterior to the first part of the rectus femoris and not very well separable from it. Origin, femur.

The rectus femoris and the three vastus muscles together constitute the great *quadriceps femoris* muscle. The origins of its several parts have already been given. Insertion, on the tibia and the patella and the tendon which extends over the patella; action, powerful extensor of the shank.

i) *Sartorius.* The very long, narrow muscle extending like a band along the middle of the medial surface of the thigh. Origin, inguinal ligament; insertion, tibia; action, adductor of the thigh, rotator of the thigh, extensor of the shank. The sartorius lies between the vastus medialis and the next muscle to be described, and covers some large blood vessels.

j) *Gracilis.* A large, thin muscle over the posterior half of the medial surface of the thigh. Origin, pubic symphysis; insertion, fascia of the distal portion of the thigh and proximal portion of the shank: action, adductor of the leg.

k) *Adductor longus* and *adductor magnus.* Cut through the middle of the gracilis and find beneath it two stout muscles, their fibers running from the median ventral line to the femur. The anterior muscle is the adductor longus; the posterior one, the adductor magnus. Origin, various parts of the ischium; insertion, femur; action, adductors of the thigh.

l) *Semitendinosus.* Split open the adductor magnus and find inside of it a cylindrical muscle, the semitendinosus. Origin, ischial tuberosity; insertion, medial condyle of the tibia; action, flexor of the shank.

m) *Semimembranosus.* This is the muscle which forms the posterior margin of the thigh, between the biceps femoris and the adductor magnus. Origin, fascia over the biceps and ischial tuberosity; insertion, with the gracilis in the fascia of the shank; action, flexor of the shank.

Cat:

a) *Tensor fasciae latae.* Examine the lateral (outer) surface of the thigh. The anterior part of this is covered by a tough fascia, the *fascia lata.* In the dorsal part of this is a muscle, the *tensor fasciae latae,* a thick triangular muscle. Origin, ilium and neighboring fascia; insertion, fascia lata; action, tightens the fascia lata.

b) *Biceps femoris.* This is the large muscle on the lateral surface of the thigh posterior to the fascia lata and covering more than half of the surface of the thigh. It has but one head in the cat. Origin, tuberosity of the ischium; insertion, patella and tibia by a tendon, and the fascia of the shank; action, abductor of the thigh and flexor of the shank.

c) *Caudofemoralis.* Clean away the fascia from the back in front of the base of the tail, as far forward as the anterior end of the pelvic girdle. Muscles will be found between the median dorsal line and the thigh. The most posterior of these is the narrow flat caudofemoralis, passing from the side of the root of the tail toward the dorsal end of the biceps femoris. Origin, transverse processes of the second and third caudal vertebrae; insertion, the muscle passes ventrally, concealed by the anterior margin of the biceps femoris. This should be lifted up and the caudofemoralis followed to its tendon; the latter is very long and passes to the patella, on which it is inserted. Action, abductor of the thigh, extensor of the shank.

d) *Gluteus maximus.* A rather thin, flat muscle immediately anterior to the preceding. It is imbedded in the fascia and is continuous with the tensor fasciae latae anteriorly. Origin, from the fascia and from the transverse processes of the last sacral and first caudal vertebrae; insertion, fascia lata and to a slight extent on the greater trochanter; action, in common with the next.

e) *Gluteus medius.* The very large, triangular muscle immediately in front of the preceding and partly covered by it. The gluteus maximus should be cut across to see it. Origin, adjacent fascia, crest of the ilium, and lateral surface of the ilium, and transverse processes of the last sacral and first caudal vertebrae; insertion, by a strong tendon on the greater trochanter of the femur; action, abductor of the thigh. Along the anterior border of this muscle the origin of the tensor fasciae latae passes internally toward the ilium.

f) *Sartorius.* This muscle forms the anterior margin from the lateral view of the thigh. It is folded over the margin and, on following it to the medial or inner surface of the thigh, will be found to cover the anterior half of the medial surface. Origin, crest and ventral border of the ilium; insertion, proximal end of the tibia and the patella and the fascia and ligaments between; action, adductor and rotator of the thigh and extensor of the shank.

g) *Vastus lateralis.* Cut through the fascia lata by a longitudinal slit extending to the patella. Separate well the sartorius from underlying parts. The tensor fasciae latae is now well exposed. The large, stout muscle which was covered by the fascia lata is the vastus lateralis. Origin, greater trochanter and surface of the femur.

h) *Rectus femoris.* At its anterior margin the vastus lateralis will be found partly separable from a stout muscle lying on its medial side and covered externally by the sartorius. The sartorius may be cut across the middle. The muscle in question is the rectus femoris. Origin, ilium near the acetabulum.

i) *Vastus medialis.* This is on the medial side of the thigh posterior to the rectus femoris, which its anterior margin partly covers. It also is covered externally by the sartorius. Origin, femur.

j) *Vastus intermedius.* On widely separating the rectus femoris from the vastus lateralis a muscle will be seen, deep down, next to the shaft of the femur. Origin, surface of the femur.

The rectus femoris and the three vastus muscles are more or less united to each other and constitute the great *quadriceps femoris* muscle. Its origins have been described; all its parts are inserted on the patella and adjacent ligaments; action, extensor of the shank.

k) *Gracilis.* This is the large, flat muscle forming the posterior half of the medial surface of the thigh. Origin, ischial and pubic symphyses; insertion by an aponeurosis which passes to the tibia; action, adductor of the leg.

l) *Adductor longus* and *adductor femoris.* Cut through the middle of the gracilis and separate each half from the underlying muscles. The latter consist of three muscles passing from the median ventral line to the femur. The most anterior of the three is quite small; this is the adductor longus. Origin, pubis; insertion, femur; action, with the next. The middle muscular mass is the large adductor femoris (corresponding to adductor magnus and brevis of other mammals). Origin, pubis; insertion, femur; action, adductor of the thigh.

m) *Semimembranosus.* The large posterior part of the mass which was covered by the gracilis. Origin, ischium; insertion, medial epicondyle of the femur and proximal end of the tibia; action, extensor of the thigh. The muscle is more or less divisible into two parts.

n) *Semitendinosus.* The most posterior muscle of the thigh, posterior to the preceding. Origin, ischial tuberosity; insertion, tibia; action, flexor of the shank.

o) *Tenuissimus.* Turn to the lateral surface of the thigh. Cut through the middle of the biceps femoris. Beneath it will be noted a very narrow, long muscle, the tenuissimus. Origin, transverse process of the second caudal vertebrae, in common with the caudofemoralis; insertion, on the same fascia as the insertion of the biceps.

On separating the biceps from the underlying muscles they will be revealed as extensions of muscles already identified on the medial surface. The adductor femoris is seen in contact with the femur posterior to the vastus lateralis; the semimembranosus comes next, and the semitendinosus is again the most caudal of the thigh muscles.

The thigh musculature of mammals, like that of reptiles, is divisible into the long muscles from the girdle to the shank and the short muscles from the girdle to the femur. On the dorsal or extensor side there is the same extensor or quadriceps femoris mass in mammals as in reptiles, comprising the rectus femoris (derived from the extensor iliotibialis of reptiles) and the vastus muscles (corresponding to the femorotibialis); to this group also belongs the sartorius, derived from the ambiens of reptiles. The biceps femoris is believed by Romer (1922) to come from the iliofibularis, but Appleton thinks the tenuissimus represents the iliofibularis. The short extensor muscles include the glutei and the tensor fasciae, considered by Romer derivatives of the iliofemoralis. Other mammalian muscles of this group, which were not seen in the dissection, are given the following reptilian correspondences by Romer: pectineus and iliopsoas with the pubo-ischio-femoralis internus, and obturator and gemellus with the pubo-ischio-femoralis posterior. On the flexor side, the gracilis, semitendinosus, and semimembranosus are long flexors corresponding to a similar group in reptiles. The adductors of the cat belong to the short flexor series.

10. The muscles of the shank.—

Rabbit:

a) *Tibialis anterior.* The lateral (outer, dorsal) surface of the shank is covered by the distal end of the biceps femoris and fascia. These should be removed. The most anterior of the muscles of the lateral surface is the tibialis anterior. Origin, lateral condyle and tuberosity of the tibia; insertion, second metatarsal; action, flexor of the foot.

b) *Peroneus*. Next dorsal to the preceding on the lateral surface, consisting of a group of several more or less fused muscles. Origin, tibia and fibula; insertion, metatarsals; action, flexor of the foot.

c) *Gastrocnemius*. This is the thin but broad muscle forming the caudal surface of the shank, divisible into two nearly separate portions, one of which is on the lateral, one on the medial, surface of the shank. Origin, lateral and medial condyles of the femur and tibia; insertion, by a strong tendon, the tendon of Achilles, which passes over the heel (calcaneus), on which it is inserted; action, extensor of the foot.

d) *Soleus*. This is the muscle just internal to that part of the gastrocnemius which is on the lateral surface of the thigh. Origin, head of the fibula; insertion and action with the preceding.

e) *Plantaris*. This is situated internal to that part of the gastrocnemius which is medial. Origin, lateral condyle of the femur; insertion and action with the preceding.

f) *Other muscles of the shank* (optional). There are three more muscles of the shank; they lie in contact with the tibia. They are: the *extensor hallucis longus*, exposed on the medial surface of the tibia; *extensor digitorum longus*, covered by the tibialis anterior; and *flexor digitorum longus*, between the tibia and the soleus and plantaris. As their names imply, these muscles are inserted by long slender tendons on the digits and act to flex and extend the digits.

It should be noted that the muscles named extensor in the shank and foot are really flexor in their action and that those named flexor are extensor in their action, following a custom borrowed from human anatomy. The custom arises from a desire to retain the names applied to the muscles of the forelimb for the muscles in the corresponding positions in the hind limb. Thus, in the supine position of the forearm, the extensors face anteriorly and the flexors posteriorly. Similarly the muscles on the anterior side of the leg are designated extensors, although they really flex the foot, and those on the posterior side, flexors, although they extend the foot. In describing the action the terms flexor and extensor are used with reference to the movement produced and not with reference to the position of the muscle on the limb.

Cat:

a) *Tibialis anterior*. Clean away the tough fascia of the shank and also the insertions of the biceps and the gracilis. Examine the lateral (outer) surface of the shank. The most ventral muscle, whose ventral border is in contact with the tibia, is the tibialis anterior. Origin, proximal parts of tibia and fibula; insertion by a strong tendon, which should be traced into the foot, where it will be found to pass obliquely to the medial side of the foot to be inserted on the first metatarsal; action, flexor of the foot.

b) *Extensor digitorum longus.* This is the muscle next dorsal to the preceding on the lateral surface of the shank. It is so closely placed to the preceding as to appear as part of it, but the line of separation will be found by a little searching. Origin, lateral epicondyle of the femur; insertion, by a stout tendon, which if followed into the foot is found to diverge into four tendons, one of which is inserted on each digit; action, extensors of the digits.

c) *Peroneus* muscles. These are next dorsal to the preceding, originating on the fibula. There are three of them, more or less fused to each other. The three end each in a tendon; the three tendons pass over the lateral surface of the lateral malleolus of the tibia and over the calcaneus and are inserted on the metatarsals and digits. Action, extensors and flexors of the foot.

d) *Gastrocnemius.* This is the large muscle forming the posterior or caudal surface of the shank. It is divisible into two large portions, one on the medial surface, the other on the lateral surface of the shank. The lateral head is subdivisible into four heads. Origins, from the surface fascia, the femur, and the tendon and fascia of the plantaris muscle (see below); insertion, by a strong tendon, which passes to the heel bone (calcaneus) on which it is inserted. Action, extensor of the foot.

e) *Soleus.* On carefully separating the lateral head of the gastrocnemius, a muscle, the soleus, will be found internal to it. It is a flat muscle in contact with the peroneus muscles ventrally; it tapers abruptly to a tendon which joins the tendon of the gastrocnemius. Origin, fibula; insertion, calcaneus; action, with the gastrocnemius, of which it is sometimes considered a part.

f) *Plantaris.* On carefully separating the medial head of the gastrocnemius a large muscle will be found internal to it, lying between the two heads of the gastrocnemius, which practically inclose it. It is fused, to a considerable extent, to the lateral head but is quite separable from the medial head, being covered on the medial side by a shining aponeurosis. Origin, patella and femur; insertion, by a thick tendon which passes in the middle of a sort of tube formed by the tendon of the gastrocnemius and soleus onto the ventral surface of the calcaneus. Here it broadens and finally divides into four slips, each attached to a digit. Action, flexor of the digits.

g) *Flexor digitorum longus.* On turning to the medial side of the shank and clearing away the surface fascia the following may be identified. Most ventrally will be seen the tibialis anterior; next comes the exposed surface of the tibia. Immediately dorsal to the bone is the flexor digitorum longus, which consists of two parts, somewhat separated. The other part is more lateral in contact with the peroneus muscles. Separate the part of the flexor which appears on the medial surface from the tibia by a cut, and lift it up. Internal to it is seen a long tendon, and on the other side of this tendon is the other part of the flexor, this part corresponding to the flexor hallucis longus of man. Both parts of the flexor terminate in slender tendons which unite distally into a broad tendon, which eventually divides into four tendons inserted on the digits. Origin, tibia, fibula, and adjacent fascia; action, flexor of the digits.

h) *Tibialis posterior*. The long tendon between the two parts of the preceding muscle was noted above. It is the tendon of the tibialis posterior, and on following this tendon proximally the belly of the muscle will be located. Origin, fibula, tibia, and fascia; insertion, scaphoid and medial cuneiform of the ankle; action, extensor of the foot.

F. SUMMARY

1. The voluntary muscles are of two kinds: the somatic or parietal muscles, derived from the epimeres, which constitute most of the musculature; and the branchial or visceral muscles, derived from the hypomere, which are limited to the gill region. Primitively, the branchial muscles operate the gill arches, but in tetrapods, they may spread over a considerable area. They come from the entire hypomere, since in the gill region the hypomere does not divide into somatic and splanchnic layers.

2. In primitive vertebrates the somatic muscles exist in the form of an axial succession of muscle segments or myotomes, separated by connective tissue partitions, the myosepta. Each myotome extends from middorsal to midventral line.

3. Beginning with fishes, the myotomes are divided into dorsal or epaxial halves and ventral or hypaxial halves by the horizontal skeletogenous septum. Although the division of the musculature remains, the septum becomes indistinct in amniotes.

4. The girdles and paired appendages interrupt the series of myotomes. The musculature of the girdles and appendages is derived in lower vertebrates from muscle buds sent out by the adjacent myotomes. Although this mode of origin can no longer be seen in the embryos of higher vertebrates, their girdle and limb musculature must be regarded as of myotomic origin phylogenetically.

5. In the evolution of tetrapods the muscles of the girdles and limbs increase in size and importance and spread over the segmented musculature. The latter also more or less loses its segmental arrangement.

6. The epaxial musculature tends throughout vertebrates to split up into longitudinal systems—some long, some short and segmental. In tetrapods there are typically three longitudinal systems: the transversospinalis (divisible into the spinalis and semispinalis systems), the longissimus, and the iliocostalis, from the middorsal line downward. These extend in general from the sacrum to the back of the skull and also have continuations into the tail; they have attachments on vertebrae and ribs. The short segmental muscles include several systems between various parts of the vertebrae and ribs and each member extends over only one body segment. Short muscles of the epaxial system extending from the first vertebrae to the skull are termed the occipital muscles.

7. The hypaxial trunk musculature in tetrapods is divisible into the oblique system on the body sides and the rectus system near the midventral line. The former splits into several layers; the latter forms the rectus abdominis in the trunk, rectus cervicis in the neck, and geniohyoid in front of the hyoid apparatus.

8. The appendicular muscles consist in fishes of a dorsal or abductor mass and a ventral or adductor mass. In tetrapods these differentiate into a number of muscles from girdle to limb

and between the sections of the limb. The accompanying table shows the reptilian and mammalian derivatives of the dorsal and ventral masses according to Romer (1922), with some corrections from Appleton (1928).

Reptiles	Mammals
FORELIMB DERIVATIVES OF PRIMITIVE DORSAL MUSCLE MASS	
Latissimus dorsi, including teres major	Latissimus dorsi, including teres major
Dorsalis scapulae (=deltoides scapularis), deltoides clavicularis	Two or three deltoids
Scapulohumeralis anterior	Teres minor
Subcoracoscapularis and scapulohumeralis posterior	Subscapularis
Anconeus	Anconeus plus epitrochlearis
Brachio- or humeroradialis	Brachioradialis
Forearm extensors and supinators	Forearm extensors and supinators
FORELIMB DERIVATIVES OF PRIMITIVE VENTRAL MUSCLE MASS	
Pectoralis	Pectoralis
Supracoracoideus	Supraspinatus, infraspinatus
Coracobrachialis	Coracobrachialis
Biceps	Biceps
Brachialis	Brachialis
Forearm flexors and pronators	Forearm flexors and pronators
HIND-LIMB DERIVATIVES OF PRIMITIVE DORSAL MUSCLE MASS	
Extensor iliotibialis	Rectus femoris
Ambiens	Sartorius
Femorotibialis	Vastus group
Iliofibularis	Tenuissimus
Pubo-ischio-femoralis internus	Iliacus, psoas, pectineus
Iliofemoralis	Gluteus group, tensor fasciae latae
Pubo-ischio-femoralis posterior	Obturator internus, gemelli
Tibialis anticus	Tibialis anterior
Peroneus group	Peroneus group
Other extensors	Other extensors
HIND-LIMB DERIVATIVES OF PRIMITIVE VENTRAL MUSCLE MASS	
Pubo-ischio-tibialis	Gracilis
Pubotibialis, flexor tibialis externus and internus	Semimembranosus, semitendinosus, biceps femoris
Adductors	Adductors
Pubo-ischio-femoralis externus	Obturator externus, quadratus femoris
Gastrocnemius	Gastrocnemius, plantaris
Tibialis posticus	Tibialis posterior
Other flexors	Other flexors

9. The branchial musculature, derived from the hypomere and innervated by cranial nerves, originally functions to move the gill arches for respiratory purposes. Very early in vertebrate history, however, the first or mandibular arch became the upper and lower jaws, and its musculature took on masticatory functions, retaining this condition throughout vertebrates. The musculature of the remaining gill arches, with the loss of gill respiration in tetrapods, became reduced and altered but continues to move those structures derived from gill arches, namely, the hyoid apparatus and the laryngeal cartilages. The following attempts to give the main points in the history of the branchial musculature, but many things are still controversial.

a) General: Primitively, probably each gill arch had a levator and a constrictor muscle. In present vertebrates there are usually present in addition to the levator and constrictor series, the following ventral series: coracobranchials, subarcuals, and ventral transversals (transversi ventrales). The last two series are lacking in elasmobranchs, and their coracobranchials are usually considered to be hypobranchial muscles, not homologous to the coracobranchials of other fishes. Their levator series is also imperfect. The ventral branchial series do not occur in connection with the mandibular and hyoid arches.

b) Mandibular arch muscles: The dorsal part of the mandibular arch constrictor is generally represented in vertebrates by the dorsal constrictor (craniomaxillaris) and the adductor mandibulae. The former persists in reptiles, where it may also give rise to certain eye muscles; the latter is the chief masticatory muscle of vertebrates, acting to close the lower jaw. The dorsal constrictor fails to persist in the adults of Dipnoi, Holocephali, Amphibia, and Mammalia, where the dorsal mandibular constrictor is then represented by one muscle mass, the adductor mandibulae. This appears in mammalian embryos and differentiates into one group of masticatory muscles, comprising the temporal, masseter, external pterygoid, and zygomaticomandibularis (lacking in the cat); and another group comprising the internal pterygoid, tensor tympani, and tensor palati. This second group, however, is thought by others to be the homologue of the first levator. The tensor tympani and tensor palati are small muscles associated with the middle ear, a result of the reduction of quadrate and articular to middle-ear ossicles.

The ventral part of the mandibular constrictor is the intermandibularis, termed mylohyoid in mammals; other derivatives are the mandibularis, when present, and the anterior belly of the mammalian digastric muscle.

The mandibular arch muscles through all their evolutionary changes are innervated by the fifth (trigeminus) cranial nerve.

c) Hyoid arch muscles: The hyoid constrictor sheet forms in tetrapods chiefly the interhyoideus and the constrictor colli. The interhyoideus becomes in mammals the posterior belly of the digastric. The digastric is the opener or depressor of the jaw in most mammals. The constrictor colli is highly developed in reptiles and birds and is usually considered to develop into the platysma and other muscles of facial expression of mammals.[6] The second or hyoid levator typically becomes the depressor mandibulae, but in some tetrapods the depressor may come from the constrictor sheet. In monotremes the hyoid constrictor is present as an interhyoideus muscle, and the hyoid levator occurs as such, running from the ear region to the remains of the hyoid arch. In other mammals the interhyoideus becomes, at least in

[6] But Edgeworth flatly disagrees with this view, maintaining that the platysma has no exact homologue in nonmammalian tetrapods and is a new formation. It must be remarked that, although Edgeworth's book is impressively erudite, it disagrees with other students of vertebrate anatomy in so many particulars that one is left by it in a very confused state of mind.

part, the posterior belly of the digastric; and the hyoid levator becomes the stapedius, a small muscle of the middle ear. This results from the reduction of the upper end of the hyoid arch to the stapes.

The hyoid arch muscles are innervated throughout all their changes by the seventh (facial) cranial nerve.

d) Other branchial arch muscles: The levators persist in gill-bearing urodeles like *Necturus* and also occur in larval stages of other amphibians. The cucullaris belongs to the levator series but has acquired attachments to the pectoral girdle. In higher tetrapods it subdivides into the trapezius and sternocleidomastoid or some variant thereof. The branchial constrictors appear to have no representatives in tetrapods. The coracobranchials are limited to fishes. The subarcual series occurs in urodeles and anuran larvae; and the first subarcual is found in reptiles, birds, and mammals as the branchiohyoid or branchiomandibularis muscle. Turtles have a second subarcual in its typical form, and the third subarcual occurs in monotremes.

The constrictor esophagi, the laryngeal muscles, and the striated pharyngeal muscles of mammals are probably derivatives of some of the ventral branchial series. This, however, is denied by Edgeworth, who considers them new formations, albeit of hypomeric origin.[7]

10. The hypobranchial musculature: This is the ventral musculature below the gill region and is formed by forward growth from the ventral ends of the myotomes behind the branchial muscles. It may be regarded as in series with the midventral systems of the hypaxial trunk musculature. It includes the rectus cervicis (sternohyoid, sternothyroid, thyrohyoid, omohyoid) in the neck, the geniohyoids in the throat, the muscles from skull, mandible, and hyoid apparatus into the tongue (genioglossus, styloglossus, hyoglossus), the intrinsic tongue musculature, and muscles of the epiglottis.

REFERENCES

Adams, L. A. 1919. A memoir on the phylogeny of the jaw muscles in recent and fossil vertebrates. Ann. N.Y. Acad. Sci., **28**.

Appleton, A. B. 1928. The muscles and nerves of the post-axial region of the tetrapod thigh. Jour. Anat., **62**.

Brock, G. T. 1938. The cranial muscles of the gecko, a general account with a comparison of the muscles in other gnathostomes. Proc. Zoöl. Soc. London, **108, B**.

Byerly, T. C. 1926. The myology of *Sphenodon*. Univ. Iowa Stud. Nat. Hist., **11**.

Camp, C. L. 1923. Classification of the lizards. Bull. Amer. Mus. Nat. Hist., **48**.

Davidson, P. 1918. The musculature of *Heptanchus maculatus*. Univ. Calif. Pub. Zoöl., **18**.

Drüner, L. 1903. Ueber die Muskulatur des Visceralskelettes der Urodelen. Anat. Anz., **23**.

Eaton, T. H. 1936. The myology of salamanders with particular reference to Dicamptoden. Jour. Morph., **60**.

———. 1937. The gularis muscle in Urodela. *Ibid.*

Edgeworth, F. H. 1935. The Cranial Muscles of Vertebrates.

Fürbringer, M. 1874. Zur vergleichenden Anatomie der Schultermuskeln. Jena Zeitschr. Naturwiss., **8**.

———. 1876. Zur vergleichenden Anatomie der Schultermuskeln. Morph. Jahrb., **1**.

———. 1900. Zur vergleichenden Anatomie des Brustschulterapparates und der Schultermuskeln. Jena Zeitschr. Naturwiss., **34**.

[7] The author has found G. Brock's article (1938) very useful in constructing the above summary and as a check on Edgeworth's views.

GADOW, H. 1882. Untersuchungen über die Bauchmuskeln der Krokodile, Eidechsen, und Schildkröten. Morph. Jahrb., **7.**

———. 1882. Beiträge zur Myologie der hinteren Extremitäten der Reptilien. *Ibid.*

GREGORY, W. K., and CAMP, C. L. 1918. Studies in comparative myology and osteology. No. III. Bull. Amer. Mus. Nat. Hist., **38.**

HAINES, R. W. 1935. Some muscular changes in the tail and thigh of reptiles and mammals. Jour. Morph., **58.**

———. 1939. A revision of the extensor muscles of the forearm in tetrapods. Jour. Anat., **73.**

HOWELL, A. B. 1933. The architecture of the pectoral appendage of the dogfish. Jour. Morph., **54.**

———. 1936. Phylogeny of the distal musculature of the pectoral appendage. *Ibid.*, **60.**

———. 1936. The phylogenetic arrangement of the muscular system. Anat. Rec., **66.**

———. 1937. Morphologenesis of the shoulder architecture. Auk, **54.**

HUBER, E. 1931. Evolution of Facial Musculature and Facial Expression.

LIGHTOLLER, G. H. S. 1939. Probable homologues. A study of the comparative anatomy of the mandibular arches and their musculature. Trans. Zoöl. Soc. London, **24.**

———. 1940. The comparative myology of the platysma: a comparative study of the sphincter colli profundus and the trachelo-platysma. Jour. Anat., **74.**

MARION, G. E. 1905. Mandibular and pharyngeal muscles of *Acanthias* and *Raia*. Amer. Nat., **39.**

MINER, R. W. 1925. The pectoral limb of *Eryops* and other primitive tetrapods. Bull. Amer. Mus. Nat. Hist., **51.**

MÜLLER, E. 1911. Untersuchungen über die Muskeln und Nerven der Brustflosse und der Körperwand bei *Acanthias vulgaris*. Anat. Hefte, **43.**

NISHI, S. 1916. Zur vergleichenden Anatomie der eigentlichen (genuinen) Rückenmuskeln. Morph. Jahrb., **50.**

OLSON, E. C. 1936. The dorsal axial musculature of certain primitive Permian tetrapods. Jour. Morph., **59.**

OSAWA, G. 1898. Beiträge zur Anatomie der *Hatteria punctata*. Arch. f. mikr. Anat., **51.**

PIATT, J. 1938. Morphogenesis of the cranial muscles in *Ambystoma*. Jour. Morph., **63.**

———. 1939. Correct terminology in salamander myology. I. Intrinsic gill musculature. Copeia.

———. 1940. Correct terminology in salamander myology. II. Transverse ventral throat musculature. *Ibid.*

ROMER, A. S. 1922. The locomotor apparatus of certain primitive and mammal-like reptiles. Bull. Amer. Mus. Nat. Hist., **46.**

———. 1923. Crocodilian pelvic muscles and their avian and reptilian homologues. *Ibid.*, **48.**

———. 1924. Pectoral limb musculature and shoulder girdle structure in fish and tetrapods. Anat. Rec., **27.**

SHANN, E. W. 1920. The comparative myology of the shoulder girdle and pectoral fin of fishes. Trans. Roy. Soc. Edinburgh, **52.**

SIEGLBAUER, F. 1909. Zur Anatomie der Schildkrötenextremität. Arch. Anat. u. Entw'-gesch.

VALLOIS, H. V. 1922. Les Transformations de la musculature de l'épisome chez les Vertébrés. Arch. morph. gén. et expér., fasc. **13.**

WILDER, H. H. 1912. The appendicular muscles of *Necturus*. Zool. Jahrb. Suppl. 15.

XI. THE COMPARATIVE ANATOMY OF THE COELOM AND OF THE DIGESTIVE AND RESPIRATORY SYSTEMS

A. THE ORIGIN AND PARTS OF THE COELOM AND THE MESENTERIES

1. Origin.—The *coelom* or *body cavity* of vertebrates is the cavity of the hypomere. It is never at any stage segmented. The outer wall of the hypomere comes in contact with the inner surface of the layer of voluntary muscles and forms the lining of the body wall. This lining is known as the *parietal peritoneum.* The inner walls of the hypomeres of the two sides come in contact in the median plane, folding around the intestine and inclosing the intestine between their two walls. The inner walls of the hypomere thus become the covering layer of the intestine and other viscera and are then named the *visceral peritoneum* or *serosa.* Above and below the intestine the two walls of the hypomere are in contact and form a double-walled membrane, which is designated as a *mesentery.* That portion of the mesentery between the dorsal wall of the coelom and the intestine is called the *dorsal* mesentery; that between the ventral wall and the intestine is the *ventral* mesentery. Different portions of these mesenteries receive special names, which will be given in the course of the dissection. The dorsal mesentery is intact for its entire length in most vertebrates, but the ventral mesentery very early disappears except in certain regions, which will be noted later. These matters have already been described in the chapter on chordate development, in which pages 74, 77 should be re-read.

In *Amphioxus* the mesoderm pouches off from the archenteron as a paired axial series of sacs whose cavities later combine to form the definitive coelom. Although the mesoderm of vertebrates no longer arises in this fashion (traces of such origin occur in urodeles), this *enterocoelous* manner of origin of the mesoderm and the coelom is commonly accepted as ancestral for vertebrates. This links the vertebrates to other groups having an enterocoelous type of coelom, namely the Chaetognatha, the Echinodermata, and the Hemichordata.

2. Divisions of the coelom.—At first the coelom consists of a continuous cavity extending the entire length of the trunk region, divided into two longitudinal halves by the dorsal and ventral mesenteries. With the partial disappearance of the ventral mesentery, the two halves of the coelom are connected ventral to the intestine (Fig. 83*C*). In the adults of all vertebrates the coelom is divided into at least two compartments by the formation of a partition. This partition, called the *transverse septum,* develops at the posterior end of the heart and cuts off the heart from all of the other viscera. The transverse septum thus divides the coelom into a small anterior compartment, the *pericardial cavity,* which contains only the heart, and a very large posterior compartment, the *pleuroperitoneal cavity,* which contains all of the other viscera (see Fig. 84). The pericardial cavity in fishes and urodeles is *anterior* to the pleuroperitoneal cavity, and the transverse septum in those groups passes *transversely* across the body (Fig. 84*A*, *B*). In the Anura and all vertebrates above Anura the pericardial cavity has descended posteriorly, so that it comes to lie *ventral* to the anterior part of the pleuroperitoneal cavity; the transverse septum then assumes an *oblique* position (Fig. 84*C*). In that portion of the pleuroperitoneal cavity which, in consequence of the descent of the pericardial cavity, lies dorsal to the pericardial cavity, the lungs are situated. This condition of the coelom, as in Figure 84*C*, is found in Anura and in most reptiles. In some reptiles, especially the Crocodilia, and in birds and mammals the pleuroperitoneal cavity is divided into anterior and posterior parts by various fusions between the transverse septum and other coelomic folds; in mammals the closure is accomplished chiefly by the *pleuroperitoneal* or *nephric fold*, which descends from the dorsal body wall and fuses with the transverse septum (Fig. 84*D*, *E*). The partition so formed is known as the *oblique septum* in birds and the *diaphragm* in mammals. In birds it is but slightly muscular, whereas in mammals it is extensively muscularized from the cervical myotomes.

The oblique septum or diaphragm forms immediately posterior to the lungs. That portion of the pleuroperitoneal cavity which is cut off anterior to the oblique septum or diaphragm

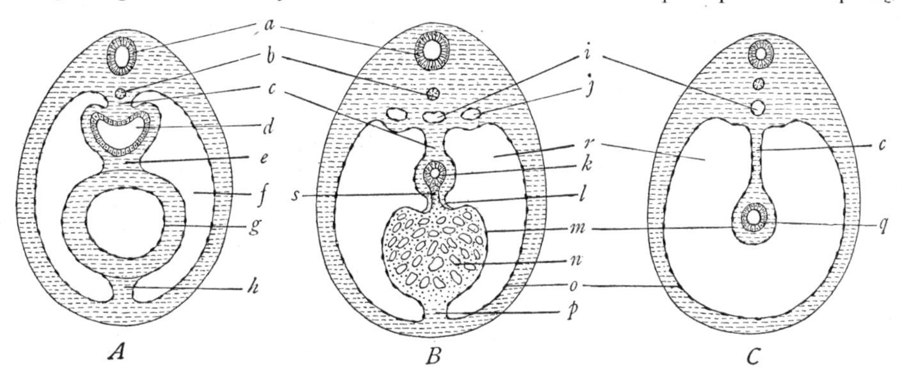

FIG. 83.—Diagrams to show the relations of certain viscera to the mesenteries. *A*, showing intestine (*d*), supported by the dorsal mesentery (*c*), and the heart (*g*), inclosed in the ventral mesentery (*e* and *h*). *B*, showing the liver (*n*), inclosed in the ventral mesentery, part of which, the lesser omentum (*l*), extends between the intestine (*k*) and the liver, and part of which, the falciform ligament (*p*), extends between the liver and the ventral body wall. *C*, showing relation of the intestine (*q*) to the dorsal mesentery (*c*). *a*, neural tube; *b*, notochord; *c*, dorsal mesentery of the digestive tract; *d*, esophagus; *e*, dorsal mesentery of the heart or dorsal mesocardium; *f*, pericardial cavity; *g*, heart; *h*, ventral mesentery of the heart or ventral mesocardium; *i*, dorsal aorta; *j*, posterior cardinal vein; *k*, duodenum; *l*, lesser omentum or hepatoduodenal ligament; *m*, serosa or visceral peritoneum; *n*, liver; *o*, parietal peritoneum; *p*, falciform ligament of the liver; *q*, small intestine; *r*, peritoneal cavity; *s*, bile duct. In *A*, *e* and *h*, and in *B*, *l* and *p* form the ventral mesentery of the digestive tract wihch incloses the heart and liver; in *C* the ventral mesentery is absent. (From Prentiss and Arey's *Textbook of Embryology*, courtesy of the W. B. Saunders Co.)

consequently contains the lungs. It consists of the two *pleural cavities* or *pleural sacs*, each inclosing a lung. The two pleural cavities are completely separated from each other; the pericardial cavity containing the heart is situated in the median line between their ventral portions. That part of the pleuroperitoneal cavity cut off posterior to the oblique septum or diaphragm is called the *peritoneal* or *abdominal* cavity; it incloses the greater part of the digestive tract and the urogenital system. It will be seen from this account that in birds and mammals the coelom is divided into *four* compartments—the pericardial cavity, the two pleural cavities, and the peritoneal cavity. This arrangement greatly increases the efficiency of lung respiration.

It is convenient to speak of the viscera as being inclosed in the coelomic cavities. This is not, however, a correct expression, for the viscera are covered by the visceral peritoneum and therefore are not really inside the coelom in the same sense that a chair could be said to be inside a room. They are outside it and have the same relations to it as if they were pushed into the coelom carrying the coelomic wall before them. To illustrate further, one cannot get into the inside of a tent or a balloon by pushing against the wall; one carries the tent or balloon wall before him and always remains in reality on the outside of the tent or balloon. Similarly, the viscera are outside the coelom, although they appear to be contained within its cavity.

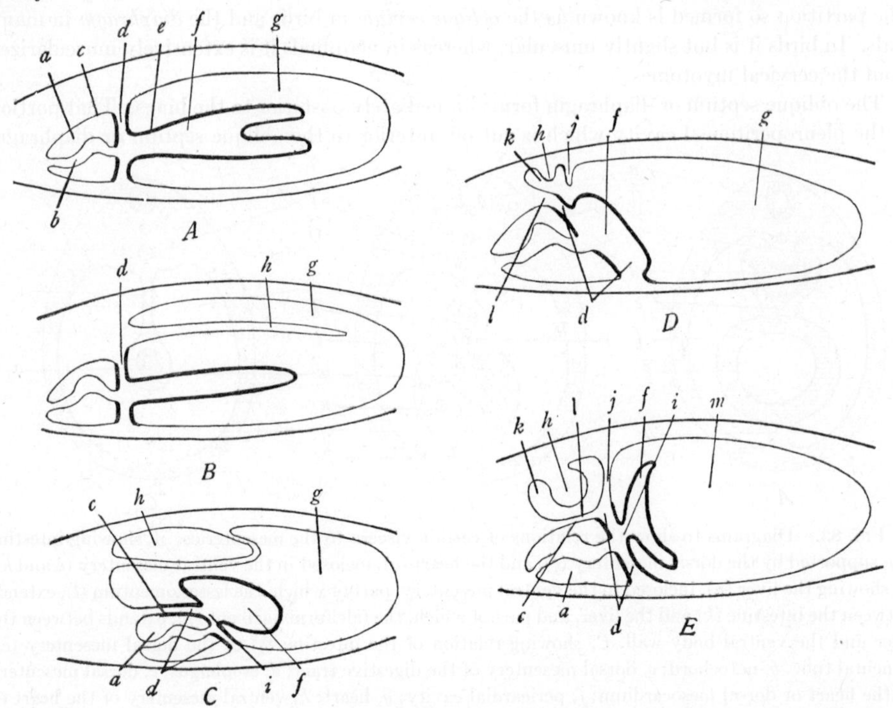

FIG. 84.—Diagrams to illustrate the divisions of the coelom in the various vertebrate classes. The transverse septum and its derivatives are indicated by thick lines. *A*, fishes, showing the division of the coelom into pericardial cavity (*a*) and pleuroperitoneal cavity (*g*) by means of the transverse septum (*d*). *B*, urodeles, similar to fishes, with the addition of the lung (*h*), which projects into the pleuroperitoneal cavity (*g*). *C*, turtle; the pericardial cavity (*a*) has descended posteriorly until it lies ventral to the anterior part of the pleuroperitoneal cavity (*g*); the anterior face of the transverse septum (*d*) has now become part of the wall of the pericardial sac; the lung (*h*) is retroperitoneal. *D*, early stage of mammals, showing the beginning of the coelomic fold (pleuroperitoneal membrane) (*j*), descending from the dorsal body wall, and the liver (*f*) inclosed within the transverse septum (*d*). *E*, later stage of mammals, showing union of the coelomic fold (*j*) with the transverse septum (*d*), the two together forming the diaphragm, which separates the pleural cavity (*k*) from the peritoneal cavity (*m*); the liver has constricted from the main part of the transverse septum, the constriction becoming the coronary ligament (*i*). *a*, pericardial cavity; *b*, heart; *c*, parietal pericardium or pericardial sac; *d*, transverse septum; *e*, serosa of the liver, this being a part of the transverse septum originally; *f*, liver; *g*, pleuroperitoneal cavity; *h*, lung; *i*, coronary ligament of the liver; *j*, coelomic fold which forms part of the diaphragm; *k*, pleural cavity; *l*, pleuropericardial membrane or anterior continuation of the transverse septum; *m*, peritoneal cavity.

B. THE DIGESTIVE TRACT AND ITS DERIVATIVES

1. The origin of the digestive tract.—The primitive intestine or archenteron, as we learned in the section on development, is produced by the process of invagination or other processes in the gastrula stage of the embryo. It is at first a simple tube of entoderm with one opening to the exterior, the blastopore, which is situated at the future posterior end of the embryo. This entodermal tube persists as the lining of the adult digestive tract and of all of its derivatives; to it there are added other layers (connective tissue and muscular layers) derived from the splanchnic mesoderm of the hypomere, which, it will be recalled, folds around the archenteron. The adult digestive tract thus consists of a thick-walled tube, composed of both entodermal and mesodermal elements, of which the latter predominates. The anterior and posterior ends of the digestive tube are lined by ectoderm as a result of invagination processes. The anterior invagination is called the *stomodaeum*, the posterior one, the *proctodaeum;* in the adult no trace of the boundary of these invaginations remains.

2. **Homology of the mouth.**—Various highly speculative theories have been advanced in the past concerning the homology of the vertebrate mouth. In *Amphioxus* the mouth originates on the left side, in series with the gill slits, and only later moves to a median position. This mode of origin has been taken to lend support to the idea that the vertebrate mouth arose by the fusion of the first pair of gill slits. However, it seems more probable that various asymmetries in the development of *Amphioxus* are peculiar to that animal and not of phylogenetic significance. During the years of acceptance of an annelid-arthropod ancestry for vertebrates, with its necessary corollary that a vertebrate is an annelid turned on its back, it was supposed that the original annelid mouth and stomodaeum are represented by the pituitary and its opening and that the present vertebrate mouth is a new formation. However, the annelid-arthropod theory of vertebrate ancestry is now discredited in favor of an origin from the echinoderm-hemichordate line. The vertebrate mouth is therefore the same as that of its immediate invertebrate ancestors and is not a new formation.

3. **Parts and outgrowths of the digestive tube.**—Along its course the digestive tube soon differentiates into various regions with different functions; a number of outgrowths also arise from the tube at various levels (Fig. 85). The regions of the digestive tube, beginning at the anterior end are: *oral* or *buccal cavity, pharynx, esophagus, stomach, small intestine, large intestine* or *colon, cloaca.* The principal outgrowths are: *oral glands, Rathke's pouch, thyroid gland, gill pouches* and *gills, tympanic cavity, thymus* and other glands derived from the gill pouches, *trachea* and *lungs, swim bladder, liver, pancreas, yolk sac,* and *urinary bladder.*

a) *Oral or buccal cavity:* This is bounded by the jaws in front and on the sides, the palate above, and the buccal floor bearing the tongue below. Lips and cheeks differentiate in higher forms. It is customary to distinguish the space between lips and teeth as the *vestibule.* In *Amphioxus* and *Ammocoetes* the oral cavity consists of the space inclosed by the oral hood and is definitely bounded from the pharynx by the velum. The position of the velum corresponds to the boundary between stomodaeum and pharynx of gnathostomes; but this boundary has no landmarks in the adult, where the oral cavity embraces not only the stomodaeum but also part of the entoderm-lined region, and is also extended forward beyond the original stomodaeal invagination by outgrowths involved in the formation of the face.

In Dipnoi, Crossopterygii, and tetrapods the nasal cavities open into the roof of the oral cavity by the *internal nares* or *choanae.* These are primitively anteriorly placed but tend to move backward, and in Crocodilia, birds, and mammals assume a far posterior position through the formation of the secondary palate, described on page 178. A part of what was originally the oral cavity thus becomes inclosed between the secondary palate and the original palate or buccal roof, and this inclosure constitutes the *nasopharyngeal passages* by which the nasal cavities are prolonged posteriorly to open by the choanae now situated near the beginning of the pharynx. The secondary palate is imperfectly formed in birds. It reaches its best development in mammals, where it is continued backward by a fleshy fold, the *soft palate,* so that the nasal passages are still further prolonged posteriorly. The purpose of the backward-shoving of the choanae is to separate the food and respiratory passages.

The *primary tongue*, found in fishes, is simply a part of the buccal floor demarcated anteriorly by a fold. It sheaths the anterior end of the hyobranchial apparatus and contains no muscles or glands. In amphibians a crescentic *gland field* develops in front of the primary tongue and fuses with it to form the *definitive* tongue. In amniotes the same process occurs (the gland field is now termed *tuberculum impar*), and in addition a pair of *lateral lingual swellings* join the tuberculum impar, so that the definitive amniote tongue is compounded of four swellings (Fig. 86). It becomes invaded by voluntary musculature, derived from the hypobranchial musculature, innervated by the twelfth (hypoglossal) cranial nerve, and is richly furnished with glands and taste organs; it also develops intrinsic muscle fibers. The tongue exhibits numerous adaptive variations in form and use. Very long, mobile tongues are seen in some lizards, in snakes, woodpeckers, and ant-eating mammals. The mechanism of protrusion of the woodpecker tongue is chiefly skeletal, for the supporting hyoid apparatus is very long and slender and is coiled, at rest, at either side of the skull. It is common for the tongue to be armed with a horny covering or with spines and thorns.

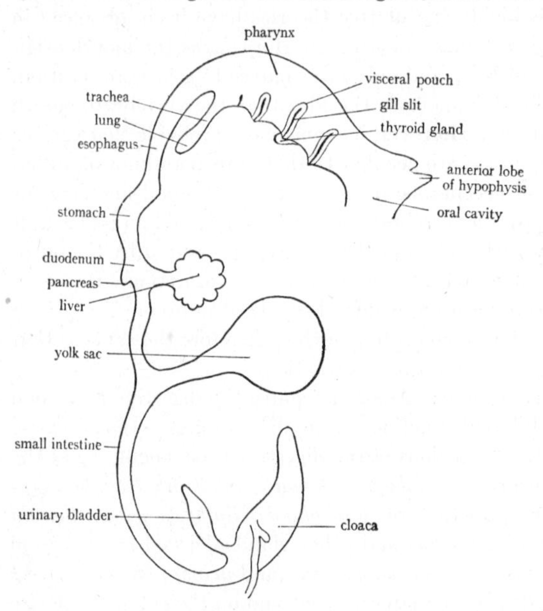

FIG. 85.—Diagram to illustrate the chief derivatives of the digestive tract. (From McMurrich's *Development of the Human Body*, after His, copyright by P. Blakiston's Son & Co.)

The teeth are structures of the oral cavity but were considered previously. Here it may be recalled that in lower vertebrates teeth may occur on various bones of the palate and jaws but that in higher ones they become limited to a single row along the jaw margin.

b) *Oral glands:* Numerous multicellular glands imbedded in the walls of the oral cavity and opening into it occur in tetrapods, but it is not possible to homologize them between different groups. They are of two general kinds—mucous-secreting glands and serous glands secreting enzymes, poisons, etc.—or are of mixed nature. They are usually named according to their location. Those of the lips (labial glands) and teeth (dental glands) appear to be modified skin glands, while those in the more internal parts of the oral cavity appear to be new formations. The oral glands are few in amphibians but increase in reptiles, where *lingual* (tongue), *sublingual, palatine,* and *labial* (lip) groups of glands occur. The poison glands of snakes are enlarged, modified upper labial glands; the association of glands with teeth is, however, nothing new, occurring in amphibians. (For the fangs of poisonous snakes, see p. 181.) Most birds are well equipped with oral glands, grouped somewhat as in reptiles, although fish-eating birds may lack them. In mammals, in correlation with the habit of chewing food, there are present enlarged oral glands, the *salivary glands,* opening into the oral cavity by one or more long ducts. The principal salivary glands are the *parotid, sublingual, submaxillary,* and *infraorbital.* The parotid gland is an enlarged cheek gland; the sublingual occurs either as one large gland or as a group of smaller glands each with its own duct and opening. The salivary secretion is employed chiefly for moistening the food and contains digestive enzymes in only a few mammals. Besides the large salivary glands, mammals have groups of small glands similar to those of reptiles, as the labial, cheek or buccal, lingual, palatine glands, etc.

c) Rathke's pouch: From the roof of the buccal cavity an epithelial evagination occurs in the embryo, producing a blind pouch, known as *Rathke's pouch*, which extends toward, comes in contact with, and finally fuses with the brain wall. The compound structure so formed is called the *hypophysis* or *pituitary body*, an important gland of internal secretion. The part de-

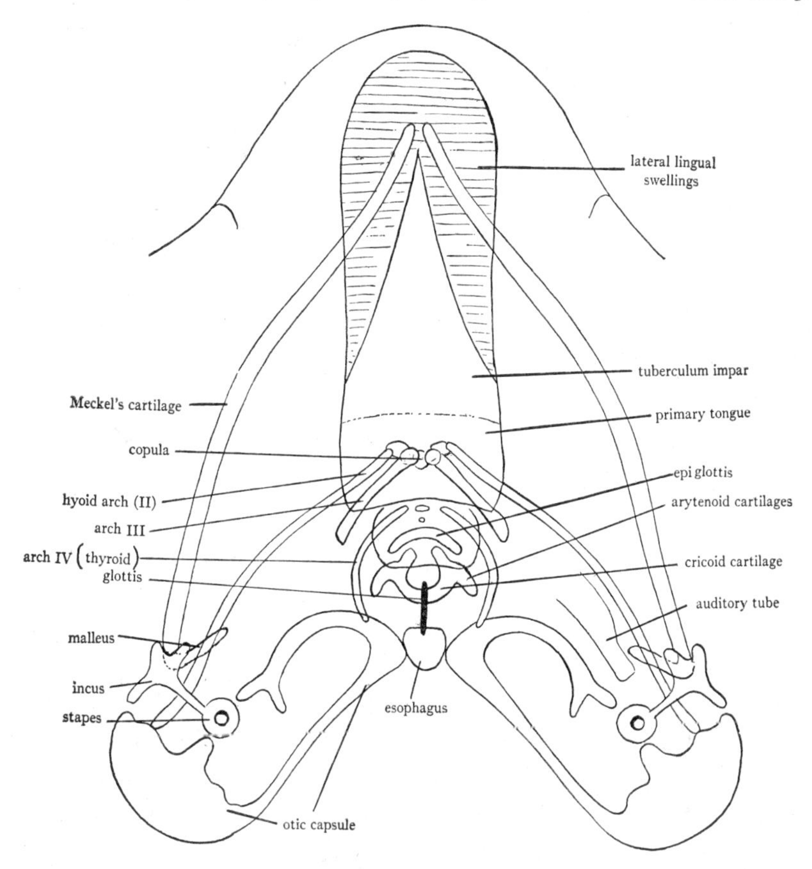

FIG. 86.—Tongue and gill arches of a pig embryo, showing areas of the tongue and relation of the gill arches to tongue, larynx, and ears. (Combined after Kallius, 1910.)

rived from the oral epithelium is termed the *adenohypophysis*.[1] Hatschek's pit and groove in *Amphioxus* are often considered forerunners of the adenohypophysis.

d) Pharynx: The part of the digestive tube immediately caudad to the oral cavity, characterized by the gill slits through its wall in adult or embryonic chordates, is termed the pharynx. In *Amphioxus* and *Ammocoetes* the pharynx is very exactly delimited anteriorly by the velum; in gnathostomes its anterior boundary is indefinite but may be set just anterior to the first gill slit or its representative (opening of the auditory tube). The gill slits are passages from the pharynx to the exterior; the pharyngeal openings are the *internal gill slits*, the

[1] Usually called anterior lobe of the hypophysis; but this term is used in different senses by different authors and now tends to refer to only a part of the adenohypophysis.

exterior openings, the *external gill slits;* and the passage from one to the other in the pharyngeal wall is a *gill pouch* or *visceral pouch.* The tissue between successive gill slits is termed the *visceral arch* or, better (to avoid embryological terms), the *branchial bar* and incloses the skeletal gill arch, a blood vessel termed an aortic arch, and a cranial nerve. From the mesodermal tissue of the branchial bar the branchial musculature, discussed in the preceding chapter, originates.

In the development of this apparatus the pharyngeal wall puts out paired evaginations, the *visceral pouches,* typically six in number (Fig. 87). Opposite each pouch the ectoderm invaginates as a *visceral furrow.* The pouches meet the furrows, and the fusion area breaks through as a *gill slit* or *gill cleft,* which persists as the external gill slit. The number of gill slits was probably large in ancestral chordates, but in present vertebrates the greatest number is

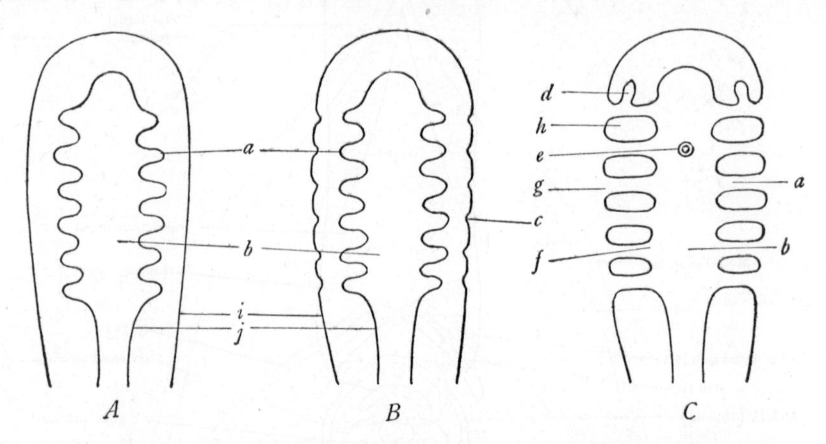

FIG. 87.—Diagrams to illustrate the formation of the visceral pouches, furrows, arches, and clefts. *A*, early stage, showing the evaginations (*a*) from the wall of the pharynx (*b*), which form the visceral pouches. *B*, later stage, illustrating formation of the visceral furrows (*c*) by invagination of the surface ectoderm. *C*, later stage, showing formation of the external gill slits (*g*) by the union of the visceral pouches and furrows and breaking-through of the points of union. *a*, visceral pouch; *b*, pharyngeal cavity; *c*, visceral furrow; *d*, tympanic cavity; *e*, thyroid gland; *f*, internal gill slit; *g*, external gill slit; *h*, visceral arch; *i*, ectoderm; *j*, entoderm.

seen in hagfishes (up to fourteen). Six is believed to be the typical gnathostome number; but the first one, between the mandibular and hyoid arches, is generally modified or reduced (forming the spiracle). Above fishes the number of gill slits appearing in the embryo gradually reduces, and often the openings fail to break through. There are, of course, no gill slits in adult vertebrates above urodeles, for they close over during development; among amniotes gills fail to appear even in the embryo.

This apparatus appears to have functioned originally as a food-catching device; as described for tunicates and *Amphioxus*, minute organisms in the water current are trapped by mucous secretions and conveyed into the esophagus. The pharynx lost this function when vertebrates increased in size and took to eating larger objects, and it then became respiratory with the aid of gills.

Gills or *branchiae* are thin-walled projections, richly supplied with blood vessels and presenting a large surface for the exchange of respiratory gases. They may be *lamellar* (platelike) or *filamentous*. In vertebrates there are two sorts of gills: the *external* gills, borne on the outer surface of the branchial bars and found only in the larvae of Crossopterygii and Dipnoi and in the larvae and some adults of amphibians; and the *internal* gills, borne on the side walls of the branchial bars, i.e., facing the gill pouches. (In some fishes the internal gills protrude through the gill slits during larval life, but these are obviously not true external gills.) In regard to internal gills, the gill on one side of a branchial bar is called a *demibranch* or *half-gill,* and the two gills on the two sides of the bar form a *whole gill* or *holobranch.* It seems probable that the external type of gill is the more ancient. Gills are absent from all stages of amniotes.

There has been much argument as to whether the lining of pharynx and gill pouches and the epithelium of the internal gills is entodermal or not. (The epithelium of external gills is definitely known to be ectodermal.) The fact that teeth often occur on the inner surfaces of the branchial bars in fishes is puzzling, since, if the lining is entodermal, one would be forced to conclude that either ectoderm or entoderm can secrete enamel. Several workers have maintained that the epithelium of the parts in question is ectodermal, for they find the oral ectoderm creeping over the original entodermal lining of the pharynx. Others deny this, and the question must be regarded as still open. If the epithelium of the internal gills is ectodermal, then it would follow that external and internal gills are variants of the same organ.

e) Thyroid gland: This is an epithelial evagination from the buccal or pharyngeal floor at the level of the hyoid arch, between the primary tongue and the tuberculum impar. It soon separates from the floor, so that it becomes ductless and proliferates into a considerable mass of follicles, usually becoming bilateral. The thyroid is a vital endocrine gland whose secretion is necessary for normal growth and sexual development. It is usually considered to be of entodermal origin, but the same argument exists here as with other pharyngeal structures mentioned above. For the phylogenetic history of the thyroid gland, see page 34.

f) Pharyngeal glands: From the dorsal and ventral ends of the gill pouches there proliferate during development epithelial masses, the *branchial buds,* variable in number in different vertebrates. The most complete set is found in lampreys, where all the pouches give off dorsal and ventral buds. The *true* or *palatine tonsils* are lymphoid masses proliferated from the second pouches. Two to four of the dorsal buds unite to form the *thymus,* a lymph gland of uncertain function. The ventral buds give rise to the *parathyroids,* very small glands concerned in calcium metabolism, and the *epithelial bodies,* glandular masses of unknown function. In mammals, contrary to other vertebrates, the thymus comes from ventral buds, the parathyroids from dorsal buds. Close behind the last gill slits there arises on one or both sides an outgrowth which separates from the pharynx and produces a number of follicles, termed the *postbranchial* or *ultimobranchial body,* of unknown significance.

g) Tympanic cavity and external auditory meatus: Beginning with amphibians, the first gill pouch puts out toward the internal ear an outgrowth whose end expands into the *tympanic cavity* or cavity of the middle ear, while the stalk of the outgrowth remains as the *auditory* or *Eustachian tube,* connecting the middle ear to the pharynx. In some reptiles and in birds and mammals an invagination corresponding to the position of the first external gill slit occurs, and the bottom of this meets the wall of the tympanic cavity, forming a double-walled membrane, the tympanic membrane or eardrum. The passage formed by the invagination is the external auditory meatus, which thus marks the position on the side of the head of the first gill slit.

h) Swim bladder, lungs, larynx, and trachea: When the vertebrates adopted the land habitat, the gill slits and gills disappeared in the adult, and their physiological role was taken over by a pharyngeal outgrowth which already existed in crossopterygian, dipnoan, and bony fishes, namely, the *swim bladder.* In *Polypterus* this arises as a ventral evagination from the pharynx behind the gill region. The evagination soon divides into two sacs, which grow backward and retain throughout life their ventral position and ventral connection with the pharynx, although the left one is smaller than the right. A similar origin is seen in Dipnoi. where the single or bilobed sac is, however, displaced to the dorsal side, although retaining its ventral opening into the pharynx. In *Polypterus* and Dipnoi these sacs, generally termed lungs, are employed in air respiration, so that these fishes can live out of water for considerable periods. Among the teleosts there is usually but a single sac, and this is displaced dorsally through a rotation process. In most teleosts the connecting duct or *pneumatic duct* remains, although also shifted to the dorsal side of the pharynx, but in others the duct degenerates, The sac, termed swim bladder, serves primarily in teleosts to alter the specific gravity of the fish as it changes level in the water and so keep it in equilibrium with external pressures. This is accomplished by alterations in the gas content of the bladder. The gas does not have the same composition as the air, containing more oxygen and carbon dioxide and less nitrogen. Oxygen can be secreted from the blood into the bladder by a gland in the latter, which contains a complicated network of capillaries. The gases in the swim bladder may also be utilized by the fish under some circumstances.

The lungs of tetrapods arise embryologically in the same way as the "lungs" of *Polypterus*, to which they are undoubtedly homologous. The pair of sacs so formed grows backward into the pleuroperitoneal cavity, and mesodermal tissue is added to the original entodermal lining. The walls of the lungs become subdivided internally into air spaces, which become smaller and smaller and more and more numerous in the tetrapod series until finally in mammals the lungs are a spongy mass of minute air pockets or *alveoli*. It is quite common among tetrapods for the left lung to be smaller than the right, as in *Polypterus*. Mammalian lungs are commonly subdivided into a few large lobes. Snakes usually have but one lung, the right one; and in some salamanders the lungs have been altogether lost. The lungs of the Crocodilia approach most nearly among reptiles to those of mammals. There is a theory to the effect that the lungs are homologous to a pair of gill slits, but supporting evidence is lacking.

The duct connecting the lungs with the pharynx is termed the *trachea* or *windpipe*, and the opening into the pharynx is the *glottis*. The trachea is very short in amphibians but of considerable length in amniotes, because of the general backward retreat of the viscera. Its wall is stiffened by complete or incomplete cartilaginous rings. On reaching the lungs the amniote trachea divides into the two *bronchi*, and these in mammals subdivide repeatedly inside the lungs forming the *bronchial tree;* the final twigs, termed *bronchioles*, terminate in little chambers encircled by alveoli.

The respiratory system of birds presents some remarkable peculiarities. There are a number of large, thin-walled *air sacs* scattered throughout the body, among the viscera and muscles and in the interior of the bones. These are connected to the lungs by bronchial tubes of rather large caliber. Inside the lung there are anastomoses between the larger bronchial tubes, and the terminal bronchioles also anastome with each other, forming a complicated network. There are, therefore, no closed alveoli at the ends of the bronchioles, but the alveoli occur along the sides of the bronchioles. The functioning of this system is not thoroughly understood, but it appears that the air passes *through* the lungs in and out of the air sacs, which therefore act like bellows, so that the lungs are completely ventilated at each breath, whereas in other vertebrates considerable used air remains behind in the lungs. The air sacs also lessen the specific gravity of the bird and may have other advantages for flight.

The beginning of the trachea at the glottis is differentiated in tetrapods into a chamber termed the *larynx*, whose walls are supported by cartilages representing reduced gill arches. In the simplest case, as in some urodeles, there is a pair of *lateral* cartilages derived from the last gill arches (Fig. 88*A*). In other amphibians these subdivide into a pair of dorsally placed cartilages, the *arytenoids*, and a cartilaginous ring, the *cricoid*. When the trachea elongates, the same *aryteno-cricoid primordium* also furnishes the cartilaginous rings of the trachea. This condition persists through reptiles (Fig. 88*B*) and birds, but in mammals there is added anterior to the cricoid a large, ventrally placed cartilage, the *thyroid* cartilage, which probably comes from the fourth and fifth gill arches plus their copula. Its compound origin is clearly seen in monotremes (Fig. 88*C*). In birds and mammals the middorsal part of the cricoid frequently separates as a *procricoid*. The larynx is very closely associated with the hyoid apparatus, being, in fact, situated immediately behind the basihyal. The muscles of the larynx are necessarily branchial muscles, since its cartilages are gill arches. The function of the larynx is the production of the voice, but in birds the voice-producing mechanism, called the *syrinx*, is at the forking of the bronchi.

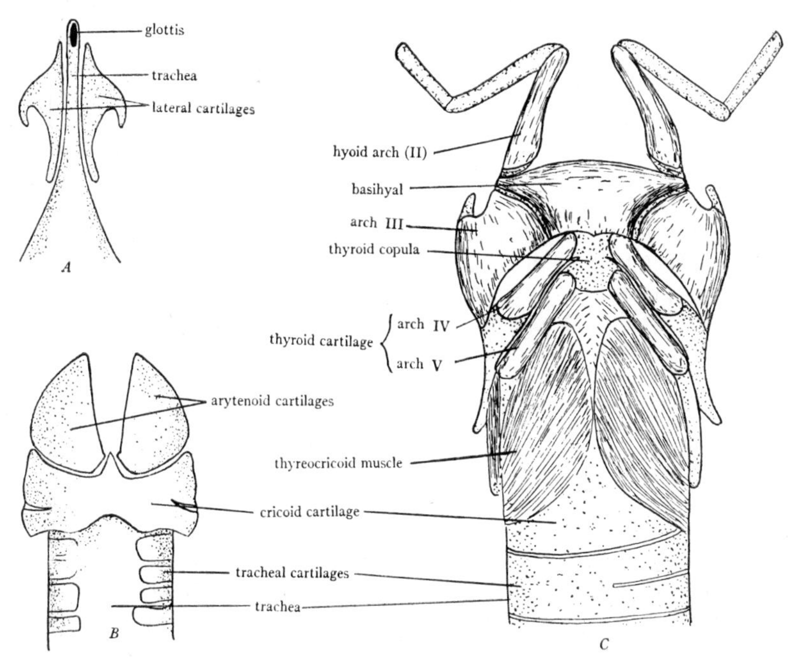

FIG. 88.—The cartilages of the larynx compared in *A*, *Necturus;* *B*, *Sphenodon;* and *C*, *Echidna* (after Göppert, 1901). In *Echidna* the thyroid cartilage is seen to originate from the fourth and fifth gill arches and their copula. Cartilage stippled in *C*.

The *epiglottis*, a cartilage-supported projection protecting the glottis, first appears in reptiles and is found throughout mammals. It is improbable that its cartilage is a modified gill arch.

i) *Esophagus*. Posterior to the glottis the digestive tract narrows to the esophagus, generally short in fishes and amphibians, elongated in amniotes because of the posterior descent of the viscera. It is lined by a stratified epithelium similar to that of the buccal cavity and is provided, for all or part of its length, by striated musculature. Any esophageal enlargement is termed a *crop* or *ingluvies*. A crop occurs in many birds, especially grain-eating ones, where it may form a large sac-shaped appendage. It serves for storage or preliminary digestion; in the pigeon tribe, the food in the crop becomes mixed with "milk," a secretion produced by fatty degeneration of the crop epithelium, and the mixture is fed by both parents to the young by regurgitation. In cud-chewing mammals the lower end of the esophagus contributes crop-like sacs to the complicated "stomach" (see later).

j) Stomach or ventriculus: The stomach follows the esophagus as a spindle-shaped to sacciform enlargement provided with enzyme- and acid-secreting glands and a thick muscular wall of smooth muscles. However, in lower vertebrates there is often little or no external demarcation between esophagus and stomach, and the proximal limit of the stomach can then be determined only by the presence of the gastric glands. The region of the stomach next the esophagus is called the *cardiac region;* the opposite end, leading into the intestine, is the *pyloric region* and terminates by a strong sphincter muscle, the *pylorus,* embracing the entrance into the intestine. A definite stomach is lacking in *Amphioxus* and cyclostomes. In lower fishes the stomach is a straight or J- or V-shaped tube, whereas teleosts frequently lack a stomach. In amphibians and reptiles there is a tubular, not much enlarged stomach. The stomach of birds is divided into the *proventriculus,* which bears the gastric glands, and the *gizzard,* with a thick muscular wall and hard cornified lining. The gizzard is best developed in grain- and seed-eating birds and contains gravel to assist in grinding up the hard food. The stomach of mammals presents many variations considered later.

k) Small intestine: This follows the pylorus and usually consists of a more or less coiled tube of considerable length. Its beginning beyond the pylorus, termed the *duodenum,* is characterized by the fact that it receives the ducts of the liver and pancreas. In cyclostomes there is a low fold in the intestinal wall, and this is apparently the forerunner of the *spiral valve,* a wide, spirally coiled fold filling the greater part of the small intestine in elasmobranchs, dipnoans, and many ganoids. The part of the intestine having the spiral valve is often called *valvular* intestine. The spiral valve is presumably a device for increasing the absorptive surface of the intestine. Teleosts lack a spiral valve and instead have one to many (up to nine hundred) diverticula, the *pyloric caeca,* projecting from the proximal part of the small intestine. These have the same histological structure as the intestine and presumably are also a device for increasing the absorptive surface. The small intestine tends, in general, to be longer in herbivorous than in carnivorous animals.

l) Liver: This extremely important gland, having no counterpart in invertebrates, is seen in simple form in *Amphioxus* as a hollow diverticulum from the intestine. In vertebrates in general it forms a large, lobed organ suspended from the transverse septum or its representative. It arises in vertebrates as a single or double hollow or solid proliferation from the intestinal epithelium at the level of the transverse septum, into which it grows. The proliferations soon branch extensively; and the branches usually anastomose, so that the adult liver is a complicated network of epithelial columns of entodermal origin with blood spaces filling the meshes of the net. The stalk of the liver outgrowth becomes the *bile duct;* and a sac, the *gall bladder,* usually develops from the liver substance along the course of the bile duct. The *Ammocoetes* larva has both gall bladder and bile duct, but these degenerate during metamorphosis. Fishes, amphibians, and reptiles generally have a gall bladder, but this organ is lacking in many birds, including pigeons, and in a number of mammals (Cetacea, Perissodactyla, many Artiodactyla, and others).

The liver has many vital functions, of which the most obvious is the secretion of the bile. This is secreted by the liver cells and poured into minute ducts, which unite to form the larger *hepatic* ducts. These join the *cystic* duct, or duct from the gall bladder; and the *common bile duct* or *choledochal duct,* so formed, empties into the small intestine not far beyond the pylorus, either with or without joining the pancreatic duct. The bile assists in the digestion of fat. Some other functions of the liver are: storage as glycogen of excess sugar in the digested food; production of urea and other nitrogenous wastes; control of food and other substances in the blood.

m) Pancreas: This gland proliferates from the intestinal epithelium, shortly beyond the liver proliferations, usually as one dorsal and two ventral outgrowths, which unite to form a single gland. This grows out into adjacent mesenteries and may form a compact or more or less diffuse organ. All three stalks of the outgrowths may persist as pancreatic ducts, but usually the number is reduced to one or two. These open into the duodenum separately from, or after joining, the common bile duct. *Amphioxus* has no pancreas, but one is present throughout vertebrates, although in many teleosts it is so diffuse that it cannot be recognized macroscopically. The pancreas consists of two kinds of cells: the regular gland cells, which secrete into the duodenum the enzyme-containing pancreatic juice, the most important digestive fluid of vertebrates; and small scattered nests of cells, termed islets of Langerhans, which produce a hormone (insulin) necessary for the metabolism of carbohydrates. In some fishes the islets are gathered together into a small body near the main pancreas.

n) Yolk sac: This is a baglike extension of the ventral wall of the small intestine found in the embryos of those vertebrates which develop by meroblastic cleavage. It is filled with yolk, gradually utilized by the embryo as food, for which purpose the yolk sac early becomes provided with numerous blood vessels. When the yolk has been used up, the diminished yolk sac is withdrawn into the body and becomes part of the intestinal wall, and the body wall closes over the aperture through which the yolk sac protruded. Although mammals do not have meroblastic cleavage, they possess during embryonic stages a vestigial yolk sac as a reminiscence of their reptilian ancestry.

o) Caecum: At the junction of small and large intestine there is generally a diverticulum, the *colic caecum.* It is first seen in a few amphibians and is usually present throughout amniotes. In reptiles it is of small size; birds generally have two colic caeca; mammals usually have a conspicuous caecum, which may in herbivorous forms be very long and large. A few mammals have two caeca. In primates and some other mammals the end of the caecum has degenerated into a slender appendage, termed the *vermiform appendix.* The function of the colic caecum is uncertain except in those animals with a very large caecum, where it assists in digestion and absorption.

p) Large intestine or colon: The large intestine is usually well delimited from the small intestine by its increased diameter and also by the presence of one or two caeca as just described. However, in fishes the large intestine is usually quite short and not distinctly marked off from the small intestine. In elasmobranchs it bears a slender projection, the *digitiform* (or *rectal*) gland, which apparently secretes mucus. This does not seem to be a colic caecum, since it is not located at the beginning of the large intestine. In amphibians and birds the large intestine is also short; it reaches a considerable length in mammals. The terminal part of the large intestine is called cloaca when it receives the urinary and genital ducts, as is the case in most vertebrate groups. There is no cloaca in Holocephali, ganoids, teleosts, and mammals except monotremes. In the absence of a cloaca the end section of the large intestine is called *rectum.*

q) Urinary bladder: In most tetrapods, a sac, the urinary bladder, serving as a reservoir for the urine, is present as a ventral outgrowth of the cloaca or, in the absence of the latter, of the urogenital canal. In amniote embryos the enlarged urinary bladder, termed the *allantois,* projects far beyond the embryo and serves as a respiratory organ. The adult bladder is but a portion of the allantois. Birds have no urinary bladder in the adult state, although their embryos are provided with a large allantois.

r) Anus: The digestive tract opens to the exterior by the anus, which, characteristically in vertebrates, is not at the posterior end of the body but is shifted to the ventral surface at the base of the tail region. This happens through overgrowth of the tail.

s) Postanal gut: In most vertebrate embryos there is a solid or hollow extension of the gut into the tail, forming the postanal gut. This later vanishes. Its occurrence would seem to indicate an originally more posterior position of the anus.

C. THE COELOM AND THE DIGESTIVE AND RESPIRATORY SYSTEMS OF ELASMOBRANCHS

Elasmobranchs illustrate an early generalized condition of the coelom and the respiratory system. The heart is far forward beneath the gill apparatus, the gill apparatus is typical, and the transverse septum has its original transverse position, cutting off the pericardial from the pleuroperitoneal cavity.

The directions apply to the spiny dogfishes and the skate.

1. The body wall and the pleuroperitoneal cavity.—Make an incision from the left side of the cloaca forward through the pelvic girdle slightly to the left of the midventral line up to the pectoral girdle. The incision will probably cut through the skin first and should then be extended through the muscle layer. To assist in exposing the interior a transverse incision may be made in the middle of the lateral body wall on each side. In the skate cut along the left side of the cloaca and then along both lateral borders of the body cavity but not anteriorly, leaving the flap of body wall adhering to the pectoral girdle. The large internal cavity is the *pleuroperitoneal cavity* and constitutes the greater part of the coelom. It is lined by a smooth shining membrane, the *parietal peritoneum*, which adheres closely to the inside of the body wall. The body wall is seen to be composed of three layers: skin, muscles, and parietal peritoneum.

2. The viscera of the pleuroperitoneal cavity.—Within[2] the cavity are a number of *organs* or *viscera*, most of which belong to the digestive tract. At the anterior end of the cavity is the large, brownish or grayish *liver*. This consists in the spiny dogfish of long *left* and *right lobes* and a small *median lobe* in which is located the long, greenish *gall bladder*. In the skate the liver is composed of *right, median*, and *left lobes* of equal length and size, and the *gall bladder* is situated in the angles between the right and median lobes. Dorsal to the liver on the left side is the large J-shaped esophagus-stomach, often distended with food. (In some specimens this is everted into the oral cavity and should be pulled back into the pleuroperitoneal cavity by exerting a gentle traction upon it.) There is no external demarcation between esophagus and stomach, but the anterior part of the organ is esophagus. The stomach continues straight backward from the esophagus to a point somewhat posterior to the caudal ends of the liver lobes; it then makes a sharp bend, decreasing considerably in diameter, and extends anteriorly, terminating in a constriction, the pylorus. Along the posterior margin of the bend of the stomach (or in the skate on the dorsal side of the bend) is a dark-colored organ, the *spleen*, a part of the lymphatic system. From the pylorus the short *intestine* extends to the anus. The first part of the intestine beyond the pylorus is called the *duodenum*. It extends for a short distance to the right

[2] It has already been explained that the organs are not, in reality, inside the coelom.

and then curves posteriorly. The *bile duct*, a long stout duct, is easily seen descending from the gall bladder to enter the duodenum shortly caudad of the bend. The bile duct, accompanied by some blood vessels, runs in a strip of mesentery. It passes to the dorsal side of the duodenal wall and runs for a short distance caudad, imbedded in the wall, before it penetrates into the cavity of the duodenum. In the curve of the duodenum reposes the *ventral* lobe of a white gland, the *pancreas*. The *dorsal* lobe of the pancreas, which is long and slender in the spiny dogfish, reaching to the spleen, should be located by raising the stomach and duodenum and looking dorsal to them. The *duct* of the pancreas is somewhat difficult to find in the dogfishes, less difficult in the skate; it lies imbedded in the tissue of the pancreas near the posterior margin of the ventral lobe and may be exposed by picking away the pancreas tissue in this region. Beyond the duodenum the intestine widens considerably, and its surface is marked by parallel rings. These rings are the lines of attachment of a spiral fold, the *spiral valve*, which occupies the interior of the intestine. (A portion of the intestine often protrudes through the anus and should be pulled back into the coelom by grasping the portion in the cavity and exerting a gentle pull.) The part of the small intestine occupied by the spiral valve is called the valvular intestine. Caudal to this, the narrower large intestine or colon proceeds to the anus. Attached to the colon by a duct is a small cylindrical body, the rectal or digitiform gland, which seems to secrete mucus. The terminal chamber of the colon is the cloaca, opening by the anus.

Cut open the esophagus-stomach and wash out its contents. Partly disintegrated squids and fish are commonly found in the organ. The anterior part, having projecting *papillae*, is the esophagus; the remainder, with lengthwise folds or *rugae*, is the stomach. Cut open the small intestine along one side midway between the large blood vessels which run lengthwise along its wall and observe the spiral valve in the interior. It consists of a fold of the intestinal wall spirally coiled so as to make a series of overlapping cones; it serves to increase the digestive and absorptive surface of the intestine.

The reproductive organs and their ducts may, in part, also be identified at this time. The gonads are a pair of soft bodies against the dorsal wall of the anterior part of the pleuroperitoneal cavity; the liver lobes must be raised to see them. In mature females the oviducts are noticeable as stout white tubes, one to each side against the dorsal coelomic wall.

3. The mesenteries.—The viscera are held in place by delicate membranes, the *mesenteries*, whose mode of origin was explained in the introduction to this section. In studying them, lift and spread each organ as it is mentioned. The dorsal mesentery extends from the median dorsal line of the coelom to the digestive tract but is not complete in the animals under consideration, as a gap is present in the region of the small intestine. That part of the dorsal mesentery supporting the stomach is called the *mesogaster;* in the skate it is limited to the anterior part of the stomach. The mesogaster incloses the spleen between its two walls, and that portion of the mesogaster from the spleen to the stomach is the *gastrosplenic* ligament. That portion of the dorsal mesentery which supports the small intestine is called the *mesentery,* in the limited sense. This is absent in the skate. In the spiny dogfish there is a fusion between the mesentery and the mesogaster, so that a sort of pocket is formed dorsal to the bend of the stomach. In the dorsal wall of this pocket is located the greater part of the pancreas, which is thus in the dorsal mesentery. The dorsal mesentery begins again in the region of the rectal gland, this portion of the mesentery being named the *mesorectum.*

The ventral mesentery is represented in these animals, as in all vertebrates, by remnants only. Such a remnant is the *gastro-hepato-duodenal ligament* extending from the right side of the stomach to the liver and duodenum, also called the *lesser omentum.* It may be roughly divided into two portions: the *hepatoduodenal* ligament, extending from the liver to the duodenum and containing the bile duct and blood vessels; and the *gastrohepatic* ligament, extending from the stomach to the liver and duodenum and in the dogfishes occupying also the angle formed by the bend of the stomach. Another remnant of the ventral mesentery is the *suspensory* or *falciform* ligament of the liver. This will be found at the anterior end of the liver, extending from the midventral surface of the liver to the midventral line of the body wall. In mature females the *mouth of the oviduct* will be noticed in the falciform ligament as a funnel-shaped aperture. After seeing the falciform ligament, the flap of body wall left in the skate may be cut off.

Each gonad has a mesentery which is a special fold arising from the dorsal wall very near the origin of the dorsal mesentery. The mesentery of the ovary is called the *mesovarium;* of the testis, the *mesorchium.* In the case of mature females, each oviduct has also a mesentery, the *mesotubarium.*

The anterior end of the pleuroperitoneal cavity will be found closed by a partition, the *transverse septum,* the posterior face of which is clothed by the parietal peritoneum. The liver is attached to the septum by the strong *coronary* ligament, which is, in fact, a portion of the septum. In its early development the liver is inclosed in the transverse septum, and subsequently, because of increased size, projects posteriorly from the septum which then narrows around the anterior end of the liver and forms the coronary ligament (see Fig. 89).

The pleuroperitoneal cavity communicates with the exterior by means of the *abdominal pores.* These will be found one on each side of the anal opening (in the skate posterior to the anus), somewhat concealed by a fold of skin.

Probe into them and note that they lead into the pleuroperitoneal cavity. Their purpose is obscure.

Draw the contents of the pleuroperitoneal cavity. Make a diagram of an imaginary section through the anterior end of the pleuroperitoneal cavity, showing gonads, stomach, and liver and the relations of the pleuroperitoneum to them and to the body wall.

4. The pericardial cavity.—Gently strip off with a forceps the layers of hypobranchial muscle on the ventral side of the throat in front of the pectoral girdle until you have exposed a membrane. This membrane is the parietal pericardium. Slit it open and see that it incloses a cavity, the pericar-

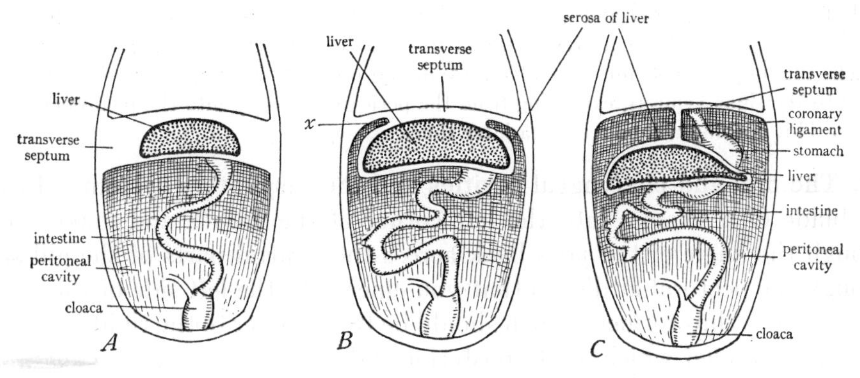

FIG. 89.—Diagrams to show the separation of the liver from the transverse septum and formation of the coronary ligament. *A*, liver inclosed in the transverse septum. *B*, beginning of constriction of the liver from the septum at *x*. *C*, completion of the constriction, leaving the liver suspended from the septum by the coronary ligament. The falciform ligament is also formed at the same time by the same constriction. (*B* and *C* after McMurrich's *Development of the Human Body*, copyright by P. Blakiston's Son & Co.)

dial cavity, in which the heart is situated. To reveal this cavity more fully, cut laterally along the anterior face of the girdle, keeping your instrument in contact with the girdle. Portions of the girdle may be sliced away, but the heart must not be injured. The pericardial cavity is thus revealed as a conical cavity lined by the parietal pericardium and containing the heart. By gently lifting up the heart note that it is attached only at its anterior and posterior ends. At these places the pericardial lining is deflected from the walls of the pericardial cavity and passes over the surface of the heart as a covering layer, the *visceral pericardium*, which is indistinguishably fused with the heart wall. With the heart lifted, note that the posterior end of the heart is a fan-shaped chamber, the *sinus venosus*, whose walls are continuous with the partition that forms the posterior wall of the pericardial cavity. This partition is the *transverse septum*, whose posterior face we have already seen. The septum is thus seen to be a partition whose anterior wall is composed of the parietal pericardium and whose posterior wall, of the parietal peritoneum. The wings of the sinus venosus are buried in the transverse septum; they constitute large venous channels through which the venous blood is returned to the heart.

We may now explain the formation of the transverse septum. Since the heart is situated in the ventral part of the body, it is necessary, in order that the blood from the dorsal body wall may reach the heart, that a bridge be formed passing from the dorsal to the ventral side. In early embryonic stages a bridge or cylinder of mesoderm develops on each side of the posterior end of the heart, connecting the splanchnic mesoderm surrounding the sinus venosus with the somatic mesoderm of the dorsal body wall. In these bridges the main venous channels pass from the dorsal body wall into the sinus venosus. Later, the bridges enlarge and finally fuse with each other and with the body wall laterally, forming a partition, the transverse septum, which thus cuts the heart off from the remainder of the coelom. In elasmobranchs the fusion is not quite complete, leaving an opening, the *pericardioperitoneal canal*, dorsal to the sinus venosus, extending from the pericardial cavity into the pleuroperitoneal cavity. This opening will be seen at a later time.

5. The oral and pharyngeal cavities and the respiratory system.—Insert one blade of a scissors into the left corner of the mouth and make a cut through the angle of the jaws back across the ventral parts of the gill slits through the pectoral girdle, so that you emerge to the left side of the stomach. A flap is thus formed, which should be turned over to the right. A large cavity is revealed which at its posterior end converges into the *esophagus*, which may be slightly slit to aid in opening the flap.

The anterior part of the cavity inclosed by the jaws and gill arches is the *buccal or oral cavity*. It is bounded in front by the upper and lower jaws, provided with teeth. The upper and lower jaws are the two halves of the first or mandibular gill arch, the cross-sections of which should be identified in the cut surface. On the floor of the mouth back of the teeth is the tongue, a flat, slight, practically immovable projection; it is a primary tongue (see p. 256). It is supported by the second or hyoid gill arch, which should be felt within it and identified on the cut surfaces. The skate lacks a tongue because of the reduction of the hyoid arch.

The posterior and greater part of the cavity under consideration is the pharynx, the wall of which is pierced by six internal gill slits. The first of these, the spiracle, is a rounded opening in the roof of the pharynx immediately posterior to the mandibular arch. The remaining five gill slits are elongated. The internal gill slits open into large cavities, the gill pouches, which in turn open to the exterior by the external gill slits. The tissue between successive gill pouches is the branchial bar (also called visceral arch). Where the gill region was cut through, examine the section of a branchial bar. Its inner, thickened part contains the cross-section of the cartilaginous gill arch and, external to this, the fine cartilages of the gill rays. The branchial bar continues to the body surface as a thin partition, the *interbranchial septum*, which bears on each face a series of low, thin folds or plates, the branchial or gill lamellae. The set of lamellae on one face of a septum constitutes a half-gill or *demibranch*, and the two demibranchs of both faces of a septum form a *holobranch* or gill. By examining all the septa determine how many demibranchs are present and where they are missing. Does the spiracle have gills?

Close to the cross-section of the gill arch cartilage may be seen the main blood vessels of the gills; to its outer side is the afferent branchial blood vessel, bringing venous blood to the gills, and to each side of the cartilage an efferent branchial vessel (injected with a colored solution) which carries the aerated blood away from the gills. Note the fine branches of these vessels in the gill lamellae. The lamellae are the respiratory mechanism of the animal in which the blood obtains oxygen and gives up carbon dioxide. The oxygen is obtained from the surrounding water, which is kept flowing through the gill pouches by the movements of the gill apparatus.

Draw the oral cavity and pharynx. Draw one branchial bar and all of its parts in cross-section.

D. THE RESPIRATORY AND DIGESTIVE SYSTEMS OF A BONY FISH

Although the bony fishes are out of the line of ascent from fish to mammals, they show interesting adaptive modifications of fish structure, especially as regards the gill apparatus.

1. Contents of the pleuroperitoneal cavity.—The coelom of bony fishes, like that of elasmobranchs, is divided into an anterior pericardial cavity and a posterior pleuroperitoneal cavity; and the two are separated by a transverse septum. Open the pleuroperitoneal cavity by a lengthwise slit from anus to pectoral girdle and enlarge, if necessary, by lateral cuts. Identify: liver, esophagus, stomach, pylorus. Esophagus and stomach are not externally delimited. The stomach varies in form in different teleosts and may simply be U-shaped or often has a saclike posterior pouch. Note finger-like projections, the *pyloric caeca*, peculiar to bony fishes, encircling the duodenum. There is no definite division of the intestine into small and large intestine, but the greater part of it, more or less looped, undoubtedly represents the small intestine. There is no cloaca, so that the terminal part of the intestine is a rectum. Observe apparent absence of a pancreas; it is present as diffuse tissue recognizable only by microscopic inspection.

The entire dorsal wall of the pleuroperitoneal cavity is occupied by a large gas-filled sac, the *swim bladder*, readily detected by pressing on the region indicated. Note that it is outside the parietal peritoneum. Slit open the swim bladder; the interior is divided by partitions in some fishes. The swim bladder is connected by a duct to the dorsal wall of the pharynx or esophagus in some teleosts but lacks such a duct in others.

2. Oral cavity, pharynx, and respiratory system.—As noted in the study of the external anatomy, the bony fishes lack external gill slits; instead the gill apparatus is covered by a backward fold, the gill cover or operculum, supported by thin bones. Below the operculum is the branchiostegal membrane supported by bony rays. Cut away one operculum and note the gill-bearing branchial bars. How many are there? Note that only the part of the branchial bar which contains the bony gill arch remains and that the thinner septum, which in elasmobranchs extends to the surface to bound the external gill slits, has been lost; hence there are no external gill slits. Note on each bar the gill rakers, supported by bone, projecting to the interior, and the gills, borne on the outer surface of the bars. The gills are here filamentous, not lamellar; and there is a double row of filaments, forming a holobranch, on each bar. Each filament is supported by a bony gill ray.

Open the mouth and cut through the angle of the jaws on one side and across the middle of the branchial bars, thus opening up the oral cavity and the pharynx. In the roof of the oral cavity study the distribution of teeth, usually borne on the premaxillae, maxillae, vomer, and dentary. At the tip of both jaws behind the teeth there is present a thin transverse membrane, the *oral valve*, which prevents water from flowing out of the mouth during respiration. Note the inflexible tongue, which is of the primary type. In the pharynx the notable feature is the gill-bearing bars, of which the first (really the second), is the hyoid arch, extending obliquely backward on each side from the tongue to the operculum. Teeth may be present on the dorsal parts of the bars.

E. THE COELOM AND DIGESTIVE AND RESPIRATORY SYSTEMS OF NECTURUS

In *Necturus* these structures are practically at the same stage as in the dogfish, except for the presence of lungs. The pericardial cavity is anteriorly placed, and the transverse septum has a transverse orientation.

Obtain a specimen and place in a wax-bottomed dissecting pan, fastening it ventral side up by pins through the legs.

1. The viscera of the pleuroperitoneal cavity.—Make a longitudinal incision through the body wall a little to the left of the midventral line from the left side of the anus through the pelvic girdle to the pectoral girdle. Spread the incision apart and look within. The large cavity is, as in fishes, the pleuroperitoneal cavity, lined by the pleuroperitoneum. The body wall consists of skin, muscle, and peritoneum, as can be seen on the cut surface. In the midventral line on the inner surface of the abdominal wall runs a large vein, the *ventral abdominal vein*.

Examine the viscera. The *liver* is the large greenish or brownish organ occupying the anterior half of the pleuroperitoneal cavity. Its margins are divided into several scallop-like *lobes* by shallow indentations. It is united to the median ventral line by a mesentery, which should not be disturbed at present. On raising the left side of the liver the elongated *stomach* will be seen dorsal to the liver. Along the left side of the stomach is situated the dark-colored *spleen*. On raising the spleen there will be seen, dorsal to it and lying along the left side of the stomach, the left *lung*, a very long slender tubular structure which terminates some distance posterior to the liver. Follow the stomach posteriorly. It is a straight tube, somewhat shorter than the liver, terminating at a constriction, the *pylorus*. From the pylorus the *small intestine* begins and makes an abrupt right-angled bend to the right. In this bend rests a white gland, the *pancreas*, which also extends onto the dorsal surface of the liver. That part of the small intestine in contact with the pancreas is known as the *duodenum*. On raising the duodenum and looking on its dorsal side the pancreas will be seen to send one tail toward the spleen and another posteriorly along the small intestine. The small intestine proceeds posteriorly, somewhat coiled. In the case of females it will be found coiled on the ventral surface of the large *ovaries*, on the surface of which the eggs will be noted. (The size of the ovaries varies with the sexual state of the animal.) To each side and dorsal to the ovaries is a large, white, much-coiled tube, the *oviduct*. Trace the intestine, posteriorly pressing the ovaries away from the median line. The small intestine widens near the anus into the short *large intestine*. It lies in female specimens between the posterior terminations of the two oviducts. At the posterior end of the pleuroperitoneal cavity ventral to the large intestine will be found a sac, generally collapsed and shriveled, the *urinary bladder*. Note the stalk by which it is attached to the ventral side of the intestine. That part of the intestine to which the urinary bladder is attached (and into which the ducts of the kidneys and gonads also open) is the *cloaca*. It terminates at the anus.

The female gonads and ducts have already been noted. The left male gonad will be found dorsal to the intestine and posterior to the spleen. Dorsal to the gonad will be seen the left kidney with its duct attached to its left margin.

2. The mesenteries.—The digestive tract is attached for most of its length to the median dorsal line of the coelom by the *dorsal* mesentery. This should be noted by pressing other organs away from the median line. It is missing in the pyloric region of the stomach. That portion of the dorsal mesentery supporting the stomach is the *mesogaster*. The spleen is inclosed in the mesogaster; that portion of the mesogaster which extends from the spleen to the stomach is designated the *gastrosplenic* ligament. The lung is also attached to the mesogaster by a short mesentery. That part of the dorsal mesentery supporting the small intestine is the *mesentery*, in the limited sense; that part supporting the large intestine is the *mesorectum*.

The *ventral* mesentery is present only in the region of the liver and urinary bladder. One part of it forms the long mesentery extending between the median ventral line of the body wall and the median line of the ventral face of the liver. This is the *falciform* ligament of the liver. It contains a number of blood vessels which pass from the ventral body wall into the substance of the liver (where they join the hepatic portal vein). In the free posterior margin of the falciform ligament the ventral abdominal vein crosses from the body wall to the liver. On raising the liver the *gastrohepatic* ligament will be seen extending from the anterior part of the stomach to the dorsal face of the liver. In the region of the pancreas the *hepatoduodenal* ligament joins the duodenum and liver and incloses the greater part of the pancreas. The tails of the pancreas, however, are situated in the mesentery of the small intestine. Both of the ligaments just mentioned are parts of the ventral mesentery. The last part of this mesentery is found extending from the urinary bladder to the midventral line of the body wall; this is the *median* ligament of the bladder.

Each gonad has a mesentery: *mesovarium* in the female, *mesorchium* in the male. The mesentery of the oviduct is the *mesotubarium*. These should be located by lifting up the structures in question.

The falciform ligament should now be severed, without, however, cutting through the ventral abdominal vein. The numerous lobes of the liver appearing as scallops of its margin may now be seen more clearly. On raising the right side of the liver the right lung may be identified dorsal to it. Is it of the same length as the left lung? The small *gall bladder* will be seen on the dorsal surface of the right side of the liver. Its duct, surrounded by pancreas tissue, may be readily traced to the duodenum. The pancreas is said to open into the duodenum by a number of fine ducts.

The anterior end of the pleuroperitoneal cavity is closed by a membrane, the *transverse septum*. The liver is attached to this by the *coronary* ligament, which is continuous posteriorly with the falciform ligament. The mode of formation of the septum and the coronary ligament was described in connection with the dogfish.

3. The pericardial cavity.—Remove the hypobranchial muscles bit by bit until you have exposed a membrane, the *parietal pericardium*. Cut through this membrane. The *pericardial cavity*, in which the heart is situated, is thus exposed. Widen the opening into the cavity by cutting laterally along the anterior margin of the pectoral girdle. The muscles between the pericardial cavity and the forelimbs may also be split. The pericardial cavity is a conical cavity lined by the parietal pericardium. On gently raising the heart the posterior wall of the cavity is seen to be formed by the transverse septum. The transverse septum is pierced by two veins (hepatic sinuses), which extend forward and enter the sinus venosus, the most dorsal chamber of the heart.

4. The oral cavity and the pharynx.—Open the mouth and cut through the angle of the jaws on each side so that the jaws can be spread widely. Carry your cuts back *to* the gill arches. The cavity thus exposed consists of an anterior *oral cavity* and a posterior *pharynx*.

The oral cavity is bounded externally by the well-developed lips. Internal to the lips are the small conical teeth. There are two rows of teeth on the roof of the mouth, the posterior row being the longer. External to the last teeth of the posterior row on each side is a slit, one of the *posterior nares* or internal openings of the nasal passages. Probe into one of the anterior nares and note emergence of the probe through the posterior naris. The floor of the mouth cavity bears a single row of teeth, which on closing the mouth will be found to fit between the two rows on the roof. Posterior to the teeth is the *tongue* supported by the strongly developed hyoid arch palpable inside it. The tongue of *Necturus* is at the same stage as that of fishes.

The pharynx walls are pierced by two pairs of gill slits; probe through them and note emergence of the probe between the external gills. Note the cartilaginous gill arches supporting the bars between successive gill slits. The gill pouches, unlike the condition in elasmobranchs, are devoid of gills, since *Necturus* has only external gills. The pharyngeal cavity narrows posteriorly into the esophagus, and on passing a probe into the latter it will be found to run dorsal to the pericardial cavity and to continue into the stomach.

5. The larynx and lungs.—Although *Necturus* resembles fishes in retaining the gill apparatus, it also has in its typical primitive state the air-breathing apparatus characteristic of land vertebrates. In the floor of the pharynx, midway between the second gill slits, will be found a short slit, the *glottis*. Its walls, as should be determined by feeling them with a fine forceps, are stiffened by a pair of delicate cartilages, the *lateral cartilages*, representing reduced gill arches (Fig. 88*A*). These are the first of the laryngeal cartilages to appear in the phylogenetic series, and they in later forms subdivide to furnish the arytenoid and cricoid cartilages of the larynx and the ring cartilages of the trachea. The small cavity into which the glottis leads and which is inclosed between the two lateral cartilages is the larynx. Cut across the gill slits of the left side, so that the pharyngeal cavity can be opened more widely, and slit the glottis posteriorly. The larynx is thus seen to lead into a narrow, flattened passage, the trachea. The posterior end of this is widened and receives two openings; probe into each and note emergence of the probe into a lung. The trachea is thus seen to lead into the lungs. The air passage in primitive air-breathing vertebrates takes the following course: anterior nares, nasal cavities, posterior nares, oral cavity, pharyngeal cavity, glottis, trachea, lungs. Slit open one of the lungs and note the smooth interior, not subdivided into air spaces.

F. THE COELOM AND THE DIGESTIVE AND RESPIRATORY SYSTEMS OF THE TURTLE

Conditions in the turtle show a marked advance over those in *Necturus*. No trace remains in the adult of gills and gill slits, although remnants of the gill arches persist. The heart and other viscera have descended posteriorly, so that the esophagus and trachea are elongated, the transverse septum takes an oblique position, and a considerable part of the pleuroperitoneal cavity is located dorsal to the heart. This cavity, however, remains undivided, as in fishes and amphibians. The large lungs have become spongy through the partitioning of their walls into air spaces.

Obtain a specimen and place in a dissecting pan. Specimens which have not been injected should be employed. Remove the plastron. This is done by sawing through the bridges on each side, lifting up the plastron and separating it with a scalpel from the surrounding skin and underlying membrane.

1. The divisions and relations of the coelom.—The removal of the plastron exposes a membrane, the *parietal peritoneum*, which covers and conceals the viscera. Note that the muscle layer which is normally present between the skin and the peritoneum is completely lacking in the ventral body wall of the turtle, because of the presence of the plastron. The ventral body wall in turtles therefore consists of but two layers: the skin with its contained exoskeleton, and the peritoneum. Because of this circumstance the parietal peritoneum can be easily separated from the inside of the body wall, a procedure which is difficult or impossible in other vertebrates. Note, however, the usual muscles in connection with the girdles and limbs.

In the median line in the anterior part of the parietal peritoneum shortly posterior to the pectoral girdle is situated a triangular membranous sac, the *pericardial sac*, which incloses the heart. It will be noticed that the heart is much more posterior in position than is the case in the fishes and *Necturus;* in fact, there has occurred a posterior descent of the heart (and of other viscera as well). The membranous sac covering the heart is, as in the dogfish, the *parietal pericardium*. Here it takes the form of an isolated sac, the *pericardial sac*, while in fishes and *Necturus* it formed the lining of a chamber surrounded by the body wall. The space between the pericardial sac and the heart is the *pericardial cavity*, a portion of the coelom. The ventral face of the pericardial sac rests in the natural position against the internal surface of the plastron, while its dorsal face is fused, as we shall see, to the parietal peritoneum. Cut into the ventral wall of the pericardial sac, thus exposing the pericardial cavity and the contained heart.

Two conspicuous veins, the *ventral abdominal veins*, run longitudinally in the parietal peritoneum between the pericardial sac and the pelvic girdle. Cut through the peritoneum halfway between the heart and pelvic girdle by a transverse cut which severs both of the abdominal veins. The large cavity thus exposed is the *pleuroperitoneal* cavity, whose walls are lined by the parietal peritoneum.

The coelom of the turtle, like that of the fishes and *Necturus*, consists of two parts, a small pericardial cavity and a much larger pleuroperitoneal cavity. We note, however, that whereas in the lower forms the pericardial cavity is anterior to the pleuroperitoneal cavity and separated from the latter by the transverse septum, in the turtle the pericardial cavity is ventral to the pleuroperitoneal cavity, and the transverse septum seems to have disappeared. We may explain this change as follows. (See also Fig. 84, p. 254.) In its posterior descent the heart must necessarily carry with it the transverse septum and the parietal pericardium. The latter, in order to move posteriorly, must separate from the body wall to which it is attached in lower vertebrates. It does this and so becomes an independent sac, the pericardial sac. This process of the splitting of the pericardial sac from the body wall is aided by the invasion forward of the pleuroperitoneal cavity. The heart contained in the pericardial sac descends posteriorly, the pleuroperitoneal cavity at the same time advancing anteriorly. The pericardial sac may be thought of as sliding posteriorly ventral to the ventral wall of the pleuroperitoneal cavity. The pericardial sac thus comes to lie ventral to the anterior part of the pleuroperitoneal cavity. The posterior wall of the pericardial sac is still the anterior face of the transverse septum, the posterior face of the latter, as in lower forms, being placed between the pericardial sac and the liver. Thus the transverse septum in the turtle forms part of the partition between the pericardial and the pleuroperitoneal cavities, and the remainder of the partition is composed of the rest of the parietal pericardium, which is now the pericardial sac. These matters will be better understood by reference to Figure 84*A*, *C*.

2. The viscera and their mesenteries.—With the bone scissors cut away the margins of the carapace on each side between fore and hind limbs so as to gain easy access to the pleuroperitoneal cavity. Masses of fat, greenish-yellow material, will be found in various places and may be removed. Lift up the edges of the cut already made in the peritoneum, widening this if necessary, and look inside. Identify in the anterior part of the pleuroperitoneal cavity the large brown *liver* lying on each side of the heart. Posterior to the liver are the coils of the *intestine*. In female specimens the *ovaries*, containing eggs of various sizes, are conspicuous objects in the lateral and posterior part of the pleuroperitoneal cavity. Running alongside each ovary is the coiled *oviduct.* Just in front of the pelvic girdle is the large bilobed *urinary bladder.*

The liver consists of *right* and *left* lobes whose lateral margins curve dorsally to fit the curves of the carapace. The pericardial sac rests in a depression between the two lobes. The latter are united by a narrow *bridge* passing dorsal to the heart. Posterior to the heart the liver is united to the parietal peritoneum by very short mesenteries, corresponding to the falciform ligament of other vertebrates. In these mesenteries the ventral abdominal veins leave the peritoneum and pass into the liver. Trace the parietal peritoneum anteriorly from this region. It passes along the dorsal face of the pericardial sac, to which it is inseparably fused. This compound membrane between the heart and liver is the transverse septum, which has assumed an oblique position, owing to the descent of the heart (Fig. 84). The ventral (original anterior) face of the septum is, as before, part of the wall of the pericardial cavity; the dorsal (original posterior) face forms part of the parietal peritoneum. The liver is, as usual, attached to the transverse septum by the *coronary* ligament. Continue to trace the parietal peritoneum to the anterior end of the pleuroperitoneal cavity. On the posterior face of the pectoral girdle it turns dorsally and passes to the carapace, of which it forms the inner lining. Similarly trace the parietal peritoneum posteriorly by lifting the posterior cut edge of the membrane. It curves dorsally, following along the anterior surface of the pelvic girdle, and passes to the inner surface of the carapace.

Press both lobes of the liver forward against the pectoral girdle and look on the dorsal surface of the liver. The elongated *stomach* will be found curving dorsal to the lateral border of the left liver lobe. On following the stomach anteriorly the narrow *esophagus* will be found entering the stomach. The stomach passes along the dorsal surface of the left lobe of the liver, to the middle of which it is attached along its entire length by the short *gastrohepatic* ligament. About opposite the bridge connecting the two lobes of the liver the stomach passes insensibly into the small *intestine,* the first part of which is the *duodenum.* The duodenum is united to the middle of the dorsal surface of the right lobe of the liver by the *hepatoduodenal* ligament. In this ligament is situated a long white gland, the *pancreas.* About one-quarter of an inch back of the right end of the pancreas a *pancreatic duct* passes from the pancreas into the duodenum and may be revealed by picking away the substance of the pancreas at this point. On the dorsal surface of the right lobe of the liver near its lateral border is the large *gall bladder,* which is connected to the duodenum by a short but stout *bile duct.* Beyond the entrance of the bile duct the small intestine turns sharply posteriorly and is then thrown into a number of coils. In the case of female specimens it will generally be necessary to remove one of the large egg-bearing ovaries at this point before the intestine can be conveniently traced farther. By lifting the coils of the small intestine note the *dorsal mesentery* which attaches it to the median dorsal line of the coelom; this part of the dorsal mesentery is the *mesentery* proper. Follow the dorsal mesentery forward and note the portions of it which support the duodenum and the stomach, named *mesoduodenum* and *mesogaster*, respectively. The mesoduodenum is fused to the hepatoduodenal ligament, so that the two appear as one; but the mesogaster is distinct from the gastro-hepatic ligament. Trace the small intestine posteriorly, noting the coiling of the mesentery corresponding to the coils of the intestine. Find on the right side the entrance of the small intestine into the *large intestine* or *colon.* At the junction of the small and large intestine is a slight projection, the *colic caecum.* The colon generally crosses the pleuroperitoneal cavity transversely and then turns posteriorly and runs straight caudad to the cloaca. Note the *mesocolon* supporting the colon. In the transverse part of the colon it is fused to the mesogaster. In the mesocolon on the dorsal side of the colon shortly beyond the caecum is a rounded red body, the *spleen.* Trace the colon to the place where it disappears dorsal to the pelvic girdle. At this point ventral to the colon will be found the large, thin-walled, bilobed *urinary bladder.* It is generally greatly distended with urine, but in some specimens may be contracted to a small mass. The bladder has no ligaments, for the peritoneum leaves the body wall around the stalk of the bladder and passes over its surface to form its visceral investment.

Cut away the pelvic girdle by making a cut through each side of it with the bone scissors and removing a median piece. The large intestine can then be traced into a tube, the cloaca, which proceeds dorsal to the girdle to the anus. At the point of entrance of the large intestine into the cloaca the urinary bladder will be found, attached to the ventral surface of the cloaca by a stalk. On each side of the stalk of the bladder, in females, a large white *oviduct* will also be seen entering the cloaca.

The female reproductive system may be noted at this time; that of the male is so inconspicuous that it will not be described at this point. The ovaries are a pair of large saclike bodies containing in their walls eggs of various sizes. Each ovary is attached by its mesentery, the *mesovarium,* to the dorsal body wall. Lateral to each ovary runs the oviduct, a large coiled white tube. It is supported by the *mesotubarium.*

3. The respiratory system.—Pry open the jaws of the turtle and cut through the angles of the jaws, cutting nearer the lower than the upper jaw, revealing the *oral cavity* and the *pharynx.* The oral cavity is bounded by the jaws, which have no teeth but are clothed with horny *beaks* of epidermal origin. These beaks extend as plates into the mouth cavity. In the roof of the mouth cavity posterior to the plate is a pair of elongated openings, the *posterior nares.* Probe them and determine that they connect with the anterior nares by passages which run through the nasal cavities. The floor of the mouth cavity is occupied by the fleshy pointed *tongue,* which is of the definitive type (p. 256).

In the pharynx note that neither gills nor gill slits are present, although, as we shall see shortly, the gill arches are represented. Behind the base of the tongue is an elevation, the *laryngeal prominence,* in the center of which is an elongated slit, the *glottis.* Feel the pair of small *arytenoid* cartilages, one on each side of the glottis; they are derived from one of the gill arches. On each side of the roof of the pharynx, posterior to the muscles which connect the skull and lower jaw, is the opening of the *auditory* or *Eustachian tube,* a canal which leads from the pharynx to the cavity of the middle ear. (The opening may have been destroyed in cutting the jaws apart.) The auditory tube and also the cavity of the middle ear are outgrowths from the first gill pouch. Posteriorly the pharynx narrows into the *esophagus.*

Cut through the skin in the median ventral line of the neck and peel away the skin from neck and throat. Separate the muscles in the median line of the neck and find a tube stiffened by rings of cartilage. This is the *trachea* or *windpipe.* Trace it forward until it disappears into the pharynx. In front of this place note the hard *body* of the *hyoid,* and by cleaning away muscles find also two pairs of horns of the hyoid extending posteriorly. The hyoid and its horns are derivatives of the second, third, and fourth gill arches. Open the mouth and make a cut around the laryngeal prominence, freeing it from its position on the dorsal surface of the body of the hyoid. The structure thus freed is the *larynx,* an expanded chamber at the anterior end of the trachea. Find in the lateral walls of the larynx the two arytenoid cartilages, small cartilages supporting the two triangular flaps which bound the glottis. Posterior to the glottis is a ring-shaped cartilage, the cricoid, much wider ventrally than dorsally. Arytenoids, cricoids, and tracheal rings are believed to be derivatives of the lateral cartilages of urodeles.

Now trace the trachea posteriorly. Note the esophagus, a soft tube, lying dorsal to or to one side of the trachea. Find just anterior to the heart the point where the trachea bifurcates into the two *bronchi* which proceed to the lungs. Raise the right and left lobes of the liver and the stomach and find dorsal to them, against the carapace, a large spongy organ, the lung, on each side. Trace a bronchus into each lung; it is accompanied by a pulmonary artery and a pulmonary vein. Study the relation of the lung to the pleuroperitoneal cavity. Note that the lung is in contact with the inner surface of the carapace and that the parietal peritoneum passes over the ventral surface of the lung, leaving the lung outside of the membrane. Such a relation to the peritoneum is spoken of as *retroperitoneal.* The posterior end of the lung, however, projects into the pleuroperitoneal cavity and is clothed with the peritoneum. Cut open the lung and observe its extremely spongy texture; cords of connective tissue divide the interior into air spaces or *alveoli.*

The path followed by the air in respiration is: external nares, nasal cavities, internal nares, oral cavity, pharyngeal cavity, glottis, larynx, trachea, bronchi, and lungs. In the pharyngeal cavity the paths of food and air cross.

Make drawings to show the parts of the digestive and respiratory systems. Make a diagram of a cross-section through the body at the level of the heart to show the pericardial and pleuroperitoneal cavities and membranes and their relation to the viscera.

G. THE COELOM AND THE DIGESTIVE AND RESPIRATORY SYSTEMS OF THE PIGEON

The chief advance of birds over the reptilian condition is the division of the pleuroperitoneal cavity by a delicate partition into a pair of anterior pleural cavities each containing a lung and a posterior peritoneal cavity. The respiratory system of birds has a number of peculiarities (p. 260), but these are not of phylogenetic significance.

Obtain a specimen and place in a dissecting pan. The feathers must be removed. It is desirable that the air sacs should have been inflated through the trachea.

1. The oral cavity and the pharynx.—Open the mouth widely by cutting through the angles of the jaws. An anterior *oral cavity* and a posterior *pharynx* are thus revealed.

a) Oral cavity: Roof and floor of the oral cavity are bounded laterally by horny *beaks* of epidermal origin which incase the jaws. Teeth are absent, as in all living birds. The roof of the mouth cavity bears a pair of elongated *palatal* folds with free fimbriated margins. These palatal folds correspond to the secondary or hard palate of mammals but differ in that they do not meet in the median line, leaving a deep palatal fissure. The hard palate of many birds is therefore a *split palate* and is normally in the condition which in mammals is the result of imperfect development. In the roof of the mouth cavity, above the palatal folds and concealed by them, are the posterior nares, found by bending aside or cutting away the palatal folds. Probe to verify connection of anterior and posterior nares. The floor of the oral cavity is occupied by the pointed tongue, having a free fimbriated posterior border. The tongue is of the compounded definitive type but is not very muscular. The numerous glands opening into the oral cavity are too small to study in gross dissection.

b) Pharynx: Note that, as in all adult vertebrates above urodeles, gill slits are absent from the lateral walls of the pharynx. In the roof of the pharyngeal cavity just posterior to the caudal ends of the palatal folds is a median aperture, the opening of the paired *auditory tubes*. Each auditory tube extends from this opening to the cavity of the middle ear; tube and cavity represent in part an evagination from the first gill pouch. In birds, unlike other vertebrates, the two auditory tubes unite to one at the point of communication with the pharynx. Posterior to this opening, the roof of the pharynx bears a pair of folds with fimbriated borders, which hang down like a curtain into the pharyngeal cavity. These folds constitute the *soft palate*. In the floor of the pharynx, immediately posterior to the caudal end of the tongue, is a hardened elevation, the *laryngeal prominence*, bearing in its center an elongated opening, the *glottis*. The margins of the glottis are also fimbriated, and immediately posterior to the glottis on each side is a fringed fold. In the walls of the glottis the supporting *laryngeal cartilages* are readily felt.

Make a drawing of the oral and pharyngeal cavities.

2. The hyoid apparatus, the larynx, the trachea, and the esophagus.—Make a median ventral longitudinal incision in the skin of the neck from the throat to the anterior end of the sternum. Deflect the skin on each side of the incision. The *trachea* or *windpipe*, a tube with walls stiffened by rings of cartilage, is immediately exposed. Dorsal to it or to one side of it is the soft *esophagus*.

Trace the trachea forward to the glottis, cleaning away the muscles which cover its anterior end. At the same time a cut may be made to the sides of the tongue, so that the tongue may be pulled down ventrally from the mouth cavity. The hyoid apparatus may now be studied. It consists of remnants of the hyoid (second) and third gill arches. It is composed of three median elements, arranged in a longitudinal series, and two pairs of *horns* or *cornua*. The most anterior of the three median pieces is the *entoglossal* cartilage. It is situated inside of the tongue and may be revealed by dissecting off the covering membrane of the tongue. It represents the two fused *ceratohyals*. From its posterior end projects posteriorly on each side a small cartilage which occupies the caudal point of the tongue, already noted. These two cartilages constitute the *anterior horns* of the hyoid and consist of the free ends of the two ceratohyals whose anterior portions fused to form the entoglossal cartilage. Posterior to the entoglossal cartilage is a median bony piece, the *basihyal*. Posterior to this is the *basibranchial* of the third gill arch. From the point of junction of basihyal and basibranchial projects on each side the long *posterior horn* of the hyoid, consisting of portions of the third gill arch. On following the posterior horns they will be found to extend toward the ears and to be divided into a proximal longer portion, the *ceratobranchial*, and a distal shorter rod, the *epibranchial*.

The cartilages of the larynx may next be identified. Cut around the laryngeal prominence, freeing it so that it can be drawn ventrally. Also free the hyoid apparatus from the ventral surface of the larynx. The larynx is the expanded chamber thus revealed at the top of the trachea and opening into the pharyngeal cavity by way of the glottis. By dissecting in the margins of the glottis on each side, expose a slender, curved, partially ossified *arytenoid* cartilage. On the ventral side of the larynx note the enlarged triangular *cricoid* cartilage. Follow this around to the dorsal side, where it terminates by much narrowed ends. Between the two dorsal ends of the cricoid cartilage is another median cartilage, the *procricoid*, which is in contact with the posterior ends of the arytenoids and which is simply a separated piece of the cricoid. Although the larynx of birds is morphologically the same as the larynx of other vertebrates from which sounds issue, in birds the voice is not produced in the larynx but in another part of the trachea, which will be seen later.

Examine the cartilages of the trachea. They are broad, hard, and bony ventrally, but narrower, softer, and cartilaginous in composition dorsally. There is consequently a somewhat soft strip along the dorsal side of the trachea which lies against the cervical vertebrae.

Trace the esophagus posteriorly. Shortly in front of the sternum it widens into an enormous bilobed sac, the *crop*, in which the food, swallowed whole, is detained for a time and may be subject to muscular and enzymatic action. It is passed on to the stomach in small quantities; and in some birds, as in pigeons, is regurgitated from the crop and fed to the young. The crop is best developed in grain-eating birds and is small or absent in many birds. The crop should be carefully loosened on all sides.

3. The anterior air sacs and the pectoral muscles.—The respiratory system of birds is the most remarkable among vertebrates. It consists not only of the lungs but also of a number of air sacs located among the viscera and of air spaces in the bones. These air sacs and air spaces communicate with the lungs by means of branches of the bronchi. This system not only aids in decreasing the specific gravity of the bird but also insures a more complete exposure of the lung tissue to the air; for the residual air is retained in the air sacs and not in the lungs as in other vertebrates, and the air in the lungs is consequently completely renewed at each inspiration. Because of the delicacy of the air sacs the student may not be able to locate all of those mentioned below, but some of them will be seen. They are best studied in freshly killed specimens in which they have been inflated through the trachea.

Dorsal to the crop in the angle formed by the two halves of the furcula or wishbone is situated the *interclavicular air sac*. Its delicate ventral wall is in contact with the dorsal wall of the crop. It consists of two lobes, one on each side of the median line; in the embryo these lobes are separate. Puncture the interclavicular air sac and find, dorsal to it on each side, another sac, the *cervical* air sac.

Extend the median ventral incision in the skin to the anus. Separate the skin from the underlying muscles on each side of chest and abdomen. The great pectoral muscles are revealed immediately internal to the skin and occupying the angle between the keel and the body of the sternum. The *pectoralis major* is the great muscle covering the entire sternum and extending to the humerus. It takes origin from the keel of the sternum, the surface of the body of the sternum, and the furcula, which will be found imbedded in its anterior border; its fibers converge toward the humerus and, passing over the shoulder, are inserted on the outer and dorsal surface of the humerus. The muscle should be followed to its insertion. Action, depresses the wing. Now carefully cut through the pectoralis major slightly to the *right* of the keel of the sternum and along the posterior margin of the furcula. The muscle can then be deflected and separates easily from the underlying *pectoralis minor*. The large *pectoral arteries* and *veins* will probably be noticed emerging between the pectoral muscles which they supply. The pectoralis minor originates from the body of the sternum and converges toward the humerus. On following the muscle laterally there will be found between it and the pectoralis major another air sac, the *axillary* sac. Cut into the axillary sac. The anterior wall of this sac is in contact with the coracoid bone; and, laterally, on looking into the sac the tuberosities of the humerus will be seen. A large opening into the humerus, the *pneumatic foramen*, is readily noticed; on probing this it will be found to lead into the interior of the humerus. It is the entrance to the air space of the humerus which communicates with the axillary air sac. The axillary air sac communicates in front with the interclavicular air sac. The pectoralis minor may now be followed to its insertion. It converges to a tendon which passes ventral to the posterior end of the cervical air sac and beneath the shoulder to the dorsal side of the humerus on which it is inserted. To see the insertion turn the bird dorsal side up and dissect away the superficial muscles of the dorsal side of the shoulder. The tendon of the pectoralis muscle, because of its mode of insertion, has a pulley-like action which enables the muscle to raise the wing. Whereas in mammals all of the pectoral muscles act together to adduct the forelimb, in birds the actions of the pectoralis major and minor are opposed to each other, the one depressing, the other raising, the limb. This arrangement eliminates all powerful muscles from the back and enables all of the wing muscles to take their origin from the firm and strong sternum.

4. The divisions of the coelom and the posterior air sacs.—Cut through the ventral abdominal wall to the *right* of the median line. Beneath the skin are the thin layers of abdominal muscles corresponding to those of mammals, and internal to this the parietal peritoneum generally impregnated with streaks of fat. Cut through this and extend the incision anteriorly, cutting through the sternum slightly to the right of the keel, keeping the scissors in contact with the bone so as to avoid injuring internal parts. Spread apart the cut edges and look within.

The small cavity posterior to the sternum is the *peritoneal* cavity. Note in it the *liver* dorsal to the posterior end of the sternum, the closely coiled *intestine*, and to the left the large firm *gizzard*. From the gizzard a mesentery extends to the ventral body wall to the left of the median line. This may be designated the *ventral ligament* of the gizzard.[3] It is continuous anteriorly with the *falciform ligament* of the liver, which extends from the median ventral region of the liver to the midventral line of the body wall and inner surface of the sternum. The falciform ligament and ventral ligament of the gizzard together constitute a partition which divides the peritoneal cavity into a large right portion and smaller left portion. This division is not found in other vertebrates. In the partition courses a small vein,[4] extending from the mesenteries in question to the liver.

Deflect the pectoralis major muscle on the left side of the sternum and make a cut through the left side of the sternum slightly to the left of the keel. Remove and discard the median piece of sternum containing the keel.

Immediately dorsal to the sternum is situated the delicate *pericardial sac* containing the *heart*. The ventral wall of the pericardial sac will probably have been opened in cutting through the sternum. The heart, as in the turtle, has descended posteriorly; and a pericardial sac has been formed of the anterior face of the transverse septum and the parietal pericardium, as described in connection with the turtle. The space between the pericardial sac and the heart is, as before, the *pericardial cavity*, a portion of the coelom. The pericardial sac is in contact on its ventral surface with the inner surface of the sternum, and anteriorly and laterally is also in contact with the inner surface of the body wall. Hence, only the posterior part of the pericardial sac is freed from the body wall. From the points where the pericardial sac meets the lateral body wall, a membranous partition extends obliquely posteriorly on each side. This partition is called the *oblique septum*. It contains a large air sac. It stretches across from the lateral body wall to that part of the

[3] This is commonly called the greater omentum in texts and manuals; but, since it is not at all homologous to the structure so named in mammals, it is desirable that the name be dropped. The ligament of the gizzard is a mesentery peculiar to birds and arises as a secondary outgrowth from the serosa of the gizzard to the ventral body wall. It is probably due to the need for additional support for the heavy gizzard.

[4] This vein is named in manuals the ventral abdominal vein, but it does not appear to be homologous to the vein of that name in other vertebrates.

pericardial sac which is derived from the transverse septum, and thus divides the pleuroperitoneal cavity into anterior and posterior portions. That part of the original pleuroperitoneal cavity left anterior to the oblique septum consists of the two *pleural* cavities, one on each side of the pericardial cavity. That part of the pleuroperitoneal cavity posterior to the oblique septum is the *peritoneal* cavity, already mentioned.

Inside the oblique septum, inclosed between its anterior and posterior walls, is a large air sac, the *posterior intermediate* air sac. Immediately anterior to this, lying to each side of the heart, is the small *anterior intermediate* air sac.

In the peritoneal cavity cut through the falciform ligament and ligament of the gizzard at their line of attachment to the ventral body wall. On either side of the viscera and slightly dorsal to them find the large *abdominal* air sac.

From the foregoing account it is seen that the coelom of birds is divided into four compartments: the pericardial cavity, the two pleural cavities, and the peritoneal cavity.

5. The peritoneal cavity and its contents.—This cavity has already been mentioned. As in other vertebrates, it is lined by the parietal peritoneum, which is deflected at certain points to form mesenteries and which continues over the surface of the viscera as the visceral peritoneum.

The viscera of the peritoneal cavity may now be studied in more detail. At the anterior end is the large *liver,* consisting of right and left lobes, the former the larger. The pericardial sac rests between the two lobes of the liver. The liver is attached to the pericardial sac (that portion of it derived from the transverse septum) by the *coronary* ligament. The falciform ligament of the liver was already noted and severed. To the left and slightly covered by the left lobe of the liver is the *gizzard.* On raising the left lobe of the liver the *gastrohepatic* ligament will be noted passing between the gizzard and the liver. The *mesogaster* connects the gizzard with the dorsal body wall. The ventral ligament of the gizzard was already noted and cut. On breaking through the gastrohepatic ligament the soft *proventriculus* will be found extending anteriorly from the gizzard dorsal to the liver. Proventriculus and gizzard together correspond to the stomach of other vertebrates; the proventriculus contains the gastric glands and performs the enzymatic part of stomachic digestion, and the gizzard has the function of triturating the food, since birds have no teeth and swallow their food whole. From the stomach, at the junction of proventriculus and gizzard, the small intestine arises. The duodenum makes a long U-shaped loop posteriorly, and its beginning is attached to the right lobe of the liver by the hepatoduodenal ligament. Between the two sides of the duodenal loop stretches the *mesoduodenum,* a portion of the mesentery of the intestine. In this is situated the *pancreas,* lying between the two limbs of the loop. From a deep depression in the dorsal surface of the right lobe of the liver, the two *bile ducts* (there is no gall bladder) emerge and pass into the duodenum. The left bile duct is the shorter and stouter of the two and enters the left limb of the duodenum about half an inch beyond the gizzard. The more slender right bile duct passes to the right limb of the duodenal loop. There are three *pancreatic ducts,* all of which pass from the right side of the pancreas into the right limb of the duodenal loop. One of these arises from the anterior part of the pancreas and passes obliquely forward, entering the duodenum near the anterior termination of the right limb of the loop. The other two ducts emerge from the middle of the pancreas and pass across to the right limb of the duodenum. The ducts are generally easily seen by spreading out the mesentery.

Trace the small intestine posteriorly from the duodenum. It is much coiled and is supported by the *mesentery,* which, owing to the small space into which the intestine is packed, is fused in many places. Near its termination the small intestine turns toward the median line, widens slightly, and then runs straight caudad in the median line. At about the middle of the peritoneal cavity it passes without enlargement into the *large intestine.* The point of junction of large and small intestine is marked by a pair of small lateral diverticula, the *colic caeca.* The short large intestine soon passes into the *cloaca.* Because of the absence of pubic and ischial symphyses in birds, the cloaca does not pass through the ring of the pelvic girdle but may be traced directly to the anus. A urinary bladder is absent, although present in the embryo as an allantois. In females the single left oviduct will probably be noticed entering the left side of the cloaca. The single ovary (left one) is situated in the anterior part of the peritoneal cavity, dorsal to the gizzard.

Make a drawing of the digestive tract.

The gizzard and proventriculus may now be freed from the adjacent air sacs and the mesenteries. On turning the gizzard far forward there will be found, between the proventriculus and the anterior end of the right limb of the duodenal loop, a rounded red body, the spleen. The gizzard may now be cut open along its posterior margin. The interior contains small stones and probably partially digested food. Note the extremely thick muscular walls and hard horny lining of the gizzard. Cut from the gizzard into the proventriculus and note the soft glandular walls of the latter. The gizzard compensates for the absence of teeth in birds and grinds up the food into small pieces; hence it is best developed in birds which eat grains and other hard foods and not so definitely set off from the proventriculus in flesh-eating birds.

6. The pleural cavities and their contents.—The posterior intermediate air sac, situated in the oblique septum, may now be punctured if this has not already been done. The two walls of the septum are now more evident. The anterior intermediate air sac may also be punctured. Against the dorsal wall of the pleural cavity of each side will be found a reddish, spongy, flattened organ, the lung. The openings of some of the air sacs into the lungs will probably be noted on some specimens. On cutting into the lung the organ will be found to be solid, not hollow as in the preceding animals.

The cavity in which each lung is contained is, as already explained, a *pleural* cavity. It is lined by a coelomic membrane, the *pleura*. As the lungs are flattened against the dorsal wall of the pleural cavity, the pleura passes over their ventral faces, leaving them outside, so to speak. The pleura, furthermore, passes over the surface of the pericardial sac and lines the inner surface of the body wall.

7. The syrinx.—Examine the posterior part of the trachea. Two slender muscles, the *sternotracheal muscles*, diverge from their insertion on the ventral surface of the trachea to their origin on the sternum. These muscles should be severed. The trachea disappears dorsal to the heart and the great blood vessels which enter and leave the heart. These blood vessels must not be injured. Loosen the trachea and pull it forward. The bifurcation of the trachea into the two *bronchi* can then be seen dorsal to the heart. Cut across the bronchi with a fine scissors and draw the trachea forward. At the point where the trachea forks into the two bronchi an expanded chamber, the *syrinx*, is present. The voice of birds issues from the syrinx, not from the larynx. Along each side of the trachea extending from the point of insertion of the sternotracheal muscles to the lateral walls of the syrinx is a muscle, the *intrinsic syringeal* muscle. The walls of the syrinx are supported by the last tracheal rings and the first bronchial half-rings. The last two tracheal rings are widely separated from each other but are connected in the midventral line by a narrow bridge of bone; the thin tracheal wall to either side of this bridge forms the *external tympaniform membranes*. On the inner side of the bronchi at their junction with the trachea are similar thin *internal tympaniform membranes*. Make a slit in the ventral wall of the syrinx and spread the cut edges apart. In the dorsal wall of the interior is a slight vertical fold, the *semilunar membrane*, supported by a bony ridge, the *pessulus*. The voice is produced by the vibrations of the pessulus and the tympaniform membranes. Large thickenings in the lateral walls of the syrinx also play a role, and the syringeal and sternotracheal muscles aid by changing the shape of the syrinx.

H. THE COELOM AND THE DIGESTIVE AND RESPIRATORY SYSTEMS OF MAMMALS

In mammals the separation of the pleural cavities from the peritoneal cavity is achieved by a muscular partition, the *diaphragm*, which consists partly of the transverse septum and which plays an important role in respiratory movements. Larynx and lungs also reach a high degree of differentiation, and the digestive tract shows many variations in correlation with various modes of life.

The directions apply to the cat and rabbit, and any differences between the two animals will be specifically mentioned.

1. Oral cavity and pharynx.—

a) *The salivary glands:* These are large masses of gland tissue which are outgrowths of the oral epithelium; the stalk of each outgrowth remains as a *salivary duct*. The glands are situated among the muscles of the head and throat. They should be located according to the following descriptions and their ducts followed as far as practicable. There are four pairs of salivary glands in the rabbit, five in the cat. The dissection should be carried out on the same side of the head as that on which the muscles were dissected.

The *parotid* gland is located ventrad and craniad of the base of the pinna of the ear, just under the skin. Remove the skin from this region and find the pinkish gland spread out under the skin anterior and ventral to the ear. Its duct passes across the external surface of the masseter muscle and penetrates the upper lip. The *submaxillary* gland has already been noted as a roundish mass at the angle of the jaw near the posterior margin of the masseter. Loosen it and find the duct springing from the internal surface. In the **cat** the beginning of this duct is surrounded by the elongated *sublingual* gland. Trace the submaxillary duct forward; it is accompanied in the cat by the sublingual duct. The duct will be found to pass internal to the digastric muscle. This muscle should be severed. The duct (or two ducts in the cat) will then be seen to pass internal to the mylohyoid muscle. This in turn should be cut and the duct traced forward. In the **rabbit** the small flattened *sublingual* gland will soon be noted lying in the path of the submaxillary duct. The submaxillary duct (accompanied by the sublingual duct in the cat), situated just external to the lining of the mouth cavity, runs forward nearly to the symphysis of the mandible and then penetrates the lining. In the **rabbit** the sublingual gland opens into the mouth cavity by several short ducts which are impractical to find. The *molar* gland, present in the **cat** only, is situated between the skin and the external surface of the mandible, just in front of the masseter muscle. It will be found by deflecting the skin at this place. It opens onto the inside of the cheek by several small ducts, impractical to locate. The *infraorbital* gland in both cat and rabbit lies in the floor of the orbit and will be seen later when the eye is dissected.

b) *The oral cavity:* Cut through the skin at the corners of the mouth and see that the skin is well cleared away over the angles of the jaws. Cut through the masseter and other muscles attached to the lower jaw at the angle of the jaws. It should then be possible to pull the lower jaw down. Pry open the mouth, grasp the lower jaw, and exert a strong traction. The jaw will generally yield; but if it does not, the ramus of the mandible may be cut through with the bone scissors. The anterior part of the cavity thus revealed is the *oral cavity*, bounded by the *lips* and *cheeks*. That part of the oral cavity lying between the teeth and lips is called the *vestibule* of the mouth. The teeth were described in connection with the skull.

The anterior portion of the roof of the oral cavity is occupied by the *hard palate*, the posterior part by the *soft palate*, which is very long in the rabbit. The difference between the hard and soft palate should be determined by feeling. The hard palate is supported by the premaxillary, maxillary, and palatine bones, as should be recalled from the study of the skull. The soft palate lacks bony support. The mucous membrane of the hard palate is thrown into a number of roughened transverse ridges. At the anterior end of the hard palate just behind the incisor teeth will be found a pair of openings, the openings of the *nasopalatine ducts* which connect the mouth and nasal cavities by way of the incisive foramina of the maxillary bones. The opening of the duct of the parotid gland may be sought for on the inside of the cheek opposite the second upper premolar tooth in the rabbit, opposite the last cusp of the third upper premolar of the cat, in which animal it is situated on a slight ridge: the openings are difficult to identify with certainty, and not much time should be spent in looking for them.

The floor of the oral cavity is occupied by the *tongue*, a fleshy muscular organ, more mobile in mammals than in most other vertebrates. It is formed by the fusion of four components (p. 256) and has a more complicated intrinsic and extrinsic musculature than the tongue of lower vertebrates. The anterior margin of the attachment of the tongue to the floor of the oral cavity has the form of a vertical fold, the *frenulum*. Halfway between the lower incisors and the frenulum will be found in the **rabbit** the two small, slitlike openings of the ducts of the submaxillary glands, the two being about an eighth of an inch apart. **In the cat** a fold runs forward from the frenulum on each side just within the teeth, and terminates anteriorly in a well-marked flattened papilla which bears the openings of the ducts of the submaxillary and sublingual glands.

Cut through the floor of the mouth on each side, keeping the scalpel next to the mandible. The tongue can now be pulled down and out between the two halves of the lower jaw. The cuts may be continued on each side at the base of the tongue back to the level of the submaxillary glands so that the tongue can be pulled well down. The surface of the tongue may now be examined in detail. In the **rabbit** the tongue is divisible into two portions: an anterior softer portion, covered with minute pointed elevations, the *fungiform papillae;* and a posterior, elevated, smoother, and harder portion. At the posterior end of the latter on each side is situated a *vallate* papilla, consisting of a round elevation set into a pit. In front of each vallate papilla on the side of the tongue is an oval area of considerable size marked by numerous fine parallel ridges, the *foliate* papilla. In the **cat** the anterior part of the tongue is covered with the *filiform* papillae, many of which are hard and spinelike, pointed posteriorly; the remainder of the tongue is provided with *fungiform* papillae; among the fungiform papillae are four to six *vallate* papillae arranged in a V-shaped row, each consisting of a round elevation set into a pit. At the sides of the vallate papillae are some very large fungiform papillae. The papillae are provided with microscopic taste buds.

c) *The pharynx:* The *pharynx* is that portion of the cavity lying posterior and dorsal to the soft palate. Pull the tongue well forward and examine the soft palate. It descends like a curtain across the posterior end of the oral cavity. Find its free posterior margin, arching above the base of the tongue. (The margin may be concealed by a leaf-shaped structure, the epiglottis, which projects from the base of the tongue. If so, the epiglottis should be pressed out of the way.) The opening formed by the free border of the palate is known as the *isthmus of the fauces*. This opening leads into the cavity of the pharynx. Shortly anterior to the free border of the soft palate on each side is a pit, the *tonsillar fossa*, which contains a small mass of lymphoid tissue, the *palatine tonsil*. The tonsillar fossa is bounded in front and behind by low folds, an anterior *glossopalatine arch* and a posterior *pharyngopalatine arch*. Now slit the soft palate forward along its median line. A cavity, the *nasopharynx*, a part of the pharynx, is revealed dorsal to the soft palate. At the anterior end of the nasopharynx are the two *posterior nares* or *choanae*, the internal ends of the nasal passages. Posterior to them on the lateral wall of the nasopharynx will be noted a pair of oblique slits; they are the openings of the *auditory* or *Eustachian tubes*, canals which connect the pharynx with the cavity of the middle ear.

The pharynx narrows posteriorly into the *esophagus*. Anterior to the entrance into the esophagus is situated the entrance into the respiratory tract. This entrance is guarded by a projecting process, the *epiglottis*, which if not already identified will be seen on pulling the tongue well forward. In the pharynx the paths for food and air are crossed (as is the case in all of the air-breathing vertebrates). However, it will be noted that, because of the formation of the palate and the consequent posterior migration of the posterior nares, the air no longer enters the oral cavity as is the case in Amphibia and most reptiles but proceeds directly into the pharynx.

2. Hyoid apparatus, larynx, trachea, and esophagus.—Press the tongue dorsally against the lower jaw and find on its external surface at its base a bone, the body of the hyoid. This is a stout bone in the rabbit, a narrow bar in the cat. Clear away muscles from its surface so as to reveal it and the two *horns* or *cornua* which extend from its sides. In the **rabbit** the horns are short processes which are connected by slender, tendinous muscles with the jugular process of the occipital bone. In the **cat** the anterior horn is long and slender and consists of a chain of four bony pieces, the last of which articulates with the tympanic bulla; the posterior horn is short and is united to the larynx. The hyoid and its horns are derived in mammals from the second and third gill arches. The hyoid supports the base of the tongue and serves for the origin and insertion of muscles.

In the median ventral line posterior to the body of the hyoid is a chamber with cartilaginous walls, the *larynx* or voice box, which constitutes the projection in the throat popularly known as Adam's apple. By making a cut through the base of the tongue and gently severing the muscle attachments the larynx may be freed and lifted forward. At the top of the larynx is a large opening, the *glottis*, from whose ventral margin the epiglottis projects. Dorsal to the glottis and bound with it by muscles is another opening, generally collapsed and concealed from view by portions of the larynx. This opening should be located by probing; the probe will be found to enter a soft tube which proceeds posteriorly dorsal to the larynx. This tube is the *esophagus*.

The structure of the larynx should now be examined in detail. The ventral wall of the larynx is supported by a large shield-shaped cartilage, the *thyroid* cartilage. A short distance posterior to this is the *cricoid* cartilage, which forms a ring around the larynx. The dorsal rim of the glottis between the glottis and the opening to the esophagus is supported by a pair of projecting cartilages, the *arytenoids*. On looking into the glottis a pair of folds, the *vocal cords*, will be seen extending from the arytenoid cartilages to the thyroid cartilage. They nearly occlude the opening. In the cat, in addition to these *true* vocal cords, there is a pair of *false* vocal cords, situated lateral to the former and extending from the tips of the arytenoid cartilages to the base of the epiglottis. It will be noted that the vocal cords are not cords but folds of the lateral wall of the larynx. On dissecting away the esophagus from the dorsal side of the larynx the dorsal side of the cricoid cartilage will be exposed. It is much broader than the ventral side. By cleaning away the mucous membrane covering it, the two arytenoid cartilages which rest on the anterior extremity of the dorsal part of the cricoid will be exposed.

From the larynx the trachea or windpipe proceeds posteriorly. Its walls are stiffened by cartilaginous rings, which are incomplete dorsally, leaving a soft strip in the dorsal wall of the trachea into which the esophagus fits. On each side of the trachea, lying against the trachea and internal to the muscles, is a flattened elongated body, one of the lobes of the *thyroid* gland. The anterior end of each lobe is at a level with the cricoid cartilage. The caudal ends of the two lobes are connected by a median portion, the *isthmus*, which crosses the ventral side of the trachea. The trachea is not to be traced farther posteriorly at this time.

3. The pleural and pericardial cavities.—The trunk of mammals is divided into an anterior thoracic region and a posterior abdominal region. Each of these regions contains cavities which are portions of the coelom. The thoracic region has three coelomic cavities: the two *pleural* cavities, laterally located, and the median *pericardial* cavity, situated between the two pleural cavities.

With the bone scissors make a cut through the ribs one-half inch to the left of the sternum, extending the cut the length of the sternum. At each end of this, cut laterally and dorsally between two adjacent ribs at right angles to the first cut. In this way a flap is formed in the chest wall. Open the flap and bend it dorsally so that you can look within. The cavity thus revealed is the left *pleural cavity* or *pleural sac*, as it is often called; a similar sac exists on the right side. The pleural sac contains the soft spongy *lung*. In the median region under the sternum lies the large heart. Note the delicate partition which stretches from the heart to the ventral median line. This partition is called the *mediastinal septum*. It consists of the two medial walls of the right and left pleural sacs in contact with each other. At the level of the heart the two walls separate so that the heart and its pericardial sac are inclosed between them. This space between the two walls of the mediastinal septum is called the *mediastinum*. The posterior wall of the pleural sac is formed by a muscular dome-shaped partition, the *diaphragm*. The pleural sac is lined by a smooth moist membrane, the *pleura*. The pleura is divided into *parietal* and *visceral* parts. The parietal pleura lines the inside of the pleural cavity, covers the anterior face of the diaphragm, and together with the medial wall of the other pleural sac forms the mediastinal septum. The visceral pleura is that part of the pleura which passes over the surface of the lung to which it is indistinguishably fused. Examine the left lung. It is a soft spongy organ divided into three lobes: a smaller *anterior*, and larger *middle* and *posterior lobes*. The anterior lobe is quite small in the rabbit. The large posterior lobe fits very neatly on the convex surface of the diaphragm. Cut into the lung; it appears solid but is really composed of innumerable minute air-cells (*alveoli*).

Now carefully cut through the mediastinal septum ventral to the heart and look into the right pleural cavity. The diaphragm may be slit along its left side so as to facilitate the spreading-apart of the thoracic walls. The right pleural cavity is similar to the left cavity. It contains the right lung. The right lung is somewhat larger than the left lung. It is divided into anterior, middle, and posterior lobes. The large posterior lobe is subdivided into two *lobules*, a *medial* and a *lateral*. The medial lobule projects into a pocket formed by a special, dorsally directed fold of the mediastinal septum. This fold, the *caval* fold, has the function of supporting a large vein, the *postcaval* vein, which ascends from the liver to the heart and will be found inclosed in the free dorsal margin of the caval fold.

Examine the heart and the pericardial sac. The pericardial sac or parietal pericardium is a sac of thin tissue inclosing the heart but not attached to it except at the anterior end, where the great vessels enter and leave the heart. The heart is freely movable inside of the pericardial sac. The narrow space between the pericardial sac and the heart is the *pericardial cavity*, a portion of the coelom. Cut through the pericardial sac so as to expose the heart. The surface of the heart is invested by a thin membrane, the *visceral pericardium*, inseparably adherent to the heart wall. The visceral pericardium is continuous with the pericardial sac at the anterior end, where the blood vessels enter and leave the heart. As the heart with its pericardial sac is situated in the mediastinum, it is evident that there are three coelomic layers surrounding the heart: the visceral pericardium closely adherent to the heart wall, the parietal pericardium or pericardial sac separated from the heart by the pericardial cavity, and the parietal pleura of the mediastinal septum, which is closely fused to the pericardial sac (Fig. 90*B*).

In the mediastinum, in the median line ventral to the anterior part of the heart and extending forward, will be found a mass of gland tissue, the *thymus*. It is larger the younger the specimen. In searching for it do not injure the large blood vessels occurring in this region. The thymus is possibly one of the glands of internal secretion and is derived from the entodermal lining of certain of the gill pouches of the embryo.

Now press the heart and the left lung over to the right. The lung will be found attached by a narrow region, the *radix* or *root* of the lung. An artery, a vein, and a bronchus or air tube pass to the lung and veins from the lung in the root; but these structures are better investigated at a later time. In the cat, furthermore, the lung is attached along most of its length to the dorsal thoracic wall by the *pulmonary ligament*, a fold of the pleura. Note that dorsal to the root of the lung the pleura continues onto the dorsal and lateral surfaces of the pleural cavity and that certain structures can be seen that lie outside the pleura. These structures lie between the dorsal portions of the two walls of the mediastinal septum and consequently are situated in the mediastinum. The most conspicuous of these structures lying in the mediastinum is the *dorsal aorta*, a very large vessel injected with a colored solution which arches away from the heart to the left and descends toward the diaphragm. About one-half an inch ventral to the aorta is another tube, the *esophagus*, also lying in the mediastinum. Trace it posteriorly to the place where it penetrates the diaphragm.

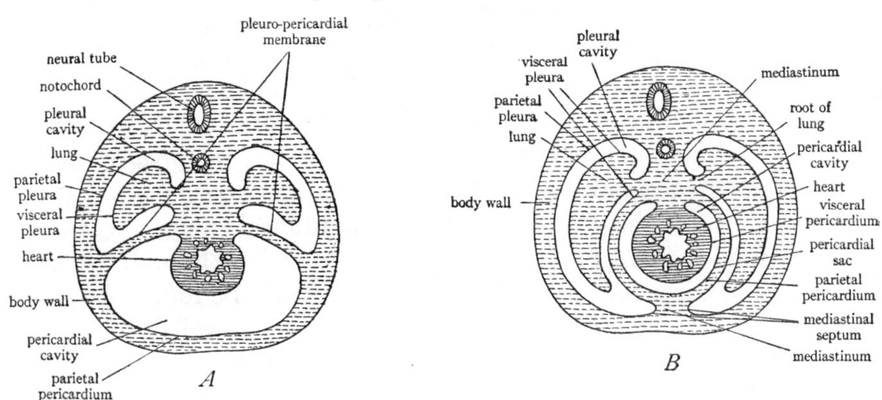

FIG. 90.—Diagrams to show the separation of the pericardial sac in mammals. *A*, early stage in which the parietal pericardium forms the lining of the body wall; the pericardial and pleural cavities are separated by the pleuropericardial membrane, which is the anterior continuation of the transverse septum. *B*, later stage, showing how the ventral extension of the two pleural cavities splits the parietal pericardium from the body wall and gives rise to the mediastinal septum; the parietal pericardium then becomes the pericardial sac. (*B* from Arey's *Developmental Anatomy*, courtesy of the W. B. Saunders Co.)

The diaphragm is a curved sheet forming the posterior wall of the thoracic cavity and completely separating it from the abdominal cavity. The center of the diaphragm is seen to consist of connective tissue forming a circular tendon, the *central tendon* of the diaphragm. The remainder of the diaphragm is muscular. The diaphragm takes origin from the ribs, sternum, and vertebrae, and is inserted on the central tendon. It is an important respiratory muscle. When contracted, it flattens, thus lengthening the pleural cavities posteriorly and causing air to rush into the lungs. The diaphragm is pierced at several points to allow important structures to pass through; the chief ones which penetrate the diaphragm were already noted, i.e., the esophagus, the aorta, and the postcaval vein. The diaphragm is a structure peculiar to mammals. It is formed in part of the transverse septum and in part of other coelomic membranes; it then becomes invaded by muscle buds from the adjacent cervical myotomes.

Make a diagram of a cross-section through the thorax showing the pleural and pericardial cavities and the relation of their linings to the thoracic wall, lungs, and heart.

4. The peritoneal cavity and its contents.—Make a longitudinal slit through the abdominal wall, a little to the left of the median ventral line from the inguinal region up to the diaphragm. Widen the opening by a transverse slit in the middle of the left abdominal wall. A large cavity, the *abdominal* or *peritoneal cavity*, is exposed. Its anterior wall is formed by the concavely arched diaphragm which completely separates the peritoneal from the pleural cavities. Posterior to the diaphragm and shaped so as to fit the concave surface of the diaphragm is the large, lobed *liver*, generally grayish brown in preserved specimens. Posterior to the liver the peritoneal cavity is filled by the coils of the intestine. In the **cat** the intestine is covered ventrally by a thin membrane impregnated with streaks of fat, the *greater omentum*. This membrane is present also in the **rabbit,** but is very much smaller and less conspicuous. At the posterior end of the peritoneal cavity may be noted the pear-shaped *urinary bladder*, generally distended with fluid. On raising the liver and looking dorsally and to the left of it will be found the *stomach* with the *spleen* attached to its left border. On the dorsal wall of the peritoneal cavity at about the level of the posterior ends of the liver lobes are the *kidneys*, round organs; to see them, gently lift the coils of the intestine. In female specimens, especially those which are pregnant, the horns of the *uterus* will be noted as a tube on each side in the posterior part of the peritoneal cavity.

The peritoneal cavity is lined by a membrane, the *peritoneum*. As in all coelomate animals, that portion of the membrane on the inside of the body wall is the *parietal* peritoneum. In both dorsal and ventral regions the peritoneum is deflected from the body wall and passes over the surface of the viscera, forming a covering layer, the *visceral* peritoneum or *serosa*, for all of the viscera. In passing to and from the body wall to the viscera the peritoneum forms double-walled membranes, the *mesenteries* or *ligaments*. The *dorsal* mesentery is present intact in mammals, and the *ventral* mesentery persists in the region of the liver and urinary bladder as in other vertebrates.

Examine the stomach first, by raising the liver and pressing it craniad. The exposure of the stomach is facilitated by slitting the diaphragm on the left side. The stomach is a large and rounded organ in the rabbit, smaller and more elongated in the cat. Find where the esophagus emerges from the diaphragm and enters the anterior surface of the stomach. The area of junction of the stomach and esophagus is called the *cardia;* and the region of the stomach adjacent to the junction, the *cardiac end* of the stomach. The shorter, slightly concave anterior surface of the stomach from the cardia to the pylorus is the *lesser curvature;* the larger, convex posterior surface, the *greater curvature.* The saclike bulge of the stomach to the left of the cardia is known as the *fundus;* the remainder of the stomach, the *body.* At the right the stomach passes into the small intestine; the point of junction, known as the *pylorus,* is marked by a constriction, beyond which the small intestine makes an abrupt bend. Along the left side of the stomach lies the *spleen,* a rather large organ in the cat, but smaller in the rabbit.

The relations of the stomach to the peritoneum are somewhat complicated. Raise the fundus and note the *mesogaster* extending from the dorsal wall to

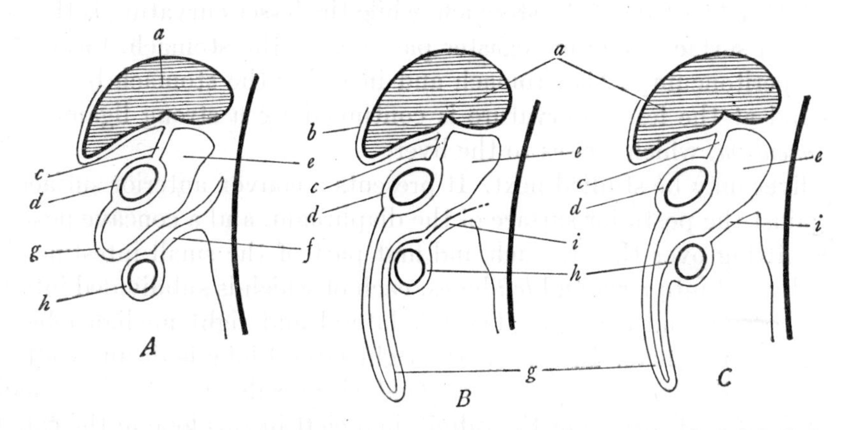

FIG. 91.—Diagrams to show the formation of the greater omentum in mammals and the fusion of the mesogaster and the mesocolon. *A*, early stage, in which the mesogaster is beginning to form a bag at *g*. *B*, the mesogaster is drawn posteriorly into a long bag (*g*), which is the greater omentum; the mesogaster and mesocolon are fusing at *i*. *C*, completion of the fusion of mesogaster and mesocolon at *i*. *a*, liver; *b*, serosa of the liver; *c*, lesser omentum or gastro-hepato-duodenal ligament; *d*, stomach; *e*, lesser peritoneal sac or cavity of the greater omentum; *f*, mesocolon; *g*, portion of the mesogaster which forms the greater omentum; *h*, intestine; *i*, fusion of the mesogaster and mesocolon. (From Arey's *Developmental Anatomy*, after Hertwig, courtesy of the W. B. Saunders Co.)

the stomach. Only a small portion of the mesogaster passes directly to the stomach; the greater part of it first descends posteriorly, forming a bag, the *greater omentum.* This is a very large and extensive sheet in the cat, covering the intestine ventrally as noted above. In the rabbit it is a short membrane dependent from the greater curvature of the stomach. The greater omentum is to be thought of as formed in the following way. Suppose one should grasp the mesogaster and pull it posteriorly, drawing it into a sac. Such a sac would have two walls, each double, i.e., composed of the two layers of the mesogaster; the sac would contain a cavity which is known as the *lesser peritoneal sac* and would open anteriorly (this opening will be seen later) (see Fig. 91). By manipulating the omentum determine that it consists of two separate walls. Having formed the greater omentum, the mesogaster returns to the stomach wall and passes onto the stomach along the greater curvature. The spleen is inclosed in the ventral wall of the great omentum just before the latter passes to the stomach. The portion of the great omentum between the spleen and the stomach is called the *gastrosplenic* (or *gastrolienal*) ligament. Posterior to the spleen, near the left kidney, a secondary fusion, the *gastrocolic* ligament, has formed between the mesogaster and the mesentery of the intestine (see Fig. 91).

The greater omentum owes its origin in part to the rotation of the stomach. The line of attachment of the omentum to the greater curvature is the original dorsal surface of the stomach, while the lesser curvature is the original ventral surface. The mesogaster passes over the stomach, forming the visceral peritoneum of the stomach and inclosing the stomach between its walls, and at the lesser curvature is continued by a strong ligament, the *lesser omentum*, which passes to the liver.

The liver may be studied next. It presents a convex anterior surface, fitting against the posterior surface of the diaphragm, and a concave posterior surface, fitting over the stomach and first part of the small intestine. The liver is divided into *right* and *left* lobes, each of which is subdivided into two lobes, a *median* and a *lateral*. The left lateral and right median lobes are larger than the others. In the cat the right lateral lobe is deeply cleft into two lobules. The large elongated *gall bladder* is imbedded in the right median lobe, on its dorsal surface in the **rabbit**, in a cleft in this lobe in the **cat**. On raising the liver and looking between the liver and the stomach another small lobe, the *caudate* lobe, will be seen. It is situated between the two layers of the lesser omentum. The *lesser omentum*, or *gastro-hepato-duodenal ligament*, is the ligament passing from the lesser curvature of the stomach to the posterior surface of the liver. It is divisible into two portions: the *gastrohepatic* ligament from the lesser curvature to the liver, and the *hepatoduodenal* ligament from the liver to the first part of the small intestine. That portion of the gastrohepatic ligament which contains the caudate lobe of the liver forms a sac which continues anteriorly the cavity of the greater omentum. In the hepatoduodenal ligament runs the *bile duct*, which should be traced from the gall bladder by gently dissecting in the ligament. Note the *cystic* duct from the gall bladder and the *hepatic* ducts from the lobes of the liver, that from the right lateral lobe being especially large. The cystic and hepatic ducts unite to form the *common bile duct*, which passes to the intestine in the hepatoduodenal ligament. It should be traced to the duodenum by cleaning away the connective tissue around it. Note to the right and dorsal to the bile duct, lying also in the hepatoduodenal ligament, the large *hepatic portal* vein. This must not be injured. Immediately dorsal to this vein, posterior to its branch into the right lateral lobe of the liver, the hepatoduodenal ligament has a free border which forms the ventral rim of an opening or slit of some size, the *foramen epiploicum* or entrance into the cavity of the omentum. It can be identified with certainty by making a slit into the cavity of the omentum and probing through the slit toward the right, toward the spot just described.

The lesser omentum extends to the middle of the posterior face of the liver, where it becomes the serosa of the liver; here its two walls part and, inclosing the liver between them, pass to the anterior face of the liver, where they again unite to form ligaments. The *falciform* ligament extends from between the two median lobes of the liver to the median ventral line; it is a thin sheet with a concave posterior border. Anteriorly and dorsally it is continuous with the *coronary* ligament, a stout ligament which attaches the liver to the central tendon of the diaphragm. The coronary ligament is circular in form, and its ring of attachment to the liver bounds a small space on the anterior face of the liver which is free from serosa.

Now trace the intestine from the pylorus. Its first portion, the *duodenum*, is bound to the liver by the hepatoduodenal ligament. The duodenum curves abruptly caudad. In the **rabbit** it is very long and forms a loop. The part of this loop which descends posteriorly is named the *descending limb;* the short turn at the most posterior part of the loop is the *transverse limb;* and the part which ascends anteriorly toward the stomach is called the *ascending limb.* The duodenum of the **cat** descends caudad for about two inches and then turns to the left. The duodenum is supported by a part of the dorsal mesentery, the *mesoduodenum*. It is also attached to the right kidney by a mesenterial fold, the *duodenorenal* ligament. Located in the mesoduodenum is the *pancreas.* It will be seen by spreading the mesentery. In the **rabbit** the pancreas consists of streaks of gland tissue scattered in the mesentery and situated chiefly along the courses of blood vessels. In the **cat** the pancreas is a definite, compact, pinkish gland which extends to the left into the dorsal wall of the greater omentum, dorsal to the greater curvature of the stomach. In the **rabbit** the *pancreatic duct* enters the duodenum about an inch or an inch and one-half anterior to the beginning of the ascending limb of the duodenum. This location of the pancreatic duct is unusual in mammals. In the **cat** there are two *pancreatic ducts.* The principal one joins the common bile duct at the point where the latter enters the duodenum. On picking away the substance of the pancreas at this point the duct is readily located. The common, slightly swollen chamber where bile and pancreatic ducts unite is known as the *ampulla of Vater*. The second or *accessory* pancreatic duct in the cat enters the duodenum about three-quarters of an inch caudad of the principal duct but is not easy to find.

From the duodenum trace the coils of the remainder of the small intestine.

It is supported by a part of the dorsal mesentery, the mesentery proper. The first portion of the small intestine beyond the duodenum is called the *jejunum*, and the remainder, the *ileum*, but there is no definite boundary between the two. Note the coils of the *mesentery* accompanying the intestine and the frequent fusions which occur (especially in the rabbit) between these coils. In the cat it will be necessary to withdraw the greater omentum from the coils of the intestine; the omentum may then be cut across near the spleen and discarded. Follow the small intestine to its enlargement into the *large* intestine. In doing this it may be necessary to tear slightly the fusions of the mesentery, but the structures should be kept as intact as practicable.

Rabbit: At the point of juncture of the large and small intestine there is an enlargement, the *sacculus rotundus*. From the sacculus rotundus extends an enormous blind sac, about a foot and a half in length. The first foot of this is very large and is known as the *caecum;* the last five or six inches is reduced in diameter and constitutes the *vermiform appendix*. Both caecum and appendix are very much longer in the rabbit than in most mammals, owing probably to the habit of the animal of ingesting large quantities of vegetable food. The wall of the caecum is marked by a spiral line which denotes the position of an internal spiral fold of the lining membrane of the caecum. From the sacculus rotundus trace the large intestine after it has given off the caecum. The large intestine beyond the caecum is named the *colon*. The colon is supported by a part of the dorsal mesentery called the *mesocolon*. The first part of the colon, the *ascending* colon, is rather long and pursues a winding course. At first its wall bears three longitudinal muscular bands, the *bands* or *taeniae* of the colon. Between the taeniae the wall of the colon is greatly puckered, forming little sacculations, the *haustra*. Beyond the ascending colon, the colon runs for a short distance transversely across the peritoneal cavity from right to left and is then named the *transverse* colon. At the left it turns abruptly posteriorly as the *descending* colon. At this turn the mesocolon is fused to the mesogaster. The descending colon passes straight posteriorly and disappears dorsal to the urogenital organs.

Cat: The junction of small and large intestine is marked by a slight projection, the *caecum*, a vermiform appendix being practically absent. The *large intestine* or *colon* passes forward as the *ascending* colon; then turns and extends across the peritoneal cavity from right to left as the *transverse* colon; and turns abruptly at the left and proceeds straight posteriorly as the *descending* colon. The mesentery of the colon is named the *mesocolon*. At the left where the transverse colon turns caudad, the mesocolon is fused secondarily to the mesogaster.

The *urinary bladder* is a sac occupying the posterior end of the peritoneal cavity, immediately internal to the body wall and ventral to the large intestine. From the ventral surface of the bladder a mesentery, the *median ligament* of the bladder (*median umbilical fold* of human anatomy) extends to the median ventral line and here continues forward for some distance. Near the exit of the bladder from the peritoneal cavity there is on each side a slightly developed ligament, the *lateral ligament* of the bladder.

The terminal portion of the descending colon is the *rectum*. Both the rectum and the duct of the bladder pass to the exterior through the ring formed by the pelvic girdle and vertebral column. They will be followed at a later time in connection with the urogenital system. For the present it may be stated that the rectum and urogenital ducts are completely separated from each other. A cloaca is therefore absent.

The small bodies which may have been noted in the mesentery, usually buried in fat, are *lymph glands*, parts of the lymphatic system. Small portions of lymphatic tissue called *lymph nodules* are also abundantly present in the wall of the intestine. Aggregations of lymph nodules are known as *Peyer's patches*. Peyer's patches are thickened oval spots on the surface of the intestine, of slightly different color from the rest of the intestinal wall. Look for them. They occur in the rabbit along the entire small intestine on the side opposite that attached to the mesentery; there is a larger patch at the place of junction of sacculus rotundus and caecum. The walls of the sacculus rotundus, caecum, and vermiform appendix are composed almost entirely of lymph nodules. In the cat, Peyer's patches occur as oval light-colored spots along the colon but are best seen from the inside.

Slit open various parts of the digestive tract along the side opposite the attachment of the mesentery. Wash out the interior. In the cat note the marked ridges or *rugae* in the wall of the stomach; these are very slight in the rabbit. Cut through the pylorus and note the thickened ridge or *pyloric valve* at this place. In the wall of the small intestine observe the velvety appearance due to the *villi* (finger-like projections of the mucous membrane). Find also the depressions marking the positions of the lymph nodules and Peyer's patches. In the rabbit slit open the sacculus rotundus, caecum, and appendix and note the spotted appearance of the interior, owing to the lymph nodules composing the walls. In the interior of the caecum is a spiral ridge. Cut through the junction of large and small intestine and note in both animals an elevation, the *ileocolic valve*, projecting into the ileum.

5. Variations of the ungulate stomach.—The stomach of mammals presents many variations in form and subdivisions, of which the most pronounced are seen among ungulates, especially the cud-chewing groups. In the pig there is a small, and in the horse and tapir a larger, area at the cardiac end of the apparently simple sacciform stomach, which is lined by stratified epithelium and hence is actually a portion of the esophagus (Fig. 92*A*). In the hippopotamus this region or forestomach is definitely separated from the true stomach and has a pair of large diverticula (Fig. 92*B*). In the camels the forestomach has become divided into a large pouch, the *rumen* or *paunch*, and a smaller one, the *reticulum* or *honeycomb;* in both a part of the wall is partitioned into cavities, the water cells, where water may be stored (Fig. 92*C*). Finally, in the typical ruminants, such as the cattle and sheep, the forestomach has three compartments, to be regarded as expansions of the lower end of the esophagus, namely, the large *rumen* or *paunch*, the *honeycomb* or *reticulum*, and the *manyplies* or *omasum* (Fig. 92*D*). The food is swallowed into the rumen; and, when this is filled, the animal ceases

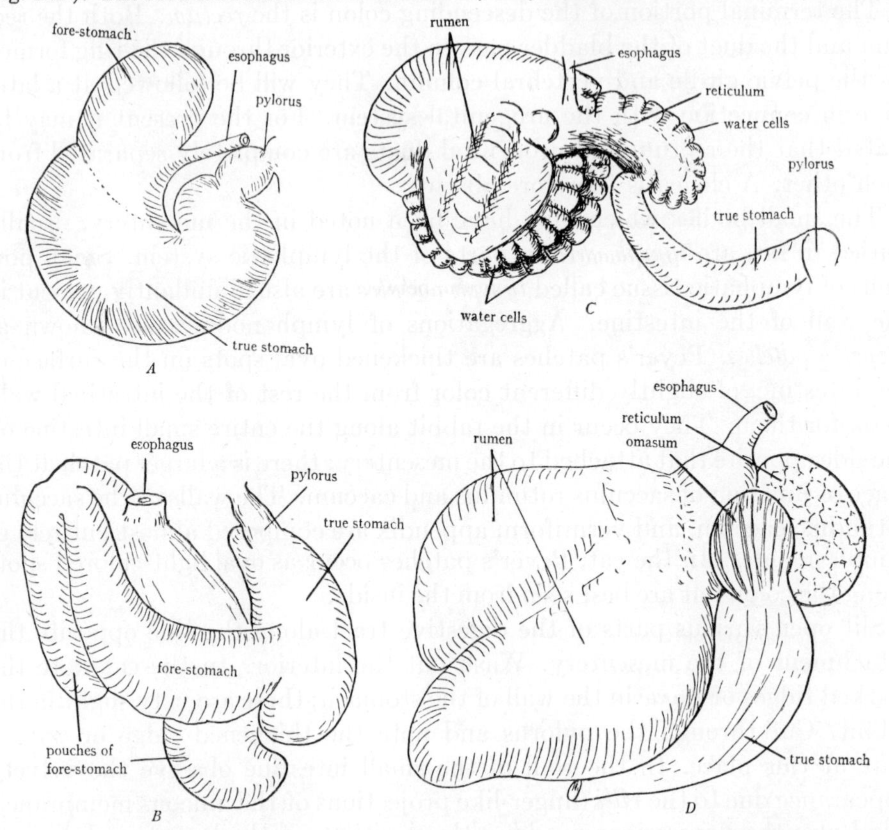

FIG. 92.—Compartments of the stomach in *A*, horse; *B*, hippopotamus; *C*, camel; and *D*, cow. In *A* the dotted line shows limit of the stratified epithelium, i.e., the part of the stomach belonging to the esophagus.

feeding and "chews its cud." The food passes into the reticulum, whose walls are subdivided into cavities in which the food is made up into small masses or cuds, and these are regurgitated into the oral cavity, to be chewed at leisure. When then swallowed, the food passes into the omasum, the walls of which bear numerous folds, like the pages of a book. This communicates directly with the true stomach or *abomasum*, which bears the gastric glands.

I. THE COMPARATIVE ANATOMY OF THE COELOM AND THE MESENTERIES

Since the development and comparative anatomy of the coelom and mesenteries are extremely difficult and complicated subjects, it seems advisable, at the risk of repetition, to make some concluding general remarks concerning them. Only a simplified account can be given here.

The coelom of fishes and urodeles is divided into two compartments, a small anterior pericardial cavity and a much larger posterior pleuroperitoneal cavity. The former is situated entirely anterior to the latter and far anterior in the body, at the level of the pharynx. The partition between the two coelomic cavities in these groups is the transverse septum. This septum is formed, as previously explained, for the purpose of conveying the great veins from the dorsal body wall into the posterior end of the heart. It arises first as a column of mesenchyme on each side, conveying a vein into the heart; later these columns enlarge and fuse to form a partition. The transverse septum is a rather thick membrane with anterior and posterior walls, which inclose between them the great veins and the posterior end of the heart. The septum in fishes and urodeles lies, as the name implies, in a plane transverse to the body axis (Fig. 84*A*, *B*, p. 254).

The relations of the transverse septum to the liver are somewhat complicated and require explanation. The liver is a diverticulum from the small intestine. It happens that at the point where the liver diverticulum grows out ventrally from the intestine, the transverse septum is situated. Consequently, the liver is compelled to grow out into the septum. It lies at first within the mesenchyme of the septum, between the two walls of the septum. Because of the fact, as already explained, that the great veins enter the heart by way of the septum, the liver also acquires important relations to these veins, as will be discussed in the section on the circulatory system. The liver rapidly increases in size, so that it can no longer be contained within the limits of the septum. It consequently bulges posteriorly, carrying with it the posterior wall of the septum, which thus becomes the peritoneal covering or serosa of the liver. Later, the region where the liver bulges from the septum narrows down on dorsal and lateral sides, leaving anterior and ventral connections between the liver and the septum (Fig. 89, p. 267). The anterior connection is named the coronary ligament and is, as the name indicates, a circular ligament, by means of which the liver is permanently in all vertebrates suspended from the transverse septum or its derivative. The ventral partition formed by the constriction of the liver from the septum is the falciform ligament of the liver, which is also, by virtue of its location, a part of the ventral mesentery (Fig. 83*B*, p. 253). We thus see that the transverse septum has the following parts: the posterior wall of the pericardial cavity, the anterior wall of the pleuroperitoneal cavity, the falciform and coronary ligaments of the liver, the serosa of the liver, and the mesoderm tissue of the liver.

In Anura and some reptiles the coelom is divided, in the same way as in fishes and urodeles, into two compartments—the pericardial and pleuroperitoneal cavities. In these groups, however, the pericardial cavity is no longer situated anterior to the pleuroperitoneal cavity but has descended posteriorly until it lies ventral to the anterior part of the latter cavity. This descent occurs during embryonic development. In this descent the pericardial cavity necessarily carries with it the transverse septum, since this septum formed the posterior wall of the cavity. Furthermore, since the parietal pericardium or wall of the pericardial cavity in the lower vertebrates lines the inside of the body wall, it is necessary that this membrane become split, at least in part, from the body wall before the descent can take place. This happens mainly by the invasion of the pleuroperitoneal cavity forward. As a result of these processes, the parietal pericardium is separated from the body wall, at least in part, and forms a delicate sac about the heart, the pericardial sac. The separation of the pericardial sac from the body wall is complete in the Anura but in the turtle is incomplete ventrally, so that the ventral wall of the pericardial cavity is still in contact ventrally with the body wall (see Fig. 84*C*).

In the groups under consideration the relations of the transverse septum remain the same as before; but its position is now oblique, indeed almost frontal, while formerly it was transverse (Fig. 84*C*). It still presents anterior and posterior walls (which are now nearly ventral and dorsal in position) and still incloses the great veins and the posterior end of the heart between its walls. Its anterior (ventral) face forms part of the pericardial sac, while its posterior (dorsal) face forms part of the lining of the pleuroperitoneal cavity, as previously. To this latter portion the liver is attached, as before, by the coronary and falciform ligaments.

The posterior descent of the pericardial cavity brings a portion of the pleuroperitoneal cavity dorsal to the pericardial cavity. Into this portion of the pleuroperitoneal cavity which lies above the heart the lungs grow out, and it later becomes the pleural cavities.

The separation of the pleural cavities containing the lungs from the peritoneal cavity is begun in reptiles and completed in crocodilians, birds, and mammals. The process is very complicated, however, involving a number of peritoneal folds and further does not appear to correspond in the different groups. In these forms there are then four compartments to the coelom: the two pleural cavities, the pericardial cavity, and the peritoneal cavity.

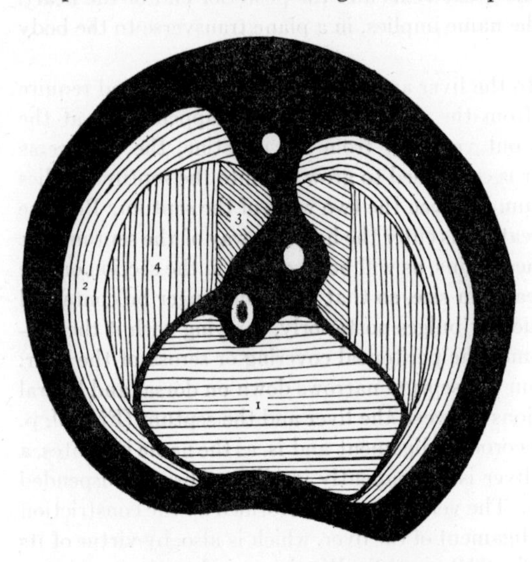

FIG. 93.—Diagram of the mammalian diaphragm showing various coelomic folds participating in its formation (after Broman, 1911). *1*, septum transversum; *2*, lateral folds from parietal peritoneum; *3*, folds from dorsal mesentery and pulmonary ligament; *4*, pleuroperitoneal membranes. The three circles are from above down the dorsal aorta, the esophagus, and the postcaval vein.

In mammals the diaphragm extends across the body in a transverse plane, cutting off the pericardial and two pleural cavities in front from the peritoneal cavity behind. The pericardial sac of mammals is free on all sides, because of the extension of the two pleural cavities ventrally (Fig. 90). The two pleural cavities, at first dorsal in position, grow ventrally in mammals, split the pericardium from the ventral body wall, and push in between the pericardial sac and ventral wall so that their medial walls meet below the pericardial sac to form the mediastinal septum.

The formation of the mammalian diaphragm is a complicated process. Its median ventral portion consists of the transverse septum; and to this there become fused several mesenterial folds, completing the partition. The principal of these folds is a pair of pleuroperitoneal membranes which descend from the dorsal body and fuse with the transverse septum (Fig. 93). Other contributions come from the dorsal mesentery and the pulmonary ligaments supporting the lungs, while the sides of the diaphragm appear to come from inward extensions of the parietal peritoneum of that region (Fig. 93). The periphery of the diaphragm becomes muscularized by invasion of muscle buds from the cervical myotomes.

In mammals, as in birds, the great veins have been gradually drawn out of the transverse septum so as to enter the anterior end of the heart, which is thus free except at its anterior end.

The coelomic linings—pericardium, pleuroperitoneum, or pleura and peritoneum—present the same relations in all of the vertebrates. Each has a parietal portion lining the body wall, a visceral portion covering the viscera, and mesenteries or ligaments connecting the parietal and visceral portions.

The relations of the mesenteries are very similar in all of the vertebrates. The dorsal mesentery extends, usually unbroken, from the dorsal median line to the digestive tract.

Each portion of it bears a special name, according to the part of the digestive tract which it supports. In mammals the mesentery of the stomach is drawn out posteriorly into a bag, the greater omentum, which apparently serves to protect the abdominal viscera (Fig. 91, p. 295). The ventral mesentery extends from the digestive tract to the ventral body wall but is reduced to remnants in the adults of all vertebrates. These remnants are found in the region of the liver and the bladder, where they form the gastrohepatic and hepatoduodenal ligaments, the falciform ligament, and the median ligament of the bladder. All viscera of vertebrates either are retroperitoneal, i.e., situated external to the coelomic lining, or they are situated between the two layers of a mesentery. The liver, the heart, and the urinary bladder are situated in the ventral mesentery; the pancreas is generally in the dorsal mesentery but may be in the ventral mesentery as well; the digestive tube may be regarded as situated in the dorsal mesentery.

J. SUMMARY

1. The digestive tract shows much variation but little directed evolution among vertebrates; in general, the intestine is longer and more coiled in the higher forms. Adaptive changes correlated with feeding habits and mode of life are common in the digestive tract.

2. The oral cavity is provided with multicellular glands in land vertebrates, and some of these are enlarged in mammals to form the salivary glands.

3. The nasal cavities do not communicate with the oral cavity in most fishes. In Dipnoi, Crossopterygii, and land vertebrates they open into the mouth for respiratory purposes. At first such posterior nares open into the anterior end of the oral cavity; but, beginning with reptiles, they shift posteriorly through the development of a hard palate. The hard palate is completed in crocodilians and mammals. In birds and mammals the nasal passages are extended still further posteriorly by the development of the soft palate; the air then passes directly into the pharyngeal cavity without first passing through the oral cavity.

4. The primary tongue is an immobile flat elevation in the floor of the oral cavity; beginning with amphibians, there is added to this a gland field developed anterior to it, and finally among amniotes a pair of lateral swellings fuse with the preceding to form the definitive tongue.

5. In fishes the wall of the pharynx is pierced by gill slits, typically six in number, situated between the skeletal gill arches. The passage from the internal to the external slit is termed the gill pouch; and the tissue between gill pouches, the branchial bar. These bars bear the internal gills, which may be lamellar or filamentous. External gills may occur on the outer surface of the bars. In bony fishes the branchial bars become much reduced, and the gill slits are covered over by the operculum.

6. Gill slits and external or internal gills occur in amphibian larvae and some adult urodeles. In amniotes gill slits appear in the embryo, but there is no trace of gills and the slits close over during development.

7. The tympanic cavity and auditory tube connecting it with the pharynx develop as an outgrowth from the first gill pouch. The external auditory meatus is a depression in the position of the first gill slit.

8. The entodermal lining of the gill pouches persists as certain glands or glandlike bodies, such as the tonsils, thymus, parathyroids, and epithelial bodies. The thyroid gland originates as an evagination of the pharyngeal floor between the second gill arches.

9. The lungs arise as an evagination from the pharyngeal floor behind the last gill slits. The evagination bifurcates into the two lungs, and in higher vertebrates the stalk of the outgrowth lengthens as the trachea. A similar development is seen in the swim bladder or "lungs" of some fishes, and there can be little doubt of the homology of swim bladder and lungs.

10. The beginning of the trachea is expanded into a chamber, the larynx, the walls of which are supported by the modified gill arches. The larynx functions for the production of the voice, except in birds, where the voice comes from the lower end of the trachea (syrinx).

11. Esophagus and trachea become elongated in amniotes because of a general posterior descent of the viscera.

12. The lungs are at first simple, smooth-walled sacs; but their walls become more and more subdivided into smaller and smaller air spaces as one ascends the vertebrate classes. In birds the lungs connect with large air sacs in the viscera and with air spaces in the bones.

13. The esophagus may present one or more enlargements for food storage.

14. Stomach and esophagus are not distinctly separated in lower vertebrates. In birds the stomach is subdivided into proventriculus and gizzard. Complicated stomachs, in which the true stomach is combined with esophageal expansions, occur in mammals, particularly ruminants.

15. The vertebrate intestine is provided with two glands—the liver and the pancreas—attached by ducts to the duodenum.

16. The division of the intestine into small and large parts is not very evident in fishes. A spiral valve in the small intestine is characteristic of some groups of fishes, especially elasmobranchs. One or two caeca at the beginning of the large intestine is characteristic of amniotes.

17. When the terminal part of the large intestine receives the urogenital ducts, it is termed the cloaca. A cloaca is generally present in vertebrates. When the urogenital ducts have their own openings, the terminal part of the intestine is the rectum; a rectum is present in cyclostomes, holocephalans, teleostomes, and placental mammals.

18. In most vertebrates there is a sacciform outgrowth, the urinary bladder, from the ventral wall of the cloaca or, in the absence of a cloaca, from the urogenital canal.

19. In all vertebrates except birds and mammals the coelom is divided into two compartments: the pericardial cavity, containing only the heart, and the pleuroperitoneal cavity, containing the other viscera. The two cavities are separated by a partition, the transverse septum.

20. In cyclostomes, fishes, and urodeles the pericardial cavity is anterior to the pleuroperitoneal cavity. From Anura on, the pericardial cavity is ventral to the anterior part of the pleuroperitoneal cavity. This change is due to a posterior descent of the heart and pericardial cavity, carrying the transverse septum with them. As a consequence of the descent the wall of the pericardial cavity, together with the transverse septum, forms a sac, the pericardial sac, around the heart. That portion of the pleuroperitoneal cavity dorsal to the heart is destined to form the pleural cavities.

21. In birds and mammals (and some reptiles) the pleuroperitoneal cavity is divided into anterior and posterior compartments by the formation of folds which join the transverse septum. The partition so produced is called the oblique septum in birds and the diaphragm in mammals. Anterior to the partition are the two pleural cavities, each containing a lung; posterior to the partition is the peritoneal cavity, containing the digestive and urogenital systems. The diaphragm differs from the oblique septum in that it contains a large amount of striated muscle derived from the cervical myotomes. The coelom in birds and mammals is thus divided into four compartments: the pericardial, the two pleural, and the peritoneal cavities.

22. A dorsal mesentery supports the digestive tract in all vertebrates and remains practically complete throughout. The ventral mesentery of the digestive tract is absent in the adult except in the regions of the liver and the bladder. There are special mesenteries for the gonads and their ducts. In mammals the mesentery of the stomach develops a special posterior prolongation called the greater omentum.

REFERENCES

Broman, I. 1911. Entwicklung der membranösen Pericardiums der Zwerchfells. Ergebn. d. Anat. u. Entw'gesch., **20**.

Göppert, E. 1901. Beiträge zur vergleichenden Anatomie des Kehlkopfs und seiner Umgebung mit besonderer Berücksichtigung der Monotremen. Semon's Zool. Forschungsreisen in Australien, **3**, Part 2.

Graham, J. D. P. 1939. The air stream in the lung of the fowl. Jour. Physiol., **97**.

Johnson, C. E. 1922. Branchial derivatives in the turtle. Jour. Morph., **36**.

Kallius, E. 1901–10. Beiträge zur Entwicklung der Zunge. Anat. Hefte, **16**, **28**, **31**, **41**.

Okita, Y. K. 1936. Studies on the physiology of the swim bladder. Jour. Fac. Sci. Tokyo Imper. Univ., Sec. IV (Zoölogy), **4**.

Pixell, Helen. 1908. On the morphology and physiology of the appendix digitiformis in elasmobranchs. Anat. Anz., **32**.

Safford, V. 1940. Asphyxiation of marine fish with and without CO_2 and its effect on the gas content of the swim bladder. Jour. Cell. Comp. Physiol., **16**.

Sewertzoff, A. N. 1927. Structure primitive de l'appareil visceral des Elasmobranches. Pubbl. Staz. zool. Napoli, **8**.

XII. THE COMPARATIVE ANATOMY OF THE CIRCULATORY SYSTEM

A. GENERAL CONSIDERATIONS

1. The parts of the circulatory system.—The circulatory system of vertebrates comprises two systems of branching tubes inclosing circulating fluids, the *blood-vascular* system and the *lymphatic* system. The former is the larger and more conspicuous of the two and is the one referred to when the expression circulatory system is employed without qualification.

The blood-vascular system is a *closed* system; i.e., it consists of a set of branching tubes, the *blood vessels,* which are continuous with one another, unconnected with other systems (except the lymphatic system), and in which the inclosed fluid travels in a circuit. The parts of the blood-vascular system are the *heart,* the *arteries,* the *veins,* and the *capillaries.* The heart is a contractile muscular organ situated in the median ventral region in the anterior part of the body. It has essentially the form of an S-shaped tube subdivided into chambers. The arteries are the vessels which *leave* the heart and in which the contained fluid flows *away from* the heart. The veins are vessels in which the contained fluid flows *toward* the heart. The student should particularly note that arteries and veins *cannot* be defined on the basis of the kind of blood which they contain. The veins of the vertebrate body fall into three classes: the *systemic* veins, which flow directly into the right side of the heart; the *pulmonary* veins, which flow from the lungs into the left side of the heart; and the *portal* veins or portal systems, in which the blood does not return directly to the heart but passes into a system of capillaries from which it is re-collected into systemic veins. The capillaries are the minute microscopic vessels which connect the ends of the arteries with the beginnings of the veins and through which the circulation is completed. All the tissues of the body are permeated with networks of capillaries through the walls of which the gaseous and other exchange between the blood and the body cells takes place.

The blood-vascular system incloses a fluid, the *blood,* which in vertebrates is colored red. The blood consists of a colorless fluid, the *plasma,* in which float microscopic cells, the *corpuscles.* The latter are of two general kinds, red and white; the former give the red color to the blood. Study of the blood lies outside of the limits of this course.

The lymphatic system is an *open* system; that is, it consists not only of branching tubes, the *lymph* vessels, but of large spaces, the *lymph sinuses;* and it is further in communication with the coelomic spaces of the body. Lymph sinuses occur beneath the skin (the student may recall the large subcutaneous lymph sinuses in the frog), between the muscles, in the mesenteries, in the walls of the digestive tract, around the central nervous system, etc. From these sinuses the fluid passes into more or less definite lymph vessels, which eventually open into the veins of the blood-vascular system. In the lower vertebrates contractile *lymph hearts* are placed in the course of the lymphatic vessels, to aid the flow, but these are absent in mammals. The lymphatic system further differs from the blood-vascular system in that nodules of tissue, the *lymph glands,* are placed in the path of the lymph vessels. The lymph glands consist of a network of connective tissue in which are imbedded masses of cells, known as lymphocytes. Lymphocytes are a variety of white blood corpuscles. The function of the lymph glands appears to be to destroy foreign particles, bacteria, etc., and to add white blood corpuscles to the circulation. The tonsils, the thymus, and the spleen belong to the category of lymph glands.

The lymphatic system contains a colorless fluid, the *lymph*, whose composition is similar to the plasma of the blood. The lymph contains white blood corpuscles, but red ones are absent. The lymph fills all of the spaces of the body and bathes all of the cells, and all exchange between the tissues and the blood must occur by way of the lymph.

Because of the delicate nature of the lymph vessels and the general diffuse character of the lymphatic system, very little of this system can be made out in an ordinary dissection. Our study of the circulatory system will therefore be confined almost entirely to the blood-vascular system.

2. The origin of the blood and of the blood vessels.—The blood vessels and the blood arise from the mesoderm, from mesenchyme cells. In the mesenchyme little patches of cells form; the central cells of these patches become modified into blood corpuscles; the peripheral cells arrange themselves to form tubes, the blood vessels. The blood vessels which arise in the somatic mesoderm are *somatic* vessels; those in the splanchnic mesoderm are *visceral* or *splanchnic* vessels.

3. The origin of the heart.—The heart arises in the ventral mesentery in the anterior part of the embryo. In cases where the embryo is closed below from the beginning, a tubular cavity appears in the ventral mesentery, and in the walls of this cavity the heart tissue differentiates. In the majority of vertebrates, with meroblastic development, the embryo is at first open below on the yolk sac (see Fig. 15*C*, *D*, p. 75). In the splanchnic mesoderm of the hypomere there arises on each side a tubular cavity, of the same nature as the blood vessels. As the hypomere closes below in the median ventral line, the two cavities are brought together and fuse to form the heart (Fig. 94). By either method of formation the heart necessarily lies in the ventral mesentery of the gut and is therefore provided with dorsal and ventral mesenteries, known as the *dorsal* and *ventral mesocardia* (see Fig. 83*A*). These, however, very rapidly disappear; the ventral mesocardium vanishes as soon as formed. The heart then swings free in the coelom, attached only at its two ends. The portion of the coelom in which the heart lies is at first continuous with the general cavity of the hypomeres but is soon cut off from the rest of the coelom by the formation of the transverse septum and is thereupon named the pericardial cavity.

4. The chief embryonic blood vessels.—The earliest vessels to form in the vertebrate embryo are the *vitelline* veins. These veins develop on the surface of the yolk sac (or in the absence of a yolk sac, along the intestine) in the splanchnic mesoderm of the hypomere. They pass to the embryo in the mesentery of the gut and enter the posterior end of the heart. At the anterior end of the heart a blood vessel, the *ventral aorta*, arises and connects with the heart. The ventral aorta extends forward to the anterior end of the pharynx. Here it divides in two, and the two branches turn dorsally, one on either side of the pharynx. This pair of branches encircling the pharynx is the first pair of *aortic arches*. Each vessel lies in the center of the first or mandibular visceral arch. On reaching the dorsal side of the pharynx the two vessels turn posteriorly and, as the *dorsal aortae*, proceed backward, situated in the median dorsal region of the body wall. Each dorsal aorta on reaching the region of the yolk sac sends out a *vitelline* artery over the surface of the yolk sac. The early embryonic circulation is thus completed (Fig. 95*A*).

There subsequently develop around the pharynx additional aortic arches connecting the ventral and dorsal aortae. In typical vertebrates six such pairs of aortic arches appear, one to each pair of visceral arches.[1] The aortic arch runs in the center of the visceral arch in front of the corresponding gill slit. The aortic arches develop in an anteroposterior sequence, from

[1] Visceral arch is the embryological term for the structure called in this book branchial bar in the adult.

in front backward. Behind the pharyngeal region the two dorsal aortae soon fuse to form a single *dorsal aorta*, which continues into the tail as the *caudal* artery; and from the aorta, as development proceeds, both segmentally arranged paired and nonsegmental unpaired branches arise to supply various body parts.

In cyclostomes and fishes the originally continuous aortic arches through the branchial bars become interrupted during development by the interpolation of a set of smaller vessels and capillaries permeating the gills. This happens in various ways in different fishes: the

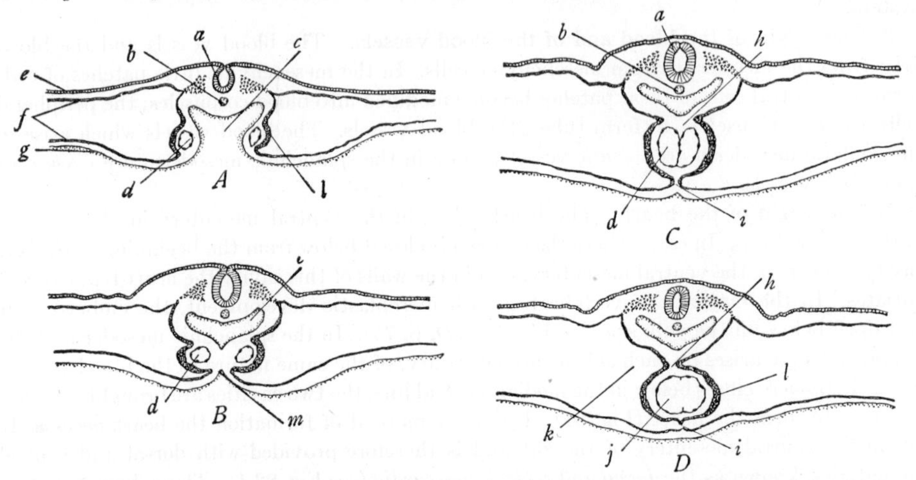

FIG. 94.—Drawings of cross-sections through four successive stages of development of the chick embryo, between twenty-five and twenty-nine hours of incubation, to show the formation of the heart. *A*, early stage, showing the open intestine at *c*, the vitelline veins (*d*) in the splanchnic mesoderm (*f*), and the thickening (*l*) in the splanchnic mesoderm covering the veins. *B*, next stage, in which the splanchnopleure has fused together at *m*, closing the intestine (*c*) and bringing the two vitelline veins (*d*) closer together. *C*, later, showing the two vitelline veins (*d*) in contact. *D*, completion of the heart by the fusion of the vitelline veins; the thickened mesoderm (*l*) becomes the muscular wall of the heart. *a*, neural tube; *b*, notochord; *c*, anterior part of the digestive tract; *d*, vitelline vein; *e*, ectoderm; *f*, somatic and splanchnic layers of the mesoderm; *g*, entoderm; *h*, dorsal mesentery of the heart or dorsal mesocardium; *i*, ventral mesentery of the heart, or ventral mesocardium (it is disappearing in *D*); *j*, lining of the heart formed by the union of the vitelline veins; *k*, pericardial cavity; *l*, muscle layer of the heart formed by the thickening of the splanchnic mesoderm; *m*, point of fusion of the splanchnopleure of the two sides. (After Patten's *Early Embryology of the Chick*, copyright by P. Blakiston's Son & Co.)

interruption in the aortic arch may be near the dorsal or near the ventral aorta, and either the ventral or the dorsal stump or both may become prolonged by a newly formed vessel, extending along the branchial bar. The vessels which lead from the ventral aorta into the branchial bars are called the *afferent branchial vessels*, and those from the branchial bars into the dorsal aorta are the *efferent branchial* or *epibranchial* vessels. As just noted, these branchial vessels consist in part of portions of the original aortic arches, in part of new formations. There is but one afferent and one efferent vessel in each bar in teleosts, but in elasmobranchs and dipnoans there are two efferents to each bar (see later, p. 324, also Fig. 100).

The early circulatory system so far described (Fig. 95*B*) does not arrange for the return of venous blood to the heart from the body of the embryo. For this purpose two pairs of veins arise in the somatic mesoderm of the body wall. These are a pair of *anterior cardinal* veins returning blood from the anterior part of the body and a pair of *posterior cardinal* veins returning blood from the posterior part of the body. In order that these veins, situated in the dorsal body wall, may reach the heart, a bridge of mesoderm is formed on each side, extending between the somatic and splanchnic mesoderm. As already learned, this pair of bridges is the beginning of the transverse septum, and by the union and extension of the bridges the septum is completed. At the level of the rear end of the heart the anterior and posterior cardinal veins of each side unite to a common vessel, the *duct of Cuvier* or *common cardinal* vein. The common cardinal vein then passes to the heart from each side by way of the transverse septum between the two walls of which this vein is inclosed. In addition to the cardinal veins there ap-

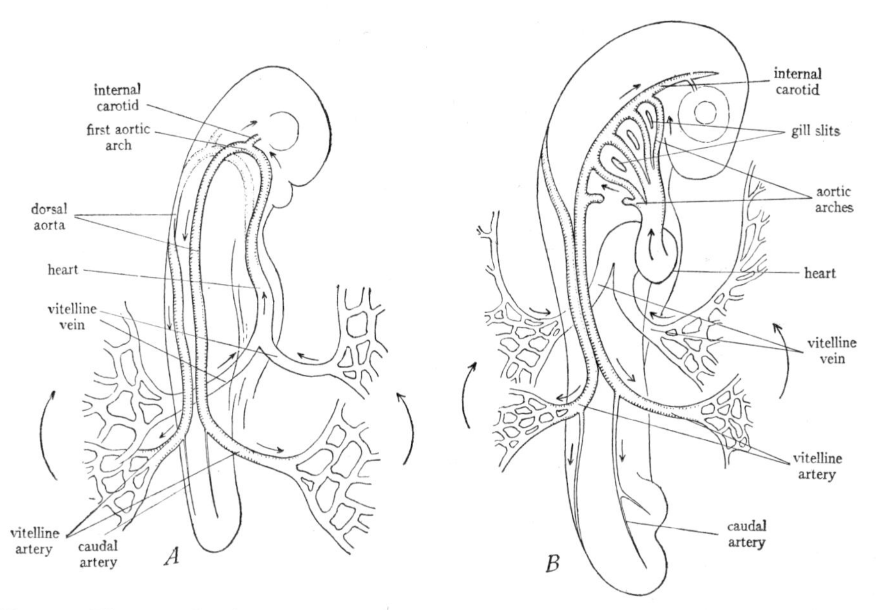

FIG. 95.—Diagrams of early vertebrate embryos to show the development of the main blood vessels. *A*, earliest stage, in which the circulatory system consists of the vitelline veins, heart, ventral aorta, first aortic arch, dorsal aortae, and vitelline arteries. *B*, later stage, showing the development of successive aortic arches following the first one. The method of formation of the aortic arches by buds from the ventral and dorsal aortae is illustrated by the last aortic arch in *B*. (From Wilder's *History of the Human Body*, courtesy of Henry Holt & Co.)

pears in vertebrate embryos a pair of veins in the lateral or ventral abdominal walls, known as the *lateral* or *ventral abdominal* veins, which enter the heart along with the common cardinal veins. They are also called the *umbilical* or *allantoic* veins in the embryos of amniotes. The main veins of the embryo at this time are illustrated in Figure 96.

The two vitelline veins are soon extended posteriorly in the embryo by means of a tributary, the *subintestinal* vein, which courses in the mesentery of the gut and constitutes the chief vein of the digestive tract. It is shown in Figure 96. It continues into the tail as the *caudal* vein, making a loop around the anus. In embryos without a yolk sac the subintestinal and vitelline veins appear as one continuous vein, called by the former name.

The arteries and veins thus far described are the chief longitudinal trunks of the embryo. From them, branches, usually segmentally arranged, extend to various parts of the body. The branches of the main longitudinal vessels are classified into three kinds: the *median visceral* or *splanchnic* branches, unpaired branches to and from the digestive tract; the *lateral vis-*

ceral branches, paired branches to and from the urogenital organs; and the *parietal* or *somatic* branches, paired vessels to and from the body wall. This arrangement is most obvious in the arteries (Fig. 97). It is more or less persistent in the adult, chiefly in the posterior part of the body. The vessels to the paired appendages are simply enlarged somatic branches.

5. The origin of the portal systems.—It has already been explained that a portal system is a portion of the venous system, the constituent veins of which, instead of entering the heart, pass into a network of capillaries, from which the blood is re-collected by a systemic vein. In other words, in a portal system a network of capillaries is interposed in the path of a vein or veins. There are two portal systems: the *hepatic portal* system, in which the interposed capillaries are located in the liver, and the *renal portal* system, in which they are in the kidney. The origin of each of these systems may be given briefly.

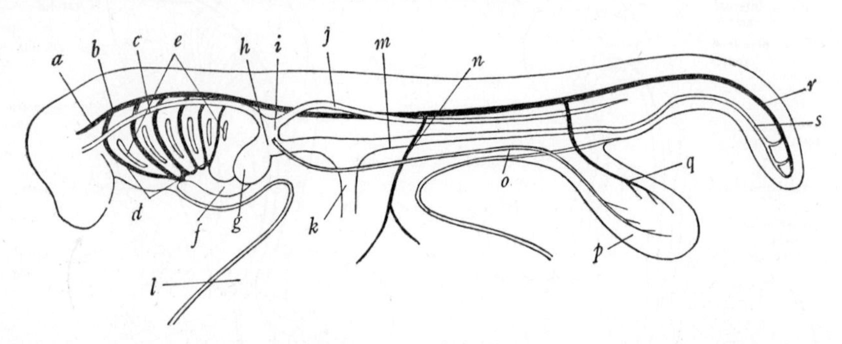

Fig. 96.—Later stage of the development of the circulatory system, showing the chief veins. Veins blank; arteries black. *a*, internal carotid; *b*, dorsal aorta; *c*, anterior cardinal vein; *d*, the six aortic arches; *e*, the six gill slits; *f*, the conus arteriosus of the heart continuing into the ventral aorta; *g*, main part of the heart; *h*, sinus venosus of the heart; *i*, duct of Cuvier or common cardinal vein; *j*, posterior cardinal vein; *k*, vitelline vein; *l*, yolk sac; *m*, subintestinal vein; *n*, vitelline artery; *o*, lateral abdominal or umbilical vein; *p*, allantois (urinary bladder); *q*, allantoic or umbilical artery; *r*, caudal artery; *s*, caudal vein. (Slightly modified from Vialleton's *Eléments de morphologie des Vertébrés.*)

a) The hepatic portal system: This arises as follows: When the transverse septum develops at the posterior end of the heart, the two vitelline veins must naturally pierce the septum on their way to the heart. At this region the liver buds out from the small intestine and extends into the transverse septum. As it grows, the liver gradually uses up the substance of the transverse septum and fills the available space in the septum. The liver substance thus comes to surround the proximal ends of the two vitelline veins (Fig. 98*A*, *B*). At first the vitelline veins pass through the liver into the heart, but soon they begin to break up into smaller and smaller vessels (Fig. 98*C*) in the liver, until the liver is occupied by a network of capillaries (of the kind known as sinusoids) which permeate the liver substance. Thus, the circulation in the vitelline veins passes from the yolk sac and digestive tract to the liver, passes through a capillary network in the liver and from the liver to the heart in the remaining proximal portions of the vitelline veins (Fig. 98*D*). The latter are now called the *hepatic* veins. Posterior to the liver by ringlike unions between the two vitelline veins (Fig. 98*C*, *D*), a single vessel, the *hepatic portal* vein, is formed; and caudad of these unions the right vein disappears, leaving the left vein with its tributary, the subintestinal vein, now named the *mesenteric* vein, to form the chief vein of the digestive tract. This arrangement insures that all of the venous blood from the digestive tract caudad of the cardia must pass through a capillary network in the liver before it can reach the heart.

b) *The renal portal system:* This develops as follows: At first the caudal vein opens into the subintestinal vein, forming a loop around the anus as in Figure 98*C*, *D*. Later the posterior cardinal veins grow posteriorly and connect with the caudal vein. The union of the caudal vein with the subintestinal vein is then broken, as in Figure 98*E*, the latter vein then becoming a tributary of the hepatic portal system. There next develops between the kidneys a vein, at first single, later paired—the *subcardinal* vein (Fig. 98*F*). This connects with the caudal vein and posterior ends of the posterior cardinal veins. The blood flows from the tail into the subcardinal veins and through the kidneys into the posterior cardinal veins. There next occurs a break between the anterior and posterior parts of the posterior cardinal veins. The anterior parts form a connection with the subcardinal veins. The posterior parts retain their connection with the caudal vein. There is then a reversal of flow through the kidneys, since the blood now passes from the caudal vein into the posterior parts of the posterior cardinal veins, now called the *renal portal* veins, through the kidneys into the subcardinal veins, and from them into the anterior portions of the posterior cardinal veins (Fig. 98*G*). At first the

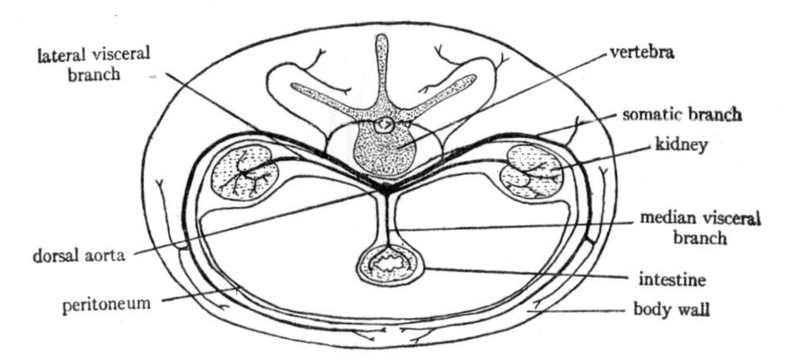

FIG. 97.—Diagram of a cross-section through a vertebrate embryo, to show the segmental branches of the aorta. (After McMurrich's *Development of the Human Body*, copyright by P. Blakiston's Son & Co.)

channels through the kidneys are direct connections between the renal portal and subcardinal veins, but later they break up into a capillary network. In this way the renal portal system is established. The blood from the tail must pass through a capillary network in the kidneys.

The foregoing account carries the circulatory system to the elasmobranch stage. The further evolution of this system will be described in connection with the dissection of the series of specimens.

6. The evolution of the heart.—A definite heart is lacking in *Amphioxus*, but there is a corresponding contractile region at the junction of the hepatic vein, ducts of Cuvier, and ventral aorta (Fig. 4). The heart, as already learned, originates in the same manner as a blood vessel and is, in fact, only an enlarged, highly muscularized blood vessel occupying a strategic position in the midventral anterior part of the trunk, close to the gills. Originally the heart was a straight tube and passes through such a stage in development but grows rapidly in length and diameter; and, since its two ends are fixed, it soon exhibits a curvature to the right and finally assumes an S shape. It becomes subdivided by constrictions into four chambers, which from the venous to the arterial end are: *sinus venosus*, *atrium*, *ventricle*, and *conus arteriosus* (Fig. 99). The sinus venosus is a thin-walled chamber, originally at the posterior end of the heart, buried in the transverse septum, where it receives the great systemic veins. The also thin-walled atrium bulges laterally. The thick-walled muscular ventricle continues into the tubular moderately muscular conus arteriosus, in which there appear during development four longitudinal ridges. The ventral ridge disappears, but the other three each differentiate into a lengthwise row of valves like little pockets, and these valves constitute the dis-

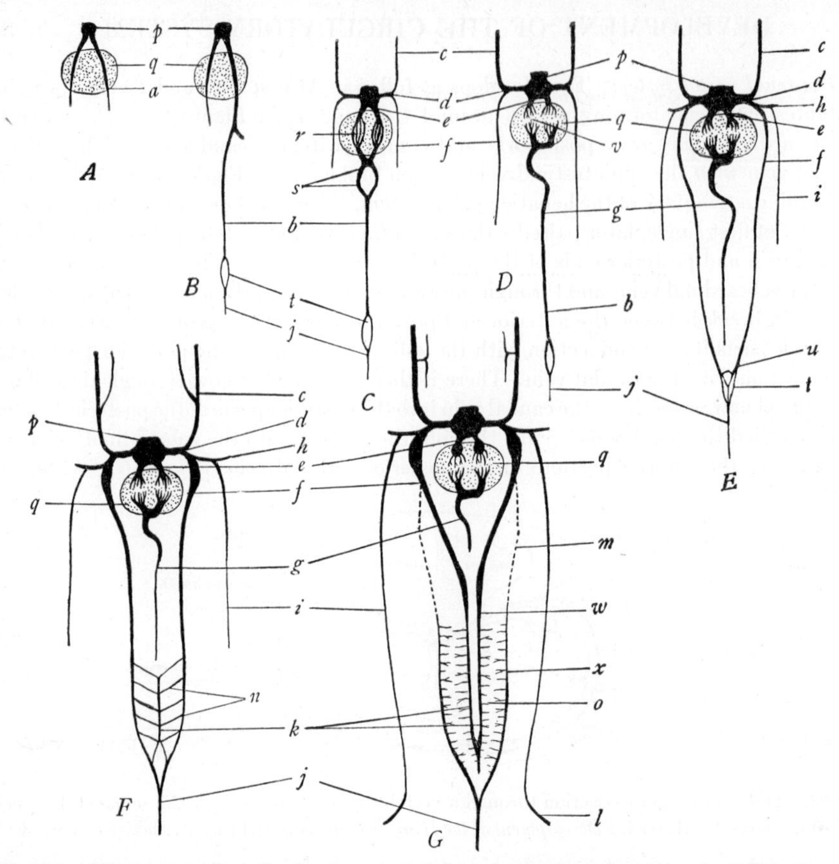

FIG. 98.—Diagrams to show the development of the veins in elasmobranchs. *A*, earliest stage, showing the two vitelline veins (*a*) passing through the liver (*q*) into the sinus venosus (*p*). *B*, the subintestinal vein (*b*) has appeared; it connects with one of the vitellines, makes a loop (*t*) around the anus, and continues into the tail as the caudal vein (*j*). *C*, the anterior and posterior cardinal veins (*c* and *f*) have appeared and connect with the sinus by way of the common cardinal vein (*d*); the vitelline veins are breaking up in the liver at *r* and caudal to the liver are connected by ring-shaped anastomoses at *s*. *D*, the vitelline veins have broken up into a network of capillaries in the liver at *v;* their proximal portions remain as the hepatic veins (*e*); their distal portions have formed the hepatic portal vein (*g*), which is continuous with the subintestinal (*b*). *E*, the posterior cardinals (*f*) have extended posteriorly and at *u* have joined the loop (*t*) formed by the caudal vein (*j*) around the anus; the subintestinal has severed its connection with the caudal; the lateral vein (*i*) and its tributary, the subclavian vein (*h*), have appeared. *F*, a new vein, the subcardinal vein (*k*), has appeared between the kidneys and connects with the posterior parts of the posterior cardinals (*f*) by means of cross-vessels (*n*) and also connects with the caudal vein (*j*). *G*, the posterior cardinal veins have joined the subcardinals (*k*) at *w*, their intermediate portions (*m*) disappearing; the posterior parts of the posterior cardinals persist as the renal portal veins (*x*), which flow into a network of capillaries (*o*) in the kidneys; the lateral abdominal veins have grown posteriorly and developed iliac tributaries (*l*) from the pelvic fins. *a*, vitelline vein; *b*, subintestinal vein; *c*, anterior cardinal vein; *d*, duct of Cuvier or common cardinal vein; *e*, hepatic vein; *f*, posterior cardinal vein; *g*, hepatic portal vein; *h*, subclavian vein; *i*, lateral abdom.nal vein; *j*, caudal vein; *k*, subcardinal vein; *l*, iliac vein; *m*, obliterated portion of the posterior cardinals; *n*, communications between subcardinals and renal portals; *o*, capillary network in kidneys; *p*, sinus venosus; *q*, liver; *r*, branching of vitelline veins in the liver; *s*, rings between the two vitellines; *t*, loop around the anus; *u*, union of posterior cardinals with the caudal vein; *v*, capillary network between hepatic portal and hepatic veins; *w*, union of posterior cardinals with subcardinals; *x*, renal portal veins. (Slightly modified after Hochstetter in Hertwig's *Handbuch der vergleichenden und experimentellen Entwickelungslehre der Wirbeltiere.*)

tinguishing characteristic of the conus. The conus continues anteriorly into the ventral aorta. The S curvature brings the ventricle near the sinus and the atrium near the conus. The heart remains in this condition in most fishes, with the chambers fairly distinct and the interior an undivided continuous cavity, through which the blood, venous in nature (i.e., carrying little oxygen and a large load of carbon dioxide), flows in one single stream in a simple route from sinus to conus. It then flows through the gills, where it is aerated.

In tetrapods the chambers of the heart compact together; the conus becomes embraced by the bulges of the atrium, and the ventricle takes on the characteristic heart shape by the close fusion of the two limbs of one bend of the S. The atrium becomes divided by an interauricular septum into right and left atria (or auricles), and the sinus is moved somewhat anteriorly, being attached to the right auricle. The blood now goes to the lungs for aeration (also to the gills in a few urodeles), and this necessitates a double circulation through the heart. The venous blood from the body returns, as before, to the sinus venosus by way of the great systemic veins and then passes into the right auricle and the ventricle. From the ventricle it exits through the pulmonary arteries, which are offshoots of the sixth aortic arches, to the lungs for aeration, returning by way of the newly formed pulmonary veins to the left auricle. It then again passes into the ventricle and out of the conus into the ventral aorta. It is seen that the aerated and nonaerated blood is flowing simultaneously through the heart in two streams, thus forming a double circulation. The two streams of blood are completely separated in the auricles but imperfectly separated in the ventricle and conus. In the conus of anural amphibians one of the longitudinal ridges remains as such, forming a sinuous fold, which helps to keep the two blood streams separated.

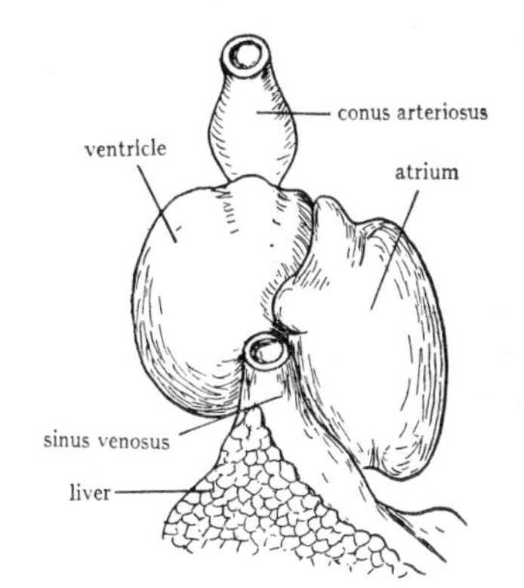

FIG. 99.—Heart of *Ammocoetes*, showing primitive arrangement of the four chambers. (After Daniel, 1934.)

It is of great interest to note that the double circulation with attendant changes in the heart already exists in the Dipnoi in about the same condition as in anurans. The atrium is almost completely divided into two chambers; the ventricle is also partially divided; and the pulmonary arteries, homologous to those of tetrapods, convey venous blood to the swim bladder.

In reptiles the heart is still more compact; and the sinus venosus, beginning to be reduced, has moved still farther anteriorly, so that the terminations of the great veins have come close to the origin of the great arteries. The two auricles are completely separated in all amniotes. The division of the ventricle begins in reptiles by the formation of an *interventricular septum* growing from the posterior end (apex) of the heart forward. This remains incomplete in all reptiles except the Crocodilia, where it forms a complete separation of the ventricle into right and left compartments. The conus arteriosus of reptiles becomes divided by somewhat spiral lines of fusion into three arterial trunks (as will be seen in the dissection of the turtle), and the characteristic pocket or *semilunar* valves are retained where these trunks leave the ventricle. The circulation follows the same course as in amphibians, but the two streams of blood are better separated because of the interventricular septum, as well as by complicated valve arrangements.

In birds and mammals the interventricular septum completely divides the interior of the ventricle into right and left compartments. The sinus venosus has vanished or, perhaps more correctly speaking, has become incorporated into the wall of the right auricle, into which, therefore, the great systemic veins open directly. The conus arteriosus is split into two trunks, the systemic aorta and the pulmonary trunk (Fig. 107). The former connects with the left ventricle, the latter with the right ventricle. The two streams of blood, venous and arterial, are wholly separated. The venous blood, passing from the right auricle into the right ventricle, exits through the pulmonary trunk to the lungs. It returns as aerated blood via the pulmonary veins to the left auricle, passes to the left ventricle, and out the systemic aorta, to be distributed throughout the body. The body of birds and mammals thus obtains pure aerated blood, and this permits the high activity characteristic of these two groups.

The heart beats at regular intervals, depending on temperature. The beat is a wave of muscular contraction which begins in the sinus venosus (or its representative in birds and mammals, called the *sinauricular node*) and sweeps over the heart to the arterial end. The hardest work is done by the ventricle, since it has to push the blood through the body, and hence the ventricle has very much thicker muscular walls than other parts of the heart. When the ventricle is divided, the left ventricle is larger and stronger than the right, since the latter operates only the short lung circuit, whereas the left ventricle forces the blood throughout the body.

7. Phylogeny of the vertebrate circulatory system.—It was formerly customary to derive the vertebrate circulatory system from that of annelids. In the latter there is a longitudinal middorsal vessel in which the blood flows forward, a longitudinal midventral vessel in which it flows backward, and segmental connecting vessels shaped like half-loops between the two longitudinal trunks. Because the direction of flow of the blood in the main longitudinal trunks of annelids is the reverse of that of vertebrates, it was assumed that somewhere in the evolution of annelid into vertebrates the transitional forms had turned over, so that the annelid dorsal side became the vertebrate ventral side. However, an annelid ancestry for vertebrates is now considered improbable, and hence one must search elsewhere for the ancestral condition of the circulatory system. The principal features of the vertebrate circulatory system are already present in *Amphioxus* (Fig. 4), i.e., the ventral aorta sending aortic arches between the gill slits, the paired dorsal aortae uniting to one behind the gill region, the anterior and posterior cardinal veins and the ducts of Cuvier, the subintestinal vein, and the hepatic portal system. There is at present no evidence for carrying the circulatory system farther back than *Amphioxus*. Hence, a more primitive condition of the circulation than that shown in Figure 96 is unknown, unless one is to assume that the developmental stages described on pages 307–9 represent ancestral conditions. It seems probable that the vertebrate circulatory system was originally more segmental than it is at present.

B. THE CIRCULATORY SYSTEM OF ELASMOBRANCHS

The circulatory system of elasmobranchs is broadly at the stage shown in Figure 96, but the first and second aortic arches have been modified as in all vertebrates. Changes have occurred in the posterior veins as indicated in Figure 98. The heart remains a one-way tube with the four typical chambers.

The account applies to the spiny dogfishes and the skate. The terminology of the blood vessels follows that of O'Donoghue (1928).

1. The chambers of the heart.—The heart, when first formed, is a simple straight tube, but it soon becomes bent upon itself in the shape of the letter S, and its wall becomes differentiated into a number of chambers. The heart of the elasmobranchs is in this condition, consisting of four chambers. The pericardial cavity has already been exposed in the preceding dissection. Spread its walls apart. Identify the chambers of the heart as follows: On raising the heart a triangular chamber will be seen extending from the heart to the transverse septum, its two corners buried in the septum. This is the *sinus venosus*, the most posterior chamber of the heart. Each corner of the sinus venosus is continuous with a large vein, the *duct of Cuvier or common cardinal* vein, which is inclosed in the transverse septum and will be seen later. Anterior to the sinus venosus is the *atrium*, a large, thin-walled chamber expanded on each side of the heart and appearing as if paired. Between the two sides of the atrium rests the *ventricle*, a thick-walled, heart-shaped chamber, the most conspicuous portion of the heart from ventral view. The pointed posterior end of the ventricle is known as the *apex;* the broad anterior end, the *base*. From the base of the ventricle a thick-walled tube runs forward and penetrates the anterior wall of the pericardial cavity. This is the *conus arteriosus*, the fourth and most anterior chamber of the heart. The blood circulates through the chambers of the heart in the following order: sinus venosus, atrium, ventricle, conus arteriosus.

2. The systemic veins.—Systemic veins have already been defined as those veins which enter the heart. All systemic veins in vertebrates open into the sinus venosus or its equivalent; i.e., they enter the phylogenetically posterior end of the heart. Because of differences between them, the dogfishes and skate will be described separately.

In dissecting the veins they are followed away from the heart, and it is often convenient to speak of them as if they proceeded from the heart to body structures. The student must, however, always bear in mind the fact that they convey the blood *from* the parts of the body *to* the heart.

Dogfish: Insert one blade of a fine scissors in the sinus wall and slit the ventral wall of the sinus venosus open in a crosswise direction. The cavity of the sinus is thus exposed and should be washed out thoroughly under a stream of running water. All of the systemic veins open into the cavity of the sinus, and the openings may now be identified, with the cut edges of the sinus wall spread well apart. Each lateral wing of the sinus which lies buried in the transverse septum receives a very large opening, the entrance of the *duct of Cuvier* or *common cardinal vein*. The natural relations of this entrance are best observed on the intact right side. On the left side carry the slit in the sinus laterally to meet the incision previously made across the gill slits. The entrance of the common cardinal vein into the sinus is thus slit open. The following may then be noted. Just medial to the main opening of the common cardinal vein into the sinus are several small apertures, most of which appear to be subdivisions of the chief opening. The most anterior of these small apertures is, however, the opening of the *inferior jugular* vein. Probe into the opening and note that the vein comes from the floor of the mouth and pharyngeal cavities, where it runs alongside the ventral ends of the gill arches. By turning back the flap, previously formed, of the floor of the mouth and pharyngeal cavities, the course of the vein will be more readily followed. Now look into the posterior part of the wall of the sinus, putting this on a stretch. In the median line of the posterior wall will be noted a white fold, and on each side of this is an opening. Probe into both openings and note that your probe passes internal to the coronary ligament and into the liver. Follow your probe into the liver by slitting the substance of the liver, and note the cavity in the right and left lobes of the liver thus revealed. These cavities, which extend nearly the entire length of the liver lobes, are the *hepatic sinuses*.[2] The two hepatic sinuses are the persistent proximal parts of the vitelline veins of the embryo.

Now probe posteriorly into one of the common cardinal veins. Raise the viscera in the anterior part of the pleuroperitoneal cavity and observe that your probe has entered a large bluish sac located in the dorsolateral wall of the pleuroperitoneal cavity. Follow this sac and its fellow of the opposite side posteriorly. Each bends toward the median region, narrowing considerably; and on pressing the viscera to one side, each may be traced posteriorly as a narrow tube lying immediately to each side of the attachment of the dorsal mesentery to the median dorsal line. These two vessels are the *posterior cardinal sinuses;* they are the chief somatic veins of the trunk. At the level of the anterior part of the liver the two posterior cardinal sinuses communicate with each other by a broad connection, which may be found by probing into the left vein and directing the probe toward the right one. In this same region each vein has on its ventral surface an extensive communication with a large blood sinus, the *genital* sinus, surrounding each gonad. Each posterior cardinal sinus also receives numerous segmentally arranged branches from the body wall (*parietal* veins) and from the kidneys (*renal* veins). The kidneys are the long, slender, flat organs, brownish in color, which lie immediately lateral to the posterior cardinal sinuses, extending the entire length of the pleuroperitoneal cavity. The parietal and renal veins are readily identifiable in those specimens in which they happen to be filled with blood, but they are impossible to see when empty.

Now turn the animal dorsal side up and locate the lateral line. Make a longitudinal incision above the gill slits on the left side along the lateral line and deepen the incision until you break into a large cavity with a smooth lining, the anterior cardinal sinus. (The anterior cardinal sinus was already exposed in the study of the branchial musculature; and, if this study was carried out, the above incision will be unnecessary.) Probe anteriorly into the anterior cardinal sinus and follow your probe by an incision. The cardi-

[2] Because the veins of elasmobranchs are not definite vessels but spaces in the tissues without definite walls, they are more correctly designated sinuses.

nal sinus can thus be traced forward above the spiracle to the eye, where it connects with an *orbital sinus* surrounding the eyeball. At the level of the posterior end of the eyeball is situated an opening in the ventral wall of the anterior cardinal sinus. On probing into this it will be found to extend medially into the skull. It is the opening of the *interorbital* sinus, which connects the two orbital sinuses. Locate the hyoid arch. In the floor of the anterior cardinal sinus, between the hyoid and third gill arches, is an opening. On probing this it will be found to lead into a vessel which extends ventrally along the outer surface of the hyoid arch. This vessel is the *hyoidean* sinus, and it connects with the inferior jugular vein. Next trace the anterior cardinal sinus posteriorly. It turns abruptly ventrally and joins the posterior cardinal sinus. On probing into the anterior cardinal sinus at the turn the probe will be found to emerge into the posterior cardinal sinus. The union of the two sinuses forms the common cardinal vein already described.

Only that part between the duct of Cuvier and the entrance of the hyoidean vein comes from the embryonic anterior cardinal vein and should properly be called by this name. The orbital sinuses and more anterior parts of the anterior cardinal system are embryologically distinct veins which have joined the anterior cardinal. Since the internal jugular vein of gnathostomes in general is similarly compounded, O'Donoghue proposes that the anterior cardinal system of elasmobranchs should also be called the internal jugular vein.

Running along the lateral walls of the pleuroperitoneal cavity, immediately external to the pleuroperitoneum, on each side is a conspicuous vein, the *lateral abdominal* vein. Note the *parietal* branches which it receives segmentally from between the myotomes. Trace the right vein anteriorly and find its entrance into the duct of Cuvier, just in front of the baglike expansion of the posterior cardinal sinus. Slit open the lateral abdominal vein at this entrance and find here, immediately posterior to the pectoral girdle, the opening of the *brachial* vein which drains the pectoral fin. The very short common stem formed by the union of the lateral abdominal vein with the brachial forms the *subclavian* vein, opening, as just seen, into the duct of Cuvier. The brachial vein passes along the posterior surface of the pectoral girdle in contact with the cartilage and may be picked up easily on the left side, where the girdle has been cut across. The vein may also be found by cutting across the base of the fin, where it forms an opening dorsal to the fin rays in the posterior half of the fin. Next trace the lateral abdominal vein posteriorly. It passes along the inner surface of the pelvic girdle and then continues posteriorly along the lateral margin of the cloacal aperture as the *cloacal* vein. At about the middle of the base of the pelvic fin it receives a *femoral* vein from the fin, whose opening into the lateral abdominal vein will be found by slitting open the latter. The femoral vein is a short vessel situated just under the dorsal skin of the fin.

In the midventral line there is a *ventral cutaneous* vein, which anteriorly bifurcates and disappears into the musculature. (There are also dorsal and lateral cutaneous veins which lie just beneath the skin and are not described in the present dissection. These cutaneous veins are connected by segmental branches.)

Draw the systemic veins.

Skate: The sinus venosus consists of a tube on each side, the central portion being reduced in size and attached to the transverse septum by a sheet of connective tissue, which may be broken. Each side of the sinus is buried in the transverse septum. Follow out the right side to the point where it disappears dorsal to the cartilage of the pectoral girdle. Carefully shave away the cartilage and surrounding tissues until the sinus venosus can be followed laterally. Insert one blade of a fine scissors in the ventral wall of the sinus and slit it open in a crosswise direction. The sinus is seen to be continuous on each side with a tube or chamber, the *common cardinal* vein or *duct of Cuvier*, which turns dorsally. All of the systemic veins open into the common cardinal vein, and their openings may now be identified. The junction of common cardinal vein with the sinus venosus is marked by a slight fold. In the anterior wall of the common cardinal vein concealed by this fold is the small opening of the *inferior jugular* vein. This opening is so small that the probe can probably not be passed into it. The inferior jugular vein drains the walls of the pericardial cavity and the floor of the mouth and pharyngeal cavities. In the posterior wall of the common cardinal vein, at its junction with the sinus venosus, is the opening of the *right hepatic* sinus. Probe posteriorly into this. It leads into the right hepatic sinus, a space situated between the anterior margin of the right part of the liver and the transverse septum; in females the sinus lies dorsal to the beginning of the oviduct, which is inclosed in the falciform ligament. In locating the hepatic sinus press the liver caudad away from the transverse septum; the sinus forms a small bag between liver and septum ventral and somewhat to the side of the esophagus. A similar left hepatic sinus exists on the left side, and the two are connected in some specimens. Cut into the right hepatic sinus and look in its posterior wall for the small openings of the *hepatic* veins that drain the liver. For the pericardioperitoneal canals, see p. 330.

Now probe into the main cavity of the common cardinal vein in a dorsal and posterior direction. The probe enters the *posterior cardinal* sinus, which is a broad, thin-walled tube lying against the dorsal wall of the pleuroperitoneal cavity. In females it is on the dorsal side of the oviduct; in males, dorsal to the testis. Follow the posterior cardinal sinus posteriorly. It swerves toward the median line, where it soon meets its fellow of the opposite side to form a single large median sinus. This sinus communicates on its ventral surface with the large *genital* sinus within each mesogonad. More posteriorly the posterior cardinal sinus separates again into two veins which proceed caudad on the medial side of the kidneys. The kidneys are rounded lobes at the posterior end of the pleuroperitoneal cavity against the dorsal wall. To see them, remove the pleuroperitoneum from the dorsal wall on each side of the cloaca. Do not injure the ducts from the kidneys. After exposure of the kidneys the posterior cardinal veins will be found on the medial side of the kidneys. They connect with each other between the kidneys.

Return to the common cardinal vein and probe into it in a dorsal direction. Turn the animal dorsal side up and locate the end of your probe. Make an incision into the spot indicated by the probe and extend the incision longitudinally forward to the eye. On carefully deepening the incision an elongated cavity with smooth walls, the *anterior cardinal* sinus, is exposed. It is situated just medial to the dorsal ends of the gill arches and visceral pouches. It may be followed forward with the aid of the probe. It turns laterally in front of the first visceral pouch, follows along the anterior border of this pouch, and then turns anteriorly again.

With the dorsal side of the animal still facing you, locate, by feeling, the chief anterior cartilage (propterygium) of the pectoral fin. It forms a crescentic ridge lateral to the gill region nearly halfway from the middorsal line to the margin. Make an incision along the medial face of this cartilage on the left side. A vein will be exposed running along the cartilage here. It is one of the *brachial* veins. Follow it posteriorly. It will be found to enter the common cardinal vein.

Turn the animal ventral side up again. Along the lateral wall of the pleuroperitoneal cavity runs the *lateral abdominal* vein. Note the *parietal* branches which it receives from the body wall at each myoseptum. Trace the lateral abdominal vein anteriorly. It passes along the internal surface of the cartilages of the pectoral fin and pectoral girdle and enters the common cardinal vein. Cut into the vein where it passes the cartilages. Immediately on the posterior side of the cartilage of the pectoral girdle a *brachial* vein will be found entering the lateral vein. Immediately posterior to this is another cartilage, and on the caudal side of that another *brachial* vein joins the lateral vein. A third brachial vein was mentioned in the preceding paragraph. The lateral vein after it has received the brachial veins is termed *subclavian* vein; but this is extremely short, entering the common cardinal vein almost immediately. Trace the lateral abdominal vein posteriorly and find its origin in a network of small vessels on the sides of the large intestine and cloaca. It passes on the inner surface of the cartilages of the pelvic girdle and pelvic fin. Slit the vein open along the surface of the cartilages; *femoral* veins will be found emerging from between the cartilages and entering the lateral vein. The largest of the femoral veins is located along the posterior side of the puboischiac bar. Probe into the femoral veins and note their distribution in the pelvic fin.

Draw the systemic veins.

3. The hepatic portal system.—A portal system is a system of veins which flows into a network of capillaries in some organ. The hepatic portal system consists of veins which collect the venous blood from the digestive tract and spleen and pour it into a network of capillaries in the liver. Locate the bile duct. Lying in the hepatoduodenal ligament alongside the bile duct is a large vein, the hepatic portal vein. Trace it posteriorly and identify the branches which it receives from the digestive tract.[3]

Dogfishes: On tracing the hepatic portal vein posteriorly it will be found first to receive a very small *choledochal* vein, which runs along the bile duct. Posterior to this, the hepatic portal vein is seen to be formed by the union of three large branches. The left branch, the *gastric* vein, passes at once to the stomach, where it is formed by the union of the *dorsal* and *ventral* gastric veins, which branch on the dorsal and ventral surfaces of the stomach (there are a dorsal and ventral anterior and a dorsal and ventral medial branches). The middle of the three branches of the hepatic portal is the *lienomesenteric* vein.[4] This passes posteriorly dorsal to the duodenum and is imbedded in the substance of the dorsal lobe of the pancreas, from which it receives small *pancreatic* veins. At the posterior end of the pancreas the vein is seen to be formed by the union of two branches: one, the *posterior lienogastric* vein, from the spleen, and the other, the *posterior intestinal* vein, which comes from the left side of the valvular intestine. The posterior lienogastric vein collects branches from the spleen and adjacent wall of the stomach. Note the numerous branches into the posterior intestinal vein from along the lines of attachment of the turns of the spiral valve. The right branch of the three that form the hepatic portal vein is the *gastrointestinal* (also called *pancreaticomesenteric*). It passes dorsal to the pylorus, from which it collects a *pyloric* vein, also receiving an *intra-intestinal* vein from the interior of the spiral valve, and then lies imbedded in the substance of the ventral lobe of the pancreas. Here it receives the *anterior lienogastric* vein from the spleen and adjacent parts of the pyloric portion of the stomach; and then its main trunk, the *anterior intestinal* vein, continues along the right side of the intes-

[3] These branches are generally filled with blood and are therefore easily traced. If they are empty, they may be readily injected through the hepatic portal vein, even in specimens which have been preserved for a long time.

[4] Considered by O'Donoghue to be the main trunk of the hepatic portal; this interpretation does not appear acceptable.

tine, from which it receives branches along the lines of attachment of the spiral valve.

Skate: The hepatic portal vein is soon seen to be formed by the union of three tributaries: a *gastric* vein from the left, a *lienomesenteric* vein from the middle, and a *pancreaticomesenteric* from the right. Follow each of these. The gastric vein passes to the right margin of the stomach and there receives the *dorsal* and *ventral* gastric veins from the dorsal and ventral surfaces of the stomach. The dorsal gastric vein receives tributaries from the spleen. The lienomesenteric vein receives a *lienogastric* branch from the spleen and adjacent stomach. Its main tributary, the *posterior intestinal* vein, runs along the left side of the intestine, beginning in the tip of the rectal gland; in its course along the intestine it receives branches along the lines of attachment of the spiral valve. It also collects from the pancreas. The pancreaticomesenteric vein collects from the pancreas and, as the *anterior intestinal* vein, from the duodenal region. It also receives a *posterior gastric* vein from the narrow portion of the stomach between the pylorus and the bend.

Trace the hepatic portal vein anteriorly in all three forms. It reaches the dorsal surface of the liver and here divides into branches which penetrate the substance of the liver. In the liver the branches fork into smaller and smaller veins and finally pass into capillaries. From these capillaries originate other veins which empty into the hepatic sinuses, which, as already seen, open into the sinus venosus (dogfish) or common cardinal vein (skate). As already explained (see Fig. 98, p. 312), the hepatic sinuses are the persistent proximal portions of the vitelline veins of the embryo, while the hepatic portal vein and its branches develop from one of the vitelline veins and the subintestinal vein of the embryo.

Draw an outline of the digestive tract and place on this outline the hepatic portal vein and its tributaries from the various parts of the digestive tract.

4. The renal portal system.—In the renal portal system the venous blood passes into a network of capillaries in the kidneys. Cut across the tail just posterior to the anal opening. In the cross-section locate the caudal blood vessels, inclosed in the haemal arch. The caudal artery is dorsal; the caudal vein, immediately ventral to the artery.

Dogfish: Probe into the caudal vein. Observe that the probe can be passed either to the right or to the left, showing that the vein forks at the anus. The two forks are the *renal portal* veins. Leaving your probe in one of the renal portal veins, locate the kidney in the pleuroperitoneal cavity. It is a long, brown organ situated against the dorsal body wall, one on each side of the middorsal line external to the pleuroperitoneum. Slit the pleuroperitoneum along the lateral border near the posterior end of the kidney on the side where your probe is inserted and gently lift the kidney away from the body wall. A space will be found between the kidney and the body wall; into this space your probe has passed. This space is the *renal portal vein* or sinus. It branches into the kidney and also receives tributaries from the body wall.

Skate: The kidney has already been exposed. Look along the medial side of the posterior part of the kidney for a vein coming from the vertebral column. Do not injure any ducts on the ventral surface of the kidney. In males the vein in question lies immediately to the dorsal side of the male duct, which will be seen passing along the ventral surface of the kidney to the cloaca. The duct may be lifted from the kidney surface and bent to one side. The vein in question is the *renal portal* vein. At first it lies along the medial border of the kidney, but soon turns onto the ventral surface of that organ, giving off branches into its substance and receiving tributaries from the body wall lateral to the kidney. The renal portal veins are continuations of the caudal vein; the latter forks at the anus giving rise to the two renal portal veins. The forking is, however, difficult to trace in the skate.

Reference to Figure 98, page 312, will show that the renal portal veins are the posterior parts of the posterior cardinal veins and that the apparent posterior parts of the posterior cardinal veins of the adult are, in reality, the subcardinal veins. Whereas in the embryo the blood flows from the subcardinals into the posterior cardinals, in the adult the direction of flow is reversed. The renal portal system provides that the blood from the tail must pass into a capillary system in the kidneys from which the blood is re-collected into the subcardinal veins. The purpose of this arrangement is obscure; it seems to have been disadvantageous, for the vertebrates later shunted part of this blood into another system and finally abandoned the renal portal system altogether.

Draw the renal portal system.

5. The ventral aorta and the afferent branchial vessels.—Turn once more to the pericardial cavity of the specimen. The conus arteriosus runs forward and penetrates the anterior wall of the pericardial cavity. Carefully pick away muscles and connective tissue from the region extending from the anterior end of the pericardial cavity to the lower jaw. In the median ventral line will be revealed a large vessel, the *ventral aorta*, which continues forward from the conus arteriosus. By dissecting carefully to the left side find the branches of the ventral aorta. They are as follows:

Dogfishes: There are three main pairs of branches, two of which subdivide into two. The most posterior pair of branches arises just at the point where the conus arteriosus passes into the ventral aorta. (At this point note the pair of *coronary* arteries passing along the conus arteriosus onto the surface of the ventricle and to the walls of the pericardial cavity. They should be preserved as far as possible.) Follow the most posterior branch of the ventral aorta. It very shortly divides in two, the posterior branch penetrating the interbranchial septum of the sixth branchial bar, the anterior branch the septum of the fifth. The middle branch of the ventral aorta arises shortly in front of the third branch and passes without division into the interbranchial septum of the fourth branchial bar. After giving off this branch the ventral aorta proceeds forward, without branching, to a point just posterior to the lower jaw. Here it forks to form its anterior pair of branches. Trace the left branch laterally. After some distance it forks, supplying the second and third branchial bars. Trace any one of the branches of the ventral aorta out into the interbranchial septum, slitting the septum. Note the small branches from the artery into the gill lamellae. The five pairs of branches of the ventral aorta are named the *afferent branchial* arteries. How many demibranchs does each supply?

Skate: There are two main pairs of branches from the ventral aorta. The posterior pair arises where the conus arteriosus passes into the ventral aorta. (At this point note the *coronary* artery, paired, passing to the conus arteriosus and to the branchial muscles. Preserve it as well as possible.) Follow out the posterior branch of the ventral aorta. After some distance it subdivides into three branches, which pass to the fourth, fifth, and sixth branchial bars, penetrating the interbranchial septa. Trace them into the septum by slitting the septum, and note the branches from each to the gill lamellae of both demibranchs of the septum. Follow the ventral aorta forward beyond the pair of posterior branches. It passes without branching for some distance and then forks into right and left branches. Follow the left branch. After a considerable distance it forks into two vessels, which penetrate the interbranchial septa of the second and third branchial bars. The five pairs of branches of the ventral aorta are named the *afferent branchial* arteries. How many demibranchs does each supply?

In front of the anterior fork of the ventral aorta will be found some soft brownish diffuse material, composing the *thyroid* gland.

6. The efferent branchial arteries and the dorsal aorta.—Open the flap already formed on the floor of the mouth and pharyngeal cavities; turn it outward and fasten it in that position. The esophagus may be cut to facilitate this procedure. With a forceps strip off the mucous membrane from the roof of the mouth and pharyngeal cavities. In the roof there will now be seen four pairs (dogfishes) or three pairs (skate) of large blood vessels extending from the angles of the gill slits obliquely caudad. They are the *efferent branchial* or *epibranchial* arteries and represent the dorsal parts of the aortic arches. Clean away the connective tissue from these arteries so that they are clearly exposed and trace them to the left, noting that they disappear dorsal to the gill cartilages. Remove these cartilages carefully. This is best done by cutting across them, grasping the cut end, and loosening the cartilage toward the middorsal line. After removing the cartilages trace each epibranchial artery toward the gills. Note that each is formed at the dorsal angle of the gill slit by the union of two vessels, a smaller *pretrematic* branch which comes from the demibranch on the anterior face of the gill pouch and a much larger *post-trematic* branch from the demibranch on the posterior wall of the gill pouch. Note the small vessel which runs from each gill lamella into the pre- and post-trematic branches. In the skate the first two epibranchial arteries unite to one, so that there are but three pairs of main vessels in the roof of the pharyngeal cavity.

Next, dissect on the right side, which has been kept intact, in order to see the full course of the epibranchial arteries. Remove the mucous membrane from the floor of the mouth and pharyngeal cavities, exposing the ventral portions of the gill arches. Remove these cartilages carefully without disturbing any of the arteries, and also remove the cartilage from the full length of the branchial bars. It will now be seen that the pre- and post-trematic branches are united at their ventral ends so that they form a complete loop, the *efferent collector loop*, around each gill cleft. Note further that the post-trematic branch on the anterior wall of each branchial bar is connected with the pretrematic branch on the posterior face of the same bar by means of cross-branches, about three to five in number in the dogfishes, one in the skate. Thus all the efferent collector loops are interconnected.

The efferent system develops in the following way (Sewertzoff, 1924). Two efferent vessels develop in each branchial bar, and these join the dorsal stump of the aortic arch, for meantime the aortic arch has broken across near its dorsal end (Fig. 100*A*). The anterior of the two efferent vessels of each bar makes a union with the posterior efferent of the bar in front and then loses connection with its own partner (Fig. 100*B*, *C*); these connections occur at both ends of the gill slit. In this way there is formed a loop around each of gill slits 2–5.

Vessels springing from the ventral ends of the collector loops will now be described. These are the *external carotid* artery and the *hypobranchial* system. The external carotid (dogfish) springs from the pretrematic branch of the first collector loop near its ventral end and runs forward along the lateral part of the lower jaw. The hypobranchial system, which is very variable, springs from the ventral ends of the second to fourth collector loops and supplies the wall of the heart, walls of the pericardial cavity, and hypobranchial musculature. **Dogfishes:** The main vessel, *commissural* artery, arises on each side from the ventral end of the second collector loop, gives off branches not followed here, and may or may not receive tributaries from the third and fourth collector loops (or these tributaries may first join a longitudinal vessel, the *lateral hypobranchial* artery). The commissural arteries then enter the anterior end of the pericardial cavity, now being named the *coronary* arteries, and are here connected by anastomosis along the dorsal surface of the conus. From this anastomosis the coronary artery of each side forks into one branch, passing along the side of the conus arteriosus and spreading out on the ventricle; and a more dorsal branch, which runs backward in the wall of the pericardial cavity, then passes to the pharyngeal floor and is finally distributed to the esophagus. Connection occurs between the hypobranchial

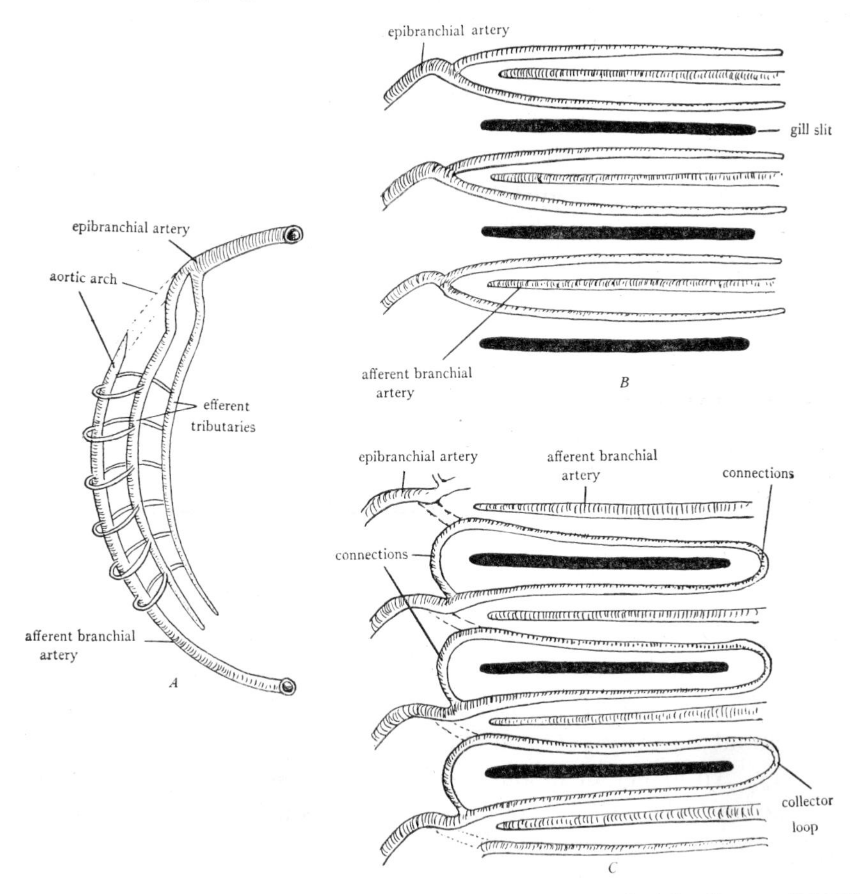

FIG. 100.—Development of the efferent collector loops of elasmobranchs (after Sewertzoff, 1924). *A*, break in aortic arch and development of paired efferent tributaries. *B*, side view of afferent and efferent branches at same stage as *A*. *C*, later stage, showing formation of the collector loops by cross-connections.

system and the subclavian artery. Skate: Contributions from the ventral ends of the collector loops, especially the second, form a longitudinal vessel, the *lateral hypobranchial* artery that anteriorly continues as the *external carotid* artery into the lower jaw, and posteriorly enters the pericardial cavity as the *anterior coronary* artery, proceeding along the conus arteriosus to the ventricle. In the posterior part of the pericardial cavity are the *posterior coronary* arteries (originating from the subclavian artery), running along the sinus venosus.

Turn again to the roof of the oral cavity. From the dorsal end of the collector loop around the second gill slit, a vessel, the second or *hyoidean* epibranchial artery, runs forward and after a short distance bends toward the median line. At this bend it gives off a branch, the *stapedial* artery, which will be found by gently shaving away the cartilage at this point. (The stapedial artery enters the orbit, where it supplies eye muscles and other structures and then proceeds to the snout.) Beyond the stapedial branch the main vessel, now known as the *internal carotid* artery, passes to the median line and joins its fellow of the opposite side; the main internal carotid artery so formed passes into the cranial cavity, where it branches to the brain, eye, and internal ear. In the dogfishes a pair of slender vessels, which represent the paired anterior parts of the dorsal aorta, connect the hyoidean epibranchial artery with the first (really epibranchial 3) of the four epibranchial arteries.

Clear away all tissue from the pretrematic part of the first collector loop. From about the middle of this pretrematic part an artery, the first afferent or *spiracular* artery, arises, passes forward, then very soon turns sharply dorsally and disappears. Turn the animal dorsal side up and remove the skin around and posterior to the spiracle. Pick up the spiracular artery again below the spiracle to the inner side of a white band, the hyomandibular nerve. Follow the spiracular artery to the walls of the spiracle and note its branches to the rudimentary gill (pseudobranch) in the spiracle. (In the skate there are numerous branches to adjacent muscles and only very small branches to the spiracular walls.) Now turn the animal ventral side up, remove the mucous membrane between the spiracle and the upper jaw, and shave away the cartilage about halfway between the upper jaw and the hyoidean epibranchial artery. An artery of moderate size will be revealed and, when traced toward the spiracle, will be found to be the continuation of the afferent spiracular artery already seen, formed by the reunion of the branches to the pseudobranch. It is the efferent or epibranchial spiracular artery; it passes into the skull ventral to the stapedial artery. When traced by scraping away the cartilage, it will be found to join the internal carotid artery.

The four epibranchial arteries (third to sixth) pass to the middorsal line of the roof of the pharyngeal cavity. Here they join in pairs in the middorsal line to form a large trunk, the *dorsal aorta*, which passes into the pleuroperitoneal cavity. In the skate the first and second epibranchials (really third and fourth) unite, and shortly posterior to this junction a vertebral artery arises on each side and passes into the cartilage of the skull, where it is distributed to the brain and spinal cord.

Draw, showing efferent branchial arteries, their branches, and their distribution to the gills.

The aortic arches of elasmobranchs may now be compared with the primitive vertebrate plan. The afferent branchial arteries are the ventral parts, and the epibranchials, the dorsal parts, of the embryonic aortic arches. As already explained, in fishes the aortic arches become broken up by the interposition of a set of small vessels running through the gills. In elasmobranchs the break occurs near the dorsal end of the aortic arches, and the efferent collector loops are new formations (Fig. 100). In vertebrate embryos there are six aortic arches. Adult elasmobranchs have preserved five ventrally and four dorsally. The five afferent branchials represent the second to sixth embryonic aortic arches. The first is imperfect in the adult, contributing to the external carotid and the afferent spiracular arteries. The four epibranchials come from the third to sixth embryonic aortic arches. The first efferent is represented by the spiracular epibranchial, and the second by the hyoidean epibranchial. In all vertebrates the arteries supplying the head are termed the carotid arteries. Usually there is a common carotid, coming from the ventral aorta (Fig. 104), and this divides into an internal carotid (compounded of the third aortic arch and dorsal aorta) and an external carotid (forward continuation of the ventral aorta). The internal carotids enter the cranial cavity and are distributed to the brain and eye; the external carotids supply the thyroid gland, tongue, and lower jaw in general. In elasmobranchs the ventral aorta has not yet assumed those relations to the external and internal carotids which make of it the common carotid; but the internal carotid, as in all vertebrates, is continuous with the forward extensions of the dorsal aortae and so with the third aortic arches (first epibranchials).

7. The dorsal aorta and its branches.—Separate the esophagus from the body wall on the left side and follow the dorsal aorta posteriorly. From the dorsal aorta between the points where the third and fourth pairs of efferent branchial arteries unite with it a *subclavian* artery is given off on each side. Trace the left one into the pectoral fin. It proceeds obliquely caudad and laterad, passing on the dorsal wall of the large bag formed by the posterior cardinal sinus. In the skate it gives off the posterior coronary artery and passes internal to a large white band, the nerve of the pectoral fin. This may be cut through.) At the lateral boundary of the posterior cardinal sinus the subclavian artery (brachial in the skate) gives rise to the small *lateral* artery, which branches into the body wall and usually proceeds posteriorly along the body wall in a position on a level with the lateral line. Farther laterally, at the point where the lateral vein enters the duct of Cuvier, the subclavian artery gives rise to the *ventrolateral* artery, which proceeds caudad halfway between the lateral abdominal vein and the midventral line, giving off segmental branches into the body wall. At the posterior end of the pleuroperitoneal cavity it anastomoses with the vessels supplying the pelvic fins. After giving off these branches into the body wall the subclavian artery, now named the *brachial* artery, proceeds into the pectoral fin.

The *dorsal aorta* is a very large vessel situated in the middorsal line of the pleuroperitoneal cavity. It has median unpaired visceral or splanchnic branches to the viscera, lateral visceral branches to the urogenital system, and somatic branches to the body wall. The median visceral branches are as follows. Work on the left side, turning the viscera to the right.

Dogfishes: Just after it has penetrated the pleuroperitoneal cavity, the dorsal aorta gives rise to the large *coeliac* artery, which distributes blood to the gonads, stomach, and liver. Near its origin the coeliac artery gives off small branches into the adjacent gonads, esophagus, and cardiac end of the stomach. It then runs posteriorly for a considerable distance without branching; it enters the gastrohepatic ligament and gives rise to three branches: the *gastric*, the *hepatic*, and the *pancreaticomesenteric* (or *intestinopyloric*) artery. The gastric artery passes to the stomach and divides into *dorsal* and *ventral* gastric arteries, which branch on the surface of the stomach and penetrate its walls. The hepatic artery turns anteriorly, runs alongside the bile duct, and enters the substance of the liver. The pancreaticomesenteric artery passes dorsal to the pylorus, gives off small branches into the pyloric portion of the stomach and the ventral lobe of the pancreas, a moderately large *duodenal* artery into the duodenum, and a large *anterior intestinal* artery along the right side of the small intestine, to which it gives off branches at the rings of attachment of the spiral valve. The dorsal aorta, after giving rise to the coeliac artery, runs without further visceral branches to the free edge of the dorsal mesentery. Here it gives off two arteries, which course in the border of the mesentery. One of these, the *gastrosplenic* or *lienogastric* artery, passes to the spleen and bend of the stomach. The other vessel, the *anterior mesenteric* artery, passes to the valvular intestine, where it runs caudad along the left side with branches at the turns of the spiral valve. Beyond the gap in the dorsal mesentery the dorsal aorta gives off the *posterior mesenteric* artery, running along the free anterior border of the mesorectum to the rectal gland.

Skate: Shortly after entering the pleuroperitoneal cavity the dorsal aorta gives off the coeliac artery, which supplies a number of organs. It has: a *hepatic* branch to the liver; an *anterior gastric* branch, which divides into *dorsal* and *ventral gastric* arteries to the stomach wall; *splenic* branches to the *spleen;* and a *gastroduodenal* branch, from which arise a *posterior gastric* artery to the posterior part of the stomach, pancreatic branches to the pancreas, and a *duodenal* branch to the pylorus and duodenum. Shortly posterior to the origin of the coeliac artery, the dorsal aorta gives rise to the *anterior mesenteric* artery, which, after some small branches to the pancreas and spleen, proceeds posteriorly along the small intestine, to which it supplies branches at the lines of attachment of the turns of the spiral valve. Shortly caudad of the origin of the superior mesenteric artery, the *posterior mesenteric* artery branches from the dorsal aorta. It sends *genital* arteries to the gonads and their ducts and then passes in the mesentery to the rectal gland.

The lateral visceral and somatic branches of the dorsal aorta are similar in the three forms under consideration. The former consist of the *genital* arteries already noted (but completely developed only in mature specimens) and the *renal* arteries into the kidneys. The latter are seen by loosening the kidney from the dorsal body wall and looking on the dorsal surface of the organ. The somatic branches consist of paired *parietal* arteries to the body wall, passing out along the myosepta. The arteries to the paired fins are merely enlarged parietal vessels. The subclavian to the pectoral fin was already seen. The paired *iliac* arteries to the pelvic fins arise from the dorsal aorta shortly in front of the cloaca. They course along the body wall, resembling the parietal arteries, and after giving off a network of branches into the walls of the cloaca and anastomosing anteriorly with the posterior end of the ventrolateral artery they enter the pelvic fins. The dorsal aorta continues into the tail as the *caudal* artery, which is situated in the haemal canal immediately ventral to the centra of the vertebrae.

Draw the dorsal aorta and its branches.

8. The structure of the heart.—The heart of elasmobranchs is a tube bent into an S shape and differentiated into four chambers. These chambers have already been named. They were originally arranged in a straight line, but the bending of the heart tube brings the ventricle in contact with the sinus venosus and the atrium in contact with the conus arteriosus. The sinus venosus has already been examined. It is a thin-walled chamber, triangular in form in the dogfishes, tubular in the skate. Cut across the connections of the sinus venosus with the transverse septum and also across the base of the ventral aorta and remove the heart from the body. Look into the previously opened sinus venosus and find the large *sinatrial aperture* which leads into the atrium. It is guarded by a pair of valves formed of the smooth free edges of the sinus wall. Note the shape of the atrium. It is a broad, thin-walled chamber with large lateral expansions on each side of the ventricle. Slit open the atrium and wash out the contained blood clots. Note the folds in its wall. Find the atrioventricular opening into the ventricle. It is guarded by two valves. Each of these is a pocket of thin tissue, the opening of the pocket being directed into the ventricle. Cut off the ventral half of the ventricle and also slit open the conus arteriosus by a longitudinal ventral incision. Note the small U-shaped cavity of the ventricle and its very thick, spongy walls forming numerous cavities and crevices in which the blood is held. Examine the atrioventricular valve from the ventricular side and note the two pockets of which it is composed and the attachment of the pockets to the ventricular wall. On the wall of the conus arteriosus note the pocket-shaped *semilunar* valves, the pockets opening anteriorly. In the dogfishes there are nine valves in three circles of three each; two circles are near the posterior end of the conus, while the third circle, composed of larger and stronger valves, is near the junction of the conus with the ventral aorta. The conus arteriosus of the skate bears three longitudinal rows of valves with five valves in each row. To distinguish the valves, run the point of a probe along the conus wall from the anterior end backward, thus opening the pockets.

Make a drawing to show the structure of the heart.

The heart of elasmobranchs contains *only venous blood.* This enters the sinus venosus from the systemic veins and passes in turn through the atrium, ventricle, and conus arteriosus, and out into the ventral aorta, which distributes the blood to the gills by way of the afferent branchial arteries. On passing through the gill filaments the blood is aerated and becomes arterial blood. It is then collected by the efferent branchial arteries and passed into the dorsal aorta, which distributes it to all parts of the body. The venous blood is returned to the heart by means of the systemic veins and the two portal systems. There is thus in the elasmobranchs (and all fishes) a *single* circulation through the heart. Since the heart contains only venous blood, the heart must obtain arterial blood from some outside source. This is accomplished by means of the coronary artery, which arises from the efferent branchial vessels and returns to the heart. As we shall see, the coronary artery originates in a different manner in fish than in other vertebrates.

9. The pericardioperitoneal canals.—Inspect the posterior wall of the pericardial cavity after the removal of the heart. In the dogfishes a large opening will be found in the posterior wall dorsal to the previous attachment of the sinus venosus. On probing into it, it will be found to lead into a canal, the *pericardioperitoneal* canal, situated along the ventral wall of the esophagus internal to the visceral peritoneum of the esophagus. Eventually the canal opens into the pleuroperitoneal cavity by a small slit. In the skate there is an opening of moderate size in the center of the posterior wall of the sinus venosus. On probing into this it will be found to lead into a canal, the *pericardioperitoneal* canal, which passes along the dorsal wall of the hepatic sinus. On probing into this canal it will be found to fork into two canals, lying on the ventral wall of the esophagus internal to its serosa. They open into the pleuroperitoneal cavity by minute slits. The pericardioperitoneal canals serve to connect the pericardial and pleuroperitoneal cavities and arise through the failure of the transverse septum to close completely across the coelom.

C. THE CIRCULATORY SYSTEM OF NECTURUS

In amphibians the heart has become more compact, although the same four chambers are present as in elasmobranchs. The atrium is completely divided by a partition into right and left auricles. Although gills persist throughout life in some urodeles, such as *Necturus,* there is, nevertheless, present a pair of lungs functional in respiration. The circulation through the gills is similar to that of fishes. The first and second aortic arches are absent in all adult tetrapods, disappearing during development. Three or four aortic arches connecting with networks in the gills are present in those urodeles with persistent gills. These are the third to sixth; or, when three arches are present, the fifth is missing. The fifth aortic arch quickly dwindles and disappears during the development of all tetrapods except a few urodeles. The arteries to the lungs, termed the pulmonary arteries, branch from the sixth aortic arches; and the portion of the sixth arches beyond the origin of the pulmonary arteries is called the *ductus of Botallus* or *ductus arteriosus*.

In the venous system certain changes have occurred, notably the formation of a new vein, the *postcaval* vein, which eventually replaces functionally the posterior cardinals (Fig. 101; further explanations on p. 337). The lateral abdominal veins shift their connections, anteriorly joining the hepatic portal system, posteriorly, the renal portal system.

There is a double circulation through the heart in *Necturus* and all tetrapods; i.e., a venous and an arterial stream course simultaneously through the heart. The venous stream is on the right side, and the arterial stream on the left. They are completely separated in the auricles but imperfectly separated in the ventricle in amphibians.

1. The chambers of the heart.—The pericardial cavity has already been exposed; if a new specimen is provided, the pericardial and pleuroperitoneal cavities are to be opened as before. The parts of the heart visible in ventral view are the *ventricle*, the *auricles*, and the *conus arteriosus*. The ventricle is thick walled and conical in form. Anterior to the ventricle on either side is a thin-walled auricle. Springing from the base of the ventricle and passing forward between the two auricles is the tubular conus arteriosus. Anteriorly the conus passes into the enlarged beginning of the ventral aorta; this muscular expansion of the base of the ventral aorta is named the *bulbus arteriosus*. Lift the apex of the ventricle and note the *sinus venosus* situated dorsad to the ventricle. The sinus venosus receives two large venous channels lying in the dorsal wall of the pericardial cavity. These are formed by the union of the two *hepatic sinuses* with the two *common cardinal* veins. The hepatic sinuses are the two large veins which emerge from the transverse septum and pass into the sinus. The common cardinal veins join the hepatic sinuses on their lateral surfaces.

2. The hepatic portal system and the ventral abdominal vein.—In the median ventral line of the body wall, posterior to the liver, inclosed in the falciform ligament, is situated the *ventral abdominal* vein. It is homologous to the lateral abdominal veins of the elasmobranchs. It receives parietal branches from the body wall. At the level of the posterior end of the liver it leaves the body wall and passes into the liver, where it is situated on the dorsal surface. After a short course it joins the hepatic portal vein at the place of attachment of the hepatoduodenal ligament.

The *hepatic portal* vein is the vein which collects the blood from the digestive tract. It is formed by the union of branches from the intestine, pancreas, spleen, and stomach. Stretch out the dorsal mesentery of the small intestine. In this mesentery, about halfway between the body wall and the intestine, runs a conspicuous vein, the *mesenteric* vein. Trace it posteriorly and note its beginning in the wall of the large intestine. As it passes forward in the mesentery, it receives numerous *intestinal* veins from the small intestine. It then passes into the substance of the pancreas, receiving small *pancreatic* veins from that organ. At the level of the pancreas the large *gastrosplenic* vein also joins the mesenteric from the left. On following the gastrosplenic to the spleen it will be found to be formed of *splenic* branches from the spleen and *gastric* veins from the stomach. The union of the mesenteric, gastrosplenic, and pancreatic veins produces a large vessel, the *hepatic portal* vein, which lies along the center of the dorsal face of the liver. It also receives the ventral abdominal vein, as already noted. Follow it along the surface of the liver. It branches into the liver substance and in its course also receives additional gastric veins from the stomach and also veins from the ventral body wall which pass into the liver by way of the falciform ligament. (These have probably been destroyed in the study of the digestive tract.) The hepatic portal vein subdivides in the substance of the liver and eventually passes into capillaries.

Draw the hepatic portal system.

3. **The renal portal system.**—Trace the ventral abdominal vein posteriorly. It soon receives some *vesical* veins from the bladder. Shortly anterior to the hind limbs the abdominal vein is seen to be formed by the union of the two *pelvic* veins which run along the inner surface of the lateral body wall just in front of the pelvic girdle. Follow one of the pelvic veins. It is joined by the *femoral* vein from the hind limb. The vein formed by this union is the *renal portal* vein. It passes at once to the dorsal surface of the adjacent kidney. In male specimens the kidney is a brownish organ of considerable size situated at the side of the intestine. In female specimens the kidney is much smaller and more slender and is situated at the common point of attachment of the mesovarium and mesotubarium. It will be located by laying the ovary to one side and the oviduct to the other side. At the posterior end of the pleuroperitoneal cavity the kidney in the female lies between the intestine and the oviduct. Having located the kidney, identify the renal portal vein near the lateral margin of its dorsal surface. At the place where the renal portal vein passes from the body wall to the surface of the kidney, the vein receives the *caudal* vein, which ascends from the tail, forks, and passes to the surface of the posterior end of each kidney. The renal portal vein runs forward along the surface of the kidney, into the substance of which it sends numerous branches. It also receives branches from the body wall.

Draw the renal portal system.

It will be seen that the ventral abdominal vein forms a connection between the renal portal and hepatic portal systems. Blood from the hind legs and tail may pass into the abdominal vein and so into the hepatic portal vein, or may pass into the renal portal vein. This arrangement appears to be a device to assist the return of the blood from the posterior regions of the body. In the elasmobranchs there is no such connection between the two portal systems (although connections have been reported in some specimens as individual variations). Further, in elasmobranchs the lateral abdominal veins enter the cardinal system of veins, while in *Necturus* their homologue, the ventral abdominal vein, empties into the hepatic portal system (Fig. 101).

4. The systemic veins.—

a) The anterior systemic veins: It has already been noted that the common cardinal vein joins the hepatic sinus on each side in the pericardial cavity. Turn to the pericardial cavity and locate the common cardinal veins. Trace one of them laterally, removing the muscles between the pericardial cavity and the base of the forelimb. Just outside of the pericardial cavity the common cardinal receives the *jugular* and *subclavian* veins. The latter lies ventral and anterior to the former. Trace the subclavian into the forelimb by removing the skin from the outer surface of the limb. The subclavian is seen to be formed at the shoulder by the union of the *cutaneous* vein from the skin and the *brachial* vein which runs along the surface of the limb muscles. The jugular vein is homologous to the anterior cardinal vein of the elasmobranchs. Follow it forward. It is formed by the union of the *external* and *internal jugular* veins. The external jugular vein first receives branches from the floor of the mouth. The main vein then passes dorsally immediately behind the gills. It may be picked up here by removing the skin behind the last gill. The external jugular may then be traced forward above the gills, where it enlarges, forming the *jugular sinus*. This sinus receives tributaries from the head and jaws. The internal jugular vein is a small vein which joins the external jugular posterior to the jugular sinus. It is difficult to find, and part of it will be seen later in the roof of the mouth.

The common cardinal vein also receives a *lateral* vein from the body wall. Remove the skin from the lateral line shortly posterior to the forelimb. Cut through the shoulder muscles so as to reveal the partition (horizontal skeletogenous septum) between the epaxial and hypaxial muscles. The lateral vein will be found situated along this partition and can be followed forward into the common cardinal vein.

b) The postcaval vein: Turn to the pleuroperitoneal cavity. Examine the dorsal mesentery of the small intestine at its junction with the dorsal body wall (in female specimens spread the ovaries apart, laying one to each side). In the mesentery runs a large vein, the *postcaval* vein. It passes forward, receiving numerous *genital* veins from the adjacent gonads and *renal* veins from the kidneys. Trace it forward. At about the level of the spleen it turns ventrally and enters the dorsal surface of the right side of the liver. It is best seen by laying the stomach to the left and the liver to the right. It passes forward, imbedded in the liver substance, and should be followed by picking away the liver tissue. It receives several *hepatic* veins from the liver; one of the larger of these lies along the midventral line of the liver and joins the postcaval near the anterior end of the liver. At the anterior end of the liver the postcaval vein emerges as a very large vessel situated in the coronary ligament. It pierces the transverse septum and forks into the two hepatic sinuses, which, after being joined by the common cardinal veins, enter the sinus venosus. The origin of the postcaval vein is discussed below.

c) *The posterior cardinal veins:* At the place where the postcaval vein turns ventrally toward the liver, it is connected with a pair of veins, the *posterior cardinal* veins. Trace these anteriorly. They lie very near the middorsal line of the anterior half of the pleuroperitoneal cavity, one to either side of the dorsal aorta. In females they are situated in the mesotubarium, along the line where this unites with the dorsal wall. Trace the posterior cardinals posteriorly and note connections between them and the renal portal veins (which are, of course, the original posterior ends of the posterior cardinals, as shown in Fig. 101). Note, also, the *parietal* veins which enter the posterior cardinals in their course along the body wall. The posterior cardinals may be traced to the transverse septum. Shortly before reaching this they diverge from each other and, penetrating the lateral portions of the septum, enter the common cardinal vein practically at the same point as the entrance of the jugular and the subclavian.

5. The pulmonary veins.—The *pulmonary* vein is a large vessel situated along the ventral side of each lung, i.e., the side opposite that which is attached to the dorsal wall. The two pulmonary veins run forward in the walls of the lungs and, shortly caudad of the transverse septum, converge and at the septum unite to one vessel. This passes through the transverse septum and, running forward in the dorsal wall of the left hepatic sinus, enters the left auricle.

6. The ventral aorta and the aortic arches.—The conus arteriosus passes anteriorly into the ventral aorta. The greater part of the ventral aorta lies within the pericardial cavity; and, because of the fact that this portion of the aorta is expanded and possesses thickened muscular walls, it is named the *bulbus arteriosus.*[5] Trace the ventral aorta forward out of the pericardial chamber by dissecting away the anterior wall of the chamber. The ventral aorta very soon forks into two vessels, which pass to the right and left. Trace the right one, since the branchial bars have been left intact on that side. Follow it toward the gill arches. It soon divides into two vessels; and sub-

[5] The term bulbus arteriosus is very ambiguously used in many texts of vertebrate anatomy and embryology. The term should be applied only to the expanded muscular base of the ventral aorta. The bulbus arteriosus does not take part in the heart beat and is not a chamber of the heart but a portion of the ventral aorta. Very few groups of vertebrates have a bulbus arteriosus; the chief group possessing it is the Teleostei. The term truncus arteriosus is another ambiguous name. It should probably be used as synonymous with ventral aorta.

sequently the posterior one again divides in two, making a total of three *afferent branchial arteries*, one to each of the gills. Trace each one into the gill, removing the skin from the gill. At the entrance into the gill the first afferent branchial artery gives off an *external carotid* artery, which turns medially, running beside the branchial artery, and then branches into the floor of the mouth. Within the gill each branchial artery sends up a loop, which branches among the gill filaments, from which other branches collect into a loop on the other side of the gill, this loop joining the branchial artery again. In addition to the two loops, a short connecting branch runs through the base of each gill.

Next turn back the flap, previously formed, of the floor of the mouth and pharyngeal cavities. Extend the incision posteriorly along the left side of the esophagus. Strip off the mucous membrane from the roof of the mouth and pharyngeal cavities. In the roof will be seen a pair of large vessels, the *roots* or *radices of the aorta*, which pass obliquely posteriorly and unite. Trace the right root laterally into the branchial bars, dissecting off the mucous membrane from the latter. Locate from the inside the branchial arteries exposed in the preceding paragraph and note their emergence from the dorsal ends of the branchial bars as the *efferent* branchial arteries. The second and third efferent branchial arteries unite as they emerge from the branchial bars, thus forming two efferent branchial arteries on each side. From the common vessel formed by the union of the second and third efferent branchial arteries an artery arises which passes posteriorly. Trace it into the pleuroperitoneal cavity and note that it courses along the dorsal or attached side of the lung. It is the *pulmonary* artery. From the first efferent branchial artery very near the place where it joins the second and third arises the *internal carotid* artery, which passes forward on the roof of the mouth. Accompanying the internal carotid artery is the *internal jugular* (anterior cardinal) vein. The efferent branchial arteries medial to the origin of the pulmonary and internal carotid arteries unite to form the root of the aorta on each side. From this springs the *vertebral* artery, which passes posteriorly along the vertebral column with dorsal branches into the skull. The two roots of the aorta then pass obliquely caudad and unite in the middorsal line to form the large *dorsal aorta.*

Draw the afferent and efferent branchial arteries and their branches.

As in elasmobranchs, the main portions of the afferent and efferent branchial vessels come from the ventral and dorsal parts of the embryonic aortic arches, respectively. The loops through the gills and the direct connection through the gill bases are new formations. The three aortic arches present in *Necturus* are the third, fourth, and sixth; the fifth is lacking in all tetrapod adults except a few salamanders. Note the origin of the external carotid from the ventral part, and of the internal carotid from the dorsal part, of the third aortic arch, and of a new vessel, the pulmonary artery, from the last or sixth arch (see Fig. 104). As already remarked, pulmonary arteries from the sixth arch make their first appearance in the Dipnoi.

7. The branches of the dorsal aorta.—Trace the dorsal aorta posteriorly into the pleuroperitoneal cavity, where it runs in the middorsal line. Immediately beyond its origin it gives off a *subclavian* artery on each side. Trace one of them. The subclavian artery passes laterally, giving off a conspicuous *cutaneous* artery that lies on the inner surface of the pectoral girdle and branches to the skin and near-by muscles. The subclavian then emits an artery to the shoulder and, as the *brachial* artery, passes into the forelimb, where it branches extensively.

In its course along the pleuroperitoneal cavity the dorsal aorta gives off both visceral and somatic branches. The first visceral branch of the dorsal aorta is the *gastric* artery. It passes to the stomach and forks into the *dorsal* and *ventral gastric* arteries supplying the corresponding walls of the stomach. The ventral gastric artery also furnishes a few small branches to the spleen. Some distance posterior to the origin of the gastric artery, the *coeliacomesenteric* artery springs from the aorta. It passes ventrally in the mesentery, giving rise to some *mesenteric* branches to the beginning of the small intestine, and then proceeds to the region of the hepatoduodenal ligament, where it branches into a *splenic* artery to the spleen, a *pancreaticoduodenal* artery to the pancreas, duodenum, and pyloric region of the stomach, and a *hepatic* artery, which runs along the dorsal surface of the liver in contact with the hepatic portal vein and supplies numerous branches to the liver substance. Posterior to the point of origin of the coeliacomesenteric vessel, the dorsal aorta gives off a number of *mesenteric* arteries into the intestine.

The lateral visceral branches of the dorsal aorta consist of numerous *genital* arteries to the testes in the male, and ovaries and oviducts in the female, and of *renal* arteries to the kidneys. The somatic branches of the aorta consist of the *parietal* or *intercostal* arteries. These arise from the dorsal side of the aorta at segmental intervals; they pass dorsally and divide in two, one branch going to each side of the body. These branches pass laterally along the internal surface of the body wall and supply the body musculature.

Near the posterior end of the pleuroperitoneal cavity the aorta gives off on each side an *iliac* artery, which passes laterally alongside the femoral vein toward the hind limb. It gives off an *epigastric* artery which runs anteriorly along the body wall, a *hypogastric* branch to the urinary bladder and cloaca, and, as the *femoral*, enters the hind limb, into which it should be traced. It runs along the medial side of the leg and at the knee gives rise to a number of branches. The dorsal aorta proceeds into the tail as the *caudal* artery, giving off a pair of *cloacal* arteries into the cloaca as it passes that region.

Draw the branches of the dorsal aorta.

8. The chambers of the heart.—Remove the heart from the pericardial cavity by cutting across both ends. The chambers of the heart were previously named. The sinus venosus is a chamber with very thin, delicate walls. It receives from behind the two large trunks formed by the union of the common cardinal vein and hepatic sinus on each side. Anteriorly the sinus passes into the auricle. The sinus connects chiefly with the right auricle and is slightly displaced to the right. By cutting open the sinus locate the *sinauricular* opening guarded by a pair of valves. Cut into one of the auricles and wash out its contents. Looking into the auricle, note the *interauricular septum* which separates the two auricles; the septum is very incomplete, being perforated by a number of openings. Remove the ventral half of the ventricle and also slit open the conus and bulbus arteriosus by a longitudinal incision. Note the thick, spongy walls of the ventricle and the numerous muscle strands in the interior. Locate the single *auriculoventricular* opening between the auricles and ventricle. It is on the left side, guarded by a pair of valves. In the base of the opened conus arteriosus note the transverse row of three *semilunar* valves. In the bulbus arteriosus is a longitudinal partition, dividing the interior into right and left channels.

The heart of *Necturus* and of all tetrapods receives both arterial and venous blood, and consequently there is a double circulation through the heart. Blood from the systemic veins enters the sinus venosus and is passed on chiefly to the right auricle. The blood from the pulmonary veins, aerated in the lungs, returns to the left auricle. Because of the incomplete nature of the interauricular septum, there is a slight mixing of arterial and venous blood. Blood from both auricles passes into the single ventricle, where some further mixing occurs, and the mixed blood exits by way of the conus arteriosus and ventral aorta. The blood next passes through the gills, where it is aerated. Since the blood which reaches the pulmonary arteries has already been aerated in the gills, it is evident that the lungs are but slightly functional in gilled salamanders.

9. Comparison of the circulatory system of *Necturus* and elasmobranchs.—It is evident that the circulatory system of *Necturus* has been somewhat modified from the elasmobranch condition. In the arterial system the chief changes concern the aortic arches. The dogfish has four complete aortic arches and considerable portions of the first and second, whereas in *Necturus* only the third, fourth, and sixth aortic arches are present and the first, second, and fifth have disappeared. In the venous system the changes are more pronounced. The anterior cardinal still remains the chief head vein, but the posterior cardinal vein is much decreased in importance, for its function has been taken over by a new vein, the postcaval vein, absent in fishes (except Dipnoi). The development of the postcaval vein is somewhat complicated; but from a study of its relations in *Necturus* it is evident that its anterior part is derived from the hepatic veins or sinuses of elasmobranchs (and these, in turn, are the proximal portions of the vitelline veins) and that the posterior part of the postcaval is formed from the subcardinal veins, chiefly the right one (Fig. 101). The posterior portions of the posterior cardinals continue to function as the renal portal veins, as in fishes. The anterior portions of the posterior cardinals are diminished in importance but have the same relations as in fishes; posteriorly they are connected with the subcardinals (postcaval), as also in elasmobranchs (Fig. 98). It will also be noted that the renal portal veins (posterior cardinals) have increased their posterior connections. Whereas in fishes they collect only from the tail, in *Necturus* they collect from both tail and posterior appendages. This is due to a union between the renal portal and abdominal veins (Fig. 101). Meantime the ventral abdominal vein (same as the lateral abdominal veins of elasmobranchs) has shifted its anterior connections. Whereas in fishes it opens into the common cardinal vein, in *Amphibia* it passes into the hepatic portal system. The abdominal vein thus becomes a connection between the renal and hepatic portal systems and provides two outlets for the blood from legs and tail. The pulmonary veins are new formations.

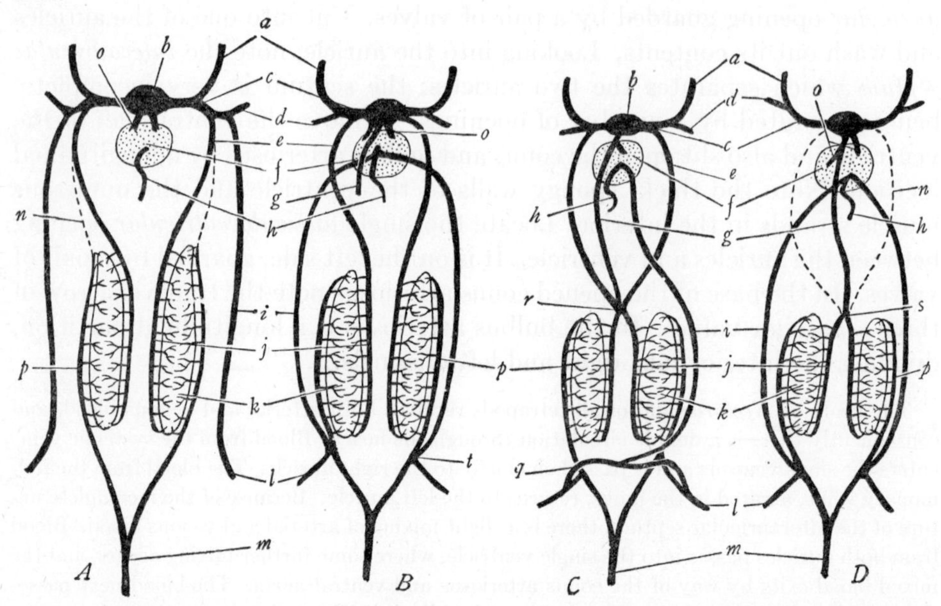

FIG. 101.—Diagrams to show the origin of the postcaval vein and the changes in the abdominal vein in amphibians and reptiles. *A*, elasmobranch stage, same as in Figure 98*G;* the lateral abdominal veins (*i*) enter the common cardinal veins (*c*) and are not connected with the renal portal veins (*p*). *B*, the lateral abdominals (*i*) have joined the renal portals at *t* posteriorly, and anteriorly pass into the liver (*f*), where they unite with the hepatic portal vein (*h*); a new vein, the postcaval vein (*g*), is seen growing caudad from the liver (*f*), where it arises from the hepatic veins (*o*). *C*, condition in the adults of urodele amphibians; the postcaval vein (*g*) has reached and fused with the posterior cardinals (*e*) and the subcardinals (*j*) at the point *r;* the two lateral abdominal veins have united to form the ventral abdominal vein (*i*), which empties into the hepatic portal (*h*). *D*, condition in adult reptiles; the anterior portions of the posterior cardinal veins (*n*) are obliterated, leaving the postcaval vein (*g*) as the sole drainage for the subcardinals (*j*) and the kidneys (*k*); the two lateral abdominal veins remain separate, as in elasmobranchs. *a*, anterior cardinal vein; *b*, sinus venosus; *c*, common cardinal vein; *d*, subclavian vein; *e*, posterior cardinal vein; *f*, liver; *g*, postcaval vein; *h*, hepatic portal vein; *i*, lateral (or in *C* ventral) abdominal vein; *j*, subcardinal vein; *k*, kidney; *l*, iliac or femoral vein; *m*, caudal vein; *n*, obliterated part of the posterior cardinals; *o*, hepatic veins; *p*, renal portal veins; *q*, pelvic veins; *r*, union of postcaval, posterior cardinals, and subcardinals; *s*, union of postcaval and subcardinals; *t*, union of abdominal vein with renal portal system.

D. THE CIRCULATORY SYSTEM OF THE TURTLE

In reptiles the heart is still more compact; the sinus venosus is still present, attached to the right auricle. As an advance over amphibians, the ventricle has become partially divided into right and left chambers by a partition, which is complete in the Crocodilia. As another advance over amphibians, the conus arteriosus and ventral aorta have disappeared as such. The conus has become subdivided into three arterial trunks by the fusion of the ridges which in lower forms give rise to the conus valves (Fig. 107). Here valves form only at the junction of the arterial trunks with the heart. All the aortic arches except the fourth have disappeared in reptiles except for portions of the third and sixth. In the venous system there are paired ventral abdominal veins, in place of the single one of amphibians; but these have the same relations to the two portal systems as in amphibians. The chief change here is that the anterior parts of the posterior cardinal veins disappear during development (Fig. 101), except in *Sphenodon*, see below.

Specimens for the study of this system should have been doubly injected, i.e., into the arterial system, and in both directions into one ventral abdominal vein. Remove the plastron. With the bone scissors cut away the sides of the carapace between fore and hind limbs.

1. The chambers of the heart.—The heart of the turtle possesses but three different chambers, in contrast to the four chambers present in the forms already considered; one of the chambers is divided into two completely separate halves. Examine the heart in the pericardial cavity, removing the ventral wall of the pericardial sac if this has not already been done. From the ventral view the visible parts of the heart are the *ventricle*—the thick-walled, conical posterior part—and the *auricles*—thin-walled chambers, one on each side anterior to the ventricle. The two auricles are entirely separate from each other, as will be seen later. The ventricle is attached to the posterior pericardial wall by a ligament, which is apparently a remnant of the ventral mesentery or ventral mesocardium of the heart, a structure which was considered in the general discussion of the coelom. Cut through this ligament, raise the ventricle, and press it forward. A large chamber, the *sinus venosus*, is revealed dorsal to the auricles and attached to the right auricle. The large bases of the systemic veins will be seen entering the sinus. Put the ventricle back in place. Observe the large vessels, arteries, which spring directly from the base of the ventricle. A conus arteriosus is apparently lacking but in fact is present as the bases of these large arteries. The conus becomes divided into three trunks in reptiles by lengthwise fusions along the longitudinal ridges from which in elasmobranchs the conus valves originate. As this process is general throughout the amniotes, all of the latter lack an intact conus arteriosus and a ventral aorta.

2. The ventral abdominal veins and the renal portal system.—Running in the ventral pleuroperitoneum from the pelvic girdle up to the heart are two large veins, the *ventral abdominal* veins. They are homologous to the lateral abdominal veins of elasmobranchs. The two veins are generally connected just anterior to the pelvic girdle by a cross-branch. Trace the veins forward. They receive *pericardial* veins from the pericardial sac, and then each turns dorsally to enter the liver. Just at this turn, each vein receives a *pectoral* branch from the pectoral muscles of that side. Trace the pectoral vein into the muscle. Slit the pleuroperitoneum alongside each abdominal vein and, by lifting the cut edges, find the places where the vein of each side penetrates the lobe of the liver.

Trace the ventral abdominal veins posteriorly. Make a longitudinal slit in the pleuroperitoneum midway between the two veins and, separating the cut edges, look within and locate the urinary bladder. Note the small *vesical* vein passing from the bladder into each abdominal vein. Continue to trace the abdominal veins posteriorly. Each passes to one side of the pointed anterior extremity of the pelvic girdle and at the same time gradually turns laterally. As it turns, it receives a *pelvic* vein which runs over the ventral surface of the muscles of the pelvic girdle. The left pelvic vein seems to be usually larger than the right one.

In their course between the heart and pelvic girdle each vein gives off laterally one or more small branches which pass to the borders of the carapace, where they join the *marginocostal* vein, to be described later.

Draw the two abdominal veins with their branches thus far noted.

Continue to trace the abdominal veins in the posterior direction. As both have identical branches, it is necessary to follow only one, selecting the one which has been most successfully injected. It passes along the dorsal surface of the pelvic girdle near the anterior margin of the latter; the girdle should be pulled toward the student in order to follow the vein. Grasp the hind leg on the side on which you are dissecting and work it back and forth until it is freely movable. Press the leg away from the carapace of that side and cut through the skin between the leg and the carapace back to the end of the tail. Remove the skin from leg and tail. Now trace the abdominal vein laterally along the base of the leg. Just beyond the pelvic vein a small *crural* from thigh muscles and a larger vein from fat enter the abdominal. About an inch and one-half lateral to this the large *femoral* vein emerges from the leg and joins the abdominal vein, now designated the *external iliac* vein. The femoral vein should be followed into the leg by separating the muscles. The external iliac vein is now situated alongside a conspicuous artery, the *epigastric* artery, both being imbedded in the abdominal wall, from which small veins pass into the external iliac vein. After a short distance the external iliac receives the *epigastric* vein, which accompanies the artery of the same name anteriorly along the curve of the carapace. It then turns abruptly posteriorly and runs between the base of the leg and the carapace, deeply imbedded in some loose tissue, which should be cleared away. The vein receives branches from the carapace and, near the posterior part of the thigh, a well-marked *ischiadic* vein from the thigh. Posterior to this point it receives several small branches from the leg and, as the *caudal* vein, passes along the side of the tail, receiving at the base of the tail a *cloacal* branch from the anal region.

Return to the point where the epigastric vein enters the external iliac vein. At this place a large vein continues forward from the anterior and dorsal surface of the external iliac. This vein, the *renal portal* vein, runs forward and dorsally, penetrating the pleuroperitoneum. Cut the pleuroperitoneum transversely halfway between the heart and pelvic girdle, cutting across both abdominal veins. Cut also into the pleuroperitoneum at the place where the renal portal vein passes through it. A layer of muscle will be found outside the peritoneum at this place. Both muscles and membrane should be slit ventrally to meet the transverse incision across the pleuroperitoneum. In this way free access is gained to the pleuroperitoneal cavity. With the left hand carefully press all of the viscera forward. It is usually necessary to detach the lung from the dorsal wall and push it forward also. With the right hand press the pelvic girdle caudad. A space cleared of viscera is thus left dorsal to, and in front of, the pelvic girdle. Look into this place near the median dorsal line for a somewhat flattened organ, the *kidney*, situated against the median dorsal wall. The kidney is retroperitoneal, i.e., dorsal to the pleuroperitoneum. This latter membrane should be stripped off from the ventral face of the kidney. (In male specimens the rounded yellow testis and black coiled epididymis will be noted attached to the ventral surface of the kidney.) The renal portal vein may now be followed from the point where it leaves the iliac through the pleuroperitoneum toward the kidney. Before reaching the kidney it receives a vein from the carapace. At about the middle of the lateral border of the kidney is a fissure; the renal portal vein enters this fissure and passes onto the ventral face of the kidney, where it immediately forks. One of its branches, the *vertebral* vein, runs forward and may be traced in well-injected specimens by separating the lung from the carapace and raising the lung and also stripping off the pleuroperitoneum from the dorsal wall. The vertebral vein passes anteriorly dorsal to the arches of the ribs and receives laterally an *intercostal* branch at each suture between the costal plates of the carapace. The intercostal veins anastomose with each other in the curve of the carapace by means of a longitudinal vessel, the *marginocostal* vein, which is the anterior continuation of the epigastric vein previously noted. The marginocostal vein also has connections with the abdominal veins. The posterior branch of the renal portal vein passes posteriorly over the ventral face of the kidney and, as the *internal iliac* or *hypogastric* vein, receives branches from the reproductive organs (male), bladder, cloaca, etc. The renal portal vein in its passage along the ventral face of the kidney gives off branches into that organ. The renal portal vein is the posterior part of the posterior cardinal vein (Figs. 98, 101). Both the vertebral and the marginocostal vein are formed by the longitudinal anastomoses of transverse segmental branches of the posterior cardinal veins of the embryo.

Draw these veins as far as you have seen them, adding them to the drawing of the abdominal veins already made.

The student should consider at this point the differences between the connections of the ventral abdominal veins of the turtle and those of their homologues, the lateral abdominal veins of elasmobranchs. The condition in the turtle is the same as in *Necturus*. The veins have formed a connection with the renal portal system posteriorly, and anteriorly they enter the liver instead of the cardinal system.

3. The hepatic portal system.—Lift up the lobes of the liver, separating them gently from the stomach and duodenum, and find on their dorsal surfaces at the place where the gastro-hepato-duodenal ligament is attached to the liver a large vein, the *hepatic portal* vein. It runs completely across the liver imbedded in its wall; and at the right, at the point where the bile duct enters the duodenum, it turns abruptly posteriorly, penetrating the mesentery. On the left note the numerous *gastric* veins entering the hepatic portal vein from the stomach. Just to the right of the bridge connecting the two lobes of the liver, two or three *anterior pancreatic* veins pass from the pancreas into the hepatic portal vein. Near the bile duct it receives *cystic* veins from the bile duct, *posterior pancreatic* veins from the right end of the pancreas, and a long *duodenal* branch from the first part of the small intestine. The hepatic portal vein should be followed posteriorly; it is imbedded in the pancreas and at the bend of the duodenum penetrates the mesentery and emerges to the left of the duodenum. Liver and duodenum must be pressed forward to follow it. The vein next passes to the posterior side of the adjacent loop of the small intestine, which should also be pressed forward. The vein will then be found to pass on the left side of the spleen in contact with that organ and to receive numerous *splenic* tributaries from it. Shortly posterior to the spleen, the hepatic portal vein reaches the central point of the mesentery, where the mesentery is thrown into a coil. At this place the numerous *mesenteric* veins, accompanied by arteries, will be seen passing in the mesentery from all parts of the intestine into the hepatic portal vein.

Draw the hepatic portal vein and its branches.

Now, by dissecting away the liver substance, trace the anterior portions of the ventral abdominal veins into the liver and find their union with the hepatic portal vein. Note how the hepatic portal vein breaks up into many branches in the liver substance. As in other vertebrates, the direction of flow in the hepatic portal vein is from the digestive tract into the liver.

Add to the drawing the connections of the ventral abdominal vein in the liver.

4. The systemic veins.—Four large systemic veins enter the sinus venosus. Turn the ventricle forward so as to obtain a clear view of the sinus. As already noted, this is not symmetrically placed but is displaced slightly to the right, connecting with the right auricle. A large vein enters the left wall of the sinus, passing around the border of the left auricle. This is the left *precaval* vein (also called anterior vena cava and descending vena cava). Another vein, the *left hepatic* vein, emerges from the bridge of the liver and enters the left angle of the posterior wall of the sinus. The very large vein which passes into the right angle of the posterior wall of the sinus is the *postcaval* vein (also named posterior vena cava and ascending vena cava); it emerges from the right lobe of the liver. Just in front of the entrance of the postcaval vein and best seen by pressing the heart to the left, the *right precaval* vein passes into the right anterior angle of the sinus venosus.

a) *The branches of the precavals:* Each precaval enters the pericardial cavity by passing through the anterior wall of the pericardial sac. From this point it may be followed forward. As both have identical branches, it is necessary to follow only one. The one whose branches appear to be filled with blood should be selected; the left one is usually easier to follow. In specimens which have been preserved for a long time the dissection of the branches of the precaval veins is generally unsatisfactory because the branches are often empty; but, of those named below, as many as the condition of the specimen permits should be identified. Be very careful not to injure the adjacent arteries springing from the ventricle. Trace the precaval forward out of the pericardial sac. Shortly anterior to the place where the precaval penetrates the pericardial sac the vein receives practically simultaneously four tributaries—three small and one large one. The most medial branch is the small *thyreoscapular* vein, which collects a branch from the thyroid gland (the gland situated in the fork of the large arteries) and then passes to the inner surface of the shoulder, where it collects from several muscles. Lateral to this vein is the slightly larger *internal jugular* vein. This runs anteriorly along the side of the neck in contact with a white nerve (vagus or tenth cranial nerve). It receives medially an extensive network of branches from the esophagus. It may be traced anteriorly to the base of the skull from which it issues, making an anastomosis with the external jugular vein, soon to be described. The third tributary of the precaval vein is the large *subclavian* vein, by far the largest of the four branches which enter the precaval. It passes along the side of the neck and, as the *axillary* vein, turns toward the shoulder. Here it is seen to be formed by the union of two large branches, the *external jugular* vein from the neck and the *brachial* from the forelimb. The external jugular lies along the side of the neck, lateral and dorsal to the internal jugular. It collects from the head and, in its passage posteriorly along the neck, has at regular intervals *vertebral* veins passing into it from between the vertebrae. Near its junction with the brachial it receives the last of the vertebral veins, which descends from the junction between last cervical and first trunk vertebra, where it connects with the anterior end of the vertebral vein described with the renal portal system. The external jugular vein also receives branches from the skin and muscles of the shoulder region. The fourth and most lateral and dorsal of the tributaries of the precaval is a small *scapular* vein which comes from the muscles covering the scapula.

Draw the branches of the precaval as far as you have found them.

b) *The left hepatic vein:* The left hepatic vein should be traced into the left lobe of the liver, from which it collects venous blood. To do this, clear away the intervening posterior wall of the pericardial sac and pleuroperitoneum.

c) *The postcaval vein:* Trace the postcaval vein posteriorly into the right lobe of the liver. Its course may be followed by making a slight hole in the vein where it enters the sinus and probing posteriorly into the hole, dissecting away the liver substance along the probe. Note the numerous hepatic veins which enter the postcaval during its passage through the liver. Find where the postcaval enters the liver from behind to the right of the hepatic portal vein. At this point the serosa of the liver is fused to the pleuroperitoneal membrane over the ventral face of the lung. This fusion should be broken and the postcaval vein freed. On following it posteriorly it will be found to swerve toward the median line, where it runs alongside a large artery (dorsal aorta). The postcaval may be traced to the posterior end of the pleuroperitoneal cavity. Its relations there will be described later.

Add the left hepatic and postcaval veins to your drawing of the precaval vein.

5. The pulmonary veins.—A *pulmonary* vein passes from each lung to the left auricle of the heart. It is situated posterior to the bronchus, where it should be identified. Follow it toward the heart. It passes dorsal to the precaval vein. The right pulmonary runs in the dorsal wall of the pericardial sac anterior to the sinus venosus and joins the left vein at the entrance of both into the left auricle. The point of entrance is near the left precaval vein.

6. The aortic arches and their branches.—From the ventricle three large arterial trunks extend forward. Together they constitute the conus arteriosus, which must be conceived of as having split into three trunks. Clean away the connective tissue from these arteries and separate them from each other. The trunk farthest to the left is the *pulmonary* artery; the vessel next to it is the *left aorta;* the third and right-hand trunk is the *right aorta*, but it is concealed from view by the large branch, the *brachiocephalic* (innominate) artery, which it gives off immediately on leaving the heart. Note the small *coronary* arteries springing from the base of the brachiocephalic artery and branching over the surface of the heart. The brachiocephalic artery lies in the median line and forks at once into large branches. In the angle of the fork lies a reddish body, the *thyroid gland.*

a) *The branches of the brachiocephalic artery:* We shall follow this vessel first; it divides at once into four trunks: the large medial ones are the right and left *subclavian* arteries, the much smaller lateral ones are the right and left *carotid* arteries. Clean away connective tissue from these vessels and follow their courses. The two subclavians embrace the thyroid gland between their bases and supply small *thyroid* arteries into this gland. Each subclavian next gives off branches to the ventral side of the neck and to the trachea, of which the chief one is the *ventral cervical* artery, a vessel arising from the subclavian about one-half inch beyond the thyroid gland and branching profusely into the esophagus, trachea, muscles of the neck, and thymus gland. The *thymus* gland is a yellowish mass lateral to the ventral cervical artery and receiving branches from it. The subclavian artery, now named the *axillary,* turns laterally and passes to the inner surface of the pectoral girdle, where a large branch arises and branches extensively into the pectoral and shoulder muscles. The axillary then turns abruptly posteriorly and about an inch beyond the turn gives off the small *dorsal cervical* into the neck, the *marginocostal* laterally, and the *vertebral* caudally. The marginocostal runs laterally and then turns posteriorly, coursing along the curve of the carapace. The vertebral passes backward along the vertebral column dorsal to the ribs alongside the vertebral vein and gives off at thc sutures of the costal plates the *intercostal* arteries, which run laterally into the marginocostal artery. At the point where the marginocostal and vertebral arteries arise from it, the axillary bends sharply laterally and, as the *brachial* artery, passes into the forelimb alongside the scapular vein.

Each carotid artery passes forward along the ventral side of the neck, soon crossing dorsal to the subclavian and then coming to lie medial to the subclavian. In specimens in which the neck is drawn into the shell the carotids usually make loops in the neck region. As the carotid artery passes the thymus gland, it gives branches into the gland. It then proceeds, without branching, the entire length of the neck in contact with the internal jugular vein and the vagus nerve, and enters the skull by a foramen in front of the auditory region.

Draw the brachiocephalic artery and its branches.

b) *The pulmonary arteries:* The pulmonary artery is the one farthest to the left of the three arterial trunks which spring from the ventricle. It divides immediately into *right* and *left* pulmonary arteries. To see this division, lift the pulmonary trunk and look on its dorsal side. Trace the left pulmonary first. It proceeds laterally posterior to the left aorta, to which it is more or less bound by connective tissue, marking the site of the embryonic *arterial ligament* or *ligament of Botallus* (significance explained later), rarely persisting in the adult. The pulmonary proceeds directly to the left lung in company with the left bronchus and left pulmonary vein. Trace the right pulmonary in the same way; it is similarly bound to the right aorta.

Add these to the preceding drawing.

c) *The right and left aortae:* Trace both of these arteries away from the heart. Each makes a curve as it leaves the heart and turns posteriorly, passing dorsal to the precaval vein, the bronchi, and the pulmonary vessels, and disappearing dorsal to the lobes of the liver. Vessels already studied may be cut to follow the aortae posteriorly. Trace the left aorta first. Grasp the stomach and left lobe of the liver and press them to the right, separating the cardiac end of the stomach from the lung. The left aorta will be found passing to the left of the esophagus and dorsal to the stomach. It gives off simultaneously three large branches. One of these is the *gastric* artery, which passes to the stomach in the cardiac region and follows the curve of the stomach along the length of this organ. After a short distance it forks into *anterior* and *posterior* gastric arteries, which supply the lesser and greater curvatures of the stomach, respectively. Another branch from the left aorta is the *coeliac* artery. It soon forks into *anterior* and *posterior pancreaticoduodenal* arteries. The anterior pancreaticoduodenal artery passes to the left end of the pancreas, there gives off branches into the pyloric end of the stomach and to the liver, then turns to the right and runs along the pancreas, supplying the liver, pancreas, and duodenum with many small branches. The posterior pancreaticoduodenal artery enters the right end of the pancreas and, passing along the pancreas, supplies branches to the liver, pancreas, duodenum, and gall bladder. The third branch of the left aorta is the *superior mesenteric* artery. It runs posteriorly in the mesentery; trace it, tearing the mesentery, to the center of the coils of the mesentery. At this point the artery breaks up in a fanlike manner into many radiating branches which traverse the mesentery to all parts of the small intestine. One branch, the *inferior mesenteric*, passes to the large intestine and accompanies it to the cloaca.

Now follow the left aorta posterior to the point where it gives rise to the superior mesenteric artery. It becomes smaller and very soon meets another vessel coming from the right. The two join in a V-shaped manner and form one vessel, the *dorsal aorta*, which continues posteriorly in the median dorsal line. Follow the vessel which meets the left aorta anteriorly, to discover its identity. Separate the right lobe of the liver from the right lung, and turn the liver and duodenum to the left. The vessel in question can then be traced anteriorly dorsal to the right bronchus and pulmonary vessels to the heart.

It is therefore the *right aorta.* Immediately beyond its origin from the heart the right aorta gives rise to the large brachiocephalic artery whose branches were followed above. It has no other branches.

Draw the right and left aortae and the branches of the latter.

7. The dorsal aorta and the postcaval vein.—The digestive tract may now be removed, except the large intestine, which is to be left in place. Follow the dorsal aorta posteriorly. It runs in the median line ventral to some long muscles and in company with the postcaval vein, which courses at first to its right and later comes to lie ventral to the aorta. We shall study the branches of both vessels. The postcaval vein is seen to be formed by two vessels running along the medial side of the kidneys. Each of these receives numerous *renal* and *genital* veins from the kidneys and reproductive organs, respectively. The postcaval vein is thus seen to originate between the kidneys. After adding these branches to your drawing of the postcaval vein, the vein may be removed and the dorsal aorta studied. The aorta gives off a number of small branches into the muscles on which it rests and then passes between the two kidneys. Hold the large intestine backward and clear away the connective tissue from between the two kidneys. The dorsal aorta is seen to give numerous *renal* arteries into the kidneys and *genital* arteries to the reproductive system. At the posterior end of the kidneys it forks into the *right* and *left common iliac* arteries.

Separate one kidney from the carapace and press it and the reproductive organs to the other side. Two large arteries will be seen emerging dorsal to the kidney. The anterior one is the *epigastric* artery; the posterior one, the common iliac mentioned in the preceding paragraph. Follow the epigastric. (If it was injured in the dissection of the renal portal system, try the other side.) It runs laterally to the point where the renal portal vein enters the pleuroperitoneal cavity. At this point it divides. The anterior branch continues to the carapace and runs forward along the curve of the carapace, supplying the fat bodies and becoming continuous with the marginocostal artery described above. The posterior branch turns and passes medially parallel to the ventral abdominal vein. It supplies the base of the leg and the pelvic muscles and terminates on the ventral surface of the pelvis.

Next follow the common iliac artery of the same side. It divides at once before it has emerged from above the kidney into an *internal iliac* and an *external iliac* artery. The external iliac forks after a short distance. The ventral and larger branch supplies the muscles of the pelvis and, as the *femoral* artery, enters the thigh. The smaller and dorsal branch passes deep dorsally to the point where the ilium is articulated to the sacral ribs; here it passes dorsal to a nerve and turns ventrally, as the *sciatic* artery, into the hind leg, running along the medial surface of the ilium. The internal iliac is best followed by replacing the kidney against the dorsal wall, pulling the large intestine backward and locating the point of origin of the internal iliacs from the common iliac. The chief branch of the internal iliac is the *haemorrhoidal* artery, which passes forward along the side of the large intestine; in addition, there are branches to the accessory bladders, the lower ends of the oviducts, and the pelvic region in general.

Draw the dorsal aorta and its branches.

8. The structure of the heart.—Separate the heart of the turtle by cutting across the great vessels and remove it from the body. The posterior chamber of the heart is the *sinus venosus*, which receives the four great systemic veins. Clean out the blood from the sinus. It is a thin-walled chamber attached to the right auricle, into which it opens by the *sinauricular opening*, guarded by a pair of thin valves. Open each auricle by making a slit in the margin and washing out the blood clots. The walls of the auricles are somewhat spongy. Look into the left auricle and note the thin *interauricular septum* which completely separates the cavity of the left auricle from that of the right one. Find the opening of the pulmonary veins into the dorsal wall of the left auricle near the septum. Find on each side the large *auriculoventricular opening* between each auricle and the ventricle. Make a cut all of the way around the margin of the ventricle, so as to make dorsal and ventral flaps of the ventricle. Spread apart the two flaps, cautiously extending your cut inward until the two flaps are attached only along the base of the ventricle. Note the exceedingly thick walls of the ventricle and the muscular columns projecting into the interior. The cavity of the ventricle is a broad but flattened cavity, usually containing a spongy network, which may be cleaned out. Spreading the two flaps widely, note in the base of the ventricle a band passing across from one side to the other. On each side of this band is an auriculoventricular opening. The band is a continuation of the interauricular septum and forms a fold or valve on each side, which partially occludes the auriculoventricular opening. The right valve continues ventrally into a ridge which is on the ventral flap of the specimen. This ridge is the incomplete *interventricular septum*. On bringing the two flaps of the specimen together, it will be seen that the interventricular septum was connected with the muscular wall of the dorsal flap and that a space is left dorsal to the septum by which the right and left ventricles communicate with each other. The right ventricle to the right of the septum is very small, while the left ventricle is much larger and communicates with the cavity of both auricles because of the incomplete character of the interventricular septum. Spread the flaps of the specimen again and pass a probe ventral to the interventricular septum. The probe emerges in the pulmonary artery. Probe into the other arterial trunks and find their openings into the ventricle. The opening of the left aorta is to the right of the interventricular septum, into the small right ventricle, while that of the right aorta is to the left of the septum, into the left ventricle; however, because of the gap dorsal to the septum, the left aorta can also obtain blood from the left ventricle. By slitting open the arterial trunks find the little pocket-like *semilunar* valves which guard their exits from the ventricle. Their presence shows that the bases of the arterial trunks represent the conus arteriosus.

9. **Circulation through the turtle heart.**—We may now attempt to explain the somewhat complicated course of the circulation through the turtle's heart. It was noted that all of the venous blood returns to the sinus venosus, which in turn connects with the right auricle; from the latter the blood passes into the right side of the ventricle. Although this is imperfectly separated from the left side of the ventricle, the venous blood is well retained in the right side because of the spongy nature of the ventricular walls. Meantime, the two pulmonary veins have returned the blood from the lungs to the left auricle. Since the function of the lungs is to aerate the blood, this blood is arterial. From the structure and relations of the heart, the right auricle always contains venous blood, and the left auricle arterial blood. The left auricle passes the arterial blood into the left side of the ventricle. There is some slight mixture of venous and arterial blood in the ventricle. As the ventricle contracts, both kinds of blood are moved toward the arterial trunks. We have noted that the pulmonary artery springs from the small right ventricle and that the opening into this artery is to the right of, and somewhat concealed by, the interventricular septum. When the ventricle contracts, the pressure practically closes the septum, so that most of the venous blood passes out into the pulmonary artery. Simultaneously, the nearly pure arterial blood in the left ventricle passes into the base of the right aorta, since that is connected with the left ventricle and since the communication with the left aorta is closed temporarily by the interventricular septum. Toward the end of the contraction the remaining blood in both ventricles passes into the left aorta, as the diminished pressure again opens up the gap in the septum. It thus happens that the brachiocephalic artery passing to the anterior part of the body contains nearly pure arterial blood, while the left aorta contains mixed blood. On account of the fusion of right and left aortae, the dorsal aorta also carries mixed blood. It is universally true among vertebrates that the arrangement of the circulatory system is such that the purest blood is received by the head; and this is no doubt correlated with the greater oxygen requirements of the nervous parts of the head.

Whereas in fishes there is a single circulation through the heart and the heart carries only venous blood, in amphibians and reptiles there is a double circulation. One half of the heart (always the left) conveys arterial blood, and one half (always the right) carries venous blood. This change is correlated with the development of the lung method of respiration, with the consequent shifting of the sinus venosus to the right side and the division of the atrium into two separate chambers. In fishes much of the force of the heart beat is lost because the blood must pass through the gill capillaries before reaching the general body parts. In tetrapods the system has become more efficient, since the blood, after a passage through the capillaries of the lungs for aeration, returns to the heart and gets driven through the body by the direct force of the heart beat. This efficiency is the purpose of the double circulation. But the apparatus as found in amphibians and reptiles is still imperfect in that the arterial and venous streams become somewhat mixed in the heart. This situation cannot be remedied merely by completing the interventricular septum (as has, indeed, happened in the Crocodilia), because the left aorta opens from the right ventricle and would continue to receive venous blood. The difficulty is caused in amphibians by the persistence of the conus arteriosus and the ventral aorta and in reptiles by the splitting of these into three arterial trunks. One can see that, if the conus and ventral aorta were to split into two trunks, one of which (pulmonary) is connected with the right side of the heart and the other (aorta) with the left side, and if, further, the interventricular septum would be completed, the difficulty would be overcome and no venous blood could get into the arterial system. This is precisely what has happened in birds and mammals, in which the double circulation is complete and perfect. This, however, has been accomplished along different lines in the two groups. In birds it has been achieved by the obliteration of the left aorta altogether; birds thus reveal their close relationship to reptiles. In mammals the conus arteriosus splits into only two arterial trunks, related to the heart as above indicated; thus, mammals must have come from a reptilian line in which the splitting into three trunks had not yet occurred.

10. Comparison of the circulatory system of the turtle with preceding forms.—Changes from the elasmobranch circulatory system noted for amphibians have continued to progress in reptiles. In the arterial system profound changes have occurred in the aortic arches (Fig. 104). Of the six aortic arches characteristic of vertebrate embryos, four complete ones (3–6) persist in elasmobranchs and some urodeles, and the first and second are imperfect in elasmobranchs, missing in amphibians; *Necturus* also lacks the fifth pair of arches. In the turtle we found that only one pair of arches remains; this pair is the fourth, which unites to form the dorsal aorta, and is usually referred to as *the* aortic arch, since there is but one. The third pair of arches is represented in the turtle by the bases of the carotid arteries (Fig. 104) and is quite disconnected from the dorsal aorta. The fifth pair is absent as in all amniotes. The bases of the sixth pair persist as the bases of the pulmonary arteries (Fig. 104); the remainder of the sixth pair connecting with the dorsal aorta is present in tetrapod embryos as the ductus of Botallus or arterial duct; but, after birth or hatching, this closes up and degenerates into a band of connective tissue, the arterial ligament noted in the dissection (Fig. 104). In reptiles conus arteriosus and ventral aorta are split into three trunks, as already noted, and do not exist as such. The branches of the dorsal aorta are similar to those of the animals already considered.

In the venous system the anterior cardinal veins persist as the precaval veins; and the bases of these, where they enter the heart, are the ducts of Cuvier. The renal portal system is like that of *Necturus* in that it collects not only from the tail but also from the hind limbs. The renal portal veins are the persistent posterior parts of the posterior cardinal veins; and the posterior cardinal system differs from that of *Necturus* only in that the anterior parts of the posterior cardinals, persistent in *Necturus*, have disappeared in adult reptiles, being functionally replaced by the vertebral veins, formed by a longitudinal anastomosis between the segmental branches of the embryonic posterior cardinal. The ventral abdominal veins as in *Necturus* have shifted their connections away from the cardinal system and join the renal portal system posteriorly, the hepatic portal vein anteriorly (Fig. 101). Thus in amphibians and reptiles the blood from the hind limbs and tail can pass either into the renal portal system or into the ventral abdominal veins and so into the hepatic portal system. This arrangement appears to be an attempt to prevent the stagnation of blood from the posterior part of the body in the kidneys. The renal portal veins pass into a capillary system in the kidneys, from which blood is re-collected into the postcaval vein. It has been reported that in some turtles there are direct channels through the kidney by which the renal portals empty into the postcaval. This marks the beginning of the retrogression of the renal portal system.

The postcaval vein, with the loss of the anterior parts of the posterior cardinal veins, becomes the chief vein of the posterior part of the body. Its mode of origin in amphibians and reptiles is shown in Figures 101, 102. Its anterior part arises from the hepatic veins (which are the vitelline veins of the embryo). Its posterior part between the kidneys is formed of the two subcardinal veins, which in the embryo are continuous with the posterior cardinals. The middle part of the postcaval between these two regions is formed of the right subcardinal and a new outgrowth from the hepatic portion.

11. The circulatory system of *Sphenodon*.—Whereas the circulatory system of other reptiles corresponds in general to that of the turtle, this system in *Sphenodon* presents a number of primitive features. There is a very short persistent conus arteriosus from which the three main arterial trunks spring. The third or carotid aortic arch persists in its entirety; and the sixth or pulmonary arch is also complete, since the two ducti arteriosi remain open, connecting

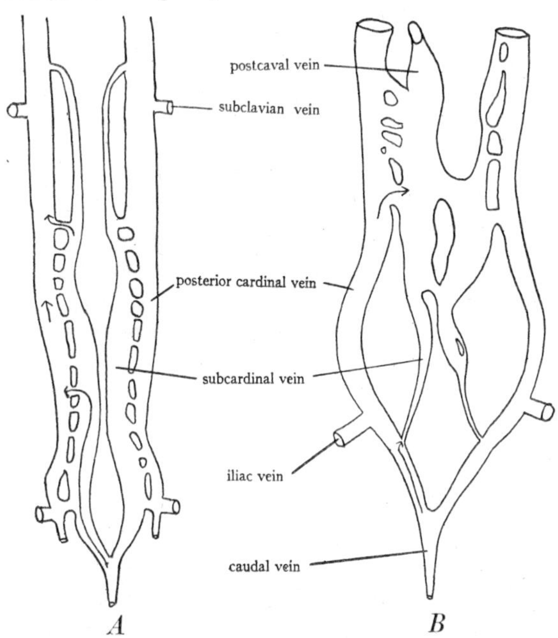

FIG. 102.—Development of the renal portal and postcaval systems of the turtle (after Stromsten, 1905). *A*, caudal vein opens into the subcardinals from which blood flows into the posterior cardinals. *B*, caudal vein has shifted to the posterior cardinals, and blood now flows from them into the subcardinals; postcaval vein forming from the right subcardinal.

the pulmonary arteries to the dorsal aorta (Fig. 103). Thus *Sphenodon* differs from all other amniotes and resembles *Necturus*, in that there are present three complete aortic arches—the third, fourth, and sixth. As a result of the persistence of the carotid arch, there is no common carotid artery, but the external and internal carotids spring separately from the arch (Fig. 103). The dorsal aorta also has a full set of segmental branches. In the venous system *Sphenodon* differs from all other reptiles (and resembles mammals) in retaining the anterior parts of the posterior cardinal veins as the vertebral veins, which give off segmental branches into the vertebral column and intercostal veins along the ribs. The renal portal and postcaval systems are similar to those of the turtle except that there is but one ventral abdominal vein. The interventricular septum of the heart is but slightly developed.

E. THE CIRCULATORY SYSTEM OF THE PIGEON

In birds the consolidation of the heart is completed. Through the incorporation of the greatly reduced sinus venosus into the right auricle and the splitting of the conus arteriosus to form the bases of the two great arterial trunks the heart comes to consist only of the two auricles and the two ventricles (now completely separated), and hence is often referred to as the "four-chambered" heart. From the comparative point of view, it would more correctly be termed the "two-chambered" heart, as only two of the original and embryonic four chambers remain as such in the adult. The entrances of the great veins, with the reduction of the sinus venosus, have shifted anteriorly. The left (fourth) aortic arch is obliterated, and only the right arch remains as the conduit to the dorsal aorta. This means that only two great arterial trunks spring from the heart, the aorta and the pulmonary; and the double circulation is now perfected. The general circulatory system is reptilian in character; the principal change is the direct connection of the postcaval vein with the renal portal system and the great reduction in the portal circulation through the kidneys.

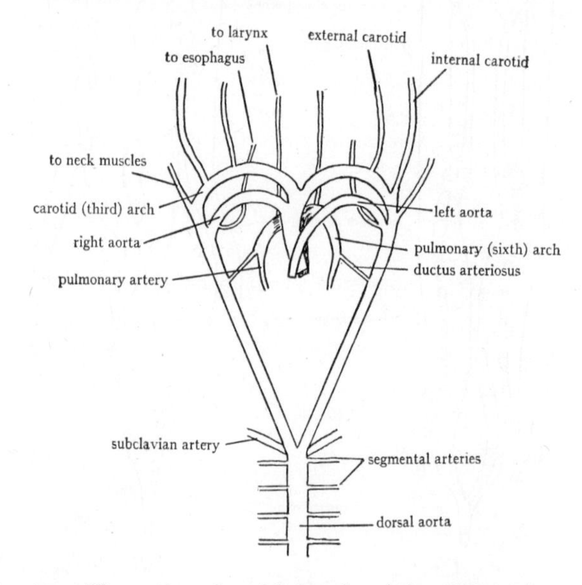

FIG. 103.—The aortic arches of *Sphenodon*. (After O'Donoghue, 1921.)

In case a fresh specimen is provided for this work, it should be opened, as before, by deflecting the pectoral muscles from either side of the keel of the sternum, then cutting through the sternum on each side of the keel and removing a median portion of the sternum including the keel. The peritoneal cavity is to be opened, as before, by a longitudinal incision. The specimen should have been injected through the pectoral artery.

1. The chambers of the heart.—The heart is relatively large and more compact than in the forms previously studied. The chambers are more closely knit together than in the lower vertebrates. The major portion of the heart is formed of the *right* and *left ventricles*, which together constitute a muscular, thick-walled cone, having a pointed *apex* directed posteriorly and a broad *base* directed anteriorly. The two ventricles are completely separated from each other, but the division between them is indistinct externally.

This division passes obliquely from the left side of the base to about the middle of the right side of the heart; the left ventricle is therefore much the larger of the two and includes the whole of the apex of the heart. Anterior to the ventricles are the two much smaller *auricles*, thin-walled chambers. The division between auricles and ventricles is generally concealed by a line of fat, which should be removed. From the anterior end of the heart between the auricles the great arteries spring without the intervention of a conus arteriosus. On raising the ventricles the dorsal portions of the auricles become visible. There is no sinus venosus, and the great veins open directly into the right auricle. There are three of these veins—the two *precavals* and the *postcaval*. The postcaval enters the right auricle from behind, emerging from the liver. The precavals come from the anterior part of the body, one on each side, and, curving toward the heart at the level of the auricles, enter the right auricle. The *pulmonary* veins may be noticed opening into the left auricle.

2. **The hepatic portal system.**—Turn to the peritoneal cavity. Cut across the falciform ligament of the liver near the gizzard, noting first the small vein passing from the ventral ligament of the gizzard in the falciform ligament to the liver. The lobes of the liver may now be turned forward. Running along the dorsal surface of the liver and branching into its substance is the large *hepatic portal* vein. The main part of the vein enters the right lobe of the liver, coursing between the two bile ducts. The remainder of it lies along the dorsal surface of the left lobe of the liver, sending branches into the liver, and at the left receives the *left* and *median gastric* veins from the margin and left side of the gizzard and from the proventriculus. Follow posteriorly that part of the hepatic portal which lies between the two bile ducts. It is soon seen to be formed by the union of three veins: a *superior mesenteric*, a *gastroduodenal*, and an *inferior mesenteric*. The superior mesenteric collects from the greater part of the small intestine. The gastroduodenal receives the *right gastric* vein from the right side of the gizzard; the *pancreaticoduodenal* vein, which runs along the duodenal loop collecting from duodenum and pancreas; and the *mesenteric* vein from the last loop of the small intestine. The inferior mesenteric vein runs along the large intestine, from which it collects many branches. At its posterior end it turns dorsally and joins the renal portal system, where it will be followed later.

Draw the branches of the hepatic portal system.

3. **The systemic veins.**—As already stated, these consist of two *precavals* and one *postcaval*.

a) The branches of the precaval veins: As both veins have identical branches, only one need be followed. Find the vein on each side lateral to the auricle and trace each into the right auricle, lifting the heart. The left precaval passes around the left auricle to enter the right auricle. The right precaval is much shorter and enters the right auricle directly.

Follow one precaval forward. It lies just posterior to a large artery and is there seen to be formed by the union of three large veins—laterally the *pectoral* vein, slightly anterior and dorsal to this the *subclavian* vein, and anteriorly the *jugular* vein. Each of these veins should be followed. The pectoral vein at its union with the other receives the *internal mammary* vein, ascending from the inner surface of the ribs, and has also a tributary from the sternum and coracoid. The main vein is formed laterally by the union of two veins emerging from the pectoral muscles. These may be followed into the muscles, from which they are seen to collect many branches. The subclavian vein passes deep dorsally ventral to a group of nerves (brachial plexus) and somewhat concealed by arteries, which should not be injured. As the *brachial* vein, it emerges from the wing and then receives a branch from the shoulder muscles. The jugular vein passes anteriorly on the dorsal side of the large arteries. On tracing the jugular forward it will be found to receive the following veins, named in order from the heart forward: on the medial side, some small and then a large branch from the crop (at the point of entrance of these into the jugular is situated a small reddish body, the *cervical lymph gland*); on the lateral side, a vein from the shoulder and, at the same level, the *vertebral* vein from the vertebral column; medially, another branch from the crop; laterally a large vein from a plexus of blood vessels in the skin of the neck; then small veins from the esophagus and trachea. On freeing the anterior end of the esophagus (also trachea) and cutting across it the jugular vein can be followed to the soft palate, where it joins its fellow of the opposite side. Posterior to this union, each receives a plexus of veins from the skin of the face. On dissecting away the soft palate from the anastomosis of the two jugular veins branches from the skull will be found passing into the anastomosis.

Draw the branches of the precaval vein as far as found.

b) *The postcaval vein:* Raise the ventricles of the heart and note once more the large postcaval vein emerging from the liver and entering the right auricle between the two precaval veins. Note the large *hepatic* veins which it receives from the liver. The left one of these hepatic veins receives the small vein of the falciform ligament mentioned previously. Follow the postcaval vein into the peritoneal cavity, turning all of the viscera to the left. The postcaval will be picked up again at the posterior margin of the right lobe of the liver in contact with the dorsal body wall. The postcaval is here seen to be formed by the union of two veins, the *iliac* veins. In males the two oval testes will be noted at this point of junction. In females the single ovary and oviduct will be noted to the left, concealing the left iliac vein. Each iliac runs along the ventral face of a three-lobed organ, the *kidney*, which is set close against the dorsal body wall. Follow the right iliac vein. From between the first and second lobes of the kidney it receives the large *femoral* vein, emerging from the leg. The femoral vein receives a small branch from the body wall. Posterior to the entrance of the femoral vein the iliac vein corresponds to the *renal portal* vein of reptiles and amphibians and may be so named. From between the second and third lobes of the kidney it receives the *ischiadic* vein, which also comes from the thigh. At the posterior end of the kidneys the two renal portal veins have an anastomosis with each other. From this anastomosis rises the *inferior mesenteric* vein, already noted as a branch of the hepatic portal system. It runs in the mesorectum. The anastomosis of the renal portals also receives in the median line a small *caudal* vein from the tail and on each side an *internal iliac* vein from the roof of the pelvic region. The left iliac and renal portal veins are the same as the right, except that in the female the left veins receive *genital* veins from the ovary and oviduct. These may be seen by turning the oviduct to the right. In their course over the kidneys the renal portal veins give off branches into the kidney, as in lower forms. There is probably some portal circulation in the kidneys, but most of the blood from the renal portals passes directly into the iliac veins. The iliac vein receives *renal* veins from the kidney, of which there is one chief renal vein, which runs along the medial side of the kidney but is so imbedded in the kidney substance as to be difficult to identify. Between the two renal portal and iliac veins runs the dorsal aorta.

Turn the animal dorsal side up and remove the skin over the thigh. By separating the muscles pick up the femoral and ischiadic veins and trace them into the leg. The ischiadic vein accompanies the large sciatic nerve and soon turns forward to run parallel to the femoral vein. Both veins are accompanied by arteries of the same name.

Draw the branches of the postcaval vein and the renal portal system.

4. The pulmonary veins.—The *pulmonary* veins emerge on each side from the lung and pass toward the heart immediately posterior to the precaval veins. There is usually one pulmonary vein from each lung, but there may be two. Note the branches collected by each vein from the lung. The veins pass to the dorsal side of the bases of the precavals and enter the left auricle. Their entrance into the auricle is best seen later, when the heart is dissected.

5. The arterial system.—It has already been noted that the great arteries spring directly from the ventricle. They are situated between the two auricles. Separate their bases from the auricles. It will be then be found that there are two arterial trunks. The larger, medially located one is the *aorta*. The smaller one, passing to the left and dorsal to the aorta, is the *pulmonary* artery. The ventral aorta in birds is split into these two vessels.

a) The anterior branches of the aorta: Follow the aorta away from the heart. The aorta immediately gives rise in the median line to two large arteries, the *brachiocephalic* (*innominate*) arteries. The aorta then turns to the right and disappears dorsally. It will be followed at a later time. Identify the branches of the brachiocephalic arteries; as both have identical branches, follow only one. Each proceeds laterally and slightly anteriorly and forks into two branches—an anterior *common carotid* artery and a lateral *subclavian* artery. The subclavian artery soon gives rise to a number of branches: the small *internal mammary* artery passing posteriorly along the inner surface of the ribs, the two *pectoral* arteries to the pectoral muscles along with the veins of the same name, and the *axillary* artery to the wing. The axillary artery runs anteriorly and, after giving off a branch into the shoulder, enters the wing as the *brachial* artery. The two common carotid arteries pass forward, and at the level of the cervical lymph gland each gives rise to a *vertebral* artery, which passes dorsally into the vertebrarterial canal of the vertebral column. The common carotid arteries then approach the median line and penetrate the muscles on the ventral surface of the vertebral column. On separating these muscles in the midventral line, the two arteries may be followed forward. They pass anteriorly side by side. Shortly before they reach the head, they diverge; and at the angle of the jaws each divides into an *external carotid,* from which branches may be traced to the esophagus, palate, and head generally, and into a more deeply situated *internal carotid,* which passes through the skull to the brain.

Draw the branches of the brachiocephalic artery.

b) The pulmonary arteries: The pulmonary artery passes to the left side of the aorta and immediately forks into *right* and *left* pulmonary arteries. The left artery goes directly to the left lung. The right artery turns and, passing on the dorsal side of the brachiocephalic arteries and posterior to the turn of the aorta, enters the right lung. It will be better seen in the next paragraph.

c) The aorta: The aorta turns to the right, forming what is called the *arch* of the aorta. This may be followed by cutting across the right precaval vein and the right brachiocephalic artery. The arch of the aorta curves to the dorsal side of the right pulmonary artery, which can now be traced into the lung, and turns caudad. Follow it by dissecting away the tissue between the heart and the right lung and by breaking through the oblique septum. Turn the viscera to the left. Cut through the postcaval vein. The aorta, now called the *dorsal aorta,* lies in the median dorsal line between the two lungs. It gives off small branches to the esophagus and body wall in its passage along the pleural cavities. At the entrance to the peritoneal cavity the large *coeliac* artery arises from the aorta. This runs posteriorly along the proventriculus, to which it branches. The coeliac artery then gives rise to the relatively small *left gastric* artery, which passes to the left side of the gizzard, branching to this side and the edge of the gizzard. The coeliac artery then passes by the spleen, to which it gives small *splenic* arteries; and just beyond the spleen it gives rise to the *hepatoduodenal* branch. This sends a *hepatic* branch into the liver and then, as the *anterior pancreaticoduodenal* artery, runs along the duodenal loop supplying duodenum and pancreas. The coeliac artery continues as the *right gastric*, which spreads out over the right surface of the gizzard. The right gastric sends a large *posterior pancreaticoduodenal* branch to the duodenal loop and pancreas and a *mesenteric* branch to a loop of the small intestine.

Very shortly posterior to the origin of the coeliac artery, the *superior mesenteric* artery arises from the dorsal aorta and branches to the small intestine. One of these branches passes along the large intestine and anastomoses with the *inferior mesenteric* artery, described below.

The dorsal aorta now passes between the two kidneys. It gives off on each side the *renolumbar* artery, which supplies the anterior lobe of the kidney and then passes to the body wall and some muscles of the thigh. In female specimens *genital* arteries are given off to the ovary and oviduct from the renolumbar artery and the renofemoral artery. The *renofemoral* artery is the large vessel arising next from the dorsal aorta; it supplies the middle and posterior lobes of the kidney, then proceeds to the lateral body wall, and, as the *femoral* artery, supplies the leg. It may be followed by turning the animal dorsal side up and looking between the muscles of the thigh along the course of the femoral vein previously identified. The femoral artery accompanies the large sciatic nerve and branches into the leg muscles. Returning to the peritoneal cavity, trace the dorsal aorta further. As it passes between the kidneys, it gives off *lumbar* arteries into the dorsal body wall. At the posterior end of the kidneys it forks. At the point of forking arise the *inferior mesenteric* artery, which runs anteriorly in the mesorectum and anastomoses with a branch of the superior mesenteric artery, and the *caudal* artery, which proceeds straight posteriorly in the median line to the tail. The two forks of the dorsal aorta are named the *internal iliac* arteries. They pass posteriorly along the roof of the pelvic region. The left one gives off branches into the oviduct.

Draw the branches of the dorsal aorta.

6. The structure of the heart.—Free the heart by cutting across the great vessels. Note that all veins enter the apparently anterior end of the heart, having shifted forward out of the transverse septum. As already stated, there is neither sinus venosus nor conus arteriosus in the bird's heart. Only two of the original four chambers of the heart have persisted, namely, atrium and ventricle; but each of these is subdivided into two completely separate halves, the right and left auricles and ventricles. The auricles are small, thin-walled chambers anterior to the ventricles. In the walls of the right auricle identify the openings of the systemic veins and in the left auricle of the pulmonary veins. Slit open the right auricle and, looking within, note the thin *interauricular* septum separating it from the left auricle. A fold extends from this septum to the entrance of the postcaval vein, partly concealing the entrance. Note the deep cleft, the *auriculoventricular* opening, through which the right auricle opens into the right ventricle. Similarly open the left auricle and find the left auriculoventricular opening. Cut across the apex of the ventricles. Observe the crescentic form and relatively thin walls of the right ventricle and circular section and enormously thickened wall of the left ventricle. What appears to be the internal wall of the ventricles is the *interventricular septum*, completely separating the cavities of the two ventricles. Open the right ventricle by a slit extending from the previously cut apex to the base. Note the single *valve* which guards the right auriculoventricular opening; it is a muscular band extending from the ventricular wall to the auriculoventricular opening. In the left side of the anterior end of the right ventricle find the opening of the pulmonary artery, or probe into the base of the pulmonary artery and note emergence of the probe into the right ventricle. At the base of the pulmonary artery are three pocket-like *semilunar* valves. Cut into the left ventricle and note the two thin membranous valves which guard the auriculoventricular opening. They are called the *mitral* valve. Each is attached by delicate cords, the *chordae tendinae*, to the wall of the ventricle. The wall of the ventricle has several muscular ridges which project into the cavity; they are called the *columnae carnae*. Find, to the medial side of the mitral valve, the opening of the left ventricle into the aorta. Probe into this and satisfy yourself that it leads into the aorta. Note the three *semilunar* valves at the beginning of the aorta.

The removal of the heart permits the tracing of the esophagus into the proventriculus and of the bronchi into the lungs. The form and extent of the lungs can also be observed to advantage at this time.

7. The circulation through the heart and the comparison of the circulatory system of bird and reptile.—In birds the heart is completely divided into right and left auricles and ventricles. The venous blood enters the right auricle from the systemic veins, passes into the right ventricle and out into the pulmonary arteries, which convey it to the lungs. After aeration in the lungs the blood returns by way of the pulmonary veins to the left auricle, from which it flows into the left ventricle and out of the aorta. Thus, the right side of the heart contains only venous blood and the left side only arterial blood. There is a *perfect double circulation*, both kinds of blood flowing simultaneously through the heart, the two streams completely separated from each other. Since the aorta is connected only with left ventricle, the arterial system receives pure arterial blood. The perfection of the double circulation is achieved from the reptilian condition by the obliteration of the left aorta. The conus arteriosus is split into two trunks—aorta and pulmonary—which spring from the left and right sides of the heart, respectively. The sinus venosus is apparently absent but in reality is reduced and incorporated into the right auricle. The semilunar valves of the aorta and pulmonary show that the basal parts of these trunks represent the conus arteriosus.

The aortic arches are further modified from the condition seen in reptiles. As in the turtle, the bases of the common carotid arteries represent the third aortic arches. The union of the subclavians with these is secondary. The arch of the aorta is the right fourth aortic arch, the left fourth arch having vanished during embryonic development. There is consequently in birds no complete aortic arch, as in the preceding forms, but only, so to speak, one-half an arch, the persistent half being the right one. The pulmonary arteries represent the sixth aortic arch, separated as in reptiles from the aorta (Fig. 104 *F*, p. 380).

The venous system is reptilian in character. The two precaval veins are similar to those of the turtle and are homologous to the anterior cardinal veins of lower forms (the internal jugular branch being the original anterior cardinal). The bases of the precavals entering the sinus venosus are the common cardinal veins. The posterior cardinal veins are, as in the turtle, represented only by their posterior portions, which are named in the adult the renal portal veins. These veins have, as in reptiles and amphibians, absorbed the veins of the legs and tail. In birds it is very interesting to note the union which is in progress between the renal portal system (posterior cardinals) and the postcaval vein. As we shall see, this union is complete in mammals. In birds the renal portal system is probably to a slight degree functional as a portal system through the kidneys, blood passing from the renal portal vein into the kidneys and re-collecting into the renal veins tributary to the postcaval vein. Most of the blood, however, passes directly from the renal portal veins into the postcaval.

The origin of the postcaval vein in birds is the same as that given for reptiles. That part of it which is situated between the kidneys is formed of the subcardinal veins, chiefly the right one. The part through and anterior to the liver comes from the vitelline veins. The middle region of the vein is a new formation. The postcaval is seen to be usurping the renal portal system and thus extending itself posteriorly. This process is completed in mammals.

The hepatic portal system is similar to that in the forms already discussed. The renal portal system is identical with that of reptiles. The inferior mesenteric vein which connects the two portal systems is probably homologous to the ventral abdominal veins of reptiles. It is, however, of decreased importance as a channel between the two portal systems, because of the junction of the renal portal system with the postcaval vein.

F. THE CIRCULATORY SYSTEM OF MAMMALS

The mammalian heart has arrived at the same stage of evolution as that of birds, although from a different reptilian stem. Sinus venosus and conus arteriosus are absent as such, and the compact heart consists only of the two auricles and the two ventricles. These are completely separated by septa, and hence there is a perfect double circulation through the heart. The conus arteriosus is split into two trunks, the aorta and the pulmonary. The aortic trunk consists of the left half of the fourth aortic arch, whereas in birds it is the right half. In the circulatory system the chief change is the fusion of the renal portal system with the postcaval vein, with the complete elimination of a portal circulation through the kidneys. Mammals thus have but one portal circulation, that through the liver; and the hepatic portal vein is consequently called simply *the* portal vein.

The specimen should have been injected in the arterial system.

1. The chambers of the heart.—The heart is relatively large and compact, the chambers closely united with each other. The pericardial sac should be removed if this has not been done previously. In case the thymus gland is well developed, it will be necessary to dissect this away from the anterior part of the heart. The greater portion of the heart consists of the two *ventricles*. These constitute a firm thick-walled cone, having a posterior pointed *apex* and a broad anterior *base*. This cone consists of two completely separated ventricles, the right and left ventricles; the division between them is marked externally by an indistinct line or groove extending from the left side of the base obliquely to the right, and terminating to the right of the apex. The groove contains branches of the coronary artery and vein which will be found ramifying over the surface of the ventricles. The left ventricle is much the larger of the two and includes the apex. Anterior to the base of each ventricle is a much smaller, thin-walled, generally dark-colored chamber, the *auricle*. Each auricle in the contracted state presents a lobe, the *auricular appendage*, projecting medially and slightly posteriorly over the ventricle; in the cat (and man) this lobe has a scalloped margin and is shaped something like the human ear (hence the name auricle, meaning little ear).[6] Extending anteriorly from the middle of the ventricular base forward between the two auricles is a large artery, the pulmonary artery. This makes an arch to the left and disappears. Dorsal to the pulmonary is another arterial trunk, the aorta. These two trunks are generally imbedded in fat, which should be removed. They represent the split conus arteriosus, and thus it seems as if in mammals the great arterial trunks spring directly from the ventricles. Grasp the apex of the heart, turn the heart forward, and note the bases of the great veins (pulmonary and systemic veins) entering the auricles. A sinus venosus is lacking as a distinct chamber, for it is greatly reduced and absorbed into the right auricle, where it can be located by physiological experiments as a small spot (*sinauricular node*) at the entrance of the systemic veins into the right auricle. From the morphological point of view, the mammalian heart is then reduced to two of the four original chambers, although each of these two has been partitioned into two parts.

2. The hepatic portal system.—Turn to the peritoneal cavity. Press the lobes of the liver forward and the other viscera to the left. Put the hepatoduodenal ligament on a stretch by widely separating the stomach and liver, without, however, tearing the ligament. In the ligament lying dorsal to the common bile duct is the large *hepatic portal* vein (commonly called simply the *portal* vein in mammals, since mammals have but one portal system). Free it by carefully cleaning connective tissue from its surface. Follow it anteriorly and note how it branches into the liver substance. Follow it posteriorly,

[6] In human anatomy only the ear-shaped lobe is termed the auricle, and the whole chamber is called atrium. There is no justification for this usage in comparative anatomy, where it is customary to call the undivided chamber atrium and its two halves the auricles.

ripping away fat and connective tissue from its surface with the dull point of a probe. Note the large branch it sends into the right lateral lobe of the liver. The branches received by the portal vein from the digestive tract are slightly different in the rabbit and cat. In preserved specimens the branches are not always easy to follow, and the student should identify as many as possible. The arteries accompanying the veins must not be injured.

Rabbit: Immediately posterior to the branch into the right lateral lobe of the liver, the portal vein receives on the right side the *gastroduodenal* vein. This vein is soon seen to be formed by the union of two veins, a larger *anterior pancreaticoduodenal* vein, which appears as a continuation of the main vein, and the smaller *right gastroepiploic* vein. The first-named vessel runs in the tissue of the pancreas alongside the first part of the duodenal loop, collecting tributaries from both pancreas and duodenum. The right gastroepiploic vein comes from the pyloric region of the stomach and receives also branches from the great omentum. Shortly posterior to the entrance of the gastroduodenal vein into the portal the portal receives on the left side the larger *gastrosplenic* vein. This vein is seen to be formed a short distance from the portal by the union of the *splenic* and *coronary* veins. The latter comes from the lesser curvature of the stomach, where it is seen to be formed by numerous branches collecting from both surfaces of the stomach. The splenic vein is a large vessel running in the great omentum past the spleen and extending as far as the left end of the stomach. In its course it collects numerous splenic branches from the spleen and the *left gastroepiploic* veins from the stomach and omentum. Some distance posterior to the entrance of the gastrosplenic vein, the portal receives the *posterior pancreaticoduodenal* vein, which runs in the mesentery of the duodenal loop, collecting from pancreas and duodenum, and anastomosing with the anterior pancreaticoduodenal vein. At the same level as the entrance of this vein the portal receives, on the opposite side, the *inferior mesenteric* vein. This may be traced alongside the descending colon and rectum, from which it receives many branches, as well as some from part of the transverse colon. The main trunk of the hepatic portal posterior to this point is now named the *superior mesenteric* vein. It collects from all parts of the intestine not already mentioned. In tracing its branches tear the mesenteries which bind together the coils of the intestine as far as necessary and also strip off fat and lymph glands. The *intestinal* vein is the large vessel collecting from the greater part of the small intestine. It runs in the middle of the mesentery, receiving many tributaries in its course. The branches from the jejunum immediately beyond the duodenum, however, enter the posterior pancreaticoduodenal vein. The very large *ileocaecocolic* vein collects from the ileum, appendix, caecum, and ascending and transverse colons. Chief among its tributaries are: the *appendicular* vein from the appendix; the *anterior ileocaecal* vein from the sacculus rotundus, proximal part of the caecum, adjacent ileum, and ascending colon; and the *posterior ileocaecal* vein from the distal part of the caecum, adjacent ileum, and ascending colon.

Draw, showing the portal system.

Cat: On following the portal away from the liver three small veins are found to enter it: the *coronary*, the *anterior pancreaticoduodenal*, and the *right gastroepiploic*. These may enter separately or may unite in any combination; usually the last two unite before joining the portal vein. The coronary vein comes from the stomach and lies in the curve between pylorus and stomach, formed at the lesser curvature by the union of many branches from both sides of the stomach. The anterior pancreaticoduodenal collects from the pancreas and duodenum. The right gastroepiploic vein comes from the pyloric region, greater curvature of the stomach, and adjacent greater omentum. Beyond the entrance of these three veins the hepatic portal receives a large tributary, the **gastrosplenic vein**. This passes to the left in the substance of the pancreas, receiving one or more small *middle gastroepiploic* veins from the stomach wall and omentum and a *pancreatic* vein from the pancreas. Beyond these tributaries the gastrosplenic is formed by the union of two main branches, the *right* and *left splenic* veins. The left splenic vein passes in the gastrosplenic ligament along the spleen, receiving branches from the spleen, the greater omentum, and several *left gastroepiploic* veins from the omentum and stomach. The right splenic vein comes from the right end of the spleen, receiving also tributaries from the omentum and stomach wall. Beyond the entrance of the gastrosplenic vein the portal is known as the *superior mesenteric* vein. This soon receives a small *posterior pancreaticoduodenal* vein from the pancreas and distal part of the duodenum; next, the *inferior mesenteric* vein from the descending colon and rectum; and then is seen to be formed by numerous converging *intestinal* branches from the small intestine, caecum, and ascending colon. The lymph glands lying along the superior mesenteric vein, as well as fat, should be removed in tracing the branches.

Draw the hepatic portal system.

It will be seen that the relations of the hepatic portal system are the same in mammals as in all other vertebrates. The system conveys all of the venous blood from the digestive tract into the capillaries of the liver. The purpose of this arrangement is that the liver cells may remove from the blood the digested food materials. There is no renal portal system in mammals, it having been completely usurped by the postcaval vein.

3. The systemic veins.—There are three systemic veins in the rabbit—two precavals and one postcaval—and two in the cat—one precaval and one postcaval. The condition in the cat is due simply to the union of the two precavals anterior to the heart (Fig. 106, p. 382). Although the branches are similar in the two animals, they will be described separately.

a) *The branches of the precaval vein:* This vein is also called the anterior vena cava and descending vena cava.

Rabbit: Turn the apex of the heart forward and examine the great veins which enter the right auricle. The *left precaval* vein comes from the left and, passing around the left auricle, enters the left side of the right auricle. It receives small *coronary* veins from the heart wall. The *right precaval* passes directly into the right anterior part of the right auricle. Note additional coronary[7] veins entering the right auricle directly.

Carefully trace the right precaval forward, clearing away connective tissue and muscle from about its course, and follow it away from the heart. At the point of entrance into the right auricle it receives, from behind, the *azygos* vein. Press the lungs to the left and follow the azygos posteriorly along the dorsal thoracic wall near the median line. Note the *intercostal* veins which enter it at segmental intervals; they course along the posterior margin of each rib. Entering the precaval immediately anterior to the entrance of the azygos is the *superior intercostal* vein, the first of the series of intercostal veins. Shortly anterior to this the *internal mammary* vein enters the precaval. This vein ascends on the internal surface of the chest very near the midventral line. Trace it posteriorly, noting branches from the intercostal muscles. It continues posteriorly on the abdominal wall as the *superior epigastric* vein. The next tributary of the precaval is the *vertebral* vein. It enters the medial side of the precaval at about the same level as the internal mammary joins the lateral side. It may be traced deep dorsally to the cervical vertebrae, from which it emerges, receiving a *costocervical* tributary from the neck. Beyond this point the precaval receives the large subclavian vein from the forelimb. Follow this laterally. It passes between the first and second ribs into the axilla and is then known as the *axillary* vein. Expose the axilla by cutting down through the pectoral muscles near the midventral line and at their insertion on the humerus. The pectoral muscles should then be separated from the underlying serratus ventralis but should not be removed. The large, stout, white cords seen crossing the axilla are the nerves of the brachial plexus and are not to be injured. Lymph glands—small rounded masses—will also be noted in the axilla.

In the axilla the axillary vein receives the following branches: the *long thoracic* vein, the *subscapular* vein, and the *cephalic* vein. The long thoracic vein runs caudad on the thoracic wall in the serratus muscle; it then passes to

[7] The term coronary (meaning literally a crown or wreath) is in mammals unfortunately applied to two vessels, those of the heart wall and a vessel of the stomach wall having radiating branches. To avoid confusion the latter is referred to as the coronary vessel of the stomach (coronaria ventriculi).

the inner surface of the skin and extends the entire length of the abdominal wall, being especially prominent in females, where as the *external mammary* it collects from the mammary glands. (The greater part of this vein was probably removed with the skin.) The subscapular vein enters the axillary dorsal to the preceding. It collects a conspicuous branch (*thoracodorsal* vein) from the latissimus dorsi and cutaneous maximus muscles; it then passes through the teres major muscle to the external surface of the shoulder, where it collects from various muscles. The cephalic[8] vein is the chief superficial vein of the arm. It can best be picked up on the outer surface of the upper arm; near the distal end of the upper arm it penetrates deep between muscles and, passing between the teres major and subscapularis muscles, emerges on the internal surface of the shoulder and enters the axillary vein at the same place as, or in common with, the subscapular vein. Immediately beyond the entrance of these tributaries, the axillary vein becomes the *brachial* vein of the arm. This proceeds along the inner surface of the upper arm in company with an artery and a nerve.

Return to the precaval vein. At the point of entrance of the subclavian vein the precaval vein receives from the neck the *external* and *internal jugular* veins. The external jugular vein is the large vein which extends forward in the depressor conchae posterior muscle (most superficial muscle of the ventral surface of the neck). It appears as the anterior portion of the precaval. The internal jugular vein is a very small vein which runs alongside the trachea, passing the thyroid gland, and accompanying the carotid artery and the vagus nerve. The place of entrance of the internal jugular, as well as its general relations, is highly variable; it may enter the precaval after the latter has received the subclavian, but it usually enters with the external jugular. The precaval vein may thus be said to be formed by the union of the subclavian, external jugular, and internal jugular veins. Follow the external jugular. Shortly anterior to its union with the subclavian it receives the *transverse scapular* vein from the ventral end of the shoulder, and near the same level it has a cross-connection (*transverse jugular* vein) with its fellow of the opposite side (this union was probably destroyed in the previous dissection). Along the neck it receives various small tributaries from muscles and about one inch posterior to the angle of the jaws is seen to be formed by the union of two veins, the *anterior* and *posterior facial* veins. The anterior facial vein proceeds to the angle of the jaws, where it is seen to be formed by the union of veins from the anterior part of the face and jaws. Its main tributaries are the *angular* vein, which passes over the ventral part of the masseter muscle and then turns to the region in front of the eye, and the *deep facial* vein, which emerges between the masseter and digastric muscles and passes along

[8] So named because the corresponding vein in man was formerly thought to connect with the head.

the surface of the masseter. Other tributaries of the anterior facial vein come from the near-by lymph and salivary glands. The posterior facial vein may next be followed. It passes to the parotid gland, where it receives a superficial vein, the *posterior auricular* vein, from the back of the ear and head. The main vein beyond the entrance of this branch lies imbedded in the parotid gland, which may be dissected from it. The vein is accompanied by the facial nerve. At the base of the ear it is formed by the *inferior ophthalmic* vein from the orbit, the *temporal* veins from the temporal region, and the *anterior auricular* vein from the region in front of the ear.

The internal jugular vein extends the length of the neck, receiving but few small branches, of which the chief ones are those from the thyroid gland. It may be traced to the occipital region of the skull, from which it emerges by way of the jugular foramen; it collects part of the blood from the brain. As already stated, its size and place of junction with the external jugular are highly variable.

The left precaval vein is identical in its tributaries with the right, except that there is no azygos vein on the left side.

Draw the branches of the precaval as far as found.

Cat: Turn the apex of the heart to the left and note the large vein which enters the anterior margin of the right auricle. This is the *precaval* vein. Note there is no such vein on the left side. Instead there is a vein called the *coronary sinus*, which runs along the dorsal surface of the heart in the groove between the auricles and ventricles. The coronary sinus will be found by cleaning out the fat from this groove. Note the numerous *coronary* veins which come from the heart wall and enter the sinus. The sinus itself opens into the left posterior corner of the right auricle. It represents the reduced proximal part of a former left precaval vein. The distal part of this vein is still present and, as we shall see shortly, is united with the right precaval. Again pressing the heart to the left, clean the base of the precaval and note the large vein which passes in front of the root of the right lung and joins the precaval as the latter enters the auricle. This tributary of the precaval is the *azygos* vein. Trace it posteriorly, pressing the right lung to the left. It passes along the dorsal thoracic wall near the middorsal line and receives at regular intervals the *intercostal* veins. These course along the posterior borders of the ribs. The most anterior of the intercostal veins join into a common trunk which enters the azygos shortly caudad to the entrance of the latter into the precaval. The azygos also receives small branches from the esophagus and bronchi.

Trace the precaval anteriorly. It receives small branches from the thymus gland and then receives a tributary of moderate size, the common stem of the *internal mammary* veins, which comes from the midventral wall of the chest.

On following this posteriorly it is soon seen to be formed by the union of two veins, the internal mammary veins, which run posteriorly in the chest wall one to each side of the midventral line and are extended onto the abdomen as the *superior epigastric* veins. In their course the two internal mammary veins receive branches from the diaphragm, chest wall, pericardium, etc. The precaval vein next receives small branches from the thymus glands and adjacent muscles, and at a level between the first and second ribs is seen to be formed by the union of two large veins. These are the *brachiocephalic* or *innominate* veins. They are the two precaval veins of embryonic stages which later unite to form the single precaval vein of adult anatomy by the crossing-over of the left vein to join the right one (Fig. 106). The branches of the two brachiocephalic veins are identical, and only one need be followed, preferably the right one, since the right side has not been touched in the previous dissection. The places of entrance of the various tributaries are, however, somewhat variable.

Immediately anterior to the junction of the two brachiocephalics, opposite the first rib, each of them receives on the dorsal side a large tributary. This is located by dissecting on the dorsal side of the vein and lifting the vein. The main part of the tributary can be traced into the cervical vertebrae; it is the *vertebral* vein and courses in the vertebrarterial canal, collecting from the brain and spinal cord. Before it enters the brachiocephalic, the vertebral is joined by the *costocervical* vein, which comes from the muscles of the back. and receives branches also from the chest wall on the inner surface of the first two ribs. The costocervical vein may be picked up by turning the animal dorsal side up and, on the side where the muscles were dissected, dissecting in the serratus ventralis and the epaxial muscles. The communication of the vertebral and costocervical veins with the brachiocephalic and with each other is variable and may not be as described here.

The brachiocephalic at the same place as the entrance of the veins just described is seen to be formed by the union of two large veins, a lateral *subclavian* and an anterior *external jugular*. The subclavian will be followed first. It passes laterally in front of the first rib into the axilla, where it is known as the *axillary* vein. Expose the axilla by cutting through the pectoral muscles near the midventral line and at their insertion on the humerus. The pectoral muscles should then be separated from the underlying serratus ventralis but should not be removed. The stout white cords crossing the axilla are the nerves of the brachial plexus and are not to be injured. Lymph glands will also be noted in the axilla. The most medial tributary of the axillary vein is the large *subscapular* vein which passes through the proximal part of the upper arm to the dorsal side of the humerus and collects from various muscles of the upper arm and shoulder, receiving also the *posterior circumflex* vein from the external surface of the upper arm. The beginnings of the subscapular vein will be found in the trapezius muscles. The axillary vein lateral to the entrance of the subscapular receives the small *ventral thoracic* vein from the medial portions of the pectoral muscles. Lateral to this it receives the *long thoracic* vein, which runs caudad along the inner surface of the pectoral muscles; and the *thoracodorsal* vein, which courses parallel to the preceding but dorsal to it and collects chiefly from the latissimus dorsi muscle. There is a broad connection between the thoracodorsal and subscapular veins. Lateral to these branches the axillary vein is known as the *brachial* vein. It runs along the inner surface of the upper arm in company with nerves and the brachial artery. These structures will be found by separating the muscles on this surface of the upper arm.

Return now to the external jugular vein. It soon receives on its medial side the very small *internal jugular* vein which passes forward in the neck alongside the trachea in company with the carotid artery and vagus nerve. The much larger external jugular vein assumes a more superficial position and in addition to small branches from adjacent muscles receives the large *transverse scapular* vein from the shoulder. This passes laterally in front of the shoulder and anastomoses with the *cephalic* vein of the arm. The cephalic vein is the superficial vein of the forelimb and will be found on the external or lateral surface of the upper arm. It also connects with the posterior circumflex vein described above. The external jugular anterior to the entrance of the transverse scapular vein is situated in the sternomastoid muscle. On following it forward it is seen to be formed at the angle of the jaw by the union of the *anterior* and *posterior facial* veins. At their point of union they are connected across the ventral side of the throat by the *transverse* vein which has probably been destroyed. The anterior facial vein collects from the face and jaws and submaxillary and lymph glands, its main tributary being the *angular* vein from the region of the eye. The posterior facial vein emerges from the parotid gland and at the place of emergence receives the *posterior auricular* vein from the pinna and back of the head. The main vein then lies imbedded in the parotid gland and may be followed by dissecting away the gland. It is then seen to be formed by the union of veins from the temporal region and region anterior to the ear.

Draw the branches of the precaval as far as found.

b) The branches of the postcaval: The following description applies to both the rabbit and the cat. Turn the apex of the heart forward and note the large vein which enters the right auricle from behind. This is the *postcaval vein* (also called vena cava posterior or inferior and ascending vena cava). It passes posteriorly in the thorax, lying slightly to the right of the median line, inclosed in the free dorsal border of the caval fold of the pleura. Follow it caudad. It passes through the diaphragm, from which it receives several *phrenic* veins. In the **rabbit** it then lies against the dorsal wall of the peritoneal cavity slightly to the right of the median line, dorsal to the right median lobe of the liver and in contact with the hepatic portal vein. It then passes into the right lateral lobe of the liver, from which it emerges near the right kidney. In the **cat** the postcaval vein passes into the right median lobe of the liver and, inclosed in the liver substance, traverses the length of the liver, emerging from the posterior lobule of the right lateral lobe. Note the large *hepatic* veins which flow from the liver into the postcaval vein. These are best seen by dissecting in the substance of the liver. Follow the postcaval posteriorly, carefully cleaning away connective tissue and fat from it and its tributaries. It runs slightly to the right of the middorsal line of the peritoneal cavity alongside the dorsal aorta, which must not be injured. The first tributary of the postcaval is the *right adrenolumbar* vein. This passes along the posterior surface of a small gland, the *adrenal* gland, which lies anterior to the kidney in contact with the postcaval vein and will be found by dissecting in the fat in this location. The adrenolumbar vein receives branches for the adrenal gland (which is one of the glands of internal secretion) and also collects from the adjacent body wall. Immediately posterior to this vein the large *right renal* vein passes from the kidney into the postcaval. Next, by turning the viscera to the right, locate the left adrenal gland and kidney and find the *left* adrenolumbar and renal veins. They are situated posterior to the right ones. The left adrenolumbar and renal veins generally unite to a common stem before they enter the postcaval. Into the left renal vein opens the vein of the left gonad. In male specimens this is the *left internal spermatic* vein; it may be traced posteriorly (in contact with the postcaval in the rabbit) to the scrotum. In female specimens it is the *left ovarian* vein which comes from the ovary, a small oval body lying about the middle of the lateral wall of the peritoneal cavity. The *right* internal spermatic or ovarian vein enters the postcaval directly, in the cat shortly posterior to the right kidney, in the rabbit much farther caudad. The postcaval vein in its course along the body wall receives at regular intervals the paired *lumbar* veins from the wall; these are seen by loosening the vein, raising it slightly, and looking on its dorsal surface. The lumbar veins are then seen passing ventrally in the median groove between muscle masses. Near the posterior end of the peritoneal cavity the postcaval receives a pair of *iliolumbar* veins. Each of these, in company with an artery, extends laterally along the body wall and receives an anterior branch from the neighborhood of the kidney. Sometimes the left ovarian vein enters the left iliolumbar. Posterior to this point the dorsal aorta comes to lie ventral to the postcaval, concealing the latter. The dissection of the remainder of the postcaval will therefore be deferred until the aorta is studied.

Variations in the postcaval and its branches from the foregoing account are common and result from the persistence of parts of the complicated embryonic conditions (Fig. 106). An apparent splitting (really a failure of fusion) of the postcaval into two main trunks caudad of the kidneys is a common variation. For an account of these variations see Huntington and McClure (1920).

4. The pulmonary veins.—Examine the roots of the lungs and note numerous veins, several on each side, entering the left auricle from the lungs. These are the *pulmonary* veins. They lie to either side of the postcaval vein; those of the right side pass dorsal to the postcaval, and in the rabbit those of the left side dorsad of the left precaval. They convey the aerated blood from the lungs into the left auricle.

5. The pulmonary artery.—The pulmonary artery is the conspicuous vessel extending from the base of the right ventricle forward between the auricles; it soon curves to the left. Its base is generally surrounded by fat, which should be cleaned away. It divides in two at the turn into *right* and *left pulmonary* arteries. The division may be found by dissecting along the pulmonary artery immediately in front of the left auricle. Press the heart to the right and follow the left pulmonary artery into the left lung. In the rabbit it passes to the dorsal side of the left precaval vein, which may now be severed. The left pulmonary artery courses parallel to, and anterior to, the most anterior of the pulmonary veins. Now turn the heart to the left and similarly find the right pulmonary artery, proceeding to the right lung; to trace it, sever the precaval vein. It lies immediately anterior to the foremost pulmonary vein. Dorsal to the right pulmonary artery lies the trachea.

6. The aorta and its branches.—Springing from the base of the left ventricle to the left of, and dorsal to, the pulmonary artery is a very large trunk, the aorta. Right and left *coronary* arteries spring from the base of the aorta, where it leaves the ventricle. The left coronary artery lies between the pulmonary artery and the left auricle and branches over the ventral and left side of the heart. The right coronary artery lies along the groove between the right auricle and right ventricle and branches to the right and dorsal surfaces of the heart.

Follow the aorta forward, cleaning away tissue from its surface. It soon describes a curve, known as the *arch of the aorta*, to the left. From the arch of the aorta spring the large arteries of the neck, head, and forelimbs. These are two in number in the cat, three in the rabbit. Beginning at the right, they are in the rabbit: the *brachiocephalic* or *innominate* artery; the *left common carotid;* and the *left subclavian*. In the cat the branches are the *brachiocephalic* or *innominate* artery to the right and the left *subclavian* to the left. The difference is due to the fact that in the cat, not in man, the left common carotid branches from the brachiocephalic; and this may also occur in the rabbit as a variation.

Trace the brachiocephalic artery forward. The precaval vein and its branches may be removed. The artery gives off small branches into the thymus gland and trachea lying dorsal to it and then divides into two branches in the rabbit and three in the cat. These are: *right subclavian* and *right common carotid* in the rabbit; and *right subclavian, right* and *left common carotids* in the cat. Each of these will be traced separately.

a) Subclavian artery: Trace the right subclavian; both have identical branches.

Rabbit: From the posterior surface of the subclavian arises the *internal mammary* artery, which follows the vein previously described along the ventral chest wall and continues on the abdomen as the *superior epigastric* artery. At the same level from the posterior surface of the subclavian, practically in common with the preceding, the *supreme intercostal* artery arises. It runs posteriorly on the dorsal wall of the thorax and receives the first *intercostal* arteries. On its anterior surface at about the same level as these the subclavian artery gives rise to the *vertebral* artery, which passes immediately dorsad toward the cervical vertebrae, where it enters the vertebrarterial canal; and to the *superficial cervical* artery, which ascends in the lateral part of the neck, supplying various muscles, its main branch (*ascending cervical*) accompanying the external jugular vein. The *transverse artery of the neck* leaves the subclavian at the same place or in common with the supreme intercostal artery. It passes dorsally in front of the first rib, through a loop formed by two nerves, and emerges on the medial side of the serratus ventralis muscle. It is best found by looking on this muscle and then tracing the artery toward the subclavian. After giving off the foregoing branches the subclavian passes in front of the first rib into the axilla, where it is named the *axillary* artery. This lies between two of the stout nerves belonging to the brachial plexus. Its branches are similar to those of the axillary vein and accompany the veins. After giving rise to the small *thoracoacromial* artery to the pectoral and deltoid muscles, the axillary gives off the *long thoracic* and *subscapular* arteries, accompanying the veins previously described. The former runs posteriorly along the serratus muscle and then, as the *external mammary* artery, passes to the under surface of the skin of the lateral abdominal wall, being especially conspicuous in females. (Most of this vessel was destroyed in removing the skin.) The subscapular has a conspicuous branch (*thoracodorsal artery*) passing caudad to the latissimus dorsi and cutaneous maximus muscles; it then turns dorsally and, perforating the teres major, emerges on the outer surface of the shoulder, supplying various muscles. Near the point of origin of the subscapular the *deep artery* of the arm arises and, after giving off branches into the subscapular muscle, passes between this muscle and the teres major to the dorsal part of the arm, where it runs in company with one branch of the cephalic vein and a nerve, all three situated internal to the lateral head of the triceps, which should be deflected. The axillary artery now passes to the upper arm, where, as the *brachial* artery, it courses along the inner surface of the limb in company with the brachial vein and nerves.

Draw the branches of the subclavian.

Cat: At the level of the first rib the subclavian has four branches: *internal mammary, vertebral, costocervical axis,* and *thyrocervical axis.* The *internal mammary* springs from the ventral surface of the subclavian, accompanies the corresponding vein along the chest wall, and passes on to the abdominal wall as the *superior epigastric* artery. The *vertebral* artery arises from the dorsal surface of the subclavian and passes dorsally into the vertebrarterial canal, giving off small branches into the neck muscles. The *costocervical axis* divides in two almost at once. One branch, the *superior intercostal* artery, passes posteriorly near the middorsal line of the thorax, giving off *intercostal* branches and then supplying the deep muscles of the back. The other branch of the costocervical axis leaves the thoracic cavity, passing deep dorsally in front of the first rib, and divides into the *transverse artery of the neck,* supplying the serratus ventralis and rhomboideus muscles, and the *deep cervical* artery to the epaxial muscles of the neck. These branches are best found by looking among the muscles in question and tracing the vessels toward the subclavian. The *thyrocervical axis* generally arises anterior to the other branches. It passes forward near the carotid artery and, after branching to the muscles of the dorsal side of the neck, turns laterally in front of the shoulder, being then named the *transverse scapular* artery; it accompanies the external jugular vein for a short distance and supplies many muscles of the shoulder and neck.

The subclavian artery now passes in front of the first rib into the axilla, where it is named the *axillary* artery. This gives off: the *ventral thoracic* artery, passing medially to the medial ends of the pectoral muscles; the *long thoracic* artery, passing posteriorly along the middle region of the pectoral muscles and then to the latissimus dorsi; and, near the arm, the large *subscapular* artery. This gives off the *thoracodorsal* artery, lying parallel but more dorsal to the long thoracic artery and supplying the latissimus dorsi; the subscapular then turns dorsally, passes through the proximal part of the upper arm dorsal to the humerus, and branches to the muscles of the upper arm and muscles of the back and shoulder.

The axillary artery then proceeds as the *brachial* to the medial surface of the forelimb, where it accompanies the brachial vein and some nerves, and branches into the limb.

Draw the branches of the subclavian.

b) *Common carotid artery:* The two common carotid arteries arise in the cat from the brachiocephalic and immediately diverge; in the rabbit the right one arises in common with the right subclavian, while the left usually springs independently from the arch of the aorta. Trace the common carotids forward. Their branches are similar in the two animals. They pass anteriorly in the neck, one to each side of the trachea, to which they give small branches. At the level of the anterior end of the thyroid gland each supplies a *superior thyroid* artery to the gland. At the level of the larynx there are branches into the larynx and adjacent parts (probably destroyed) and an *occipital* branch into the dorsal muscles of the neck. The common carotid at about this same level gives off the *internal carotid* artery. In the rabbit this artery arises at the place where the carotid passes to the dorsal side of the shining ligament of the digastric. In the cat it is a degenerated cord and rises slightly caudad to the occipital artery. In both animals the internal carotid passes dorsally in company with nerves and enters the skull by a foramen through the tympanic bulla. It need not be followed. The artery beyond this point is called the *external carotid* artery. At the angle of the jaw it branches to all parts of the head. Its chief branches are: the *lingual* artery into the tongue, and the *external maxillary* running along the ventral border of the masseter muscle and branching to the upper and lower lips and jaws. The main artery then passes along the posterior border of the masseter muscle. It receives *auricular* and *temporal* branches from the pinna and temporal regions and then, as the *internal maxillary* artery, turns internal to the masseter muscle and is lost to view. It need not be followed farther.

Draw the branches of the common carotid artery.

c) *The thoracic aorta:* After having given rise to the subclavians and the carotids, the aorta arches to the left. Note, as it passes the left pulmonary, the strong fibrous band which connects the two vessels. This is the *arterial ligament* or *ligament of Botallus* and is the remnant of the embryonic connection between the aorta and the pulmonary (Fig. 104). Follow the aorta posteriorly, pressing the left lung to the right. It descends posteriorly, lying against the dorsal wall of the thorax to the left of the median line. It is situated within the mediastinum; the mediastinal wall may be cleared away. The aorta in its course along the thorax is named the *thoracic aorta*. Its chief branches are the paired *intercostal* arteries which arise from the aorta at segmental intervals and run along the thoracic wall along the posterior margin of the ribs. The aorta also has small *bronchial* arteries to the bronchi and *esophageal* arteries to the esophagus. Along the dorsal surface of the aorta on its left side runs a delicate tube, resembling a streak of fat. This is the *thoracic duct,* the *main lymphatic channel* for the posterior part of the body. Trace it forward; its connection with the jugular vein, generally at the point of union with the subclavian, may be found.

The aorta penetrates the diaphragm, to which in the rabbit it gives *superior phrenic* arteries and passes into the peritoneal cavity, where it is known as the abdominal aorta.

d) The abdominal aorta: Turn the digestive tract to the right and locate the dorsal aorta after it has passed the diaphragm. It will be found against the dorsal wall in the median dorsal line. Clear away the mesogaster and clean the surface of the aorta. Follow it identifying its branches. These branches consist of unpaired *median visceral* branches to the digestive tract, paired *lateral visceral* branches to the kidneys and reproductive organs, and paired *somatic* branches to the body wall.

Shortly posterior to the diaphragm the aorta gives rise to two large unpaired visceral arteries, the *coeliac* and the *superior mesenteric* arteries. In the cat the second is shortly posterior to the first, while in the rabbit the superior mesenteric artery lies one-half inch posterior to the coeliac. As the branches of these two vessels are different in the two animals, because of the differences in their digestive tracts, it will be necessary to describe them separately.

Rabbit: The coeliac artery near its origin from the aorta gives rise to the small *inferior phrenic* arteries to the diaphragm. Beyond this point the *splenic* artery arises from its posterior surface. This vessel passes in the mesogaster to the spleen, where it runs in the gastrosplenic ligament. In its course to the spleen it provides the *short gastric* arteries to the left end of the stomach; along the spleen it supplies *splenic* branches to the spleen; beyond the spleen it branches into the omentum; at about the middle of the spleen a large branch, the *left gastroepiploic* artery, arises from the splenic and passes to the greater curvature of the stomach. The coeliac artery beyond the splenic passes to the lesser curvature of the stomach, where it may best be followed by turning the stomach forward. Here it gives off a group of vessels, the *left gastric* (or *coronary*) arteries, which radiate to the stomach wall on both sides of the lesser curvature and also send small branches to the esophagus. Shortly beyond this point the coeliac artery is known as the *hepatic* artery, which passes along the right end of the lesser curvature, very shortly giving rise to the *gastroduodenal* artery. This runs to the pyloric region and there branches into the *anterior pancreaticoduodenal* artery to the pancreas and first part of the duodenum and the *right gastroepiploic* artery which returns to the stomach wall by way of the great omentum. The hepatic artery now passes to the dorsal side of the pylorus and enters the hepatoduodenal ligament. After giving off the small *right gastric* artery to the pylorus it proceeds to the liver, lying to the right of the bile duct.

The superior mesenteric artery is the chief artery of the intestine and has many and complicated branches in the rabbit, these branches following, for the most part, the branches of the hepatic portal vein. Clean the surface of the vessel and follow it; it runs alongside the superior mesenteric vein. The first branch is the small *middle colic* artery, arising from the ventral wall of the superior mesenteric and passing to the transverse colon and beginning of the descending colon. At the same level, but from the dorsal side, arises the *posterior pancreaticoduodenal* artery, which passes to the duodenal loop and pancreas. The superior mesenteric artery then forks into the *intestinal* artery, which runs in the mesentery of the small intestine and gives off numerous branches ventrally into the intestine, and into the large *ileocaecocolic* artery. This last has many branches to the ileum, the caecum, the appendix, and the ascending colon. Its branches are: small arteries to the terminal part of the ascending colon; the *anterior right colic* artery, which forks several times, supplying the greater part of that portion of the ascending colon which bears the haustra; the *posterior right colic* artery, arising near the preceding and supplying the remainder of the haustra-bearing region of the ascending colon, its end joining one end of the preceding; the *appendicular* artery, arising with the preceding and running along the appendix and that part of the ileum adjacent to the appendix; the large *posterior ileocaecal* artery, passing to the greater part of the caecum and to that portion of the ileum lying between the caecum and the ascending colon; the much smaller *anterior ileocaecal* artery to the more distal part of the caecum and adjacent ileum, and running toward the preceding, with which it anastomoses; and the *caecal* artery or arteries to that portion of the caecum which adjoins the appendix.

Cat: The coeliac artery passes toward the stomach and very soon divides, at about the same level, into three branches. The most cranial one is the *hepatic;* the next one, the *left gastric;* and most caudal and largest is the *splenic.* Trace the splenic artery. It courses in the great omentum toward the spleen and forks. One branch goes to the left end of the spleen and also sends branches into the pancreas and the *short gastric* arteries to the stomach. The other branch passes to the right end of the spleen and supplies also branches to the pancreas, the omentum, and the *left gastroepiploic* arteries to the greater curvature. The left gastric or *coronary* artery passes to the lesser curvature, where it splits into many branches, supplying both sides of the stomach. The hepatic artery passes along the border of the left end of the pancreas and to the dorsal side of the lesser curvature and enters the hepatoduodenal ligament. It is best found by separating the stomach and liver. It lies to the left side of the hepatic portal vein. As it passes the pylorus, it gives off the large *gastroduodenal* branch. This branches into the *anterior pancreaticoduodenal* artery, descending along the beginning of the duodenum and supplying also the pancreas; the *right gastroepiploic*, passing from the pylorus along the greater curvature of the stomach to the left; and the small *pyloric* artery to the pyloric region (this may also arise independently from the hepatic). The hepatic artery proceeds into the liver, sending a *cystic* artery to the gall bladder.

The superior mesenteric artery supplies the greater part of the intestine. It passes toward the intestine. Follow it, cleaning away fat and lymph glands from its surface. It first gives rise to the *middle colic* artery, which passes to the transverse and descending parts of the colon. A little farther on the superior mesenteric gives rise simultaneously to the *posterior pancreatico-duodenal* artery, which ascends along the duodenum, supplying it and the pancreas and anastomosing with the anterior pancreaticoduodenal; and to the *ileocolic* artery to the caecum and terminal portion of the ileum, and sending also a *right colic* branch to the ascending colon (this last may arise independently from the superior mesenteric). The superior mesenteric then divides into numerous *intestinal* branches to the small intestine.

Draw the branches of the coeliac and superior mesenteric.

Return now to the dorsal aorta. Its next branches are the paired *adreno-lumbar* and *renal* arteries. In the rabbit the adrenolumbars are branches of the renals, but in the cat they arise independently. They pass close to the adrenal gland, to which they give an adrenal branch, and then course along the dorsal body wall. In the cat each sends a *phrenic* artery anteriorly to the diaphragm. The renal arteries are large vessels passing into the kidneys. The aorta posterior to the kidneys gives rise to the paired arteries to the gonads (these may, however, branch from the renals). They are the *internal spermatic* arteries in the case of the male and run posteriorly on the dorsal wall to the scrotum. In the female the corresponding *ovarian* arteries are larger, and in the cat convoluted, and pass laterally to the ovaries. In its passage along the middorsal line the aorta gives off paired *lumbar* arteries at segmental intervals. These are found by loosening the aorta and looking on its dorsal surface. Posterior to the genital arteries the *inferior mesenteric* arises as an unpaired visceral branch and passes to the descending colon and rectum, running in the mesocolon. In the mesocolon it forks into the *left colic* artery, passing craniad along the descending colon, and the *superior haemorrhoidal* artery, passing caudad to the posterior part of the descending colon and the rectum.

Add these vessels to the drawing of the aorta.

The digestive tract may now be removed and discarded, leaving the end of the large intestine in place. Hold the stump of the colon, together with the urinary bladder and in female specimens the uterus (the forked coiled tube at the posterior end of the peritoneal cavity), back against the pubes and follow the aorta farther. Near the end of the peritoneal cavity it forks into the two *common iliac* arteries in the rabbit; in the cat it gives off a pair of *external iliac* arteries, followed shortly by a pair of *internal iliac* arteries. Anterior to this place in the cat, or in the rabbit at the level of the fork or from the common iliac arteries, a pair of *iliolumbar* arteries arises and passes laterally along the body wall. The iliolumbar artery divides into an anterior branch, which passes forward toward the kidney, and a posterior branch, which extends to the thigh.

The two common iliac arteries in the rabbit soon fork into an anterior *external iliac* and a posterior *internal iliac*. In the cat the external and internal iliacs arise separately from the aorta, the latter immediately posterior to the former. After giving rise to the iliacs the aorta continues in the middorsal line as the small *median sacral* or *caudal* artery, lying halfway between the two internal iliacs. In the cat this vessel arises from the fork of the internal iliacs. In the rabbit it springs anterior to the forking of the aorta, from the dorsal surface of the latter; its origin is concealed by the postcaval vein and will be seen later. The sacral artery supplies the sacral region and the tail.

Follow the external iliac. It passes laterocaudad out of the peritoneal cavity, in the rabbit to the dorsal side of the inguinal ligament. As it passes through the abdominal wall or shortly beyond the wall, it gives rise to the *deep femoral* artery (cat) or the *inferior epigastric* (rabbit). In the cat this vessel gives off branches into the thigh, while these are lacking in the rabbit. In both animals the following branches are present: branches into the mass of fat between the thighs and into the external genital organs, of which one branch, in male specimens, constitutes the *external spermatic* artery; and the main vessel then, as the *inferior epigastric* artery, turns craniad and ascends in the abdominal wall, running along the inner surface of the rectus abdominis muscle. It anastomoses with the superior epigastric artery. In the rabbit there arises, either from the inferior epigastric at the origin of the latter from the external iliac or from the external iliac itself near by, the *superficial epigastric* artery which extends forward on the inner surface of the skin of the abdominal wall and anastomoses with the external mammary, a branch of the long thoracic. These vessels are particularly prominent in females, but the greater part of their course is destroyed in removing the skin. The external iliac, now named the *femoral*, proceeds along the center of the medial surface of the thigh, giving branches into the leg muscles.

Follow the internal iliacs, being careful not to injure the end of the postcaval vein lying in contact with them or any parts of the urogenital system (in males do not injure the male ducts curving around the base of the urinary bladder). The internal iliacs lie against the dorsal wall. At their origin from the common iliac (rabbit) or posterior to their origin from the dorsal aorta (cat) each gives rise to an *umbilical* artery, which passes to the bladder or in female rabbits to the uterus, first with a branch to the bladder. The internal iliacs then pass to the dorsal side of the postcaval vein. To follow them dissect as deeply as possible between the rectum and the base of the thigh. The internal iliacs give some branches to the pelvis, and then each gives off a *middle haemorrhoidal* artery to the rectum. This accompanies the rectum to the anus but cannot be followed at this time. In female cats the *uterine* artery arises from the middle haemorrhoidal and passes anteriorly again to the uterus. The internal iliacs cannot be followed farther conveniently. They give branches to the tail and thigh.

Draw the branches of the iliacs, adding them to the drawing of the dorsal aorta already made.

7. The posterior portion of the postcaval vein.—The postcaval may now be followed caudad from the point where it was previously left, by removing the arteries which cover it. Its tributaries should be traced, as far as practicable, to the posterior end of the peritoneal cavity, dissecting deeply dorsally as before.

Rabbit: The postcaval receives, at the same level as the forking of the dorsal aorta, the two large *external iliac* veins. It then continues in the mid-dorsal line for a short distance caudad to this point, this portion often receiving the name of *common internal iliac* vein, and is then seen to be formed by the union of the two *internal iliac* or *hypogastric* veins. Trace the external iliac; its branches are similar to and accompany those of the artery of the same name. It soon receives the *vesical* vein from the bladder: this vein in females also collects from the uterus. At the place where it passes through the abdominal wall, the external iliac receives the *inferior epigastric* vein. The main part of this runs forward along the internal surface of the rectus abdominis muscle and anastomoses anteriorly with the superior epigastric. The inferior epigastric near its place of entrance into the external iliac also receives tributaries from the fat between the bases of the thighs and the external genital region and sends a *superficial epigastric* vein along the inner surface of the skin of the lateral abdominal wall. This last-named vessel is particularly conspicuous in females but is destroyed in removing the skin; it anastomoses with the external mammary, a tributary of the long thoracic. The external iliac passes to the dorsal side of the inguinal ligament and, as the *femoral* vein, continues along the medial side of the leg, in company with the femoral artery. Follow the internal iliacs. After a short distance the *sacral* or *caudal* vein enters one of them, usually the left one; it accompanies the caudal artery. Caudad of this, each internal iliac receives the *middle haemorrhoidal* vein, which ascends from the anus and lies along the side of the rectum. The tributaries of the internal iliac cannot be followed farther conveniently; they come chiefly from the gluteal region.

Draw the branches of the postcaval, adding them to the drawing previously made. The iliacs and postcaval may then be removed and the origin of the caudal artery from the aorta traced.

Cat: The postcaval is formed dorsal to the forking of the aorta by the union of the two large *common iliac* veins. One of them, usually the left one, receives the small *sacral* or *caudal vein*, which lies parallel to the artery of the same name. About one inch posterior to its junction with the postcaval each common iliac is formed by the union of the *internal iliac* (*hypogastric*) and the *external iliac*. The former receives branches from the gluteal region and receives the *middle haemorrhoidal* vein, which runs along the sides of the rectum from the anus forward, and also collects from the bladder. The external iliac passes out of the abdominal cavity. At the place of exit it receives on its medial side the *deep femoral* vein, which collects from the thigh, from the fat between the thighs, from the external genital region (receiving in males the *external spermatic* vein from the testes), and also receives the *inferior epigastric* vein from the inner surface of the rectus abdominis muscle. The branches from the thigh may enter the external iliac separately. The external iliac, now known as the *femoral* vein, passes along the thigh, receiving branches from the leg muscles.

Draw these branches as far as found, adding them to the drawing of the postcaval vein previously made.

8. The structure of the heart.—Remove the heart from the body, cutting across the bases of the great vessels. Identify the systemic veins entering the right auricle (three in the rabbit, two in the cat) and the pulmonary veins entering the left auricle. Cut into the wall of each auricle by a transverse slit and wash out the clotted blood which generally fills the interior. Note the thick ridged walls of the auricular appendages and the thinner smoother walls of the remainder of the auricle. Note the *interauricular septum* extending dorsally between the two auricles and completely separating them. Find the large *auriculoventricular openings* in the floor of the auricles. In the cat find, near the dorsal edge of the interauricular septum, the opening of the coronary sinus into the right auricle, noting the valve—the *valve of the coronary sinus*—which guards the opening. The coronary sinus is the remnant of the original left precaval vein. Cut off the apex of the heart and note the thick walls and rounded form of the left ventricle and the smaller size, thinner walls, and crescentic form of the right ventricle. Cut open the right ventricle by an oblique cut, beginning at the cut surface already made and extending out through the pulmonary artery, slitting open this artery. Wash out the right ventricle. Its cavity is rather small, the walls being deeply cleft by muscular ridges, the *trabeculae carnae*. From the walls project a number of pointed finger-like muscles, the *papillary* muscles, which are connected by slender fibers, the *chordae tendinae*, to a thin membrane. This membrane consists of three flaps and is called the *tricuspid valve*. Two of the flaps can be stretched by pulling on the cut surfaces of the ventricle, while the third lies collapsed against the interventricular septum, to which it is fastened without the intervention of the papillary muscles. The tricuspid valve guards the right auriculoventricular opening and prevents the blood from flowing back from the ventricle into the auricle. In the base of the pulmonary artery note the three pocket-shaped *semilunar* valves. The pulmonary artery is the sole exit from the right ventricle. Similarly cut open the left ventricle by a longitudinal slit from apex to base. Wash out the interior. The cavity of the left ventricle is considerably larger than that of the right, and its walls thicker. The two ventricles are completely separated by the *interventricular septum*, which appears as the common internal wall of both ventricles. Note in the left ventricle the trabeculae carnae, the papillary muscles, and the chordae tendinae. The latter are attached to the membranous *bicuspid* or *mitral* valve, which consists of but two flaps. This guards the left auriculoventricular opening and prevents the regurgitation of the blood from the ventricle back into the auricle. By probing, find the sole exit of the left ventricle, its opening into the aorta. Follow the probe by a cut and note the three semilunar valves at the base of the aorta.

Make drawings to illustrate the structure of the heart.

The removal of the heart permits a clearer view of some of the structures of the pleural cavity. The student should examine carefully the forking of the trachea into the bronchi, the form of the lungs and their relation to the pleural cavity, and the pulmonary arteries and veins.

9. Comparison of the mammalian heart and circulatory system with those of the preceding animals.—In mammals, as in birds, only two of the original four chambers of the heart, found in fishes and amphibians, persist, namely, the atrium and the ventricle. Each is, however, divided into right and left chambers by a septum, which is complete in mammals in both chambers. The division of the atrium into two auricles begins in the amphibians and is complete in anurans and all amniotes. The division of the ventricle, already seen in the dipnoan fishes, is partial in reptiles and is completed in the crocodilians, birds, and mammals. In reptiles the sinus venosus is still present as a distinct chamber; but in birds and mammals it is reduced to a small area, not grossly detectable, in the wall of the right auricle. In all amniotes the conus arteriosus has become subdivided among the bases of the great arteries, which therefore appear to spring directly from the ventricles. The valves of the conus, so conspicuous in the dogfish, are present only at the junction of these arteries with the ventricles. In reptiles it was seen that the conus arteriosus is divided into three trunks and that this arrangement is impractical since it allows mixing of venous and arterial blood. Birds and mammals, with their constant warm body temperature and high activity, require a more perfect circulation. In them the conus arteriosus is subdivided into two trunks, the aorta and the pulmonary. The pulmonary artery leaves the right ventricle, the aorta the left ventricle. All of the systemic veins open into the right auricle and the pulmonary veins into the left auricle. The venous blood from the body passes into the right auricle, thence into the right ventricle, and from this into the pulmonary artery, which conveys it to the lungs for aeration. The aerated or arterial blood returns to the left auricle by way of the pulmonary veins, passes into the left ventricle and out through the aorta, which distributes it to all parts of the body. A complete and perfect double circulation is maintained through the heart. The right half of

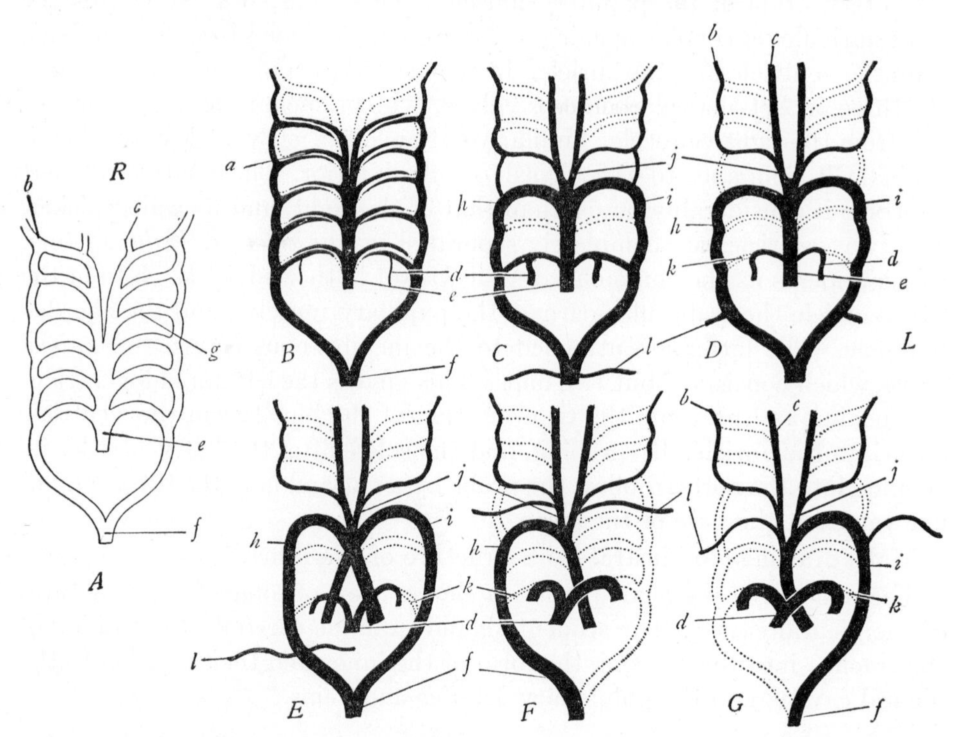

FIG. 104.—Diagrams to show the evolution of the aortic arches. *A*, primitive condition with six aortic arches. *B*, fishes, the first aortic arch missing. *C*, some urodeles, the first, second, and fifth arches missing. *D*, anurans, with the connection (*k*) between the pulmonary arteries (*d*) and the aorta obliterated. *E*, reptiles, showing the ventral aorta split into three trunks and the fourth aortic arch (*h* and *i*) persistent on both sides. *F*, birds, the ventral aorta split into two trunks, and the fourth aortic arch (*h*) persistent on the right side only. *G*, mammals, the ventral aorta split into two trunks and the fourth aortic arch (*i*) persistent on the left side only. *a*, interruption of the aortic arches by the gill capillaries in fishes; *b*, internal carotid; *c*, external carotid; *d*, pulmonary, developed from the sixth arch; *e*, ventral aorta; *f*, dorsal aorta; *g*, aortic arch; *h*, right fourth aortic arch, called the right aorta above urodeles; *i*, left fourth aortic arch, called the left aorta above urodeles; *j*, common carotid; *k*, arterial ligament or obliterated vessel originally connecting pulmonary and aorta; *l*, subclavian. (Slightly modified from Wilder's *History of the Human Body*, courtesy of Henry Holt & Co.)

the heart is thus venous, the left half arterial; and, as both interauricular and interventricular septa are complete, there is no mixing of arterial and venous blood in the heart, nor can any pass between the great arteries, because of the obliteration of all such connections.

Little trace is left in the adult mammal of the original system of aortic arches passing around the pharynx, although these arches occur in a fishlike condition in the mammalian embryo (Fig. 104). The first and second aortic arches disappear, the third or carotid arch contributes to the common carotid and base of the internal carotid (Fig. 104), the fourth or aortic arch remains on the left side as the arch of the aorta, the fifth is completely degenerate, and the sixth or pulmonary arch forms the bases of the pulmonary arteries. The embryonic connection or arterial duct between the aorta and the pulmonary, which represents the dorsal half of the sixth aortic arch, becomes shriveled at birth, to form the arterial ligament.

There has been much argument in the literature, started by Lewis (1906), as to whether the mammalian embryo actually has any trace of a fifth aortic arch and if not, whether, therefore, the pulmonary arteries are really derivatives of the sixth arch. Various investigations on this point have shown that in some mammalian embryos it is impossible to find any definite fifth aortic arch; however, in others, such as the cat (Fig. 105), the diminished fifth arch is certainly present. There does not seem to be any reason to doubt that the story of the aortic arches is the same for mammals as for other vertebrates; the fifth arch, it will be remembered, begins to degenerate in amphibians, and it is not surprising that in mammals it has become almost or quite erased from even the embryo.

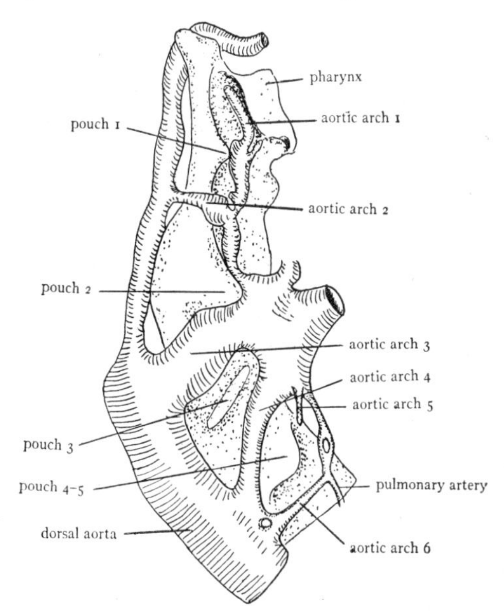

Fig. 105.—The aortic arches in the cat embryo at a stage of 5 mm., showing fifth arch; first arch degenerating. (After Coulter, 1909.)

The venous system is obviously built on the reptilian plan and passes through reptilian stages in its development (Fig. 106). There are two precaval veins, as in the turtle and other reptiles; or in some mammals the left precaval joins the right one by a cross-connection in front of the heart, forming a single precaval (Fig. 106). The proximal end of the left precaval then becomes the coronary sinus of the heart. The coronary sinus and bases of the precavals are the common cardinal veins of the dogfish and of the embryo. The internal jugular branch is the original anterior cardinal vein, but in the adult mammal it is often exceeded in importance by its branch, the external jugular.

The proximal portion of the right posterior cardinal persists in mammals as a small part of the azygos vein. However, the greater part of the azygos vein comes from the supracardinal system. This system in mammals appears to be homologous to the vertebral veins of the turtle, which, it will be recalled, are formed by the lengthwise anastomosis of segmental branches of the posterior cardinals. The supracardinal system is first seen in reptiles as the vertebral veins and persists throughout birds and mammals. In monotremes and some marsupials there is a pair of azygos veins (supracardinals), each of which may open directly into the corresponding precaval; or the left one may open into the right one.

The main change from the reptilian condition is seen in the postcaval vein, which has now taken over the entire renal portal system. Whereas in the turtle the postcaval system extends only to the rear end of the kidneys and has only capillary connections through the kidneys with the renal portal system, in mammals it collects from the whole body, caudad of the liver. This has been accomplished by the union of the postcaval vein with the rear parts of the renal

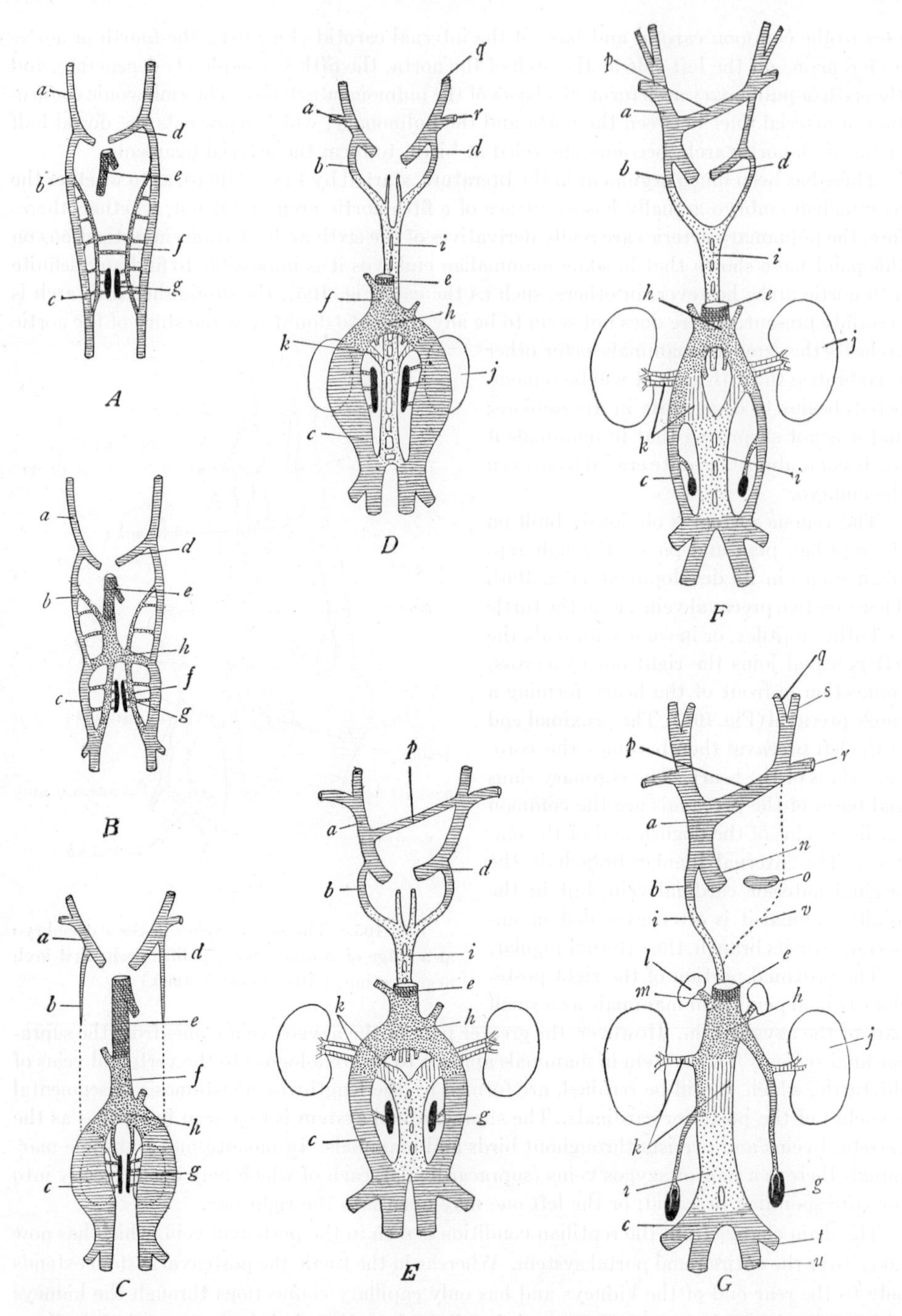

FIG. 106.—Diagrams to show the development of the postcaval vein in the cat. The *cardinal* system of veins is crosshatched, the *subcardinal* veins closely stippled, the *hepatic* veins indicated by cross, vertical, and oblique hatching combined, the *supracardinal* veins by open stippling, and the *renal collar* by

portal system (Figs. 101, 106) by way of the supracardinal veins, homologous, as stated above, to the vertebral veins of reptiles. The renal portal veins then vanish during development, and in the adult mammal no trace remains of a portal circulation through the kidneys. The anterior part of the postcaval vein is formed, as in other amniotes, of the hepatic (vitelline) veins and the right subcardinal; the rear part of the postcaval, from the posterior cardinal system (renal portal system); while between these two parts the postcaval comes chiefly from the supracardinal system (Fig. 106). It is noteworthy that in monotremes, as in reptiles, the postcaval is double back of the level of the kidneys (persistence of both subcardinals), and this condition is not infrequent in other mammals as a retention of an embryonic stage, as will be understood by reference to Figure 106. The complicated embryological history of the mammalian postcaval and its origin from so many different sources also explains numerous other variations commonly found in this vein in mammals.

In conclusion, it may be stated that the embryology of the mammalian circulatory system furnishes a beautiful and striking example of the repetition of evolutionary stages. In its development the mammalian circulatory system passes successively through each of the stages which we have found to persist as the adult condition in the types we have studied, and the evolution of this system can be determined equally well either by studying its development in the mammal or by studying and comparing its form in the adults of the different classes of vertebrates which were ancestral to the mammal.

(FIG. 106.—*Continued*)

vertical hatching. *A*, early stage, showing the anterior and posterior cardinal veins (*a*, *b*, *c*), the common cardinal vein (*d*), the subcardinal veins (*f*), and the outgrowth (*e*) from the hepatic veins of the liver. *B*, next stage, showing the union of the hepatic outgrowth (*e*) with the subcardinal veins (*f*), to form the proximal part of the postcaval vein; the two subcardinals have united with each other at *h*. *C*, the anterior part of the posterior cardinal vein (*b*) has separated from the posterior part (*c*), *c* now being the renal portal vein; the postcaval vein is seen to be formed of the hepatic vein (*e*) and the right subcardinal (*f*), and to be united by means of the two subcardinals below *h* with the renal portals (*c*). *D*, the supracardinal system of veins (*i*), represented by open stippling, has appeared and has united anteriorly with the anterior parts of the posterior cardinals (*b*), medially with the subcardinals by an anastomosis (*k*), named the renal collar, and posteriorly with the renal portals (*c*). *E*, union of the two anterior cardinals by a cross-connection (*p*), and development of the renal veins from the renal collar (*k*); the supracardinal veins have separated into anterior parts connected with the posterior cardinals (*b*) and posterior parts connected with the subcardinals and renal portals (*c*). *F*, continuation of *E*. *G*, adult stage; the left anterior cardinal joins the right by means of the cross-vein (*p*), which is the left innominate vein; the common stem (*a*), which is the right anterior cardinal, enters the heart by way of *n*, which is the right common cardinal vein; the left common cardinal vein persists as the coronary sinus (*o*); the right anterior parts of the posterior cardinal vein and supracardinal form the azygos vein (*b* and *i*), while on the left side these are obliterated at *v*; the postcaval vein is now complete and is seen to be composed of the hepatic vein (*e*), the right subcardinal, the anastomosis between the two subcardinals at *h*, the right renal collar (*k*), the posterior part of the supracardinal vein (*i*), and the posterior parts of the renal portals (posterior cardinals) (*c*): the left subcardinal and posterior cardinal contribute to the vein of the left gonad, hence the asymmetrical arrangement of the genital veins of mammals. *a*, anterior cardinal; *b*, anterior part of the posterior cardinal; *c*, posterior part of posterior cardinal or renal portal; *d*, common cardinal; *e*, hepatic portion of the postcaval (this is partly removed in Figs. *D*–*G*); *f*, subcardinal; *g*, gonad; *h*, union between the two subcardinals; *i*, supracardinal; *j*, kidney (metanephros); *k*, renal collar or union between subcardinals and supracardinals; *l*, adrenal gland; *m*, vein to adrenal gland; *n*, base of the precaval vein or right common cardinal; *o*, coronary sinus or left common cardinal; *p*, left innominate or connection between the two anterior cardinals; *q*, internal jugular; *r*, subclavian; *s*, external jugular; *t*, external iliac; *u*, internal iliac. (After Huntington and McClure in the *Anatomical Record*, Vol. 20.)

G. SUMMARY OF THE CIRCULATORY SYSTEM

1. The entire circulatory system is derived from the mesoderm.

2. The first blood vessels are the vitelline (omphalomesenteric) veins. These course along the intestine and are continued posteriorly as the subintestinal vein. In forms with yolk sacs they are the veins of the yolk sac.

3. As the walls of the hypomere fuse on the ventral side of the embryo, the two vitelline veins unite to form the heart. The heart lies in the median ventral part of the body, inclosed in the ventral mesentery.

4. The anterior end of the heart continues forward as the ventral aorta.

5. From the ventral aorta a series of vessels, shaped like half-loops, pass around the sides of the pharynx and join the dorsal aortae, at first double, fusing posteriorly to a single vessel. The half-loops, termed the aortic arches, form by sprouts from both dorsal and ventral aortae. Typically in vertebrates there are six aortic arches, although primitively there were presumably a greater number. The aortic arches course through the branchial bars, in front of the corresponding gill slit.

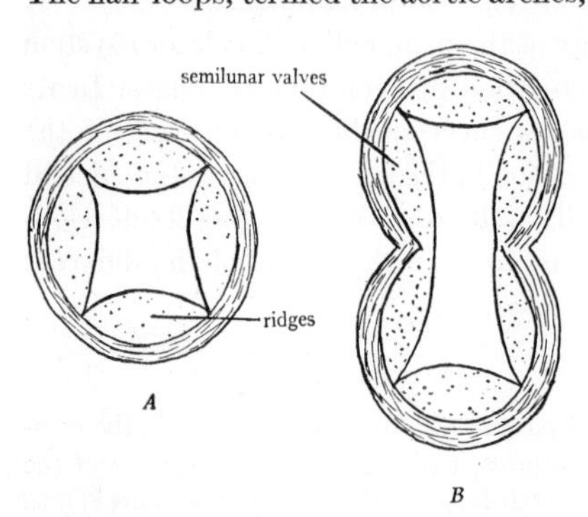

FIG. 107.—The process of splitting of the conus arteriosus into two trunks (birds and mammals) and the retention of the ridges as the semilunar valves, shown in diagrammatic cross-section.

6. The dorsal aorta proceeds posteriorly along the middorsal line of the coelom, supplying branches to all parts of the body below the heart.

7. The chief somatic veins at first are the anterior and posterior cardinal veins, uniting at the level of the heart to a common cardinal vein (duct of Cuvier), on each side which enters the sinus venosus; the subcardinal veins extending along the kidneys; and the vein of the lateral body wall, the abdominal or umbilical vein, opening into the common cardinal.

8. Both arteries and veins are provided with paired segmental and unpaired nonsegmental branches. The former are of two kinds: the somatic vessels to the products of the epimere, and the lateral visceral branches to the products of the mesomere. The unpaired branches consist of visceral vessels to the digestive tube. The vessels of the paired appendages—subclavians and iliacs—may be regarded as enlarged somatic vessels.

9. In all vertebrates the vitelline and subintestinal veins of the embryo become converted into the hepatic portal system (Fig. 98). The proximal portions between the liver and the heart form the hepatic veins; within the liver a network of capillaries arises; and posterior to the liver one of the vitellines with the subintestinal becomes the hepatic portal vein.

10. Originally (and in vertebrate embryos) the caudal vein continued into the subcardinal veins, and the blood then passed through the kidneys into the posterior parts of the posterior cardinal veins (Fig. 98). But in all present vertebrates the caudal vein during development shifts so as to open into the posterior parts of the posterior cardinal veins, and there is then a reversal of flow through the kidneys, from the posterior cardinals into the subcardinals. The posterior parts of the posterior cardinal veins then detach from the anterior portions and are known as the renal portal veins, and the subcardinal veins then establish connections with the anterior portions of the posterior cardinals (Fig. 101). There is thus formed the renal portal system characteristic of lower vertebrates, with blood flowing into the renal portal veins (posterior parts of the posterior cardinals) and through a capillary network in the kidneys into the subcardinal veins.

11. The venous system of fishes is in essentially the stage reached to this point, with a renal portal and a hepatic portal system, with the cardinal veins as the chief somatic veins, aided by an inferior jugular in the ventral head region and by lateral abdominal veins.

12. The arterial system of fishes is also in a relatively primitive condition. There is a ventral aorta, connected by aortic arches with a dorsal aorta; the latter supplies the body caudad to the heart. Of the original six pairs of aortic arches, the first and second are represented by remnants, and the third to sixth are present entire. In fishes, however, there is a break in each aortic arch, with an interpolated new portion consisting of smaller branches and capillaries in the gills. The parts of the aortic arches below the break become the afferent branchial arteries; those above it, the efferent branchial or epibranchial arteries.

13. The primitive vertebrate heart is a bent tube consisting of four chambers, named from the posterior end forward: sinus venosus, atrium, ventricle, and conus arteriosus. There is a single circulation, composed of venous blood, through the heart from sinus through the chambers in the order named. This condition is permanent in fishes (except Dipnoi, which have a double circulation similar to that of urodeles).

14. The anterior cardinal veins persist in all vertebrates as the internal jugular veins, which, after receiving other veins of the anterior parts of the body, form large trunks, the precaval veins. The proximal portions of the precavals are the common cardinal veins (ducts of Cuvier). In some mammals the left precaval joins the right precaval in front of the heart, so that there is but a single precaval stem (cat, man). In such a case the reduced detached base of the left precaval (which is the left common cardinal) remains as the coronary sinus, a vein in the heart wall.

15. The anterior portions of the posterior cardinal veins lose their importance in land vertebrates and disappear above urodeles, except that in mammals the right one contributes slightly to the unpaired azygos vein.

16. The posterior parts of the posterior cardinal veins are the renal portal veins of all vertebrates. In fishes they collect from the tail only, but above fishes they usurp the veins from the legs (at first tributaries of the abdominal veins). The renal portal veins in fishes, amphibians, and reptiles conduct the blood from tail, legs, and general pelvic region into a capillary net in the kidneys, from which the blood is re-collected into the subcardinal veins running between the two kidneys.

17. Simultaneously with these changes in the posterior cardinals there appears a new vein, the postcaval vein, found in a few fishes and in all tetrapods. It is formed in lower vertebrates by the union of a hepatic vein (derived from the vitellines) with the subcardinal veins, chiefly the right subcardinal. The subcardinals then become parts of the postcaval vein. In mammals the supracardinal veins also contribute to the postcaval, forming the part between the subcardinals and the renal portal veins. The supracardinal veins first appear in reptiles, where they are the vertebral veins of the adult. Their anterior parts persist as the paired azygos veins of monotremes and some marsupials; the right one, as the single azygos vein of other mammals.

18. Simultaneously with these changes the abdominal veins, which become connected with the renal portal veins when the latter usurp the veins from the legs, change their anterior connections. Whereas they originally entered the common cardinal veins, they now enter the hepatic portal vein. The abdominal veins thus come to constitute a connection between the renal portal and the hepatic portal systems.

19. In amphibians and reptiles the postcaval vein extends only to the posterior end of the kidneys.

20. In birds the postcaval vein establishes direct connections with the renal portal veins; the renal portal circulation is thus greatly reduced. In mammals the connection is completed,

so that the veins from the legs, tail, and adjacent regions pass directly into the postcaval. The renal portal circulation then vanishes.

21. With the changes outlined in paragraph 20 the abdominal vein loses its function. It is probably present in birds but has established different connections, and it is lacking in mammals except monotremes. In embryonic stages, however, this vein (or veins) is of great importance as the veins of the allantois or embryonic respiratory organ (umbilical veins).

22. The hepatic portal system remains unchanged throughout.

23. The ventral aorta is present in fishes and amphibians. It is absent in amniotes, for in these the conus arteriosus splits into either three trunks, right and left aortae and pulmonary (reptiles), or into two trunks, aorta and pulmonary (birds and mammals).

24. The aortic arches are more or less modified in land vertebrates. The first two vanish during development in all tetrapods. In a few urodeles the third to sixth arches are present, but in all other tetrapods the fifth arches atrophy early in development. The remaining urodeles, gymnophionans, and *Sphenodon* retain the third, fourth, and sixth arches in their entirety. In anurans, reptiles (except *Sphenodon*), birds, and mammals the third and sixth arches lose their connection with the dorsal aorta, and only the fourth arches retain this connection. The third or carotid arches persist as the common carotids, and the dorsal and ventral aortae anterior to the third arches contribute to the internal and external carotids, respectively (Fig. 104). The fourth arches form the aorta. In anurans and reptiles both fourth arches persist, forming right and left aortic arches which unite dorsally to produce the dorsal aorta; but in birds the left fourth arch and in mammals the right fourth arch disappear during development. The sixth aortic arches give rise, at about their middle, to the pulmonary arteries; and then the parts of the sixth arches dorsad to this point (known as the ductus arteriosus or duct of Botallus, open during embryonic stages) degenerate to a ligament, so that the ventral parts of the sixth arches become the proximal parts of the pulmonary arteries.

25. The dorsal aorta is the chief artery of the body caudad of the heart in all vertebrates, and has much the same branches throughout.

26. The pulmonary veins appear as new structures in air-breathing vertebrates (including Dipnoi). They enter the left auricle, and simultaneously there occurs a change in the heart and the double circulation is initiated.

27. In amphibians the four chambers of the heart are retained, as in fishes, but the atrium is partially or completely divided into right and left auricles. The sinus venosus is then attached to the right auricle, and the pulmonary veins enter the left auricle. The right side of the heart consequently contains venous blood and the left side arterial blood, but these become somewhat mixed in the ventricle and also in passing through the conus arteriosus. A double circulation is thus present in an imperfect condition.

28. In amniotes the conus arteriosus becomes split into two or three arterial trunks and thus disappears as such. The great arteries then appear to spring directly from the ventricles. In all amniotes there are two completely separated auricles.

29. In reptiles the sinus venosus is retained; the ventricle is incompletely separated into right and left chambers except in the crocodilians, where the interventricular septum is complete; and the conus arteriosus is subdivided into three trunks. The double circulation is improved over that of amphibians but is still imperfect because of the incompleteness of the ventricular septum and the presence of three arterial trunks.

30. In birds and mammals the sinus venosus is reduced to a mere node in the wall of the right auricle, and the systemic veins then enter the right auricle directly. The ventricle is completely divided by a partition into right and left ventricles; and the conus arteriosus is split into two trunks, the aorta leading from the left ventricle, and the pulmonary from the

right. The right side of the heart is venous, the left arterial, and the double circulation is complete and perfect; only pure arterial blood passes into the aorta for distribution to the body.

REFERENCES

COLES, ESTHER. 1928. The segmental arteries in *Squalus sucklii.* Univ. Calif. Pub. Zoöl., **31.**

COULTER, C. B. 1909. Early development of the aortic arches of the cat with especial reference to the presence of a fifth arch. Anat. Rec., **3.**

DANIEL, J. F. 1926. The lateral blood supply of primitive elasmobranch fish. Univ. Calif. Pub. Zoöl., **29.**

———. 1934. The circulation of the blood in *Ammocoetes. Ibid.*, **39.**

DANIEL, J. F., and CURLIN, A. R. 1928. The circulation of blood in the larva of *Triturus torosus.* Univ. Calif. Pub. Zoöl., **31.**

DANIEL, J. F., and STOKER, EDITH. 1927. The relation and nature of the cutaneous vessels in selachian fishes. Univ. Calif. Pub. Zoöl., **31.**

DE RYKE, W. 1926. The vascular structure of the kidney in *Chrysemys* and *Chelydra.* Anat. Rec., **33.**

GELDEREN, C. VON. 1927. Zur vergleichenden Anatomie der Vv. cardinales posteriores, der V. cava inferior und der Vv. azygos (vertebrales). Anat. Anz., **63.**

HAMMOND, W. S. 1937. The developmental transformations of the aortic arches in the calf. Amer. Jour. Anat., **62.**

HUNTINGTON, G. E., and MCCLURE, C. F. 1920. The development of the veins in the domestic cat. Anat. Rec., **20.**

KIMBALL, PAULINE. 1923. A contribution to the anatomy and development of the arterial and venous systems in turtles. Anat. Rec., **25.**

———. 1928. A comparative study of the vas subintestinale in the vertebrates. Amer. Jour. Anat., **42.**

LEWIS, F. T. 1906. Fifth and sixth aortic arches and related pharyngeal pouches in pig and rabbit. Anat. Anz., **28.**

MATHUR, P. N. 1940. The venous system of the pond turtle *Lissemys.* Proc. Indian Acad., **11,** Sec. B.

MILLER, W. S. 1900. Contributions from the anatomical laboratory of the University of Wisconsin (*Necturus*). Bull. Univ. Wis., **33,** Sci. Ser. 2.

O'DONOGHUE, C. H. 1921. The blood vascular system of the tuatara, *Sphenodon.* Phil. Trans. Roy. Soc. London, B, **210.**

O'DONOGHUE, C. H., and ABBOTT, EILEEN. 1928. The blood vascular system of the spiny dogfish, *Squalus acanthias* and *Squalus sucklii.* Trans. Roy. Soc. Edinburgh, **55,** Part III.

ROBINSON, B. L. 1918. Concerning the renal portal system in *Chrysemys marginata.* Anat. Rec., **14.**

SEWERTZOFF, A. N. 1933. Entwicklung der Kiemen und Kiemenbogengefässe. Zeitschr. f. wiss. Zool., **121.**

SHANER, R. F. 1921. The development of the pharynx and aortic arches of the turtle. Amer. Jour. Anat., **29.**

STROMSTEN, F. A. 1905. A contribution to the anatomy and development of the venous system of *Chelonia.* Amer. Jour. Anat., **4.**

XIII. THE COMPARATIVE ANATOMY OF THE UROGENITAL SYSTEM

The urogenital system of vertebrates comprises the *excretory* or *urinary* system and the *reproductive* system. The former consists of a pair of *kidneys* and their ducts and functions for the elimination of nongaseous waste metabolic products, especially those resulting from the breakdown of proteins. The reproductive system consists of a pair of sex glands or *gonads* and their ducts. Because of the close association anatomically of these two systems it is customary to consider them together as the *urogenital* system. The association consists of the utilization of the excretory ducts by the sex glands for the discharge of the sex cells to the exterior. The evidence indicates that this relation is primary among the vertebrates and that there is a tendency toward the complete separation of the two systems in the evolution of the vertebrate classes. Such separation is, however, achieved only in the females of the primate mammals (and teleosts).

A. PHYLOGENY, EMBRYOLOGY, AND STRUCTURAL PLAN OF THE UROGENITAL SYSTEM

1. Phylogeny of the nephridial tubules.—The excretory organs of animals in general are termed *nephridia;* and they consist of tubules, the *nephridial tubules.* These are of two kinds: the *protonephridia,* whose inner ends are closed by ciliated cells (*flame cells*) or flagellated cells (*solenocytes,* Fig. 108*B*); and the *metanephridia,* whose inner ends open into the coelom by a ciliated funnel. Protonephridia are probably the more primitive and are characteristic of noncoelomate animals, such as flatworms. Metanephridia first appear in annelids, and the open funnel seems to be an adaptation for removing coelomic fluid with its contained wastes. So far, the matter seems relatively simple; but there has been endless argument about the nature of the coelomic funnel, and consequently the phylogeny of the nephridia remains one of the most confused and unsettled questions in the whole of comparative anatomy. According to the work of Goodrich (1895–99), there are two kinds of coelomic funnels on the metanephridia. One kind, found in oligochaete annelids, is really a part of the nephridial tubule and alone merits the name *nephrostome.* In the other kind the tubule has appropriated the coelomic funnel belonging to the sex duct or *gonoduct* and hence properly called *gonostome* or *coelomostone.* According to Goodrich, each segment of a primitive segmented animal was provided with a pair of protonephridia with external pores (nephridiopores) and a pair of sex ducts or *gonoducts* opening into the coelom by a funnel, the gonostome or coelomostome. Not infrequently, so it seems, the coelomostome fused onto and adopted the protonephridial tubule, whereupon the latter's solenocytes and also the gonoducts degenerated, and the compounded tubule (called by Goodrich *nephromixium,* although *mixonephridia* would seem to be more suitable) with its acquired funnel remained, serving the combined function of excretory tubule and genital duct. In this way there arose the close association of the excretory and genital systems. It does not seem to be a very sensible arrangement, since it surely cannot be healthy for the sex cells to be subjected to the concentrated nitrogenous wastes in the urine. However, it may be supposed that in the annelids, where this arrangement originated, the fluid conveyed by the nephridia was dilute and not very different from the coelomic fluid. In vertebrates, however, the situation is different, for there the concentrated urine is toxic. Hence, as already remarked, there is seen throughout the vertebrates a striving to separate the urinary and genital ducts.

Now, it is generally agreed among vertebrate anatomists that the nephridial tubules of vertebrates are of the "mixed" type; i.e., they are, or primitively were, provided with coelomic funnels which are really genital funnels or coelomostomes, although there is no direct evidence of their genital nature. But it is an astounding fact, glossed over by phylogeneticists, that *Amphioxus* has segmentally arranged protonephridia with solenocytes (Fig. 108); closely similar nephridia occur only in polychaete annelids. How does it happen that *Amphioxus*, so close to the vertebrates in many anatomical features, has nephridia of a type wholly different from theirs? The stock answer is, of course, that *Amphioxus* must have

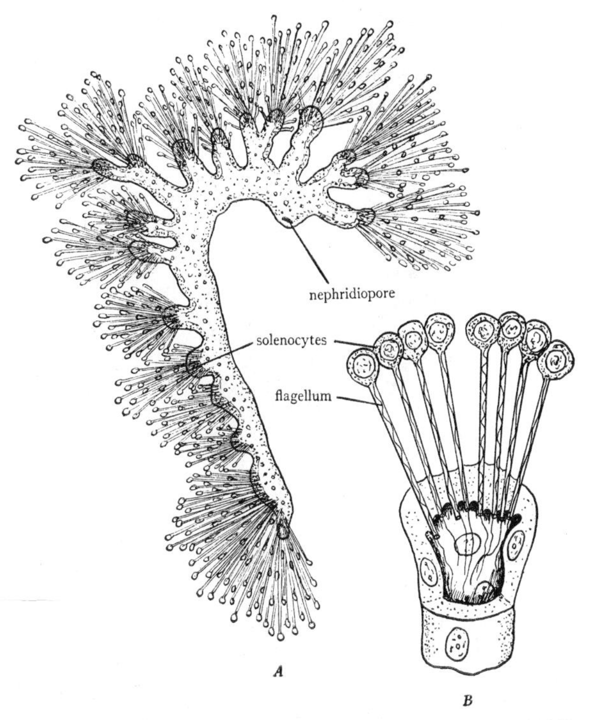

FIG. 108.—Nephridium of *Amphioxus* (after Goodrich, 1902). *A*, entire nephridium, showing diverticula bearing clusters of solenocytes. *B*, group of solenocytes enlarged, nephridium cut open.

separated from the vertebrate line very far back, although this view is difficult to reconcile with its many vertebrate resemblances. Again, it is now generally believed that the chordates stem from the echinoderm-hemichordate line and have not even a remote relationship with annelids. How, then, did *Amphioxus* get its annelid type of solenocytes? No explanation is available at present. The echinoderms lack an excretory system, and that of hemichordates is limited to a structure similar to the glomeruli of the vertebrate kidney (see below). Hence, the vertebrate nephridial tubules with their coelomic funnels must have originated independently of those of annelids. It therefore is questionable whether the work of Goodrich, done on annelids, applies to the vertebrate kidney. In the author's opinion, work on annelids has, on present views, no bearing on the homology of vertebrate nephridia, and hence the interpretation of these nephridia must be based on their embryology.

2. Development of the vertebrate kidney.—The kidneys of vertebrates, as already intimated, consist of a mass of tubules, the *uriniferous tubules*. These tubules are wholly of mesodermal origin. They form from the tissue of the mesomere, also called nephrotome and intermediate cell mass. The mesomere, it will be recalled, is that part of the mesoderm immediately lateral to the epimere (Fig. 15, p. 75), from which it soon becomes completely separated. It is more or less segmented and has coelomic cavities, the *nephrocoels*, continuous with the general coelom of the hypomere. The primitive nephridial tubules develop by outgrowth of the outer wall of the mesomere, one pair of tubules to each body segment. The inner end of the tubule opens into the nephrocoel of that segment; the outer ends of the tubules turn backward and fuse to form a duct, which grows caudad and opens into the cloaca. The opening of the nephridial tubule into its nephrocoel is considered to be the nephrostome.[1] The originally wide communications of the nephrocoels with the general coelom usually narrow, forming ciliated funnel-like coelomic openings, the *peritoneal funnels*.[2] A relation is early established between the tubules and the adjacent blood vessels, which send

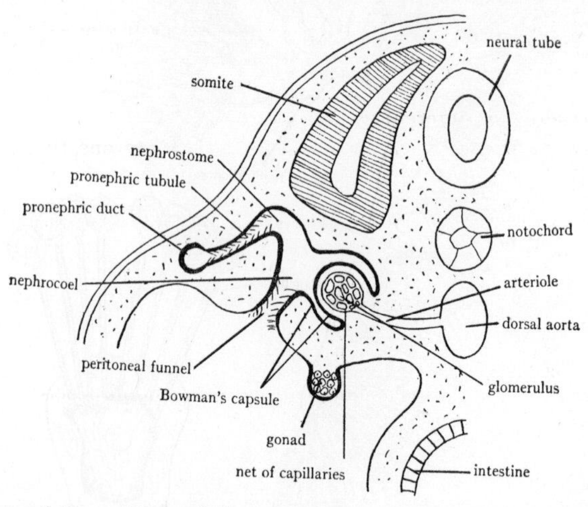

FIG. 109.—Diagram of the structure of the pronephros and its relations to other structures; seen in cross-section; glomerulus of the internal type.

a spherical network of capillaries, called a *glomerulus*, toward the nephrocoel (Fig. 109). An arteriole from the dorsal aorta leads into each glomerulus, and the venule leaving the glomerulus empties into the cardinal system, which permeates the kidneys. The glomerulus pushes in the wall of the nephrocoel, which is then termed *Bowman's capsule;* Bowman's capsule, plus the glomerulus, constitutes a *Malpighian body*, also called *renal corpuscle*. The presence of renal corpuscles is the most characteristic histological feature of vertebrate kidneys, although these bodies are lacking in some bony fishes, whose kidneys are therefore referred to as *aglomerular*.

The embryological findings lead to the belief that the ancestral vertebrate had a pair of kidneys extending the entire length of the coelom and composed of segmentally arranged tubules (Fig. 109), one per segment to each kidney, each opening into the coelom by a peri-

[1] But Goodrich objects altogether to the use of the term nephrostome with regard to the vertebrate kidney and terms the opening in question nephrocoelostome. There does not appear to be any need for such niceties of terminology, especially in view of the state of confusion of the whole subject.

[2] Usually called nephrostomes; the author is unable to see any objection to this, for the opening into the general coelom would seem to correspond better with the annelid nephrostome than the opening into the nephrocoel. However, Goodrich and others are very insistent upon the above terminology.

toneal funnel, provided near the funnel with a glomerulus, and opening laterally into a duct connecting with the cloaca. Such a kidney is termed an *archinephros* or *holonephros*, and its duct the *archinephric duct.* Among existing vertebrates a holonephros is found only in the embryos of the myxinoid cyclostomes. This extends over about seventy segments and consists of segmental tubules, beginning in nephrocoels open to the coelom and ending in a duct running along the length of the holonephros and opening into the cloaca; only the more posterior tubules are provided with glomeruli. In other vertebrates the kidney tubules develop in two or three separated groups, in an anteroposterior succession. These three groups are called the *pronephros*, the *mesonephros*, and the *metanephros* (Fig. 111).

a) *Pronephros:* The pronephros develops from the most anterior part of the mesomere over a limited, usually small, number of segments. Its tubules and its duct, the pronephric duct, develop as described above for the archinephros and archinephric duct. It varies considerably in development and structure in different vertebrates, and its one constant and distinguishing character is that it has approximately only one tubule per segment. Other distinctions previously listed have been found not generally valid, for complete pronephric tubules with a nephrocoel, internal glomerulus, and peritoneal funnel—in short, tubules like the primary mesonephric tubules—occur in the embryos of gymnophionans and some fish. It is common for the pronephros to lack peritoneal funnels and typical renal corpuscles. The glomeruli may project directly into the general coelom, so that they lack a Bowman's capsule; such glomeruli are termed *external.* This is usually caused by the nephrocoels becoming widely open into the general coelom, so that peritoneal funnels are also lacking. In other cases the nephrocoels may coalesce to a large chamber, which becomes more or less closed off from the general coelom, so that peritoneal funnels are again lacking; the glomeruli project into this pronephric chamber and may also fuse to a compound glomerulus, often called *glomus.* The pronephric tubules open by their nephrostomes into such a chamber.

The pronephric duct, formed anteriorly by the fusion of successive tubules, extends itself backward beyond the rear limits of the pronephros by utilizing mesomere tissue and opens into the cloaca.

A pair of pronephroi appear during embryonic stages of all vertebrates but are well developed and functional only in free-living larvae, such as those of cyclostomes, many fishes, and amphibians. The *Ammocoetes* larva has pronephroi composed of from three to six tubules, each with a nephrostome opening into the pericardial cavity, but there is only one glomus (Fig. 110). Well-developed pronephroi are also seen in frog tadpoles and in the larvae of *Polypterus* and of lungfishes. Among amniote embryos the pronephroi are usually very transient structures. In all vertebrates except a few bony fish and the myxinoid cyclostomes the pronephroi degenerate during development, but the pronephric ducts persist and become the ducts of the mesonephroi. The myx-

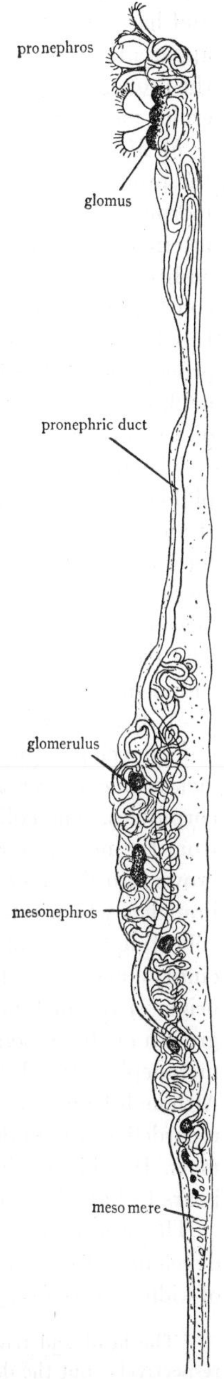

FIG. 110.—Kidney of the *Ammocoetes* larva of *Petromyzon* (after Wheeler, 1900), consisting of pronephros and mesonephros; the former has four peritoneal funnels and a glomus.

inoids have an embryonic holonephros, of which the anterior and posterior parts persist while the intermediate region degenerates. The anterior part or *head* kidney is a mass of branched tubules opening by many nephrostomes into the pericardial cavity and provided with one glomus; the long rear part or *trunk* kidney retains the segmental arrangement of the tubules and has glomeruli but has lost the peritoneal funnels.[3] In most bony fishes the pronephroi apparently remain, but the kidney tissue has been replaced by lymphoid tissue; however, the pronephroi are persistent and functional in several of those bony fishes which have aglomerular mesonephroi.

b) *Mesonephros:* The mesonephroi develop in the mesomere immediately or shortly behind the pronephroi. At first they consist of segmental pairs of tubules whose development and morphology are typical; i.e., each has a peritoneal funnel and a renal corpuscle, always of the internal sort, and each opens at its outer end into the pronephric duct. Subsequently, however, additional generations of tubules develop, so that the segmental arrangement is lost, and this constitutes the only constant difference between pro- and mesonephros. The later mesonephric tubules lack peritoneal funnels. The mesonephric tubules connect with the already present pronephric duct, which thereupon is renamed the *mesonephric* or *Wolffian duct.* The mesonephros, also called *Wolffian body,* forms a conspicuous organ, especially in embryos, composed of a confusion of coiled tubules.

The mesonephros is the functional kidney of the adults of lamphreys, fishes, and amphibians. The peritoneal funnels are retained only in some selachians (including *Acanthias*), *Amia*, and amphibians, but they do not necessarily open into the kidney tubules. In Anura, for instance, they lose their connection with the tubules and open secondarily into veins. The mesonephroi appear as prominent and functional organs in the embryos of amniotes but degenerate except for those parts which become associated with the reproductive system. The mesonephroi of amniote embryos lack peritoneal funnels or have very transient and rudimentary ones, except in monotremes, where there are well-developed funnels.

c) *Metanephros:* This is the third kidney of vertebrates, found only in amniotes. Its tubules develop in the remaining, most posterior part of the mesomere, behind the mesonephros, lack segmental arrangement, have no peritoneal funnels, but are well provided with renal corpuscles. The collecting system of the metanephros develops from an outgrowth put out from the mesonephric duct near its entrance into the cloaca (Fig. 111*C*). This evagination grows into the mesomere, where the metanephridial tubules are developing, and divides into branches which establish connections with these tubules. These branches develop into the collecting tubules of the metanephros and the *calices* (spaces into which the collecting tubules empty), and the stalk of the evagination becomes the *metanephric duct* or *ureter.*[4] This double origin of the metanephros and its possession of a separate duct (which, however, is an outgrowth of the mesonephric duct) are the distinctive characteristics of the metanephros. The metanephric tubules become very long and complicated, with several loops and convolutions. Although traces of metanephric development occur in anamniotes, the metanephros is to be regarded as an amniote structure and is the kidney of the adults of reptiles, birds, and mammals. It achieves that isolation of the urinary function from the genital function which appears to have been the trend in the evolution of the urogenital system.

The occurrence of three kidneys in anteroposterior succession in the mesomere during the development of amniotes may be taken to indicate that the vertebrates originally had a pair of kidneys extending the whole length of the coelom, the archinephroi or holonephroi already

[3] The head and trunk kidneys of myxinoids are usually regarded as pronephroi and mesonephroi, respectively; but the development shows that they are persistent parts of the holonephros, i.e., both are pronephros.

[4] It is desirable that the term ureter be reserved for the duct of the metanephros.

mentioned. It seems that only the pronephros retains the characteristic structure of the holonephros; mesonephros and metanephros are more recently evolved modifications of the

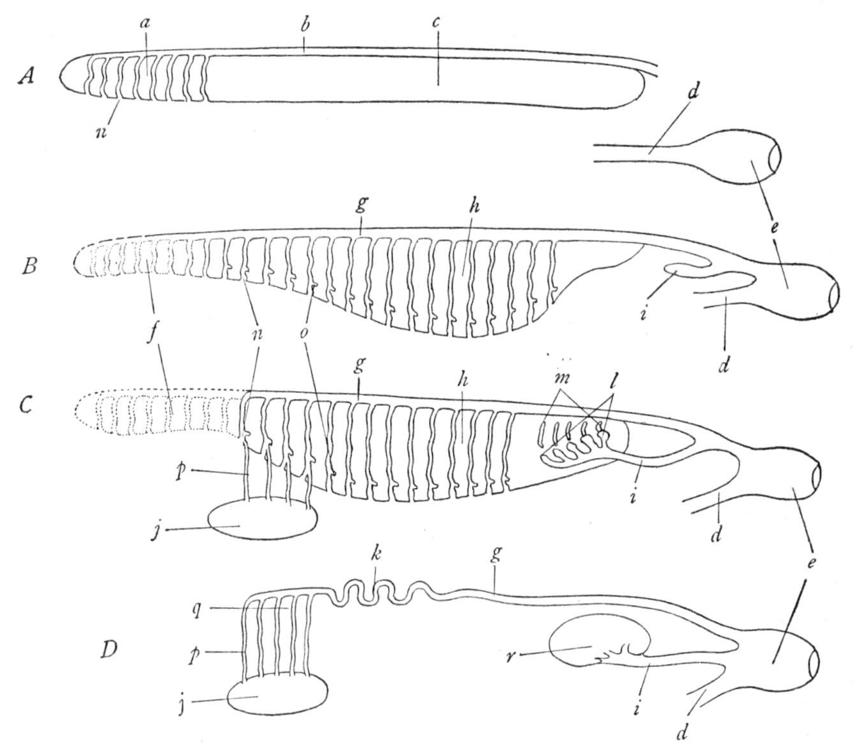

FIG. 111.—Diagrams to show the development of the three kidneys and their ducts and their relation to the male gonad. *A*, early stage showing the pronephros (*a*) developing from the anterior end of the mesomere (*c*), and the pronephric duct (*b*), which has not yet reached the cloaca (*e*). *B*, next stage, illustrating the degeneration of the pronephros at *f*, the development of the mesonephros (*h*) from the middle portion of the mesomere, the junction of the pronephric duct, now the mesonephric duct (*g*), with the cloaca, and the beginning of the metanephric evagination (*i*) from the mesonephric duct. *C*, later stage, showing connection between certain tubules of the mesonephros and the testis (*j*) by means of tubules, the efferent ductules (*p*), which grow out from the mesonephros; and the penetration of the metanephric evagination into the posterior end of the mesomere, where it is subdividing to form the collecting apparatus (*l*), which becomes associated with secretory metanephric tubules (*m*), developed in the mesomere. *D*, final stage, in which the mesonephros has disappeared except for the remnant (*q*) which connects with the testis (*j*) by means of the efferent ductules (*p*); the mesonephric duct (*g*) persists as the vas deferens; the two parts of the metanephros shown in *C* have united to form a single organ (*r*). *a*, pronephros; *b*, pronephric duct; *c*, mesomere or nephrotome; *d*, intestine; *e*, cloaca; *f*, degenerating pronephros; *g*, mesonephric or Wolffian duct; *h*, mesonephros or Wolffian body; *i*, metanephric evagination from the Wolffian duct in *B*, ureter in *C* and *D*; *j*, testis; *k*, coiled portion of the deferent duct forming part of the epididymis; *l*, collecting part of the metanephros derived from the Wolffian duct; *m*, excretory tubules of the metanephros derived from the mesomere; *n*, nephrostome; *o*, renal corpuscle or Malpighian body; *p*, efferent ductules; *q*, remnant of the mesonephros, forming part of the epididymis; *r*, metanephros.

original kidney, making for greater efficiency. It may be supposed that draining the coelomic fluid was an important function of the primitive kidney, but with increasing size and activity of early vertebrates this would be a most inadequate method of ridding the body of excretory substances. Relation to the blood stream became more and more important, and hence the peritoneal funnels disappear and more and more tubules and renal corpuscles appear. The amniote kidney, of course, obtains all the urine constituents from the blood.

3. Concept of the opisthonephros.—In anamniotes the so-called mesonephros, although corresponding anatomically to the mesonephros of amniotes, typically extends the length of the coelom behind the pronephros and hence has used up the mesomere tissue from which in amniotes both mesonephros and metanephros come. It therefore is not exactly equivalent to the amniote mesonephros but topographically represents both meso- and metanephros. For this reason, Kerr[5] has suggested that it be called by a different name and has proposed the term *opisthonephros*. This concept is here adopted, and hence the kidney of adult selachians and *Necturus* will be called an opisthonephros.

4. The urinary bladder.—The urinary bladder is, in all forms above fishes, a sacciform evagination from the ventral cloacal wall, which serves as a reservoir for the urine. In fishes the bladder, when present, is formed chiefly by the enlargement of the terminal portions of the mesonephric ducts. In the embryos of amniotes there is an enormous cloacal evagination, the *allantois*, which may be regarded as a greatly expanded urinary bladder; it does act as a reservoir for kidney excretions, but its chief function in the amniote embryo is respiratory. It spreads out over the inner surface of the egg shell in bird and reptile embryos or the uterine wall in mammals, and its blood vessels are the chief site of gaseous exchange. The adult bladder develops at the base of the allantoic stalk, except in birds, which lack a urinary bladder. The kidney ducts generally do not open into the bladder directly except in mammals.

5. The gonads and their ducts.—The gonads or sex glands consist of a pair of *testes* in the male, *ovaries* in the female. The gonads develop from a pair of elongated ridges, the *genital ridges*, which arise to the medial side of the kidneys. The length of these ridges suggests that the gonads originally extended the length of the pleuroperitoneal cavity, but in present vertebrates they are usually much shorter than this. The peritoneum covering the developing gonad thickens to become the *germinal epithelium*, which is the source of the epithelial elements of the gonad, while the mesoderm tissue of the genital ridge forms the other histological elements of the gonads. The gonads come to project into the coelom and, when mature, are usually provided with mesenteries.

Vertebrates have a pair of gonads, but in *Amphioxus* there are numerous segmentally arranged gonads. These are sacs which lie along the ventral ends of the myotomes close against the wall of the atrium, into which they discharge by rupture, for *Amphioxus* lacks genital ducts. In cyclostomes the paired embryonic gonads unite to a single one in the adult, but this is of typical vertebrate construction. Cyclostomes, however, also lack sexual ducts, and the sex cells are discharged by a pair of abdominal pores piercing the body wall from the coelom to the exterior. The gnathostome vertebrates have genital ducts for conveying the sex cells to the exterior; but, except in the bony fishes, these ducts are kidney ducts appropriated by the gonads. The aberrant and puzzling conditions in teleosts will not be considered here.

a) *Male:* The testis consists of a mass of sperm-forming or *seminiferous tubules*, which develop by proliferation from the germinal epithelium. These tubules connect with the mesonephric tubules of the adjacent region of the mesonephros or opisthonephros by a set of tubules, termed the *testicular network*. This varies in form and development in different vertebrates. Inside the testis the seminiferous tubules join a longitudinal canal or a network; from these, small tubules, termed the *efferent ductules* (*ductuli efferentes*, old name, *vasa efferentia*), cross in the mesorchium into the mesonephros or opisthonephros, where they open into the

[5] *Text-Book of Embryology*, Vol. II: *Vertebrata with the Exception of Mammalia* (1919).

cavities of the renal corpuscles, from which mesonephric tubules lead to the mesonephric duct. Thus the passage for the sperm from the seminiferous tubules to the exterior comprises a canal or network in the testis, the efferent ductules in the mesorchium, certain renal corpuscles and mesonephric tubules of the region of the kidney adjacent to the testis, the mesonephric duct, and the cloaca. Frequently the efferent ductules are connected inside the mesonephros or opisthonephros by a longitudinal anastomosis. There has been much uncertainty regarding the nature of the efferent ductules. In some vertebrates they grow out from the renal corpuscles into the testis, and in others they proliferate from the testis into the kidney. They are usually interpreted to be closed-over coelomic grooves from the peritoneal funnels to the testis along the peritoneum, but there does not appear to be any good evidence for this view.

That portion of the mesonephros or opisthonephros which thus comes to serve the male genital system is termed the *epididymis*. It loses its urinary function; its peritoneal funnels and renal corpuscles degenerate; its tubules, which convey the sperm, are known as the *epididymidal tubules;* and that part of the mesonephric duct involved in the epididymis is called the *epididymidal duct.*

In elasmobranchs and amphibians, in which the opisthonephros is the functional adult kidney, the anterior part of the opisthonephros connected to the testis by the efferent ductules becomes an epididymis, and only its posterior part acts as a urinary organ (Fig. 112). The mesonephric duct in the males of these groups serves both genital and urinary functions, although in many elasmobranchs, special *accessory urinary ducts* arise, so that the mesonephric duct is here largely genital (see further, p. 404). In male amniotes, in which the metanephros is the functional adult kidney, the persistent mesonephric duct has only genital functions and is then called, in the adult, the *deferent duct* (*ductus deferens;* old name, *vas deferens*). In amniotes the mesonephroi disappear except for that portion whose tubules convey the sperm, which persists as a more or less distinct body, the *epididymis* (Fig. 112). The proximal part of the deferent duct, included in the epididymis, is usually much coiled, forming the *epididymidal duct.*

b) *Female:* The ovaries are masses of connective tissue containing various sizes of eggs, each of which is surrounded by a layer of nutritive cells forming a *follicle.* The ovaries, unlike the testes, *never* have any connection with the kidneys. Whereas the sperm, as related above, travel in a closed system of tubes from testis to cloaca, the ovaries of vertebrates have no connection with their ducts, and the ripe eggs are discharged into the coelom (Fig. 112*B*, *D*). The ducts of the ovaries are called the *oviducts* or *Müllerian ducts.* In elasmobranchs they originate by the lengthwise splitting of the pronephric ducts, half remaining as the mesonephric duct, half becoming the oviduct, which appropriates one or more peritoneal funnels to serve as its coelomic opening, termed the *ostium.* This developmental history does not apply to tetrapods, where the oviduct and its funnel develop directly from the mesomere; but the findings in elasmobranchs are usually taken to indicate the phylogenetic origin of the vertebrate oviduct.

The oviducts are never directly connected to the ovaries, but each opens near the corresponding ovary by the fringed and ciliated funnel or ostium. The ripe eggs, discharged into the coelom by rupture of their follicles, are conveyed to the ostia of the oviducts chiefly by ciliated tracts of the coelomic wall. However, in a few mammals, as rats, the ovary is inclosed in a peritoneal sac which also includes the ostium so that the eggs cannot escape into the coelom.

The oviducts in the majority of vertebrates remain as two separate tubes opening into the cloaca (Fig. 113*A*). In mammals each oviduct is differentiated into a narrower anterior portion, called the *uterine* or *Fallopian tube,* which bears the ostium, and a wider more muscular

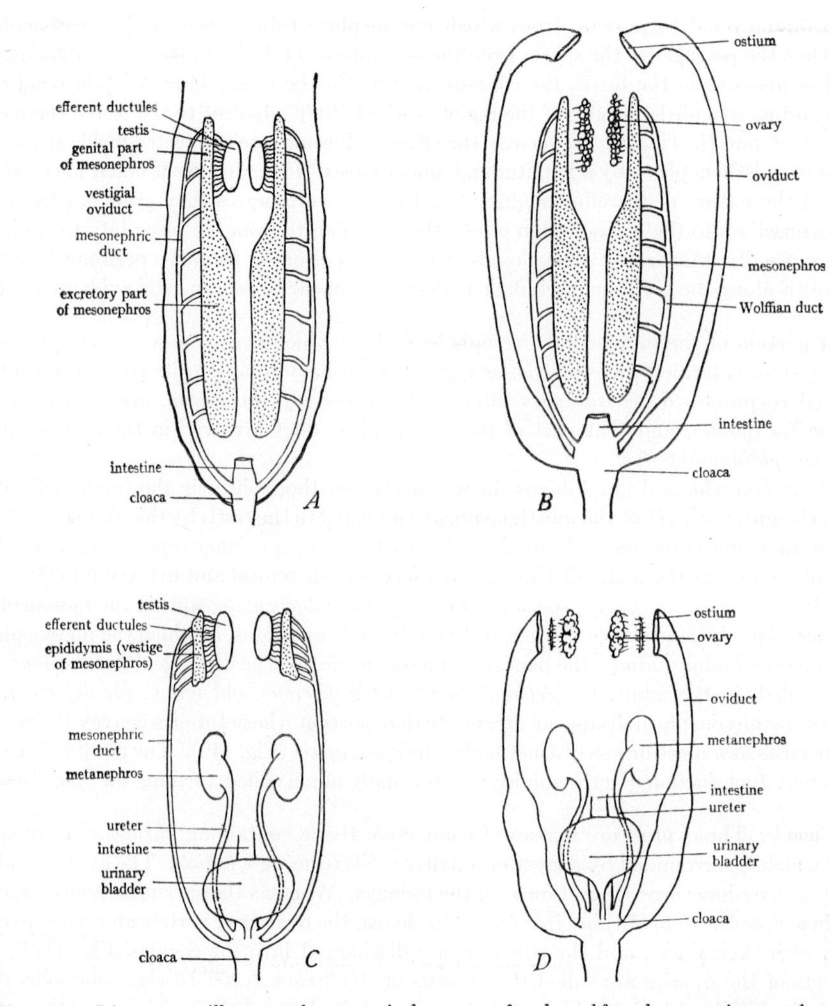

FIG. 112.—Diagrams to illustrate the urogenital systems of male and female anamniotes and amniotes. *A*, male elasmobranch or amphibian; the mesonephros is differentiated into anterior genital and posterior excretory portions; the genital part is connected with the testis by means of the efferent ductules, which are outgrowths from the mesonephros; the mesonephric or mesonephric duct serves as both genital and excretory duct; the oviduct or Müllerian duct is vestigial. *B*, female elasmobranch or amphibian; the ovary is not connected with the mesonephros; the mesonephros and mesonephric duct serve only excretory functions; the oviduct is well developed and opens into the coelom by the ostium near the ovary. *C*, male reptile, bird, or mammal; the excretory part of the mesonephros has disappeared, but the genital part persists as the epididymis (in part), which is connected, as in anamniotes, with the testis by means of the efferent ductules; the mesonephric duct is purely genital and is renamed the deferent duct; the excretory function is served by metanephroi and ureters. *D*, female reptile, bird, or mammal; the mesonephros and mesonephric duct have entirely vanished; the condition of the ovary and oviduct is the same as in anamniotes; the excretory function is served by the metanephroi and ureters, exactly as in the male. (The changes in the relation of the urogenital ducts and cloaca which occur in mammals are not indicated in these figures but are shown in Figures 113 and 115.) (Slightly modified from Wilder's *History of the Human Bodu*. courtesy of Henry Holt & Co.)

posterior portion, the *uterus*. In the monotremes, or egg-laying mammals, each uterus opens separately into the cloaca. In the marsupials the terminal portion of the uterus is differentiated as a *vagina* to receive the penis. In mammals above marsupials the two vaginae fuse to a single vagina (hence the name Monodelphia). There is also generally more or less fusion of the two uteri (Fig. 113). When only the posterior portions of uteri are fused, the fused portion is called the *body* of the uterus, and the separate portions the *horns* of the uterus. In man and

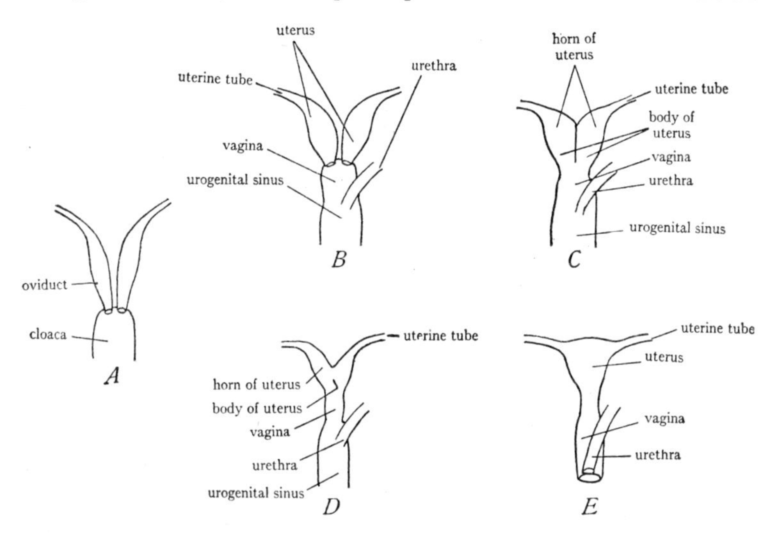

FIG. 113.—Diagrams to show the various types of mammalian oviducts. *A*, condition found in the majority of female vertebrates; the two oviducts are completely separate and open independently into the cloaca. *B–E*, various conditions found in mammals, showing differentiation of the oviducts into uterine tube, uterus, and vagina, and progressive fusion of the lower parts of the oviducts: *B*, *duplex* type, found in rodents, in which the two vaginae are fused; *C*, *bipartite* type, occurring in carnivores; not only are the vaginae fused but the lower parts of the two uteri are fused to form a single *body*, divided in two by a partition which represents the fused walls of the two uteri; the upper parts of the two uteri remain separate as the *horns; D*, *bicornuate* type, found in many ungulates, similar to *C* except that the partition has disappeared; *E*, *simplex* type, occurring in man and the apes, in which both vaginae and uteri are fused along their entire lengths, leaving only the uterine tubes separate. Note further that in *B–D* the urethra joins the vagina to form the urogenital sinus, which opens to the exterior, while in *E* the urethra and vagina are wholly separate and open independently to the exterior. (From Wiedersheim's *Comparative Anatomy of Vertebrates*, courtesy of the Macmillan Co.)

other primates the uteri are fused along their entire length, producing the single *uterus* or *womb*. The young of the placental mammals develop only in the uterine part of the oviducts; in those forms with partially fused uteri, only in the horns.

c) *Bisexuality of the vertebrates:* The developmental history of the vertebrate reproductive system shows that each embryo has, at some stage, the structural features of both sexes and develops into one sex by the degeneration, or failure to differentiate, of the structures characteristic of the other sex. Which sex develops depends upon genetic factors and hormonal influences. According to present genetic theory, the sex of the individual is fixed at the time of the fertilization of the egg from which it comes, but can be wholly or partially altered in the direction of the other sex by the application of the proper hormones at labile stages. Persistence of embryonic conditions explains various anomalies of the reproductive system

Change of sex during early life is not uncommon in the lower vertebrates. Thus the myxinoids are hermaphroditic in youth, with the anterior part of the gonad female, the rear part male, and become one sex in the adult by the degeneration of the part of the gonad of the opposite sex. It is quite common for genetically male frogs to pass through a female phase in which the gonads have the structure of ovaries; later the gonads transform into testes. Among birds, transformation of a functional female into an anatomical male has been repeatedly observed or experimentally produced. On removal of the left ovary of hens, the rudimentary right ovary enlarges into a testis, and the bird takes on the plumage and behavior of a rooster. Various grades of transformation of gonads and ducts in the direction of the opposite sex, forms termed *intersexes*, can be elicited in juvenile or even adult vertebrates by injection of appropriate hormones or by grafting hormone-producing parts of the genital system.

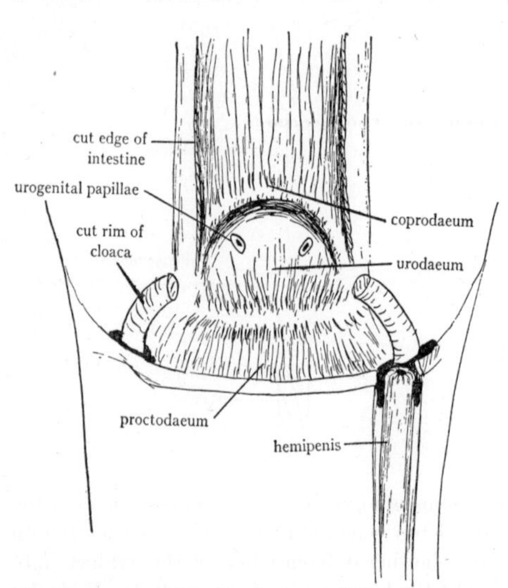

FIG. 114.—Cloaca of a male lizard (*Varanus*), cut open to show the divisions of the cloaca (from a dissection by H. Raven); the body wall on the left side behind the cloaca has been slit to show the left hemipenis.

6. Phylogeny of the urogenital ducts.—According to the embryological findings, the kidney tubules and the urogenital ducts are coelomoducts, i.e., they are wholly of mesodermal origin and connect the coelom to the exterior. It does not necessarily follow, however, that they are nephromixia, compounded of nephridial tubules and of coelomic funnels originally belonging to the gonads. There is nothing in the development to indicate a separate origin of the coelomic funnels. As long as the annelids were considered ancestral to the vertebrates, it was natural to attempt to derive the nephridial tubules and ducts from those of annelids. But the matter takes on a different aspect if the chordates stem from the echinoderm-hemichordate groups, for none of these forms have nephridial tubules, and neither do they have any genital ducts, except the very short direct connections of the gonad sacs to the exterior. Nor have *Amphioxus* and the cyclostomes any genital ducts. It would therefore appear that in the vertebrates it is the gonads which have appropriated the nephridial ducts to serve as genital ducts, rather than the reverse, as usually supposed. The facts seem to indicate that the vertebrate gonads never had any ducts of their own and that they borrowed the conveniently adjacent kidney ducts. This brought about the degeneration of the anterior part of the holonephros and may, indeed, have been the cause of the peculiar embryonic history of the vertebrate kidneys, with its succession of pro-, meso-, and metanephros. These may represent repeated attempts to escape utilization by the gonads.

7. History of the cloaca.—A cloaca occurs in all vertebrates except cyclostomes, holocephalans, teleostomes, and placental mammals. The cloaca is the common chamber into which open the intestine and urogenital ducts. In many vertebrates, especially reptiles and birds, the cloaca is more or less distinctly divided by folds into the *coprodaeum*, or part into which the intestine opens; the *urodaeum* or *urogenital sinus*, into which the urinary and genital ducts open; and the *proctodaeum* or terminal part, formed of the ectodermal proctodaeal invagination and opening by the anus. These divisions are well evidenced in the cloaca of reptiles (Fig. 114). Whether the ancestral vertebrate stock had a cloaca or not, and whether its absence in cyclostomes is primitive or derived cannot be stated; but, on the general supposition that the kidney ducts originally opened directly to the exterior, a cloaca is not an original vertebrate characteristic. In cyclostomes the mesonephric ducts open on a urogenital papilla behind the anus.

In the males of selachians and amphibians the cloaca receives only the mesonephric (Wolffian) ducts, whereas in the females of these groups both oviducts and mesonephric ducts enter it; however, males often have rudimentary oviducts (Fig. 112). In the males of reptiles and birds the cloaca receives the mesonephric ducts (ducti deferentes) and the ureters; in females, the oviducts and the ureters (Fig. 112). In addition, in many fishes and in amphibians, reptiles, and the embryos of birds and mammals the urinary bladder (or, embryonically, the allantois) opens into the urodaeal part of the cloaca.

Only in mammals does the cloaca undergo progressive change. A cloaca is present in monotremes, but already a considerable fold marks off the urodaeal from the coprodaeal

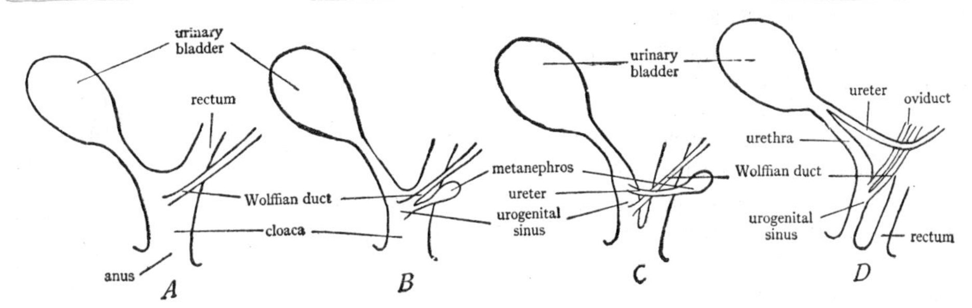

FIG. 115.—Diagrams to illustrate the changes in the cloaca in mammals during development. *A*, early embryonic stage, showing the cloaca receiving the urinary bladder, the rectum, and the Wolffian duct, as in the lower vertebrates. *B*, later stage, showing the beginning of the fold which divides the cloaca into a ventral urogenital sinus, which receives the urinary bladder, Wolffian ducts, and ureters, and into a dorsal part, which receives the rectum. *C*, further progress of the fold, dividing the cloaca into urogenital sinus and rectum; the ureter has separated from the Wolffian duct and is shifting anteriorly. *D*, completion of the fold, showing complete separation of the cloaca into ventral urogenital sinus and dorsal rectum. Note in *D* that the ureter has shifted farther, so that it opens into the urinary bladder.

region (Fig. 121*A*). The completion of such a fold in placental mammals separates the cloaca into two parts, each with its own opening to the exterior. The dorsal part includes only the intestine, here termed rectum, opening by the anus; the ventral part is the *urogenital sinus* or *urogenital canal*, which receives the stalk of the bladder (*urethra*) and the excretory and genital ducts. The division of the cloaca is incomplete in many marsupials and some other mammals, where a shallow cloaca, corresponding to the proctodaeum, remains. The ureters (metanephric ducts) open into the urogenital sinus in the embryos of placental mammals but subsequently shift so as to open into the urinary bladder (Fig. 115). Thus finally the ureters open into the bladder, while the deferent ducts (mesonephric ducts) in males or the vagina in females unite with the urethra (stalk of the bladder) to form a common tube or chamber, the urogenital sinus or canal, which opens to the exterior in front of the anus by a urogenital aperture. Throughout the placental mammals the male ducts join the urethra, and hence the sperm must, for a short distance, use a passage in common with the urine. Although this passage is very much shorter in amniotes than in anamniotes, it may be said that the male vertebrate never achieves complete separation of urinary and genital paths. This has, however, been accomplished by the females of the highest placental mammals (order Primates, including man, apes, and monkeys), in which by another fold the urethra separates from the vagina and each has an external opening. In primate females, then, there are three openings in the perineum: the anus, the mouth of the vagina, and the mouth of the urethra.

8. The adrenal gland.—This important gland of internal secretion is topographically close to the kidneys or the gonads and is conveniently considered with them, although functionally it is not related to the urogenital system. In mammals the *adrenal* or *suprarenal glands* are a pair of definite bodies anterior to the kidneys. Each consists of two histologically different parts, the *cortex* and the *medulla*, which also have different embryonic origins. The cortex or peripheral tissue is of mesodermal origin, formed by ingrowth of the peritoneum; the medulla or central tissue comes from the sympathetic ganglia of the nervous system. Although both cortical and medullary tissue exist in all vertebrates, they do not have the topographical relation which occurs in mammals, and hence more general terms for them are necessary. The cortical tissue is called *interrenal* tissue; and the medullary tissue, because it stains brown with chromic salts, is termed the *chromaffine* tissue. In cyclostomes small groups of chromaffine and interrenal cells occur along blood vessels. In selachians there is present between the kidneys an interrenal organ composed of paired more or less discontinuous strands of interrenal tissue; the chromaffine tissue consists of a series of small masses imbedded in the kidneys along segmental branches of the dorsal aorta. These are often called *suprarenal* bodies or, erroneously, interrenal bodies. In amphibians and reptiles the adrenal gland, consisting of a mesh of interrenal tissue with the chromaffine cells in the interstices, forms elongated bodies lying imbedded in the kidneys or situated close to the gonads. The adrenals of birds are also close to the gonads and have the same structure as those of reptiles.

B. THE UROGENITAL SYSTEM OF SELACHIANS

The directions apply to the spiny dogfishes and the skate. The dogfishes supplied for dissection are usually immature, and it is therefore difficult or impossible to locate in them all of the parts of the urogenital system. At least a few mature males and females should be on hand for demonstration. Skates are sexually mature while still relatively small.

Remove the digestive tract except cloaca and liver.

1. Female urogenital system.—

Dogfish: The ovaries are a pair of soft bodies, oval in form, situated dorsal to the liver, each with a mesentery, the mesovarium. In mature specimens the ovaries contain large eggs, consisting chiefly of yolk.

The kidneys are long, slender, brown bodies lying against the dorsal body wall, one to each side of the dorsal aorta; they are retroperitoneal. Free their lateral borders by slitting the pleuroperitoneum and note thickness of the organ at various levels. The thinner anterior portion has lost its urinary function and is degenerate in females; the broader, thicker posterior part performs the work of excretion. Between the two kidneys is a tough, shining ligament which should not be mistaken for a duct. Both components of the adrenal complex are present in dogfishes but are not very practical to find. The chromaffine masses, called suprarenal bodies, may be noticed as a longitudinal series of light spots near the medial border.

The kidneys of elasmobranchs, usually considered mesonephroi, are better termed opisthonephroi. In *Squalus* and in a number of other selachians there are persistent peritoneal funnels, but these do not connect with the kidney tubules, becoming closed off from these during embryonic stages; they open into short blind tubes.

The *oviducts* in *immature* females are slender tubes running along the ventral face of the kidneys, without mesenteries. In *mature* females they are very large tubes which spring free from the kidneys by means of well-developed mesenteries, the *mesotubaria*. Trace the oviducts forward. They pass forward along the dorsal coelomic wall, curve around the anterior border of the liver, and enter the falciform ligament. Here the two oviducts are united to a common opening, the *ostium*, a wide funnel-like aperture lying in the falciform ligament, with the opening facing posteriorly into the pleuroperitoneal cavity. As already indicated, the ostium is formed by the fusion of peritoneal funnels of the pronephros. To find the opening it is generally necessary to separate the walls of the ostium, as these tend to adhere. Trace the oviducts posteriorly. They are narrow at first, but in mature specimens they soon present a slight enlargement, the *shell gland* or *nidamental gland*. This in *Squalus* secretes a thin membrane in which several eggs become inclosed. Posterior to this gland, the oviduct narrows again; and in mature females it enlarges greatly to form the *uterus*, which swings free by means of the mesotubarium. In immature females there is no uterine enlargement or any mesotubarium, but the oviducts widen slightly as they proceed posteriorly along the ventral faces of the kidneys.

Trace the oviducts to the cloaca. Cut open the cloaca by a median slit which opens up the intestine. Note the opening of the intestine into the ventral part of the cloaca, which is thus the *coprodaeum*, and the slight fold which separates this from the dorsal urogenital region of the cloaca or *urodaeum*. Note the *urinary papilla* in the middorsal wall of the urodaeum. In mature specimens the large openings of the oviducts are readily seen to each side of the urinary papilla; in immature females they are in the same position but quite small and are best found by cutting into the posterior ends of the oviducts and probing toward the cloaca.

The kidney ducts, the mesonephric or Wolffian ducts, are somewhat difficult to find in females. The duct lies along the ventral face of the kidney exactly dorsal to the oviduct in immature females, along the line of attachment of the mesotubarium in mature ones. Locate it in immature specimens by carefully stripping off the oviduct and also freeing the peritoneum from the ventral face of the kidney. The mesonephric duct is a slender tube proceeding directly to the cloaca, where the two ducts join to open by the terminal pore of the urinary papilla.

Skate: The ovaries, a pair of elongated, soft bodies containing large yellow eggs, are situated dorsally in the anterior half of the pleuroperitoneal cavity. The large oviducts pass dorsal to them. Follow one oviduct forward; its narrow anterior portion passes along the dorsal coelomic wall, curves around the anterior margin of the liver, and, after entering the falciform ligament, unites with its fellow to a single common opening, the *ostium*. This is a wide, funnel-like aperture situated in the ligament and facing caudad. Trace the oviducts caudad. After a short distance they widen greatly to a conspicuous bilobed swelling, the *shell* or *nidamental gland*, which secretes the horny case in which the eggs are laid. The wide *uterus* continues from the shell gland and proceeds to the cloaca, supported by a thickened mesotubarium. Cut open the cloaca in the midventral line, also slitting the intestine. Note the opening of the intestine into the ventral part of the cloaca, which thus constitutes a *coprodaeum*, and the conspicuous horizontal fold which separates this from the dorsal urogenital part, or *urodaeum*. Cut into the latter by cutting forward through this fold. The urodaeum is greatly extended and thickened craniad. Find the oviducal openings, one to each side of this thickened part of the cloaca, and the urogenital opening in the middorsal wall between them. The terminal part of the cloaca leading to the anus is the *proctodaeum*.

The kidney is an opisthonephros; its main part, the *caudal* opisthonephros, consists, in female skates, of a thick rounded lobe lying dorsad at each side of the cloaca, best revealed by stripping off the thick peritoneum which covers its ventral surface. The anterior or cranial part of the opisthonephros is nearly degenerate in females but will be found as diffuse brownish tissue extending forward ventral to the dorsal aorta. From the medial surface of the caudal opisthonephros several ducts, the *accessory urinary ducts*, pass anteriorly and medially in contact with the posterior cardinal vein and open into a small chamber, the *urinary sinus*, situated on the dorsal surface of the anterior end of the cloaca. The two urinary sinuses of the two sides unite into a common chamber, which is sometimes called the *urinary bladder*. It does not correspond to the bladder of higher forms, since it consists of the enlarged terminations of the mesonephric ducts. Cut into this, note the entrance into it of the two urinary sinuses, and find the opening in its middorsal wall by which it opens into the cloaca. The *Wolffian ducts* or *ducts* of the cranial mesonephros are slender tubes extending anteriorly from the urinary bladder, lying on the dorsal surface of the strong white portions of the mesotubaria.

Draw, showing kidneys, gonads, and their ducts, and the opened cloaca.

2. The male urogenital system.—The *testes* are a pair of soft bodies dorsally situated, each with a mesorchium. In the dogfishes they are located in the anterior part of the pleuroperitoneal cavity, dorsal to the liver. In the skate they are broad, flat bodies against the dorsal wall.

In the dogfishes the kidneys are identical in both sexes and should be next examined according to the directions given under females. In the male skate, however, the cranial part of the kidney is very much better developed than in the female and extends forward as a firm cylindrical body on either side of the middorsal line.

As explained in the introduction (p. 395), the *male ducts* in the majority of vertebrates are the *mesonephric* or *Wolffian ducts*. In mature males these ducts are consequently much larger than in females. The Wolffian ducts run posteriorly along the ventral face of the kidneys. In immature specimens each is a slender, straight tube, similar to that of the female, but in mature males it is greatly coiled. Each testis is connected with the corresponding region of the opisthonephros by delicate ducts, the *efferent ductules*, which run in the mesorchium and can sometimes be seen by holding the mesorchium up to the light. They vary in number in different selachians (said to be from four to seven in *Squalus acanthias*, one in skates and rays). They connect with the tubules of the cranial part of the opisthonephros, which has practically lost its urinary function and serves as part of the male genital system, conveying the sperm into the Wolffian duct. The part of the opisthonephros penetrated by the efferent ductules is therefore in male selachians an *epididymis*. From the tubules of the epididymis, the sperm pass into the Wolffian duct, now called male duct or ductus deferens, which forms a greatly coiled tube (*epididymidal duct*) on the vental surface of the epididymis and on the ventral surface of *Leydig's gland*. Leydig's gland, believed to secrete a fluid beneficial to the sperm, is the part of the opisthonephros behind the epididymis. Leydig's gland has peritoneal funnels (lacking in the epididymis), although these do not connect with the ductus deferens, but lacks Malpighian bodies. As the ductus deferens approaches the caudal opisthonephros, it becomes less coiled and enlarges upon the surface of the latter as a wide, straight tube, the *seminal vesicle*. Trace this to the cloaca by removing the peritoneum from the ventral face of the kidney. At its posterior end on the sides of the cloaca the seminal vesicle terminates in a sac, the *sperm sac*, which projects craniad as a blind sac lying against the ventral surface of the seminal vesicle.

Cut open the cloaca as directed under the female and identify its parts as directed there. There is no difference in the cloaca of the dogfish between the males and females, but in the male skate the cloaca is very much smaller than in the female and is not divided into intestinal and urogenital parts. In the median dorsal line there is in the male skate a *urogenital papilla*.

The sperm sacs should now be cut open and the *papillae*, where the seminal vesicles open into them, identified. The two sperm sacs unite at their posterior ends to form a *urogenital sinus*, which opens at the tip of the urinary papilla.

The seminal vesicle of male *Squalus* does not receive any kidney tubules from the opisthonephros on which it lies. Instead, this part of the opisthonephros has an *accessory urinary duct* separate from the mesonephric duct (here seminal vesicle). This accessory duct will be found by lifting out the seminal vesicle (note lack of tubules connecting this with the opisthonephros) and gently picking in the kidney tissue on the medial side of the groove where the vesicle lay. The accessory urinary duct is a delicate, white tube which will be seen to receive collecting tubes from the kidney substance at intervals. The accessory ducts are difficult to trace; they enter the sperm sacs where the minute openings occur to the medial side of the openings of the seminal vesicles. In male skates, conditions are the same as in the female.

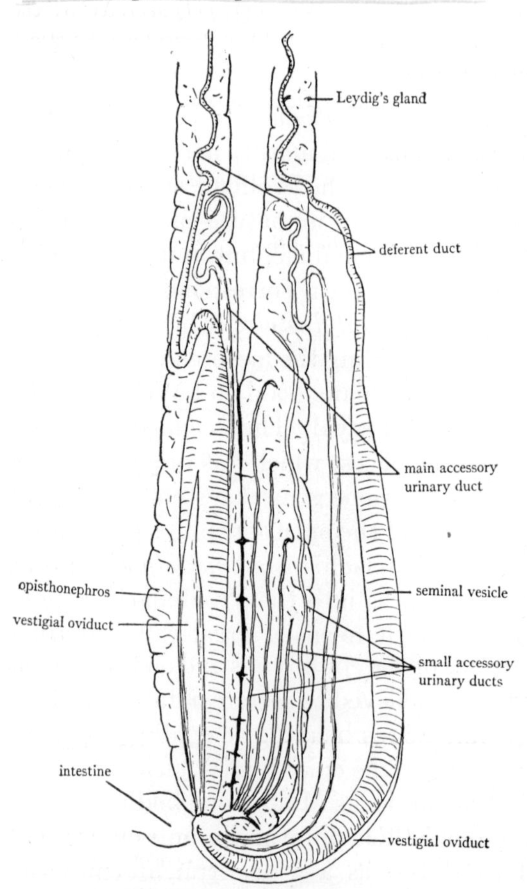

Fig. 116.—Urogenital system of a male shark (*Galeus*), to show mainly accessory urinary ducts (after Borcea, 1906). On the left side the seminal vesicle has been pulled out of place; it is shown *in situ* on the right side.

Female *Squalus* lack an accessory urinary duct; instead the collecting tubules from the caudal opisthonephros pass directly into the mesonephric duct. The method of discharge of the functional caudal part of the opisthonephros varies greatly in different selachians. It is usually provided with one or more accessory urinary ducts (Fig. 116), which arise by the fusion of the enlarged collecting tubules of this part of the kidney. In some cases there is a very large accessory duct in both sexes; in others, a slender one in one or both sexes. An accessory duct may be present in the male, while in the female, the collecting tubules open directly into the mesonephric duct without forming any accessory duct. In a number of forms, especially the skates and rays, the enlarged collecting tubules run backward and unite only when near the cloaca, so that the accessory duct is very short; or the tubules may unite in groups, so that several short ducts are produced.

Draw, showing gonads, kidneys, their ducts, and the opened cloaca.

Look in male dogfishes for vestiges of the ostium and oviducts in the vicinity of the liver.

3. Breeding habits of selachians.—The eggs of selachians are very large, containing an immense amount of yolk, and have the meroblastic type of development. In many forms, especially the skates and rays, the eggs are laid in horny cases (Fig. 117) secreted by the oviducal gland, which in this event is very large and of complicated histological construction. The horny cases are usually of flat, rectangular shape with strings at the corners for attachment to objects; but may be spirally coiled, also with strings. There are one to several eggs in each case, and these require several months to hatch. Other selachians are ovoviviparous, i.e., the eggs develop inside the uterus into fully formed young fish. At first, as in *Squalus*, several eggs may be inclosed in a thin, horny membrane ("candle") formed by the oviducal gland; but this soon disappears, and the developing fish ("pups") lie in the uterus. The embryo gradually separates from the yolk mass, which forms a sac hanging from its belly. This yolk sac may form very intimate relations with the uterine wall, with an exchange of materials. The uterine wall may put out thin folds or papillae which come in contact with the yolk sac, and the latter may also send out elevations against the uterine wall. The uterine development of selachians is usually of long duration, up to two years (twenty-one months or more in *S. acanthias*, according to Ford, 1919).

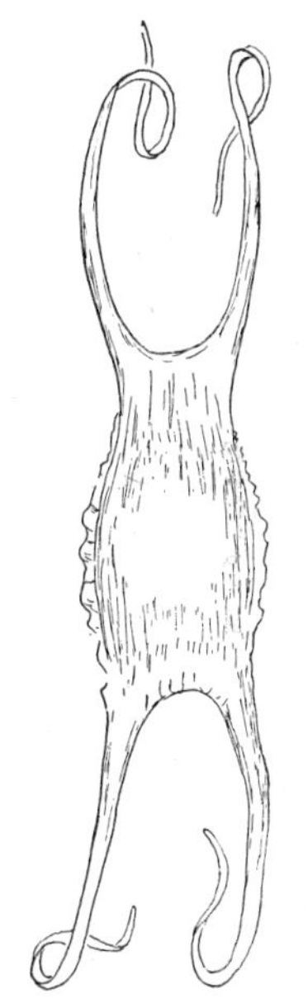

FIG. 117.—Egg case of a skate.

C. THE UROGENITAL SYSTEM OF NECTURUS

The urogenital system of *Necturus* is morphologically of the same grade as that of selachians; the kidneys are opisthonephroi, and in males the mesonephric duct serves both urinary and genital functions.

1. The female urogenital system.—The *ovaries* have already been noted as elongated saclike bodies bearing eggs of various sizes. Note the *mesovarium*. Lateral to each ovary, running along the dorsal body wall, is the *oviduct*, a thick, white, coiled tube supported by the mesotubarium. Follow it anteriorly. At the anterior end of the pleuroperitoneal cavity it becomes of a more delicate texture and is fastened to the lateral wall. Here it opens by a funnel-shaped opening, the *ostium;* the dorsal rim of this is fastened to the body wall, but the ventral rim is free and can be lifted to expose the opening. Trace the oviducts posteriorly to the cloaca. They enter this, one to each side of the large intestine. Cut the cloaca open by a lateral slit extending up into the intestine. Note the *papilla* by which each oviduct opens into the cloaca. A transverse fold separates the coprodaeum from the urodaeum.

The kidneys or **opisthonephroi** are long, slender organs extending from the cloaca forward to the medial side of the oviducts and inclosed in the mesotubarium. The kidneys of *Necturus* are thus covered on both sides by peritoneum, a rather unusual condition in vertebrates. The female kidney is divisible into a slender, anterior genital part and a thicker, caudal part. The adrenal glands consist of small bright flecks and patches along the sides of the postcaval vein. The mesonephric duct lies along the lateral border of each kidney but is very delicate in females and difficult to locate. Seventy to eighty delicate collecting ducts cross from the kidney in the mesentery into the mesonephric duct. The latter proceeds to the cloaca, into which it opens to the dorsal side of the oviduct, becoming imbedded in the medial wall of the oviduct. It opens into the cloaca by a minute pore on the dorsal side of the oviducal papilla. In tracing it, make a cut along one side of the cloaca, freeing the cloaca from the body wall. Note the urinary bladder extending from the midventral region of the cloaca and find its opening into the cloaca.

Draw, showing ovaries, kidneys, their ducts, and the opened cloaca.

2. The male urogenital system.—The testes are a pair of elongated bodies situated to the sides of the small intestine, each supported by a mesorchium. Dorsal and lateral to each testis is the long brown kidney or opisthonephros, inclosed, as in the female, in the same mesentery that supports the mesonephric duct. The kidney, as in the female, is divisible into anterior genital and posterior urinary parts; but the former, although thin and flat, is much wider in the male. The genital part is an epididymis. Along the lateral border of the kidney runs the conspicuous coiled mesonephric duct, which, as in male vertebrates, in general also acts as the male duct or ductus deferens. Two to four delicate efferent ductules cross the mesorchium from the anterior part of the testis into the genital kidney, from which about twenty-six collecting tubules run transversely into the mesonephric duct. The genital kidney lacks peritoneal funnels but has renal corpuscles. Trace the mesonephric duct to the cloaca; along the caudal functional part of the kidney, the duct is narrower, loses its convolutions, and receives fifty to sixty collecting tubules from the kidney. Open the cloaca by a slit to one side of the midventral line, carrying the slit into the large intestine. Note transverse fold dividing cloaca into coprodaeum and urodaeum. The small openings of the mesonephric ducts, difficult to find, are in the dorsolateral wall of the cloaca, just caudad of the fold. Note the urinary bladder and its opening into the ventral cloacal wall.

Draw, showing testes, kidneys, mesonephric ducts, and opened cloaca.

The adrenal glands are the same in the male as in the female.

In *Necturus* the collecting ducts of the opisthonephric tubules pass directly laterally into the mesonephric duct, a primitive condition. In many salamanders, however, the same process has occurred as in selachians. In either the male or in both sexes the more posterior or all of the collecting ducts turn posteriorly, become elongated, and open directly into the cloaca as a bundle of accessory urinary ducts.

3. Breeding habits of salamanders.—The breeding habits of salamanders are somewhat peculiar. Breeding usually occurs in the spring, when the male may show heightened colors and develop a crest. Aquatic and some terrestrial species breed in water, but some terrestrial salamanders breed on land. There is a prolonged courtship, in which the male stimulates the female by various contacts, involving the release of hedonic substances from his special skin glands, and in some species the male clasps the female after the manner of frogs. In the final phase of mating, the male walks away, closely followed by the female, and deposits spermatophores in front of her, which are picked up by her cloacal lips. These spermatophores are gelatinous masses bearing a ball of sperm and are secreted by the cloacal glands of the male. The sperm proceed into tubules in the roof of the female cloaca, where they remain until needed. In a few forms, direct transfer of the spermatophore by cloacal contact obtains. The eggs are laid singly or in gelatinous masses, in the water or under objects on land, and in water hatch into a larva which soon develops external gills. Most salamanders undergo a process of metamorphosis, in which these external gills are lost and other morphological changes occur; but *Necturus* and a few others retain the larval aspect throughout life. In land-breeding species, the gilled stage is passed inside the egg mass, and the young, at hatching, have much the adult aspect. The breeding habits of *Necturus* are imperfectly known, but presumably the male deposits a spermatophore which is picked up by the female. This occurs in the fall, but the eggs are not laid until the following spring. The female hollows out the sand under a log or rock in the water, then lays the eggs singly on the under side of this object by turning her body over. She remains in this "nest," guarding the eggs.

D. THE UROGENITAL SYSTEM OF THE TURTLE

In turtles and other reptiles the kidneys are metanephroi with their own ducts, the ureters, running to the cloaca. Only that part of the mesonephric system remains which functions in the male as epididymis and ductus deferens, and these are degenerate or rudimentary in females. They are homologous to the parts of the same name in male selachians and amphibians (Fig. 112).

Remove the digestive tract, if not already done, leaving the large intestine in place.

1. The female urogenital system.—This consists as usual of a pair of *ovaries* and a pair of *Müllerian ducts* or *oviducts*. The ovaries have already been noted as large baglike bodies in the posterior part of the pleuroperitoneal cavity. They usually contain yellow eggs in various states of development. Each ovary is supported by a mesentery, the *mesovarium*. Along the posterior border of each ovary runs the *oviduct*, a large white coiled tube, supported by the mesotubarium. Trace the oviduct forward and find the *ostium;* this lies in the mesentery and has winglike borders which are generally closed together and should be spread apart to see the opening. Trace each oviduct to the cloaca. Each opens into the side of the anterior end of the cloaca, ventral to the opening of the intestine. The stalk of the large bilobed *urinary bladder* joins the cloaca midway between the two oviducts.

The cloaca has already been exposed. (If not, do so by cutting through the pelvic girdle on each side and removing the median portion of the girdle.) Clear away connective tissue from around the cloaca. Attached to each side of the cloaca posterior to the oviducts are two elongated sacs, the *accessory urinary bladders*. Their function is uncertain; but in females, at least, they carry water, employed in softening the soil while digging a nest. A dark structure visible through the cloacal wall is the *clitoris*, homologous to the male penis and without function in the female.

Now cut open the cloaca to one side of the clitoris, extending the cut in the median ventral line up to the stalk of the bladder. Look into the cloaca. Observe that the clitoris consists simply of thickenings in the ventral wall. Find the large openings of the accessory bladders. Next note the opening of the large intestine. This is the most dorsal of the openings and is somewhat separated by a fold from the urogenital openings, so that the cloaca is divisible into coprodaeum and urodaeum. Ventral to the opening of the intestine are the openings of the oviducts on thickened papillae. They are best found by cutting into the oviduct and probing posteriorly into the cloaca. Between and ventral to the oviducal openings is the opening of the urinary bladder.

The kidneys of the turtle are *metanephroi*. They have already been identified as flattened lobed organs fitting snugly against the posterior end of the pleuroperitoneal cavity. The renal portal vein and its tributary, the internal iliac, run along the ventral face of each kidney. Dissect off this vein; directly dorsal to it is a tube, the *metanephric duct* or *ureter*, extending from the middle of the kidney to the cloaca. It enters the cloaca at the base of the oviduct. By making a slit in it and passing a probe into it, its opening into the cloaca will be found just anterior to the thickening caused by the oviducal entrance.

Draw the female urogenital system with opened cloaca.

2. The male urogenital system.—The male system consists of the paired *testes* and their *ducts*. The ducts of the testes are mesonephric or Wolffian ducts, now termed deferent ducts.

Expose the cloaca as directed in the female and find the two accessory bladders attached to its lateral walls. Note the place of attachment of the rectum to the cloaca and ventral to this the attachment of the urinary bladder. The dark mass seen through the ventral wall of the cloaca is the *penis* or organ of copulation, which is inserted into the female cloaca at mating so that the sperm are injected directly into the female system. It is to be noted that a penis is first met with in reptiles. A pair of rounded masses projects from the anterior wall of the cloaca to either side of the stalk of the bladder; these are the *bulbs of the corpora cavernosa*, part of the penis. Muscles which retract the penis will be seen attached to the ventral wall of the cloaca.

The kidneys were previously identified as flattened, lobed bodies fitting against the posterior wall of the pleuroperitoneal cavity. Each *testis* is a yellow spherical body attached to the ventral face of the kidney by the mesorchium. Lateral and posterior to the testis is an elongated, dark-colored coiled body, the *epididymis*. The testis is connected to the anterior part of the epididymis by the minute *efferent ductules* which run in the mesorchium. The efferent ductules and epididymis are remnants of the mesonephros. The male duct or deferent duct begins as a greatly coiled tube on the surface of the epididymis, here termed epididymidal duct. Remove the peritoneal covering of the epididymis, uncoil the deferent duct, and trace it to the cloaca, which it enters craniad of, and at the base of, the bulb of the corpora cavernosa.

Next cut open the cloaca, inserting the blade of the scissors into one corner of the anus and cutting far to one side to avoid injuring the penis. Spread apart the cloacal walls and study the penis. It consists of two spongy ridges, the *corpora cavernosa* or *cavernous bodies*, in the ventral wall of the cloaca. Between these folds in the midventral line is a deep groove, the *urethral groove*, which in the natural condition is practically converted into a tube by the approximation of the cavernous bodies. The urethral groove terminates caudad at the base of a heart-shaped projection, the *glans* of the penis. The anterior ends of the cavernous bodies form the *bulbs*, already noted, which project forward into the coelom at the sides of the stalk of the bladder. The bulbs are filled with blood, which they receive from the internal iliac vein. All parts of the penis are highly spongy and vascular. In the sexual act the blood from the bulbs rushes into the spongy spaces of the cavernous bodies and the glans, erecting them and causing the cavernous bodies to come in contact above the urethral groove, converting the latter into a canal for the passage of the sperm.

The kidneys are *metanephroi*, and their ducts the *metanephric ducts* or *ureters*. The ureter will be found immediately to the dorsal side of the epididymis, which should be removed. The ureter is a short, straight tube proceeding to the cloaca, into which it opens just anterior to the opening of the ductus deferens. The two openings will be found at the sides of the anterior beginning of the urethral groove.

Find the openings of the accessory bladders, the urinary bladder, and the rectum into the cloaca. The latter is dorsal to the urogenital openings.

Draw the male urogenital system with open cloaca.

3. Breeding habits of reptiles.—In all reptiles the sperm are injected into the female by an act of sexual union (*copulation*), usually preceded by a courtship behavior. *Sphenodon* lacks a copulatory organ, and copulation occurs by cloacal contact. The copulatory apparatus of snakes and lizards is unique among vertebrates, consisting of a pair of sacs, the *hemipenes*, often armed with spines or teeth, lying beneath the skin behind the cloaca (Fig. 114). At copulation one hemipenis turns inside out to the exterior and is inserted into the female cloaca, acting to guide the sperm. Crocodilians have a penis similar to that of turtles. All reptiles produce large-yolked, shelled eggs similar to those of birds; but in some snakes and lizards these are retained in the oviducts until they hatch, so that the young are born alive. In some of these viviparous forms, intimate nutritive relations may become established between embryo and uterus in the form of uterine ridges or folds in contact with modified areas of the embryonic membranes. A degeneration of the intervening epithelia may bring maternal and embryonic blood vessels in close contact. Other reptiles lay eggs, numbering a few to one hundred, in a crude nest, dug out in sand or earth, or under objects or composed of a mound of vegetable material. There is generally no or little parental care, and the eggs are incubated by the heat of the sun plus the heat of fermentation of the decaying vegetable debris. The common pond turtles breed in summer, at that time leaving the pond and ascending a near-by slope, where the female excavates a hole with her feet, moistening the ground by water from the accessory bladders. A batch of eggs, mostly five to fifteen, is laid in the hole, which is then filled in again. The incubation period in reptiles lasts a few to several weeks (two months in the alligator).

E. THE UROGENITAL SYSTEM OF THE PIGEON

The urogenital system of birds is very similar to that of reptiles except for the peculiarity that the right ovary and oviduct are degenerate in birds with a few exceptions. The large size of the eggs is usually cited as the cause of this degeneration, since there would not be room in the compact pelvis for two functional ovaries and ducts. However, both ovaries and ducts are functional in some birds, especially birds of prey.

1. The female urogenital system.—Remove the digestive tract, leaving the large intestine in place. In adult birds there is a single *ovary* and *oviduct* on the *left* side. The right ovary and duct are present in the embryo but almost entirely disappear before hatching. The ovary is a mass containing eggs of various sizes, situated at the anterior end of the left kidney, attached by a short mesovarium. Posterior to the ovary the coiled left oviduct proceeds to the cloaca, being supported by the mesotubarium. The *ostium* is situated in the mesotubarium near the ovary; it is a wide opening with winglike borders fastened to the mesotubarium. A small remnant of the right oviduct is attached to the right side of the cloaca.

The kidneys are *metanephroi*. Each is a flattened, three-lobed organ situated against the dorsal wall. The *ureters* or *metanephric ducts* are located just dorsal to the renal portal veins, which should be stripped from the face of the kidney. The ureter begins on each side at the groove between the anterior and middle lobes of the kidney and extends straight posteriorly to the cloaca. The left ureter is concealed by the oviduct.

The cloaca is an expanded chamber receiving the rectum on its median ventral surface, the left oviduct to the left, the very small right oviduct to the right, and the ureters dorsal to the oviducts. Cut into the cloaca to the right of the rectum. Note that the cavity of the cloaca is subdivided. There is a large ventral portion (*coprodaeum*) into which the rectum opens. Dorsal to this and separated from it by a fold is the *urodaeum*, into which open the oviducts and ureters. The opening of the left oviduct is readily found here; the openings of the ureters are more medial and smaller. The most dorsal compartment of the cloaca is the *proctodaeum*, a small chamber with a raised rim, which opens to the anus. In the anterior wall of the proctodaeum dorsal to the rim an opening may be noted; it leads into a small pouch, the *bursa of Fabricius*, which seems to have some function in young birds but degenerates with maturity.

Draw the urogenital system and cloaca.

2. **The male urogenital system.**—The testes are a pair of oval organs at the anterior ends of the kidneys; their size varies considerably with the season. They lack definite mesorchia. The kidneys and the ureters should be studied according to the directions given for the female. The male ducts or deferent ducts spring from the medial surface of the testes near their posterior ends, with the intervention of an epididymis too small to be identified macroscopically. The deferent ducts are slender, convoluted tubes which pass caudad parallel to the ureters. Trace both ducts to the cloaca.

The cloaca is smaller than in the female, and the lips of the anus more protruding. The rectum enters medially and ventrally, the urogenital ducts laterally. Cut into the cloaca as directed under female and identify its chambers as there described. They are the same in the two sexes, except that the male urodaeum is smaller and receives the two deferent ducts instead of the oviducts. Ureters and deferent ducts open on small papillae in the lateral walls of the urodaeum. Draw.

A copulatory organ is lacking in most birds, which copulate by cloacal contact. However, a penis occurs in ratite birds (ostrich, emu, cassowary, apteryx) and also in ducks and a few others. Apparently its absence in most birds is a secondary loss.

F. THE UROGENITAL SYSTEM OF MAMMALS

The urogenital system of primitive mammals is reptilian in character (see below) but is modified in later forms—in males by the process known as the descent of the testes, and in females by the great reduction in the size of the eggs and the adaptation of the female ducts for intrauterine development of the young.

Remove the digestive tract, leaving the rectum in place.

1. **The kidneys and their ducts.**—The kidneys of mammals are metanephroi, and their ducts the metanephric ducts or ureters. The kidneys are large oval organs situated against the dorsal wall of the peritoneal cavity; they are retroperitoneal. The right kidney is usually considerably anterior to the left one. Clear away fat and connective tissue from about the kidneys and note their characteristic bean shape, convex laterally, concave on their medial faces. The concavity is termed the *hilus;* and from it a white tube, the *ureter*, passes out, turning posteriorly. Follow the ureters caudad, clearing away fat, and note their entrance into the urinary bladder. In females the ureters pass dorsal to the horns of the uterus; in males dorsal to a white cord, the male duct or ductus deferens, which loops over the ureter and disappears dorsal to the bladder.

Remove by a cut the ventral half of a kidney. Within the hilus there is a cavity, the *renal sinus*, occupied by the renal artery and vein and by the expanded beginning of the ureter, termed the *renal pelvis*. Into this pelvis the substance of the kidney projects as the *renal papilla*, on which are situated the microscopic openings of the collecting tubules. The kidney substance is readily divisible into two areas, a peripheral *cortex* and a central *medulla*. The cortex contains the renal corpuscles and the convoluted and looped portions of the kidney tubules. The medulla is marked by lines which converge to the renal papilla; these lines are the collecting tubules. It will be recalled that the collecting tubules, the pelvis, and the ureter arise by outgrowth from the mesonephric duct. The collecting tubules and renal papilla together form a *renal pyramid*, of which there is but one in the rabbit and cat but about twelve in the human kidney.

Draw the section of the kidney.

The *urinary bladder* is a pear-shaped sac at the posterior end of the peritoneal cavity. It is ventral to the rectum in the male, ventral to both rectum and uterus in the female. The free anterior end of the bladder is named the *apex* or *vertex*, the posterior portion the *fundus*. The fundus continues posteriorly as a narrowed stalk, the *urethra*[6] (also called *neck of the bladder*). The bladder is covered by the peritoneum, which is continuous with that of the abdominal wall by means of the median and lateral ligaments previously noted. The pouch between the bladder and rectum (male) or bladder and uterus (female) is named the *rectovesical* or *vesicouterine pouch*, respectively.

Draw the excretory system.

2. The female reproductive system.—This consists of a pair of *ovaries* and their ducts. The ovaries are very small oval bodies located at the sides of the peritoneal cavity at the anterior end of the coils of the uterus. Each will be seen to bear little clear vesicles, the *Graafian follicles*, each of which contains an egg or ovum; in pregnant females the ovary also bears little hard lumps, the *corpora lutea*, which represent follicles from which the eggs of the pregnancy were discharged. The ovary is suspended by the *mesovarium*, which extends forward to the kidney and is continuous posteriorly with the ligament of the uterus.

The ducts of the ovaries are, as in other vertebrates, the *Müllerian ducts* or *oviducts*, but they are differentiated into several distinct parts in mammals. The uppermost portion of the oviducts is a slender, convoluted tube which passes lateral to the ovary and curves over its anterior end; its mesentery,

[6] The term urethra is in much confusion in comparative anatomy, owing to the differences between the urogenital systems of various mammals. Although in the embryo the urethra is the same as the urogenital sinus, this is not the case in the adults of most mammals, and consequently the use of urethra as synonymous with urogenital sinus appears to be inadvisable. Urethra is therefore here employed in the same sense as in human anatomy, i.e., as the name of the duct leading from the bladder to the exterior.

the *mesosalpinx*, forms a sort of hood, partly inclosing the ovary. This portion of the oviduct is the *uterine* or *Fallopian tube*. It opens in front of the ovary (rabbit) or to the lateral side of it (cat) by the *ostium* having fringed borders, the *fimbriae*. On tracing the uterine tube posteriorly it is found to widen suddenly into a thick-walled tube, the *uterus* (rabbit) or *horn of the uterus* (cat). The size of this depends on whether the animal is pregnant or not; in pregnant animals the uteri or horns are greatly enlarged and exhibit a series of swellings, each of which contains an embryo (these will be examined later). The strong fold of peritoneum supporting the uteri or horns is the *mesometrium*. Mesovarium, mesosalpinx, and mesometrium together are called the *broad ligament* of the uterus in mammals. The *round* ligament of the uterus is the fold extending from the beginning of the uterus or horn posteriorly to the body wall; it is continuous with, but at right angles to, the broad ligament. In the **cat** the two horns of the uterus unite in the median line, dorsal to the bladder, to a single tube, the *body of the uterus*. Body and horns together constitute the *uterus or womb*, but the young develop only in the horns. In the **rabbit** the two uteri are separate along their entire lengths, and consequently there is no division into body and horns. In the **cat** the body of the uterus continues posteriorly as the *vagina;* in the rabbit the two uteri join the *vagina;* the vagina is a tube situated in the median line between the bladder and the rectum. It exits through the ring formed by the pelvic girdle and vertebral column.

The external genital parts or *external genitalia* were described with the external anatomy. Review this (p. 55). Then in the **rabbit** make an incision through the skin forward from the vulva. In the median line beneath the skin is a hardened body, the *clitoris*, homologous to the penis of the male. Its anterior end is attached by ligaments to the ischium and pubic symphysis. Cut across the clitoris and note the two *cavernous bodies* of which it is composed. In the cat the clitoris is minute.

Now cut through the pubic and ischial symphyses and spread the legs well apart. Trace the urethra, the vagina, and the rectum posteriorly. At first the urethra lies on the ventral face of the vagina, to which it is bound by tissue; it then unites with the vagina to form a common tube, the *urogenital canal* or *urogenital sinus*. Dissect this free, lift it out, and follow it to the urogenital aperture. Cut open the urogenital aperture and note the free posterior end, or *glans*, of the clitoris projecting into the cavity in the **rabbit**. Free the rectum from the urogenital canal and follow it to the anus. Along its sides in the rabbit are a pair of elongated *anal glands;* in the **cat**, the rounded *anal glands* or *sacs* occur to either side of the rectum close to the anus. The secretions of the anal glands are strongly odoriferous and presumably of sexual nature.

Draw the female urogenital system.

Cut open the vagina. In the rabbit note the external uterine orifice with raised fringed lips by which each uterus opens into the vagina. The rabbit uterus is of the *duplex* type (Fig. 113*B*). In the cat the body of the uterus is divided into lateral halves by a median partition, and the horns open to either side of this partition. The cat uterus is of the *bipartite* type (Fig. 113*C*). The lower end of the uterus, called the *cervix*, projects into the vagina by a fold. The opening of the body of the uterus into the vagina is the external uterine orifice.

3. **The male reproductive system.**—Review the description of the external genitalia on page 55. The two testes are lodged in the *scrotum*, divided into two compartments by the internal partition. Cut through the skin ventral and in front of one testis, exposing the testis as an oval, white body. Clear away tissue anterior to the testis and find a white cord, the *spermatic cord* (**cat**), or ductus deferens (**rabbit**), passing forward and entering the peritoneal cavity through a canal, the *inguinal canal*. The two ends of this canal are the *external* and *internal inguinal rings* (**cat**); these are indefinite in the **rabbit**. The spermatic cord contains a white duct, the ductus deferens, and blood vessels and nerves. Trace the spermatic cord (**cat**) or ductus deferens (**rabbit**) by cutting open the inguinal canal. The two deferent ducts turn toward the median line, loop over the ventral surfaces of the ureters, and disappear on the dorsal surface of the urethra.

Now cut through the pubic and ischial symphyses and spread the legs apart. Trace the deferent ducts and the urethra caudad, separating them from the rectum. The ureters may be cut through and the bladder held caudad. The deferent ducts pass along the dorsal surface of the urethra. In the **rabbit** they enlarge and enter an expanded sac, the *seminal vesicle*, which pouches forward between the rectum and the bladder. Cut into the seminal vesicle and note the openings into its ventral wall of the two deferent ducts, and in its posterior dorsal wall the thickening caused by the *prostate* gland. Find the union of the seminal vesicle and urethra to form a common tube, the *urogenital sinus*. In the **cat** the two deferent ducts join the urethra without the formation of a seminal vesicle, the point of junction being surrounded by a slight enlargement, the *prostate gland*. The common tube thus formed is the *urogenital canal* or *sinus*. The *bulbourethral glands* or *Cowper's glands* are small swellings situated on the urogenital canal shortly posterior to the prostate gland in the rabbit, about an inch posterior in the cat. The terminal inch of the urogenital canal is inclosed in the penis. Cut into the prepuce and note the pointed projection within it, called the *glans* of the penis. Note that the prepuce is simply a fold of skin around the glans. At the tip of the glans is the *urogenital opening*. The glans in the cat bears a number of minute spines. Dissect anteriorly from the glans, exposing the remainder of the penis as a hardened cylindrical structure. Find where the urogenital canal enters its anterior end. Note, also, the strong attachments of the penis to the pelvic region. Cut across the middle of the penis and note that it is composed of two cylindrical bodies, the *corpora cavernosa* or *cavernous bodies*, closely placed. The urogenital canal, here called the *cavernous urethra*, lies on the dorsal side of the penis, resting in a depression between the two cavernous bodies. At the anterior end of the penis the two cavernous bodies diverge, forming the *crura* of the penis, which are attached to the ischia. The cavernous bodies are spongy structures and in the sexual act become distended with blood, so that the penis is caused to project out of its sheath, the prepuce.

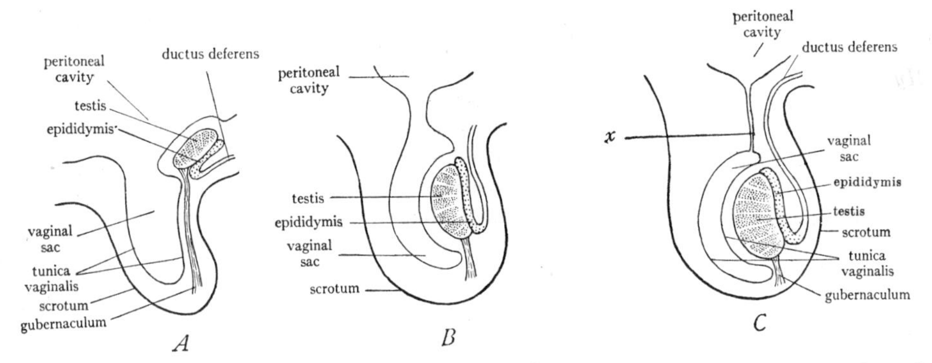

FIG. 118.—Diagrams to illustrate the descent of the testis in the male mammal. The testis descends into the scrotum, which is a sac of the body wall containing a portion of the peritoneal cavity, called the vaginal sac; in the descent the gubernaculum or ligament of the testis shortens; the passage *x*, along which the descent occurs and which at first forms a connection between the vaginal sac and the peritoneal cavity, is later completely obliterated in the higher mammals; it does not correspond to the inguinal canal. (From Arey's *Developmental Anatomy*, courtesy of the W. B. Saunders Co.)

Draw the parts of the male genital system.

Trace the rectum to the anus, following the directions given for the female.

The structure of the testis may now be investigated. Each testis is inclosed in a white fibrous sac, which is the peritoneal pouch made by the descent of the testis (Fig. 118). Cut open this sac, exposing the cavity, or *vaginal sac*, in which the testis lies; this is a part of the peritoneal cavity. The *tunica vaginalis* which lines this cavity is reflected over the surface of the testis as a covering layer. This deflection lies along the middorsal line of the testis, and a mesorchium is thus formed between testis and the wall of the vaginal sac. The posterior end of the testis is attached to the posterior scrotal wall by a short, stout ligament, the *gubernaculum*, continuous with the mesorchium and homologous to the round ligament of the uterus. The ductus deferens lies along the dorsal surface of the testis much coiled, forming the epididymidal duct. This begins at the anterior end of the testis as a coil, the *head* of the epididymis, which receives the invisible efferent ductules from the testis. The coiled epididymidal duct then passes along the dorsal surface of the testis as the *body* of the epididymis and finally, at the posterior end of the testis, makes another coiled mass, the *tail* of the epididymis, to which the gubernaculum is attached. From this the much convoluted ductus deferens proceeds anteriorly and passes into the inguinal canal, where it becomes a straight tube. The head of the epididymis corresponds to the epididymis of lower forms and is derived from the mesonephros.

Draw, showing contents of the scrotum.

4. The descent of the testes in mammals.—In the embryos of male mammals, as in the adults of other vertebrates, the testes lie inside the peritoneal cavity; but during development in many mammals they descend to a position outside the body. There is first formed a sac of the body wall, the scrotal sac, which contains all of the body wall layers—skin, muscles, and peritoneum—and incloses an extension of the coelom, the vaginal sac. The testes then descend into the scrotal sac, as shown in Figure 118, carrying with them their ducts, blood vessels, nerves, etc. The canal along which the descent occurs remains open in some mammals; but in the higher ones it is completely obliterated, and a new canal, the *inguinal canal*, is later secondarily formed around the spermatic cord. The descent of the testes explains the peculiar looping of the deferent ducts over the ureters and the course of the internal spermatic vein. The testes retain their position in the peritoneal cavity in monotremes, most marsupials, *Hyrax*, elephants, Sirenia, and some edentates. They descend to the lower part of the peritoneal cavity in cetaceans. In a number of mammals the testes descend into the scrotum at the breeding season but are withdrawn into the peritoneal cavity at other times, as in moles, bats, many rodents. A permanent inguinal position in a more or less projecting sac is seen in most ungulates, carnivores, some marsupials, and primates. The cause of the descent of the testes remains mysterious. It was independently discovered by a Japanese, Fukui, and by C. R. Moore, of the University of Chicago, that the temperature inside the scrotum is lower than that of the peritoneal cavity and that the testes of animals with a scrotum degenerate and are unable to produce mature sperm if they are moved up into the peritoneal cavity. It appears that the scrotum has a temperature-regulating function. This does not, however, explain the origin of this situation, since in many mammals and in birds the testes are normally situated in the peritoneal cavity and thus continually exposed to body temperatures. The ovaries of mammals also move caudad from their embryonic position; and, in fact, the descent of the testes appears to be the extreme case of the general phenomenon of a posterior descent of the viscera evident in the vertebrate series.

5. The evolution of the copulatory organ.—Aquatic animals, as a rule, discharge their sex cells into the water, where fertilization occurs; but internal fertilization is an obvious necessity among terrestrial forms. The transfer of the sperm directly into the body of the female is generally accomplished by a protrusible organ termed the *penis*. It is self-evident that the penis cannot be homologous among the varied groups of animals in which it occurs, but a homology does appear to exist with regard to the penis of the amniotes. A penis is lacking among the anamniotes, in the vast majority of which fertilization is external, the male shedding sperm as the female emits her eggs. In selachians and holocephalans the pelvic fins are modified as intromittent (sperm-transferring) organs, but the act of copulation appears to have been seldom witnessed; apparently, only one clasper is used at a time. An exhaustive study of the structure and mode of operation of selachian claspers has been made by Leigh-Sharpe (1920–24). Among teleost fishes which are viviparous, internal fertilization is accomplished by the modified anal fin, or direct cloacal contact may occur. Amphibians in general lack a copulatory apparatus, but in the gymnophionans the cloaca is everted (turned inside out) for this purpose. As already noted, a true penis is first found in reptiles; this is a differentiation of the ventral cloacal wall and occurs in both sexes. The female homologue of the penis in amniotes is termed the *clitoris*. The amniote penis consists primitively of a mass of tissue in the ventral cloacal wall, called the *corpus cavernosum* (or *corpus fibrosum*), having upon its dorsal surface a *urethral groove* as a sperm conduit (Fig. 119). It is usual for the penis to have at its caudal end a swelling termed the *glans*. The penis of monotremes, which also lies imbedded in the ventral cloacal wall, appears, despite certain peculiarities, to be derivable from

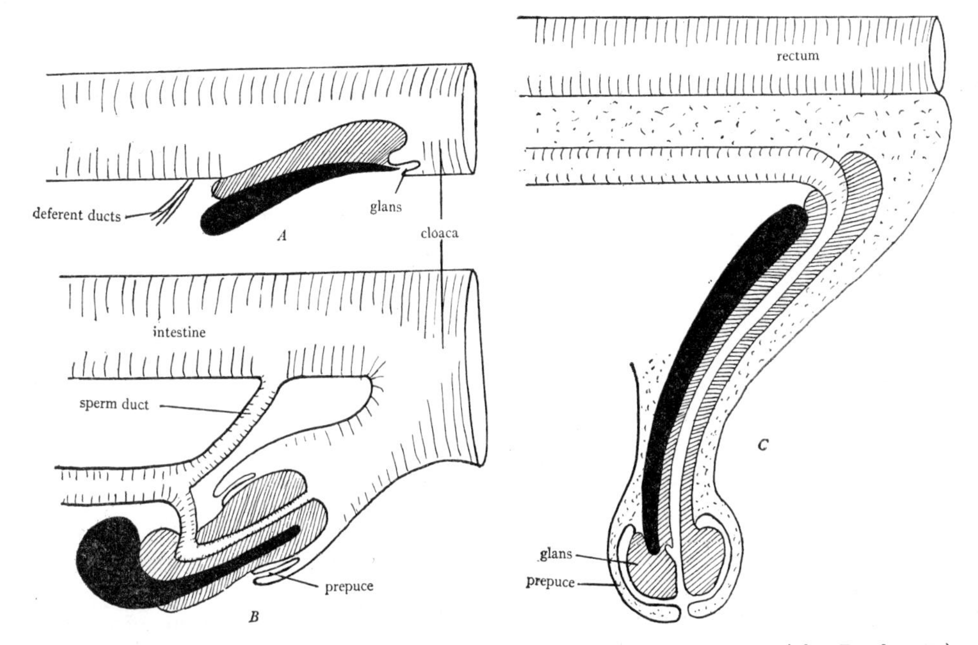

FIG. 119.—Diagrams to compare the penis of *A*, turtle; *B*, *Echidna;* and *C*, man (after Broek, 1910); cavernous body black, spongy body, crosshatched. Note in *Echidna* the forking of urogenital canal, one fork opening into the cloaca, the other traversing the penis, so that urine does not pass through the penis.

that of turtles by the closing-over of the sperm groove, so that there results a cylindrical corpus cavernosum inclosing a canal. This canal conveys only sperm, as the urine passes out through the general cloacal cavity. The tissue surrounding this sperm canal differs from that of the cavernous bodies and is called *corpus spongiosum*. The monotreme penis also shows a feature characteristic of the mammalian penis, the inclosure of the glans in a skin pocket, the *prepuce*, a new development not found in reptiles (Fig. 119*B*). Peculiar to the monotremes is the subdivision of the sperm canal into many openings on the glans, which also is subdivided (Fig. 121*C*). Above monotremes it is usual for the corpus cavernosum to subdivide into two corpora, as seen in the dissection. The canal which traverses the penis of placental mammals conveys both sperm and urine and, although called urethra, would be more correctly termed male urogenital canal. It also is inclosed in a corpus spongiosum, distinct from the cavernous bodies. The mammalian penis, originally imbedded in the ventral cloacal wall and directed backward, comes among higher mammals to take a more and more external position (coincident with the development of a scrotum) and to be directed forward (Fig. 119*C*). It is a peculiarity of many marsupials that the penis is forked, in correlation, so some believe, with the double vagina of this group (Fig. 122*C*).

6. The accessory sex glands.—These are, strictly speaking, glandular outgrowths of the genital canals, whose secretion is of importance to the vitality and functioning of the sex cells. Although occurring in lower vertebrates—as Leydig's gland and the glands of the claspers of selachians, and the cloacal glands of salamanders—they reach their greatest complexity in male mammals, where the terminology is much confused. That these glands are concerned in the male sex function is proved by their degeneration on castration (operative removal of the testes) and by their return to normal histology on the injection of appropriate sex hormones into castrated animals. The following are the principal accessory sex glands of male mammals (Fig. 120):

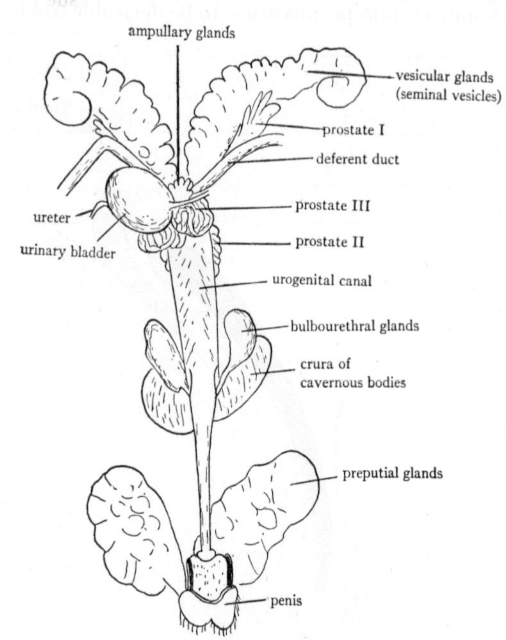

FIG. 120.—Urogenital system of the male mouse (after Rietschel, 1929), to show the accessory sex glands.

a) *Ampullary glands*, found in the wall of the widened terminal part of the ducti deferentes as in the rabbit.

b) *Vesicular glands*, usually called *seminal vesicles*. The latter name is a misnomer, as it has been proved that they do not store sperm but instead produce a secretion. The vesicular glands are a single or paired sacciform or tubular evagination, often of large size, from the deferent ducts at their junction with the urethra. They form elongated sacs in rats and guinea pigs (Fig. 120), also occur in man, but are lacking in the cat.

c) *Prostate glands*, bunches of tubular glands at the beginning of the urogenital canal. In rodents, such as rats and guinea pigs, there are three groups of prostate glands (Fig. 120); and it has been shown that the secretion of the first prostate causes the secretion of the vesicular glands (= seminal vesicles) to coagulate, producing a plug which in these animals closes the vaginal orifice after copulation.

d) *Urethral glands*, glandular differentiation in the wall of the proximal part of the urogenital canal.

e) *Bulbourethral glands* or *Cowper's glands*, more prominent swellings along the urogenital canal.

f) *Skin glands*, found around the sexual and anal orifices. Although, strictly speaking, these are not accessory sex glands, they are conveniently listed here. They include the *preputial glands* (Fig. 120), invaginated from the skin of the prepuce, the *anal glands* or *sacs* opening alongside the anus, and the *inguinal glands*.

The function of most of these glands is not understood.

7. The urogenital system of the lower mammals.—The urogenital system of monotremes, especially the female system, retains primitive reptilian characters. There is a cloaca, which receives the intestine and urogenital canal; but in the female a deep fold considerably separates the two latter (Fig. 121*A*). In both sexes, ureters, bladder, and genital ducts open into the upper end of the urogenital canal, as the ureters have not yet shifted so as to enter the bladder directly, as they do in higher mammals (Fig. 121). The oviducts open separately into the proximal part of the urogenital canal (Fig. 121*A*), as in reptiles, and are not much differentiated into uterus and vagina, although showing a distinct uterine tube. Monotremes do not gestate their young but instead lay large yolky, reptilian-like eggs having a horny shell. The testes retain the primitive abdominal position (Fig. 121*B*), near the kidneys, and have

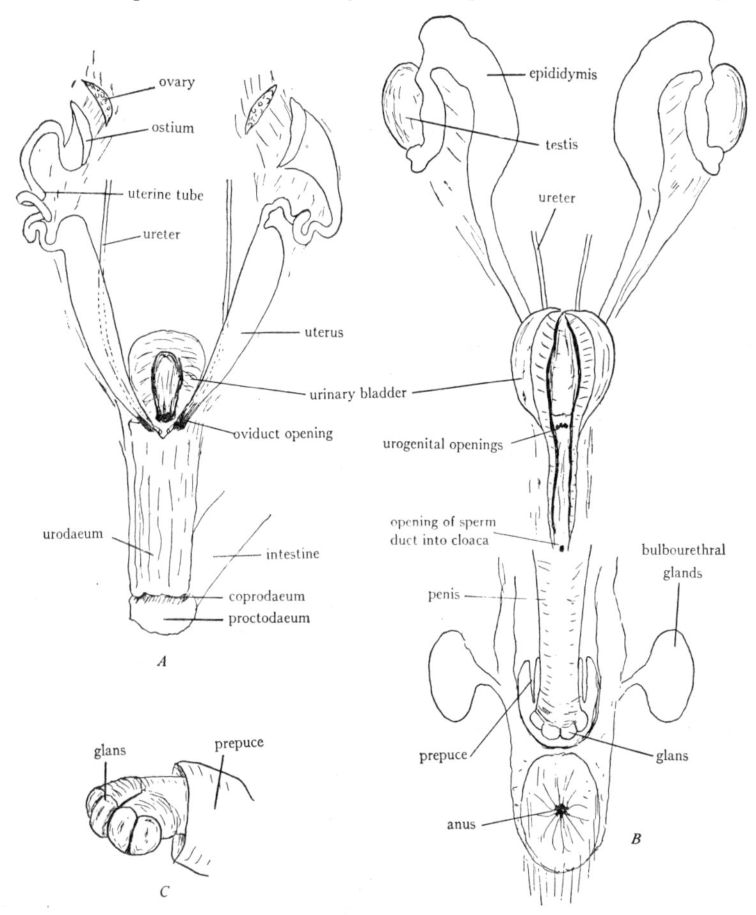

Fig. 121.—Urogenital system of the monotreme *Echidna* (from dissections made by H. Raven). *A*, female; *B*, male; *C*, penis enlarged to show four-parted glans. Urinary bladder and cloaca have been cut open.

the usual epididymis. The penis, consisting of a single cavernous body, lies imbedded in the ventral cloacal wall and has a prepuce and (subdivided) glans (Fig. 121*B*, *C*). The main urogenital canal enters the cloaca but gives off a sperm canal which traverses the penis (Fig. 119*B*); thus monotremes are unique among male mammals in that the urine does not pass through the penis. This is generally regarded as a reptilian reminiscence. The monotreme male system is mostly lacking in accessory sex glands, being provided only with a pair of bulbourethral glands (Fig. 121*B*).

In marsupials the intestine and urogenital canal are almost completely separated; but usually a shallow cloaca, representing the proctodaeum, remains. The ureters in both sexes enter the urethra or the bladder base. The oviducts are differentiated into uterine tube, uterus, and vagina. The uteri remain completely separate and are well marked off from the vaginae, into which they open on papillae. The vaginae enter the urogenital canal separately or after union into a *common vagina*, and the urethra joins the urogenital canal at the same

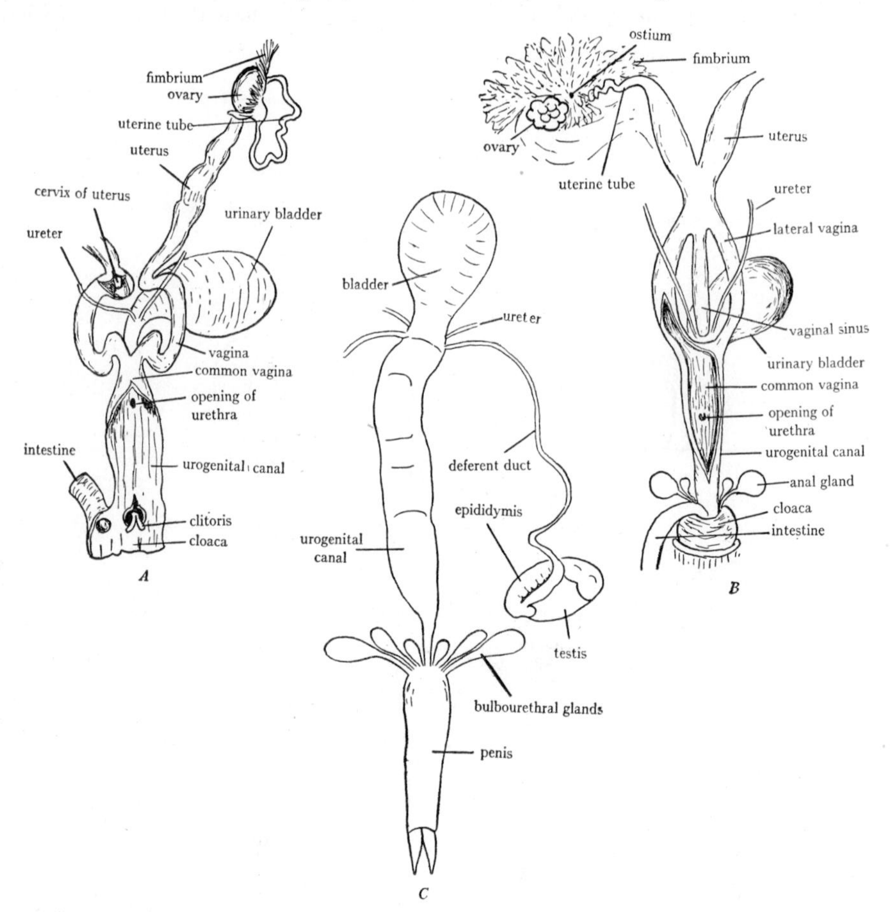

FIG. 122.—Urogenital systems of marsupials. *A*, female opossum (after Brass, 1880), showing inward bend of vagina; *B*, female wombat (after Brass, 1880), vaginae have fused at the bends, and a vaginal sinus has grown backward from the fusion; *C*, male opossum (after Moore, 1941), showing bulbourethral glands and forked penis; the thick wall of the urogenital canal (urethra) contains prostatic type of glands.

level as the vaginae or after entering the common vagina. The vaginae of marsupials are peculiar in that at their proximal ends, where each receives a uterus, they are bent toward the median line and secondarily fused there into a *vaginal sinus* (except in a few marsupials, where these outpouchings remain separate [Fig. 122*A*]). From this sinus vaginalis a tube grows backward toward the urogenital sinus (Fig. 122*B*), with which it may connect only during birth or permanently. Thus, most marsupials appear to have three vaginae (Fig. 122*B*): the two primary ones, called *lateral vaginae;* and the middle one, the vaginal sinus (not to be confused with the common vagina). The young are gestated in the uteri for only a short period and are born in a very immature condition, with functional mesonephroi; they are born by way of the vaginal sinus when this is sufficiently developed.

In male marsupials the deferent ducts open into the urogenital canal, in whose ventral wall the penis lies (Fig. 122*C*). The cavernous body of the penis is divided into a pair of bodies to varying degrees in different marsupials, and there is also present the spongy body around the cavernous urethra. As already remarked, the distal part of the marsupial penis is frequently forked. The marsupials usually have three pairs of bulbourethral glands, which come from one pair of embryonic primordia. Preputial and anal glands are also present. The testes of marsupials may be abdominal or may be contained in a scrotum at the breeding season or permanently.

G. THE EMBRYONIC MEMBRANES

1. General.—There are four embryonic membranes in vertebrates: the *yolk sac*, the *allantois*, the *amnion*, and the *chorion* (Fig. 123).

The yolk sac is simply a saclike expansion of the ventral wall of the intestine, narrowed into a *yolk stalk* near the body of the embryo. It occurs in the embryos of all vertebrates with meroblastic development, in which it is filled with yolk utilized as food by the embryo. A yolk sac also occurs in mammals as an inheritance from their meroblastic ancestors, and in monotremes is full of yolk and functions as in reptiles. In other mammals it is empty but may be large and vascular, especially in marsupials, where it functions as a placenta constituting the yolk-sac type of placenta (see p. 423). With the appearance of intrauterine development the mammalian embryo obtains nutrition from the maternal blood and requires no stored food.

The allantois is a large sacciform evagination from the floor of the cloaca and may be regarded as an embryonic urinary bladder. Besides serving to hold embryonic excretory material, it has a more important respiratory function (see below). The adult bladder of amniotes comes from the allantoic stalk.

The amnion and the chorion are simultaneously formed in the amniote embryo by a fold of the body wall (somatopleure) which rises up around the embryo, meets above it, and fuses across (Fig. 123*A*). The outer limb of the fold becomes the chorion, the inner limb the amnion. The amnion forms a sac inclosing the embryo except below (Fig. 123*C*). The chorion is the outermost membrane of the embryo and is in contact with the egg shell or uterus, as the case may be. The yolk sac and allantois are between chorion and amnion on the ventral side of the embryo (Fig. 123*D*).

The yolk sac and the allantois are lined by entoderm, outside which there is a highly vascular layer of mesoderm. The blood vessels of the yolk sac are termed the vitelline vessels, and much was said of them in the chapter on the circulatory system. The blood vessels of the allantois are termed the umbilical or allantoic vessels, and the umbilical veins are homologous to the ventral or lateral abdominal veins of lower vertebrates. The amnion is lined by ectoderm on the side facing the embryo and has an outer mesodermal layer, and the chorion consists of an outer ectoderm and an inner mesoderm. The amnion and the chorion never have any blood vessels.

A yolk sac may occur in any group of vertebrates; but the allantois, amnion, and chorion are found only in reptiles, birds, and mammals, whence these three classes are termed amniotes (Greek, *amnion*, a veil, referring to the amnion covering the embryo like a veil). In birds and reptiles the allantois expands enormously between amnion and chorion, and its outer wall comes in contact and fuses with the chorion to form the *chorioallantoic membrane*. This lies against the egg shell, through which respiratory gases diffuse, so that the blood vessels of the chorioallantoic membrane become the respiratory mechanism of the sauropsidan embryo, carrying gases to and from the embryo.

It was already learned that viviparity may occur in various groups of lower vertebrates;

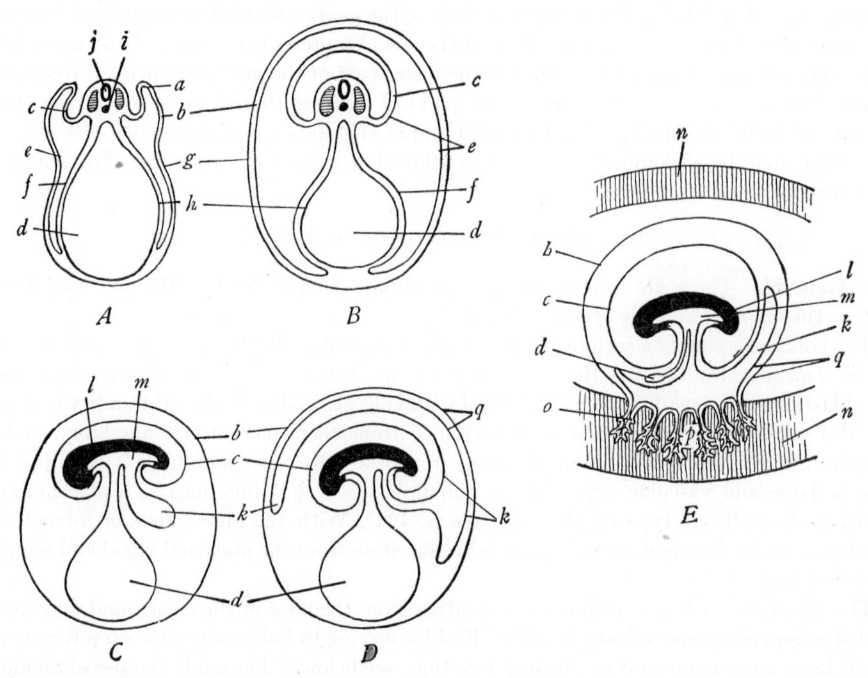

FIG. 123.—Diagrams to illustrate the mode of formation of the embryonic membranes of amniotes and of the placenta. *A*, cross-section through an early stage of the formation of the amnion and the chorion; a fold of the somatopleure is seen rising up at *a;* the outer wall of the fold (*b*) becomes the chorion, the inner wall (*c*), the amnion. *B*, later stage after completion of the process of formation of the amnion and chorion; the fold fuses across the above dorsal surface of the embryo, forming two membranes, an outer chorion (*b*), which incloses embryo and yolk sac, and in inner amnion (*c*), which incloses the embryo. *C*, sagittal section of a later stage of the embryo to show the origin of the allantois (*k*) as an evagination from the digestive tract (*m*). *D*, later stage, following *C*, showing the spreading of the allantois, between the chorion and the amnion and yolk sac; note that the outer wall of the allantois is in contact with the chorion, the two together forming the chorioallantoic membrane *q*. *E*, formation of the placenta in mammals by the penetration of the chorioallantoic membrane (*q*) into the wall of the uterus (*n*); the penetration takes the form of treelike ingrowths which are called the chorionic villi (*o*); note small size of the mammalian yolk sac (*d*); the placenta consists of the inner part of the uterine wall and the chorionic villi; the latter are generally restricted to certain areas of the chorioallantoic membrane. In *C*, *D*, and *E* the two layers of which the chorion, amnion, allantois, and yolk sac are each composed are omitted for simplicity. *a*, amniotic fold of the somatopleure; *b*, chorion; *c*, amnion; *d*, yolk sac; *e*, somatic mesoderm; *f*, splanchnic mesoderm; *g*, ectoderm; *h*, entoderm; *i*, notochord; *j*, neural tube; *k*, allantois; *l*, body of embryo, head to the left; *m*, digestive tract of embryo; *n*, wall of uterus; *o*, chorionic villi; *p*, placenta; *q*, chorioallantoic membrane. (Suggested by figures in Hertwig.)

i.e., the eggs are retained in the oviducts (or ovaries in some teleosts) and develop there into young animals ready for free existence. Such viviparity always involves eventually a nutritive relation between the embryo and the maternal tissue. Leaving out of account some very bizarre nutritive relations which may develop in certain families of viviparous teleost fishes (see articles by C. L. Turner), the nutritive arrangement typically consists of projections or folds of one or the other of the embryonic membranes which are in contact with, or dovetail with, similar projections or folds of the oviducal or uterine wall. A mutual alteration of structure of embryonic membranes and oviducal or uterine wall for the nutrition and respiration of the embryo is termed a *placenta*. It must be noted that the placenta is compounded of both embryonic and maternal tissues. Among viviparous selachians and reptiles and also in marsupials the yolk sac is the embryonic membrane involved, and the placenta is then termed a *yolk-sac placenta* or *omphaloplacenta*. In mammals above marsupials, the chorioallantoic membrane is the embryonic membrane involved, and hence their placenta is called an *allantoic* or *true* placenta. For the same reason these mammals are termed the *placental* mammals. The name placenta is derived from the round cakelike shape of the human placenta (Latin, *placenta*, a cake).

The young of all mammals except monotremes develop in the uterus, a highly modified region of the oviducts, with the aid of a placenta. In marsupials this is a yolk-sac placenta, except in one case, *Perameles*, which has an incipient allantoic placenta. In all other mammals there is an allantoic placenta which, however, shows varying degrees of development. The mammalian placenta consists, in general, of vascular finger-like projections (*villi*) of the chorioallantoic membrane which fit into depressions of the uterine wall or are imbedded in the uterine wall. When these villi occur all over the chorioallantoic membrane, the placentation is termed *diffuse* (many ungulates, whales, some primates). When they occur in separated bunches, called *cotyledons*, the placentation is cotyledonary (ruminants). A *zonary* or ring-shaped placenta is found in carnivores, elephants, and *Hyrax;* and a *discoid* placenta is found in insectivores, bats, rodents, and primates. When at birth the embryonic part of the placenta is ejected without taking with it any of the uterine tissue, the placenta is spoken of as *nondeciduate;* when the uterine part of the placenta is also shed, this is termed *deciduate*. The diffuse and cotyledonary placentas are nondeciduate, the others deciduate.

The intimacy of the relation between the chorioallantoic membrane and the uterine lining evolves gradually among the placental mammals. In the simplest cases, mostly forms with a diffuse placenta, the embryonic and uterine villi merely interdigitate, and all the cell layers of both remain intact, forming the *epitheliochorial* type of placenta. In the next or *syndesmochorial* type, seen in ruminants, the cotyledons fit into depressions of the uterine wall, and the epithelium and part of the connective tissue of the latter disappear; but at birth the cotyledons are withdrawn and do not carry uterine tissue with them. In the *endotheliochorial* placenta, found in the zonary type, embryonic and uterine tissues are inseparably intergrown, and the embryonic villi are in contact with the maternal blood vessels. Finally, in the *haemochorial* placenta, which is of the discoid type, there is an extensive breakdown of maternal tissues with destruction of blood vessels, so that the embryonic villi lie in pools of maternal blood.

2. Anamniote embryo of the dogfish.—Cut open the pregnant uterus of a dogfish and remove an embryo, or examine embryos provided. Note that the embryo is naked or *anamniote*. From the middle of its ventral wall hangs the large yolk sac, filled with yolk and attached to the embryo by the narrowed yolk stalk. The yolk sac is covered externally by a layer of the body wall and internally consists of the intestinal wall inclosing the yolk. As the yolk is used up, the yolk sac is gradually withdrawn into the body. Draw.

3. Amniote embryo of the cat.—If pregnant females are available, open one of the enlargements in the horns of the uterus. The enlargement contains an embryo. Note that the embryo is inclosed in a thin membrane, the amnion, which forms a sac around it. The placenta of the cat is of the zonary type, and on the inner surface of the uterine wall at the enlargement a thickened vascular ring of tissue will be found, which is the placenta. It will probably peel off, especially in late stages of pregnancy. Open the amnion and note the *umbilical cord* extending from the belly of the embryo to the inner surface of the amnion, where the latter is applied to the placenta. The umbilical conveys the umbilical blood vessels to and from the embryonic part of the placenta and is a connection between the embryo and its own membranes, not a connection between the embryo and its mother. There is no direct connection between embryo and mother, but substances can diffuse from the mother's blood into the embryonic blood by way of the placenta.

H. SUMMARY OF THE UROGENITAL SYSTEM

1. The urogenital system is derived from the mesomere of the embryo.

2. The urinary or excretory system consists of the paired kidneys and their ducts. Primitively, each kidney extended the length of the pleuroperitoneal cavity and was composed of segmentally arranged tubules and a collecting duct. Such a kidney is termed holonephros or archinephros, and its duct the archinephric duct.

3. The primitive kidney tubules consist of the following parts: a ciliated peritoneal funnel opening into the general coelom; a tubule leading from this to the nephrocoel (coelomic cavity of each mesomere); an opening, the nephrostome, from this into the kidney tubule proper; kidney tubule; and the opening of this into the archinephric duct. The archinephric duct is formed by the union of backward extensions of the kidney tubules.

4. A capillary network termed a glomerulus is primitively associated with each kidney tubule. It is either an external glomerulus, projecting into the coelom near the peritoneal funnel, or an internal glomerulus, pushing in the wall of the nephrocoel. The combination of glomerulus and nephrocoel wall is termed a renal corpuscle.

5. A holonephros exists only in the embryos of myxinoid cyclostomes; in vertebrate embryos, in general, the kidneys develop in two or three successive sections in an anteroposterior direction.

6. The most anterior vertebrate kidney is termed the pronephros; it is similar to the holonephros, composed of segmental tubules provided with peritoneal funnels. It appears in the embryos of all vertebrates but persists and remains functional throughout life only in the cyclostomes and a few teleost fishes. Its ducts, the pronephric ducts, grow backward and open into the cloaca.

7. As the pronephroi degenerate, there develop behind them the second kidneys, or mesonephroi (also called Wolffian bodies), which differ chiefly in having more than one tubule per segment. The pronephric ducts persist and are utilized by the mesonephroi as their ducts, being then termed the Wolffian or mesonephric ducts.

8. In amniotes the mesonephroi degenerate during development, and there appear behind them the third kidneys, or metanephroi. These develop in part from the remaining tissue of the mesomere and in part by outgrowth from the Wolffian ducts near their termination in the cloaca. The stalks of the outgrowth form the metanephric ducts or ureters. The metanephroi

are the functional kidneys of adult amniotes; they have many tubules per segment and lack peritoneal funnels.

9. In anamniotes, in general, the second kidney utilizes the tissue which in amniotes furnishes both mesonephros and metanephros. Hence, the ichthyopsidan kidney does not exactly correspond to the mesonephros of amniote embryos, although usually called so. It is here termed opisthonephros.

10. Pronephros, mesonephros, and metanephros appear to be successive alterations of the original holonephros.

11. A urinary bladder is generally present as an evagination of the ventral cloacal wall. The kidney ducts generally open directly into the cloaca, not into the bladder; but in mammals they shift so as to open into the bladder.

12. The genital system of vertebrates consists of the paired male sex glands or testes and the paired female sex glands or ovaries and their respective ducts. These ducts, however, are kidney ducts; hence the close association of urinary and genital systems. Testes and ovaries arise as mesomere thickenings which project into the coelom near the kidneys.

13. The ducts of the ovaries are termed Müllerian ducts or, better, oviducts; they have no direct connection with the ovaries but open close to them into the coelom by a funnel-like ostium believed to represent one or more pronephric peritoneal funnels. The oviducts of elasmobranchs originate by a longitudinal splitting from the pronephric ducts; and, although this method of origin does not obtain in other vertebrates, it may be regarded as the phylogenetic source of the oviducts.

14. Except in placental mammals the oviducts enter the cloaca separately. In placental mammals the oviducts are more or less fused and differentiated into uterine tube, uterus, and vagina. Fusion proceeds from the distal end proximally, involving first the vaginae, then the uteri. Partial fusion of the two uteri results in a uterus with horns, complete fusion in the single uterus of the primates. The uterine (Fallopian) tubes always remain separate.

15. The testes are located inside the abdominal cavity except in mammals, where in many cases they descend temporarily or permanently into an inguinal pouch of the body wall, called the scrotum.

16. In all vertebrates except cyclostomes and teleostomes the testes use as ducts the mesonephric ducts, which are then called deferent ducts. Hence, in males in which the mesonephros (or opisthonephros) is the functional adult kidney, the mesonephric ducts convey both urine and sperm. In amniotes, where the metanephros is the adult kidney, the mesonephric ducts have only genital functions.

17. The mesonephric duct is always connected to the corresponding testis by the intervention of a persistent part of the mesonephros, termed the epididymis, whose renal corpuscles connect with the seminiferous tubules of the testis by a varying number of tubules, called the efferent ductules.

18. The vertebrate embryo has all the structures of both sexes and is potentially hermaphroditic. In the development of a male the oviducts are suppressed; in the development of a female the mesonephric ducts are suppressed or, in anamniotes, limited to an excretory function.

19. A cloaca is characteristic of most vertebrates, but in placental mammals it becomes subdivided by a fold into a dorsal rectum and a ventral urogenital canal or sinus, in such a way that the latter gets the urogenital ducts and the bladder. In male placental mammals the deferent ducts (mesonephric ducts) join the stalk of the bladder (urethra) to form a urogenital canal, which pierces the penis and opens at its tip. In most female placental mammals the urethra and common vagina unite similarly to form a urogenital canal; but in female primates,

urethra and vagina are completely separate and open separately to the exterior, so that a urogenital canal is absent and the path of the urine is wholly distinct from the genital passage.

20. In mammals the male urogenital canal is generally provided with a variety of glands, termed the accessory sex glands, whose secretions are of importance for the vitality of the sperm.

21. A definite organ of copulation, the penis, is first seen in reptiles, occurs in some birds, and is general throughout mammals. It is a differentiation of the ventral floor of the cloaca and consists primitively of a cylinder of spongy tissue, the cavernous body, on whose surface a groove conveys the sperm. In mammals the groove closes over, becoming the cavernous urethra; and the cavernous body subdivides into two bodies.

22. The general direction of evolution in the urogenital system is toward the separation of excretory, genital, and intestinal functions.

REFERENCES

ALVERDES, K. 1928. Die Epididymis der Sauropsiden im Vergleich zu Säugetier und Mensch. Zeitschr. f. mikr. anat. Forsch., 15.

BATES, G. A. 1914. The pronephric duct in elasmobranchs. Jour. Morph., 25.

BISHOP, S. C. 1926. Notes on the habits and development of the mud puppy, *Necturus maculosus*. Bull. N.Y. State Mus., 268.

BORCEA, I. 1906. Système uro-génital des Elasmobranches. Arch. zool. expér. et générale, ser. 4, 4.

BRASS, A. 1880. Beiträge zur Kenntnis des weiblichen Urogenitalsystems der Marsupialen. Thesis, Leipzig.

BROCK, A. J. P. VON. 1910. Entwicklung und Bau des Urogenital-Apparates der Beutler und dessen Verhältnis zu diesen Organen anderer Säuger und niederer Wirbeltiere. Morph. Jahrb., 41.

CAGLE, F. R. 1937. Egg-laying habits of the slider turtle, the painted turtle, and the musk turtle. Jour. Tenn. Acad. Sci., 12.

CHASE, S. W. 1923. The mesonephros and urogenital ducts of *Necturus*. Jour. Morph., 37.

DAWSON, A. B. 1922. The cloaca and cloacal glands of the male *Necturus*. Jour. Morph., 36.

ENGLE, E. T. 1926. The copulation plug and the accessory genital glands of mammals. Jour. Mammology, 7.

EYCLESHYMER, A. C. 1900. The habits of *Necturus*. Amer. Nat., 40.

FORD, E. 1919. A contribution to our knowledge of the life histories of the dogfishes. Jour. Marine Biol. Assoc., 12.

FRASER, E. A. 1927. Observations on the development of the pronephros of the sturgeon. Quart. Jour. Mic. Sci., 71.

FUKUI, N. 1923. Action of body temperature on the testicle. Japan Med. World, 3.

GERECKE, H. 1932. Das Kopulationsorgane von *Testudo*. Jena Zeitschr. Naturwiss., 66.

GOODRICH, E. S. 1895. On the coelom, genital ducts, and nephridia. Quart. Jour. Mic. Sci., 37.

———. 1902. On the structure of the excretory organ of *Amphioxus*. *Ibid.*, 45.

———. 1934. The early development of the nephridia in *Amphioxus*. *Ibid.*, 76.

GRAY, P. 1930. The development of the mesonephros in *Rana*. Quart. Jour. Mic. Sci., 73.

KINDAHL, MARTHA. 1938. Zur Entwicklung der Exkretionsorgane von Dipnoern und Amphibien. Acta Zool., 19.

KINGSBURY, B. F. 1895. The spermatheca and methods of fertilization in some American newts and salamanders. Trans. Amer. Mic. Soc., 17.

LEIGH-SHARPE, W. H. 1920–24. The comparative morphology of the secondary sexual characters of the elasmobranch fishes. Jour. Morph., **34, 35, 36, 39.**

MAHADEVAN, G. 1940. Preliminary observations on the structure of the uterus and the placenta of a few Indian elasmobranchs. Proc. Indian Acad. Sci., B, **11.**

MASCHKOWZEFF, A. 1926. Zur Phylogenie des Urogenitalsystems der Wirbeltiere. Zool. Jahrb., Abt. Anat., **48.**

METTEN, H. 1939. Studies on the reproduction of the dogfish. Phil. Trans. Roy. Soc. London, B, **230.**

MOORE, C. R. 1924. Heat application and testicular degeneration. The function of the scrotum. Amer. Jour. Anat., **34.**

———. 1941. Role of sex hormones in sex differentiation of the opossum. Physiol. Zoöl., **14.**

PICKEL, F. W. 1899. The accessory bladders of the Testitudinata. Zool. Bull., **2.**

PRICE, G. C. 1897. Development of the excretory organs of a myxinoid. Zool. Jahrb., Abt. Anat., **10.**

———. 1904. A further study of the development of the excretory organs in *Bdellostoma.* Amer. Jour. Anat., **4.**

QUIGLEY, J. P. 1928. Life history and physiological condition of *Squalus sucklii.* Biol. Bull., **55.**

RAUTHER, MAX. 1904. Ueber den Genitalapparat einiger Nager und Insectivoren, insbesondere die akzessorischen Genitaldrüsen derselben. Jena Zeitschr. Naturwiss., **38.**

REDDY, A. R. 1938. The development of the anuran kidney. Proc. Indian Acad. Sci., **8.**

RIETSCHEL, P. E. 1929. Zur Morphologie und Histologie der Genitalausführgänge der weissen Maus. Zeitschr. f. wiss. Zool., **135.**

RISLEY, P. L. 1933. Development of the reproductive system in *Stenothernus.* Zeitschr. f. Zellforsch. u. mikr. Anat., **18.**

STROMSTEN, F. A. 1912. Nest digging and egg laying habits of Bell's turtle. Univ. Iowa Studies Nat. Hist., **10.**

TURNER, C. L. 1937——. Series of Studies on nutritive arrangements in viviparous fishes. Jour. Morph. **61, 62, 67.**

WALSCHE, L. DE. 1929. Etude sur le développement du pronéphros et du mésonéphros chez les Chéloniens. Arch. de Biol., **39.**

WEEKES, H. C. 1935. A review of placentation among reptiles. Proc. Zoöl. Soc. London.

WHEELER, W. M. 1899. The development of the urogenital organs of the lamprey. Zool. Jahrb. Abt. Anat., **13.**

XIV. THE COMPARATIVE ANATOMY OF THE NERVOUS SYSTEM AND THE SENSE ORGANS

A. GENERAL CONSIDERATIONS

This chapter will include both the nervous system proper and the sense organs; the former conducts and correlates stimuli and probably also initiates actions, and the latter are differentiated for the reception of stimuli.

1. The parts of the nervous system.—The unit of structure of the nervous system is the *nerve cell* or *neuron* (also called ganglion cell), which consists of a protoplasmic mass, the *cell body* (containing the nucleus), and one or more processes or *neurites,* which may be very long. The most primitive type of nerve cell is the *neurosensory* cell, in which the usually elongated cell body is situated at the surface and acts as a *sensory receptor* and also as a *conductor*, for its inner end continues as a nerve fiber, joining the central nervous system. Neurosensory cells are common in invertebrates but in vertebrates are limited chiefly to the olfactory epithelium and the retina of the eye (Fig. 124). The next type is the *unpolarized* nerve cell, usually with several neurites, along which impulses pass in either direction. The final type, the common type in vertebrates, is the *polarized* nerve cell,[1] which *receives* impulses along one or more processes, the *afferent* processes or *dendrites,* and *sends them out* along one process, the *efferent* process or *axon.*[2] Axon and dendrites differ histologically as well as functionally. An aggregation of nerve-cell bodies outside the central nervous system is called a *ganglion* (plural, *ganglia*). A similar group inside the central nervous system is termed a *nucleus*[3] or *center,* or less often also a ganglion.

A *nerve* is a bundle of neurites, also called *nerve fibers* in this connection. In vertebrates nerves are nearly always devoid of cell bodies, since these are situated in ganglia or in the central nervous system. Nerve fibers may or may not be each inclosed in a fatty sheath, called the *medullary* or *myelin* sheath. When so inclosed, nerve fibers are spoken of as *medullated* or *myelinated* or *white* fibers; fibers which are naked without any such sheath are known as *nonmedullated* or *unmyelinated* fibers, also *gray* fibers.

The vertebrate nervous system consists of three parts: the *central* nervous system, the *peripheral* nervous system, and the *autonomic* (also called *sympathetic* and *vegetative*) nervous system. The central nervous system comprises the *brain,* situated inside the skull, and the *spinal cord,* situated inside the neural canal formed by the arches of the vertebrae. Brain and cord are composed both of nerve-cell bodies, which form the *gray matter,* and of nerve fibers, which form the *white matter,* so called because the myelin sheaths lend a white color. The peripheral nervous system consists of the *cranial nerves,* springing from the brain, and the *spinal nerves,* springing from the spinal cord. Both kinds of nerves consist of the neurites of nerve-cell bodies found inside the brain or cord or in outside ganglia. The nerves, especially

[1] By some, only the polarized type is called neuron and only the axon is called neurite; but such restricted meanings are undesirable from a comparative viewpoint.

[2] These processes conduct in either direction, but impulses *function* in polarized nerve cells only in the direction indicated.

[3] This term is certainly most unfortunate but has become firmly established in neurology. It has, of course, nothing to do with the nucleus of a cell but is used in the sense of a center.

the spinal nerves, are markedly metameric in arrangement, there being typically a pair of nerves to each body segment; but this metamerism is probably imposed upon the nerves by the metamerism of the muscles and vertebrae. The autonomic nervous system controls and regulates in general the involuntary activities of the body and the organs which subserve those functions, such as the heart, digestive tract, smooth musculature in general, secreting glands, blood vessels, respiratory, and urogenital systems. It consists of a ganglionated cord lying to either side of the vertebral column along the dorsal coelomic wall, of ganglia in the head and among the viscera, and of connecting and distributing nerves.

2. The development of the nervous system.—The central nervous system is formed, as already learned, by the infolding of the ectoderm along the middorsal region of the embryo to form a tube. That portion of the tube situated in the head becomes the brain, and that part posterior to the head becomes the spinal cord. The originally single layer of inrolled ectoderm cells proliferates to form a thick zone of cells around the central cavity; most of these become nerve cells, but some give rise to supporting cells or neuroglia. This position of the zone of nerve-cell bodies or gray matter around the central cavity is the primitive arrangement and is retained throughout the vertebrates in the spinal cord; but in the brain of the higher and also more specialized forms among lower vertebrates there is considerable migration of gray matter to the periphery. Neurites arise by outgrowth from the embryonic nerve cells and may remain in the central nervous system to form a peripheral zone of medullated fibers, the white matter, or may grow out of it to form the nerves. The ganglia outside the brain and cord originate by migration of nerve-cell bodies, mostly from the neural crests, a pair of longitudinal strands left outside the neural tube at the time of its closure.

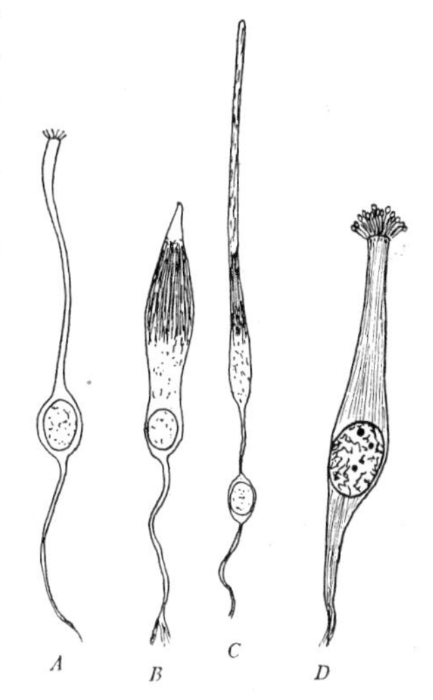

FIG. 124.—Neurosensory cells. *A*, olfactory cell of the olfactory epithelium of a mammal. *B*, cone, and *C*, rod visual cell of the mammalian retina. *D*, neurosensory cell of the vascular sac of the fish brain (after Dammerman, 1910).

In the development of the brain the original simple tube becomes marked off by a transverse ventral fold into two regions, the primitive *fore* and *hind* brain (Fig. 125*A*). This fold is situated just behind the future *infundibulum*, and this place also marks approximately the anterior end of the notochord. The anterior part of the primitive forebrain, called the *telencephalon*, puts out paired lateral swellings, which become the *cerebral hemispheres*. The region behind this is the *diencephalon* or *thalamencephalon* and soon shows a dorsal evagination, the *pineal apparatus* (Fig. 125*C*). The diencephalon becomes marked off by a constriction from the last part of the primitive forebrain, the *mesencephalon*, which develops a pair of pronounced dorsal swellings, the *optic lobes* (Fig. 125*C*, *D*). A deep constriction forms between mesencephalon and primitive hindbrain. The anterior part of the latter, called *metencephalon*, puts out a dorsal enlargement, the *cerebellum* (Fig. 125*E*); and the remainder and greater part of the hindbrain is the *myelencephalon* or *medulla oblongata*, characterized by its thin roof and the large nerve ganglia along its side walls (Fig. 125*C*). The medulla oblongata is continuous with the spinal cord, from which it is not definitely delimited. These five main regions of the embryonic brain persist as such in the adult condition, into which they develop by thickenings, folds, and outgrowths of their walls.

3. Gross features of the adult central nervous system.—The description of the nuclei and

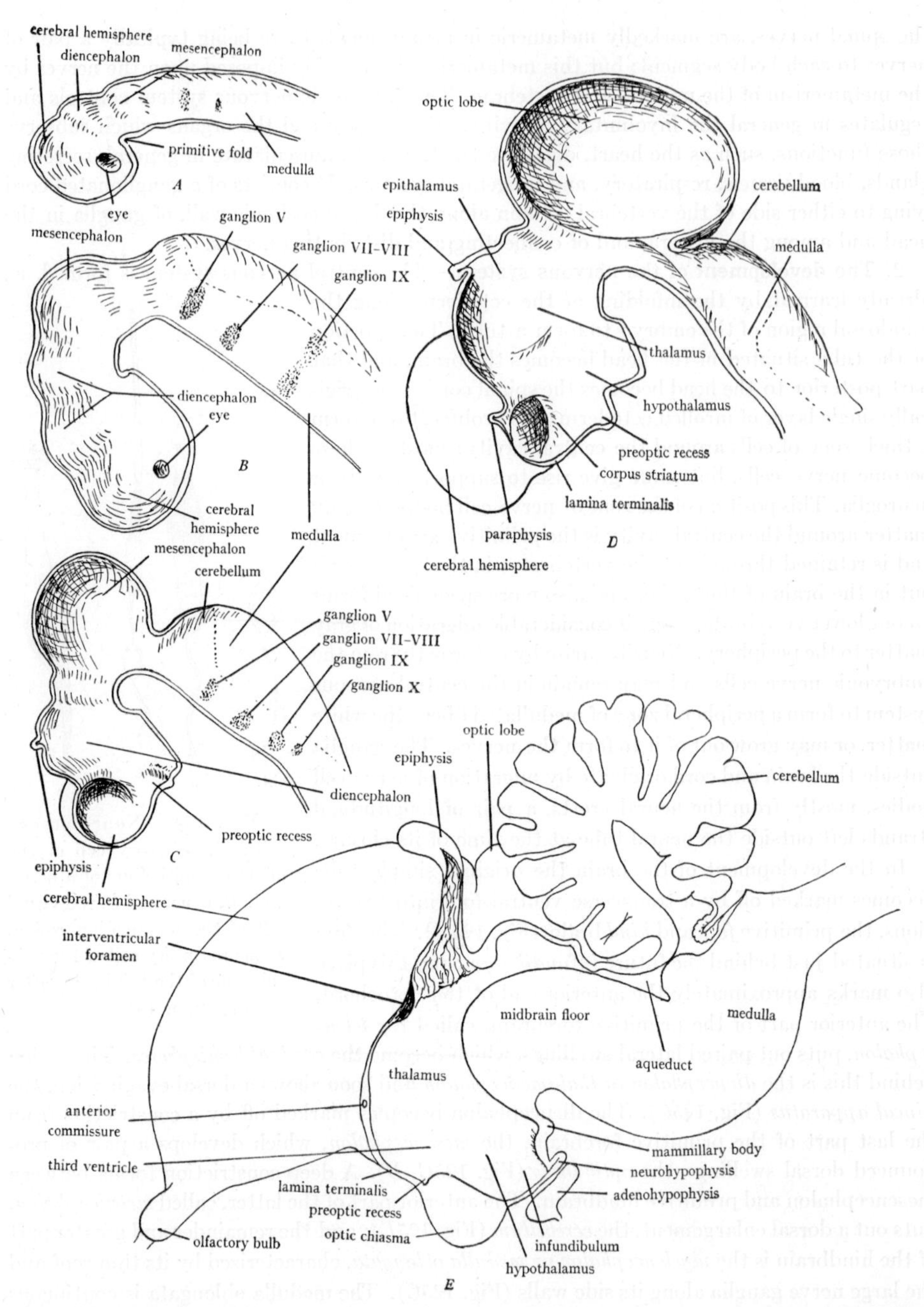

FIG. 125.—Five stages in the development of the chick brain (after Streeter, 1933). *A*, ventral fold has formed; *B*, main regions of the brain are outlined; *C*, epiphysis has appeared, cerebral hemispheres are expanding; *D*, paraphysis has appeared, cerebral hemispheres and optic lobes are much expanded; *E*, olfactory bulbs have evaginated, cerebellum is well developed, parts of the hypothalamus are shown. *C*, *D*, and *E* are sagittal sections.

tracts of the brain and cord lies outside the province of this book and will be found in textbooks of neurology.

The adult spinal cord is usually a slightly dorsoventrally flattened rod; this flattening is very pronounced in cyclostomes, where the bandlike cord (Fig. 126*B*) lies on the notochord. In tetrapods the cord presents *cervical* and *lumbar enlargements*, associated with the limbs; these are absent in fishes and limbless tetrapods such as snakes. The cord extends the full

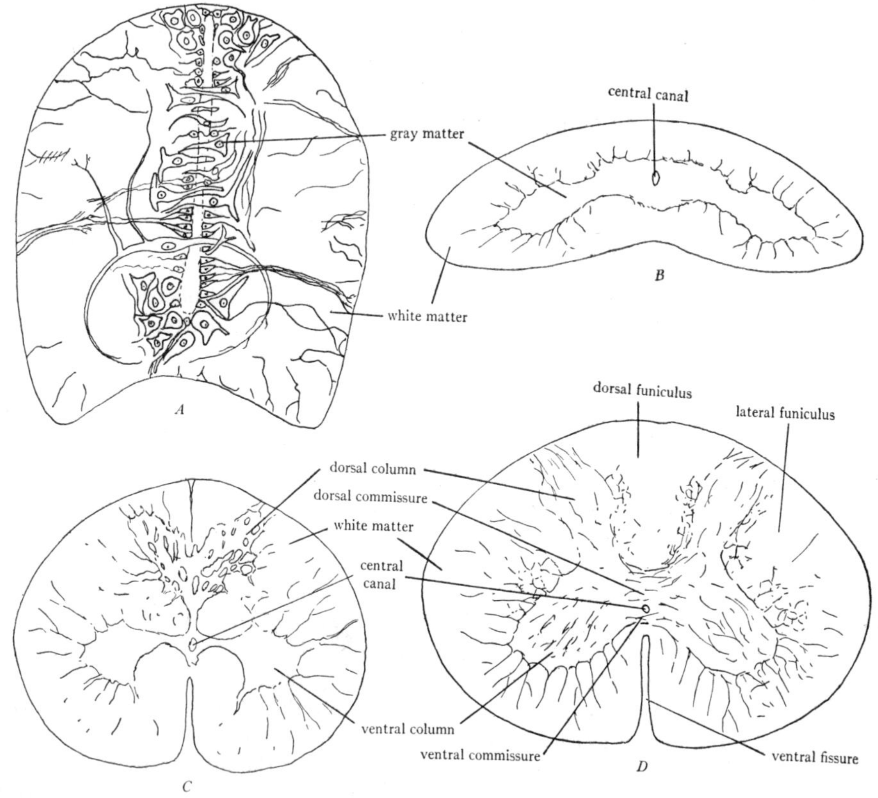

FIG. 126.—Section of the spinal cord of *A*, *Amphioxus; B*. *Ammocoetes; C*, selachian; and *D*, amniote, to show arrangement of gray and white matter. (*A* after Franz, 1923; *B*, after Tretjakoff, 1909; *C*, after Glees, 1940.)

length of the vertebral canal in primitive fishes, reptiles, and birds; but in mammals and many teleosts is much shorter than the vertebral column, tapering to a fine strand, the *terminal filament*. The cavity of the spinal cord is reduced to a very small *central canal* in the adult, and this is lined by a one-layered columnar epithelium, termed the *ependyma*. The gray matter of the cord, composed chiefly of nerve-cell bodies and nonmedullated neurites, occupies its central region and in amniotes has in cross-section the well-known H or butterfly-like shape (Fig. 126*D*), being actually a four-fluted cylinder. The two dorsal limbs of the H are known as the *dorsal columns* (old name, "posterior" horns); the two ventral limbs form the *ventral columns* ("anterior" horns); and between these there is in mammals a more or less definite *lateral* column on each side. The bar of the H incloses the central canal, passing above and below it as the *dorsal* and *ventral commissures*, respectively. In anamniotes the gray matter in section varies from a rounded or quadrangular shape to a figure somewhat like an inverted T and Y (Fig. 126*C*); in cyclostomes it forms a broad band (Fig. 126*B*). The white matter, constituted of ascending and descending bundles or tracts of medullated neurites, forms a thick peripheral zone of the cord and is divided by the gray columns into *dorsal*, *lateral*, and *ventral funiculi* (Fig. 126*D*). Columns and funiculi are, of course, less evident and defined in anamniotes; and they also differ in shape and proportions at different levels of the cord. The funiculi may be subdivided by longitudinal furrows, of which the deepest is the *median ventral fissure*; these also, except the last, are indistinct in lower forms.

The vertebrate brain forms a conspicuous lobed enlargement continuous with the anterior end of the spinal cord. It occupies the cranial cavity of the skull, and the shape of this cavity conforms to the contours of the brain. The brain rests ventrally on the chain of cartilage bones ossified in the floor of the chondrocranium. Although the brain is bent at the mesencephalon in the embryos of all vertebrates, it straightens out in line with the longitudinal body axis in the later development of the lower vertebrates; and a permanent flexure is limited to birds and mammals, being especially notable in the brain of primates. The arrangement of the gray and white matter of the cord persists into the rear part of the brain, but anteriorly the gray and white matter intermingle irregularly; in the cerebral hemispheres and cerebellum the gray matter forms a peripheral stratum over more central white matter. The white matter of the brain consists of ascending and descending tracts connecting various parts of brain and cord; the gray matter consists mostly of "nuclei" (also called centers and ganglia), i.e., of aggregates of cell bodies whose neurites form the tracts of the white matter or issue in the cranial nerves.

The fifth, most posterior part of the brain, the myelencephalon or medulla oblongata, is continuous with the spinal cord through the foramen magnum and cannot be definitely delimited from the cord, although the boundary is usually placed just anterior to the level of the first spinal nerve. The rear part of the medulla is similar to the spinal cord, having a thick wall around the very small central canal; but soon this canal widens out into a large rhomboidal or triangular cavity, termed the *fourth ventricle*, the roof of which is greatly thinned out and consists of only the ependymal epithelium. To the outside of this, however, there adheres the richly vascularized pia mater (see below, under meninges), the two together forming the *tela choroidea*. This in certain regions is invaginated into the cavity of the fourth ventricle as vascularized tufts, known as *choroid plexi*. The nervous part of the walls of the medulla consists chiefly of the nuclei of origin and of termination of the fifth to tenth (twelfth in amniotes) cranial nerves and their connections. These nuclei are arranged in primitive vertebrates according to their functions into four longitudinal columns (see below, under functional components), and this arrangement more or less persists throughout the vertebrates.

The fourth division of the brain, or metencephalon, forms dorsally the *cerebellum*. This is of slight development in cyclostomes and amphibians, consisting of a ledge in front of the tela choroidea of the medulla, but is large in fishes of active habits, as selachians and some teleosts. In the amniote series it begins of rather small size in most reptiles but quickly increases in size and complexity of structure, being highly developed in birds and mammals. It typically consists of a median portion, the *body* or *corpus*, and a pair of lateral lobes, which from their shape in selachians are called the *auricular lobes* and which contain extensions of the fourth ventricle, the *auricular recesses*. In amniotes the auricular lobes are usually called the *floccular lobes*. The body of the cerebellum is generally marked by transverse fissures, which increase in number in the amniote series, where the body seems more or less definitely divisible into anterior, middle, and posterior lobes (Fig. 127). The floccular lobes are connected with the posterior lobes; and the marked lateral areas, termed cerebellar hemispheres, characteristic of mammals, appear to be developments of the anterior and middle lobes. The cerebellum has a surface layer of gray matter known as the *cerebellar cortex*, and in birds and mammals this is underlain by a thick stratum of white matter, so that sections of the cerebellum of these groups show the familiar "arbor vitae," lobulations of white matter covered by gray. It is characteristic of the teleost cerebellum that there is a thick invaginated lobe, termed the *valvula*. The cerebellum usually contains a cavity, the *cerebellar ventricle*.

The floor of the metencephalon shows little differentiation from the medulla in lower forms; but, with increasing importance of the cerebellum, it comes to be considerably occupied by tracts to and from the cerebellum. Hence in mammals this floor, termed the *pons*, has a conspicuous fibrous structure, visible on the ventral surface in front of the medulla.

The functions of the cerebellum, so far as known, are the co-ordination of muscular movements, the maintenance of muscular tone, and the equilibration of the body in space. Consequently, it has connections with practically all levels of the central nervous system, but par-

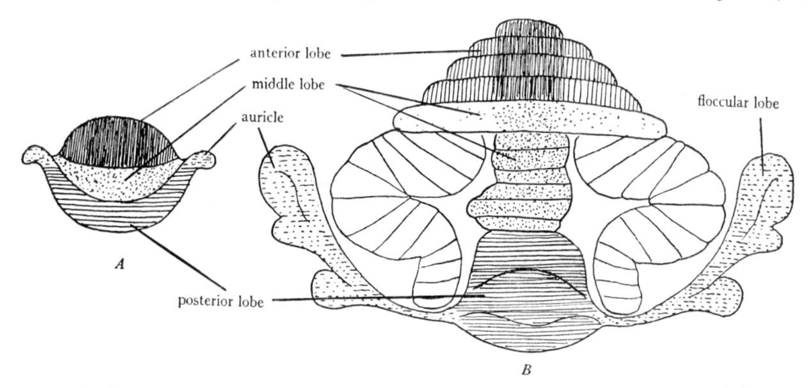

FIG. 127.—Evolution of the cerebellum from *A*, reptilian stage, to *B*, mammalian stage (after Ingvar, 1918). Vertical hatching, anterior lobe; stippling, middle lobe; crosshatching, posterior lobe; dashed lines, auricles. In the mammal the white areas and less closely lined and stippled areas are new formations, developments of anterior and middle lobes. The auricles become the floccular lobes.

ticularly with sensory impulses from muscles and joints (*proprioceptive impulses*) and with the equilibratory mechanism of the inner ear.

The *mesencephalon* or *midbrain*, the third division of the brain, is also best developed dorsally, where its thick roof, known as the *optic tectum*, has usually two curved eminences, the *optic lobes* or *corpora bigemina*. In mammals there are four such externally apparent eminences, the *corpora quadrugemina*, of which the anterior pair are also called the *superior colliculi*, the posterior pair the *inferior colliculi*. The inferior colliculi, however, are already present from fish upward but are not evident externally. Their exposure on the surface in mammals results from the retrogression of the superior colliculi. The midbrain roof is concerned chiefly with auditory and optic functions. The midbrain floor consists chiefly of tracts to and from other parts of the brain. The increase in tracts from the cerebral hemispheres results, in mammals, in a pair of conspicuous bundles, the *cerebral peduncles*, visible externally on the ventral surface of the midbrain. The midbrain is solid in mammals except for a narrow canal, the *cerebral aqueduct* (old name, *aqueduct of Sylvius*), which passes through it from the fourth ventricles. In lower vertebrates the optic lobes are usually hollow, containing cavities, termed the *optic ventricles*, which open into the aqueduct.

The *diencephalon* or *thalamencephalon*, the second division of the brain, is chiefly a relay center for the cerebral hemispheres, and consequently its size and importance parallel those of the latter. It has a central cavity, the *third ventricle*, into which the aqueduct opens, and which in mammals is much compressed laterally. The roof of the third ventricle is thinned to an ependyma and forms a tela choroidea. This has choroid plexi projecting into the interior or, in certain lower forms (petromyzonts, some ganoid fishes, selachians, some reptiles), *choroidal sacs*, often very large, bulging to the exterior. The term *parencephalon* is sometimes applied to the choroidal roof of the third ventricle. Ventrally the anterior boundary of the third ventricle is a thin membrane, the *lamina terminalis* (Fig. 125*E*), believed by many to represent the original anterior end of the embryonic brain; more dorsally the diencephalon is bounded from the telencephalon by a transverse infolding, the *transverse velum*. The diencephalon is divisible throughout the vertebrates into a dorsal *epithalamus*, a middle thick *thalamus*, and a ventral *hypothalamus* (Fig. 125*E*). The epithalamus consists chiefly of the *epiphyseal apparatus* and a pair of small masses, termed the *habenulae*. The epiphyseal apparatus apparently consisted primitively of two outgrowths of a sensory nature; and in lampreys, some ganoids, and many reptiles there are still two outgrowths—an anterior *parapineal* or *parietal body*, and a posterior *pineal body* or *epiphysis*. In petromyzonts each develops a simplified eye. In *Sphenodon* and a number of other reptiles the anterior or parietal outgrowth forms a well-differentiated eye (usually called pineal eye, but more correctly termed parietal eye), and the posterior outgrowth is the pineal body proper. In other vertebrates only the pineal outgrowth appears in the embryo; and this develops into the pineal body, with no trace of eye formation. Except in lampreys the pineal body is of glandular character and is generally believed to be an endocrine gland, although convincing proof is wanting. The parietal eye of reptiles occupies the parietal foramen of the skull, and this foramen is noticeable in the skull of primitive extinct tetrapods and their crossopterygian ancestors (Figs. 58*E*; 59*A*, *B*). The possibility that vertebrates originally had three eyes is suggested by these facts but must be left an undecided question.

The thalamus comprises the lateral walls of the diencephalon; it is a relay center for tracts passing to and from the cerebral hemispheres and hence increases in size and importance, especially its dorsal part, in higher vertebrates. It consists of numerous "nuclei," i.e., masses of gray matter. In reptiles and mammals the dorsal parts of the thalami are fused across the third ventricle by a connection of gray matter termed the *intermediate mass*.

The hypothalamus or ventral part of the diencephalon has the following main parts: *optic chiasma* (crossing of the optic nerves at their entrance into the brain), *tuber cinereum*, *infundibulum*, *hypophysis*, and *mammillary* region. The hypothalamus reaches its greatest development in fishes, where it is an important correlation center for olfactory, gustatory, and other sensory impulses. It is small in amphibians but increases again in reptiles and mammals, although there functioning somewhat differently than in fishes. The tuber cinereum is an area of gray matter behind the optic chiasma having to do with olfactory sensations. It is continued ventrally as a stalk, the *infundibulum*, whose cavity is an extension of the third ventricle and from which depends the hypophysis or pituitary body. The latter is composed partly of the oral epithelium of Rathke's pouch (*adenohypophysis*, see p. 257) and partly of nervous tissue from the infundibulum (*neurohypophysis*). The hypophysis is a very important endocrine gland, secreting a number of hormones. In fishes and amphibians the infundibulum is expanded posteriorly into a pair of swellings, the inferior lobes; and below these extends as a soft-walled, very vascular structure, the *saccus vasculosus* or *vascular sac*, which is close to the pituitary. The vascular sac is lined by an epithelium containing neurosensory cells (Fig. 124*D*); and this, together with the fact that it occurs only in aquatic vertebrates, indicates some sensory function concerned with aquatic life. The mammillary region is the most posterior part of the hypothalamus and usually has a single or paired swelling, the *mammillary bodies*.

The telencephalon or endbrain, the most anterior division of the brain, consists chiefly of the paired *olfactory bulbs* and *cerebral hemispheres.* The olfactory bulbs are anterior outgrowths of the cerebral hemispheres (Fig. 125*E*). They abut against the rear wall of the nasal sacs; and, when the latter are considerably anterior to the brain, the bulbs remain in contact with them, and the connecting region is drawn out into a more or less elongated stalk, the *olfactory stalk* or *peduncle.* The olfactory bulbs are generally hollow but are secondarily solid in mammals. Their size is correlated with the development of the sense of smell, and hence they are very reduced in birds (which have almost no sense of smell) and are relatively small in primates.

The cerebral hemispheres constitute the major part of the telencephalon. They form large lateral bulges to either side of the original median anterior brain wall, the lamina terminalis, which is thus in higher vertebrates left at the bottom of the cleft between the hemispheres (Fig. 125*E*). Not infrequently, however, the olfactory bulbs or cerebral hemispheres or both may be secondarily fused together dorsally. The lamina terminalis extends forward and upward from the *preoptic recess,* a depression in front of the optic chiasma, to the *neuroporic recess,* the point of last closure of the neural folds of the embryo. Just below the neuroporic recess the lamina terminalis contains a bundle, the *anterior commissure,* which connects olfactory regions of the two hemispheres. Above the neuroporic recess the original telencephalic roof, extending backward to the velum transversum, is thinned to form a tela choroidea, from which choroid plexi project into the cavities of the cerebral hemispheres. Immediately in front of the velum transversum the telencephalic roof has a dorsal branched or folded evagination, the *paraphysis* (Fig. 125*D*). This is evident in the embryos of all classes of vertebrates but does not persist in the adults of the higher ones; its function is unknown.

In lower vertebrates the cerebral hemispheres are smooth, rounded, or elongated lobes which are not noticeably out of proportion to the rest of the brain, but in the amniote series they increase greatly in size. This process culminates in primates, where the relatively enormous and much convoluted hemispheres cover and conceal the other divisions of the brain.

Each cerebral hemisphere contains a cavity, known as the *lateral ventricle,* and these communicate with each other and with the third ventricle by a passage (or pair of passages) termed the *interventricular foramen* (old name, *foramen of Munro*). These ventricles are somewhat reduced in reptiles and birds but well developed in mammals, where, however, they are of very irregular form because of the bulgings of the hemisphere wall.

In lower vertebrates the hemispheres are largely olfactory, and the main olfactory areas are traceable throughout the vertebrate series; but additional areas appear and expand in higher forms. The wall of the cerebral hemisphere is roughly divisible into a dorsal half, the *roof* or *pallium,* and a ventral half, or *subpallium.* The pallium primitively shows two main regions concerned with olfaction: a medial region, the *primordium of the hippocampus,* and a lateral region, the *primordium of the pyriform lobe.* Beginning with amphibians, but more clearly indicated in reptiles, there appears between these middorsally, another region, the *neopallium* (Fig. 128). The neopallium is a nonolfactory area which expands rapidly in the amniote series without increasing much in thickness. This process of expansion culminates in higher mammals, where the area of the neopallium is still further greatly increased by the formation of convolutions. The expansion of the neopallium in mammals is so great that the pyriform lobe is forced to the ventral side, and the hippocampus is caused to fold into the interior, forming the characteristic "horn of Ammon" (Fig. 128). During its evolution the pallium acquires a superficial layer of gray matter, termed the *cerebral cortex,* characterized by a zonary arrangement of certain types of nerve-cell bodies. The cortex, although indicated in reptiles and birds, is first fully differentiated in mammals, most typically in the area developed from the neopallium. The hippocampus and pyriform lobe are olfactory centers; the neopallial cortex is essentially somatic, receiving the terminations of the somatic senses (vision, hearing, skin senses) and initiating and correlating muscular responses; it is further the seat of all conscious feelings, actions, and thoughts.

The ventral half of the wall of the cerebral hemisphere is roughly divisible into a medial *septum* and a massive lateral area, which may be broadly termed the *corpus striatum* or *striatal region* (Fig. 128). This in reptiles and birds forms a conspicuous mass bulging into the lateral ventricle from below (Fig. 128*A*) but in mammals is less obvious (Fig. 128*B*). It consists of several masses of gray matter and has connections with the cerebral cortex, thalamus, and midbrain. Its functions are not very clearly known, but in mammals there is evidence that it has a stabilizing influence on automatized muscular co-ordinations, such as walking.

Beginning in some marsupials, there occurs in mammals a broad horizontal white fibrous band, the *corpus callosum*, consisting of medullated fibers connecting the neopallial cortex of

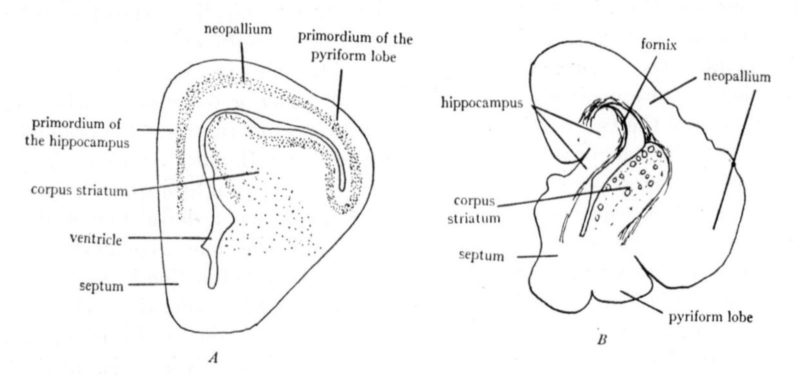

FIG. 128.—Section through the cerebral hemisphere of *A*, *Sphenodon*, and *B*, *Ornithorhynchus*, to show infolding of hippocampus and expansion of the neopallium in the mammalian brain. (*A*, after Smith, 1919; *B*, after Hill, 1894.)

the two hemispheres. It is generally believed that these fibers, gradually increasing in numbers as the cortex expands, occupy the upper part of the lamina terminalis and cause this to extend also. Below the corpus callosum is another white fiber tract, the *fornix*, which runs from the hippocampus to the mammillary body on each side. Although a fornix system (i.e., a connection between olfactory areas of the cerebral hemispheres and the hypothalamus) is present in all the vertebrate classes, the fornix (Fig. 128*B*) is a mammalian feature. Between the corpus callosum and the fornix there stretches in mammals a double-walled membrane, the *septum pellucidum*, with a cleft between the two walls.

The term *brain stem* is applied to the brain minus the cerebral hemispheres and the cerebellum. The medulla is often referred to as the *bulb*.

4. The meninges.—The central nervous system does not fill the bony canal in which it lies but is separated from it and protected by fluid and by connective tissue membranes, the *meninges* (singular, *meninx*). In cyclostomes and fishes in general there is but one membrane, the *primitive meninx*. Amphibians have two, an outer *dura mater* and an inner *secondary meninx;* and similar conditions obtain in reptiles and birds. In mammals there are three meninges—a relatively tough outer dura mater and two delicate inner ones, the *arachnoid*, and the *pia mater*, both differentiations of the secondary meninx. The pia mater is closely applied to the brain and cord, into which it may send connective tissue partitions, and is separated by a considerable *subarachnoid space* from the weblike arachnoid. Outside the dura mater there is further a fat-vascular cushioning layer. The spaces around the brain and cord and between the meninges are in communication with the central canal of the cord and the ventricles of the brain; and all are filled with a fluid, the *cerebrospinal fluid,* differing but little from the lymph.

5. The functional divisions of the nervous system and the peripheral nervous system.—It has proved convenient in discussions of the nervous system to divide the body functions into two categories, *somatic* and *visceral.* The somatic functions are those mediated by the body wall, i.e., the skin, the musculature, and the skeleton. The visceral functions are those carried on by the other systems—the digestive, respiratory, circulatory, urogenital, and endocrine systems. Each of these two categories has further two components, an *afferent* or *sensory,* and an *efferent* or *motor,* so that the nervous system is made up in general of four functional components. These are: the *somatic sensory component,* which handles the impulses from the sense organs of the skin, the special sense organs of the head (eye, ear), and the sensations from the deeper body-wall structures, such as muscles and joints (*proprioceptive* impulses); the *somatic motor component,* which handles outgoing impulses to the voluntary musculature except the branchial musculature; the *visceral sensory component,* which deals with sensations from the visceral systems, including the special senses of smell and taste; and the *visceral motor component,* which mediates impulses to the involuntary musculature of the viscera, to the branchial musculature, and to glands, etc. The visceral components involve the autonomic, as well as the central, nervous system (see below).

The dorsal half of the central nervous system (gray matter) is sensory; the ventral half, motor. This division is well marked in the spinal cord, continues into the medulla, but is somewhat distorted in higher levels of the brain. In the spinal cord the order of arrangement of the four components is, from dorsal to ventral: somatic sensory, visceral sensory, visceral motor, somatic motor (Fig. 129*B*). The dorsal gray columns of the cord are somatic sensory; the ventral gray columns are somatic motor; the gray region between is visceral and in mammals forms a distinct lateral column connected with the autonomic system. The four components continue in the same order into the medulla, where they form four longitudinal areas, which are especially clear in fishes (Fig. 129*A*). Large areas of the gray matter of brain and cord also serve for association and co-ordination of impulses.

It has been shown by the brilliant work of a succession of American neurologists that each nerve of the peripheral nervous system typically is made up of fibers belonging to all four functional components. This concept has been of great value in understanding the cranial nerves and their central connections.

The spinal nerves issue from the spinal cord at segmental intervals, emerging through the intervertebral foramina. They are named after the vertebral regions—cervical, thoracic, lumbar, sacral, caudal. In cases where the spinal cord does not reach to the rear end of the neural canal, the last pairs of spinal nerves run inside the canal before issuing at the appropriate level, and so form a bundle termed the *cauda equina* or *horse's tail.* The attachment of a nerve, often multiple, to the central nervous system is termed its *root.* Each spinal nerve has two roots: a *dorsal* or *sensory* root, which passes into the dorsal gray column, and a *ventral* or *motor* root, which issues from the ventral gray column (Fig. 130). The dorsal root is composed chiefly of somatic sensory and visceral sensory fibers. It is a rule that the nerve-cell bodies of sensory fibers are located not in the central nervous system but in outside ganglia. Each dorsal root bears a ganglion, the *dorsal* or *spinal* ganglia, which is made up of the cell bodies of the somatic sensory and visceral sensory fibers (Fig. 130). The ventral root is composed of somatic motor fibers and in higher vertebrates also contains some or all of the visceral motor fibers, which in lower vertebrates exit by the dorsal root. The cell bodies of the somatic motor fibers are in the ventral gray column, and those of the visceral motor fibers are in the lateral column or the region of the gray matter corresponding to this position. It is a rule that the cell bodies of the somatic motor system are always inside the brain and cord.

Except in lampreys the dorsal and ventral roots unite distal to the spinal ganglion to form a *spinal nerve*, which then exits through the intervertebral foramen of its appropriate body segment and divides into three branches—the *dorsal ramus*, the *ventral ramus*, and the *communicating ramus*. Dorsal and ventral rami both contain somatic sensory and somatic motor fibers (also fibers of the autonomic system) and supply the skin and voluntary muscles of dorsal and ventral regions, respectively, of the appropriate body segment. The cutaneous (skin) areas supplied by the dorsal and ventral rami of each spinal nerve can be mapped out,

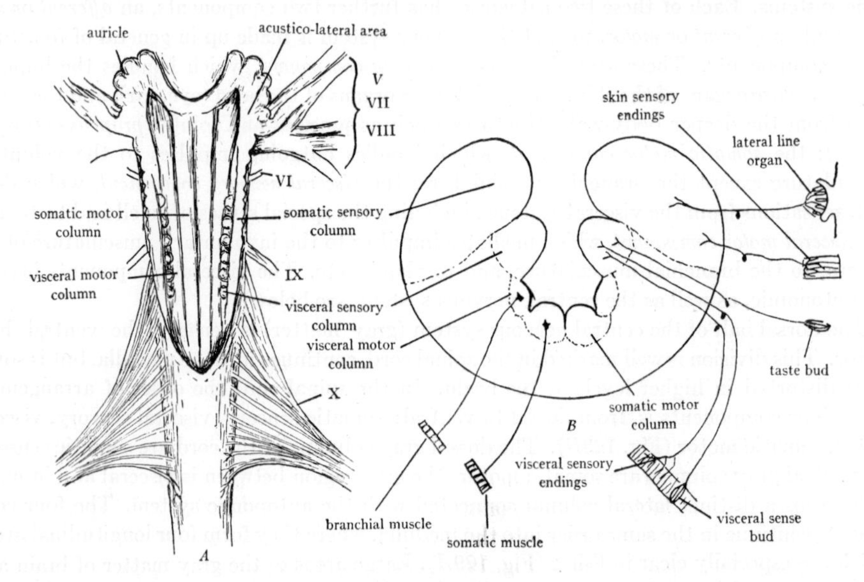

FIG. 129.—Medulla of a selachian with tela choroidea removed to show the longitudinal columns formed by the four functional components (after Gegenbaur, 1871). *B*, diagrammatic section through the medulla of a lower vertebrate to show the four functional areas and the parts they supply (after Johnston, 1902*b*).

and such maps are of great service in determining the level of injury to the spinal cord, since feelings are lost from skin areas below damage to dorsal parts of the cord. The somatic motor fibers in the dorsal ramus supply the epaxial muscles; those in the ventral ramus, the hypaxial muscles. The communicating ramus joins a ganglion of the autonomic system. For the original arrangement of the roots and branches of the spinal nerves see below, under nervous system of *Amphioxus* and cyclostomes.

In connection with the paired appendages, the ventral rami of a variable number of successive spinal nerves of the region concerned branch and anastomose to form a network or *plexus*, from which the nerves going to the appendages emerge. These plexi are the *brachial* plexus for the anterior, the *lumbar* or *lumbosacral* for the posterior, appendage. In their passage through the plexus the nerve fibers intermingle and cross over, so that the emergent nerves have no correspondence in content to those that entered the plexus. The occurrence of these plexi is taken to indicate: that the limb muscles arise from the hypaxial parts of the myotomes, since only the ventral rami enter into these plexi; that the limb muscles are derived from a number of myotomes, since each ramus supplies only the muscles derived from a single myotome and several rami are concerned in the plexi; and that the appendage muscles must have undergone considerable torsion and displacement, resulting in a crisscross arrangement of their motor nerves, since each nerve retains connection with the muscle of its body segment. The plexi also carry sensory fibers. In the neck region the *cervical* nerves may form a *cervical* plexus for the neck muscles, and this has connections with the cervical sympathetic and the spino-occipital nerves or their equivalents. It may be continuous with the brachial plexus, and a *cervicobrachial* plexus results.

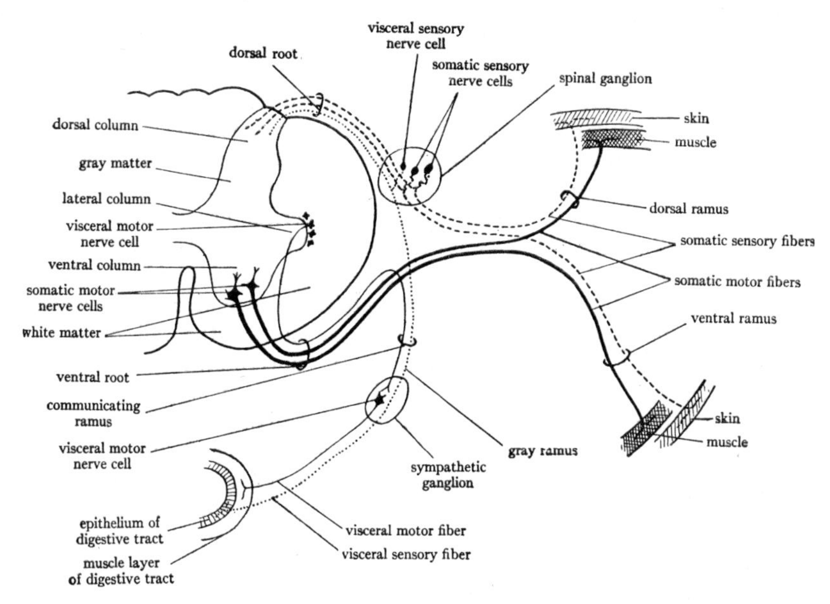

FIG. 130.—Diagram of a cross-section through the mammalian spinal cord with an attached spinal nerve and sympathetic ganglion to show the functional components; lower vertebrates differ in that some or all of the visceral motor fibers exit through the dorsal root. Broken lines, somatic sensory fibers; dotted lines, visceral sensory fibers; light continuous lines, visceral motor fibers; heavy continuous lines, somatic motor fibers. (Slightly altered from Herrick's *Introduction to Neurology;* courtesy of the W. B. Saunders Co.)

In mammals the *phrenic* nerves, which innervate the diaphragm, are derived from the more anterior spinal nerves entering into the brachial plexus. This indicates that the myotomes which contribute to the diaphragm are of cervical origin and have migrated posteriorly.

In some fishes (selachians, ganoids, Dipnoi) there is associated with the lumbosacral plexus a *collector* nerve, a longitudinal nerve which is joined by a number of spinal nerves anterior to the plexus and which then enters the plexus.

In the transitional region between medulla and spinal cord, what were originally the most anterior spinal nerves undergo some modification, losing their dorsal or sensory roots and becoming primarily of somatic motor nature. These are termed the *spino-occipital* nerves. Originally (in cyclostomes) outside the skull, the more anterior ones (generally two in number) become incorporated into the skull in fishes and amphibians because of the absorption of the more anterior vertebrae into the skull. These nerves are then termed, more exactly, the *occipital* nerves, since they then exit through the occipital region of the skull. In amniotes the more posterior ones, called *occipito-spinal* nerves, also become included in the skull; and, further, their nuclei of origin shift anteriorly, so that they are actually in the medulla. This is the reason that amniotes have twelve cranial nerves, anamniotes ten, since the last two in amniotes originate chiefly from the spino-occipital nerves. Naturally, it follows that amniotes do not have any spino-occipital nerves as such. The spino-occipital nerves vary considerably in number and occurrence among fishes and amphibians and usually have connections with the cervical plexus and the cervical sympathetic. From the unions of the spino-occipital nerves with the more anterior cervical nerves there springs a large *hypobranchial* nerve supplying the hypobranchial muscles. Spino-occipital nerves are generally lacking in bony fishes.

The cranial nerves—ten or twelve in number, as just explained—differ from the spinal nerves in that their dorsal and ventral roots do not unite and may appear to be quite separate nerves. As in the case of the spinal nerves, the visceral and somatic sensory fibers of the cranial nerves have their cells of origin in ganglia on the sensory roots close to the entrance of the latter into the brain. Because of the great changes involved in cephalization, the cranial nerves have not the regular composition of the spinal nerves, and analysis of their components has been one of the important contributions of American neurologists.

a) *Nervus terminalis:* In 1895 there was described by Pinkus in the lungfish *Protopterus* a cranial nerve in addition to the usual ten, and this nerve has since been identified throughout the gnathostomes (except birds). It issues from the telencephalon near the neuroporic recess, runs along the inner side of the olfactory stalk, and terminates in the walls of the nasal sacs. It has one or more ganglia along its course. Its nature is uncertain; often considered to be a somatic sensory nerve, it is more probably related to the autonomic system.

b) *Olfactory nerve* (*I*): The olfactory nerves differ from all other vertebrate nerves in that their fibers originate from the neurosensory cells of the olfactory epithelium of the nasal sacs. The numerous fibers from these cells constitute the olfactory nerves, which run a very short course into the olfactory bulbs, in mammals passing through the pores of the cribriform plate of the ethmoid bone. Although the nasal sacs and the olfactory epithelium are of ectodermal origin, and hence the olfactory nerves should belong to the somatic sensory division, it appears from the central connections (i.e., brain pathways) of the olfactory impulses that the whole olfactory system is part of the visceral sensory division. Since the telencephalon of lower vertebrates is concerned almost entirely with olfactory (and related gustatory) impulses, it is often called a visceral type of cerebrum. The *organ of Jacobson* or *vomeronasal organ*, a sensory area of the interior of the nose, has a more or less separate nerve, the *vomeronasal nerve*, considered a separated part of the olfactory nerve. This terminates in a somewhat distinct part of the olfactory bulb, called the *accessory olfactory bulb*.

c) *Optic nerve* (*II*): The optic nerves are not really nerves at all but brain tracts. The sensory part of the eye, the retina, is embryologically part of the brain wall and consists of several layers of nerve cells. The fibers of the last layer form the optic nerve, which exits from the inner surface of the eyeball and runs a short course into the optic chiasma of the hypothalamus. Here the optic fibers cross wholly or, in mammals, partly to the opposite side. The optic system belongs to the (special) somatic sensory division; although the fibers enter the ventral surface of the brain, they terminate in dorsal areas of thalamus and midbrain.

d) *Oculomotor nerve* (*III*): The oculomotor nerves are motor nerves to certain muscles which move the eyeball (internal, inferior, and superior rectus and inferior oblique muscles). As embryology shows that these muscles arise from regular myotomes in series with the trunk myotomes, the oculomotor nerve is primarily a somatic motor nerve (it also carries autonomic fibers). The cells of origin lie in the midbrain floor in the somatic motor area, and the roots of the oculomotor nerves emerge on the ventral surface of the midbrain.

e) *Trochlear nerve* (*IV*): This nerve supplies one of the eyeball muscles (superior oblique) and is therefore similar in content and connections to the oculomotor nerve. Although its roots emerge from the metencephalic roof just behind the optic lobes, the nuclei of origin are situated in the midbrain floor in the somatic motor area.

f) Abducens nerve (VI): This is the third of the nerves of the eyeball muscles, innervating the external rectus muscle, and its nature and relations are similar to the others. The nuclei of origin lie in the floor of the medulla, and the roots emerge on the ventral surface of the anterior end of the medulla.

The reason that there are three cranial nerves to supply six small muscles is that these muscles come from three different myotomes (see discussion of head segmentation).

g) Auditory or acoustic nerve (VIII): This is the nerve of the internal ear; and, as this is an ectodermal sense organ, the auditory nerve belongs to the somatic sensory division. As such, it has one or more ganglia located close to the internal ear, and the fibers from these ganglia constitute the nerve which enters the somatic sensory column of the medulla. As the internal ear has two functions—that of equilibration, vested in the semicircular ducts and their ampullae, and that of hearing, vested in the cochlea or its phylogenetic forerunner, the lagena—so the auditory nerve carries two kinds of impulses and in mammals is accordingly divided into a vestibular and a cochlear nerve, each with a separate ganglion. In lower vertebrates the fibers corresponding to the cochlear nerve do not form a separate nerve. The vestibular fibers of the auditory nerve have strong connections with the cerebellum, as this is concerned chiefly with equilibration.

h) The branchial nerves and the lateralis system: The remaining cranial nerves (i.e., fifth, seventh, ninth, and tenth) are closely related to the branchial bars and gill slits; and the last three are also involved in a sensory system, the *lateralis system,* peculiar to aquatic vertebrates. The lateralis system consists of sensory canals and their attendant nerves, ganglia, and central pathways. The sensory canals comprise the lateral-line canal along the side of the trunk, indicated by the lateral line, and canals distributed over the head. These canals contain sensory cells similar to those of the internal ear, and the whole lateralis system is so closely related to the acoustic system that it is customary to treat both together as the *acousticolateralis* system. The lateralis system is a skin-sense system and hence belongs to the somatic sensory division; and each of its nerves bears a sensory ganglion. These ganglia, however, differ from other sensory ganglia in that their cells come from ectodermal sensory patches, termed the *dorsolateral placodes,* which form a series in line with the *auditory placode* (sensory patch which develops into the internal ear). The lateralis ganglia lie close to the ganglia of the seventh, ninth, and tenth cranial nerves; and the lateralis nerves accompany, or appear to form, branches of these three nerves. The roots of the lateralis system enter the anterior part of the somatic sensory column of the medulla; this region, which also receives the auditory nerve, is large and well developed in aquatic vertebrates, being termed the *acousticolateral area.* It is continuous with the auricular lobe of the cerebellum. The whole lateralis system vanishes completely in land vertebrates, and in anuran amphibians it disappears when the aquatic tadpoles metamorphose into the adults.

Each of the four branchial nerves supplies a definite branchial bar and its musculature and continues to supply the same through all its phylogenetic changes. Typically, each branchial nerve forks around a particular gill slit, having a *pretrematic* branch in front of, and a *posttrematic* branch behind, the gill slit; there is, further, a *pharyngeal* branch supplying the pharyngeal lining. Since the branchial musculature is of visceral nature (see p. 199), the branchial nerves lack somatic motor fibers. The pharyngeal branch consists exclusively of visceral sensory fibers; the visceral motor fibers to the branchial musculature are carried in the posttrematic branch, and both pre- and posttrematic branches have visceral sensory fibers. Somatic sensory fibers are well developed only in the fifth nerve. The sensory fibers of the branchial nerves have their cells of origin in ganglia close to or fused with lateralis ganglia;

these branchial-nerve ganglia also differ from ordinary sensory ganglia in that their cells are derived in part from the *epibranchial placodes*, sensory patches at the upper angle of the gill slits. The roots of the branchial nerves are attached to the sides of the medulla and relate to the appropriate functional columns of the latter. The visceral motor components of the branchial nerves are also peculiar in not being mediated by way of the autonomic system.

i) *Trigeminus nerve* (*V*): The trigeminus nerve is the nerve of the first (mandibular) gill arch, i.e., of the upper and lower jaws, and forks around the mouth opening, suggesting a gill-slit origin of the latter. It has a very large somatic sensory component to the eye region (*ophthalmic* branch), upper jaw (*maxillary* or pretrematic branch), and lower jaw (*mandibular* or posttrematic branch), supplying skin, teeth, etc. The last contains the visceral motor fibers to the muscles of the mandibular arch (i.e., first constrictor and levator, adductor mandibulae, masseter, temporal, anterior belly of the digastric, some of the pterygoid muscles, etc. [see pp. 217–19]). The trigeminus is not accompanied by any lateral-line component and lacks somatic motor and visceral sensory components. The somatic sensory part bears a large ganglion, the *Gasserian* or *semilunar* ganglion, attached to the somatic sensory column of the medulla; the visceral motor part enters the medulla by a separate root, which has its nucleus of origin in the visceral motor column.

An apparent part of the trigeminus, termed the *deep ophthalmic* (*ophthalmicus profundus*) nerve, is conspicuous in fishes but in mammals is not distinct from the ophthalmic branch of the trigeminus. This was originally a separate cranial nerve (see below, metamerism of the head).

Because of the very large spread of the somatic sensory part of the trigeminus, so as to supply practically the skin of the whole head, the somatic sensory component of the remaining branchial nerves (leaving the lateralis system out of account) is greatly reduced or wanting.

j) *Facial nerve* (*VII*): The facial nerve is the nerve of the second or hyoid arch and its musculature and forks around the first gill slit (spiracle). It has only a small or no somatic sensory component (apart from the lateralis system) but has a large visceral sensory element from the taste buds and more anterior portions of the buccopharyngeal lining. The facial nerve is accompanied by large trunks of the lateralis system and has visceral motor fibers to the muscles of the hyoid arch (hyoid constrictor and levator, interhyoideus, constrictor colli, posterior belly of the digastric, depressor mandibulae, platysma and other muscles of facial expression, etc.). The sensory fibers have their cells of origin in the *geniculate* ganglion, attached to the side of the medulla. Commonly the ganglia of the fifth, seventh, and eighth cranial nerves are fused into one complicated mass, which also includes in fishes lateralis ganglia.

k) *Glossopharyngeal nerve* (*IX*): This is the nerve of the third branchial bar; it forks around the second gill slit. It, too, consists chiefly of visceral sensory fibers from the more posterior taste buds and pharyngeal lining and of visceral motor fibers to the muscles derived from the third branchial arch. It usually has a small lateralis accompaniment in fishes. The sensory ganglion, called *petrosal*, is attached to the side of the medulla in line with the facial ganglion.

l) *Vagus nerve* (*X*): This is a large nerve with a wide distribution, supplying all the remaining branchial bars and slits and most of the viscera. It is accompanied by a very large lateralis component, which, after supplying a small part of the head canals of the lateralis system, runs, as the large lateral-line nerve, along the side of the trunk under the lateral line. The vagus proper has a small somatic sensory part (with cells of origin in the *jugular* ganglion) from the skin around the ear, but its main or *visceral* trunk consists of visceral sensory and motor fibers. This runs along the upper ends of the remaining gill arches and gives off a ganglionated branch to each of these; from each ganglion there springs the usual pretrematic, posttrematic, and pharyngeal branches. With the loss of the branchial apparatus the branchial part of the vagus is much reduced; but the visceral motor fibers in the posttrematic branches continue to supply the muscular derivatives of the more posterior branchial bars, i.e., the striated muscles of the pharynx and larynx. After passing the branchial region the visceral trunk of the vagus continues posteriorly and supplies visceral sensory fibers (cells of origin in the *nodosal* ganglion) to the esophagus, larynx, trachea, and viscera in general, except the extreme posterior ones. The visceral branches also carry visceral motor fibers (*preganglionics*, see below) to the heart muscle and the smooth musculature of the same viscera, but these relay in ganglia of the autonomic system.

The only explanation for the supplying of several branchial bars by the vagus is that it must have appropriated at least parts of the nerves originally behind and separate from the vagus. Such shifts are common in the nervous system and may be explained by Kappers' theory of neurobiotaxis (see below).

m) Spinal accessory nerve (*XI*)*:* This nerve is a visceral motor nerve whose roots are in line with the ventral or motor roots of the spinal nerves. It is compounded in higher amniotes of a part of the vagus and of contributions from the spino-occipital nerves. The spinal part is a visceral motor nerve to the cucullaris muscle or its derivatives; the vagal part goes to branchial musculature and also accompanies the visceral branches of the vagus, functioning by way of the autonomic system. In lower vertebrates nerve XI is represented by a branch of the vagus. It is not well developed as such in reptiles.

n) Hypoglossal nerve (*XII*)*:* This is a pure somatic motor nerve innervating the muscles of the tongue. It will be recalled that the tongue muscles are hypobranchial muscles; i.e., they come from regular myotomes behind the branchial region which curve around the rear end of this region and grow forward in the midventral area. These myotomes are primitively innervated from the ventral roots of the regular spinal nerves. So long as the tongue remains nonmuscular, as in fishes, there is no distinct development of the hypoglossal nerve. With increasing importance and muscularity of the tongue the motor roots supplying its musculature shift forward to become the hypoglossal nerve.

6. Concept of neurobiotaxis.—Through a study of phylogenetic changes in the position of brain nuclei, especially the motor nuclei of the medulla, the distinguished Dutch neurologist, C. U. Ariens Kappers, reached the conclusion that nuclei tend to migrate in the direction from which they receive their most frequent or most important stimulation. This phenomenon he termed *neurobiotaxis*. First the dendrites grow out toward the stimulatory direction, and then the cell bodies shift, with correlated shortening of the dendrites. On the same principle, nuclei having the same function and hence receiving impulses from the same tracts tend to aggregate. The forward shifts of the anterior spinal nerves into the brain, mentioned above, are the result of attractive stimuli from the medulla.

7. The autonomic nervous system.—As this system has been adequately investigated only in mammals, the description concerns chiefly this class. The autonomic system is a complex of ganglia, nerves, and plexi through which the heart, lungs, digestive tract, other viscera, glands, including the sweat glands, walls of the blood vessels, and smooth musculature generally, including that of the feather and hair follicles, receive their motor innervation. For the autonomic system is, in effect, an elaboration of the visceral motor division of the nervous system. Although the visceral sensory fibers also run in the autonomic system, they have the same general relations as the somatic sensory fibers, having their cell bodies with the latter in the spinal and cranial ganglia, and hence require no special consideration. On the contrary, the arrangement of the visceral motor fibers differs from that of other parts of the nervous system in that their fibers relay in autonomic ganglia. These ganglia contain cell bodies whose fibers, known as *postganglionic* fibers, usually nonmedullated, terminate in smooth musculature, heart muscle, or glands. These cell bodies receive their stimulation by way of medullated *preganglionic* fibers, which originate from cell bodies located in the visceral motor columns of the spinal cord or brain and pass out with the cranial or spinal nerves. It is another peculiarity of the autonomic system that, to most of the parts which it supplies, it gives off a double innervation; and in general one set of fibers is retarding or inhibitory and the other set is stimulatory. The two sets also react differently to certain drugs. The elaboration of the autonomic system in higher vertebrates is readily understood if one considers that a nice regulation of the heart beat, respiratory movements, intestinal peristalsis, and many other important visceral functions is necessary to proper functioning, especially in relation to conditions.

It may be mentioned once more that the branchial musculature and its phylogenetic derivatives throughout the vertebrates, although part of the visceral motor division, are not innervated by way of the autonomic system. There are no pre- and postganglionic fibers and no relays in outside ganglia, in connection with these muscles.

The autonomic system of mammals is divisible into the following parts: the *cranial outflow*, the *cervical sympathetic*, the *thoracolumbar chain of ganglia*, the *sacral outflow*, the *collateral ganglia*, and the *peripheral ganglionated plexi* (Fig. 131).

a) *The cranial outflow:* The preganglionic fibers of the head autonomic system have their cells of origin in the visceral motor column of midbrain and medulla and pass out in the oculomotor, facial, glossopharyngeal, vagus, and spinal accessory nerves. The oculomotor fibers run to the *ciliary* ganglion in the orbit, where they relay; the postganglionic fibers supply the ciliary muscles of the lens, and the iris muscles which regulate the size of the pupil. The autonomic fibers of the facial nerve run to the *sphenopalatine* ganglion, in the pterygoid fossa, from which postganglionic fibers supply the lacrymal or tear glands; and to the *submaxillary* ganglion near the submaxillary gland, from which postganglionic fibers continue to the submaxillary and sublingual salivary glands, causing them to secrete. The facial branch which carries the secretory fibers to the submaxillary ganglion is termed the *chorda tympani*, because it traverses the middle ear along the inner surface of the eardrum, and is quite famous in physiology because of the numerous experiments on salivary secretion performed by means of it. The preganglionic fibers in the glossopharyngeal proceed to the *otic* ganglion, near the foramen ovale; and from this the postganglionic fibers go to the parotid salivary gland. The autonomic part of the vagus is very large and extensive; the visceral branches of the vagus consist chiefly of preganglionic visceral motor fibers, which relay in the peripheral ganglia and plexi in the thoracic and abdominal viscera as far posteriorly as the rear end of the kidneys. The autonomic part of the accessory nerve is of vagal origin and accompanies the course of some of the visceral branches of the vagus.

b) *The cervical sympathetic:* This consists of three ganglia in the neck region: the *superior*, *middle*, and *inferior cervical* ganglia and the connections between them, composed of preganglionic fibers entering the cervical ganglia from the upper thoracic spinal nerves by way of the white communicating rami (see below). The inferior cervical ganglion may be fused with the first thoracolumbar ganglion, and the resulting ganglion is termed the *stellate* ganglion. After relaying in the cervical ganglia the postganglionic fibers supply, in general, the same head parts as the cranial outflow, i.e., the head glands and the intrinsic eye musculature, further the heart and bronchi, constituting the second autonomic innervation of these parts.

c) *Thoracolumbar chain of ganglia:* This is a linear series of ganglia and connecting cords found on the dorsal coelomic wall to either side of the vertebral column. Together with the cervical ganglia, these are generally known as the *sympathetic trunks*. The number of ganglia only approximately corresponds to that of the vertebrae. The ganglia are connected to the spinal nerves by the communicating rami, each of which typically consists of two parts, a

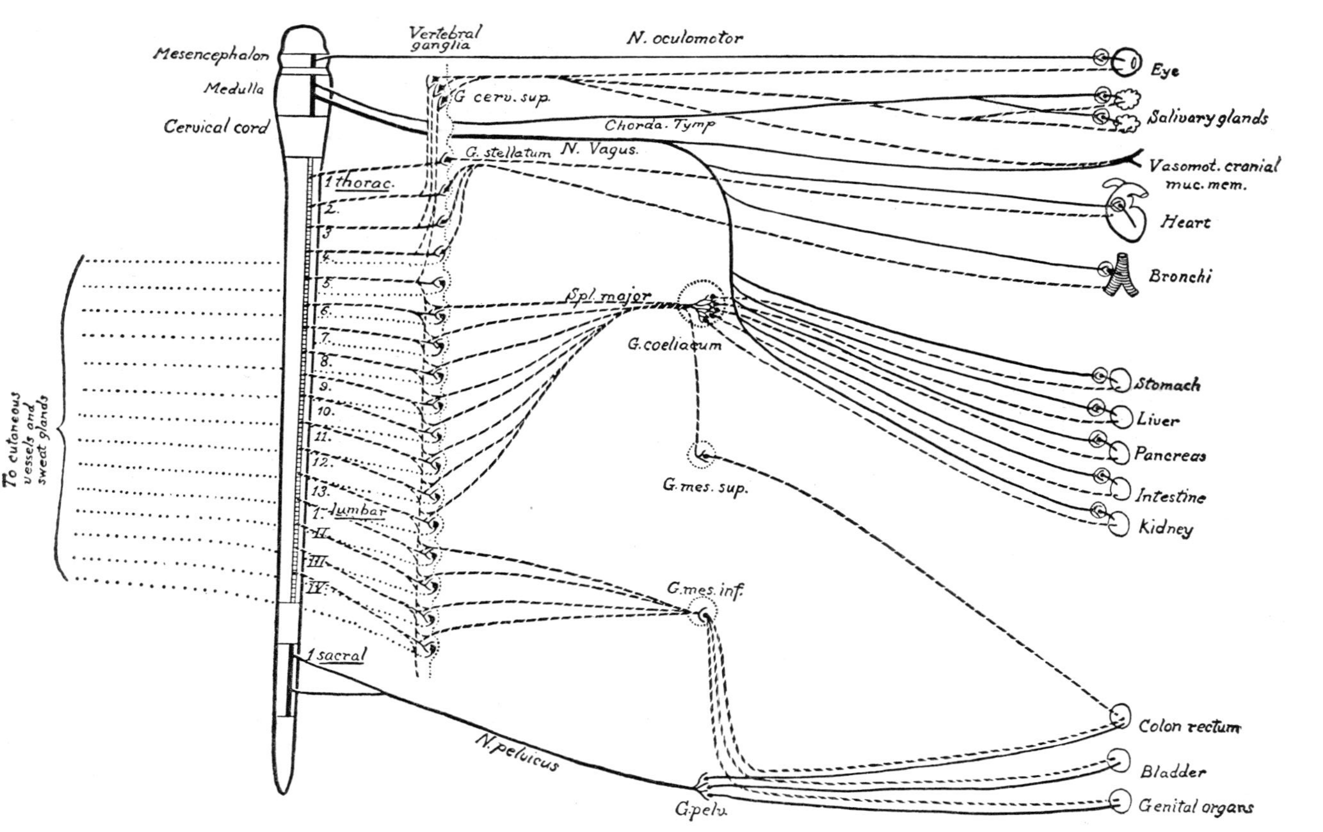

Fig. 131.—Diagram of the mammalian autonomic system, showing double innervation of the viscera. (From Ranson's *Anatomy of the Nervous System* [6th ed.], Fig. 250; courtesy of the W. B. Saunders Co.)

white and a *gray* ramus. The white ramus, made of medullated fibers, conveys the preganglionic fibers from their cells of origin in the lateral gray columns of the spinal cord to the corresponding ganglion of the thoracolumbar chain; the fibers may relay there, but more often they relay in the collateral or peripheral ganglia. The gray ramus carries postganglionic nonmedullated fibers from cells of origin in the ganglia of the thoracolumbar chain to the spinal nerves. These fibers pass out with the spinal nerves to blood vessels and skin and mostly terminate in the smooth musculature of the walls of the blood vessels and of the hair and feather follicles, and in skin glands. As already noted, there are no white rami, only gray rami, connecting the cervical ganglia with the cervical nerves; the same is the case for the sacral part of the sympathetic trunks.

The preganglionic fibers of the white rami usually do not relay in the thoracolumbar ganglia but course in the sympathetic trunks and emerge in the cervical sympathetic or in special *splanchnic* nerves from the thoracolumbar trunks to collateral and peripheral ganglia, where the relay occurs. The postganglionic fibers supply the same viscera as reached by the vagus nerve and constitute the second autonomic supply of these viscera; and the postrenal viscera, such as the colon, rectum, bladder, and genital organs, are reached from the lumbar part of the sympathetic trunks.

The sympathetic trunks continue into the tail in tailed mammals and have communicating rami with the caudal spinal nerves.

d) The sacral outflow: Preganglionic fibers from the sacral spinal nerves run directly to collateral ganglia without passing through the sympathetic trunks; the postganglionic fibers supply the same postrenal viscera mentioned above, giving these the usual second autonomic innervation.

Because of their similar morphological relations and physiological role, it has become customary in recent years to lump the cranial and sacral outflows together as the *craniosacral* outflow, also called the *parasympathetic* system. From what was said above, it is evident that the preganglionic fibers of this system go directly to collateral and peripheral ganglia and plexi without passing through the sympathetic trunks. The fibers of this system are stimulated by choline derivatives. The thoracolumbar outflow (including the cervical sympathetic) is termed the *sympathetic proper* and is stimulated by adrenalin. These two parts of the autonomic system constitute the double innervation of viscera and have antagonistic or, better, balancing actions. Thus, parasympathetic nerves to the heart by way of the vagus and accessory have a retarding effect upon the heart beat, whereas the sympathetic innervation of the heart, from the cervical sympathetics, has an accelerating action; both are necessary to regulate the heart beat in accord with body demands.

e) The collateral ganglia: This includes the four head ganglia already mentioned—the ciliary, sphenopalatine, otic, and submaxillary ganglia—and some abdominal ganglia closely associated with large arteries. The chief of the latter are the *coeliac* and *superior mesenteric* ganglia, located near the origin of the superior mesenteric artery (or coeliac axis) from the aorta, and the *inferior mesenteric* ganglion, alongside the inferior mesenteric artery. The first two are often fused together. In the collateral ganglia the preganglionic fibers from the thoracolumbar trunks (by way of the splanchnic nerves) and from the sacral outflow terminate; and the postganglionic fibers begin in the cells of these ganglia and proceed to their terminations in the viscera.

f) The peripheral ganglia and plexi: This comprises a vast and complicated system of sympathetic ganglia and networks upon or inside the viscera and blood vessels. Such ganglionated networks occur on the arch of the aorta, in the heart wall, on the esophagus, in the lungs along the bronchi, and in the stomach, adrenal glands, spleen, gonads, intestine, etc., along the arteries which supply them. There are also ganglionated plexi in the walls of the intestine, termed the plexi of Meissner and Auerbach. The large mass of plexi and ganglia associated with and including the coeliac and superior mesenteric ganglia is termed the *coeliac* or *solar* plexus. The preganglionic fibers of the parasympathetic system terminate in the peripheral ganglia and plexi, and from the cells of the latter the postganglionic fibers originate and pass into the musculature of the viscera. There is a *pelvic* or *hypogastric* plexus which receives the sacral outflow and from which postganglionic fibers reach the bladder, lower part of the intestine, external genitals, etc.

g) *Autonomic system of nonmammalian vertebrates:* Knowledge here is somewhat scanty. Cyclostomes do not appear to have any definite sympathetic trunks or collateral ganglia. In selachians there are sympathetic trunks connected to the spinal nerves by communicating rami. There is a close relation of these sympathetic ganglia to the suprarenal bodies (p. 400). Definite sympathetic trunks are seen in the Dipnoi and are well developed in the bony fishes, which also have a cranial autonomic system. Sympathetic trunks are present in amphibians, and the cranial part is more or less developed. Beginning with reptiles, the whole autonomic system approaches that of mammals. The sympathetic trunks in tailed tetrapods generally extend into the tail.

8. The sense organs.—The sense organs are specialized cells or organized groups of cells and tissues which respond to environmental changes by initiating nerve impulses; these go to the central nervous system and there evoke appropriate responses. The sense organs are in general specific; i.e., each kind responds only to a particular environmental agent as light, temperature, etc. The sensitive part may be either the nerve endings themselves or neurosensory cells (see p. 428) or sensory cells with which nerve endings make contact.

a) *General cutaneous sense organs:* This comprises the general sensory equipment of the skin, mediating temperature, touch, pain, etc. This includes various types of nerve endings, such as fine branches, networks, bulbs, and disks; a variety of encapsulated sense organs, usually called *tactile corpuscles*, in which the nerve endings are inclosed in layers of connective tissue; and sensory cells, with which nerve endings make contact. For the details of these structures textbooks of histology should be consulted.

b) *Proprioceptive sense organs:* These are found in the muscles, tendons, and joints and consist mostly of spindle-shaped bodies provided with nerve endings. They play an important role in the maintenance of body equilibrium and co-ordinated movements.

c) *Special cutaneous sense organs—neuromast system:* This system, peculiar to aquatic vertebrates, consists of sense organs in a linear arrangement. Primitively and still in cyclostomes, amphibians, and some other forms, the lines of sense organs are exposed on the body surface. More often they are sunk into canals, which may remain open as a groove (Holocephali, primitive sharks) but generally are closed over, with pores at intervals, as in most fishes. In bony fishes the canals run through the scales. These canals are the canals of the lateralis system and consist of the main lateral-line canal and four main branches on the head: a *hyomandibular* canal along the lower jaw, a *supraorbital* canal above the orbit, an *infraorbital* canal along the head below the orbit, and a *supratemporal* canal transversely across the rear part of the head. The head canals are supplied by branches accompanying the facial nerve, except that the supratemporal canal is innervated from the vagus and a short piece near the junction of supratemporal and main lateral-line canals is supplied by the glossopharyngeal. The lateralis system is found in cyclostomes and most fishes, in aquatic larvae of amphibians, and throughout life in those amphibians which lead a permanent aquatic existence. The function of the system is the detection of water vibrations.

The sense organs of the lateralis system consist of *neuromasts*, which are clusters of sensory and supporting cells (Fig. 132*A*). The sensory cells have a terminal hair projecting into the lumen of the canal and are innervated by terminal twigs of the nerves of the lateralis system. The sensory cells closely resemble those of the internal ear; and there is, as already noted, a close relation between the acoustic and lateralis systems. The neuromast lines develop from a series of placodes continuous with the auditory placode and from there grow out over the head and along the trunk.

In addition to the neuromast system there are several other related sense organs in fishes. Prominent among these are the *pit organs* and the *ampullae of Lorenzini.* The former consist of isolated neuromasts sunk in pits on the head, scattered or arranged in rows. The ampullae of Lorenzini, limited to selachians and holocephalians, are sensory vesicles at the inner end of long slime-filled tubes lying under the skin and opening on the surface by a pore. The snout

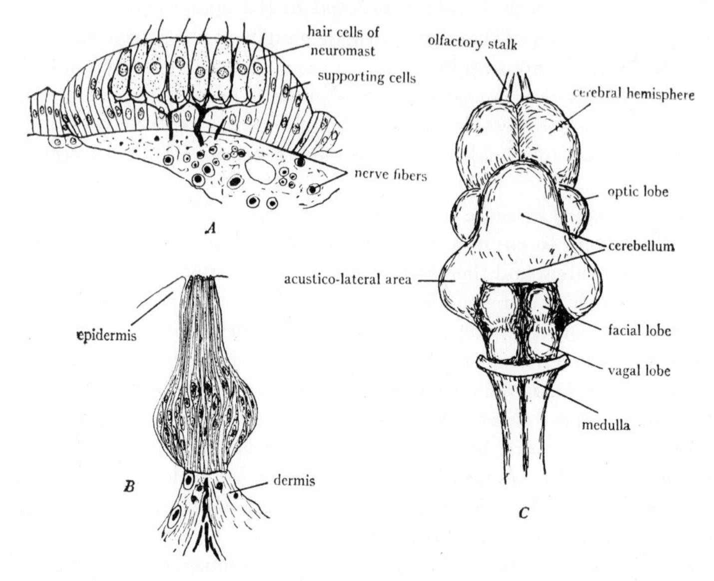

FIG. 132.—Sensory organs and centers. *A*, section through a canal of the lateralis system of the dogfish, showing a neuromast, composed of sensory hair cells supplied by nerve twigs, and of supporting cells (after Johnson, 1917). *B*, an external taste bud of a siluroid fish (after Herrick, 1901). *C*, dorsal view of the brain of a catfish, with roof of fourth ventricle removed to show the facial and vagal lobes, taste centers (after Herrick, 1905).

and jaws of selachians are full of these tubular organs, whose function is uncertain; recently it has been suggested that they are thermoreceptors (Sand, 1938). Pit organs and ampullae are innervated from the lateralis system and have similar central connections.

d) Special cutaneous sense organs—ear: The sensory part of the ear is vested in the internal ear, also called the *membranous labyrinth;* middle and external ears are accessory mechanisms for conveying sound waves to the internal ear. The latter develops from an area of thickened ectoderm, the *auditory placode,* which sinks inward, forming a sac, the *otic vesicle* or *otocyst.* The connection with the surface forms a canal, the *endolymphatic duct,* which in selachians remains open to the exterior but is closed in other vertebrates. Its blind end often expands into a sac, the *endolymphatic sac,* which in amphibians and some reptiles is very much expanded and branched. The otic vesicle becomes divided into a dorsal chamber, the *utriculus,* and a ventral one, the *sacculus.* The endolymphatic duct enters at about their junction or is attached to the sacculus. From the wall of the utriculus there become constricted off three canals, the *semicircular ducts,* two vertical and one horizontal, opening into the utriculus at both ends. One end of each duct is expanded into a chamber, the *ampulla*, containing a sensory patch or *crista ampullae*. There is a similar sensory patch or *macula* in the utriculus and in the sacculus. Already in fishes the lower part of the sacculus puts out an evagination, the *lagena*, also containing a sensory macula. The lagena becomes of increasing importance in the amniote series, where it lengthens and coils spirally, finally producing in mammals the *cochlear duct* (old name, *cochlea*). The cochlear duct is the seat of hearing in mammals; its wonderful histological construction should be learned from textbooks of histology. In lower vertebrates the lagena is supposed to be the seat of hearing. The other parts of the internal ear are concerned with spatial equilibration and therefore have strong connections with the cerebellum by way of the vestibular fibers in the auditory nerve.

The internal ear is filled with a fluid, the *endolymph*, whose movements are believed to furnish the stimulus for equilibratory adjustments. Around the internal ear the skull bone more or less follows the contours of the ear, forming the *bony labyrinth*. Between the bony and membranous labyrinths is a space, or set of spaces, filled with a fluid, the *perilymph*.

The sensory patches of the internal ear, i.e., the cristae and maculae and some other patches, are similar to neuromasts, consisting of supporting cells and hair cells. The hair cells are innervated by the fine twigs of the auditory nerve.

In some bony fishes the two internal ears are connected by a transverse canal, and this, remarkably enough, may communicate by a duct with the swim bladder; or a chain of little bones, the *Weberian ossicles*, known to be derived from the adjacent vertebrae, may extend on each side from the transverse canal to the anterior wall of the swim bladder. Apparently, the Weberian ossicles convey changes in the swim bladder to the internal ear, much as the middle-ear ossicles transmit sound waves to the eardrum.

In fishes the internal ear contains stony bodies, the *otoliths*, placed near the maculae. In selachians, where the endolymphatic ducts open to the exterior, these are of extraneous origin and consist of loose masses of rock fragments and sand grains. But in bony fishes they are definite, secreted bodies which show rings of growth and hence are utilized in determining the age of fish. Movements of the otoliths with changes of position are believed to act as stimuli to the hair cells. In land vertebrates the maculae are covered with a gelatinous material containing crystals.

e) Special cutaneous sense organs—eye: The vertebrate eye is the most remarkable of all sense organs. Its essential sensory mechanism, the *retina*, is a part of the brain wall. In the embryo the lower region of the diencephalon bulges laterally as a rounded eminence, the *optic vesicle*. The distal part of this invaginates, forming a two-walled cup, the *optic cup*. The wall which lines the cavity of the cup thickens and becomes the retina; and the other wall, facing the brain, thins and develops black pigment, becoming the *pigment layer* of the retina. Meantime the surface ectoderm over the opening of the cup thickens, invaginates, and cuts off a vesicle, the *lens vesicle*, which differentiates into a rounded or biconvex transparent body, the *lens* of the eye, having the same function as the lens of a camera. The distal rim of the optic cup does not become nervous but forms the greater part of the *ciliary body* and of the iris, a circular, colored membrane, having a central hole, the *pupil*, for the passage of light into the eye. The iris contains radial and circular smooth muscles for regulating the size of the pupils; and these muscles, unlike other smooth musculature, are of ectodermal origin. They and the ciliary muscle (derived from surrounding mesenchyme) are, as already learned, innervated by autonomic postganglionic fibers from the ciliary ganglion. The ciliary body is concerned with the curvature of the lens. In addition to the ectodermal parts of the eye, there develop from the surrounding mesenchyme two coats: an inner, vascular layer, the *choroid* coat, which fuses to the pigment layer; and an outer, tough protective layer, the *sclerotic* coat. The whole structure is termed the *eyeball*. Over the front of the eye the sclerotic coat alters to a transparent membrane, the *cornea*, to which is adherent the original surface ectoderm, here termed the *conjunctiva*. The conjunctiva turns back at the attachment of the eyelids, forming their lining.

The *retina* or nervous part of the eye consists of a layer of sensory cells and several layers of ganglion cells. The sensory cells are of the neurosensory type (Fig. 124*B*, *C*) and are of two sorts, the more elongated *rods* and the shorter *cones*. It is a peculiarity of the eye that this sensory layer faces the back of the eye, i.e., the choroid coat, rather than the direction from which light enters the eye; and the basal layer of the retina, from which the optic nerve springs, lies next the cavity of the eye. The fibers of the optic nerve traverse this surface of the retina, converging to one spot, termed the *blind spot;* here they traverse the whole thickness of the retina and the two coats of the eyeball to emerge on the medial surface of the eyeball as a bundle of fibers, the so-called optic nerve.

The cavity of the eye between lens and retina is occupied by a gelatinous material termed the *vitreous humor* or *vitreous body*. The cavity between iris and cornea contains a thin fluid, the *aqueous humor*.

As already intimated, the eyeball is operated by a set of muscles so arranged as to move it in all directions; details of these muscles will be seen in the dissections.

f) General visceral sense organs: The sensory terminations in the viscera and mesenteries are similar to those of the skin, consisting of free nerve endings, encapsulated endings, etc. From these the visceral sensory nerve fibers proceed through the trunks of the autonomic system to the cells of origin in the spinal and cranial ganglia.

g) Special visceral sense organs—olfactory organ: The olfactory organ begins as an ectodermal placode, median and faintly paired in cyclostomes, lateral and paired in gnathostomes; this invaginates, forming the olfactory sac. The lining epithelium of the olfactory sac becomes differentiated, at least in part, into olfactory epithelium, which contains the olfactory neurosensory cells (p. 428). In fishes the olfactory surface is generally increased by thin folds which nearly fill the sac, and there are two external openings for keeping up a water current through the sac. The history of the respiratory part of the nasal sacs was already given (p. 45). In gnathostomes the interior of the nasal cavities presents various folds and swellings, part of which bear olfactory epithelium while the rest are clothed with ordinary or mucus-secreting epithelium. In mammals the olfactory epithelium is borne chiefly on the folds of the ethmoid turbinates, and these are the more complicated and infolded the keener the olfactory sense of the animal in question.

In tetrapods a portion of the ventromedial wall of each nasal sac evaginates to form a rounded or elongated blind sac, known as the *vomeronasal* or *Jacobson's organ*, which generally opens into the nasal cavity. It is lacking in adult turtles, crocodilians, birds, bats, primates, and aquatic mammals, although usually appearing in the embryo. It reaches a high development in snakes and lizards, where it is cut off from the nasal cavity and instead opens into the oral cavity. Its function was for years a mystery, but it has recently been shown for lizards and snakes that Jacobson's organ assists the tongue in the recognition of food (Noble and Kumpf, 1936; Wilde, 1938).

h) Special visceral sense organs—taste: The organs of taste are the *taste buds*, little bud-like clusters (Fig. 132*B*) of supporting and sensory cells, of entodermal origin, situated in the epithelium; in mammals a canal runs from the bud to the surface. The sensory or gustatory cells are elongated cells with a terminal projection and are innervated by the terminal twigs of the gustatory nerves. The taste buds reach their widest distribution and greatest numbers in fishes, where (ganoids and teleosts) they may spread over the external surface of the head and even over the body. Scavenger fishes, such as the siluroids (catfish, bullheads, etc.), which depend largely on external taste buds for finding food in their environment, may have as many as one hundred thousand taste buds strewn over the whole body and also concentrated on special food-finding barbels. In such fishes the gustatory centers in the visceral sensory area of the medulla are greatly enlarged and may cause immense bulges, termed the *facial* and *vagal lobes* (Fig. 132*C*). In selachians and tetrapods in general the taste buds are confined to the oral cavity and pharynx. Birds have few taste buds and a corresponding poor sense of taste. The taste buds are anatomically best developed in mammals, where they are located chiefly on the papillae of the tongue but also occur on the palate, pharynx, and epiglottis. Although taste buds may be mounted on tongue papillae throughout the vertebrates, this relation is more definite in mammals. The taste buds of mammals occur principally on the foliate, fungiform, and vallate papillae (see p. 289); they vary in numbers from several hundred to ten or more thousand.

The gustatory fibers run in the seventh, ninth, and tenth cranial nerves. External taste buds, when present, and those of the anterior part of the oral cavity are supplied by the facial nerve; taste buds of the rear part of the oral cavity and of the pharynx are supplied by the

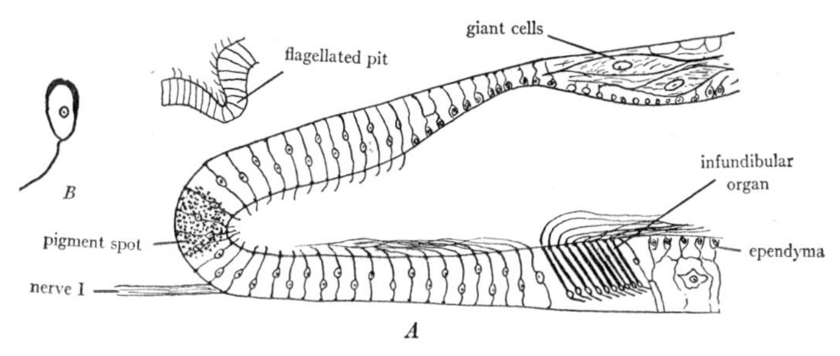

FIG. 133.—*A*, sagittal section through the brain vesicle of *Amphioxus; B*, a photoneurosensory cell from the spinal cord of *Amphioxus*. (Both after Franz, 1927.)

glossopharyngeal and vagus nerves. The cells of origin of the gustatory fibers are in the ganglia of these three cranial nerves, and the fibers enter the side of the medulla by the dorsal roots of these nerves. In mammals the facial nerve innervates the fungiform papillae (which in man lack taste buds), the glossopharyngeal innervates the vallate and foliate papillae, and the vagus supplies taste buds on the epiglottis and adjacent regions.

9. The nervous system and sense organs of *Amphioxus*.—The brain of *Amphioxus* is delimited from the rest of the neural tube by its enlarged cavity. The wall of this brain vesicle is a one-layered flagellated epithelium (Fig. 133*A*). At the anterior end there is a mass of pigment granules in the brain wall, but this has been shown to be unresponsive to light. Near this, on the left side only, is the flagellated pit (p. 24), which would seem to be a sense organ, although nervous connection with the brain vesicle is lacking. From the ventral side of the anterior end of the brain vesicle springs the first pair of nerves, considered by some to be homologous to the nervus terminalis of vertebrates. It is sensory to the tip of the head, especially to the fin there. In the floor of the rear part of the brain vesicle is an area of special neurosensory cells with very long flagella; this is termed the *infundibular organ* and from the similarity of the cells is considered by many to correspond to the region of the vascular sac of the fish infundibulum. Just behind this level springs dorsally the second brain nerve, also sensory to the tip of the head and fin, and considered by some to represent the deep ophthalmic nerve. This would mean that this region of the brain vesicle is midbrain and that the part behind with the more anterior spinal nerves would be medulla. However, the brain of *Amphioxus* is certainly degenerate, and comparisons with the vertebrate brain lack adequate foundation. The fact that the notochord extends beyond the tip of the *Amphioxus* brain would indicate that the whole of the latter is hindbrain. The lack of olfactory, optic, and oculomotor nerves and their central connections makes identification very difficult.

The spinal cord of *Amphioxus* has some resemblance to that of vertebrates, having a central region of nerve-cell bodies, many of gigantic size (Fig. 126*A*), and a peripheral region of fibers, which, however, are nonmedullated. The central canal is continued dorsally by a narrow slit. The spinal nerves show very primitive characteristics: the dorsal and ventral roots alternate, they do not unite, and the dorsal roots carry nearly all of the visceral components. The dorsal roots spring from the cord opposite the myosepta and run in these, branching to the skin as somatic sensory fibers and also giving off a visceral branch which connects with gangliated autonomic plexi in the wall of the digestive tract. The dorsal roots lack ganglia, and the sensory cells of origin appear to be inside the spinal cord. The ventral roots spring from the cord opposite the myotomes and enter these at once; they also carry autonomic fibers (visceral motor) for the blood vessel walls. The roots of the two sides of the *Amphioxus* body also alternate, but this is the result of the general right-left asymmetry of this animal and is not regarded as of phylogenetic importance.

The first five spinal nerves (i.e., the third to the seventh, inclusive, in the series of nerves) innervate the oral tentacles, oral hood, and velum and have therefore been compared to the trigeminus nerve of vertebrates. The nerves behind these, supplying the branchial region, would then correspond to the branchial nerves of vertebrates. However, the branchial region of *Amphioxus* is very long and includes about half of the sixty or so pairs of spinal nerves. If this has any significance at all, it would mean that the vertebrate brain has been greatly foreshortened and its caudal end moved far forward. However, all such ideas are highly speculative.

The "eyes" of *Amphioxus*, consisting of numerous single photo-neurosensory cells each with a pigment cup (p. 24, Fig. 133*B*), occur in the ventral wall of the spinal cord. They resemble the retinal cells of lower invertebrates.

10. The nervous system and main sense organs of cyclostomes.—The brain of cyclostomes is much compressed in the anteroposterior direction because of the great expansion of the buccal funnel and is correspondingly deep through the thalamic region, but it has the same general parts as other vertebrate brains, and its tracts and centers are retained in the vertebrate series (Fig. 134*A*). The relatively small telencephalon is divided into olfactory bulbs and cerebral hemispheres (also called olfactory lobes) of about equal size and is practically entirely devoted to the sense of smell. The diencephalon contains a large third ventricle and is divisible into the usual three regions; but the thalamus proper, especially its dorsal portion, is not well developed. In the epithalamus there are large habenulae; and the petromyzonts (but not the myxinoids) have both parts of the epiphyseal apparatus, a parapineal and a pineal body, each with a functional eye. There are large choroidal sacs near the epiphyseal apparatus. The hypothalamus is large with a typical pituitary body, infundibulum (but no vascular sac), and mammillary region. The optic lobes of the midbrain are rather small; and the cerebellum, practically absent in myxinoids, is in petromyzonts a mere shelf over the anterior end of the fourth ventricle. There are very large tela choroidea in the roof of the cyclostome brain.

The olfactory organ consists of one very large median unpaired olfactory sac which, however, was apparently originally double, since there are two olfactory nerves. The peculiar relations of the olfactory sac to the hypophysis were explained previously (p. 30). The eyes begin their development in the *Ammocoetes* larva in typical vertebrate fashion but fail to differentiate very far and continue their development only after metamorphosis. Adult petromyzonts have fairly well-developed eyes, which, however, appear to be primitive in some respects, especially in lacking an accommodatory mechanism for the lens. In myxinoids the degenerate eyes are buried under the skin. The internal ears of cyclostomes are also peculiar, having but two semicircular ducts (petromyzonts) or only one (myxinoids), but this has two ampullae, one at each end, and hence would appear to be derived from the two-duct condition. There is also no differentiation of the vestibule into sacculus and utriculus in cyclostomes. A lateralis system is present, better developed in petromyzonts than in myxinoids; it consists of groups of neuromasts in pits arranged linearly. There are external, as well as internal, taste buds.

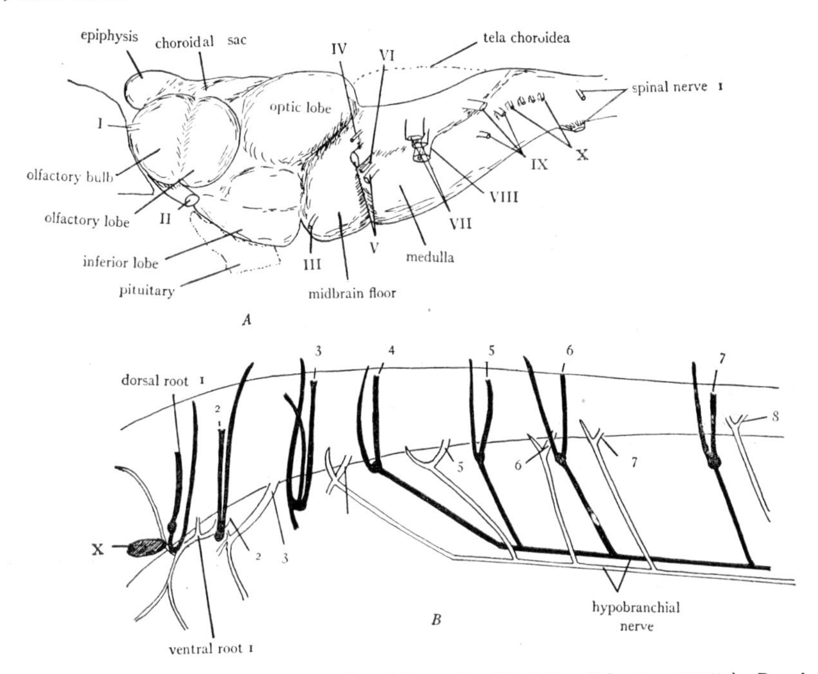

FIG. 134.—*A*, the brain of *Petromyzon*, viewed from the side (after Johnston, 1902*a*). *B*, spinal cord of *Petromyzon*, showing complete set of sensory (black) and motor roots of the spinal nerves, their alternate arrangement, failure of union of sensory and motor roots, and formation of the hypobranchial nerve (after Johnston, 1905).

The cranial nerves are typically vertebrate in number and arrangement. There is no nervus terminalis. Myxinoids have no optic nerves, and the eye-muscle nerves are also degenerated. The profundus nerve is distinct; and the trigeminus proper is large and well developed, with two main branches, the ophthalmic and the maxillo-mandibular. The lateral-line and auditory nerves are well developed, and they and their central connections show the same close relationship as in typical vertebrates. The glossopharyngeal and vagus nerves have somatic sensory components (although the glossopharyngeal tends to be greatly reduced in myxinoids). The facial, glossopharyngeal, and vagus have the same relations to the gill slits as in gnathostomes; the vagus supplies all but the first two.

The spinal nerves of petromyzonts have primitive characters similar to those of *Amphioxus:* the dorsal and ventral roots alternate on the cord, and they do not unite (Fig. 134*B*). There are, however, sensory ganglia on the dorsal roots. In myxinoids the greater part of the two roots unite, although some branches remain independent, so that this group is intermediate between petromyzonts and gnathostomes. In the anterior part of the spinal cord there is a complete set of dorsal and ventral roots (Fig. 134*B*), whereas in gnathostomes the anterior sensory roots are generally missing. The dorsal root of a given segment lies behind the ventral root, and this appears to be the primitive arrangement (Fig. 134*B*). In petromyzonts a number of the more anterior sensory and motor nerves contribute a branch to a lengthwise nerve; and the two nerves so formed, one sensory and one motor, run alongside each other posteriorly and then turn forward to innervate musculature corresponding to the hypobranchial musculature. Hence the formation of a hypobranchial (or hypoglossal) nerve is already seen in cyclostomes.

11. The segmentation of the head.—The question whether the vertebrate head was originally segmented like the trunk has engaged the attention of leading anatomists for one hundred and fifty years. From a long series of embryological researches the conclusion has been reached that the head is composed of a series of mesodermal segments which, through the process of cephalization, have become almost indistinguishably fused together. As in the case of the trunk, the segmentation is primarily myotomic; and that of the skeleton, gill slits, etc., has been imposed upon these parts by the segmental arrangement of the epimeres. General agreement has been reached that the otic capsule represents an important landmark and that there are three epimeres or somites in front of the capsule and a variable number behind it (Fig. 135). The three prootic somites are termed the premandibular, mandibular, and hyoid somites, respectively. Each is hollow, containing a coelomic space; and these, except the first (termed the anterior head cavity), are continuous with the general coelomic space of the hypomere. The epimeres, as already learned, give rise only to somatic muscle, and in the head region the three prootic somites produce the eye muscles. The premandibular somite gives rise to the four eye muscles innervated by the oculomotor nerve, the mandibular somite to the superior oblique, and the hyoid somite to the external rectus muscle. There are variants to this account of the origin of the eye muscles, and the statement made is to be regarded as only generally true.

Below and behind the ear the number of head somites (metaotic somites) varies from two or three to ten or twelve or more, according to various estimates. In cyclostomes there is a complete set of metaotic epimeres, and all give rise to somatic muscle; but, in general, one or two metaotic epimeres fail to form muscle, presumably because they are more or less crowded out of existence by the enlarging otic capsule. The remaining ones contribute to the hypobranchial musculature.

The hypomeric portions of the head segments give rise to the branchial musculature. In embryonic stages a part of the hypomeric coelom runs in each visceral arch (branchial bar).

Of great interest has been the working-out of the innervation of the head segments. Each of the original head segments was innervated by a dorsal sensory and a ventral motor nerve, which have remained separate. The oculomotor nerve is the motor nerve of the premandibular segment; the profundus nerve is the sensory nerve. The nerves of the mandibular segment are the trochlear (motor) and the trigeminus (sensory). The abducens nerve is the motor nerve; the facial, the sensory nerve of the hyoid segment and also the visceral motor nerve of the branchial musculature of this segment. Of the metaotic segments the first one or two do not form any somatic musculature, hence their somatic motor roots are absent; but the somatic sensory and visceral motor components are represented by the glossopharyngeal nerve for the fourth head segment (or first metaotic) and the vagus for the fifth head segment. From the sixth onward the myotomes contribute to the hypobranchial musculature and are supplied by the motor roots of these segments, while the other components appear to have been appropriated by the vagus nerve.

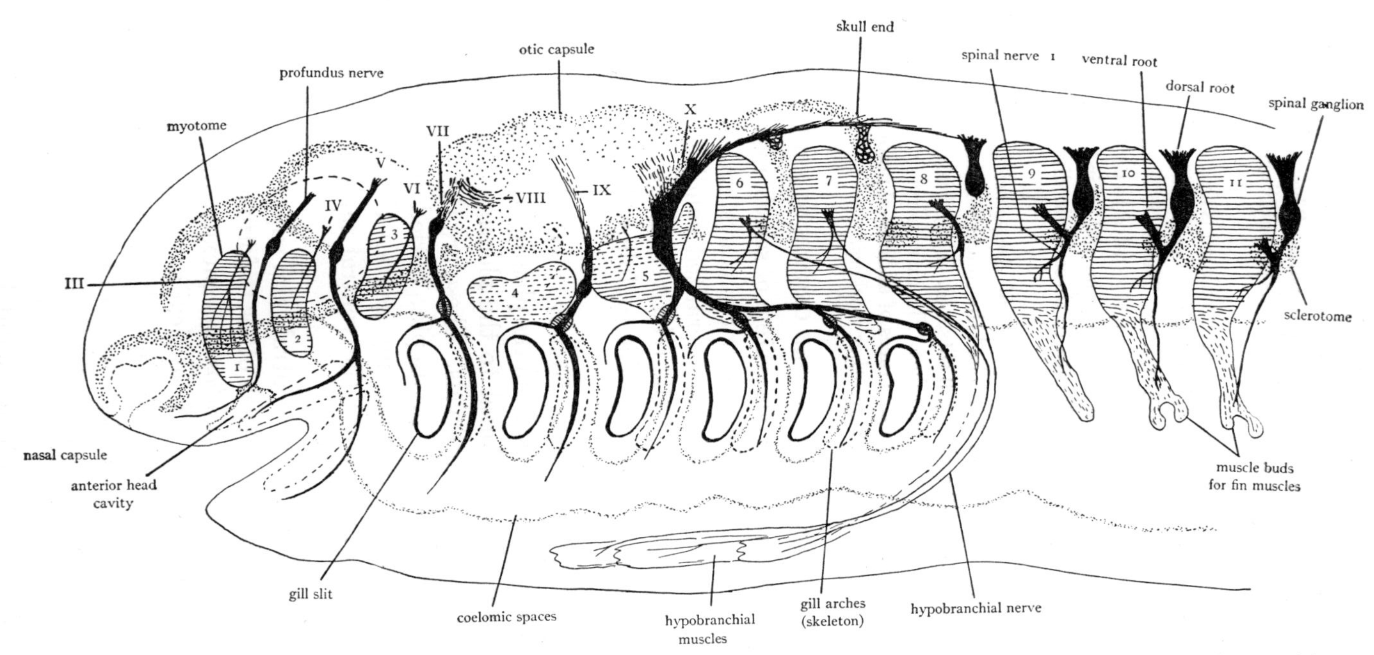

FIG. 135.—Diagram based on selachians to illustrate the segmentation of head and the segmental arrangement of the cranial nerves (after Goodrich, 1918). The somites are crosshatched and bear Arabic numerals (*1*, *2*, etc.). The Roman numerals (*III*, *IV*, etc.) designate the cranial nerves. The heavily stippled part is the outline of the chondrocranium and nasal and otic capsules, also the sclerotomes. The lightly stippled outline indicates the coelomic spaces; these run up through each branchial bar. The jaw cartilages and gill arches are in dashed outline. The motor roots of segments *4* and *5* and the sensory roots of *6* and *7* are degenerate.

The relation of the gill slits to the head segmentation has been frequently debated, and it is now generally believed that they are not primarily segmental but that a segmental arrangement has been imposed upon them by the surrounding parts. They are located at the segmental boundaries, i.e., at the myosepta. The number was probably originally greater than in present gnathostomes.

Another question on which much literature has accumulated is that of the segmentation of the brain. During the development of the brain in various vertebrates there are frequently seen regularly repeated expansions, termed neuromeres. These are regarded by a number of authors as primary nervous segments; and there are usually stated to be two neuromeres in the telencephalon, two in the diencephalon, one in the midbrain, and six in the hindbrain. Others, however, have vehemently denied any significance to these brain divisions, and this latter view may be said to prevail at present.

As a result of an extensive study of the development of *Ammocoetes*, the eminent Russian comparative anatomist Sewertzov (1916–17) reached the conclusion that there were five prootic segments with accompanying gill arches, gill slits, etc. The labial cartilages (p. 203) are regarded by him as remnants of the extra gill arches. The extra nerves were mostly originally separate parts of what is now the trigeminus. Traces of additional prootic segments besides the accepted three have also been reported by other workers.

B. THE NERVOUS SYSTEM AND SENSE ORGANS OF SELACHIANS

A thorough knowledge of the nervous system of selachians is indispensable for the understanding of the vertebrate nervous system. Not only is the nervous system of these animals in a generalized condition, but the cartilaginous nature of the skeleton and the relatively large size of the cranial nerves makes the nerve distribution easier to follow than in other vertebrates.

1. The spinal nerves and fin plexi.—Remove all of the viscera, including the kidneys and reproductive organs, from the pleuroperitoneal cavity. Note against the dorsal coelomic wall dorsal to the pleuroperitoneum the white nerves passing out at segmental intervals. These are the *ventral rami* of the spinal nerves. They lie along the myosepta, buried in the muscle, and will be revealed by cutting along the myosepta. Farther laterally they emerge to the internal surface. Trace the ventral rami into the hypaxial muscles.

In the regions of the paired fins the ventral rami supply the muscles of the fins and are more or less united with each other to form a plexus. The plexus for the posterior appendage is the *lumbosacral* plexus; for the anterior appendage, the *cervicobrachial* plexus. These plexi are as follows:

Dogfish: The lumbosacral plexus to the pelvic fin is found by cutting through the skin on the dorsal side of the base of the fin. On carefully separating the fin muscles from those of the trunk the nerves of the plexus are seen as white cords passing into the base of the fin. They are more or less imbedded in connective tissue, which should be carefully cleaned away. There are ten nerves passing into the fin, of which, however, only the last ones are united by cross-branches to form a plexus. The first of the ten is called the *collector* nerve. Trace it forward and note that it is formed by the union of branches from the ventral rami anterior to the fin.

The cervicobrachial plexus to the pectoral fin is located by cutting through the skin at the base of the fin on the ventral side. On separating the skin from the muscles of the trunk, nerves will be seen passing in the connective tissue to the pectoral fin. Proceed carefully forward, carrying your cut into the coelom at the side of the esophagus. The plexus is then seen to consist of a number of nerves (eleven in the spiny species) passing from the spinal cord into the fin. Only the first four or five of these, situated on the dorsal side of the bag formed by the posterior cardinal sinus, are united by cross-branches to form a true plexus, the posterior ones passing directly into the fin.

Skate: A large number of ventral rami supply the pectoral fin, the anterior ones uniting to a plexus. Strip off the pleuroperitoneum at the level of the subclavian artery and note there the enormous *nerve trunk* of the *brachial plexus.* It is formed by the union (within the neural canal) of a large number of ventral rami. This will be seen later. Follow out the nerve trunk to the pectoral fin. It lies along the posterior side of the curved cartilage (propterygium) which is situated in the pectoral fin about halfway from the middorsal line to the margin. Cut through skin and muscles on the dorsal side of the animal along the posterior and lateral side of this cartilage and expose the trunk. It supplies only the anterior part of the pectoral fin. The posterior part, as already noted, is supplied by direct ventral rami, not forming a plexus. The lumbosacral plexus for the pelvic fin is located as follows: Remove the skin from the base of the fin on the dorsal side. This exposes a fan-shaped layer of muscles. Cut through this, and just ventral to it will be found a number of nerves which diverge into the fin muscles.

2. The sense organs.—For the rest of the dissection a large separate head will generally be provided. In that case the specimens used up to this point may be discarded. A very careful dissection of this head, on which the student will be graded, is required.

a) The ampullae of Lorenzini: It has already been noted that the skin of the head is perforated by pores, from which mucus exudes under pressure. Note the distribution of the pores. Remove a piece of skin from a region bearing pores (in the skate from the ventral side of the head) and note that each pore leads into a canal of varying length lying beneath the skin. Each canal, named the *canal of Lorenzini,* terminates in a little bulb, the *ampulla of Lorenzini,* which is supplied by a nerve, a delicate white fiber easily seen attached to the ampulla. The function of this system is uncertain, but from the fact that it is innervated by the lateralis nerves it is usually supposed to have the same function as the lateralis system (see also p. 448).

b) The pit organs: Pit organs (p. 448) occur throughout the selachians but are not easy to see or to distinguish from the pores of the canals of Lorenzini. In the spiny dogfishes they occur in a row on the ventral side at the level of the bases of the pectoral fins, and there is also a row shortly in front of the level of the first gill slits; they also occur irregularly above the anterior part of the lateral line.

c) The lateralis system: In fishes and aquatic phases of amphibians there is present a system of sense organs, the *lateralis system*, related to the aquatic mode of life. This system, together with its nerves, is completely lost in land vertebrates. It consists of the *lateral-line canals;* the sense organs of these canals, termed *neuromasts;* and the nerves supplying these (see further, p. 447). The following dissection will trace the course of the canals of the lateralis system; the neuromasts in the canals are microscopic.

Dogfish: Along the trunk the system consists of the lateral line, which marks the position of a canal. Find the lateral line on the head. Remove the skin at this place, noting the underlying canal and the pores connecting the canal with the surface. Trace the lateral line forward, removing the skin as you proceed. At the level of the spiracles the canals of the two lateral lines are connected by the *supratemporal* canal. Anterior to this, each forks into a *supraorbital* canal, passing forward above the eye, and an *infraorbital* canal, passing ventrally between the eye and the spiracle and then forward below the eye. Trace the supraorbital canal to the end of the rostrum; here it turns and proceeds posteriorly again, parallel to its former course, and becomes continuous with the infraorbital canal. The latter gives off a *hyomandibular* branch, running posteriorly along the sides of the jaws, and turns to the ventral surface of the rostrum, passing first posterior to the nostril and then turning forward between the two nostrils. There is also a short *mandibular* canal under the skin just behind the lower jaw; it is not connected with the other canals.

Skate: The lateral-line system is more complex than in the dogfish and more difficult to follow. The lateral-line canal runs on the dorsal surface just lateral to the middorsal spines. Remove the skin at this place and identify the canal. Trace it forward, removing the skin as you proceed. At the posterior end of the cartilage (propterygium) of the anterior part of the pectoral fin it gives off two canals, which proceed posteriorly over the surface of the fin. It then proceeds above the eye as the *supraorbital* canal, apparently connecting with its fellow by a cross-union on the posterior part of the skull. The supraorbital canal passes in front of the eye and, as the *infraorbital* canal, below the eye. In the region of the eye it gives off branches over the rostrum and a long branch which proceeds posteriorly along the lateral margin of the fin. On the ventral side of the skate there is a prominent canal passing just lateral to the gill slits. Trace this forward, noting branches behind and in front of the nostril and on the ventral surface of the rostrum. On the surface of the pectoral fins, after removal of the skin, the numerous, very long canals of Lorenzini are noticeable.

Draw, showing distribution of the canals.

d) *The olfactory organs:* These consist of a pair of olfactory sacs on the ventral side of the rostrum, opening externally by the nostril or external naris, with which are associated various flaps of skin, arranged so as to permit water circulation through the sac. Dissect the skin away from one olfactory sac and cut away the flaps so that you can look into the sac. Note the numerous plates or lamellae arranged in rows inside of the sac; these are covered with olfactory epithelium, the sense of smell being quite keen in fishes. Prove to yourself that the olfactory sac is closed internally, having no communication with the oral cavity.

Draw, showing the lamellae.

e) *The eye muscles:* Remove the tissue from about the eye on the same side of the animal as under (*c*) and completely expose the eyeball. In doing this, first cut away the upper eyelid (or in the skate the skin over the eye), noting that the inner lining of the eyelid is continuous with a thin layer (*conjunctiva*) which adheres closely to the external surface of the eyeball. Next cut away very carefully the cartilage between the eye and the brain (which is seen as a white, lobed structure in the median region) and also the cartilage in front of the eye. Do not injure the brain and do not cut into the elevation dorsal to the spiracle. The stout, white bands seen in this dissection are cranial nerves. In the skate very little cutting is required. The large, somewhat spherical body exposed is the *eyeball.* In the dogfishes it is imbedded in a gelatinous material, which should be carefully cleaned out.

The eyeball reposes in a cavity, the *orbit,* to the walls of which it is attached by muscular bands, the *eye muscles.* These are voluntary muscles derived from the myotomes of the first, second, and third segments of the head. There are six of these eye muscles, which should be identified as follows: From the dorsal view four of them will be seen. The one which is attached to the anterior wall of the orbit is the *superior oblique.* The other three originate from the posterolateral angle of the orbit and are named *recti* muscles. The most anterior one is the *internal* or *medial rectus;* its insertion on the eyeball is covered dorsally by the superior oblique. The next rectus muscle is the *superior rectus,* more dorsally situated than the others. The third, the *external* (or *lateral*) *rectus,* is inserted on the posterior surface of the eyeball. Next, raise the eyeball dorsally and note that the conjunctiva or most superficial coat over the external surface of the eyeball is continuous with the lining of the lower lid. Cut through this and free the eyeball ventrally, cleaning out the gelatinous and fibrous tissue which will be found here. On lifting the eyeball the remaining two eye muscles will be seen. The *inferior oblique* originates from the anteromedial corner of the orbit, the *inferior rectus* from the posteromedial angle of the orbit; both are inserted in contact with each other on the middle of the ventral surface of the eyeball. The white cords seen among the eye muscles are nerves.

Draw the eyeball and its muscles from dorsal view, showing as many of the muscles as possible.

The eye muscles originate from the orbit and are inserted on the eyeball. Their action is to turn the eyeball in various directions. As already stated, they are derived from three prootic head segments. (See further, p. 454.)

f) The structure of the eyeball: Cut through the eye muscles at the insertions and remove the eyeball. As already noted, the outermost coat covering the front of the eyeball is the conjunctiva, which is deflected onto the inner surface of the eyelids. Note the free edge of the conjunctiva clinging to the eyeball where the eyelids were cut. The conjunctiva is the epidermis of the skin and not one of the true coats of the eye. The outermost coat of the eyeball is the *sclera* or *sclerotic* coat, a very tough membrane composed of connective tissue. The front part of the sclera is transparent and is named the *cornea;* the conjunctiva is inseparably fused to the outer surface of the cornea. Through the transparent cornea can be seen an opening, the *pupil.* Cut off the dorsal side of the eyeball so that you can look within the cavity. Place the larger piece under water. The large spherical body in the interior is the *crystalline lens.* Note that internal to the sclera is a black coat, the *choroid* coat, and internal to this a soft, often collapsed, greenish layer, the *retina.* Follow the choroid coat to the front of the eye and note that there it is separated from the cornea, forming a black curtain, the *iris,* in the center of which is an opening, the *pupil.* The iris divides the cavity of the eyeball into an external cavity, the *anterior chamber* of the eye between the iris and the cornea, and an internal chamber, the *cavity of the vitreous humor*, between the lens and the retina. The anterior chamber contains a fluid, the *aqueous humor;* the cavity of the vitreous humor contains a gelatinous material, the *vitreous humor* or *vitreous body*, collapsed in the preserved specimen. The lens *in life* is attached to the junction of cornea and sclera by a circular membrane, the *ciliary body*, marked with radiating folds and applied to the inner surface of the iris; the attachment is lost in preserved specimens.

Draw the section, showing the structures of the eye.

The retina is the nervous part of the eye containing the sensory cells (rods and cones) which are stimulated by light. The lens and the two humors focus the light upon the retina. The focus is changed in fishes by moving the lens back and forth. The pupil regulates the amount of light admitted. The coats of the eye serve for protection and to darken the interior.

In the orbit after removal of the eyeball note the origins of the six eye muscles, the *optic pedicel*, a cartilaginous stalk situated among the rectus muscles and helping support the eyeball, and the *optic nerve*, a stout white stalk located in front of the rectus muscles. The stout white band in the floor of the orbit is the *infraorbital* nerve.

g) *The internal ear:* The ear in fishes consists only of the *internal ear* or *membranous labyrinth.* This is imbedded in the otic region of the skull, in a set of channels in the cartilage termed the *cartilaginous labyrinth.* In the dogfishes and skate the internal ear is situated between the spiracle and the middorsal line in a pronounced elevation caused by the otic region of the chondrocranium. In the median line between these two elevations will be

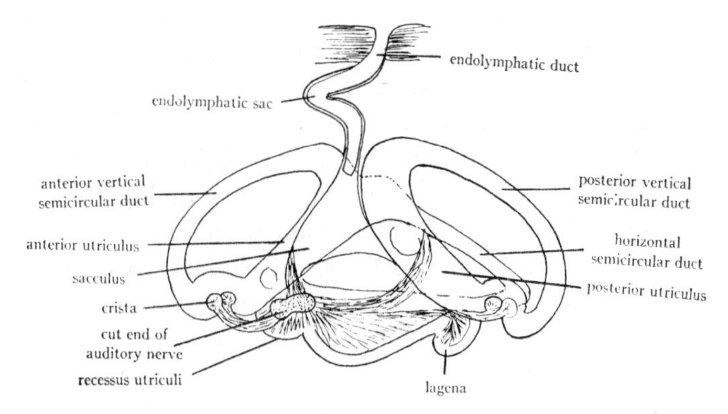

Fig. 136.—The right internal ear of the dogfish, *Squalus acanthias* (after Retzius, 1881), viewed from the inner side.

found a pair of small holes in the skin. Upon removing the skin bearing these holes the endolymphatic fossa of the chondrocranium will be found beneath it. In this fossa are the two *endolymphatic ducts*, which open on the skin by the two holes just mentioned and connect the cavity of the internal ear with the surface. Shortly below the surface each endolymphatic duct has a slight enlargement, the *endolymphatic sac.* Very carefully shave off with a scalpel the cartilage of the elevation containing the ear, working on the same side as before. There will soon be noticed a canal in the cartilage containing a delicate curved tube; this is the *anterior semicircular duct.* Continue removing the cartilage without injuring this duct. The muscles posterior to the ear may also be removed. Another tube will soon be uncovered posterior to the first one; this is the *posterior semicircular duct.* There will next be revealed the thin-walled central chambers of the ear to which these ducts are attached. Continue picking away the cartilage in small bits, leaving all parts of the internal ear in place. A third duct, the *horizontal semicircular duct*, lying below and lateral to the others, will next be exposed. When the cartilage has been removed as far as possible, the parts of the internal ear may be identified. The central part, to which the ducts are attached, consists of three delicate chambers: the *anterior utriculus*, the *posterior utriculus*, and the *sacculus*. The semicircular ducts are slender tubes, curved in a semicircle and each terminating in a rounded sac, the *ampulla*. The ampullae of the anterior and horizontal ducts are in contact, and the chamber which they enter is the anterior utriculus; these two ducts spring from the dorsal end of this chamber. The chamber to which both ends of the posterior duct are attached is the posterior utriculus.[4] The larger chamber occupying a central position and receiving the endolymphatic duct (connection difficult to find) is the *sacculus*, fitting into a rounded depression in the cartilage. In each ampulla will be seen a white sensory patch or *crista*, to which a branch of the auditory nerve is attached. Larger sensory patches, termed *maculae*, occur in the sacculus and utricular chambers. Inside the sacculus is a white mass of sand grains, collectively termed the *otolith;* the movement of these grains may be concerned in equilibration. The lower posterior wall of the sacculus near the ampulla of the posterior canal has a slight bulge, the *lagena*, considered to be the beginning of the cochlear duct of higher vertebrates; it contains a macula. Compare your dissection with Figure 136.

Draw, showing parts of the internal ear. After the drawing has been made, the sacculus may be opened and the otolith examined. The communications of the anterior and posterior utriculi with the sacculus may be sought.

The internal ear has two functions, that of hearing and that of equilibration. It is filled with a fluid, the endolymph, while the cartilaginous labyrinth is filled with perilymph. Sound waves impinging on the head or changes in the position of the head affect the endolymph, which in turn excites the sensory cells of the cristae and maculae, producing, in the first case, the sensation of hearing and, in the second, sensations of the animal's position in the water, enabling it to keep in the desired position. According to the experiments of Maxwell, both (or either) the cristae and the maculae control equilibration. The capacity to perceive sound is presumably limited to the maculae or to certain of them.

3. The dorsal aspect of the brain.—The brain is now to be exposed by carefully picking away the cartilage in small pieces from its roof. The cranial nerves, white strands passing through the cartilage, must not be injured. One side of the head has thus far been left intact for the study of the cranial nerves. This side is now to be exposed along with the brain, as far as necessary. Remove the upper eyelid, as directed, under the eye; but leave all structures intact. In removing the cartilage between the brain and the eye the following nerves will be noted: the *superficial ophthalmic* nerve, running in the wall of the orbit near the dorsal surface; the small *trochlear* nerve, passing through the back wall of the orbit to the superior oblique eye muscle; in the skate the larger *oculomotor* nerve accompanying the trochlear. Dissect

[4] Considered by Retzius and others, however, to be part of the posterior duct, in which case this duct is described as forming a circle and opening directly into the sacculus.

forward to the olfactory sacs, exposing them dorsally, leaving the ophthalmic nerve in place. Remove the skin behind the spiracle and note the *hyomandibular* nerve, passing posterior to the spiracle; this nerve is also to be preserved. To expose the posterior part of the brain the internal ears of both sides may be cut through and the mass of muscles posterior to the ear removed as much as necessary. Nerves will be seen passing through the cartilage in the ventral part of the ear but are not to be dissected out for the present. In short, the dorsal side of the brain is to be fully exposed, leaving all of the more superficial nerves intact. The dorsal aspect of the brain will then be studied first, and the cranial nerves afterward.

The brain is situated in a cavity in the chondrocranium, which it only partially fills. It is covered by a delicate membrane, the *primitive meninx*, in which the blood vessels of the brain are situated. The meninx is connected by strands with the membrane lining the cartilaginous walls of the cranial cavity. The space between brain and chondrocranium is filled in life by a fluid.

The most anterior structures of the brain are the large *olfactory bulbs*, nervous masses situated in contact with the dorsal walls of the olfactory sacs. From the olfactory sac a number of very short fibers, which together constitute the *olfactory nerve*, pass into the olfactory bulb. The olfactory bulb is spherical in the dogfishes, elongated in the skate. Each olfactory bulb is connected with the next part of the brain by a stalk, the *olfactory stalk* or *peduncle*, also called *olfactory tract*. The olfactory stalks enter the large rounded lobes, the *cerebral hemispheres*, which form the anterior (or lateral in the skate) part of the main mass of the brain. These are subdivided by a slight transverse groove into anterior and posterior lobes, of which the anterior ones are sometimes termed *olfactory lobes*, since they are of olfactory nature. All of the parts so far mentioned belong to the telencephalon.

Posterior to the cerebral hemispheres is a depressed region, the *diencephalon* or *thalamencephalon*, which has a thin and discolored roof, consisting of a *tela choroidea* (p. 432), highly vascularized. From this, choroid plexi project into the interior, forming the choroid plexi of the third ventricle. The *optic nerves* pass from the orbit into the ventral surface of the diencephalon and are easily seen in the skate; in the dogfishes they can be seen by gently pressing the diencephalon to one side. Behind the diencephalon is the *midbrain* or *mesencephalon*, consisting dorsally of two rounded lobes, the *optic lobes* or *corpora bigemina*. A pair of nerves, the *trochlear* or *fourth* cranial nerves, arises from the posterior borders of the optic lobes and passes forward to an eye muscle. By gently pressing the optic lobes to one side a nerve, the *oculomotor* or *third* cranial nerve, will be seen emerging from the ventral surface of the midbrain and passing to the orbit, on each side.

Posterior to the optic lobes and somewhat overhanging them is the large main mass or *body* of the *cerebellum*, slightly divided into four quadrants by faint longitudinal and transverse grooves; it belongs to the metencephalon. Behind this the elongated remaining section of the brain, continuous posteriorly with the spinal cord, is the *medulla oblongata* or *myelencephalon*. The greater part of the roof of the medulla is also formed by a tela choroidea with choroid plexi, roofing the cavity of the fourth ventricle. The anterior end of the medulla is continuous with the *auricles* of the cerebellum, two earlike projections at the sides of and below the body of the cerebellum. Remove the tela choroidea from the roof of auricles and medulla, thus revealing the large cavity of the fourth ventricle. On lifting the posterior end of the body of the cerebellum it will be seen that the auricles are continuous with this and with each other, thus providing the main pathway for the acousticolateral, auditory, and equilibratory impulses to reach the cerebellum. The entire dorsal rim of the medulla forms an elongated strip on each side, the *somatic sensory column;* the anterior part of this, continuous with the auricles, is the *acousticolateral area*. This somatic sensory column is the primary terminus for general cutaneous and proprioceptive impulses coming in along the sensory nerves and spinal tracts. Near the anterior end of the somatic sensory column, at the middle of the acousticolateral area, will be seen, by pressing the latter toward the middle, the roots of a number of nerves. These are the roots of the *fifth*, *seventh*, and *eighth cranial* nerves, to be studied in more detail later. At the posterior end of the somatic sensory column, just anterior to the point where the walls of the medulla close together, will be noted the stout root of the *tenth* cranial nerve. In the lateral wall of the medulla, just ventral to the somatic sensory column, is another longitudinal area marked by a row of rounded elevations; this area is the *visceral sensory column*. As its name implies, it is associated with sensations from the viscera. In fishes the gills are important visceral structures, so that a considerable portion of this column is connected with the gills; in fact, each of the little elevations is said to be a center for one branchial bar. Ventral to the visceral sensory column is the very slender *visceral motor column*, from which impulses go to the visceral muscles; in the head region, as already learned, these are the branchial muscles or their derivatives in higher vertebrates. In the floor of the fourth ventricle are two conspicuous *somatic motor columns*, separated by a median groove. These are the places of origin of impulses to the somatic muscles, derived from the myotomes; in the head these consist of the six eye muscles and the hypobranchial musculature.

The fourth ventricle narrows posteriorly and is finally roofed over by the fusion of the walls of the medulla. Shortly beyond this point the medulla is continuous with the spinal cord. The posterior end of the medulla marks the posterior end of the brain but is not sharply defined.

Draw the dorsal aspect of the brain. Place it in the middle of the page, so that the cranial nerves can be added later.

4. The cranial nerves.—There are ten cranial nerves in fishes. They are to be dissected with great care and their distribution noted and drawn. This distribution is, in general, similar in all vertebrates except that certain nerve trunks present in fishes disappear in the land vertebrates. One of the most striking examples of homology is found in this distribution of the cranial nerves, which in man still continue to supply the same parts as in the fish.

Add to your drawing of the dorsal side of the brain an outline of the head of the animal, putting in outline the olfactory sacs, eyes, ears, and gill slits. As you dissect the cranial nerves according to the directions to be given immediately, add each to this drawing, showing, as accurately as possible, the location and course of each of these nerves and the parts of the head which they supply. It is necessary to enter a nerve on one side only, and by using the two sides of the drawing for different nerves it will be possible to enter all of them.

a) *The first or olfactory nerve:* This nerve has already been noted. It arises from the olfactory cells in the lamellae of the olfactory sac and passes by very short branches into the olfactory bulb. These branches are practically invisible. From the olfactory bulb, after a relay, the olfactory impulses pass along the olfactory stalks into the cerebral hemispheres. The olfactory nerve is a pure sensory nerve, belonging to, or at least closely related to, the visceral sensory division.

b) *Nervus terminalis:* The terminal nerve (not counted as one of the ten cranial nerves) is a fine white thread which will be found along the medial side of the olfactory stalk. On tracing it caudad, it will be found to separate from the stalk and come in contact with the middle of the anterior face of the cerebral hemisphere. It then runs along this face and into the fissure between the two cerebral hemispheres. It probably belongs to the autonomic system.

c) *The second or optic nerve:* The optic nerve arises in the retina of the eye and passes through the coats of the eye, emerging ventral to the internal rectus muscle. Find it there on the intact eye by pressing this muscle against the eyeball. The nerve is a stout, white trunk which pierces the cartilage of the orbit and passes to the ventral side of the diencephalon, where it may be seen by gently raising the diencephalon. As already explained, the optic nerve is not really a nerve but a brain tract (see p. 440); it also is a pure sensory tract, carrying the visual impulses and discharging them into the diencephalon and optic lobes.

d) *The fourth or trochlear nerve:* The trochlear nerve arises in the midbrain and emerges in the groove between the optic lobes and the cerebellum. Trace it on the side where the eye is still intact. It passes forward in the cranial cavity to about the level of the cerebral hemispheres; it then turns abruptly laterally, pierces the wall of the orbit, and is distributed to the superior oblique muscle of the eyeball. It is the motor nerve of this muscle and carries only somatic motor impulses. Although it appears externally to emerge from the roof of the midbrain, the motor cells from which it originates are, in fact, in the floor of the midbrain in a forward extension of the somatic motor column.

e) The third or oculomotor nerve: The oculomotor nerve arises from the floor of the midbrain. It is readily noticed in the skate, ascending to the orbit near the preceding nerve. In the dogfishes it is deeply situated and is seen by pressing the cerebellum away from the wall of the orbit. Follow it into the orbit on both sides, getting its general relations first on the side where the eyeball was removed. It emerges into the orbit very near the insertion of the superior rectus muscle and is situated ventral to the superficial ophthalmic nerve, already noted. It should not be confused with the *deep ophthalmic* nerve, which is in contact with it as it enters the orbit; the deep ophthalmic nerve runs through the orbit in contact with the medial surface of the eyeball. This nerve will be better seen on the intact side. Observe the branches given by the oculomotor nerve, immediately after its entrance into the orbit, to the internal and superior recti muscles. On the intact side now loosen the eyeball and cut the insertions of the superior oblique and superior rectus close to the eyeball. Identify the deep ophthalmic nerve passing in the dogfishes dorsal to the internal rectus and lying against the eyeball; in the skate it passes ventral to the internal rectus. Free and preserve this nerve. Cut through the insertion of the inferior rectus and the optic nerve and, pressing the eyeball outward, note the branch of the oculomotor nerve which runs along the posterior side of the inferior rectus muscle, turns ventral to it, and then runs forward in the floor of the orbit to the interior oblique. Note, also, that the branch to the inferior rectus also gives off a branch, one of the *ciliary nerves,* which enters the eyeball in company with an artery. Along this ciliary nerve small brown masses can be noted; they are part of a ganglionated *ciliary plexus* belonging to the autonomic system. Ciliary branches also enter the eyeball from the deep ophthalmic nerve. The function of the ciliary plexus and nerves is to control the smooth muscles of the iris, regulating the size of the pupil. Throughout vertebrates it also controls the accommodation mechanism of the lens, but this is poorly developed in selachians. The oculomotor nerve is a pure somatic motor nerve, except for the visceral motor fibers of the autonomic system which accompany it.

f) The sixth or abducens nerve: The abducens originates from the somatic motor column on the ventral surface of the anterior end of the medulla. Its origin will be seen later. It penetrates the orbit at the point of origin of the external rectus muscle and passes along the ventral surface of this muscle, to which its fibers are distributed. It will be seen as a white ridge on the ventral surface of the muscle. Like the other eye-muscle nerves, it is a somatic motor nerve.

As just seen, the third, fourth, and six cranial nerves are somatic motor nerves to the muscles of the eyeball. The reasons for this arrangement are explained above (p. 454), where also the sensory parts of these nerves are indicated.

g) The fifth or trigeminus nerve: The trigeminus is a very large nerve with four main branches in elasmobranchs (three in land vertebrates). The trigeminus is attached to the medulla near the anterior end of the somatic sensory column, just behind the auricles of the cerebellum. Its roots here are inextricably mingled with the roots of the seventh and eighth nerves, the three together forming a conspicuous mass at the place stated. The trigeminus passes through the adjacent wall of the orbit and should be followed into the orbit by carefully picking away the cartilage around it. As soon as it penetrates the orbit, the trigeminus divides into four branches. The first of these, the *superficial ophthalmic* branch, is part of the superficial ophthalmic trunk, which has already been mentioned several times. This large trunk passes forward in the dorsal part of the cartilage of the medial wall of the orbit. Trace it forward. It passes out of the orbit through the ophthalmic foramen in the chondrocranium and above the olfactory bulb. Only a small part of this trunk is trigeminal; this is sensory to the skin dorsal to the orbit. The second branch of the trigeminus is the deep ophthalmic nerve. It passes through the orbit ventral to the preceding (giving off small ciliary nerves) and leaves the orbit by the orbitonasal canal. On tracing it forward through the canal it will be found to come in contact with the superficial ophthalmic trunk from which it again separates, supplying sensory fibers to the dorsal and lateral skin of the snout. Both the superficial and deep ophthalmic parts of the trigeminus are somatic sensory nerves.

The two remaining branches of the trigeminus lie in the floor of the orbit. To see them, remove the eyeball or study the side where the eyeball was previously removed. A broad white band, the *infraorbital* trunk, is seen in the floor of the orbit, passing obliquely laterally. In the **dogfishes** this trunk is composed of the mixed fibers of the *maxillary branch* of the trigeminus and the *buccal branch* of the seventh nerve (see below). In the orbit the larger and more medial portion of the trunk is the maxillary branch, but farther out this becomes inextricably mingled with the buccal nerve. In the **skate** the infraorbital trunk is divisible into three trunks, of which the outer one is the *maxillary* branch of the trigeminus, the middle one the *mandibular* branch of the trigeminus, and the inner one the *buccal* branch of the seventh nerve. As before, however, it should be remembered that there is an admixture of fibers of the fifth and seventh nerves in these trunks. Trace the maxillary branch of the trigeminus and buccal branch of the seventh out from the orbit, along the ventral surface of the rostrum. The branches pass to the region below and in front of the eye, to the medial side of the nostril (in the skate to the lateral side of the nostril also) and to the angle of the jaws. The maxillary branch is sensory to the skin of the rostrum, while the buccal nerve supplies the infraorbital lateral-line canal and near-by ampullae of Lorenzini.

The fourth branch of the trigeminus is the *mandibular* branch. In the dogfishes it separates from the infraorbital trunk where the latter enters the orbit from the brain and passes along the posterior wall of the orbit. The mandibular nerve is seen to branch to various muscles in the floor of the orbit (these are gill-arch muscles) and, on following it out of the orbit, will be seen to be distributed to muscles of the lower jaw and to send a sensory branch to the skin of the lower jaw, this branch being situated just behind the teeth. In the skate the position of the mandibular nerve was described above as between the maxillary and the buccal nerves. Follow it forward. It curves around the angle of the jaw and supplies muscles of the lower jaw and the adjacent skin.

It will be observed that all of the branches of the fifth nerve are somatic sensory nerves coming from various sensory organs of the skin, except the mandibular nerve, which also contains some motor branches to muscles. As those muscles are branchial muscles, this part of the mandibular nerve belongs to the visceral motor system. As explained above (p. 442), the deep ophthalmic or profundus nerve is not really part of the trigeminus nerve but an independent nerve with its own ganglion in cyclostomes and selachians. It is the sensory nerve of the same head segment of which the oculomotor nerve is the motor part. In higher vertebrates the profundus nerve is incorporated into the ophthalmic branch of the trigeminus. The trigeminus nerve proper is the sensory nerve of the same head segment of which the trochlear nerve is the motor part. It is important to note further that the trigeminus is the *nerve of the upper and lower jaws,* i.e., *first* or *mandibular arch.* It is the sensory nerve of this arch and the motor nerve of its branchial muscles. The sensory ganglion (Gasserian ganglion) of the trigeminus nerve and the profundus ganglion are in the mass mentioned above, including the roots of the fifth, seventh, and eighth nerves.

h) The seventh or facial nerve: This nerve is intimately related to the trigeminus. It arises in common with the latter from the anterior end of the medulla and divides into three main branches. Two of these branches pass through the orbit in common with the trigeminus. The *superficial ophthalmic* branch of the facial nerve accompanies the same branch of the trigeminus and forms the greater part of the superficial ophthalmic trunk, supplying the supraorbital lateral-line canal and adjacent ampullae of Lorenzini. The buccal branch of the facial nerve, as already noted, forms in the orbit the outer half of the infraorbital trunk and supplies the infraorbital lateral-line canal and near-by ampullae of Lorenzini. These two branches of the facial nerve are really parts of the lateralis system, hence somatic sensory, and vanish with the disappearance of this system. The third branch of the facial nerve is the large *hyomandibular* trunk, which has already been located posterior to the spiracle. Trace it inward toward the brain, cutting tissues in its path. It turns ventrally and runs through the anterior part of the ear capsule deep down. Follow it by removing the cartilage of the ear capsule in small pieces. The nerve passes ventral to some of the branches of the nerve of the ear and joins the anterior end of the medulla in common with the trigeminus root. Near the brain it has an enlargement or ganglion (*geniculate* ganglion). From this ganglion is given off the *palatine* nerve. It will be found by dissecting carefully around and on the ventral surface of the ganglion. In the skate it is easily seen. It runs forward below the orbit along the roof of the mouth, where it supplies the taste buds and the lining epithelium in general. Now trace the hyomandibular outward, past the spiracle. It turns ventrally and breaks up into branches on the side of the head. These branches supply the hyomandibular and mandibular lateral-line canals, ampullae and similar sense organs, the muscles of the hyoid arch, and the lining of the floor of the mouth cavity and tongue.

The hyomandibular is seen to be a mixed nerve with lateral-line (somatic sensory), visceral motor, and visceral sensory components. The sensory part of the facial nerve belongs to the same head segment as that of which the abducens constitutes the motor nerve. The facial nerve is the *nerve of the first gill slit* (*spiracle*) and of the *second* or *hyoid arch*, hence visceral motor to the musculature of that arch. With the loss of the lateralis parts the facial nerve of land vertebrates consists only of the palatine (=pharyngeal branch) and certain parts of the hyomandibular nerve (=posttrematic branch).

i) *The eighth or auditory nerve:* The auditory nerve is a pure somatic sensory nerve extending from the internal ear to the brain. It enters the anterior end of the medulla and is there mingled with the roots of the fifth and seventh nerves. Follow it into the internal ear, on the side opposite that on which the hyomandibular was dissected. Note its branches to each ampulla and the fanlike arrangement of the branchlets to the crista of each ampulla. The auditory nerve also collects a number of branches from the walls of the sacculus and utriculus. It carries impulses for hearing and equilibration into the acousticolateral area of the medulla, to which it will be found to be attached. There is no motor nerve corresponding to the auditory nerve, and this nerve is not considered to be one of the segmental head nerves.

j) *The ninth or glossopharyngeal nerve:* This nerve passes through the floor of the middle of the ear capsule (where it is likely to be mistaken for a part of the auditory), parallel to the hyomandibular nerve. Pare away as much of the ear capsule as is necessary to reveal it. Find its attachment to the medulla posterior to the auditory nerve. Trace it out of the ear capsule. Just before it exits from the ear capsule, it bears a swelling, the *petrosal* ganglion. Insert a knife blade into the second (first typical) gill slit (in the skate into the dorsal wall of the corresponding visceral pouch) and slit the gill cleft open dorsally. The petrosal ganglion will now be seen to be located near the upper limits of the cleft. Dissect the nerve from the ganglion toward the gill slit. It very soon divides into three branches—two smaller anterior ones, and a larger posterior one. The most anterior branch is the *pretrematic* branch; this passes to the anterior wall of the visceral pouch, to which it is a sensory nerve. The second branch is posterior to the pretrematic branch; it is named the *pharyngeal* branch and is a sensory nerve to the mouth cavity. (This branch appears to be lacking in the skate.) The third and largest is the *posttrematic* branch. It passes to the posterior wall of the visceral pouch and is both sensory and motor, its motor components supplying the muscles of the third branchial bar. The glossopharyngeal nerve is the nerve of the *second visceral pouch* and of the *third branchial bar*. The glossopharyngeal nerve also has a lateralis component in the form of a minute nerve which supplies a short stretch of the main lateral-line canal just before this is joined by the supratemporal canal.

k) *The tenth or vagus nerve:* The vagus nerve is the very large trunk passing through the posterior border of the ear capsule. It is attached to the sides of the posterior part of the medulla. Dissect it out and follow its course. It passes to the anterior cardinal sinus, the wall of which is formed in the dogfishes of a tough membrane. Open up the anterior cardinal sinus by a deep cut through the muscles above the upper ends of the gill slits. Follow the vagus nerve into the anterior cardinal sinus. In the **dogfishes** it divides into two trunks at the point where it penetrates the tough wall of the sinus. The medially situated trunk is the *lateral* branch of the vagus and passes posteriorly internal to the lateral line, whose canal it supplies. The lateral trunk is the *visceral* branch of the vagus, which continues along the anterior cardinal sinus. In the **skate** the vagus runs for a short distance in the sinus before dividing into a more dorsal *lateral* branch, which passes posteriorly internal to the lateral-line canal to which it is distributed, and a more ventral *visceral* branch. In all three forms open up the remaining visceral pouches as directed for the ninth nerve and determine the distribution of the visceral branch of the vagus. With the walls of the sinus well spread open, note the four branches crossing the floor of the sinus to the visceral pouches. Dissect out each of these and observe that each bears a ganglion, beyond which it divides into three branches: an anterior pretrematic branch, a middle pharyngeal branch, and a posterior posttrematic branch. The pharyngeal branch seems to be missing in the skate. As in the case of the ninth nerve, the pre- and posttrematic branches embrace the visceral pouch, which lies between them; all three branches have the same functions as described for the ninth nerve. We thus see that the vagus nerve supplies the remaining branchial bars, beginning with the fourth, and the remaining visceral pouches, beginning with the third. After supplying the gill apparatus the visceral branch of the vagus passes on into the pericardial and pleuroperitoneal cavities, supplying the heart, digestive tract, and other viscera (by way of the autonomic system, see p. 444).[5]

For the general relation of the cranial nerves to head segmentation and the gill apparatus, see pages 441–42.

5. The occipital, hypobranchial, and first spinal nerves.—Very carefully expose the spinal cord posterior to the medulla by shaving away the cartilage of the neural arches in thin slices. On the dorsolateral surface of the cord note the little swellings, the *dorsal* or *spinal ganglia*, attached to the cord by the *sensory* or *dorsal root*. In the skate the spinal ganglia are elongated. Between the first spinal ganglion and the root of the vagus note two or three small roots springing from the side of the medulla. These are the *occipital* nerves, which represent the ventral roots of original spinal nerves of which the dorsal roots have disappeared. They innervate some muscles of this region, also help to form the hypobranchial nerve described below, and contribute to the cervicobrachial plexus. On pressing the spinal cord to one side the *ventral* or *motor roots* of the spinal nerves will be seen arising from the ventrolateral region of the cord, at the same level as the occipital nerves. The ventral roots are formed by the union of several small rootlets coming from the cord. The most anterior ventral roots are situated anterior to the dorsal roots which belong to the same segment.

In the **dogfish** the union of dorsal and ventral roots to form a spinal nerve is not easy to follow. It may usually be seen by carefully paring down the cartilage along the side of the spinal cord. In the **skate** the union is easily followed; the roots pass through the cartilage at the side of the cord and unite as they exit from the cartilage. A large number of the most anterior spinal nerves (really the ventral rami of the spinal nerves, the dorsal rami being very slender in the skate) are then seen to unite to form the very large nerve of the brachial plexus previously noted.

The *hypobranchial* nerve is a trunk formed by contributions from the occipital nerves and the first spinal nerves. In the **dogfishes** it may be located as follows: Insert one blade of the scissors in the angle of the jaw and cut back across the gill slits through to the side of the esophagus, as was done in an early stage of the dissection (if the same specimen is still being used, this

[5] The foregoing account does not attempt to give the smaller branches of the cranial nerves. For an extended account of the branches of the cranial nerves see the important article of Norris and Hughes (1920).

cut will already have been made). Open the flap thus formed and expose the roof of the mouth. Make a longitudinal cut through the mucous membrane of the roof in the median dorsal line. Strip the membrane laterally, carrying with it the free dorsal ends of the gill cartilages, which will be readily located just lateral to the middorsal line. The visceral branch of the vagus nerve, which was already seen from the other side, is now exposed. It lies along the thin ventral wall of the anterior cardinal sinus. Emerging from the muscle now exposed in the roof of the mouth will be seen the ventral rami of the spinal nerves. In the **dogfishes** two of these (they appear as one but will be found to consist of two on dissecting them toward the median line) pass obliquely toward the visceral branch of the vagus and enter its sheath, thus appearing to join it. The trunk they form, which is the hypobranchial nerve, can, however, be readily separated from the vagus. It lies just anterior to the nerves of the cervicobrachial plexus, to which it contributes branches. After passing dorsal to the dorsal side of the visceral branch of the vagus (which may be noted proceeding to the esophagus) and behind the last visceral pouch, the hypobranchial nerve turns ventrally and courses along the floor of the oral cavity supplying hypobranchial muscles.

In the **skate** the hypobranchial nerve leaves the spinal cord in common with the trunk of the brachial plexus. Locate this trunk again in the anterior wall of the pleuroperitoneal cavity, dorsal to the pleuroperitoneum and just behind the cartilage of the pectoral girdle. The entire trunk, with the exception of one branch, passes out dorsal to the cartilage to the pectoral fin. This one branch, the hypobranchial nerve, turns forward and is distributed to the muscles of the floor of the mouth.

The occipital and hypobranchial nerves are probably the homologues of the twelfth or hypoglossal nerve of amniotes, having been taken into the cranium in those forms. The muscles supplied by the hypobranchial nerve are derived from certain metaotic myotomes and in higher forms furnish the extrinsic musculature of the tongue.

6. The ventral aspect of the brain.—Carefully free the brain from the chondrocranium; cut through the olfactory stalks and lift the anterior end of the brain. You will next see the two optic nerves entering the ventral surface of the diencephalon. Cut through them and lift the brain further. Next, pare away the wall of the orbit on one side. It will then be seen that certain structures attached to the ventral surface of the diencephalon extend ventrally into a deep pit (sella turcica) in the floor of the cranial cavity. Take especial care to lift these out intact. Then cut through the remainder of the cranial nerves, cut across the spinal cord, and lift the brain out of the cranial cavity.

Examine the ventral surface of the brain. Note the forward continuation of the internal carotid artery on the midventral line of the brain. It forks around the ventral part of the diencephalon (farther posteriorly in the skate) and passes forward to the telencephalon, distributing many branches to all parts of the brain as well as to the orbit. The ventral surface of the brain presents nothing new except as regards the diencephalon, where several additional structures are visible. The two optic nerves are seen attached to the anterior end of the ventral surface of the diencephalon; as they enter the latter, they cross, the crossed region being named the *optic chiasma.* From the chiasma a broad band, the *optic tract,* extends dorsad and caudad into the dorsal part of the diencephalon and into the optic lobes. It is readily seen, especially in the spiny dogfish, by scraping off the primitive meninx at this place. It is thus evident that the visual impulses pass into these two portions of the brain. Posterior to the optic chiasma the floor of the diencephalon bulges ventrally and posteriorly as the *infundibulum,* consisting in large part of two rounded lobes, the *inferior lobes.* From between the two inferior lobes a stalk projects caudad and widens into a soft body. The dorsal part of this has thin discolored walls, being highly vascularized; this is termed the *vascular sac* (*saccus vasculosus*) and is a part of the infundibulum. The remainder of the structure hanging from the infundibulum is the *hypophysis* or *pituitary body,* generally more or less torn in removing the brain. This is an important gland of internal secretion. Dorsal to the vascular sac the roots of the oculomotor nerves will be found springing from the floor of the midbrain. The ventral surface of the remainder of the brain presents nothing new. The roots of the cranial nerves should be identified on the medulla. On the ventral surface of the medulla will be found the roots of the sixth nerves. If not identifiable on the brain, they will usually be found adhering to the floor of the cavity from which the brain was removed.

Draw a profile view of the brain.

7. The sagittal section and the ventricles of the brain.—The brain, like the remainder of the central nervous system, is hollow. Its cavities are known as *ventricles* and are continuous with each other by means of narrow passages. The *fourth* or last ventricle of the brain has already been identified as the cavity within the medulla oblongata. The ventral portion of this ventricle is named, from its shape, the *fossa rhomboidea.* Bisect the brain by a median sagittal cut. Examine the cut surface, best under water. From the fourth ventricle a narrow passage, the *aqueduct* of the brain, extends anteriorly. It communicates with the cavity of the cerebellum, the *cerebellar ventricle,* and the cavities of the optic lobes, the *optic ventricles.* Below the aqueduct is the thick floor of the midbrain. The aqueduct opens into the cavity of the diencephalon, the *third* ventricle. The thin roof of the diencephalon forms a tela choroidea, from which a choroid plexus is folded into the cavity. The anterior part of this roof in the dogfishes extends dorsally into a sac, the *paraphysis*, resting against the telencephalon, of which it is considered a part. Posterior to the paraphysis a thin transverse partition, the *velum transversum*, is seen in the roof of the diencephalon, particularly in the dogfishes. This marks the dorsal boundary between diencephalon and telencephalon. The small thickened region of the diencephalon just in front of the anterior end of the optic lobe is the *habenula*, a smell center. From the habenula a slender process, the *pineal body*, may be seen in favorable specimens, extending dorsally, just back of the paraphysis. The entire roof of the diencephalon, including choroid plexus, habenula, and pineal body, is named the *epithalamus*. The lateral walls of the diencephalon constitute the *thalamus*, an important correlation center for various body senses. The ventral part of the diencephalon is named the *hypothalamus*. It consists of the infundibulum, including the inferior lobes, the hypophysis, and the *mammillary* bodies. The former should be identified on the section. The mammillary bodies are the thickened part of the ventral wall above the vascular sac. Note that the cavity of the third ventricle extends into all parts of the hypothalamus. The third ventricle connects by a passage, the *foramen of Monro*, or *interventricular foramen*, with the cavity in each half of the telencephalon. (In the skate, the telencephalon is solid.) These two cavities are named the *first* and *second* or *lateral* ventricles. They extend out into the olfactory bulbs through the olfactory tracts. Cut into the telencephalon to see its ventricles.

Draw the sagittal section.

8. Autonomic system.—According to the work of Young (1933), the autonomic system of selachians consists of a paired chain of ganglia along the middorsal region of the pleuroperitoneal cavity, imbedded anteriorly in the dorsal wall of the posterior cardinal sinuses and posteriorly in the kidney, close to the suprarenal bodies; and of ganglionated plexi in the viscera with which the ganglia of the chains connect. There is also a ciliary plexus in the orbit and ganglia along the gill arches, apparently for visceral sensory supply to the pharyngeal lining. There is no division into antagonistic sympathetic and parasympathetic systems. The communicating rami consist only of white rami.

C. THE NERVOUS SYSTEM AND SENSE ORGANS OF NECTURUS

1. The spinal nerves.—The spinal nerves are best found as follows: Make a longitudinal cut along the side of the body below the lateral line. Cut through the external and internal oblique muscles and separate this mass of muscle from the thin layer of transverse muscles lying next to the coelom. The *ventral rami* of the spinal nerves will now be seen running between the oblique and transverse muscles, along the myosepta, and supplying the hypaxial muscles. Trace one of these toward the vertebral column, cutting away muscles from its course. It lies just behind the rib, which may be cut away. The nerve may be traced up to the vertebra, where it is imbedded in an orange-colored material. On carefully clearing this away the *dorsal gan-*

glion of the spinal nerve will be found imbedded in it. It is a rounded, brownish body from which spring two nerves: the ventral ramus, just followed, and the smaller *dorsal ramus*, which supplies the epaxial muscles.

2. The limb plexi.—The ventral rami of the spinal nerves form a plexus for each limb, the motor nerves for the limb muscles arising from this plexus. The *brachial* plexus is located as follows: Make a cut in the ventral body wall just medial to the base of the forelimb. Separate the pectoral and shoulder muscles from the sternohyoid muscle. The brachial plexus is then easily seen running posterior to the scapula. It consists of the ventral rami of three spinal nerves (3, 4, and 5) which have cross-connections with each other. Beyond the plexus, nerves proceed into the forelimb.

The *lumbosacral* plexus is located by cutting through the skin and both layers of oblique muscles longitudinally just dorsal to the base of the hind limb. On separating the oblique from the transverse muscle layers the plexus is exposed. It consists of the ventral rami of three (sometimes four) spinal nerves. The anterior one of the three gives off a slight branch to the nerve next posterior to it, sometimes receives a contribution from the ramus anterior to it, and, as the *crural* nerve, passes into the limb. The middle of the three nerves is much the stoutest and, after receiving contributions from the nerve next posterior to it, enters the limb as the *ischiadic* (sciatic) nerve.

3. The sense organs of the head.—Because of the small size of the eye and its similarity to that of other vertebrates, it will not be investigated.

a) *Lateral-line system:* Lateral-line sense organs are present in *Necturus* but are impractical to find in gross dissection. They are situated along lines similar to those of fishes.

b) *Nose:* Probe into the external naris and follow your probe with the cut, opening the entire nasal passage to the internal naris. Note the folds or *olfactory lamellae* in the interior of the passage. Unlike the condition in fishes, the olfactory sac opens into the oral cavity and has both olfactory and respiratory functions.

c) *Ear:* The ear, as in fishes, consists of the internal ear only, imbedded in the otic region of the skull. Expose the dorsal surface of the skull by cleaning away the muscles. Locate the otic capsules, one at either side of the posterior end of the skull. Cautiously shave away the cartilage here and locate the three semicircular ducts and vestibule, as done in the dogfish. There are three of the former—*anterior vertical*, *posterior vertical*, and *horizontal*—each with an *ampulla*, all arranged exactly as in the dogfish. The vestibule is divided indistinctly into a dorsal *utriculus*, from which the ducts spring, and a ventral *sacculus*. The latter contains a crystalline mass, the *otolith*.

4. The dorsal aspect of the brain.—Remove the roof of the skull. This is best done by stripping it off in slivers with a knife. After the brain is revealed, study its dorsal surface. The brain is covered by a membrane, the *primitive meninx*, which is more or less divisible into the *pia mater*, a delicate pigmented membrane adhering to the brain, and an outer *dura mater*, which is separated from the skull by the *peridural* space.

The most anterior portion of the brain consists of the two elongated cerebral hemispheres. The olfactory bulbs are not distinct externally, being included in the cerebral hemispheres. Between and behind the posterior ends of the cerebral hemispheres is a thin roof constituting a tela choroidea, the anterior part of which forms a choroidal sac (parencephalon, see p. 434), projecting forward in the groove between the two hemispheres. Immediately behind this is another dorsally projecting process, the pineal body or epiphysis. Behind the diencephalon are the two elongated optic lobes, which are the dorsal part of the mesencephalon or midbrain. Behind this is another dark-pigmented tela choroidea, which should be removed, revealing the triangular cavity of the fourth ventricle. The anterior end of this cavity is overhung by a narrow shelf just back of the optic lobes; this shelf is the cerebellum, very undeveloped in urodeles. The remainder of the brain, inclosing the cavity of the fourth ventricle, is the medulla oblongata. Anteriorly the medulla is continuous below the optic lobes, with projections which form the auricles of the cerebellum. The walls of the medulla are divided into dorsal or *sensory* portions and ventral or *motor* portions, the latter forming a broad band on either side of the median ventral groove. The further subdivisions of these are not evident in *Necturus*.

Draw the dorsal view of the brain.

5. The cranial nerves.—

a) The first or olfactory nerve: This is a stout band passing from the olfactory sac to the anterior end of the telencephalon.

b) The second or optic nerve: This small nerve may be seen in the floor of the cranial cavity by pressing the telencephalon to one side. It passes obliquely caudad to the ventral surface of the diencephalon.

c) The eye-muscle nerves: The *third* or *oculomotor*, *fourth* or *trochlear*, and *sixth* or *abducens* nerves are so small in *Necturus* as to be scarcely discernible in gross dissection. The originate from the same regions of the brain and supply the same eye muscles as in elasmobranchs.

d) The fifth or trigeminus nerve: The fifth nerve is the large trunk arising from the side of the anterior end of the medulla. Trace it out through the skull. It passes in front of the otic capsule and immediately enters a large ganglion, the *semilunar* or *Gasserian* ganglion of the trigeminus. From the ganglion three large nerves are given off—the *ophthalmic*, *maxillary*, and *mandibular* branches. The former passes forward through the quadrate cartilage and runs anteriorly alongside the frontal bones. The maxillary nerve proceeds along the margin of the upper jaw. The mandibular nerve passes laterally to the angle of the jaws and then turns forward along the lower jaw.

e) The seventh or facial nerve: This arises just behind the trigeminus and sends a branch forward to join the latter at the semilunar ganglion. This branch of the facial passes out with the ophthalmic nerve as the *superficial ophthalmic* branch of the facial and with the maxillary nerve as the *buccal* branch. Both of these go to lateral-line organs. The greater part of the facial arises from the medulla in common with the auditory nerve ventral to the above-named branch. This common *acousticofacial* trunk of facial and auditory passes into the anterior part of the otic capsule. From here the main trunk of the facial or hyomandibular may be followed laterally. It branches to muscles, lateral-line organs, etc.

f) The eighth or auditory nerve: This arises in common with the facial and is distributed to the internal ear. Its branches are readily noted in the otic capsule.

g) The ninth or glossopharyngeal and tenth or vagus nerves: These arise together from the medulla posterior to the acousticofacial trunk, by three roots. The common trunk passes along the posterior margin of the ear capsule and enters a large ganglion. From this several nerves arise which may be traced into the external gills (these nerves are not homologous to the pre- and posttrematic branches of elasmobranchs) and to the branchial bars. The most posterior branch of the vagus gives off a *lateral* branch, which passes to the lateral line which it accompanies. The vagus also supplies the viscera.

Add the cranial nerves to your drawing of the brain, as far as seen.

6. Ventral aspect of the brain.—Remove the brain by cutting across the spinal cord, the olfactory, and other nerves. The ventral view reveals some additional parts of the diencephalon. At the anterior end of the ventral surface of the diencephalon is the small *optic chiasma,* formed by the optic nerves. Posterior to this is the large *infundibulum,* from the posterior end of which projects the hypophysis.

Draw the ventral view.

7. The autonomic system.—The urodele amphibians have a pair of sympathetic trunks alongside the dorsal aorta, extending from the vagus region to the end of the tail; and there are also the usual plexi along the large blood vessels and in the viscera. A cephalic part of the autonomic system appears to be usually absent.

D. NERVOUS SYSTEM AND SENSE ORGANS OF THE TURTLE

The brain of reptiles shows considerable advance over that of fishes and amphibians in the increased size of the cerebral hemispheres. The cerebellum in some, such as *Sphenodon* and the crocodilians, is not much advanced from the urodele stage, forming a shelf over the anterior part of the fourth ventricle, but in others, such as the turtles, forms a projecting hollow body. The entire lateralis system has vanished.

1. The spinal nerves, the autonomic system, and the spinal cord.—Remove all of the viscera from the pleuroperitoneal cavity, leaving the large neck muscles in the middorsal region undisturbed. Identify the spinal nerves as the white cords passing along the sutural lines of the costal plates. The *sympathetic chain* should be identified as a white cord or cords located on the sides of the mass of neck muscles.

a) Spinal nerves and limb plexi: Carefully expose the spinal nerves of the trunk, avoiding injury to the sympathetic system. These nerves, called the *dorsal* spinal nerves, run along the sutures between the costal plates of the carapace. In most cases each consists of two branches, a smaller *dorsal ramus* and a larger *ventral ramus*. On tracing these toward the vertebral column they will be found to come from a large ganglion, the *dorsal* or *spinal* ganglion, situated in contact with the center of the centrum.

Expose the *brachial plexus* in the depression between the neck and the dorsal end of the scapula. It is generally formed by the cross-unions between the ventral rami of the last four *cervical* spinal nerves and the first dorsal spinal nerve. The last-named nerve may be identified as the one in front of the first typical rib (really the second rib). The four cervical nerves form a complex network on the surface of the shoulder muscles. From this network the large *median* nerve proceeds along the anterior surface of these muscles and the smaller *ulnar* and *radial* nerves along the posterior surface; the radial is the most dorsally situated one. The first dorsal nerve sends a branch near its ganglion to the brachial plexus and is then distributed to the carapace just posterior to the forelimb.

The next six dorsal nerves are similar to the first description given. The ventral rami of the eighth, ninth, and tenth dorsal nerves, together with the two *sacral* nerves, form the *lumbosacral plexus* for the hind limb. This lies on the medial surface of the muscles covering the ilium and is found by carefully separating the ilium, with its muscles, from the median region. The branches from the three dorsal nerves unite to a sort of knot, from which several nerves proceed to the anterior part of the leg. The two sacral nerves, receiving also a contribution from the tenth dorsal nerve, unite to form a large trunk, the *sciatic* nerve, situated among the muscles on the posterior side of the leg.

There is a pair of *caudal* spinal nerves corresponding to each caudal vertebra, but these need not be looked for. There are nine pairs of *cervical* nerves, which will be found by looking in the neck at the same level as the level of emergence of the nerves of the brachial plexus.

b) The sympathetic system: Locate the vagus nerve in the neck. It is the conspicuous white cord running along the side of the neck. Trace it posteriorly. The sympathetic trunk is bound with it but at about the level of the first nerve of the brachial plexus separates from the vagus and enters a swelling or ganglion, the *middle cervical* ganglion. The sympathetic cord proceeds dorsally from this ganglion and lies on the ventral surface of the brachial plexus, where it presents two successive swellings, which together constitute the *inferior cervical* ganglion. Observe branches from these ganglia. The sympathetic cord passes to the ganglion of the first dorsal spinal nerve, to which its own ganglion is fused. It then proceeds as a delicate white cord across the second rib and again forms a ganglion, which is fused to the ganglion of the second dorsal nerve. The sympathetic cord then passes more ventrally, lying on the side of the long neck muscles. Follow it here and note the ganglia which it bears at intervals and the branches from these ganglia. Note particularly the branches between the sympathetic ganglia and the adjacent spinal ganglia. These branches constitute the *ramus communicans* and consist of the visceral motor and visceral sensory fibers passing between the sympathetic and central nervous systems. The ganglia and branches of the sympathetic are particularly noticeable in the urogenital region.

Draw the spinal nerves, plexi, and sympathetic system as far as seen.

2. The sense organs of the head.—

a) The nasal cavities: The external nares lead into wide chambers, the nasal cavities. Cut off the external nares and the roof of the skull posterior to them, thus revealing the nasal cavities. They are separated by a median *septum*, partly bony. From the ventral region of the septum a conspicuous fold projects into the nasal cavity. On the posterior wall of the nasal cavity is a slight projection, a *concha* or *turbinal*. Posterior to this the nasal cavity connects by a passage with the roof of the mouth cavity, the nasal cavities thus serving as respiratory passages.

b) The eye: Although the eye is small, it can be dissected with a little care. Its parts are very similar to those of the fish eye. Make an incision through the skin around one eye and with the bone scissors remove the skull dorsal to and between the eyes. The two eyes are seen to be close together, separated by a median membranous *interorbital septum*. Near this septum on each side runs an artery. On the anterior dorsal surface of the eyeball is a gland, the *Harderian* gland. Over the posterior and ventral surface of the eyeball extends the much larger *lacrimal* or *tear* gland. Remove these glands, thus exposing the surface of the eyeball and the eye muscles. Extending from the interorbital septum to the dorsal surface of the eyeball is the *superior oblique* muscle. Posterior to this and inserted in the eyeball near it is the *superior rectus*. Between and ventral to these two is the *internal rectus*. Passing above the internal rectus are two nerves, the *trochlear* to the superior oblique muscle, and the *ophthalmic* branch of the trigeminus. Loosen the eyeball ventrally and, raising it as far as possible, examine the ventral surface. The anterior part of this surface is covered by a flat muscle, the *pyramidalis*, which originates on the eyeball and passes to the eyelids and nictitating membrane. Remove this and clean the ventral surface of the eyeball. The *inferior oblique* and *inferior rectus* muscles are then seen converging to their insertions on the ventral surface of the eyeball. The *external rectus* is posterior to them.

Remove the eyeball and open it by cutting off its dorsal side. Identify the coats of the eyeball, the lens, the cavities of the eye, and the two humors as in the elasmobranch eye, as the structure is practically identical with the latter. Note, however, the difference in the shape of the lens of the turtle eye.

c) The ear: The ear consists of two parts, a *middle* ear and an *internal* ear. The former is located posterior to the angle of the jaws internal to a circular area of skin. Remove this piece of skin and find beneath it a smaller cartilaginous circular plate, the *tympanic membrane* or *eardrum.* Make a cut around the margin of this and carefully raise it. Attached to its internal surface, posterior to the center, is a rod-shaped bone, the *columella,* whose inner end is fastened to the wall of a large cavity. This cavity is the *tympanic cavity* or *cavity of the middle ear.* It is an evagination from the first visceral pouch. Ventral to the inner end of the columella is a slit bounded by raised lips. This slit is the opening of the *auditory* tube connecting the pharyngeal cavity with the cavity of the middle ear and representing the stalk of the evagination by which the latter was formed. Considerably internal to the point of attachment of the inner end of the columella lies the internal ear. It will be more definitely located later. It is similar in structure to the internal ear of elasmobranchs.

3. Dorsal aspect of the brain.—Remove the roof of the skull and expose the brain. The brain is covered by a tough membrane, the *dura mater.* On cutting carefully through this a more delicate membrane, the *secondary meninx,* will be found adhering to the brain; this is more or less pigmented and vascularized. The space between the two membranes, crossed by strands, is the *subdural space,* and that between the dura mater and the skull is the *peridural space.* Remove the dura mater from the dorsal surface of the brain.

The brain has the same divisions as in the preceding animals, but the proportions of the parts are somewhat altered. The anterior half of the brain is composed of the conspicuous enlarged *cerebral hemispheres,* relatively much larger than in fish and amphibians; at their anterior ends, separated from them by a slight groove, are the small *olfactory bulbs.* Evidently, the whole olfactory apparatus is reduced, as compared with the selachian. Between the posterior ends of the cerebral hemispheres lies the choroid roof of the diencephalon, projecting dorsally as a choroidal sac that adheres to the dura mater and is generally torn off in removing the latter. The pineal body or epiphysis lies on this sac. With the removal of the choroidal sac the diencephalon is seen as a depressed area posterior and ventral to the cerebral hemispheres. Behind it are the two rounded optic lobes, and behind these the cerebellum, smaller than in selachians but larger than in amphibians. The cerebellum overhangs the medulla, whose roof is made of the usual tela choroidea; when this is removed, the cavity of the fourth ventricle is revealed. The dorsal rims of the medulla are, as usual, the somatic sensory columns, and the anterior ends of these are continuous with the cerebellum by way of the small auricular lobes of the latter. In the floor of the medulla the two somatic motor columns are conspicuous, one to either side of the midventral groove.

Draw the dorsal aspect of the brain.

4. The cranial nerves.—The dissection of the cranial nerves is a matter of some difficulty, and the following description is consequently not complete.

a) *The olfactory nerves:* These are the two stout nerves extending from the dorsal portions of the olfactory sacs to the anterior end of the olfactory bulbs of the brain.

b) *The optic nerves:* On cutting through the olfactory nerves and raising the anterior end of the brain, the optic nerves are seen as two stout trunks situated below the cerebral hemispheres and passing out of the orbit.

c) *The trochlear nerve:* This small nerve arises on each side of the brain in the dorsolateral angle between the optic lobe and the cerebellum. It passes ventrally and forward and will be seen by pressing the cerebral hemisphere away from the skull. It lies behind the larger oculomotor nerve. To find the course of the trochlear nerve in the orbit expose the undissected eye as before, clearing away the glands. Cut through the superior oblique muscle at its point of insertion on the eyeball and find below it the trochlear nerve terminating in this muscle. Medial to the trochlear is the ophthalmic branch of the trigeminus.

d) *The oculomotor nerve:* This nerve originates from the floor of the midbrain immediately in front of the trochlear and is seen by pressing the cerebral hemisphere away from the skull. Loosen the ventral side of the eyeball and raise it. Among the loose tissues between the eyeball and the floor of the orbit, generally adhering to the eyeball, is a nerve, the *maxillary* branch of the trigeminus. Free this from the eyeball. Cut away the pyramidal muscle, and, raising the eyeball and pressing it as far medially as possible, separate the inferior and external rectus muscles and find between and above them the stout white trunk of the optic nerve. The *oculomotor* is in contact with the ventral surface of the optic nerve and branches to the same four eye muscles as in the dogfish. These branches are not readily followed.

e) The trigeminus nerve: This is a stout trunk whose origin from the anterior end of the medulla will be seen by pressing the cerebellar region of the brain away from the skull. The trunk passes laterally and enters its ganglion, the *semilunar* ganglion, which lies in a depression in the medial wall of the skull. The trigeminus has three branches—the *ophthalmic*, the *maxillary*, and the *mandibular*—distributed to the orbit and nose, the upper, and the lower jaw, respectively. Remove the eyeball on the side where it is still present, leaving the ophthalmic and maxillary nerves intact. Cut through the roots of the nerves anterior to the trigeminus so as to raise the brain, and bend it away from the side being dissected. Follow the ophthalmic nerve forward and note its distribution to the nasal sacs. Follow it posteriorly toward the root of the trigeminus. It enters the skull; runs, in company with the trochlear nerve, between the dura mater and the skull; and finally joins the semilunar ganglion. Follow the maxillary nerve posteriorly. Besides the branch below the eyeball, already noted, there is a branch in the floor of the orbit, running obliquely forward. These two branches unite to form the main trunk of the maxillary nerve at the posterior end of the orbit. Trace this nerve posteriorly among the muscles to where it pierces the skull. At this point it is joined by the *mandibular* branch of the trigeminus. Trace this laterally. After branching into adjacent muscles the mandibular nerve proceeds ventrally and enters the lower jaw. Mandibular and maxillary branches pass together through a foramen in the skull and connect with the semilunar ganglion.

f) The facial and auditory nerves: These arise together from the side of the medulla just behind the root of the trigeminus and immediately separate into an anterior facial nerve and a posterior auditory. The latter is distributed to the internal ear. This is situated in the skull opposite the acousticofacial root. This part of the skull may be broken open with the bone forceps. The semicircular ducts, ampullae, and vestibule of the internal ear will be noted. The auditory nerve will be seen branching among these structures. The facial nerve passes through the anterior part of the ear capsule and will be seen again later.

g) The glossopharyngeal nerve: This arises by a small root from the medulla immediately posterior to the acousticofacial root and passes out through the posterior part of the ear capsule.

h) The vagus, the spinal accessory, and the hypoglossal: The vagus and spinal accessory (eleventh) nerves arise together by a number of roots from the side of the medulla posterior to the preceding nerve, the more anterior roots belonging to the vagus and the posterior ones to the accessory. On cutting through these roots the more ventrally situated roots of the hypoglossal (twelfth) nerve will be seen. The three nerves pass out from the skull close together.

i) The abducens nerve: Cut through all of the nerve roots on one side of the brain and tilt the brain toward the opposite side. The abducens nerves will be seen springing from the ventral surface of the medulla at about the same level as the acousticofacial root.

The seventh, ninth, tenth, and twelfth nerves may be traced farther, as follows: Turn the head ventral side up and remove the skin and superficial muscles from the hyoid apparatus. Locate the anterior and posterior horns of the hyoid. On the side of the neck, near the dorsal end of the anterior horn and posterior to it, the hypoglossal nerve will be seen emerging. It branches into the muscles over the anterior horn and sends a branch forward into the tongue muscles. Very near the point of emergence of the hypoglossal but situated more deeply and nearer to the cartilage of the anterior horn will be found the glossopharyngeal nerve. It runs between the two horns toward the median ventral line and supplies adjacent muscles and lining of the mouth cavity. Lateral to these nerves, a branch of the facial will be found crossing the anterior horn near its dorsal end and passing into the muscles lying along the posterior border of the mandible.

Make a median longitudinal incision through the whole floor of the mouth and pharyngeal cavities and open the two flaps so that the roof of these cavities is revealed. Locate the vagus (really vagosympathetic) trunk in the neck and trace it anteriorly to its point of exit from the skull, removing the mucous membrane from the roof of the pharyngeal cavity. The vagosympathetic trunk passes to the dorsal side of the hypoglossal nerve, seen above, and there enters a ganglion, the *superior cervical* ganglion of the sympathetic. From this ganglion numerous branches pass out. Internal to the hypoglossal locate the glossopharyngeal nerve, the carotid artery being situated between the two. Slightly anterior to these will be found the facial nerve, as it exits from the skull. Its branches pass to the muscles between the anterior horn of the hyoid and the lower jaw, one of them curving over the ventral surface of the horn, as noted above. The vagus nerve proceeds posteriorly and supplies the heart and other viscera. It will be noted that the facial, glossopharyngeal, and vagus nerves are much reduced, because of the loss of the lateral-line system and the gill apparatus. Note, however, that these nerves continue to supply the remains of the branchial bars and their muscles.

Add the cranial nerves to your drawing of the brain.

5. Ventral aspect of the brain.—Remove the brain from the skull and examine the ventral surface. On the ventral surface of the diencephalon note the optic chiasma, the infundibulum just behind this, with the hypophysis projecting ventrally from the latter. Note the roots of the abducens and hypoglossal nerves arising from the ventral surface of the medulla.

6. Median sagittal section.—Make a median sagittal section and study the cut surface. Identify the *fourth ventricle* in the medulla, the *aqueduct* or passage below the cerebellum, the *optic ventricle* in the optic lobe, the *third ventricle* in the diencephalon. Note the increased size of the diencephalon, as compared with the elasmobranch, and the backward extension of the cerebral hemisphere over the diencephalon. The diencephalon is divided into *epithalamus, thalamus,* and *hypothalamus,* as in the dogfish, each including the parts previously enumerated. Note that the cerebral hemisphere presents a solid medial wall, called the *septum.* Cut into the roof or pallium of the hemisphere. Note its cavity, the *lateral ventricle,* and the large mass protruding from the floor into the ventricle; this mass is the *corpus striatum.* Draw the section.

E. THE NERVOUS SYSTEM AND SENSE ORGANS OF THE PIGEON

The brain of the pigeon and other birds contrasts with that of the preceding animals in the greatly enlarged cerebral hemispheres and cerebellum and the pronounced curvature; whereas the olfactory bulbs and other parts of the olfactory apparatus are greatly reduced.

1. The spinal nerves and the autonomic system.—Carefully remove the remaining viscera from one-half of the trunk. Note the ventral rami of the spinal nerves passing laterally along the dorsal body wall between the ribs in the trunk region. Trace them toward the vertebral column and note, at the points where they emerge from the vertebrae, the ganglia of the autonomic system lying on the spinal nerves and the delicate white cords connecting the ganglia, forming the sympathetic trunks.

a) Spinal nerves and limb plexi: In the neck the *cervical* spinal nerves will be seen by separating the vertebral column from the skin. They pass out at segmental intervals. The vagus nerve is the white cord which passes ventral to the proximal portions of the cervical nerves.

The ventral rami of the last cervical nerves, together with that of the first of the trunk, form the *brachial* plexus to the wing. This is a network, formed by the union of branches of four stout nerves, which receives a small branch from the succeeding nerve.

The next five ventral rami pass out between the ribs. Following them is the *lumbosacral* plexus, divisible into three parts: the *lumbar,* the *sacral,* and the *pudendal* plexus. The lumbar plexus is formed by three nerves; from it nerves pass into the thigh. The sacral plexus arises from the union of five nerves, the first of which is the same as the third nerve contributing to the lumbar plexus. These five unite to produce a large trunk, the *sciatic* nerve, which passes along the dorsal side of the thigh between the muscles and proceeds down the leg. It will be found by separating the muscles along the middle of the dorsal surface of the thigh. It courses alongside the femoral artery and vein.

The remaining spinal nerves posterior to the sacral plexus form the pudendal plexus and pass obliquely posteriorly to the tail and cloacal region.

b) The autonomic system: This has already been identified on the sides of the vertebral column. It consists, on each side, of a chain of two cords and segmental ganglia. One of the cords passes ventral to the head of the rib, the other dorsal to it. A sympathetic ganglion lies fused to each spinal nerve in the trunk region as the latter emerges from the vertebral column. On scraping off one of these sympathetic ganglia, the *spinal ganglion* belonging to the spinal nerve will be found dorsal to it. At about the middle of the rib-bearing region a plexus of nerves and ganglia will be seen extending ventrally from the main sympathetic cords and surrounding the dorsal aorta and its main branches to the digestive tract. This is the *coeliac* plexus. Posterior to this region the sympathetic cords are reduced and consist of a single trunk on each side. A sympathetic cord accompanies the pudendal plexus and has a ganglion in the middle of this plexus. Anteriorly the sympathetic cords pass across the ventral side of the brachial plexus, having ganglionic enlargements on the latter, and then enter the vertebrarterial canals.

2. The sense organs of the head.—

a) The nasal cavities: Open one nasal cavity by a longitudinal slit just above the margin of the upper jaw from the external naris to the head. Note the median septum between the two nasal cavities and the swellings, the *turbinals* or *conchae,* projecting from the septum into the nasal cavity. There are three turbinals in a row: the first two large and conspicuous, the third and most posterior one consisting only of a small rounded swelling on the roof of the cavity in close contact with the posterior end of the second concha. Only this third concha is provided with olfactory epithelium. Beyond the conchae the nasal passages connect with the pharyngeal cavity.

b) The eye: Cut through the skin around the eyeball and also remove the roof of the skull between the two eyes. Note the relatively large size of the eyeballs and the *interorbital septum* between them. Along the dorsal margin of the septum course the two *olfactory nerves.* Press the eyeball outwardly away from the skull. Two thin, flat muscles will be seen extending to the eyeball from the orbit; the anterior one is the *superior oblique,* the posterior one the *superior rectus.* Cut through the superior oblique at its insertion on the eyeball and press it against the orbit. The *internal rectus* will now be seen extending to the eyeball ventral to the superior oblique. The white nerve crossing the orbit against the internal surface of the superior oblique is the *ophthalmic* branch of the trigeminus. Dorsal to it the smaller *trochlear* nerve is seen terminating on the superior oblique. The thin sheet of muscles on the surface of the eyeball is the *quadrate,* a muscle of the eyelids. On the anterior surface of the eyeball ventral to the superior oblique is a white, fatlike mass, the *Harderian gland.* Press the eyeball posteriorly and find anterior to this gland, against the anterior wall of the orbit, the *inferior oblique* muscle. On pulling the eyeball forward the *external rectus* is seen extending to the posterior surface of the eyeball. Free the ventral margin of the eyeball. In the posterior ventral region, on raising the eyeball, may be seen a small gland, the *lacrimal gland.* Two muscles will be seen on the ventral surface of the eyeball. The anterior one is the *inferior rectus,* the posterior one the *external rectus.* On cutting through the inferior rectus the *pyramid,* a muscle of the eyelids, will be revealed internal to it. Cut through all of the rectus muscles and the inferior oblique at their insertions on the eyeball and remove the eyeball, severing the optic nerve. The pyramid and quadrate muscles are now more readily seen extending on the surface of the eyeball to the optic nerve; the quadrate muscle is broad and dorsally situated, the pyramid narrow and ventral. They are concerned in operating the nictitating membrane. In the orbit note the extent of the Harderian gland.

Cut off the dorsal part of the eyeball and identify the structures of the eye. Note the *sclerotic* coat, continuing as the transparent *cornea* over the exposed part of the eye; the *conjunctiva,* passing over the external surface of the cornea and continuing onto the eyelids; the black *choroid* coat internal to the sclerotic and forming the *iris* in front; the soft *retina.* Note the peculiar ridged structure, the *pecten,* projecting from the choroid coat through the retina in the medial wall of the eyeball and extending to the lens. The pecten is a structure found in the eyes of birds and reptiles. Its function is uncertain; according to Menner (1938), it acts to throw a shadow on the retina, hence makes birds very sensitive to any movement in their visual field. Loosen the lens and observe that it is encircled by a structure continuous with the choroid coat and marked by radiating ridges, the *ciliary processes.* The whole structure, which is called the *ciliary body* and contains the *ciliary* muscles, holds the lens in place and can change the curvature and position of the lens. Note the shape of the lens—flat externally, more convex internally. The chambers of the eye and the two humors are as in the dogfish. Peel the iris from the cornea and note the stiff bony ring, composed of the small *sclerotic* bones, encircling the cornea.

Draw, showing structure of the eye.

c) *The ear:* The ear of birds consists of three parts—the *external* ear, the *middle* ear, and the *internal* ear. The external ear comprises the passage, the external auditory meatus, situated below and behind the eye. Cut into this on the same side of the head on which the eye was dissected and find, at its internal end, a circular transparent membrane, the *tympanic membrane.* Through the membrane the *columella* can be seen extending from its internal surface inwardly. Remove the tympanic membrane, noting the columella adhering to its internal surface. The cavity of the middle ear is now exposed; medially and ventrally it is connected to the pharyngeal cavity by the auditory tube; posterior and slightly dorsal to it is situated the internal ear. The inner end of the columella adjoins a tiny bone, the *stapes*, which fits into an opening, the *fenestra ovalis* or *vestibuli*, which leads into the internal ear. Look for these at the inner end of the columella. Next, carefully break away in small pieces the spongy bone behind the middle ear. Three bony *semicircular canals* are revealed. Each of them contains a membranous *semicircular duct*, as will be seen by breaking open one of them. The three ducts are situated in the same planes and have the same names as in elasmobranchs. The remaining structures of the internal ear, consisting of two small chambers—the *utriculus* and the *sacculus*—are too difficult to dissect.

3. Dorsal aspect of the brain.—Expose the brain, removing the roof of the skull and the side of the skull where the sense organs were dissected, including the medial wall of the orbit. Note the *dura mater* inclosing the brain; on removing this, note the very delicate *secondary meninx* next to the brain substance. Unlike the condition in the preceding forms, the brain closely fills the cranial cavity.

The brain is short and broad and strongly curved, in correlation with the biped gait. The curvature results from flexures of the brain in three regions. The chief or *primary flexure* occurs in the region of the midbrain, with the result that the posterior part of the brain is bent nearly at right angles to the anterior part. The second or *nuchal* flexure takes place in the medulla, bending the medulla at an angle to the spinal cord. The *pontal* flexure in the region ventral to the cerebellum bends the brain in the opposite direction from the other two flexures, with the result that this region of the brain is depressed.

The anterior end of the brain consists of the two very small *olfactory lobes*. Posterior to them are the large *cerebral hemispheres*, separated by a deep *sagittal fissure*. These are so enlarged posteriorly as to completely conceal the diencephalon from dorsal view, with the exception of the delicate *pineal body*, which is seen in the posterior end of the sagittal fissure. The large *optic lobes* of the midbrain are ventral to the posterior ends of the cerebral hemispheres. Posterior to the hemispheres is the curved *cerebellum*, marked by transverse grooves. Posterior and ventral to this is the *medulla oblongata*, its anterior end depressed beneath the cerebellum. The roof of the medulla is composed, as usual, of a tela choroidea.

Draw the brain from the side.

4. The cranial nerves.—These are somewhat difficult to follow in detail. Work on the side left intact. There are twelve pairs of cranial nerves.

a) The olfactory nerves: These are two stout and elongated nerves passing from the nasal sacs along the dorsal margin of the interorbital septum to the olfactory lobes.

b) The optic nerves: On the side where the wall of the orbit was removed note the stout, white *optic tract* in front of the optic lobe. Follow this toward the orbit and find the optic nerve connected with its anterior end.

c) *The trochlear nerve:* The cranial origin of this nerve is difficult to see at the present stage of the dissection. It arises in the deep groove between the optic lobe and the cerebellum and passes ventrally between the optic lobe and medulla. It runs forward in the floor of the cranial cavity to the orbit. To find it in the orbit, expose the intact eye as before. Cut through the superior oblique at its insertion on the eyeball and, lifting it, note the trochlear nerve passing to it and spreading out on its ventral surface. The ophthalmic branch of the trigeminus runs close to the trochlear nerve.

d) *The oculomotor nerve:* The cranial origin of this nerve will be seen later. It branches to the inferior oblique, superior, inferior, and internal rectus muscles. Remove the eye which is still in place, cutting the eye muscles as near the eyeball as possible and preserving the ophthalmic nerve intact. Look for the branches of the oculomotor among the eye muscles in question. The branch to the inferior oblique in the floor of the orbit is the most conspicuous of them.

e) *The abducens nerve:* This will be found on examining the posterior surface of the external rectus muscle. This nerve also supplies the pyramid and quadrate muscles.

f) *The trigeminus nerve:* This has three branches—the *ophthalmic*, the *maxillary*, and the *mandibular*. The ophthalmic has already been noted in the dorsal part of the orbit. Follow it forward, noting its distribution to the walls of the nasal cavities. In the floor of the orbit, near its outer margin, locate the maxillary nerve. Trace it forward, noting its branches to the orbit and upper jaw. Trace the maxillary nerve posteriorly, carefully cutting away tissues in its path. Caudad of the orbit it is joined by the mandibular nerve. Trace this, noting branches to muscles and main trunk passing into the lower jaw. Trace the common trunk of the maxillary and mandibular nerve toward the skull and note that they are joined in the skull by the ophthalmic nerve. At the point of union is the *semilunar ganglion*, lying in the skull. From the ganglion the trigeminus nerve may be traced to its origin from the side of the medulla below the optic lobe.

g) *The facial nerve:* This arises from the medulla just back of the root of the trigeminus and passes through the anterior part of the ear capsule, where it will be found by scraping away the latter.

h) *The auditory nerve:* This arises close to the facial and passes out with it into the ear capsule, to the various parts of which it is distributed.

i) *The glossopharyngeal and the vagus:* These nerves arise close together just behind the ear capsule and will be found there by carefully dissecting in the muscles. The glossopharyngeal is the smaller of the two and is anterior in position. It enters a ganglion, the *petrosal* ganglion, beyond which it is distributed to the palate, pharynx, and larynx. The vagus nerve is considerably larger than the glossopharyngeal. It passes laterally parallel and posterior to the glossopharyngeal and enters its ganglion, the *jugular* ganglion, which is united with the petrosal ganglion, the two forming a mass. Beyond this the vagus turns posteriorly and passes down the neck, supplying respiratory system, heart, and other viscera. Portions of the sympathetic system are intermingled with the ninth and tenth nerves.

j) The spinal accessory and the hypoglossal: The former passes out with the vagus and is distributed to certain muscles. The hypoglossal is found just posterior to the vagus. It is distributed to certain neck muscles and sends a branch forward to the tongue.

5. Ventral aspect of the brain.—Remove the brain from the cranial cavity, preserving the roots of the cranial nerves as far as possible. Those not kept attached to the brain will be found in the cranial cavity.

Note form of the olfactory lobes and cerebral hemispheres from the ventral aspect. Between the optic lobes is the diencephalon. In the center of this is the *optic chiasma,* marked by cross-lines. From the optic chiasma the strong white *optic tracts* pass laterad and dorsad to the optic lobes and dorsal part of the diencephalon. Behind the chiasma is a depressed area, the *infundibulum,* from which the *hypophysis* extends ventrally. The latter is usually left behind in removing the brain and will be found in a deep pit, the *sella turcica,* in the floor of the cranial cavity. The infundibulum bears a central cleft where the hypophysis was torn from it. At the sides of the infundibulum are the roots of the oculomotor nerves. Posterior to the diencephalon is the depressed medulla. Between the medulla and the optic lobe is the slender root of the trochlear nerve. On the ventral surface of the medulla are the roots of the abducens nerves; they should also be sought in the floor of the cranial cavity. On the sides of the medulla look for the roots of the fifth, seventh, eighth, ninth, and tenth nerves, situated in a row. The twelfth nerve arises from the ventral surface of the medulla about on the same level as the ninth and tenth roots. The eleventh nerve arises from the spinal cord by several roots and ascends to a position immediately behind the tenth root.

Draw the ventral view of the brain.

6. Sagittal section.—Make a median sagittal section of the brain and study the cut surface. In the medulla note the fourth ventricle overhung by the cerebellum. Note the thick ventral wall of the medulla and the pontal flexure causing a ventrally directed bend in the medulla. In the cerebellum observe the small *cerebellar ventricle* and the arrangement of the gray and white matter resulting in section in a treelike appearance, called the *arbor* *vitae.* Each fold of the cerebellum consists of a central plate of white matter surrounded by a thick covering of gray matter. Anterior to the cerebellum is a region consisting dorsally of the mesencephalon and ventrally of the diencephalon. The optic lobes do not appear in the section, but the median part of the midbrain forms the dorsal part of the section. A narrow cavity, the *third ventricle,* is present in the diencephalon and extends into the infundibulum. In front of the latter appears the optic chiasma. Note how the cerebral hemisphere arches back over the diencephalon and midbrain, and note the strong connection of the diencephalon with the hemisphere. The cavity of the cerebral hemisphere is not visible in the median section. The medial wall of the hemisphere is called the *septum,* its dorsal wall the *pallium.* Cut into the latter, noting its thinness, and inside note the cavity or *lateral ventricle* of the hemisphere and the great mass, the *corpus striatum,* bulging from the floor. The function of the corpus striatum is not definitely known, but it seems to have a steadying effect on voluntary movements; and the delicacy and precision of movement necessary in flight may account for the relatively enormous size of the corpus striatum in birds.

F. THE NERVOUS SYSTEM AND SENSE ORGANS OF MAMMALS

The mammalian brain is notable for the enormous size of the cerebral hemispheres, whose surface is still further increased in many mammals by convolutions, and for the high development of the cerebellum. Although this is not evident on the surface, the thalamus is also greatly enlarged and differentiated and has extensive connections with the cerebral hemispheres. There are four lobes in the midbrain, instead of the usual two. The olfactory apparatus is usually well developed and presents the same tracts and connections as in lower vertebrates. A pronounced curvature is characteristic of the mammalian brain.

For the complete dissection of the nervous system a new specimen is necessary, but the greater part of this system can be worked out on the same specimen as used for preceding systems. If a new animal is provided, open it by a longitudinal cut from the perineum through the anterior end of the sternum. If the old specimen is used, it will not be possible to see the branches of the autonomic system and the vagus to the viscera or the peripheral distribution of some of the cranial nerves. In working on the nerves all structures other than nerves may be removed in order to expose the latter. In following one nerve, adjacent nerves must not be destroyed.

1. The spinal nerves, the autonomic system, and the vagus.—

a) Cervical portion of the sympathetic and the vagus: Locate the vagus nerve at a point near the larynx. It lies alongside the carotid artery. The nerve crossing the vagus near the larynx and giving off branches into the sternohyoid, sternothyroid, and related muscles is the *descending branch* of the twelfth or *hypoglossal* nerve.

Rabbit: The vagus nerve and the cervical part of the sympathetic trunk lie together on the dorsal surface of the carotid artery. The vagus is larger and more lateral. Toward the medial side of the sympathetic trunk, posterior to the larynx, may be separated a delicate nerve, the *cardiac branch* of the *vagus* (*depressor nerve* of the heart). Trace the sympathetic posteriorly. Just in front of the subclavian artery it enters a ganglion, the *inferior cervical* ganglion. From this ganglion cords pass to either side of the subclavian artery, forming the *ansa subclavia*, and unite again to another ganglion, the *first thoracic ganglion*, situated posterior and dorsal to the artery.

Cat: The sympathetic trunk is inseparably bound with the vagus, the two forming a large *vagosympathetic* trunk coursing lateral to the carotid artery and bound with it by a common sheath. Trace it caudad. Just in front of the first rib, branches of sympathetic origin arise from the trunk and proceed toward the esophagus. Shortly posterior to this point the sympathetic separates from the vagus and generally enters a ganglion, the *middle cervical* ganglion, which lies in contact with the vagus. From this ganglion cords pass on either side of the subclavian artery, forming the *ansa subclavia*, and, proceeding directly dorsally, unite to form a large ganglion, the *inferior cervical* ganglion, which lies against the neck muscles between the heads of the first and second ribs.

From the inferior cervical ganglion in **both animals** *cardiac* branches pass to the heart. The conspicuous nerve lying lateral to the vagus is the *phrenic nerve* or nerve of the diaphragm. The right vagus, just after passing ventral to the subclavian artery, gives off the *recurrent* or *posterior laryngeal* nerve, which runs anteriorly along the side of the trachea to the larynx. The left recurrent nerve arises much farther posteriorly from the left vagus.

b) *The anterior cervical spinal nerves:* The spinal nerves emerge from the spinal cord in pairs between successive vertebrae, passing out through the intervertebral foramina. Those of the cervical region are called the *cervical* nerves; there are eight pairs of them. The ventral rami of the first four cervical nerves are loosely united with each other to form the *cervical* plexus; the last four, together with the first thoracic, form the *brachial* plexus. As the first two are small and more or less mingled with the posterior cranial nerves, they will not be studied at this stage of the dissection.

To expose the cervical nerves, pull the muscles which are inserted on the anterior end of the sternum (sternomastoid, sternohyoid, sternothyroid) laterally or cut across them where necessary, thus exposing the musculature of the vertebral column. Look along the side of this, dorsal to the carotid artery, and note the ventral rami of the spinal nerves emerging at intervals. At about the level of the posterior end of the larynx lies the third cervical nerve in the rabbit, fourth in the cat. As already stated, the nerves thus exposed are the ventral rami only; the dorsal rami are exposed only by more radical dissection, which will not be attempted here. The dorsal rami supply the epaxial musculature. Note the branches of the exposed ventral rami to the muscles of the side of the neck.

From the ventral ramus of the fourth cervical nerve (rabbit) and fifth cervical nerve (cat) arises the *phrenic* nerve. It passes posteriorly parallel to the vagus, in the rabbit close to the vertebral musculature. It receives a branch from the fifth (rabbit) or sixth (cat) cervical nerve and then continues posteriorly into the thorax. As it passes the sympathetic ganglia, it receives contributions from them. In the thorax it lies at the side of the pericardial sac, just ventral to the root of the lung. Trace it posteriorly and note how it spreads on the surface of the diaphragm. The phrenic nerves are the motor nerves of the diaphragm; their origin from the cervical nerves shows that the muscles of the diaphragm are derived from cervical myotomes.

c) *The brachial plexus:* The ventral rami of the fourth to eighth (rabbit) or fifth to eighth (cat) cervical nerves, together with the ventral ramus of the first thoracic nerve, are united by intercommunicating branches, called *ansae,* to form the brachial plexus, which innervates the muscles of the shoulder, breast, forelimb, and diaphragm. The fourth cervical (rabbit), or fifth (cat), takes part in the plexus only through its contribution to the phrenic nerve.

To expose the brachial plexus cut through the pectoral muscles near the midventral line and separate the pectoral muscles from the underlying serratus muscle. The plexus lies in the axilla along with the axillary artery and vein. Then cut through the pectoral muscles as near as possible to their insertion on the humerus and separate them from the muscles of the upper arm. In this way the course of the nerves into the forelimb is exposed.

The connections of the nerves of the plexus are so intricate that it is impossible to describe them. The following points may be noted, however. **Rabbit:** The fifth cervical immediately sends a branch to the sixth cervical and then proceeds laterally into the neck muscles. The sixth cervical is a broad nerve which, after communicating with the seventh nerve, passes to the shoulder muscles. The seventh is smaller and, after contributing to the eighth, likewise innervates the shoulder muscles. The eighth cervical and first thoracic unite to one trunk as they emerge from the vertebral column. From this trunk arise the nerves of the limb. **Cat:** The sixth cervical has a broad connection with the seventh and then proceeds to the shoulder. The seventh and eighth cervicals and the first thoracic are very stout trunks which are intricately connected with each other and from which proceed the nerves of the forelimb.

The chief nerves from the brachial plexus are the following:

1. The *phrenic* nerve. This was described above.

2. The *suprascapular* nerve. This is the most anterior nerve arising from the sixth cervical. The main part of this nerve passes between the supraspinatus and subscapular muscles to supply the supraspinatus and infraspinatus. In the cat a branch of this nerve passes over the shoulder to more superficial parts.

3. The *ventral thoracic* nerves. These nerves supply the pectoral muscles and will be found entering the inner surface of these muscles between the two incisions made above. They are the most ventral of the nerves of the plexus. There are two of these nerves—one arising from the seventh cervical, the other from the eighth cervical and first thoracic. The former is small in the rabbit.

4. The *subscapular* nerves. There are three of these, dorsally situated and passing into the inner surface of the shoulder. The first arises from the sixth cervical and passes to the subscapular muscle; the second arises from the seventh cervical and supplies chiefly teres major; the third comes from the seventh and eighth cervicals and runs posteriorly along the internal surface of the latissimus dorsi muscle.

5. The *axillary* nerve. This nerve originates chiefly from the seventh cervical. It passes through the upper part of the upper arm, ventral to the triceps, and, emerging on the lateral surface of the upper arm, supplies chiefly the deltoid muscles.

6. The *dorsal* or *long thoracic* nerve. This nerve is best located by examining the outer surface of the serratus ventralis muscle. On tracing it anteriorly the nerve will be found to pass internal to the scalenes and to spring from the seventh cervical nerve close to the vertebral column.

7. The *musculocutaneous* nerve (cat only). This arises from the ventral surface of the sixth and seventh cervicals. It passes to the biceps muscle, forking as it approaches the muscle. The posterior branch continues along the surface of the muscle and at the elbow passes to the lateral surface of the arm and supplies the skin of the forearm.

8. The *radial* nerve. This is the largest nerve springing from the plexus. Seventh and eighth cervicals and first thoracic nerves contribute to its formation. It passes to the upper arm and, coursing between the humerus and the triceps, turns distally. It supplies many muscles of the forelimb.

9. The *median* nerve. This nerve lies posterior to the radial. It arises in the cat by branches from the last three nerves of the plexus and in the rabbit chiefly from the first thoracic. It passes to the upper arm and then turns distally running along with the brachial artery.

10. The *ulnar* nerve. This lies just posterior and parallel to the median

nerve, originating chiefly from the first thoracic nerve. The ulnar and median nerves supply the limb distal to the elbow, although in the rabbit the median nerve innervates the biceps.

11. The *medial cutaneous.* This is the small nerve which runs in contact with the ulnar nerve. It turns superficially just above the elbow and is distributed to the skin of the forearm.

Draw, showing the main parts of the brachial plexus.

d) The thoracic portions of the vagus and the sympathetic: Trace the two vagi toward the heart. They pass dorsal to the roots of the lungs. The left vagus, just caudad of the aortic arch, gives off the left recurrent laryngeal nerve, which turns cephalad, passing on the dorsal side of the aorta, and ascends along the side of the trachea. At the roots of the lungs the vagi give rise to a plexus—the *pulmonary* plexus—to the lungs. This plexus also extends to the heart as the *cardiac* plexus. The cardiac branches of the sympathetic system, noted above, join the cardiac plexus. The cardiac plexus is situated at the bases of the aorta and pulmonary arteries. In the rabbit the cardiac branches of the vagus may be traced into this plexus.

Caudad of the pulmonary plexus the two vagi in the **rabbit** continue posteriorly along the sides of the esophagus, to which they furnish small branches, and penetrate the diaphragm. In the **cat** each vagus divides just posterior to the root of the lungs into dorsal and ventral branches. The ventral branches of the two sides immediately unite into one trunk, which proceeds posteriorly, lying on the left ventrolateral surface of the esophagus. The two dorsal branches continue posteriorly, lying along the right and left sides of the esophagus; near the diaphragm on the dorsal side of the esophagus they unite to one trunk. In this manner are formed the *dorsal* and *ventral divisions* of the vagi; they pass through the diaphragm. In their course along the esophagus they furnish branches to it.

Locate again the inferior cervical ganglion. Note the communicating branches from this ganglion to the brachial plexus. In the cat a particularly stout branch extends anteriorly ventral to the bases of the sixth to eighth cervical nerves, giving branches to them. Trace the sympathetic trunk posteriorly from the inferior cervical ganglion. The contents of the pleural cavities may now be cleaned out. The sympathetic trunk is a white cord lying to each side of the vertebral column, passing ventral to the heads of the ribs. At segmental intervals, generally in the places between the ribs, it presents a ganglionic enlargement.

e) The thoracic spinal nerves: The first thoracic nerve contributes to the brachial plexus, as already learned. The ventral rami of the remaining thoracic nerves pass laterally as the *intercostal* nerves, lying along the posterior side of each rib. These nerves are readily exposed by running the point of an instrument along the posterior side of each rib, slitting open the fascia of the intercostal muscles. As each nerve emerges from the intervertebral foramen, it receives one or two *communicating branches* (*rami communicantes*) from the adjacent sympathetic ganglion. These branches are rather delicate, and the student may not be able to see them. The *dorsal rami* of the thoracic spinal nerves supply the epaxial muscles. To see them, turn the animal dorsal side up and carefully cut down through the mass of epaxial muscles close to the vertebrae. The dorsal rami will then be seen emerging from the vertebral column and penetrating the epaxial mass, accompanied by blood vessels. There are twelve or thirteen pairs of thoracic nerves.

f) The abdominal portions of the vagus and the sympathetic: Trace the vagi into the peritoneal cavity, removing the liver if not already done. In the **rabbit** the left vagus crosses the ventral surface of the esophagus obliquely to the right and is distributed to the lesser curvature and ventral surface of the stomach. The right vagus crosses the dorsal surface of the esophagus obliquely to the left and is distributed to the dorsal surface of the stomach. In the **cat** the ventral division of the vagus passes to the lesser curvature, the dorsal division to the greater curvature. In both cases the vagi form plexi on the stomach, called the *ventral* and *dorsal gastric plexi*, which also connect with the near-by sympathetic plexi, described in the next paragraph.

Locate again the posterior part of the thoracic portion of the sympathetic trunk. Expose it and note the nerve, the *greater splanchnic* nerve, which arises from the sympathetic trunk on each side and passes obliquely ventrally toward the diaphragm. In the cat this nerve is accompanied by additional smaller nerves, the *lesser splanchnic* nerves, arising from the sympathetic shortly posterior to the origin of the greater splanchnic nerve. The splanchnic nerves pass to either side of the crura of the diaphragm into the peritoneal cavity. (The crura of the diaphragm are the muscular cords which fasten the diaphragm to the lumbar vertebrae.) Turn the abdominal viscera to the right and look on the left surface of the superior mesenteric artery near its origin from the aorta. Two prominent sympathetic ganglia will be found lying on the superior mesenteric artery. These are the *coeliac* and *superior mesenteric* ganglia; the former lies in front of, or on the left surface of, the artery; the latter behind, or on the ventral surface of, the vessel. The two ganglia are bound together by a strong connection. The splanchnic nerves of both sides may be traced into the coeliac ganglion. From this ganglion a prominent *coeliac plexus* will be seen extending toward the stomach, where it connects with the gastric plexi of the vagi. This great sympathetic plexus, formed around and dorsal to the stomach, is often called the *solar* plexus. From the coeliac and superior mesenteric ganglia and adjacent plexi also arise plexi for the liver, spleen, adrenal glands, gonads, and the great blood vessels. Some of these will probably be seen. The *inferior mesenteric* ganglion of the sympathetic system lies in the mesocolon alongside the inferior mesenteric artery. It is situated in the *inferior mesenteric plexus*, from which networks extend to adjacent structures.

The main sympathetic trunk of the abdominal region should now be traced caudad from the place of origin of the splanchnic nerves. The two trunks descend deep dorsally, lying in the groove between two muscle masses. At segmental intervals they have ganglionic enlargements from which nerves pass to the ganglia and plexi, already noted. At the posterior end of the peritoneal cavity the sympathetic trunks gradually diminish and disappear.

g) The lumbar and sacral spinal nerves and the lumbosacral plexus: There are seven pairs of *lumbar* nerves and four (rabbit) or three (cat) pairs of *sacral* nerves. The ventral rami of the last four lumbar nerves form a *lumbar* plexus, those of the sacral nerves a *sacral* plexus; but since the two plexi are united with each other, they may be considered together as the *lumbosacral* plexus.

Remove all viscera from the peritoneal cavity, including the postcaval vein and aorta. In the dorsal wall is a muscular mass extending from the vertebrae to the pelvic girdle. This consists of a lateral larger muscle, the *iliopsoas*, and a smaller medial one, the *psoas minor*. In the rabbit the psoas minor is a slender muscle which occupies only the posterior part of the middorsal region; its stout shining tendon passes to the dorsal side of the inguinal ligament. In the cat the psoas minor extends nearly as far anteriorly as the iliopsoas; it narrows posteriorly to a tendon, which passes obliquely laterally on the ventral surface of the iliopsoas, which is thus exposed both medially and laterally to the tendon of the psoas minor. The psoas minor covers a part of the iliopsoas in both animals, and the greater part of the lumbar plexus is situated between the two muscles. Note the abdominal parts of the sympathetic cords between the posterior portions of these muscles.

Locate the last thoracic spinal nerve. It lies about one-half inch posterior to the last rib. The first nerve posterior to this on the dorsal wall is the ventral ramus of the first lumbar nerve. Shortly posterior to this is the second lumbar nerve. These two nerves pass to the muscles and skin of the abdominal wall; in the cat each divides into two branches. The third lumbar nerve emerges dorsal to the iliopsoas muscle and divides into a larger lateral branch to the abdominal wall and a more slender medial branch, which passes obliquely caudad, reaching and following the course of the iliolumbar artery and vein. The fourth lumbar nerve is the first of the lumbar plexus. It has two main branches, the *lateral cutaneous* nerve and the *genitofemoral* nerve. The former is the stout trunk which emerges between the iliopsoas and psoas minor muscles and accompanies the course of the iliolumbar artery and vein, passing to the thigh. The genitofemoral nerve is a long slender nerve which runs along the medial border of the psoas minor muscle, lateral to the sympathetic cords. In the posterior part of its course it accompanies the external iliac artery. It supplies the thigh and abdominal wall of, and adjacent to, the inguinal region. After locating these two branches of the fourth lumbar trace them toward the vertebral column, removing the psoas minor as far as necessary. Find the point of emergence of the fourth lumbar from the vertebral column and note the connection, very stout in the cat, between the fourth lumbar and the fifth.

The fifth lumbar contributes by means of its connection with the fourth lumbar to the lateral cutaneous branch named above and also forms a strong union with the sixth lumbar. To expose these remove the rest of the psoas minor. The common trunk, formed by the union of branches from the fifth and sixth lumbar nerves, passes laterally as the large *femoral* nerve. Trace this to the thigh. It courses along the center of the medial surface of the thigh in company with the femoral artery and vein. It innervates adjacent muscles of the thigh and then continues down the shank and foot as the *saphenous* nerve.

The *obturator* nerve arises from the connecting band between the sixth and seventh lumbar nerves and passes obliquely caudad, dorsal to the pubis, through the obturator foramen and into the gracilis and other muscles.

The seventh lumbar and the first sacral unite to form a very large trunk, the *sciatic* nerve. The sixth lumbar and second sacral also contribute small branches to this nerve. Follow the sciatic nerve. It turns dorsally, passing between the ilium and the vertebral column. Thrust an instrument through the place where it turns and dissect where the instrument emerges on the dorsal side of the animal. On separating the muscles at this place the sciatic nerve is exposed. Expose it as near to the vertebral column as possible. The *gluteal* nerves will be seen separating from the anterior side of the main trunk and passing into the gluteus muscles. (The nerve on the posterior side of the sciatic trunk is the posterior cutaneous, described below.) Follow the sciatic nerve down the leg. After giving off branches into the thigh muscles it divides shortly above the knee into a lateral branch, the *peroneal* nerve, which passes between the insertions of the biceps femoris and the gastrocnemius, and a more medial branch, the *tibial* nerve, passing between the two heads of the gastrocnemius.

The sacral nerves are united by ansae to form the sacral plexus. The first sacral also, as seen above, takes part in the formation of the sciatic nerve. The chief nerves arising from the sacral plexus are the *pudendal* nerve and the *inferior haemorrhoidal*. The latter arises in the cat from the point of union of the three sacral nerves and passes to the bladder and rectum. The pudendal nerve arises from the large trunk formed by the union of the second and third sacral nerves and may also in the cat receive a branch from the sciatic. This trunk passes laterally parallel and posterior to the sciatic. From it arises the pudendal nerve, which turns toward the rectum and urogenital organs, and the *posterior cutaneous* nerve, which continues laterally into the biceps femoris muscle. It will be found by turning the animal dorsal side up and looking where the sciatic nerve was exposed. The nerve in question lies immediately posterior to the sciatic nerve and enters the biceps femoris. The fourth sacral nerve in the rabbit is of moderate size; it passes laterally and then turns to the sides of the rectum, which it innervates in common with the pudendal nerve.

Draw, showing the lumbosacral plexus.

The foregoing nerves are all the ventral rami of the lumbar and sacral nerves. To see the small dorsal rami of the lumbar nerves, proceed as directed for the dorsal rami of the thoracic nerves. The *caudal* spinal nerves will not be considered.

2. The spinal cord and the roots of the spinal nerves.—With the bone scissors cut out a piece of the vertebral column, two or three inches long, from the posterior thoracic and anterior lumbar region. Remove the epaxial muscles from this piece so as to expose the vertebrae, and with the bone scissors cut off the neural arches of the vertebrae, thus exposing the neural canal. In this canal, but not completely filling it, lies the spinal cord. Note that the spinal cord is loosely inclosed in a tough membrane, the *dura mater*, from which strands pass to the walls of the neural canal. The space between the dura mater and the spinal cord is the *subdural space*. Slit open the dura mater. The spinal cord is closely invested by a membrane, the *pia mater*, which cannot be separated from its surface. Between these two is a delicate membrane, the *arachnoid*, which is almost impossible to identify in gross dissection. The arachnoid and pia mater of mammals together correspond to the secondary meninx of lower vertebrates. The spaces around and between these membranes are filled in life with the *cerebrospinal fluid*, which is a modified lymph.

From the sides of the spinal cord observe the roots of the spinal nerves arising in pairs at segmental intervals. They are insheathed in the dura mater, which follows them to their exit from the intervertebral foramina and is continuous with their sheaths outside of the vertebral column. Examine one of the roots in detail. Although it appears at first glance to be single, a little gentle picking in the center of the root with the point of a probe will reveal that it is composed of two parts. One of these, the *dorsal* root, is attached to the dorsolateral region of the cord and near the intervertebral foramen bears a large oval swelling, the *dorsal* or *spinal ganglion.* The dorsal root in mammals carries sensory fibers only, and the nerve cells from which the sensory fibers originate are located in the spinal ganglion. The other root, the *ventral* root, arises from the ventrolateral region of the cord by several branches, which unite to one trunk. The ventral root carries motor fibers only, arising from motor cells in the cord. The dorsal and ventral roots unite beyond the ganglion to form the spinal nerve, which then exits through the intervertebral foramen and divides into the dorsal ramus to the epaxial muscles and adjacent skin, the ventral ramus to the hypaxial muscles and adjacent skin, and the communicating rami to the sympathetic system. These rami were already seen.

Cut through the roots of the spinal nerves and remove a small section of the spinal cord for examination. Identify in the median dorsal line a groove, the *dorsal median sulcus;* in the median ventral line, another groove, the *ventral median fissure.* Lateral to the dorsal median sulcus is the *dorsolateral sulcus,* along which the dorsal roots enter the cord. The region between the dorsal median and dorsolateral sulci is called the *dorsal funiculus.* The lateral region of the cord between the dorsolateral sulcus and the line along which the ventral roots emerge is the *lateral funiculus.* Between this and the ventral median fissure is the *ventral funiculus.*

Make a diagram through the cord, showing its funiculi and the roots of the spinal nerves.

Make a clean cut across the cord and examine the cut surface. The section is divisible into a central darker material, the *gray matter*, shaped like a butterfly, in which the nerve cells of the cord are located; and a much thicker, white material, the *white matter*, surrounding the gray matter and composed of nerve fibers. The white matter is subdivisible into the funiculi named above. Each funiculus consists of a number of tracts or bundles of fibers, whose functions are known; but these tracts are not visibly differentiated from each other.

3. The peripheral distribution of the posterior cranial nerves.—In this section will be described the peripheral course of the fifth, seventh, and ninth to twelfth cranial nerves. For the complete dissection of these it is necessary to have a specimen of which the head is intact, but most of them can be found, in part at least, on the same specimen on which the previous dissections were made.

a) The eleventh or spinal accessory nerve: This nerve supplies the sternomastoid, cleidomastoid, levator scapulae ventralis, and trapezius muscles. It is a pure motor nerve and is apparently derived from the vagus.

Rabbit: Separate the sternomastoid and cleidomastoid, on the one hand, from the basioclavicularis and levator scapulae ventralis, on the other. Running near the dorsal border of the levator scapulae ventralis and parallel to it is the spinal accessory nerve. Branches of the second to fourth spinal nerves pass ventral to it and unite with it by branches. Trace it posteriorly and note its branches on the inner surface of the trapezius. Trace it anteriorly and note branches to the levator scapulae ventralis, sternomastoid, and cleidomastoid.

Cat: Cut through the clavotrapezius near its origin and deflect it ventrally, thus exposing the levator scapulae ventralis. On the inner surface of the clavotrapezius along the dorsal border of the levator scapulae ventralis runs the main part of the spinal accessory nerve. Trace it posteriorly, noting branches into the trapezius muscles and the levator scapulae ventralis. Trace it anteriorly. It passes dorsal to the second cervical nerve, to which it is connected by a network, and near this region gives branches to the sternomastoid and cleidomastoid muscles. It then passes through the cleidomastoid muscle.

b) The vagus, the sympathetic, and the hypoglossal nerves: Follow the vagus and sympathetic anteriorly. Stretch the head forward by cutting across the lateral muscles of the neck. At about the level of the posterior end of the larynx the vagus and carotid artery are crossed ventrally by the *descending branch* of the *hypoglossal* or twelfth cranial nerve. This passes obliquely caudad toward the median line and supplies the sternohyoid, sternothyroid, and thyrohyoid muscles. Continue forward. At about the place where the common carotid artery divides into external and internal carotids, a conspicuous nerve is seen crossing the ventral surface of the vagus and carotid artery and curving anteriorly. This is the main part of the *hypoglossal* nerve. Follow it forward. It passes to the dorsal side of the mylohyoid muscle, which may be cut, and innervates the muscles of the tongue.

About halfway between the descending branch and main part of the hypoglossal nerve, but deeper dorsally and passing to the dorsal side of the carotid artery, is situated the *superior laryngeal* branch of the vagus nerve. It runs obliquely caudad to the larynx, which it penetrates, passing through the fibers of the thyrohyoid muscle.

Follow the vagus and sympathetic once more. At the place where the descending branch of the hypoglossal crosses them, the two separate in the cat. Shortly anterior to this the vagus in both animals presents an elongated swelling, the *nodosal* ganglion. At about the same level, but more medial in position, the sympathetic trunk enters the *superior cervical ganglion* of the sympathetic, an elongated pinkish body. The two ganglia lie just posterior to the hypoglossal as it curves forward into the tongue.

The hypoglossal, the accessory, the vagus, and the sympathetic are all involved in a plexus in which the first cervical nerves also take part.

c) The ninth or glossopharyngeal nerve: This lies very close to the main part of the hypoglossal nerve but more deeply dorsal. Dissect directly internal to the hypoglossal, where it curves anteriorly to the tongue. The *glossopharyngeal* is a smaller nerve lying dorsal to the hypoglossal along the sides of the pharynx anterior to the larynx. It is situated between the two horns of the hyoid. It divides into two branches: a smaller *pharyngeal* branch passing medially into the pharynx and a main *lingual* branch which enters the tongue. The former is a motor nerve to muscles of the pharynx, while the lingual branch is a nerve of taste.

Follow the nerves thus far described toward the point where they emerge from the skull. They are found to converge to a point to the medial side of the tympanic bulla. Here the ninth, tenth, and eleventh nerves emerge from the brain through the jugular foramen, located on the medial side of the bulla. The twelfth nerve emerges near the others through the hypoglossal foramen (consisting of several openings in the rabbit).

d) The seventh or facial nerve: The main part of this nerve is very superficial in position. It emerges at the posterior end of the masseter muscle at the base of the ear, in a sort of depression. On carefully searching in this region it will be found as a stout white band, in contact with the main part of the external carotid artery. At this place the facial gives off a branch to the posterior part of the digastric muscle, and the *posterior auricular* nerve to the pinna. (The large nerve to the pinna, which may be noticed dorsal to this branch of the facial, is the *great auricular* nerve originating in the cervical plexus.) The facial then proceeds forward, branching over the external surface of the masseter muscle, and passes to the lips and region of the eye. It supplies the various parts of the platysma muscle, which, it may be recalled, is a dermal muscle of the head and neck, serving to move the ears, lips, eyelids, whiskers, etc. The platysma muscle is a branchial muscle originally belonging to the hyoid arch, hence its innervation by the facial nerve.

e) The fifth or trigeminus nerve: This nerve has three main branches: the *ophthalmic*, the *maxillary*, and the *mandibular*. The former is best studied with the eye, since it passes into the orbit.

To locate the mandibular branch of the trigeminus proceed as follows, freeing one half of the mandible. Cut through the attachment of the digastric to the mandible and deflect the digastric backward. Cut through the attachments of all of the muscles along the medial surface of the mandible, keeping the knife against the bone. Next, free the lateral or outer surface of the body of the mandible from muscle attachments, chiefly the masseter. Cut through the symphysis of the mandible (place of junction of the two halves of the mandible at their anterior tips). Carefully bend the half of the mandible thus freed outward, so as to expose the side of the muscular mass which forms the floor of the mouth and pharyngeal cavities. The main part of the mandibular branch of the trigeminus, the *inferior alveolar* nerve, will now be seen passing into a foramen, the mandibular foramen, situated on the medial surface of the mandible. In the rabbit the *mylohyoid* nerve, another branch of the mandibular, will be noted to the medial side of the inferior alveolar and proceeding ventrally to muscles of the floor of the mouth cavity. The inferior alveolar nerve runs in the interior of the mandible supplying the teeth and then emerges through the mental foramen on the lateral surface of the mandible at the level of the diastema. There the nerve, now named the *mental* nerve, may be found and followed into the lower lip.

Trace the inferior alveolar nerve posteriorly. It converges toward another branch of the mandibular nerve, the *lingual* nerve, which should then be followed forward. It passes into the tongue, lying close to the hypoglossal. The lingual branch of the trigeminus innervates the mucous membrane of the tongue but is not a nerve of taste.

Follow both lingual and inferior alveolar nerves centrally again. In front of the tympanic bulla, behind the point where the body of the mandible bends dorsally into the ramus of the mandible, will be seen the *auriculotemporal* branch of the mandibular nerve, joining the other two. On tracing it peripherally it is found to pass to the skin of the cranial side of the pinna and in the cat also sends branches along the side of the face, in company with the branches of the facial.

The tympanic bulla may now be exposed. Emerging from the bulla will be found a slender nerve which very soon joins the lingual branch of the mandibular. This is the *chorda tympani* (so called because it runs in the tympanic membrane), a branch of the facial nerve. Its fibers pass out with the lingual nerve and supply the taste buds on the anterior part of the tongue; it also innervates the sublingual and submaxillary salivary glands, by way of the submaxillary ganglion of the autonomic system.

Besides the branches of the mandibular nerve here named, there are branches to the muscles of mastication, namely, the temporal, the masseter, the anterior part of the digastric, and the pterygoids.

Remove the half of the mandible. This will reveal additional branches of the mandibular nerve. One of these, the *buccinator*, will probably be noticed extending to the angle of the mouth, where it supplies the masseter muscle and the lips. The main trunk of the *maxillary* nerve, the second branch of the trigeminus, may now be sought. It is a very stout trunk lying at the sides of the palate in front of, and more deeply situated than, the main trunk of the mandibular nerve. It is somewhat concealed by an artery (internal maxillary) which runs along its ventral surface and should be removed. The maxillary nerve is then revealed as a large trunk which passes forward along the side of the hard palate and disappears dorsal to the teeth. Cut away the zygomatic arch on the same side on which the half of the mandible was removed; in the rabbit cut away also the ridge which holds the molar and premolar teeth. By this operation the contents of the orbit are revealed. Note in the cat the small reddish *infraorbital* salivary gland lying close to the maxillary nerve. In the rabbit the very large reddish mass of the *Harderian* gland and the smaller yellowish mass of the *infraorbital* salivary gland anterior to it are readily noticed. The maxillary nerve should now be investigated. It is seen to divide into a large main trunk, the *infraorbital* nerves, and a small medial branch, the *sphenopalatine* nerve, which passes into the hard palate. The infraorbital nerves pass forward above the teeth, which they supply, and emerge through the infraorbital foramen, situated internal to the root of the zygomatic arch. On separating the upper lip from the teeth the foramen is readily found, and the nerve is seen emerging from it to supply the upper lip and side of the nose. Follow the sphenopalatine nerve toward the palate, cutting away the bone. It connects with a ganglion, the *sphenopalatine* ganglion of the sympathetic system. This ganglion lies near the sphenopalatine foramen. The chief branch of the sphenopalatine nerve is the *palatine* branch, which passes into the hard palate by a foramen. In the cat this nerve arises before the ganglion is reached; in the rabbit, beyond the ganglion. Other branches of the sphenopalatine nerve pass from the ganglion into the nasal cavity.

4. The sense organs of the head.—

a) The eye, the eye muscles, and the nerves of the orbit: Dissect on the other side from that on which the cranial nerves were worked out. Identify upper and lower eyelids and the nictitating membrane, a fold projecting from the anterior corner of the eye. Make a slit through the junction of upper and lower lids at the posterior corner of the eye so that the eyelids can be pulled away from the eyeball. Note that the skin passes onto the inner surface of the eyelids and continues over the exposed surface of the eyeball, thus forming the outermost covering membrane, the *conjunctiva*, for this part of the eyeball. Make an incision through the skin above the eye and deflect the skin downward toward the eye on which you are working, stretching the skin away from the head. On the skin of the inner surface of the upper eyelid note a thin sheet of muscle fibers, proceeding in a somewhat circular direction. This is the *orbicularis oculi*, a part of the platysma, and has the function of closing the eyelids.

Rabbit: Stretch the upper eyelid away from the head and clean away the connective tissue between it and the eyeball. A thin sheet of muscle will be found extending from beneath the supraorbital arch to the upper eyelid. This is the *levator palpebrae superioris*, which raises the eyelid. Repeat the foregoing directions on the lower eyelid, stretching the skin away from the eyeball. On the inner surface of the lower eyelid note the rest of the orbicularis oculi. The *depressor palpebrae inferioris* may be noted extending from the zygomatic arch to the lower eyelid; it lowers the eyelid. Remove the surrounding skin and eyelids, cutting them away from the eyeball. With the bone clippers cut away the supraorbital arch and clean away tissue between the dorsal surface of the eyeball and the orbit. A slender but strong muscle will now be seen extending from about the middle of the wall of the orbit to the dorsal surface of the eyeball; this is the *superior oblique* muscle. It separates the thin sheet of the levator palpebrae superioris into two parts, which pass on either side of it. Trace the superior oblique to the wall of the orbit. Here there will be found a tendinous cord, the *trochlea*, over which the muscle passes. Next, remove the levator palpebrae superioris and find underneath its posterior portion the thin, flat *superior rectus* muscle. The insertion of the superior oblique on the eyeball is concealed under the margin of the superior rectus.

Remove the half of the mandible and the zygomatic arch on the side on which you are working. This fully exposes the ventral side of the eyeball. Along the ventral surface of the outer part of the eyeball extends the yellowish *infraorbital* salivary gland. Medial to this, extending beneath the eyeball, is the larger *Harderian* gland, which pours its secretion onto the nictitating membrane. Remove these glands; note the white part of the Harderian gland extending far medially. The *inferior oblique* muscle is now seen extending to the eyeball from the anteroventral region of the orbit. Posterior to it is the *inferior rectus* muscle, originating from the posteroventral region of the orbit. Note the branch of the *oculomotor* nerve running along the anterior border of the inferior rectus and supplying both muscles. The nerve which runs along the posterior border of the inferior rectus, innervating the lower eyelid, is the *zygomatic* branch of the maxillary nerve. Immediately behind the inferior rectus is the *external* or *lateral rectus*. The nerve passing along the posterior margin of the external rectus is the *lacrimal* branch of the maxillary. It passes to the *lacrimal* gland and to the skin between the eye and base of the pinna. The lacrimal gland is a small, reddish body which will be found by pressing the eyeball forward and searching against the posterodorsal wall of the orbit. Two nerves pass the point of origin of the external rectus from the orbit. The larger is the *oculomotor*, the smaller the *abducens*. Cut through the insertions of the inferior oblique and inferior and external recti at the eyeball and deflect them ventrally. Above the inferior rectus the *internal* or *medial rectus* will be seen inserted on the eyeball. Look on the inner surface of the external rectus and find the abducens nerve, curving around the posterior border of the origin of this muscle and passing onto its surface. Return to the dorsal surface of the eyeball, cut through the insertion of the superior oblique at the eyeball, and press the eyeball ventrally. Two nerves will be seen on the medial wall of the orbit. The lower one is the *trochlear* nerve. Trace it to the medial surface of the superior oblique. The upper nerve is the *frontal* nerve, one of the main branches of the *ophthalmic* branch of the trigeminus. It passes to the dorsal part of the orbit and exits through the anterior supraorbital foramen, to be distributed to the upper eyelid and skin in front of the orbit. It may have been cut in removing the supraorbital arch. The white part of the Harderian gland will be noted in the anterior part of the orbit. Cut through all of the insertions of the eye muscles at the eyeball and through the optic nerve, removing the eyeball. The *optic nerve* is the stout white trunk near the superior rectus. The muscles around the optic nerve, exclusive of those already identified, belong to the *retractor bulbi.* Find the main trunk of the oculomotor nerve and trace its branches to the retractor bulbi and superior and internal recti. The main nerve curves below the optic nerve. The *nasociliary* branch of the ophthalmic nerve may be noted passing between the superior oblique and the retractor bulbi. Its main portion, the *ethmoidal* nerve, leaves the orbit by a small foramen in front of the superior oblique muscle. On tracing this nerve posteriorly, fine branches to the orbit may be seen.

Trace the nerves of the orbit to their exits from the skull. The third, fourth, and sixth nerves and the ophthalmic and maxillary branches of the trigeminus pass through the orbital fissure. The mandibular branch of the trigeminus passes through the foramen lacerum.

Cat: Remove the eyelids and the surrounding skin, cutting them away from the eyeball. Remove the half of the mandible and the zygomatic arch from the side on which you are working. Press the eyeball ventrally away from the supraorbital arch. In the anterodorsal angle of the orbit a strong fibrous connection will be found between the wall of the orbit and the eyeball. On investigating this it is found to consist of two fibrous bands which form a pulley; this is known as the *irochlea.* The tendon of the *superior oblique* muscle passes over the trochlea and is inserted on the eyeball. Its insertion is much expanded and extends caudad from the trochlea. Posterior to the insertion of the superior oblique is a thin, flat muscle, the *levator palpebrae superioris,* or elevator of the upper eyelid. This passes to the dorsoposterior surface of the eyeball. Cut through this at its insertion. Posterior to this muscle in the dorsoposterior angle of the orbit is the flattened *lacrimal* gland. Cut out the nictitating membrane and examine its internal surface. It is found to be roughened, because of the presence of the *Harderian* gland in its wall.

Turn to the ventral surface of the eyeball, exposed by the removal of the mandible, the zygomatic arch, and part of the hard palate. Identify again the small reddish *infraorbital* salivary gland, situated back of the last tooth. On clearing away connective tissue and fat the *inferior oblique* eye muscle will be seen extending from the anterior part of the orbit to the ventral surface of the eyeball. Ventral and at right angles to the inferior oblique is the *inferior rectus.* The branch of the oculomotor which innervates the inferior oblique runs along the posterior border of the inferior rectus. The *zygomatic* branch of the maxillary nerve passes along the posterior border of the inferior rectus to the lower eyelid but may have been destroyed. Posterior to the inferior rectus is the *external rectus;* between and internal to them appears one of the four parts of the *retractor bulbi* muscle. Along the posterior border of the external rectus runs the *lacrimal* branch of the maxillary nerve, supplying the lacrimal gland and adjacent skin. On detaching the eyeball from the posterior wall of the orbit another part of the *retractor bulbi* will be seen, next posterior to the external rectus; dorsal to this is the *superior rectus.* Cut through both obliques at their insertions and press the eyeball posteriorly. Note the *internal rectus* on the anterior surface of the eye and above it the remainder of the retractor bulbi.

Cut through all of the eye muscles and the optic nerve at their insertion on the eyeball and remove the latter. Note the four parts of the retractor bulbi around the optic nerve. Deflect the external rectus ventrally and note the *abducens* nerve ascending on its inner surface. Running along the ventral surface of the optic nerve, note a slender nerve, the *long ciliary* branch of the *ophthalmic* branch of the trigeminus; it accompanies the optic nerve into the eyeball. Look on the inner surface of the inferior rectus for the branch of the *oculomotor* to this muscle. Note the *ciliary ganglion* of the autonomic system near this branch and observe branches between this ganglion and the oculomotor and long ciliary nerve; further, the short ciliary nerves passing from the ganglion along the optic nerve. As explained above (p. 444), the ciliary ganglion and ciliary nerves belong to the cephalic part of the autonomic system and carry visceral motor fibers to the smooth muscles of the iris and the ciliary muscle. Find the main trunk of the oculomotor ventral to the optic nerve at the place of passage of both through the wall of the orbit and note branches of the oculomotor to the retractor bulbi and superior rectus muscles. Bend all eye muscles except the superior oblique ventrally, leaving the superior oblique against the medial wall of the orbit. Crossing the inner surface of the superior oblique obliquely forward are two nerves. They are parts of the ophthalmic branch of the trigeminus. The lower one is the *ethmoidal* nerve; it passes through a foramen into the nasal cavity. The upper one is the *infratrochlear* nerve. It goes to the anterior part of the upper eyelid. Posterior to, and parallel to, the posterior margin of the superior oblique is the *frontal* branch of the ophthalmic. It innervates the upper eyelid and integument anterior to the eyelid. The *trochlear* nerve lies slightly dorsal and medial to the proximal portions of the ethmoidal and infratrochlear nerves. It runs obliquely dorsad and anteriorly and enters the superior oblique muscle at about the middle of its posterior margin.

The structure of the eyeball may now be investigated. It is very similar to that of all vertebrates. The outer, tough *sclerotic* coat or *sclera* is continuous with the transparent *cornea* covering the exposed surface of the eye. As found above, the cornea is covered externally by the *conjunctiva.* Cut off the top or dorsal side of the eyeball and look within. The large *lens* will be observed. Internal to the sclera is the black *choroid* coat of the eye, and internal to that the greenish-gray *retina.* Between the lens and the retina is a large chamber, the *cavity of the vitreous humor,* containing a gelatinous mass, the *vitreous humor* or *vitreous body.* Remove the lens. The choroid coat terminates behind the cornea as a black curtain, the *iris,* bearing in its center a round hole, the *pupil.* The space between the cornea and the iris is called the *anterior chamber* of the eye and is filled in life with a fluid, the *aqueous humor.* The boundary between the iris and the rest of the choroid coat constitutes a ring known as the *ciliary body.* It consists of two parts: a ring of thickened processes, the *ciliary processes,* next the iris, and a ring of radially arranged ridges, the *orbiculus ciliaris,* extending to the main part of the choroid coat. Both parts of the ciliary body contain the *ciliary muscle;* this is a smooth muscle having both meridional and circular fibers. It originates on the sclera, is inserted on the walls of the ciliary body, and has the function of changing the shape of the lens. By making a new cut parallel to the first around the equator of the eyeball the relations of cornea, iris, and ciliary body will be more clearly observable. Note the marked thickening due to the ciliary body. Examine the lens. Note its biconvex form, as compared with the spherical form of the lens of the fish eye. Around the equator of the lens will be found the torn attachment of a membrane. This membrane holds the vitreous body. Where it is attached to the lens it exhibits parallel ridges, the *zonular fibers,* which in life fit into the hollows between the ciliary processes. The zonular fibers constitute the *suspensory ligament* of the lens, which passes from the lens to the ciliary processes. By means of this ligament traction can be exerted on the lens and its shape altered to some extent. The small space between the suspensory ligament and the iris is the *posterior chamber* of the eye. Peel the lens and note that it is composed of concentric coats or *lamellae,* like the coats of an onion, each lamella being composed of lens fibers.

Draw the section of the eye.

b) The nasal cavities: Detach the head of the animal at the joint between the occipital condyles and the atlas, and discard the body. Cut off the pinnae. Clear the dorsal surface of the skull down to the bone. With a saw, saw completely through the head slightly to one side of the median sagittal plane. Use the saw only for the bony parts. After having sawed through the roof of the skull, cut down through the brain with a single sliding stroke of a blunt knife like a table knife. The brain and skull should thus be cut in two, one part being slightly larger than the other. Wash the cut surfaces gently under the tap, and study the nasal cavities.

The nasal cavities are very long in the rabbit, shorter in the cat. They are divided into *right* and *left* cavities or *fossae* by a perpendicular plate, the *septum* of the nose, which is present on the larger section of the head. The septum consists of cartilage anteriorly and of thin bone posteriorly, the latter being the *perpendicular plate* of the *ethmoid* bone. On the smaller section the lateral and posterior walls of the nasal fossa are seen to be occupied by delicate scrolled and folded bones, the *turbinated bones* or *conchae*. In the **rabbit** these are readily divisible into an anterior concha, the *inferior concha* or *maxilloturbinal*, much folded and located on a separate small bone of the skull; a *middle concha* or *nasoturbinal*, a long single fold dorsal to the preceding and dependent from the nasal bone; and the *superior concha* or *ethmoturbinal*, part of the ethmoid bone. In the **cat** the turbinals are closely crowded together, but by prying them apart gently there can be distinguished a small anterior *maxilloturbinal* on the maxilla; above this a single fold, the *nasoturbinal*, dependent from the nasal bone; and a great mass of folds, the *ethmoturbinal*, filling the greater part of the nasal fossa. The ethmoturbinals are also called the *ethmoid labyrinths*, and the spaces inclosed by the bony folds are called the *ethmoid cells*. Definite passages known as the *meatuses* of the nose run between the conchae and conduct the air to the nasopharynx. They connect with the nasopharynx below the ethmoturbinals.

The posterior dorsal part of the nasal fossa is closed by the *cribriform* plate of the ethmoid, which unites with the perpendicular plate of the ethmoid bone medially and with the parts of the ethmoid which bear the labyrinths laterally. The anterior end of the brain (olfactory bulbs) is readily seen to abut against the cribriform plate, and through this plate the fibers of the olfactory nerve pass from the olfactory membrane covering the ethmoid labyrinths to the olfactory bulbs.

c) *The structure of the ear:* Carefully remove the brain from the two halves of the skull, preserving the latter. In doing this the roof of the skull may be cut away. Loosen the brain on all sides by passing a blunt instrument between the brain and the skull. The tough membrane, the *dura mater*, which covers the brain should be retained with the brain. Carefully cut the cranial nerves where they pass through the foramina of the skull, leaving their roots attached to the brain. Note the small, round, reddish body, the *pituitary* body, attached to the ventral surface of the brain, set into a depression in the floor of the skull; keep this body attached to the brain if possible. Preserve the two halves of the brain in a vessel of water or, if they are to be kept for some time, in weak formaldehyde.

After removal of the brain examine the cavities of the skull on the larger piece. Anteriorly behind the cribriform plate is the small *anterior* or *olfactory fossa* in which the olfactory bulbs are situated. Posterior to this is the large *middle* or *cerebral fossa* lodging the cerebrum. Behind this is the smaller *posterior* or *cerebellar fossa* for the cerebellum. The cerebral and cerebellar fossae are partly separated by a bony ledge, the *tentorium*, which is continued in life by the dura mater. In the floor of the cerebral fossa in the basisphenoid bone is the *sella turcica*, lodging the pituitary body. Note also the *optic foramen* in front of this and behind this, near the ventral end of the tentorium, the foramina for the passage of the third to sixth cranial nerves. In the wall of the cerebellar fossa observe an area of hard, white bone; this is the *petrous* portion of the temporal. In the center of this is a foramen for the passage of the auditory nerve into the internal ear. Above this in the rabbit is a depression, the *floccular fossa*, which lodges a part of the cerebellum called the *flocculus*. In removing the rabbit brain the flocculus is left behind in the fossa. In front of the ventral part of the petrous bone, just behind the tentorium, is the internal opening of the *facial canal* for the passage of the facial nerve. Behind the middle of the petrous bone is the *jugular foramen* for the passage of the ninth, tenth, and eleventh nerves. Behind this the twelfth nerve passes through one or more foramina.

The ear of mammals consists of three parts, the *external*, the *middle*, and the *internal* ear. The external ear includes the *pinna* or *auricle* and the *external auditory meatus* leading into the interior of the bulla; these have already been noted. The middle ear is situated in the tympanic bulla, the internal ear in the petrous portion of the temporal bone. Both are consequently in the wall of the cerebellar fossa. With the bone clippers remove this wall in one piece and discard the remainder of the skull. Clean away the muscles from its external surface, exposing the tympanic bulla.

Rabbit: With the bone clippers cut away the ventral wall of the tympanic bulla. A large cavity, the *cavity of the middle ear* or *tympanic cavity*, is revealed. In the lateral wall of this cavity is a ringlike elevation of bone across which is stretched a delicate membrane, the *tympanic membrane* or *eardrum*. By probing into the external auditory meatus determine that the meatus terminates at the eardrum, which closes its internal opening. The tympanic membrane has a nearly vertical position. Extending toward the tympanic membrane from the medial wall is a short calcareous process which supports the *chorda tympani* branch of the facial nerve as it crosses from the facial to the tympanic membrane. Anterior to the tympanic membrane is a depression in which are lodged the three little ear bones. These bones are so small and so deeply lodged in the depression that they cannot be distinctly seen; but on picking in the depression with a forceps it is usually possible to extract one or more of them. Compare them with pictures of the malleus, incus, and stapes in books.

Cat: Remove the fleshy part of the external auditory meatus down to the tympanic bulla. The meatus will be found to terminate at an oval opening with a slightly elevated rim. Across the rim is stretched the delicate *tympanic membrane* or *eardrum*. The *handle* of the *malleus* or *hammer* is visible through the eardrum attached to its internal surface. Next, remove with the bone clippers the ventral wall of the bulla. The interior is the *tympanic cavity of the middle ear*. Note that it is divided by a bony plate into a larger, posteroventral chamber and a smaller, anterodorsal chamber. The latter is the one covered above by the tympanic membrane. Break open the plate of bone, exposing this cavity, which is called the tympanum proper and which contains the ear bones. Note the membrane which lines it and the eardrum forming its anterodorsal wall. From the posterodorsal region of the cavity a calcareous process projects toward the eardrum and carries the *chorda tympani nerve*, a branch of the facial, to the eardrum. From the internal surface of the eardrum the three little ear bones are plainly seen extending into a depression in the internal wall of the tympanum. These may be extracted and examined. Compare with pictures of the malleus, incus, and stapes in books.

There now remains between the bulla and the cerebellar fossa the hard white mass of the petrous bone, already noted. This contains the internal ear. Because of the complexity of the internal ear and its small size, a dissection of it is impractical, but its main parts can be seen by breaking away the petrous bone in small fragments. The tiny, spirally coiled chamber in the bone is the *cochlea;* it contains a spiral tube, the *cochlear duct*, in which the organ of sound perception (organ of Corti) is located. In the thicker, harder part of the petrous bone are semicircular channels, the *semicircular canals*, inclosing the *semicircular ducts*.

The internal ear is thus seen to be inclosed in channels in the petrous bone, consisting of the cochlea, the semicircular canals, and the vestibule or connecting chamber; together these constitute the *bony labyrinth*. The internal ear proper, or *membranous labyrinth*, is contained in the bony labyrinth. Its parts are: the *sacculus* and *utriculus* inclosed in the vestibule; the *semicircular ducts* arising from the utriculus and situated inside of the semicircular canals, and the *cochlear duct* arising from the sacculus and inclosed in the cochlea. The cochlear duct is a mammalian feature developed from the lagena, already present in fishes; the lengthening and spiral coiling of the lagena begin in birds.

5. The structure of the brain.—

a) *The membranes or meninges of the brain:* With the two halves of the brain previously removed before you, study the membranes of the brain. The brain is covered by a tough membrane, the *dura mater*. This consists of the dura mater of lower forms fused to the internal lining (periosteum) of the skull. A considerable space, the *subdural* space, is present between the dura mater and the other membranes of the brain. The dura mater dips down between the larger divisions of the brain. The surface of the brain is covered by the delicate *pia mater,* in which the blood vessels run. The pia mater follows closely all of the folds of the brain surface. Between the pia mater and the dura mater is another membrane, the *arachnoid,* very delicate and difficult to see. It is best found covering the depressions on the surface of the brain, for the pia mater dips down into these depressions while the arachnoid passes over them. Between the arachnoid and the pia mater is the *subarachnoid* space crossed by a delicate web of tissue. All of the spaces between the meninges of the brain are filled in life with the *cerebrospinal* fluid.

b) *The dorsal aspect of the brain:* Remove the dura mater. Fit the two halves of the brain together and study the dorsal surface. At the anterior end of the brain are the two *olfactory bulbs,* relatively small, rounded masses into whose anterior surfaces the fibers of the olfactory nerve enter. Posterior to them are the enlarged pear-shaped *cerebral hemispheres.* Their surfaces are much convoluted in the cat, consisting of folds, the *gyri,* with grooves, the *sulci,* between the gyri. The two hemispheres are separated from each other by a deep median sagittal fissure, the *longitudinal cerebral fissure* (which is on the larger piece of the brain). Gently spread open the fissure and note at its bottom a thick, white mass connecting the two hemispheres. This is the *corpus callosum,* a structure characteristic of the mammalian brain, lacking, however, in monotremes and some marsupials. It is composed of nerve fibers passing between the hemispheres. At the posterior end of the longitudinal fissure is a small, reddish mass of folded tissue, which is part of the *choroid plexus* of the roof of the diencephalon. The diencephalon or region of the brain posterior to the cerebral hemispheres is in mammals completely concealed from dorsal view by the posterior extension of the hemispheres above it. The posterior ends of the cerebral hemispheres are in contact with the *cerebellum,* a large mass with a much convoluted surface. Between the cerebellum and the cerebral hemispheres is the *midbrain,* also concealed from dorsal view by the hemispheres. It is readily revealed by bending the hemispheres and the cerebellum apart. It consists of four rounded lobes or hillocks, known as the *corpora quadrugemina* or *colliculi.* The two anterior ones are named the *superior* colliculi, the two posterior ones the *inferior* colliculi. The cerebellum consists of a median lobe, the *vermis* or *worm,* and a pair of lateral lobes, the *hemispheres*. The hemispheres are, for the most part, new formations (see Fig. 127). From each hemisphere in the rabbit arises by a narrow stalk another lobe, the *flocculus*, which, as already seen, is left behind in the floccular fossa of the petrous bone when the brain is removed from the skull. Identify on the hemispheres the cut surfaces where the flocculi were attached. In the cat the floccular lobes are not definitely separated from the main mass of the cerebellum. The floccular lobes are homologous to the auricular lobes of lower forms (Fig. 127). Posterior to the cerebellum and partly overlapped by the vermis is the *medulla oblongata;* lift the vermis and note beneath it the cavity of the *fourth ventricle* in the medulla. This is normally roofed over by a membrane, the *medullary velum*, which contains a choroid plexus, probably removed in sectioning the brain. In the cat the choroid plexus projects on each side between cerebellum and medulla as a little tuft of vascular tissue. At each side of the posterior pointed end of the fourth ventricle is a tract terminating in a club-shaped area, the *clava*. Lateral to each clava is another elongated area, the *tuberculum cuneatum*. These two belong to the *somatic sensory column*. Anterior to them a white bundle passes toward the cerebellum, disappearing ventral to an elevation which lies just ventral to the hemisphere of the cerebellum. The bundle is the *restiform body* or *posterior peduncle* of the cerebellum, which conveys impulses from the medulla and spinal cord to the cerebellum. The elevation over the restiform body is the *area acustica* or *primary auditory center*. The general resemblance of these structures to those found in the dogfish should be evident.

Draw the dorsal view of the brain.

c) The ventral aspect of the brain: Note the *basilar* artery (continuation of the two vertebral arteries) running in the midventral line and forming a circle around some structures in the center of the ventral surface. This circle, the *circle of Willis*, is joined on each side by the *internal carotid* artery. Note the arteries arising from the basilar and circle of Willis and distributed over the brain, coursing in the pia mater. The arteries should be removed.

At the anterior end of the ventral surface are the two olfactory bulbs. From each one a definite white tract, the *olfactory tract*, extends obliquely caudad and terminates posteriorly in a lobe, the *pyriform lobe*, which forms the posteroventral part of the cerebral hemispheres. The fissure or sulcus which separates the pyriform lobe from the rest of the cerebral hemisphere is called the *rhinal fissure*. Inclosed between the two pyriform lobes is the ventral side of the *diencephalon*, or *thalamencephalon*. At the anterior end of this is the *optic chiasma*, from which the *optic nerves* project. The region between the optic chiasma and the olfactory tracts is called the *anterior perforated substance*. Behind the optic chiasma is a slight, rounded elevation, the *tuber cinereum*, from which the *pituitary body* or *hypophysis* depends by a stalk.

In case the pituitary body was torn off in removing the brain, a slitlike aperture will be noticed in the center of the tuber cinereum marking the place of attachment of the pituitary body. Immediately posterior to the attachment of the pituitary body is an area, the *mammillary body*, not distinctly marked off from the tuber cinereum. Posterior to this is a depressed area, the *posterior perforated substance*, from which arise the two third, or *oculomotor*, nerves. From beneath (dorsal to) the pyriform lobes a thick white bundle will be seen passing obliquely backward on each side of the posterior perforated substance. These bundles are the *cerebral peduncles*, belonging to the midbrain. The fourth or *trochlear* nerves arise on the side of the brain between the cerebellum and the inferior colliculi and pass ventrally over the outer surface of the peduncles.

The remainder of the ventral surface of the brain belongs to the hindbrain and consists of the *pons* and the medulla oblongata. The pons is the heavy band of fibers which crosses the ventral surface of the hindbrain immediately behind the posterior perforated substance. By following it around to the sides of the brain it will be seen to narrow to a white cord, the *brachium pontis* or *middle peduncle* of the cerebellum, which passes into the substance of the cerebellum. The pons is, in fact, a bridge between the two hemispheres of the cerebellum. It is the ventral part of the metencephalon, of which the cerebellum is the dorsal portion. Immediately posterior to the brachium pontis and partly concealing it is the thick root of the *trigeminus* nerve. On close examination this will be seen to consist of a large dorsal portion, the *sensory* root (*portio major*), which consists of the somatic sensory fibers of the trigeminus, and a very small ventral portion, the *motor* root (*portio minor*), which contains the visceral motor fibers for the muscles of mastication (masseter, temporal, digastric, etc.). Posterior to the pons and of about half its width is another bundle of transverse fibers, the *trapezoid* body. Close inspection will show that the trapezoid body originates from the area acustica or auditory center; it passes toward the median line but, before reaching it, turns forward and disappears dorsal to the pons. The trapezoid body is the main tract which carries the auditory impulses to the more anterior portions of the brain. Attached to the side of the area acustica is the root of the eighth or *auditory* nerve. Just ventral to this and behind the root of the trigeminus is the root of the *facial* nerve, emerging through the trapezoid body. In the median ventral line of the medulla is a groove, the *median ventral fissure.* Along each side of this runs a narrow bundle of fibers; each emerges dorsal to the posterior margin of the pons and proceeds straight posteriorly. These two tracts are the *pyramids* or *somatic motor tracts;* they convey impulses from the cerebral hemispheres to the voluntary muscles. At the place where the pyramids emerge from above the pons are the roots of the sixth or *abducens* nerves. The small root of the ninth or *glossopharyngeal* nerve will be found at the posterior boundary of the acoustic area, at the point where the restiform body passes dorsal to it and about on a line with the root of the eighth nerve. The equally small root of the tenth or *vagus* nerve lies immediately posterior to, and on a line with, the root of the ninth nerve. Posterior to the vagus are the numerous roots of the eleventh or *spinal accessory* nerve, arising in a line. The main root of the accessory ascends from the spinal cord but is probably missing in the specimen. The roots of the twelfth or *hypoglossal* nerve emerge along the lateral border of the pyramid, posterior to the preceding roots.

Draw the ventral view of the brain, including the roots of the cranial nerves as far as you have seen them.

d) The median sagittal section: Now cut the larger half of the brain along the longitudinal cerebral fissure so as to obtain an exact median sagittal section. In making such a cut use a dull knife and pass it through the brain with one sliding stroke. Examine the cut surface. The cerebral hemisphere forms a thick roof, which arches posteriorly above the diencephalon and midbrain. In the cerebral hemisphere identify the section of the *corpus callosum.* This is an obliquely placed longitudinal band of white material. Both anterior and posterior ends are enlarged, the former being named the *genu,* the latter the *splenium.* From about the middle of the corpus callosum a band of fibers, the *fornix,* curves ventrally. Between the fornix and the anterior half of the corpus callosum stretches a thin membrane, the *septum pellucidum,* consisting of two leaves. If the brain is sectioned exactly in the median sagittal plane, the section will pass between the two leaves of the septum pellucidum; but often the whole septum is left on one half; in this case a slitlike opening into a cavity, the *lateral ventricle,* will appear on the other half between the fornix and the corpus callosum. The fornix passes downward and soon turns (as the *column* of the fornix) into the interior of the brain, where it is lost to view. Immediately in front of the point where it disappears is the section of a small round bundle, the *anterior commissure.* From the anterior commissure a delicate membrane, the *lamina terminalis,* extends ventrally to the optic chiasma. The fornix, the anterior commissure, and the lamina terminalis form the anterior boundary of a deep but narrow chamber, the *third ventricle,* which lies in the middle of the diencephalon. The cavity of the third ventricle extends ventrally into the tuber cinereum and the pituitary body.

The diencephalon is the massive region extending between the fornix and lamina terminalis and midbrain. It consists of three parts: a dorsal region, the *epithalamus;* a central and lateral region, the *thalamus;* and a ventral region, the *hypothalamus.* The hypothalamus includes the optic chiasma, the tuber cinereum, the mammillary body, and the hypophysis or pituitary body, all of which should be identified in the section. The epithalamus includes the structures in the roof of the diencephalon. These are: the *tela choroidea,* a thin folded vascular membrane between the cerebral hemisphere and the diencephalon; the *pineal body,* a stalked body lying in the *tela choroidea;* the *habenula,* a small mass just in front of the attachment of the pineal body to the diencephalon; and the *posterior commissure,* a small circular area just posterior to the habenula. The *thalamus* constitutes the greater part of the diencephalon. On the cut surface it presents a large, round mass, the *intermediate mass* or *middle commissure;* this is not really a commissure but merely the cut median mass of the thalamus. The greater part of the thalamus is concealed by the overhanging cerebral hemisphere. On the smaller piece of the brain remove the cerebral hemisphere and then examine the dorsal and lateral regions of the thalamus. Three low elevations are present. The most dorsal and medial one is the *pulvinar.* Lateral to this and whiter in color is the *lateral geniculate body.* A white band, the *optic tract,* is plainly seen ascending from the optic chiasma and terminating on the lateral geniculate body. Posterior and ventral to the lateral geniculate body is a smaller swelling, the *medial geniculate body.* Behind the geniculate bodies will be recognized the corpora quadrugemina as two low hillocks. Ventral to them runs the stout cerebral peduncle, the anterior part of which is crossed externally by the optic tract.

Returning to the medial sagittal section, identify in the roof of the midbrain the two hillocks formed by the superior and inferior colliculi or corpora quadrugemina. Below them is a narrow passage, the *aqueduct* of the brain, which connects the third ventricle in the diencephalon with the fourth ventricle in the medulla. Below the aqueduct is the thick floor of the midbrain, the *tegmentum.* At the sides of this are the cerebral peduncles, not exposed in the section. In the section of the cerebellum note the curious branching treelike arrangement of the white matter, forming the arbor vitae or tree of life. This appearance is brought about by the fact that each fold of the cerebellar surface consists of gray matter or nerve cells, with a central plate of white matter or nerve fibers. The cerebellum fits into the fourth ventricle, from which, however, it is separated in the normal condition by a membrane, the *medullary velum.* Part of this velum will probably be found below the anterior part of the cerebellum. Identify in the section the mass formed by the pons. The section of the medulla has nothing of additional interest.

Draw the sagittal section.

e) *Further structure of the cerebral hemispheres:* On the intact half of the brain begin to cut away the roof of the cerebral hemisphere in thin slices. Note that the superficial substance of the roof is gray, the interior white—a reversal of the original condition. This gray outer coat of the mammalian cerebral hemispheres is called the cortex and is composed of a characteristic stratified arrangement of nerve cell bodies. It is a relatively recent evolutionary development. The white matter under the cortex consists of fibers carrying impulses to and from the cortex. In the cat and other higher mammals the cortex is much convoluted, and each convolution consists of a central core of white matter covered peripherally by a thick coat of gray matter. Continue to shave the brain ventrally until the corpus callosum is exposed as a narrow band of fibers conveying impulses between the two hemispheres. Remove the corpus callosum and the cortex lateral to it. This exposes the cavity of the cerebral hemisphere, called the lateral ventricle. It is filled by two conspicuous elevations—an anterior, smaller darker one, the *corpus striatum*, and a posterior larger one, the *hippocampus*. Remove the side of the hemisphere so as to expose the hippocampus. It is a curved body with an anterior free margin, the *fimbria*. Cut through the medial attachment of the hippocampus, raise the anterior border, and roll the hippocampus back. Observe that the part of the hippocampus still attached is continuous with the pyriform lobe. The hippocampus is a part of the original external surface of the brain, which has been invaginated into the interior (Fig. 128). The turning-back of the hippocampus reveals the thalamus. Note the thick stalk extending from the thalamus into the cerebrum, immediately in front of the pulvinar and lateral geniculate body. Scrape the surface of this and note that it consists of a great mass of fibers radiating from the thalamus into the cerebral hemisphere. This is more evident in the rabbit than in the cat, since in the cat the fibers turn dorsally. This radiating mass is called the *corona radiata*.

f) Functions of the parts of the brain: As an aid in the understanding of the anatomy of the brain, a few statements may be made concerning the functions of the parts identified in the preceding sections. The olfactory bulbs, the olfactory tracts, the pyriform lobe, the tuber cinereum, the fornix, the habenulae, the mammillary body, and the hippocampus belong to the olfactory apparatus of the mammal. The olfactory impulses come along the olfactory nerves into the olfactory bulbs, are then relayed along the olfactory tracts to the pyriform lobe, and from the pyriform lobe pass to the hippocampus. The hippocampus has extensive connections with other parts of the brain for reflex purposes. These connections occur by way of the fimbria, a mass of nerve fibers. The fimbria connects with the fornix and this in turn with the mammillary body. The habenula is also connected with the hippocampus. In the dogfish practically the whole of the telencephalon is concerned with smell, while here the olfactory functions occupy but a part of the telencephalon, the remainder having developed new connections and functions. The diencephalon is the great center of correlations in the mammalian brain. Its relation to the olfactory sense has already been noted. The lateral geniculate body is the primary optic center, in which, as we saw, the optic tracts terminate in part. From the lateral geniculate body the optic impulses pass to the cerebral hemisphere by way of the corona radiata. The pulvinar and the superior colliculus of the midbrain are also concerned in optic impulses, the latter being a reflex center for these impulses. The primary auditory center is located in the area acustica; from here the auditory impulses are carried, in part by the trapezoid body, to the inferior colliculus and the medial geniculate body, which constitute secondary and tertiary auditory centers. From the medial geniculate body the auditory impulses are carried in the corona radiata to the cerebral cortex. In a similar way other sensations, such as pain, touch, temperature, pressure, consciousness of muscle and joint movements, and position of the body in space, are carried by definite paths (which are, for the most part, invisible externally on the brain) to the thalamus, from which they are relayed to the cerebral cortex. (The clava and tuberculum cuneatum are concerned with joint and muscle sense and steadiness of body movement and position.) It will thus be seen that practically all sensations make a relay in the diencephalon from which they ascend by a new path, the corona radiata, to the cerebral cortex. The corona radiata is thus the great pathway between the diencephalon and the cerebral cortex by means of which the sensations are projected, as it were, upon the cortex. Furthermore, it is well known on which area of the cerebral cortex each sensation is projected, although this localization is not as exact and definite as formerly supposed. There remain considerable areas of the cortex to which no definite tracts from below can be traced, and it is assumed that these areas are concerned with thought, emotions, etc.

The impulses toward voluntary movements originate in a more or less definite part of the cortex, termed the *motor cortex*, pass downward in the corona radiata into the cerebral peduncles, and appear on the ventral surface of the medulla as the pyramidal tracts or pyramids, which descend the whole length of the spinal cord and make connections with the motor cells of the ventral columns of the cord. The cerebral peduncles, besides carrying the pyramidal tracts, also carry large tracts from the cortex to the pons, where they pass into the cerebellum.

The cerebellum is the great center for equilibration and motor co-ordination. Its functions are involuntary and unperceived by the conscious mind. It is connected with the rest of the brain by means of its peduncles: the posterior peduncles or restiform bodies, which connect it with the medulla and spinal cord; the brachium pontis or middle peduncle, which joins the two sides of the cerebellum and also conveys tracts between the cerebellum and cerebral cortex; and the anterior peduncles (which were not seen in the dissection), which extend between the cerebellum and midbrain and thalamus. The cerebellum has extensive connections with the area acustica, since the impulses from the ampullae of the semicircular ducts, which are concerned with equilibration, terminate in the area acustica.

G. SUMMARY OF THE NERVOUS SYSTEM AND THE SENSE ORGANS

1. The nervous system and the nervous parts of the sense organs are derived from the ectoderm.

2. The nervous system is subdivided into the central, peripheral, and autonomic nervous systems. The first includes the brain and spinal cord; the second, the cranial and spinal nerves; and the third is a system of ganglia, cords, and plexi controlling involuntary functions.

3. Functionally the nervous system is divisible into four components: the somatic sensory, handling sensory impulses from skin and other layers of the body wall; visceral sensory, handling sensations from the viscera; visceral motor, conveying motor impulses to the smooth musculature and the branchial muscles; and somatic motor, dealing with impulses to the voluntary musculature (except the branchial muscles). These components are arranged in the spinal cord and medulla from dorsal to ventral in the order named. The autonomic system is primarily an elaboration of the visceral motor component.

4. The spinal cord consists of a central gray region and a peripheral white region. The gray matter is of sensory nature dorsally, receiving the somatic and visceral sensory impulses.

The ventral half of the gray matter is of motor nature, containing the cells of origin of the somatic and visceral motor impulses. The cells of origin of sensory impulses are always outside the brain and cord in ganglia.

5. The spinal nerves arise from the spinal cord at segmental intervals by way of two attachments or roots, a dorsal or sensory root and a ventral or motor root. The sensory root bears a ganglion, the dorsal or spinal ganglion, which contains the cell bodies of the sensory fibers.

6. Primitively (*Amphioxus*, petromyzonts) the dorsal and ventral roots remain separate, but in gnathostome vertebrates they unite just beyond the spinal ganglion to form a spinal nerve. This soon divides into a dorsal ramus, which passes to the epaxial musculature and adjacent skin; a ventral ramus, to the hypaxial musculature and adjacent skin; and a communicating ramus, which connects with the autonomic system.

7. In the region of the appendages the ventral rami of the spinal nerves are intricately united by cross-connections to form plexi from which the nerves to the muscles of the appendages arise. The chief plexi are the brachial plexus to the anterior appendages and the lumbar or lumbosacral plexus to the posterior appendages. These plexi indicate that the limb muscles come from the hypaxial parts of the myotomes, that several myotomes contribute to the limb musculature, and that the limb muscles must have undergone torsion and change of position.

8. The autonomic system consists of ganglia in the head region; a pair of ganglionated cords in the cervical, thoracic, and lumbar regions; sacral nerves; and a complicated set of ganglia and ganglionated plexi among and in the viscera. The visceral sensory fibers have relations similar to those of the somatic sensory fibers, having their cells of origin in the spinal ganglia. But the visceral motor fibers of the autonomic system are peculiar in that a chain of two neurones is involved in the innervation of the smooth musculature. The first neurone in the brain or spinal cord sends a preganglionic fiber to a collateral or peripheral ganglion, and from this the second or postganglionic fiber proceeds to the musculature or to the glands.

9. The vertebrate brain consists of five main parts, named from anterior to posterior: telencephalon, diencephalon, mesencephalon or midbrain, metencephalon, and myelencephalon or medulla oblongata.

10. The telencephalon differentiates into the olfactory bulbs, the olfactory tracts, olfactory lobes, and cerebral hemispheres. The hemispheres are lateral expansions of the telencephalon. The olfactory part of the telencephalon is of paramount importance in the lower vertebrates, but later becomes subordinated to the cerebral hemispheres. These latter increase in size in the vertebrate scale until in mammals they cover most of the remaining parts of the brain. In mammals the two hemispheres are connected by the corpus callosum. A dorsal evagination, the paraphysis, is soon lost in vertebrates.

11. The diencephalon differentiates into the hypothalamus, the thalamus, and the epithalamus. The hypothalamus includes the optic chiasma, infundibulum, tuber cinereum, and mammillary bodies and is particularly well developed in fishes. The ventral part of the infundibulum unites with Rathke's pouch, growing in from the roof of the oral cavity, to form the hypophysis or pituitary. In fishes the infundibulum also includes the inferior lobes and vascular sac. The thalamus is the central mass of the diencephalon. The epithalamus or dorsal part of the diencephalon includes the epiphyseal apparatus and the habenula. The former originally comprised an anterior parapineal and a posterior pineal outgrowth, both of which originally bore an eye; but only the pineal body or epiphysis (without any eye) persists in most vertebrates.

12. The mesencephalon or midbrain is composed dorsally of the optic lobes, of which there are two in most vertebrates, four in mammals. In mammals the midbrain floor has a pair of conspicuous bundles, the cerebral peduncles.

13. The metencephalon develops the cerebellum dorsally; this is small in cyclostomes and amphibians but increases in size and importance in the amniote series.

14. The medulla oblongata undergoes little change in the vertebrate series; it contains all or part of the nuclei of origin and the terminations of the fifth to twelfth cranial nerves.

15. The brain is hollow, containing cavities termed ventricles. The first two or lateral ventricles are in the cerebral hemispheres; the third ventricle is in the diencephalon; the fourth in the medulla. They are all connected with each other. In lower vertebrates the olfactory bulbs, optic lobes, and cerebellum are also hollow. The roofs of the ventricles are thinned to a vascular membrane termed the tela choroidea, from which vascular tufts or choroid plexi project into the ventricles.

16. The brain and cord are protected by membranes termed meninges, of which there is one in fishes, two in amphibians, birds, and reptiles, and three in mammals.

17. Ten cranial nerves spring from the brain in fishes and amphibians, twelve in amniotes; but the last two are not really new formations. As in the case of the spinal nerves, the sensory fibers in the cranial nerves have their cells of origin in external ganglia on the roots of the nerves in question. In the cranial nerves, sensory and motor roots do not unite, and the cranial nerves are irregular as to their functional components.

18. The first or olfactory nerve extends from the olfactory epithelium in the nose to the olfactory bulbs. It is a pure sensory nerve.

19. The second or optic nerve extends from the retina to the diencephalon. It is not a true nerve but a tract of the brain.

20. The third or oculomotor nerve is a somatic motor nerve to the inferior oblique, superior, inferior, and internal recti and some accessory muscles of the eyeball. It originates in the midbrain.

21. The fourth or trochlear nerve is a somatic motor nerve to the superior oblique muscle of the eyeball. It arises from the midbrain.

22. The sixth or abducens nerve, originating in the floor of the medulla, is a somatic motor nerve to the external rectus muscle of the eyeball.

23. The eighth or auditory nerve is a somatic sensory nerve with its cells of origin in a ganglion or ganglia in the internal ear. It supplies the cristae and maculae of the internal ear and carries equilibratory and auditory impulses into the acousticolateral area of the medulla.

24. The fifth, seventh, ninth, and tenth cranial nerves are known as the branchial nerves, since they are definitely related to the branchial bars and gill slits. Typically, each one forks around a gill slit, sending a pretrematic branch in front of, and a posttrematic branch behind the slit; there is, further, a visceral sensory pharyngeal branch to the pharyngeal lining. The posttrematic branch carries visceral motor fibers to the branchial muscles belonging to that particular branchial bar.

25. The fifth or trigeminus nerve is the chief somatic sensory nerve of the head and the nerve of the first branchial (mandibular) arch. In all vertebrates it has three branches: the ophthalmic branch, to the orbit and nasal region (composed in fishes of two parts, the superficial and deep ophthalmic); the maxillary, to the upper jaw and roof of the oral and pharyngeal cavities; and the mandibular, to the lower jaw and floor of these cavities. The first two are pure somatic sensory nerves; the mandibular also carries visceral motor fibers to the

branchial musculature of the mandibular arch. The trigeminus is attached to the side of the medulla, and its sensory part has a ganglion.

26. The seventh or facial nerve is the nerve of the second or hyoid arch and is large in forms provided with a lateral-line system. With the disappearance of this system it becomes limited to visceral sensory branches to taste buds and the pharyngeal lining, and visceral motor fibers to the musculature of the hyoid arch. The facial springs from the side of the medulla, and its root is provided with a sensory ganglion.

27. The ninth or glossopharyngeal nerve, attached to the medulla and having a sensory ganglion, is the nerve of the third branchial bar and, like the other branchial nerves, has visceral sensory fibers to taste buds and the pharyngeal lining and visceral motor fibers to the muscles of its particular bar.

28. The tenth or vagus nerve is the nerve of the remaining branchial bars and in aquatic vertebrates has also a large lateral-line component. It appears to have appropriated the visceral components of several originally distinct spinal nerves posterior to itself. With the loss of the lateral-line system and the branchial mode of breathing the vagus is somewhat reduced but continues to supply the corresponding region of the pharynx with visceral sensory fibers and to be the visceral motor nerve of the pharyngeal and laryngeal musculature belonging to the appropriate branchial bars. In addition, the vagus extensively innervates the heart, lungs, stomach, and other viscera with preganglionic visceral motor fibers belonging to the autonomic system. The vagus is attached to the medulla and has two or more ganglia.

29. The eleventh or spinal accessory nerve of amniotes is compounded of vagal and spinal nerves; it springs from the medulla and upper spinal cord and is a visceral motor nerve to muscles of branchial origin.

30. The twelfth or hypoglossal nerve is a nerve of spinal origin which has become included into the cranial cavity. It is a somatic motor nerve to hypobranchial musculature; the tendency to form such a nerve is already seen in cyclostomes.

31. The lateralis system is a system of skin sense organs arranged in lines over the head and along the side of the trunk, limited to aquatic vertebrates. The sense organs consist of neuromasts; these are clusters of hair cells, and serve to detect vibrations in the water. The neuromasts may occur naked on the surface but are usually imbedded in canals; they are supplied by lateralis branches of the seventh, ninth, and tenth cranial nerves, and these branches disappear in land vertebrates.

32. The chief sense organs of the head are the nose, eyes, and ears.

33. The sensory part of the nose consists primitively of a pair of olfactory sacs invaginated from the surface epithelium. These at first are blind sacs, but in some fishes (Dipnoi, Crossopterygii) and in land vertebrates they establish communication with the oral cavity for respiratory purposes. Thereafter the nasal cavities have both respiratory and olfactory functions; the latter is limited to certain areas of the wall of the cavities. The nose itself is formed by the fusion of certain projections around the mouth and is supported by bones and cartilages of the skull.

34. In higher vertebrates, particularly mammals, the walls of the nasal cavities develop complex outgrowths, the turbinals or conchae, for the purpose of increasing both respiratory and olfactory surfaces.

35. The eyes are compound structures. The nervous part of the eye is formed by an evagination from the brain. The lens of the eye is an invagination from the adjacent ectoderm. The coats of the eye, sclera and choroid, are formed in the surrounding mesenchyme. The eye is moved by muscles which are very constant in arrangement in the different vertebrate classes, except that in mammals the superior oblique operates by means of a pulley.

36. The ear consists of internal, middle, and external portions. The internal ear is an invagination from the ectoderm. It differentiates into the three semicircular ducts, the sacculus, the utriculus, and the endolymphatic duct. Fishes and many urodeles possess only the internal ear. The internal ear of cyclostomes is primitive and has only one or two semicircular ducts. The internal ear of mammals is more complicated than that of other vertebrates because of the development of a spiral outgrowth, the cochlear duct, from the sacculus. The cochlear duct is the seat of hearing, and the semicircular ducts are concerned in equilibration.

37. Beginning with amphibians the middle ear is added to the internal ear. It consists of a chamber developed by outgrowth from the first gill pouch. The outer wall of this chamber comes in contact with the skin, producing a double-walled membrane, the tympanic membrane or eardrum. Within the middle ear is a chain of little bones, derived from the gill arches but differing in nature in different tetrapods.

38. Beginning with reptiles the tympanic membrane sinks into the skull, leaving a passage, the external auditory meatus, extending from the tympanic membrane to the exterior. This passage is deepened in birds and mammals, and in the latter a fold of skin, the pinna, develops around the external rim of the meatus. Pinna and meatus constitute the external ear.

39. Embryological studies show that the vertebrate head was originally composed of a number of segments each with a sensory and a motor nerve. In general there are three segments anterior to the otic capsule and a variable number behind it. The eye muscles come from the three prootic segments, the hypobranchial musculature from metaotic segments; two or three of the latter nearest the otic capsule degenerate in most vertebrates without developing any musculature. Of the three prootic segments, the profundus branch of the trigeminus, the trigeminus proper, and the facial are the sensory nerves; the oculomotor, trochlear, and abducens, the motor nerves. The glossopharyngeal and vagus are sensory nerves of metaotic segments, but the motor parts are mostly missing because of degeneration of the myotomes here.

REFERENCES

Adelmann, H. B. 1925. The development of the neural folds and cranial ganglia of the rat. Jour. Comp. Neurol., **39**.

Baumgartner, E. A. 1915. The development of the hypophysis in *Squalus acanthias*. Jour. Morph., **26**.

Christensen, K. 1927. The morphology of the brain of *Sphenodon*. Univ. Iowa Studies Nat. Hist., **12**.

Dammerman, K. W. 1910. Der Saccus vasculosus der Fische, ein Tieforgan. Zeitschr. f. wiss. Zool., **96**.

Dendy, A. 1910. On the structure, development, and morphological interpretation of the pineal organs and adjacent parts of the brain of the tuatara (*Sphenodon*). Phil. Trans. Roy. Soc. London, B, **201**.

Ewart, J. C., and Mitchell, J. C. 1892. The sensory canals of the common skate. Trans. Roy. Soc. Edinburgh, **37**.

Franz, V. 1923. Haut, Sinnesorgane, und Nervensystem der Akranier. Jena Zeitschr. Naturwiss. **59**.

———. 1924. Lichtsinnversuche am Lanzettfisch zur Ermittlung die Sinnesfunktion des Stirn- oder Gehirnbläschens. Wissensch. Meeresuntersuch., Abt. Helgoland, **15**.

Fürbringer, Max. 1896. Über die Spino-occipitalen Nerven der Selachier und Holocephalen und ihre vergleichende Morphologie. Festschrift für C. Gegenbaur, **3**.

Gegenbaur, C. 1871. Über die Kopfnerven von Hexanchus und ihr Verhältnis zum "Wirbeltheorie" des Schadels. Jena Zeitschr. Naturwiss. **6**.

Goodrich, E. S. 1915. The chorda tympani and middle ear of reptiles, birds, and mammals. Quart. Jour. Mic. Sci., **61**.

———. 1918. On the development of the segments of the head in *Scyllium*. *Ibid.*, **63**.

Herrick, C. J. 1901. The cranial nerves and cutaneous sense organs of North American siluroid fishes. Jour. Comp. Neurol., **11**.

———. 1905. The central gustatory paths in the brains of fishes. *Ibid.*, **15**.

Hill, A. 1894. The cerebrum of *Ornithorhynchus*. Phil. Trans. Roy. Soc. London, B, **184**.

Ingvar, S. 1919. Zur Phylo- und Ontogenese des Kleinhirns. Folia Neurobiol., **11**.

Johnson, S. E. 1917. Structure and development of the sense organs of the lateral canal system of selachians. Jour. Comp. Neurol., **28**.

Johnston, J. B. 1902*a*. The brain of *Petromyzon*. Jour. Comp. Neurol., **12**.

———. 1902*b*. An attempt to define the primitive functional divisions of the central nervous system. *Ibid.*

———. 1905. The cranial nerve components of *Petromyzon*. Morph. Jahrb., **34**.

———. 1906. The Nervous System of Vertebrates.

Kappers, C. U. Ariens; Huber, G. Carl; and Crosby, Elizabeth. 1936. The Comparative Anatomy of the Nervous System of Vertebrates, including Man. 2 vols.

Löwenstein, O. 1936. The equilibrium function of the vertebrate labyrinth. Biol. Rev. (Cambridge), **11**.

McKibben, Paul S. 1913. The eye-muscle nerves of *Necturus*. Jour. Comp. Neurol., **23**.

Menner, E. 1938. Die Bedeutung des Pecten im Auge des Vogels für die Wahrnehmung von Bewegungen nebst Bemerkungen über sein Ontogenie und Phylogenie. Zool. Jahrb., Abt. allg. Zool., **58**.

Neal, H. V. 1918*a*. The history of the eye muscles. Jour. Morph., **30**.

———. 1918*b*. Neuromeres and metameres. *Ibid.*, **31**.

Noble, G. K., and Kumpf, K. F. 1936. The function of Jacobson's organ in lizards. Jour. Genetic Psychol., **48**.

Norris, H. W. 1928. The parietal fossa and related structures in the plagiostome fishes. Jour. Morph., **48**.

———. 1929. The distribution and innervation of the ampullae of Lorenzini of the dogfish, *Squalus acanthias*. Jour. Comp. Neurol., **47**.

———. 1941. The Plagiostome Hypophysis, General Morphology and Types of Structure. Pub. by author.

Norris, H. W., and Buckley, M. 1911. The peripheral distribution of the cranial nerves of *Necturus*. Proc. Iowa Acad. Sci.

Norris, H. W., and Hughes, Sally P. 1920. The cranial, occipital, and anterior spinal nerves of the dogfish, *Squalus acanthias*. Jour. Comp. Neurol., **31**.

Papez, J. 1929. Comparative Neurology.

Piatt, J. 1938. Morphogenesis of the cranial muscles in *Ambystoma*. Jour. Morph., **63**.

Ranson, S. W. 1939. Anatomy of the Nervous System. 6th ed.

Reed, C. I. 1916. The epibranchial placodes of *Squalus acanthias*. Ohio Jour. Sci., **16**.

Retzius, G. 1881. Das Gehörorgan der Wirbelthiere. 2 vols.

Sand, A. 1938. The function of the ampullae of Lorenzini with some observations on the effect of temperature on sensory rhythms. Proc. Roy. Soc. London, B, **125**.

Sewertzov, A. N. 1916–17. Etudes sur l'évolution des vertébrés inférieurs. I. Morphologie du squelette et de la musculature de la tête des Cyclostomes. II. Organisation des ancêtres des vertébrés actuels. Arch. Russ. Anat. Hist. Embryol., **1**.

Shino, K. 1912. Beitrag zur Kenntnis der Gehirnnerven der Schildkröten. Anat. Heft. **47**.

Smith, G. Elliott. 1919. A preliminary note on the morphology of the corpus striatum and the origin of the neopallium. Jour. Anat., **53**.

Streeter, G. L. 1933. The status of metamerism in the central nervous system of chick embryos. Jour. Comp. Neurol., **57**.

Tretjakoff, D. 1909. Das Nervensystem von *Ammocoetes*. I. Das Rückenmark. Arch. f. mikr. Anat., **73**.

———. 1927. Das peripherische Nervensystem des Flussneunauges. Zeitschr. f. wiss. Zool.. **129**.

Warren, J. 1905. The development of the paraphysis and the pineal region in *Necturus*. Amer. Jour. Anat., **5**.

Wilde, W. S. 1938. The role of Jacobson's organ in the feeding reaction of the common garter snake. Jour. Exper. Zoöl., **77**.

APPENDIX A

PRONUNCIATION AND DERIVATION OF TECHNICAL WORDS

Some compound words and names combined of two or more words are given under their components. Some common prefixes are also given. (L, Latin; G, Greek; F, French.)

A (G prefix, without)

Ab (L prefix, away from)

Abdomen, abdominal—ab dough′ men, ab dom′ i nal (L, of uncertain origin)

Abducens—ab due′ senz (L, *ab*, from; *duco*, lead)

Acanthias—a kan′ the as (G, *acantha*, thorn)

Acentrous—a sen′ trous (G, *a*, without; *kentron*, center)

Acetabulum—ass i tab′ yu lum (L, name of a kind of cup)

Acipenser—ass i pen′ ser (L, a sturgeon)

Acoustico—a kous′ ti koh′ (G, *akoustikos*, related to hearing)

Acromion—a krow′ me on (G, *akros*, top; *omos*, shoulder)

Actinopterygii—ak′ ti nop′ ter yg′ee eye (G, *aktinos*, ray; *pterygion*, wing, fin)

Ad—(L prefix, toward, upon)

Adductor—a duck′ tore (L, *ad*, to; *duco*, lead)

Adrenal—add ree′nal (L, *ad*, upon; *renes*, kidneys)

Afferent—aff′err ent (L, *ad*, to; *fero*, bear)

Ali—al′ ee (L prefix, *ala*, wing)

Allantois—a lan′ toe iss (G, *allas*, sausage; *eidos*, form)

Alveolar, alveolus, alveoli—al vee′ oh lar, -lus, -lie (L, a little cavity)

Amia—aim′ ee ah (G, a kind of fish)

Amnion, amniota, amniote—am′ nee on, am′ nee oh′ tah, am′ nee oat (G, a membrane of the embryo)

Amphibia—am fib′ ee ah (G, *amphi*, double; *bios*, life)

Amphicoelous—am′ fee see′ lous (G, *amphi*, double; *koilos*, hollow)

Amphioxus—am′ fee ox′ us (G, *amphi*, both; *oxys*, sharp)

Amphiplatyan—am′ fee pla′ tee an (G, *amphi*, both; *platys*, flat)

Ampulla, ampullae—am pull′ ah, am pull′ ee (L, flask)

Anamnia—an am′ nee ah (G, *an*, without; *amnion*, embryonic membrane)

Anastomosis—a nass′ toh mow′ sis (G, an opening)

Anconeus—an′ ko nee′ us (L, *ancon*, the bend of the arm)

Ankylosis—an′ kee lo′ sis (G, *ankylos*, bent)

Ansa—ann′ sah (L, a handle)

Anura—a new′ rah (G, *an*, without; *oura*, tail)

Anus—ay′ nus (L, ring)

Aorta, aortic—ay or′ tah, -tik (G, *aorte*, to lift)

Aponeurosis—ap oh′ new row′ sis (G, *apo*, from; *neuron*, tendon)

Apophysis—a poff′ y sis (G, an outgrowth)

Aqueduct—ak′ we duct (L, *aqua*, water; *duco*, lead)

Aqueous—ay′ kwee us (L, *aqua*, water)

Arachnoid—a rack′ noid (G, *arachne*, spider)

Archenteron—ar ken′ ter on (G, *archos*, chief, first; *enteron*, intestine)

Arcualia—ar′ kiu ale′ ee ah (L, *arcus*, bow)

Arteriosus—ar teer′ ee oh′ sus (L, *arteria*, artery)

Arytenoid—ar′ ee tee′ noid (G, *arytaina*, pitcher, funnel)

Astragalus—ass trag′ a lus (G, an ankle bone)

Atlas—at′ lass (G, *tlao*, to bear)

Atrium—ay′ tree um (L, a court)

Auditory—aw′ di toe ry (L, *auditorius*, pertaining to hearing)

Auricle, auricular—aw′ ree kal, aw rik′ yu lar (L *auricula*, a little ear)

Aves—ay′ veez (L, birds)

Axilla, axillary—axe ill′ ah, axe′ i lay ree (L, *axilla*, a little axis)

Azygos—az′ ee gos (G, *a*, without; *zygon*, yoke)

Balanoglossus—bahl′ an oh gloss′ us (G, *balanos*, acorn; *glossa*, tongue)

Basalia—bah sail′ ee ah (L, *basis*, the base)

Basi—base′ ee (L prefix, at the base of)

Basilar—bass′ ee lar (L, *basis*, the base)

Biceps—buy′ seps (L, *bi*, two; *caput*, head)

Blastocoel—blas′ toe seal (G, *blastos*, germ; *koilos*, hollow)

Blastoderm—blas′ toe derm (G, *blastos*, germ; *derma*, skin)

Blastula—blas′ tiu lah (L, a little germ)

Brachial, brachialis, brachium—bray′ kee al, -kee ay′ lis, -kee um (L, *brachium*, arm)

Brachiocephalic—bray′ kee oh se phal′ ik (L, *brachium*, arm; *cephalicus*, pertaining to the head)

Branchia, branchiae, branchial—bran′ kee ah, -kee ee, -kee al (G, *branchia*, gills)

Branchiostegal—bran′ kee oss′ te gal (G, *branchia*, gills; *stego*, to cover)

Bronchus, bronchi, bronchial—bron′ kus, bron′ kai, bron′ kee al (G, *bronchos*, windpipe)

Buccal—buck′ al (L, *bucca*, cheek or mouth)

Buccinator—buck′ si nay′ tor (L, pertaining to blowing a trumpet)

Bulla—bull′ ah (L, a round seal or locket)

Bursa—burr′ sah (L, a purse)

Caecum, caeca—see′ kum, see′ kah (L, *caecus* blind)

Calcaneus—kal kay′ nee us (L, *calx*, heel)

Canine—kay nine′ (L, *canis*, dog)

Capillary—kap′ ee lay′ ry (L, *capillus*, hair)

Capitular, capitulum—ka pit′ yu lar, -yu lum (L, a small head)

Carapace—kair′ a pace (F, probably from L, *capa*, hood)

Cardia, cardiac—kar′ dee ah, -dee ak (G, *kardia*, heart)

Carina—ka rye′ nah (L, keel)

Carnivora, carnivore—kar niv′ oh rah, kar′ ni vore (L, *caro*, flesh; *voro*, devour)

Carotid—ka rot′ id (G, *karos*, stupor)

Carpales, carpus—kar pay′ les, kar′ pus (G, *karpos*, wrist)

Caudal—kaw′ dal (L, *cauda*, tail)

Cava, caval—kave′ ah, -al (L, *cavus*, hollow)

Cavernosa, cavernous—kav′ er no′ sah, -nous (L, *caverna*, hollow)

Centrale, centralia—sen tray′ lee, -tray′ lee ah (L, *centralis*, central)

Centrum—sen′ trum (L, center)

Cephalic—se phal′ ik (G, *kephale*, head)

Cephalization—seph′ al i za′ tion (G, *kephale*, head)

Cephalochordata—seph′ a low chor day′ tah (G, *kephale*, head; *chorde*, string)

Cerato—ser′ a toe (G prefix, *keras*, horn)

Cere—sear (L, *cera*, wax)

Cerebellum—ser′ e bell′ um (L, a little brain)

Cerebral, cerebrum—ser′ e bral, ser′ e brum (L, brain)

Cervical—ser′ vi kal (L, *cervix*, neck)

Chelonia—kee low′ nee ah (G, *chelone*, turtle)

Chiasma—kai as′ ma (G, cross-mark)

Choana, choanae—koh′ a nah, -a nee (G, funnel)

Chondrocranium—kon′ dro cray′ nee um (G, *chondros*, cartilage; *kranion*, skull)

Chondrostei—kon dross′ tee eye (G, *chondros*, cartilage; *osteon*, bone)

Chordata, chordate—kor day′ tah, kor′ date (L, *chorda*, cord or string)

Chordocentrous—kor′ dough sen′ trous (L, *chorda*, cord; *centrum*, center)

Chorion—koh′ ree on (G, membrane)

Choroid (*or* **chorioid**)—koh′ ree oid, koh′ roid (G, *chorion*, membrane)

Ciliary—sill′ ee a ry (L, *cilium*, eyelid)

Cinereum—si nee′ ree um (L, ashy)

Ciona—sigh′ oh na (G, *kion*, pillar)

Circulatory—sir′ kiu la tow′ ree (L, *circulo*, to form a circle)

Cirrus, cirri—sir′ rus, -ree (L, *cirrus*, tuft, lock of hair)

Clava—clay′ va (L, branch, club)

Clavicle—klav′ i kel (L, *claricula*, little key)

Clavo—clay′ voh (L combining form, *clavis*, key, referring to the clavicle)

Cleido—kly′ dough (G combining form, *kleis*, key, referring to the clavicle)

Cleithrum—klyth′ rum (G, *kleithron*, bar, gate)

Clitoris—kly′ to riss (G, *kleio*, to close)

Cloaca—klo ay′ kah (L, sewer)

Coccyx, coccygeal—kock′ six, -sij′ ee al (G word)

Cochlea—kock′ lee ah (L, snail)

Coeliac—see′ lee ak (G, *koilia*, stomach)

Coelom—see′ loam (G, *koilos*, hollow)

Colon—koh′ lon (G, *kolon*, member)

Columella—kol′ yu mell′ ah (L, little column)

Concha—kon′ ka (L, shell)

Condyle—kon′ dill (L, *condylus*, knuckle)

Conjunctiva—kon′ junk tie′ va (L, *conjunctus*, join together)

Copula—kop′ yu lah (L, *cum*, together; *apo*, blind)

Coracoid—kor′ a koid (G, *korakoeides*, like a crow's beak)

Corium—koh′ ree um (L, leather)

Cornea, corneum—kor′ nee ah, -um (L, *corneus*, horny)

Cornu, cornua—kor′ niu, kor′ niu ah (L, horn)

Corona, coronary—koh row′ nah, kor′ oh nay ry (L, crown or wreath)

Coronoid—kor′ oh noid (G, *korone*, crow)

Corpora, corpus—kor′ po rah, kor′ pus (L, body)

Costal—kos′ tal (L, *costa*, rib)

Cranial, craniate, cranium—kray′ nee al, -nee ate, -nee um (G, *kranion*, skull)

Cribriform—krib′ ree form (L, *cribrum*, sieve)

Cricoid—kry′ coid (G, *kriksos*, ring)

Crista—kris′ tah (L, crest)

Crocodilia—krok′ oh dill′ ee ah (L, crocodile)

Crossopterygii—cross sop′ ter yg′ ee eye (G, *krossoi*, fringe; *pteron*, wing)

Crura, crural, crus—kru′ rah, kru′ ral, kruss (L, *crus*, leg)

Ctenoid—ten′ oid (G, *ktein*, comb)

Cuneiform—kiu′ nee i form (L, *cuneus*, wedge)

Cutaneous—kiu tay′ nee us (L, *cutis*, skin)

Cuvier—kiu vyay (name of French anatomist)

Cycloid—sigh′ kloid (G, *kyklos*, circle)

Cyclostomata, cyclostome—sigh′ klo stow′ ma tah, -stome (G, *kyklos*, circle; *stoma*, mouth)

Cystic—sis′ tik (G, *kystis*, bladder)

Deferens—deff′ er enz (L, *defero*, carry away)
Deltoid—dell′ toid (G, *delta*, fourth letter of the Greek alphabet, triangular in form)
Dentary—den′ ta ree (L, *dens*, tooth)
Dentine—den′ tin (or teen) (L, *dens*, tooth)
Dermal, dermis—derr′ mal, derr′ miss (G, *derma*, skin)
Dermatome—derr′ ma tome (G, *derma*, skin; *tomos*, cutting)
Diaphragm—dye′ a framm (G, *dia*, between; *phragnymi*, to inclose)
Diastema—dye′ a stee′ mah (G, interval)
Digastric—dye gas′ trik (G, *di*, two; *gaster*, belly)
Digitigrade—dij′ i ti grade (L, *digitus*, finger, toe; *gradus*, step, walk)
Diphycercal—diff′ ee sir′ kel (G, *diphyes*, twofold; *kerkos*, tail)
Diplospondyly—dip′ low spon′ dy lee (G, *diploos*, double; *spondylos*, vertebra)
Duodenum—diu′ oh dee′ num (L, *duodeni*, twelve)
Dura—diu′ rah (L, hard)

Ect, ecto—ekt, ek′ toh (G prefix, the outer, outside)
Edentata—ee′ den tay′ tah (L, *e*, without; *dens* tooth)
Efferent—eff′ er ent (L, *effero*, to bear away from)
Efferentia—eff′ er en′ shi a (L, *effero*, to bear away from)
Elasmobranch, elasmobranchii—ee las′ mow brank, -mow bran′ kee eye (G, *elasmos*, plate; *branchia*, gills)
Encephalon—en sef′ a lon (G, *enkephalos*, brain)
End, endo—end, en′ dough (G prefix, within, inside)
Endostyle—en′ dough style (G, *endo*, within; *stylos*, column)
Entoderm—en′ toe derm (G, *endo*, within; *derma*, skin)
Ep, epi—epp, epp′ ee (G prefix, upon)
Epaxial—epp axe′ i al (G, *epi*, upon; L, *axis*)
Epidermis—epp′ ee derr′ miss (G, *epi*, upon; *derma*, skin)
Epididymis—epp′ ee did′ y miss (G, *epi*, upon; *didymos*, testis)
Epimere—epp′ ee mere (G, *epi*, upon; *meros*, part)
Epiphysis, epiphyseal—e piff′ ee sis, epp′ ee fis′ e al (G, an outgrowth)
Epiploic, epiploicum—epp′ ee plo′ ik, -plo′ ee kum (G, *epiploon*, omentum)
Epistropheus—epp′ e stro′ fee us (G, *epi*, upon; *strepho*, turn)
Esophagus, esophageal—ee soph′ a gus, ee′ so faj′ ee al (G, *oisophagos*, gullet)
Ethmoid—eth′ moid (G, *ethmoe*, sieve)
Eustachian—you stay′ kee an (after Eustachius, an anatomist)
Ex—(L prefix, out, outside)
Excretory—eks′ kree toh ry (L, *ex*, out; *cerno*, separate)

Facet—fas′ et (L, *facies*, face)
Facial—fay′ shal (L, *facies*, face)
Falciform—fal′ see form (L, *falx*, sickle)
Fallopian—fa loh′ pee an (after Fallopio, an anatomist)
Fascia—fash′ ee ah (L, a band or bandage)
Fasciculus, fasciculi—fa sik′ yu lus, -yu lye (L, a little bundle)
Fauces—fa′ sees (L, throat)
Femoral, femoris—fem′ oh ral, -oh riss (L, *femur*, thigh)
Femur—fee′ mur (L, thigh)
Fenestra—fee nes′ trah (L, window)
Fibula, fibulare—fib′ yu lah, -yu lay′ ree (L, *fibula*, clasp, buckle)
Filoplume—fill′ (or file) oh plume (L, *filum*, thread; *pluma*, feather)
Fimbria—fim′ bree ah (L, border)
Flocculus—flock′ yu lus (L, a little piece of wool)
Follicle—foll′ i kel (L, *folliculus*, a little bag)
Fontanelle—fon′ ta nell′ (F, a little fountain)
Foramen, foramina—fo ray′ men, -ram′ ee nah (L, an opening)
Fornix—for′ niks (L, an arch or vault)
Fossa—foss′ ah (L, a pit or cavity)
Frenulum—fren′ yu lum (L, a little bridle or bit)
Frontal—frun′ tal (L, *frons*, brow)
Fundus—fun′ dus (L, the bottom)
Fungiform—fun′ ji form (L, *fungus*, a mushroom)
Funiculus—fiu nik′ yu lus (L, a small rope or cord)
Furcula—fur′ kiu lah (L, a little fork)

Ganglion—gan′ glee on (G, a tumor)
Ganoid—gan′ oid (G, *ganos*, bright)
Gasserian—ga see′ ri an (from a physician Gasser)
Gastric, gastro—gas′ trik, gas′ troh (L, *gaster*, stomach)
Gastrocentrous—gas′ troh sen′ trous (L, *gaster*, stomach; *centrum*, center)
Gastrocnemius—gas′ trok nee′ me us (G, *gaster*, stomach; *kneme*, shank)
Gastrocoel—gas′ troh seal (G, *gaster*, stomach; *koilos*, hollow)
Gastrula—gas′ trew lah (L, a little stomach)
Geniculate—ji nik′ yu late (L, *genu*, knee)
Geniohyoid—ji nye′ oh high′ oid (G, *geneion*, chin; *upsilon*, a Y-shaped letter of the Greek alphabet)
Genital—jen′ i tal (L, *gigno*, to reproduce)
Genu—jee′ new (L, knee)
Germinativum—jerr′ mi nay tiv′ um (L, *germino*, to sprout)
Glans—glanz (L, an acorn)
Glenoid—glee′ noid (G, *glene*, a socket)
Glomerulus—glow mer′ yu lus (L, *glomus*, a ball of yarn)
Glossopharyngeal—gloss′ oh fa rin′ jee al (or fare′ in jee′ al) (G, *glossa*, tongue; *pharynx*, pharynx)
Glottis—glott′ iss (G, *glotta*, tongue)

Gluteus—glu tee′ us (G, *gloutos*, rump)

Gnathostomata, gnathostome—nath′ oh stow′ ma tah, nath′ oh stome (G, *gnathos*, jaw; *stoma*, mouth)

Gonad—gonn′ ad (G, *gonos*, seed)

Graafian—grahf′ ee an (after de Graaf, a Dutch physician)

Gracilis—grass′ i lis (L, slender)

Gubernaculum—giu′ ber nak′ yu lum (L, a rudder)

Gyrus, gyri—jye′ rus, -rye (G, *gyros*, round)

Habenula—ha ben′ yu la (L, a little band)

Haemal—hem′ al (G, *haima*, blood)

Hamulus—ham′ yu lus (L, a little hook)

Hemichordata—hem′ i core day′ tah (L, *hemi*, half; *chorda*, cord)

Hepatic, hepato—he pat′ ik, hep′ a toe (G, *hepar*, liver)

Hermaphroditic—her maff′ row dit ik (G, *Hermes*, Mercury; *Aphrodite*, Venus)

Heterocercal—het′ er oh ser′ kal (G, *heteros*, different; *kerkos*, tail)

Heterocoelous—het′ er oh see′ lous (G, *heteros*, different; *koilos*, hollow)

Heterodont—het′ er oh dahnt (G, *heteros*, different; *odon*, tooth)

Heteronomous—het′ er on′ oh mous (G, *heteros*, different; *nomos*, law)

Hilum—high′ lum (L, eye of the bean)

Hippocampus—hip′ poh kam′ pus (G, *hippos*, horse; *kampos*, sea monster)

Holoblastic—holl′ oh blas′ tik (G, *holos*, whole; *blastos*, germ)

Holostei—ho loss′ tee eye (G, *holos*, whole; *osteon*, bone)

Homocercal—home′ oh ser′ kal (G, *homos*, the same; *kerkos*, tail)

Homonomous—hoh mon′ oh mous (G, *homos*, the same; *nomos*, law)

Humerus—hiu′ mer us (L, the bone of the upper arm)

Hy, hyo—high, high′ oh (prefix referring to the hyoid)

Hyoid, hyal—high′ oid, high′ all (G, *upsilon*, a Y-shaped letter of the Greek alphabet)

Hyp, hypo—hipp, high′ poh (G or L prefix, below, less than)

Hypaxial—hip axe′ ee al (G, *hypo*, below; *axis*, axis)

Hypoglossal—high′ poh gloss′ al (G, *hypo*, below; *glossa*, tongue)

Hypomere—high′ poh mere (G, *hypo*, below; *meros*, part)

Hypophysis—he poff′ ee sis (G, *hypo*, below; *phyo*, to cause to grow)

Hypural—hip your′ al (G, *hypo*, below; *oura*, tail)

Ichthyopsida—ik′ thy op′ si dah (G, *ichthys*, fish; *opsis*, appearance)

Ileum—ill′ e um (G, *eilo*, twist)

Iliac—ill′ i ak (L, *ilium*, flank)

Ilium—ill′ ee um (L, flank)

Incisor, incisiva, incisive—inn sigh′ sor, -sigh′ si vah, -sigh′ sive (L, *incido*, to cut into)

Incus—inn′ kuss (L, anvil)

Infra—inn′ frah (L prefix, below)

Infundibulum—inn′ fun dib′ yu lum (L, funnel)

Inguinal—inn′ gwi nal (L, *inguen*, groin)

Innominate—inn nomm′ i nate (L, *innominatus*, without a name)

Integument—inn teg′ u ment (L, *intego*, to cover)

Inter—inn′ ter (L prefix, between)

Intercalary—inn terr′ ka lay ree (L, *intercalo* to put between)

Intestine, intestinal—inn tess′ tin, inn tess′ ti nal (L, *intestinus*, inside)

Intra—inn′ trah (L prefix, within)

Invagination—inn vaj′ i nay′ shun (L, *in*, in; *vagina*, sheath)

Iris—eye′ riss (G, rainbow)

Ischial, ischium—iss′ kee al, iss′ kee um (G, *ischion*, hip)

Isolecithal—eye′ so less′ i thal (G, *isos*, equal; *lekithos*, yolk)

Iter—eye′ ter (L, passage)

Jejunum—jee jew′ num (L, hungry)

Jugal—jew′ gal (L, *jugum*, yoke)

Jugular—jew′ giu lar (L, *jugulum*, the collar bone)

Labia, labial—lay′ bee ah, -bee al (L, *labium*, lip)

Lacertilia—lass′ err till′ ee ah (L, *lacerta*, lizard)

Lacrimal (also spelled **lacrymal**)—lack′ ree mal (L, *lacrima*, tear)

Lagena—la jee′ nah (G, *lagynos*, flask)

Lamella, lamellae—la mell′ ah, -mell′ ee (L, *lamina*, a thin sheet)

Lamina—lamm′ ee nah (L, a thin plate)

Laryngeal, larynx—la rin′ jee al (or lar′ in jee al), lar′ inks (G, *larynx*, gullet)

Latissimus—la tiss′ i muss (L, the broadest)

Lepidosteus—lepp′ i doss′ tee us (G, *lepis*, scale; *osteon*, bone)

Levator—le vay′ tor (L, a lifter)

Lienal, lieno—lie ay′ nal, lie ay′ no (L, *lien*, spleen)

Linea—linn′ ee ah (L, line)

Lingual—linn′ gwal (L, *lingua*, tongue)

Lissamphibia—liss′ am fib′ ee ah (G, *lissos*, smooth; *amphibios*, double life)

Lobule—lobb′ yule (G, *lobos*, lobe)

Longissimus—lon jiss′ ee mus (L, the longest)

Lorenzini—loh ren zee′ nee (Italian anatomist)

Lumbar—lumm′ bar (L, *lumbus*, loin)

Lutea, lutein—liu′ tee ah, -tee in (L, *luteus*, yellow)

Lymphatic—limm fat′ ik (L, *lympha*, clear water)

Lymphocytes—limph′ oh sites (L, *lympha*, clear water; G, *kytos*, hollow place)

Major—may′ jor (L, the greater)

Malar—may′ lar (L, *mala*, cheek)

Malleolus, malleoli—ma lee′ oh lus, -oh lye (L, little hammer)

Malleus—mahl′ ee us (L, hammer)

Malpighian, malpighii—mahl pig′ ee an, -ee eye (after Malpighi, an Italian biologist)

Mammal—mam′ mal (L, *mamma*, breast)

Mammalia—ma may′ lee ah (L, *mamma*, breast)

Mammary—mam′ a ree (L, *mamma*, breast)

Mammillary (or **mamillary**)—mam′ ill lay ree (L, *mamma*, breast)

Mandible, mandibular—man′ di bl, man dib′ yu lar (L, *mandibula*, jaw)

Manubrium—ma niu′ bree um (L, handle)

Manus—may′ nuss (L, hand)

Marsupial, marsupialia—marr siu′ pee al, mar siu′ pee ay′ lee ah (L, *marsupium*, pouch)

Masseter—ma see′ ter (G, *maseter*, a chew)

Mastoid—mass′ toid (G, *mastos*, breast)

Mater—may′ ter (L, mother)

Maxilla, maxillary—maks ill′ ah, maks′ i lay ree (L, *maxilla*, jawbone)

Maximus—maks′ i mus (L, the largest)

Meatus—mee ate′ us (L, passage)

Mediastinal, mediastinum—mee′ dee ass tie′ nal, -num (L, *mediastinus*, being in the middle)

Medius—mee′ di us (L, middle)

Medulla, medullary—me dull′ lah, medd′ u lay ree (L, *medulla*, marrow, pith)

Meninx, meninges—mee′ ninks, mee nin′ jees (G, membrane)

Mental—men′ tal (L, *mentum*, chin)

Meroblastic—merr′ oh blas′ tik (G, *meros*, part; *blastos*, germ)

Mes, meso—mess, mess′ oh (G prefix, the middle)

Mesenchyme—mess′ en kime (G, *mesos*, middle; *enchyma*, in a fluid)

Mesenteric, mesentery—mess′ en tare′ ik, -y (G, *mesos*, middle; *enteron*, gut)

Mesoderm—mess′ oh derm (G, *mesos*, middle; *derma*, skin)

Mesogaster—mess′ oh gas′ ter (G, *mesos*, middle; *gaster*, stomach)

Mesomere—mess′ oh mere (G, *mesos*, middle; *meros*, part)

Mesopterygium—mess′ op terr yg′ ee um (G, *mesos*, middle; *pteron*, wing)

Mesorchium—mess or′ kee um (G, *mesos*, middle; *orchis*, testis)

Mesotubarium—mess′ oh tiu bay′ ri um (G, *mesos*, middle; L, *tubus*, tube)

Mesovarium—mess oh vay′ ree um (G, *mesos*, middle; L, *ovum*, egg)

Met, meta—met, met′ ah (L or G prefix, between, after, reversely)

Metacarpal—met′ a kar′ pal (G, *meta*, after; *karpos*, wrist)

Metacromion—met′ a krow′ mee on (G, *meta*, after; *akromion*, point of the shoulder)

Metamere, metamerism—met′ a mere, me tam′ er izm (G, *meta*, after; *meros*, part)

Metapleural—met′ a plu′ ral (G, *meta*, after; *pleura*, side)

Metapterygium—met ap′ ter yg′ ee um (G, *meta*, after; *pteron*, wing)

Metatarsal—met′ a tar′ sal (G, *meta*, after; *tarsos*, a flat surface)

Minimus—min′ i mus (L, the least)

Minor—my′ nor (L, the lesser)

Mitral—my′ tral (F, *mitre*, a peaked cap)

Molar—mow′lar (L, *mola*, millstone)

Monotremata, monotreme—mon′ oh tremm′ a tah, -treem (G, *monos*, one; *trema*, hole or opening)

Müllerian—me lerr′ i an (after Müller, a German physiologist)

Multifidus—mull tiff′ i dus (L, many cleft)

Myelon—my′ e lon (G, marrow)

Mylohyoid—my′ low high′ oid (G, *myle*, mill; *upsilon*, the letter *y*)

Myocomma, myocommata—my′ o komm′ ah, -komm′ a tah (G, *mys*, muscle; *komma*, that which is cut off)

Myotome—my′ oh tome (G, *mys*, muscle; *tome*, cutting)

Naris, nares—nay′ riss, nay′ rees (L, nostril)

Necturus—nek too′ rus (G, *nektos*, swimming; *oura*, tail)

Nephros—neff′ ross (G, kidney)

Nephrostome—neff′ row stome (G, *nephros*, kidney; *stoma*, mouth)

Nephrotome—neff′ row tome (G, *nephros*, kidney; *tome*, cutting)

Neural—niu′ ral (G, *neuron*, nerve)

Neurocoel—niu′ row seal (G, *neuron*, nerve; *koilos*, hollow)

Nictitating—nik′ ti tay′ ting (L, *nicto*, wink)

Nodosal—no dose′ al (L, *nodosus*, knotty)

Notocentrous—no′ toe sen′ trous (G, *notos*, back; *kentron*, center)

Notochord—no′ toe kord (G, *notos*, back; *chorde*, string)

Nuchal—new′ kal (L, *nucha*, nape of the neck)

Oblongata—ob lon gay′ tah (L, *oblongus*, oblong)

Obturator—ob tiu ray′ ter (L, *obturo*, to close, shut)

Occipital—ok sip′ i tal (L, *occiput*, back of the head)

Oculomotor—ok′ yu loh mow′ tor (L, *oculus*, eye; *motor*, mover)

Odontoid—o don′ toid (G, *odous*, tooth)

Olecranon—oh′ lee kray′ non (G, *olene*, ulna; *kranion*, skull)

Olfactory—ol fak′ toh ree (L, *olfacere*, to smell)

Omentum—oh men′ tum (L, fat skin)

Omo, om—oh′ mow, ohm (G prefix, *omos*, shoulder)

Opercular, operculum—o per′ kiu lar, -lum (L, *operculum*, lid)

Ophidia—o fid′ ee ah (G, *ophis*, serpent)

Ophthalmic—off thal′ mik (G, *ophthalmos*, eye)

Opistho, opisth—o pis′ tho, o pist′ (G prefix, *opisthen*, behind)

Oral—oh′ ral (L, *os*, mouth)

Orbicularis—or bik′ yu lay′ riss (L, *orbiculus*, a little circle)

Orbital, orbito—or′ bi tal, or′ bi toe (L, *orbita*, orbit)

Ostium—oss′ tee um (L, mouth or entrance)

Otic—oh′ tik (G, *oticos*, pertaining to the ear)

Otolith—oh′ toe lith (G, *ous*, ear; *lithos*, stone)

Ovarian, ovary—o va′ ree an, oh′ va ree (L, *ovum*, egg)

Oviduct—oh′ vi duct (L, *ovum*, egg; *ductus*, duct)

Palatal, palate, palatine—pal′ a tal, pal′ ate, pal′ a tiyn (or -tin) (L, *palatum*, palate)

Pallium—pahl′ ee um (L, cloak)

Palpebra—pahl′ pe brah (L, eyelid)

Pancreas, pancreatic—pan′ kree ass, pan′ kree at′ ik (G, *pas*, all; *kreas*, flesh)

Papilla, papillae—pa pill′ ah, -ee (L, nipple)

Para—par′ ah (G prefix, beside, near)

Paraphysis—pa raff′ ee sis (G, *para*, beside; *phyo*, produce)

Parietal—pa rye′ ee tal (L, *paries*, wall)

Parotid—pa rot′ id (G, *para*, beside; *ous*, ear)

Patella—pa tell′ ah (L, a small pan or dish)

Pecten—pek′ ten (L, comb)

Pectineal—pek′ ti nee′ al (L, *pecten*, comb)

Pectoral, pectoralis—pek′ toe ral, -ray′ liss (L, *pectoralis*, referring to the chest)

Pellucidum—pe liu′ see dum (L, translucent)

Pelvic, pelvis—pell′ vik, -viss (L, *pelvis*, basin, the pelvis)

Penis—pee′ nis (L, male external sex organ)

Peri—perr′ ee (L or G prefix, around)

Pericardial, pericardium—per′ ee kar′ dee al, -um (G, *peri*, around; *kardia*, heart)

Perineum—perr′ i nee′ um (L, *perineon*, origin uncertain)

Peripheral—pe riff′ er al (G, *peri*, around; *phero*, bear)

Peritoneal, peritoneum—perr′ i toe nee′ al, -um (G, *peri*, around; *teino*, stretch)

Peroneal, peroneus—perr oh nee′ al, -us (G, *perone*, pin of a brooch, referring to the fibula)

Petrosal, petrous—pe trow′ sal, pee′ trous (L, *petrosus*, rocky)

Phalanges, phalanx—fay lan′ gees, fay′ lanks (G, *phalanx*, battle line)

Pharyngeal, pharynx—fa rin′ jee al (or far′ in jee′ all) far′ ynks (G, *pharynx*, throat)

Phrenic—fren′ ik (G, *phren*, diaphragm)

Phylum—fye′ lum (G, *phylon*, race, tribe)

Pia—pie′ ah (L, *pious*, kind)

Pineal—pin′ ee al (L, *pinea*, pine cone)

Pinna—pin′ ah (L, feather)

Pisces—piss′ ees (L, fish)

Pituitary—pi tiu′ i tay ree (L, *pituita*, mucus)

Placenta, placentalia—pla sen′ tah, plass′ en tay′ lee ah (L, *placenta*, cake)

Placoid—plak′ oid (G, *plax*, plate)

Plantaris—plan tay′ riss (L, *planta*, sole of the foot)

Plantigrade—plan′ ti grade (L, *planta*, sole; *gradus*, walk)

Plastron—plass′ tron (F, a breastplate)

Platysma—pla tiz′ mah (G, flat plate)

Pleura, pleural—plew′ rah, -ral (G, *pleura*, rib, side)

Plexus—pleks′ us (L, interweaving)

Plumulae—plew′ miu lee (L, little feather)

Pneumatic—niu mat′ ik (G, *pneuma*, breath, spirit)

Polyodon—po lee′ oh don (G, *polys*, many; *odous*, teeth)

Polypterus—po lip′ ter us (G, *polys*, many; *pteron*, wing)

Pons—ponz (L, bridge)

Portal—pour′ tal (L, *porta*, gate)

Post—(L prefix, behind, after)

Pre—pree (L prefix, before)

Prepuce—pree′ pyuse (F, probably from L, *pre*, before; and G, *posthion*, penis)

Primate—pry′ mate (L, *primus*, first)

Pro—proh (L or G prefix, before)

Procoelus—proh′ see lous (G, *pro*, before; *koilos*, hollow)

Proctodaeum—prok′ toe dee′ um (G, *proktos*, anus; *daio*, divide)

Prostate—pross′ tayte (G, *prostates*, in the front rank)

Proventriculus—proh′ ven trick′ yu lus (L, *pro*, before; *venter*, belly)

Pseudo—sue′ do (G prefix, *pseudes*, false)

Pterygium—te ryg′ ee um (G, *pterygion*, a little wing or fin or projection, from *pteron*, wing)

Pterygo, pterygoid—terr′ i go, -goid (G, *pteron*, wing)

Pterylae—terr′ y lee (G, *pteron*, feather, wing; *hyle*, wood)

Pubic, pubis—piu′ bik, -bis (L, *pubes*, hair, by inference, maturity)

Pulmonary—pull′ mow nay′ ree (L, *pulmon*, lung)

Pulvinar—pul vine′ ar (L, *pulvinus*, cushion)

Pygal—pye′ gal (G, *pyge*, rump)

Pygostyle—pye′ go style (G, *pyge*, rump; *stylos*, column)

Pylorus—pye loh′ rus (G, *pyloros*, gatekeeper)

Quadrate—kwad′ rate (L, *quadratus*, square)

Quadriceps—kwad′ ree seps (L, *quattuor*, four; *caput*, head)

Rachis—ray′ kis (G, spine)

Radial, radiale, radius—ray′ dee al, ray′ dee ay′ lee, -dee us (L, *radius*, ray)

Ramus, rami—ray′ mus, ray′ mee (L, a branch)

Raphe—ray′ fee (G, seam)

Rectrices—rek try′sees (L, *rectus*, straight)

Rectum, rectus—rek′ tum, -tus (L, *rectus*, straight)

Remiges—rem′ i jeez (L, *remus*, oar)

Renal—ree′ nal (L, *renes*, kidneys)

Reptilia, reptile—rep till′ ee ah, rep′ till (L, *reptilus*, reptile, from *repo*, creep)

Respiratory—re spire′ a toh′ ree (or res′ pee rah toh′ ree) (L, *respiro*, breathe back)

Rete—ree′ tee (L, net)

Retina—ret′ i nah (L, *rete*, net)

Retro—ret′ roh (L prefix, back, backward)

Rhinal—rye′ nal (G, *rhis*, nose)

Rhomb, rhombo—romb, romm′ boh (L or G prefix, referring to a geometric figure, a kind of parallelogram)

Rhomboideus—rom boy′ dee us (L, *rhombus*, or G, *rhombos*, rhomboid in form)

Rhyncocephalia—rin′ koh see fay′ lee ah (G, *rhynchos*, snout; *kephale*, head)

Rodent, rodentia—row′ dent, row den′ she ah (L, *rodens*, gnawing)

Rostral, rostrum—ross′ tral, -trum (L, *rostrum*, beak)

Sacculus—sak′ yu lus (L, little sac)

Sacral, sacrum—say′ cral, -krum (L, *sacer*, sacred)

Sagittal—saj′ i tal (L, *sagitta*, arrow)

Salivary—sall′ i vay′ ree (L, *saliva*, spit)

Sartorius—sar toe′ ree us (L, *sartor*, tailor)

Sauropsida—sah ropp′ si dah (G, *sauros*, lizard; *opsis*, appearance)

Scalene, scalenes—skay′ lean, skay′ leans (G, *skalenos*, uneven)

Scapula—skap′ yu lah (L, shoulder blade)

Sciatic—sigh at′ ik (L, *sciaticus*, originally *ischiadicus*, from G, *ischion*, hip)

Sclera—sklay′ rah (G, *skleros*, hard)

Sclerotic—skle rot′ ik (G, *skleros*, hard)

Sclerotome—skle′ roh tome (G, *skleros*, hard; *tome*, cutting)

Scrotal, scrotum—skroh′ tal, -tum (L, uncertain origin)

Scute—skiut (L, *scutum*, shield)

Sella—sell′ ah (L, a seat or saddle)

Semilunar—semm′ i liu′ nar (L, *semi*, half; *luna*, moon)

Seminal—semm′ i nal (L, *semen*, seed)

Seminiferous—semm′ i niff′ er ous (L, *semen*, seed; *fero*, to bear)

Septum—sepp′ tum (L, fence, wall)

Serosa—see row′ sah (L, *serum*, serum)

Serratus—se rate′ us (L, *serra*, saw)

Sesamoid—sess′ a moid (G, *sesamon*, a plant, referring to the shape of the seeds)

Sinus—sigh′ nus (L, *sinus*, curve)

Skeletogenous—skell′ e toj′ ee nous (G, *skeleton*, from *skello*, make dry)

Solar—soh′ lar (L, *sol*, the sun)

Soleus—soh′ lee us (L, *solea*, sole of the foot)

Somatic—soh mat′ ik (G, *soma*, body)

Somatopleure—soh′ ma toe plure (G, *soma*, body; *pleura*, side)

Somite—soh′ might (G, *soma*, body)

Spermatic—sperr mat′ ik (L, *sperma*, sperm)

Sphenodon—sfee′ noh don (G, *sphen*, wedge; *odon*, tooth)

Sphenoid—sfee′ noid (G, *sphen*, wedge)

Spiracle—spear′ ah kel (or spire′ ah kel) (L, *spiraculum*, air hole)

Splanchnic—splank′ nik (G, *splanchnon*, one of the viscera)

Splanchnopleure—splank′ noh plure (G, *splanchnon*, a viscus; *pleura*, side)

Splenial, splenic—splee′ nee al, splenn′ ik (G, *splen*, spleen)

Splenius—splee′ nee us (G, *splenion*, bandage)

Squamata—skwa may′ tah (L, *squama*, scale)

Squamosal—skwa moh′ sal (L, *squama*, scale)

Stapes—stay′ peez (L, stirrup)

Stegocephala—steg′ oh seff′ ah lah (G, *stego*, cover; *kephale*, head)

Sternebra, sternebrae—stir′ ne brah, -bree (G, *sternon*, breastbone; L, *vertebra*, joint)

Sterno—stir′ noh (G combining word, *sternon*, breastbone)

Sternum—stir′ numm (G, *sternon*, breastbone)

Stomodaeum—stow′ moh dee um (G, *stoma*, mouth; *daio*, divide)

Stratum—stray′ tum (L, a spread or cover)

Striatum—stry ay′ tum (L, *stria*, furrow)

Stylo—sty′ loh (G combining word, *stylos*, column)

Sub—subb (L prefix, under)

Subclavian—subb clay′ vee an (L, *sub*, under; *clavis*, key, referring to the clavicle)

Sulcus, sulci—sull′ kuss, sull′ sigh (L, a furrow or groove)

Supine—sue′ pine (L, *supino*, to put on the back)

Supra—siu′ prah (L prefix, above)

Suture—siu′ chur (L, *suo*, to sew)

Symphysis—simm′ fee sis (G, union)

Syn—sin (L or G prefix, together, with)

Synotic—sy not′ ik (G, *syn*, together; *ous*, ear)

Syrinx—sir′ inks (G, pipe)

Tarsal, tarsalia, tarsus—tar′ sal, tar say′ lee ah, tar′ sus (G, *tarsos*, a flat surface)
Tectum—tek′ tum (L, *tego*, to cover)
Tegmentum—teg men′ tum (L, *tego*, to cover)
Tel, tele—tell, tell′ ee (G, *tele*, far)
Teleost, teleostei—tell′ ee ost, tell′ ee os′ tee eye (G, *teleos*, whole, perfect; *osteon*, bone)
Teleostome, teleostomi—tell′ ee oh stome, tell′ ee oss′ toe my (G, *teleos*, perfect; *stoma*, mouth)
Telolecithal—tell′ oh less′ i thal (G, *telos*, end; *lekithos*, yolk)
Temporal—tem′ poh ral (L, *tempus*, temple)
Tendon—ten′ don (L, *tendo*, to stretch)
Tentorium—ten toh′ ree um (L, *tendo*, to stretch)
Teres—tee′ reez (L, round)
Testes, testis—tess′ teez, tess′ tiss (L, testis)
Thalamus—thal′ a mus (L, chamber)
Thecodont—thee′ koh dont (G, *theke*, sheath; *odon*, tooth)
Thoracic, thoraco—thoh ras′ ik, thoh′ ra koh (L, *thorax*, thorax)
Thorax—thoh′ raks (L, thorax)
Thymus—thy′ mus (G, *thymon*, thyme)
Thyroid (or **thyreoid**)—thy′ roid, thy′ ree oid (G, *thyreos*, shield)
Tibia, tibiale, tibialis—tibb′ ee ah, tibb′ ee ay′ lee, -ay′ lis (L, *tibia*, the shin bone)
Tonsil—ton′ sill (L, *tonsilla*, tonsil)
Trabecula, trabeculae—tra bek′ yu lah, -lee (L, little beam)
Trachea—tray′ kee ah (G, *trachys*, rough)
Trapezius—tra pee′ zee us (G, *trapeza*, table)
Trapezoid—tra′ pe zoid (G, *trapeza*, table)
Trematic—tree mat′ ik (G, *trema*, hole, opening)
Triceps—try′ seps (L, *tres*, three; *caput*, head)
Tricuspid—try kuss′ pid (L, *tres*, three; *cuspis*, point)
Trigeminus—try jem′ i nus (L, three at a birth, triple)
Trochanter—troh kan′ ter (G, *trochos*, wheel)
Trochlea, trochlear—trock′ lee ah, -lee ar (L, *trochlea*, pulley)
Tuber—tiu′ ber (L, a swelling)
Tubercular, tuberculum—tiu burr′ kiu lar, -lum (L, *tuberculum*, a little swelling)
Tunicata, tunicate—tiu′ ni kay′ tah, tiu′ ni kate (L, *tunica*, tunic)
Turbinal—turr′ bi nal (L, *turbo*, anything that whirls)
Turcica—turr′ see kah (L, turkish)
Tympanic, tympanum—tim pan′ ik, tim′ pa num (L, *tympanum*, drum)

Ulna, ulnar, ulnare—ull′ nah, ull′ nar, ull nay′ re (L, *ulna*, elbow)
Umbilical, umbilicus—um bill′ i kel, um bi lye′ kus (L, *umbilicus*, navel)
Uncinate—unn′ see nate (L, *uncus*, hook)
Unguligrade—unn giu′ li grade (L, *ungula*, hoof; *gradus*, walk)
Ureter—yu ree′ ter (G, *ouron*, urine)
Urethra—yu ree′ thrah (G, *ouron*, urine)
Urinary—yu ri nay′ ree (L, *urina*, urine)
Urodela, urodele—yu′ row dee′ lah, yu′ row deal (G, *oura*, tail; *delos*, evident)
Urogenital—yu′ row jen′ i tal (G, *ouron*, urine; L, *genitalis*, genital)
Uropygium, uropygial—yu′ row pij′ i um, -i al (G, *oura*, tail; *pyge*, rump)
Urostyle—yu′ row style (G, *oura*, tail; *stylos*, column)
Uterus, uterine—yu′ ter us, yu′ ter inn (or -ein) (L, *uterus*, womb)
Utriculus—yu trik′ yu lus (L, a little bag)

Vagina—va jye′ nah (L, a sheath)
Vagus—vay′ guss (L, wandering)
Vallate—vall′ ate (L, *vallo*, to surround with a wall)
Vas, vasa—vass, vay′ sah (L, a duct)
Vastus—vass′ tus (L, vast)
Velum—vee′ lum (L, veil)
Venosus—vee noh′ sus (L, *vena*, vein)
Venous—vee′ nous (L, *vena*, vein)
Ventricle—ven′ tree kel (L, *venter*, belly)
Vermis—verr′ mis (L, worm)
Vertebra, vertebrae—verr′ te brah, -bree (L, a joint)
Vertebrata, vertebrate—verr′ te bray′ tah, -brate (L, *vertebra*, joint)
Vesical—vess′ i kal (L, *vesica*, bladder)
Vibrissa, vibrissae—vie briss′ ah, -ee (L, hairs in the nostrils)
Villi, villus—vill′ eye, -us (L, shaggy hair)
Viscera, visceral—viss′ err ah, -al (L, *viscus*, internal organ)
Viscus—viss′ kus (L, internal organ)
Vitelline—vi (or vie) tell′ inn (L, *vitellus*, yolk)
Vitreous—vit′ ree ous (L, glassy)
Vomer—voh′ mer (L, plowshare)
Vulva—vull′ vah (L, covering)

Wolffian—wolf′ ee an (after K. F. Wolff, German anatomist)

Xiph, xiphi—ziff, ziff′ ee (G, *xiphos*, sword, used in combining words)

Zyg, zygo—zigg, zye′ go (or zigg′ oh) (G combining word, *zygon*, yoke)
Zygapophysis—zye′ (or zigg) ga poff′ ee sis (G, *zygon*, yoke; *apophysis*, process)
Zygomatic—zye′ go (or zigg′ oh) mat′ ik (G, *zygoma*, yoke or bar)

APPENDIX B

PREPARATION OF MATERIALS

1. Killing the specimens.—*Necturus* is best killed by placing in hot water; turtles by injection of ether or chloroform into the cloaca or, better, the trachea; birds and mammals by inclosing them in a tightly closed vessel with a wad of cotton soaked in ether or chloroform. In handling turtles pull the head forward by inserting a stout hook behind the jaw and pry open the mouth.

2. Preparation of skeletons.—Skeletons are best prepared from fresh materials or, in the case of marine forms, from those that have been preserved in brine. It is difficult to prepare skeletons from specimens that have been preserved in formalin. To prepare a skeleton remove the skin, all of the viscera, and as much of the muscles as possible and soak the specimen in water. The remaining flesh will decay and may be removed with a stiff brush or forceps. This process of maceration in cold water takes some time. The process may be much shortened, and tough specimens are more easily prepared by immersing the specimen for a few hours in hot or boiling water to which Gold Dust or the following soap solution has been added. Kingsley gives the following formula for the soap solution:

75 gm. hard soap
12 gm. potassium nitrate (saltpeter)
150 cc. strong ammonia
2,000 cc. distilled or soft water

Mix thoroughly. In using, take one part of the soap solution to three or four parts of water. The length of time required before the flesh will separate easily from the bones varies with different animals and is shorter for small or young specimens than for large and old ones. Davison gives the time for the cat as two to four hours in boiling water or, better, three to six hours in water kept at 75°–90° C. He also states that if the bones are heated for only one to two hours at a temperature not above 85° C., the ligaments will be preserved, and that, when the skeleton is dried, the ligaments will harden and hold the bones together. Skeletons containing a considerable amount of cartilage should not, of course, be boiled but should be treated only with moderately hot water. For fishes a few minutes' treatment with hot water is generally sufficient. Cartilaginous skeletons should be preserved in weak formalin or in 70 per cent alcohol.

3. Injection of the circulatory system.—To render the blood vessels conspicuous and more easily followed, it is advisable and, in fact, practically necessary that the arteries, at least, be injected with a colored solution. Injection syringes for this purpose may be obtained from dealers in laboratory supplies, or an ordinary rubber atomizer bulb may be used. A glass cannula is inserted in the vessel to be injected. A cannula is simply a piece of glass tubing drawn out in the flame at one end to a size suitable for the blood vessel into which it is to be inserted. It is also desirable that the end to be inserted in the vessel be slightly enlarged, as it will then hold more securely in the blood vessel. Loosen the vessel to be injected from the surrounding tissues, pass a cord under it, and tie the cord in a loose single knot above the vessel. With a fine scissors make a V-shaped cut into the blood vessel, having the cut extend not more than halfway through the vessel, and immediately insert the cannula into the vessel. Tighten the knot around the cannula and blood vessel by pulling on the two ends of the cord, but make only a single knot. The cannula is connected with the injection syringe or rubber bulb by a piece of rubber tubing. The whole system should be filled with the injection fluid in advance, in order to avoid forcing air into the blood vessels. After everything is ready, inject the solution into the blood vessel by a steady but not too forceful pressure on the syringe or bulb. The success of the injection should be determined by examining the small vessels of the skin or in the intestinal walls. When these are deeply colored, the injection is complete, and the cannula is then withdrawn, the cord being immediately tightened around the blood vessel and tied with a double knot to prevent leakage. In case only the arterial system is injected, as is usually the case, the blood should not be let out of the animal; the injection will force the blood around into the veins and distend them. If the systemic veins are to be injected, the blood should be drained from the animal by opening the vein, at the place where it is to be injected.

Two kinds of injection fluids are in general use: the starch suspension and the latex emulsion.

a) *Starch injection method:* The following formula is given by Davison:

100 cc. water
20 cc. glycerin
20 cc. concentrated formalin
85 gm. cornstarch
Sufficient coloring matter to give a deep color

Stir thoroughly and strain through a fine cheesecloth to remove all lumps. Stir the mixture immediately before using, as the starch has a tendency to settle to the bottom on standing. The coloring matter used is: for red color, vermilion (mercuric sulphide) or powdered carmine; for yellow, chrome yellow (lead chromate); for blue, Prussian blue. This medium will keep indefinitely in a tightly covered jar.

b) *Latex* (*liquid rubber*) *injection method:* This method has come into use in recent years. It has the advantage that blood vessels so injected do not break but merely stretch when pulled, but otherwise is not particularly better than the starch method and is more troublesome and expensive to employ. The following directions have been kindly furnished by the General Biological Supply House, Chicago. The rubber emulsion may be purchased from this or other supply-houses. It has the consistency of cream and comes ready to use in red, blue, and yellow colors. It remains liquid at average temperatures and should *not* be stirred, as stirring traps air bubbles, and this must be avoided. Liquid latex should be kept in tightly covered glass containers in a cool, dark place. It does not keep indefinitely, and hence only limited amounts should be purchased at a time. If the latex should become too thick, plain water or, better, ammonia water should be added. When of the consistency of cream, the latex will flow freely into the smaller vessels. Use an ordinary syringe with glass barrel and rubber tip plunger of any desired capacity and a regular needle of the proper size.

After the injection the latex must be caused to set, and this is done by exposure to dilute acids. Smaller specimens, such as frogs and *Necturus*, are placed in a bath of 5 per cent acetic acid for twenty-four hours and then transferred to formalin or any other suitable preservative. Or acetic acid may be added to the preservative, thus eliminating the necessity of an acid bath. Embalmed specimens need not be treated with an acid bath, as the carbolic acid in the embalming fluid is sufficient to cause the latex to set. Acetic acid should *not* be added to the embalming fluid, as this may cause the latex to set prematurely.

Injection is usually made into the following vessels: in elasmobranchs the arteries are injected by way of the caudal artery, exposed by cutting across the tail; the hepatic portal system is injected by way of the hepatic portal vein near the liver backward or in both directions or into the posterior mesenteric vein forward. Injection of the systemic veins is more difficult and is usually not done; a method is given by Rand in the *American Naturalist*, Volume **39** (1905). In *Necturus* the arteries are most easily injected by way of the bulbus arteriosus. The hepatic portal system should be injected in both directions into the hepatic portal vein just before it enters the liver. The postcaval can be injected, if desired, through the large hepatic vein seen on the ventral surface of the liver just behind the transverse septum. The arterial system of the turtle is injected either backward by way of one of the carotids or into the left aorta away from the heart. It is practically impossible to work out the renal portal and hepatic portal systems on preserved turtles unless they are injected. Injection is very easily made of both of these systems by way of the abdominal vein. The plastron is, of course, to be removed by sawing across the bridges. Insert the cannula into one of the abdominal veins about halfway between the heart and pelvis and inject forward. Injection of the arterial system of the pigeon can be done by way of the pectoral artery toward the heart. According to Parker, the systemic veins and the hepatic portal system may be injected by way of the coccygeo-mesenteric vein in both directions. The arterial system of mammals is injected by way of either the carotid or the femoral artery, the cannula being directed toward the heart. The veins are usually not injected, but it is stated that they may be injected by way of the external jugular toward the heart. The hepatic portal system may be injected by way of the hepatic portal vein near its entrance into the liver. In injecting arteries and veins in the same specimen, differently colored injection fluids should, of course, be used.

4. Preservation of specimens.—Specimens are usually preserved in 5 per cent formalin (5 cc. of commercial formalin plus 95 cc. of water). A slit should be made in the body cavity, or the formalin should be injected into the body cavity through a small opening to insure preservation of the viscera. A portion of the roof of the skull should be removed in the smaller specimens in order that the brain may harden. In the case of mammals additional measures are necessary in order to insure preservation. Mammals should be embalmed. This is done by injecting an embalming fluid into the blood vessels before the injection fluid is sent in. The embalming fluid may consist of 5 per cent formalin or, better, 5 per cent formalin plus one-sixth its volume of glycerin. A still better but more expensive embalming fluid is one used for human bodies, with the following formula:

	Parts by Volume
Formalin	1.5
Carbolic acid (melted crystals)	2.5
Glycerin	10.0
Water	86.0

The embalming fluid is injected preferably into the femoral artery through a cannula in the same way as already described for the injection fluid. Bensley recommends that the injection of the embalming fluid should be done not with a syringe but by attaching the cannula to a receptacle containing the embalming fluid elevated about three feet and allowing the fluid to run into the vessel under this pressure for about two hours. The animal should be arranged in a position suitable for dissection, with the limbs spread well apart and the head tilted backward. After the embalming, the injection fluid is run into the same cannula by attaching the syringe to it; Bensley recommends that twenty-four hours elapse between the embalming and the injection with the colored fluid.

If the animal has been thoroughly embalmed, it will keep without being immersed in a preserving fluid. It should be prevented from becoming dry by being placed in airtight receptacles or wrapped in cloths saturated with the formalin-glycerin solution. It is best to sponge the hair with a mixture of alcohol and water containing 2 per cent formalin. The animals may, however, if preferred, be immersed in 1 per cent formalin solution. During the dissection the animals should be kept wrapped in cloths moistened with formalin-glycerin.

Further details on these matters will be found in: Bensley, *Practical Anatomy of the Rabbit* (University of Toronto Press); Reighard and Jennings, *Anatomy of the Cat* (Henry Holt & Co.); and Davison, *Mammalian Anatomy* (P. Blakiston's Son & Co.).

INDEX

PRINTED IN U.S.A.